Advances in Intelligent Systems and Computing

Volume 886

Series editor

Janusz Kacprzyk, Systems Research Institute, Polish Academy of Sciences,
Warsaw, Poland
e-mail: kacprzyk@ibspan.waw.pl

The series "Advances in Intelligent Systems and Computing" contains publications on theory, applications, and design methods of Intelligent Systems and Intelligent Computing. Virtually all disciplines such as engineering, natural sciences, computer and information science, ICT, economics, business, e-commerce, environment, healthcare, life science are covered. The list of topics spans all the areas of modern intelligent systems and computing such as: computational intelligence, soft computing including neural networks, fuzzy systems, evolutionary computing and the fusion of these paradigms, social intelligence, ambient intelligence, computational neuroscience, artificial life, virtual worlds and society, cognitive science and systems, Perception and Vision, DNA and immune based systems, self-organizing and adaptive systems, e-Learning and teaching, human-centered and human-centric computing, recommender systems, intelligent control, robotics and mechatronics including human-machine teaming, knowledge-based paradigms, learning paradigms, machine ethics, intelligent data analysis, knowledge management, intelligent agents, intelligent decision making and support, intelligent network security, trust management, interactive entertainment, Web intelligence and multimedia.

The publications within "Advances in Intelligent Systems and Computing" are primarily proceedings of important conferences, symposia and congresses. They cover significant recent developments in the field, both of a foundational and applicable character. An important characteristic feature of the series is the short publication time and world-wide distribution. This permits a rapid and broad dissemination of research results.

More information about this series at http://www.springer.com/series/11156

Kohei Arai · Supriya Kapoor
Rahul Bhatia
Editors

Advances in Information and Communication Networks

Proceedings of the 2018 Future of Information and Communication Conference (FICC), Vol. 1

Editors
Kohei Arai
Faculty of Science and Engineering
Saga University
Saga, Japan

Rahul Bhatia
The Science and Information
(SAI) Organization
Bradford, UK

Supriya Kapoor
The Science and Information
(SAI) Organization
London, UK

ISSN 2194-5357 ISSN 2194-5365 (electronic)
Advances in Intelligent Systems and Computing
ISBN 978-3-030-03401-6 ISBN 978-3-030-03402-3 (eBook)
https://doi.org/10.1007/978-3-030-03402-3

Library of Congress Control Number: 2018959425

This Springer imprint is published by the registered company Springer Nature Switzerland AG
The registered company address is: Gewerbestrasse 11, 6330 Cham, Switzerland

Preface

On behalf of the organizing committee of the Future of Information and Communication Conference (FICC), it is an honour and a great pleasure to welcome you to the FICC 2018 which is held from 5 to 6 April 2018 in Singapore.

The conference is organized by the SAI Conferences, a group of annual conferences produced by The Science and Information (SAI) Organization, based in the UK.

The digital services nowadays are changing the lives of people across the globe. So is the information and communication which plays a significant role in our society. The Future of Information and Communication Conference (FICC 2018) focuses on opportunities for researchers from all over the world. Bringing together experts from both industry and academia, this information and communication conference delivers programs on latest research contributions and future vision (inspired by the issues of the day) in the field and potential impact across industries.

FICC 2018 attracted a total of 361 submissions from many academic pioneering researchers, scientists, industrial engineers and students from all around the world. These submissions underwent a double-blind peer review process. Of those 361 submissions, 104 submissions (including nine poster papers) have been selected to be included in this proceedings. It covers several hot topics which include ambient intelligence, communication, computing, data science, intelligent systems, Internet of things, machine learning, networking, security and privacy. The conference held over two days hosted paper presentations, poster presentations as well as project demonstrations.

Many thanks go to the keynote speakers for sharing their knowledge and expertise with us and to all the authors who have spent their time and effort to contribute significantly to this conference. We are also indebted to the organizing committee for their great efforts in ensuring the successful implementation of the conference. In particular, we would like to thank the technical committee for their constructive and enlightening reviews on the manuscripts.

We are pleased to present the first proceedings of this conference as its published record. Our sincere thanks to all the sponsors, press, print and electronic media for their excellent coverage of this conference.

Hope to see you in 2019 in our next Future of Information and Communication Conference but with the same amplitude, focus and determination.

Kohei Arai

Contents

End to End Deep Neural Network Frequency Demodulation of Speech Signals

Dan Elbaz(✉) and Michael Zibulevsky

Department of Computer Science, Technion Israel Institute of Technology, 32000 Haifa, Israel
elbazdan@gmail.com, mzib@cs.technion.ac.il

Abstract. Frequency modulation (FM) is a form of radio broadcasting which is widely used nowadays and has been for almost a century. We suggest a software-defined-radio (SDR) receiver for FM demodulation that adopts an end-to-end learning based approach and utilizes the prior information of transmitted speech message in the demodulation process. The receiver detects and enhances speech from the in-phase and quadrature components of its base band version. The new system yields high performance detection for both acoustical disturbances, and communication channel noise and is foreseen to out-perform the established methods for low signal to noise ratio (SNR) conditions.

Keywords: Frequency Modulation (FM)
Long Short-Term Memory (LSTM) · Software-Defined-Radio (SDR)
Deep learning · End-to-end learning · Amplitude noise · Phase noise

1 Introduction

Frequency modulation (FM) is a nonlinear encoding of information on a carrier wave. It can be used for interferometric, seismic prospecting, telemetry and many more applications, each with its own statistics, dominated by the underlying generating process. However, its widest use is for radio broadcasting, which is commonly used for transmitting audio signal representing voice.

Communication transmission channel is subject to various distortions, noise conditions and other impairments. Those impairments severely degrade FM demodulator performance when a critical level is exceeded. As a result thereof, the intelligibility and quality of the detected speech decreases significantly. This phenomenon is known as the Threshold Effect.

Long Short-Term Memory (LSTM) recurrent neural networks [8] are powerful models that can capture long range dependencies and non-linear dynamics. In

This research was supported by the Intel Collaborative Research Institute for Computational Intelligence (ICRI-CI).

K. Arai et al. (Eds.): FICC 2018, AISC 886, pp. 1–11, 2019.
https://doi.org/10.1007/978-3-030-03402-3_1

many signal estimation tasks, the advantage of recurrent neural network becomes significant only when there is a statistical dependency between the examples. This paper introduces FM demodulator based on Long Short-Term Memory recurrent neural network.

The main contributions of this work are as follows:

- Utilizing the LSTM abilities to capture the temporal dynamics of speech signals and take advantage of the prior statistics of the speech to overcome transmission channel disturbances.
- Taking an end-to-end learning based approach for filtering both acoustical disturbances, modeled as phase noise and transmission channel disturbances, modeled as amplitude noise. In this approach, the LSTM learn to map directly from the modulated baseband signal to the modulating audio that had been applied at the transmitter, thus creating a baseband to speech mapping.

We demonstrate this method by applying it to Frequency modulation (FM) decoding in varying levels of amplitude and phase noise and show it has a superior performance over legacy reception systems in low SNR conditions.

2 Problem Formulation and Related Work

Traditionally radio transmission decoding and speech enhancement are considered as two separate problems. However, optimal signal estimation algorithms are usually constructed on the basis of statistical properties of a measurement process and prior statistical or deterministic model of the reconstructed signals. This is often a difficult problem with no analytic solution that can be only approximately solved based on simplistic models of noise and signal. On the other hand, any signal estimation can be considered as a non-linear mapping from input data to the desired output. Having a universal function approximation tool in hand, we can learn such mapping using a set of training examples, pairs of input modulated baseband signals and the desired audio output signals.

Except for the traditional methods for radio transmission decoding and for minimum mean square error (MMSE) based speech enhancement techniques, several, though not many, neural networks based methods have been proposed for each of the two problems separately. For example the radio transmission decoding, and for channel noise estimation [1,12] and recently [2], however, these works deal with digital communication for which bit-streams are mapped to symbols, moreover, traditionally the symbols are precoded and scrambled before transmitted, therefore effectively the coded data stream is uncorrelated from time-sample to time-sample [21], and use of the prior speech data to overcome the noise in the transmission channel is not possible.

The fact that the modulating input is proportional only to the instantaneous frequency of the received FM signal has driven the development of traditional FM demodulators to rely on very short time frame processing in order to extract the modulating signal, disregarding long range dependencies that are present in the transmitted voice.

In [14], author addressed the analog FM problem, but the approach taken was to imitate the way a conventional FM demodulator works by implementing different neural network for each building block separately. It used memoryless (or very short memory) feed-forward neural network with only one input at some intermediate blocks, and therefore it did not take into account the prior knowledge of the transmitted speech. Moreover, demodulation was performed directly on the passband signal so the input of the neural network needed to be sampled with very high sampling rate in order to detect the change in the input, resulting in several samples, most of which are redundant, and a very large network for actual sampling rates that is very difficult to train and not suitable for practical use.

As for the problem of speech enhancement, several neural networks based solutions were suggested and shown to give good performance, for example [4, 10, 22] and in [9] an LSTM based model was suggested.

While the listed works have applied neural networks to the task of radio demodulation and for speech enhancement separately, neither of these works suggested the task of radio transmission decoding with the prior information of transmitted speech messages. In this sense, our project is entirely novel as our network exploits the prior knowledge of the speech signal to overcome both acoustical disturbances and noise in the communication channel and performs audio reconstruction by directly operating on the baseband representation of the modulated data.

In [20], demodulation is viewed as a problem of inference and learning and it was suggested to use a demodulation process that can be shaped by user-specific prior information. We adopt this approach in our work and suggest a neural network based solution for this problem.

3 Background

3.1 Signal Modulation

Frequency Modulation is the process of modulating a sine wave with an information message $x_m(t)$ in the following manner:

$$y(t) = A_c cos\Big(2\pi f_c t + 2\pi f_\Delta \int_0^t x_m(\tau)\, d\tau\Big)$$

where $x_m(t)$ is the data signal, which is typically a speech signal, $x_c(t) = A_c \cos(2\pi f_c t)$ is the sinusoidal carrier, f_c is the base frequency of the carrier, A_c is the amplitude of the carrier and f_Δ is the frequency deviation, which represents the maximum shift away he carrier's base frequency.

Sinusoid with frequency modulation can be decomposed into two amplitude-modulated sinusoids that are offset in phase by one-quarter cycle ($\pi/2$ rad). The amplitude modulated sinusoids are known as in-phase and quadrature components or the I/Q components. By using simple trigonometric identities the general expression representing the transmitted signal can be expressed as follows:

$$y(t) = A_c cos(2\pi f_c t) cos\left(2\pi f_\Delta \int_0^t x_m(\tau) d\tau\right)$$
$$- A_c sin(2\pi f_c t) sin\left(2\pi f_\Delta \int_0^t x_m(\tau) d\tau\right)$$

The I/Q components can be defined in the following way:

$$I(t) = A_c cos\left(2\pi f_\Delta \int_0^t x_m(\tau) d\tau\right)$$
$$Q(t) = A_c sin\left(2\pi f_\Delta \int_0^t x_m(\tau) d\tau\right)$$

and we can represent the modulated signal with its I/Q components in the following way:

$$y(t) = I(t) cos(2\pi f_c t) - Q(t) sin(2\pi f_c t)$$

This signal has a bandpass spectrum centered around the carrier frequency f_c. It is common to analyze communication systems by using the low pass equivalents, also referred to as baseband (or I/Q components) of the original band pass signals.

3.2 Noise Model

During the process of transmission and reception, the signal is subject to several impairments. On the receiving side, the modulator's role is to reconstruct the original signal from the received signal with maximal level of reliability, overcoming the impairments introduced by the transmission and reception phases.

As mentioned, the message signal undergoes several distortions, the signal impairments due to those distortions can be divided into two categories:

(1) Phase noise: Impairments due to environmental conditions such as audio distortions and the operation of frequency modulation, those original audio additive impairments are translated to the phase to become phase noise.

$$r(t) = A_c cos\left(2\pi f_c t + 2\pi f_\Delta \int_0^t (x_m(\tau) + n(\tau)) d\tau\right)$$

(2) Amplitude noise: Impairments due to communication channel distortions such as convolution with the communication channel, multi-path, additive noise due to propagation characteristics of the channel environment, etc. those impairments are translated to additive amplitude noise, $r(t) = y(t) + n(t)$.

In communication systems, the statistical model for each of the above noise models is usually assumed to be white Gaussian noise. For clarity a Fig. 1 presents a diagram depicting the communication system and its elements.

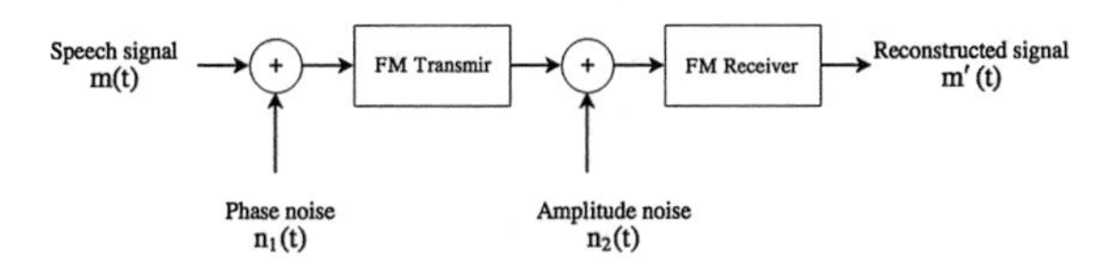

Fig. 1. Communication system with amplitude and phase noise sources.

3.3 Long Short-Term Memory Networks

Long short-term memory (LSTM) is a recurrent neural network architecture, that achieve excellent performance on a general sequence to sequence learning problems [18]. Due to the vanishing and exploding gradient [13] the training of recurrent neural networks is a challenging task. To address this issue, the LSTM cell has been introduced by [8]. We are using the common LSTM version [6], with the following update equations:

$$\begin{aligned}
i_t &= \sigma_i \left(W_{xi}x_t + W_{hi}h_{t-1} + W_{ci}c_{t-1} + b_i\right)\\
f_t &= \sigma_f \left(W_{xf}x_t + W_{hf}h_{t-1} + W_{cf}c_{t-1} + b_f\right)\\
c_t &= f_t c_{t-1} + i_t tanh \left(W_{xc}x_t + W_{hc}h_{t-1} + b_c\right)\\
o_t &= \sigma_o \left(W_{xo}x_t + W_{ho}h_{t-1} + W_{co}c_t + b_o\right)\\
h_t &= o_t tanh \left(c_t\right)
\end{aligned}$$

where, σ is the logistic sigmoid function, i, f, o and c are respectively the input gate, forget gate, output gate and cell activation vector cells at time t. x_t is the input feature vector, h_t is hidden output vector, b_i, b_f and b_o are the bias terms and $W_{hi}, W_{hf}, W_{ho}, W_{xi}, W_{xf}$ and W_{xo} are the weight matrices connecting the different inputs and gates with the memory cells.

3.4 Natural Speech Structure

Natural speech is composed of many timescale features, generated by anatomic processes that control sound production. A typical segment of speech can be decomposed to sentences or words that are of a typical time scale of one second. On a smaller time scale, words can be decomposed into phonemes, which are one of the units of sound that distinguish one word from another. Usually phonemes last a duration which is smaller than 10^{-1} s. We can look on an even smaller time scale, such as pitch 10^{-2} and formants 10^{-3}. For an optimal reconstruction to take place, all those timescales need to be accounted for in the reconstruction task.

4 Method Description

4.1 Dataset and Training Procedure

In order to support high quality audio transmissions broadcast, FM stations use large values of frequency deviation. The FM broadcast standards in the United States specify a value of 75 kHz of peak deviation and 240 kHz sampling frequency of the output signal. The default value of the modulating audio signal is 48 kHz. For the above reasons, the training set was generated using Matlab FM modulation [7] with the above stated standard specifications. The above system constraints dictates the number of baseband samples the modulator produces for each audio sample on its input (five in-phase and five quadrature). In order to avoid manipulating FM passband signal directly, we assume that conversion to baseband from intermediate frequency will be performed by another digital or analog hardware. This conversion process is known as synchronous detection or heterodyning the signal down to baseband and it is usually performed in the analog front end. Converting the high frequency signal to baseband signal, enables more convenient processing in a lower sampling rate than the original carrier frequency and alleviates the demodulator (either standard or DNN based) computational demands. The audio waveforms used in our experiments were downloaded from TIMIT Acoustic-Phonetic Continuous Speech Corpus [3]. The TIMIT corpus includes 16-bit, 16 kHz speech waveform file for each utterance. We used male speakers from New England dialect region. The speech material in the TIMIT corpus is subdivided into portions for training and testing. The criteria for the subdivision has no relation to the data distributed, and can by found in the corpus documentation. For the input of the neural network we used two features, samples of the in-phase and quadrature components of the baseband signal. For compatibility with standard United States specifications described above the waveforms were up-sampled to 48 kHz.

4.2 Architecture

We utilize the abilities of the LSTM network to capture the temporal dynamics of speech signals for the problem of source signal estimation from noisy frequency-modulated measurements. As dictated by the underlying generating speech, future samples are also related to current samples. To exploit this dependency we introduce a small delay of 100 samples, this makes the system slightly non causal. However, it enables us to use bidirectional LSTMs [16], which are trained using input information from the past as well as from the future of a specific time frame. This is achieved by processing the data in both directions with two separate hidden layers. For combining multiple levels of representations of the modulated speech signal we use deep architectures. Deep RNNs can be created by stacking multiple LSTM layers on top of each other, with the output sequence of one layer forming the input sequence for the next. The stacking of multiple recurrent hidden layers have proven to give state-of-the-art performance for acoustic modeling [5,11]. For the above stated reasons we have decided to adopt Deep bidirectional LSTM architecture based on the architecture proposed in [5]. For regularization we added a dropout layer [17]. We unrolled the network to length of 100 time steps, using backpropagation through time [15] in the training phase. Long term dependencies were accounted for by preserving the network state between batches, as the last state of a batch was used as the initial state for the following batch. The entire system was optimized with MSE loss function and RmsProp [19] optimization method.

5 Experimental Results

In order to evaluate the performance of the DNN demodulator we used both the Mean Squared Error (MSE) objective measure. We compare the performance of proposed LSTM demodulator against the performance of conventional demodulator implementation from Matlab communication toolbox, which is based on [7]. In both cases, DNN and conventional demodulator, the modulated signal sample rate is set to 240 kHz and the frequency deviation is set to 75 kHz (United States standard). In order to boost the performance of the conventional FM receiver and compensate FM characteristic in that it amplifies high frequency noise and degrades the overall signal-to-noise ratio, we used matlab FM broadcasters. FM broadcasters insert a pre-emphasis filter prior to FM modulation to amplify the high-frequency content. The FM receiver has a reciprocal de-emphasis filter after the FM demodulator to attenuate high-frequency noise and restore a flat signal spectrum. For clarity the full FM broadcast system is depicted in Fig. 2. We start with the noise free case, i.e. neither phase nor amplitude noise were added to the modulated signal. For the noise free case we get output SNR of 36.56 dB, these results indicate high quality reconstruction. Figure 3 shows the spectrogram of the original audio and spectrogram of the LSTM demodulator reconstruction for noise free case. We investigated the performance of the proposed receiver, by comparing the audio reconstruction quality in experiments employing FM modulation in the presence of various levels of additive white Gaussian (AWGN) noise in both the amplitude and phase.

The comparison between the proposed LSTM demodulator and conventional demodulator is presented in Figs. 4 and 5. Figure 4 show the speech reconstruction MSE for various levels of AWGN amplitude noise. Figure 5 show the speech reconstruction MSE for various levels of AWGN phase noise.

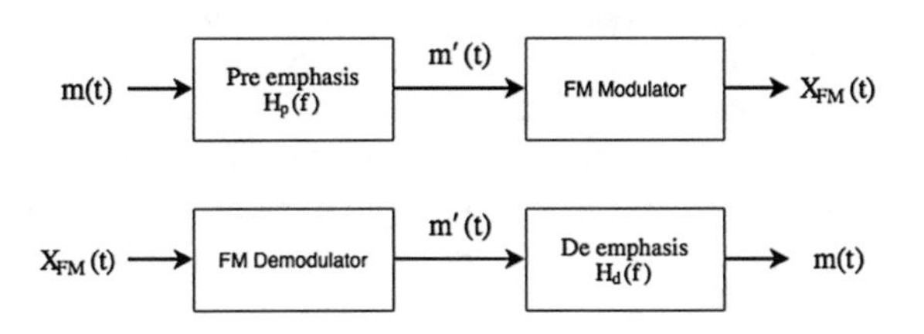

Fig. 2. Full broadcast system, used for performance comparison.

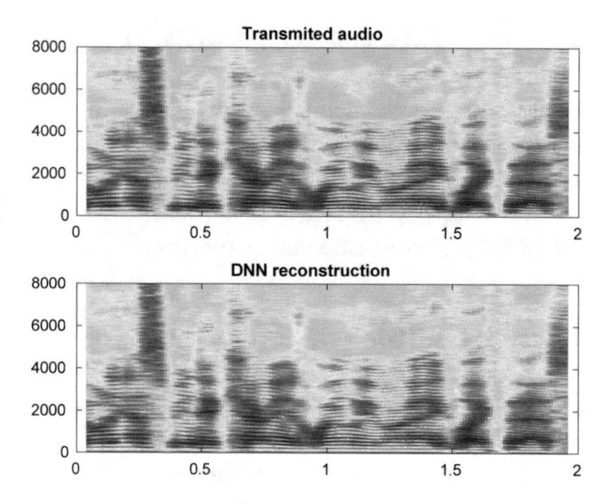

Fig. 3. Spectrogram of the original audio signal and DNN demodulator reconstruction.

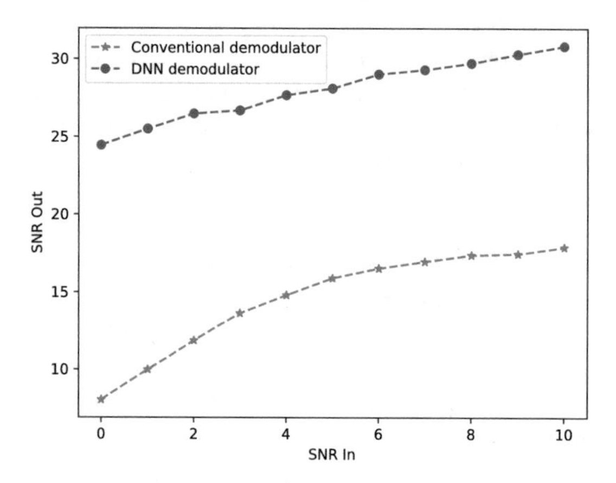

Fig. 4. Audio reconstruction quality comparison- Conventional Vs. DNN based demodulator for various levels of amplitude noise.

The conducted experiments show that the proposed receiver has a clear advantage over the conventional receiver in noise conditions, this is mostly due to the fact that the proposed LSTM demodulator takes advantage of the statistics of the generating speech signal. We prove this point by limiting the memory of the network to only one time step, as in theory we can map the FM signal back to audio with almost no memory. Though we were able to reconstruct the audio for noise free case with rather small reconstruction error of SNR 17.76 dB, however for low SNR conditions, of 0 dB amplitude noise, reconstruction was not possible without using memory, and the demodulation failed. This experiment shows that indeed in order for quality reconstruction to take place under noise conditions the statistics of the generating speech signal must be accounted for. Next we compare the reconstruction quality of the proposed LSTM demodulator

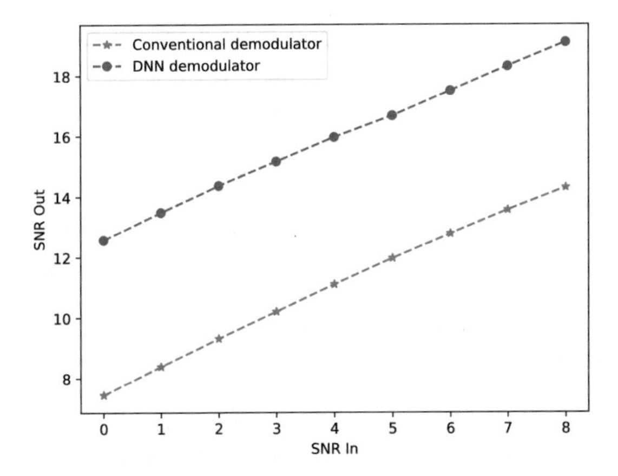

Fig. 5. Audio reconstruction quality comparison- Conventional Vs. DNN based demodulator for various levels of phase noise.

and conventional demodulator, in the presence of phase noise. This is done by adding AWGN noise both to the modulating speech signal and to the frequency modulated signal separately. We add phase noise by adding AWGN with SNR of 0 dB with respect to the speech signal, We also add 0 dB AWGN amplitude noise to the modulated FM signal. The experiment showed the SNR of the LSTM demodulator reconstructed signal is 12.32 dB, were as for conventional demodulator we got 4.54 dB for the reconstructed signal. Again a clear performance advantage for the LSTM demodulator in this case as well. As seen the LSTM demodulator outperforms the conventional demodulator in the case of both amplitude and phase noise, thus creating an end-to-end radio receiver that can overcome both communication channel disturbances and acoustical disturbances, modeled as phase noise.

6 Conclusion

We have presented a new approach to decode FM transmission of audio speech signals based on bidirectional stacked LSTM. In this approach we utilize the statistics of the information message, more specifically long and short time-scale temporal structure in speech. As a result, the proposed receiver has a clear advantage over the conventional receiver as it yields much higher reconstruction quality and can overcome both distortions in the information message and distortions in the transmission channel. With the availability of sufficient computation power, which became practical with the appearance of powerful graphical processing units (GPU) and corresponding software, the proposed receiver can be used as an extremely robust radio receiver.

References

1. Amini, M., Balarastaghi, E.: Universal neural network demodulator for software defined radio. Int. J. Mach. Learn. Comput. **1**(3), 305–310 (2011)
2. Fan, M., Wu, L.: 2017 International Conference on Communication, Control, Computing and Electronics Engineering (ICCCCEE) (2017)
3. Garofolo, J.S., Lamel, L.F., Fischer, W.M., Fiscus, J.G., Pallett, D.S., Dahlgren, N.L.: DARPA TIMIT acoustic-phonetic continuous speech corpus CD-ROM. NASA STI/Recon Technical report N, 0, pp. 1–94, January 1993
4. Goehring, T., Bolner, F., Monaghan, J.J.M., van Dijk, B., Zarowski, A., Bleeck, S.: Speech enhancement based on neural networks improves speech intelligibility in noise for cochlear implant users. Hear. Res. **344**, 183–194 (2016)
5. Graves, A., Mohamed, A., Hinton, G.: Speech recognition with deep recurrent neural networks. In: ICASSP, no. 3, pp. 6645–6649 (2013)
6. Graves, A.: Generating sequences with recurrent neural networks. preprint. arXiv:1308.0850 (2013)
7. Hatai, I., Chakrabarti, I.: A new high-performance digital FM modulator and demodulator for software-defined radio and its FPGA implementation. Int. J. Reconfigurable Comput. **2011** (2011)
8. Hochreiter, S., Schmidhuber, J.U.: Long short-term memory. Neural Comput. **9**(8), 1735–1780 (1997)
9. Kolbaek, M., Tan, Z.-H., Jensen, J.: Speech enhancement using long short-term memory based recurrent neural networks for noise robust speaker verification. In: IEEE Workshop on Spoken Language Technology (SLT), no. 1, pp. 305–311 (2016)
10. Kumar, A., Florêncio, D.: Speech Enhancement In Multiple-Noise Conditions using Deep Neural Networks. CoRR, abs/1605.0 (2016)
11. Li, X., Wu, X.: Constructing long short-term memory based deep recurrent neural networks for large vocabulary speech recognition. In: 2015 IEEE International Conference on Acoustics, Speech and Signal Processing (ICASSP), pp. 4520–4524 (2014)
12. Önder, M., Akan, A., Doğan, H.: Advanced neural network receiver design to combat multiple channel impairments. Turkish J. Electr. Eng. Comput. Sci. **24**(4), 3066–3077 (2016)
13. Pascanu, R., Mikolov, T., Bengio, Y.: On the difficulty of training recurrent neural networks. JMLR.org (2013)
14. Rohani, K., Manry, M.T.: The design of multi-layer perceptrons using building blocks (1991)
15. Rumelhart, D.E., Hinton, G.E., Williams, R.J.: Learning representations by back-propagating errors. Nature **323**(6088), 533–536 (1986)
16. Schuster, M., Paliwal, K.K.: Bidirectional recurrent neural networks. IEEE Trans. Sig. Process. **45**(11), 2673–2681 (1997)
17. Srivastava, N., Hinton, G., Krizhevsky, A., Sutskever, I., Salakhutdinov, R.: Dropout: a simple way to prevent neural networks from overfitting. J. Mach. Learn. Res. **15**, 1929–1958 (2014)
18. Sutskever, I., Vinyals, O., Le, Q.V.: Sequence to sequence learning with neural networks. In: Advances in Neural Information Processing Systems (NIPS), pp. 3104–3112 (2014)
19. Tieleman, T., Hinton, G.: Lecture 6.5-rmsprop: Divide the gradient by a running average of its recent magnitude. In: COURSERA: Neural Networks for Machine Learning (2012)

20. Turner, R.E., Sahani, M.: Demodulation as probabilistic inference. IEEE Trans. Audio Speech Lang. Process. **19**(8), 2398–2411 (2011)
21. Wornell, G.W.: Efficient symbol-spreading strategies for wireless communication. Research Laboratory of Electronics, Massachusetts Institute of Technology (1994)
22. Xu, Y., Du, J., Dai, L.-R., Lee, C.-H.: A regression approach to speech enhancement based on deep neural networks. IEEE/ACM Trans. Audio Speech Lang. Process. **23**(1), 7–19 (2015)

Performance Enhancement of MIMO – MGSTC Using a New Detection and Decoding Technique

Karim Hamidian[1(✉)] and Wurod Qasim Mohamed[2]

[1] Electrical Engineering Department, California State University Fullerton, Fullerton, CA, USA
khamidian@exchange.fullerton.edu

[2] Communications Engineering Department, Diyala University, Diyala, Iraq
Wurod89@csu.fullerton.edu

Abstract. In this paper, the performance of Multi – Group Space Time Codes (MGSTC) which is one of the Multiple Input Multiple Output (MIMO) communication methods, was investigated and improved using a new signal detection and decoding technique called parallel decoding algorithm. It is shown that the new technique reduces the overall signal detection time, thus it increases the speed of signal processing at the receiver. In addition, the new technique prevents possible error propagation that may be present in serial detection methods. Simulation results demonstrate the advantages of using parallel decoding technique.

Keywords: Multiple Input Multiple Output (MIMO) · Multi – Group Space Time Codes (MGSTCs) · New detection technique · Parallel decoding · Error propagation · V-BLAST

1 Introduction

The key challenge of the current wireless communications technology is to continue improving the performance of Multiple Input Multiple Output (MIMO) techniques. This is achieved by a combination of signal processing techniques that improve the performance of wireless communication techniques through combating and exploiting the multipath scattering between multiple transmit and multiple receive antennas [1]. In the beginning of 1990s, two new techniques were developed [2]. One of these new techniques uses multiple transmit antennas to achieve transmit diversity and reduce the effect of fading [3]. In 1998, a paper published by Alamouti [4] developed another way to achieve transmit diversity using a simple signal processing technique at the receiver.

The two main classes of MIMO communication techniques are spatial diversity and spatial multiplexing, which are achieved by the space time coding (STC) and spatial demultiplexing methods, respectively [5]. Table 1 summarizes the main characteristics of the MIMO categories [6].

Space time coding technique is a coding method that enables transmission of the replicas of a transmitted signal using two or more transmit antennas to achieve full

K. Arai et al. (Eds.): FICC 2018, AISC 886, pp. 12–25, 2019.
https://doi.org/10.1007/978-3-030-03402-3_2

Table 1. The MIMO categories

MIMO technique	Purpose	Approach	Method
Spatial diversity	Improve reliability	Combat multipath	Space time coding
Spatial multiplexing	Increase capacity	Exploit multipath	Spatial demultiplexing

diversity [7]. This implies increasing the transmission reliability of the MIMO system with lower computational complexity. This is achieved without any bandwidth expansion under assumption that the channel state information is available at the receiver (CSIR) only [8].

The key purpose of the spatial multiplexing is to increase the data transmission rate without requiring the bandwidth expansion by transmitting multiple independent data streams over multipath channels. This technique exploits the multipath propagation between the transmitter and the receiver [1, 9]. In general, the basic concept of a spatial multiplexing is layered space time (LST) coding, where a layered indicates a data stream from a single transmit antenna.

In the following sections, the performance of Multi – Group Space Time Codes MGSTC-MIMO communications system is investigated using either the standard decoding algorithm or the proposed parallel decoding algorithm. The results show that the new decoding algorithm enhances the performance of this system.

2 MIMO – MGSTCS Method

Multi – Group Space Time Codes (MGSTCs) is a MIMO communications method that achieves both spatial diversity and spatial multiplexing simultaneously. In 1999, the first example of MGSTCs was published by Tarokh et al. [10].

2.1 Encoder of a MIMO – MGSTCs

Under assumption of CSIR only, the number of the data streams that can be transmitted simultaneously is equal to the smallest number between the number of transmit and receive antennas. However; the number of the receive antennas should be equal or greater than the number of the transmit antennas to enable the receiver to detect all the independent data streams that are transmitted [11]. Therefore, the maximum number of the data streams that can be transmitted is

$$N_{stream} = N_t \tag{1}$$

These N_{stream} data streams split into q groups. Each group consists of $B_1, B_2, \ldots, B_q$ bits, respectively as shown in Fig. 1.

Then each of the q group bits is mapped into the specific group of the transmit antennas by space time coding, which is called a component code. Each component code can differ from one another within the same encoder and can use either STBCs or

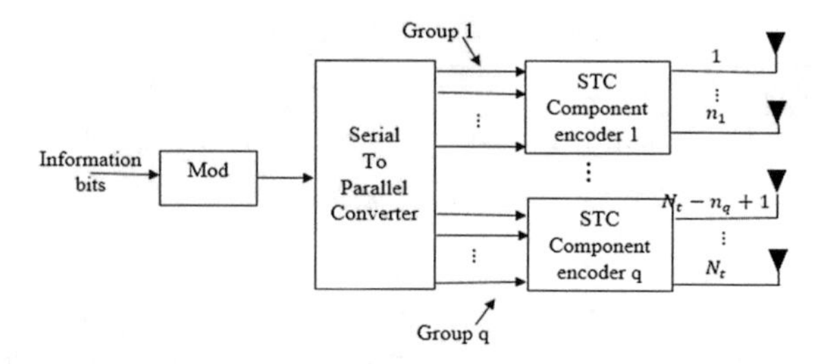

Fig. 1. Multi-group space time coding MGSTCs encoder.

STTCs. Let the output of the component codes be shown as $S_{C_1}, S_{C_2}, \ldots, S_{C_q}$, respectively [6]. Therefore, the transmit signal matrix S can be denoted as

$$S = \begin{bmatrix} S_{C_1} \\ S_{C_2} \\ \vdots \\ S_{C_q} \end{bmatrix} \tag{2}$$

Where, the first component code (S_{C_1}) transmitted signal is

$$S_{C_1} \triangleq \begin{bmatrix} s_1(1) & \ldots & s_1(p) \\ \vdots & \ddots & \vdots \\ s_{n_1}(1) & \ldots & s_{n_1}(p) \end{bmatrix} \tag{3}$$

2.2 Performance Analysis of a MIMO – MGSTCs Decoder

At the receiver, each received component code has interferences from all other component codes. Thus, there is an inter-group interference [12]. Inter-group interference can be suppressed through decoding processes. MGSTCs encoding and decoding processes are performed under assumption of channel state information at the receiver (CSIR) only [13]. For simplicity, the channel matrix $H(N_r \times N_t)$ is decomposed into components as following equation shows:

$$H = \begin{bmatrix} H_{C_1} & H_{C_2} & \ldots & H_{C_q} \end{bmatrix} \tag{4}$$

Where $H_{C_1}(N_r \times n_1)$ refers to the channel component that is associated with transmitted component code S_{C_1}. When this channel component is removed, then the received interference from the component code S_{C_1} is truncated and the resulting channel matrix $H_{C-C_1}(N_r \times (N_t - n_1))$ is

$$H_{C-C_1} = \begin{bmatrix} H_{C_2} & \ldots & H_{C_q} \end{bmatrix} \tag{5}$$

To decode the first space time component code (S_{C_1}) that is transmitted over the n_1 transmit antennas, and to suppress all the remaining signals which constitute interference and that are associated with H_{C-C_1}, it is necessary that the number of receive antennas satisfies $N_r \geq N_t - n_1 + 1$. This interference suppression can be achieved through pre-multiplying the received matrix R. By the matrix Θ_{C-C_1} as following equation shows:

$$\widetilde{R}_{C_1} \triangleq \Theta_{C-C_1} R \tag{6}$$

Where $\widetilde{R}_{C_1}$ is the estimated received signal after interference suppression and Θ_{C-C_1} consists of the set of null space vector of H_{C-C_1}. Each row of the Θ_{C-C_1} matrix is an orthonormal basis vector of the null space of H_{C-C_1}, as the following equation shows

$$\Theta_{C-C_1} H_{C-C_1} = 0_{[(N_r - N_t + n_1)\times(N_t - n_1)]} \tag{7}$$

Assume Rayleigh distributions, thus the receive matrix is $R = \sqrt{\rho} HS + Z$ and using the previous equations, the $\widetilde{R}_{C_1}$ can be written as

$$\widetilde{R}_{C_1} = \sqrt{\rho}\,\Theta_{C-C_1} H_{C_1} S_{C_1} + \Theta_{C-C_1} Z \tag{8}$$

Clearly, (8) shows that the energy of $\widetilde{R}_{C_1}$ is only from S_{C_1} component code. This demonstrates the interferences from all other component codes are vanished. Therefore, the previous equation can be rewritten as follows:

$$\widetilde{\mathrm{R}}_{\mathrm{C}_1} = \sqrt{\rho}\,\mathrm{H}_1^{(\mathrm{eff})} \mathrm{S}_{\mathrm{C}_1} + \widetilde{\mathrm{Z}}_{\mathrm{C-C}_1} \tag{9}$$

Where $H_1^{(eff)}$ refers to the effective channel matrix for decoding the first component code and it represents the following expression

$$H_1^{(eff)} = \Theta_{C-C_1} H_{C_1} \tag{10}$$

Where $\widetilde{Z}_{C-C_1}$ refers to the noise part that is associated with transmitting the first component code which can be denoted as

$$\widetilde{\mathrm{Z}}_{\mathrm{C-C}_1} = \Theta_{\mathrm{C-C}_1} \mathrm{Z} \tag{11}$$

Hence, the S_{C_1} component code can be decoded by applying the maximum likelihood decoding to the (9), where $\widehat{S}_{C_1}$ refers to the estimate of S_{C_1}, and is computed as the following equation shows [13]:

$$\widehat{S}_{C_1} = \arg\min_{\{S_{C_1}\}} \left\{ \left\| \widetilde{R}_{C_1} - \sqrt{\rho} H_1^{(eff)} S_{C_1} \right\|_F^2 \right\} \tag{12}$$

The decoding each of the remaining component codes is similar to that which is used to decode the first component code, of course after subtracting the previous decoded component codes from the overall received signal. For example, to decode S_{C_2} component code, the received signal is the overall received signal after canceling the interference that is associated with the S_{C_1} component code and exploiting the result from the previous process as following equation shows:

$$R_{C_1} \triangleq R - \sqrt{\rho} H_{C_1} \widehat{S}_{C_1} \tag{13}$$

Then applying the pre-multiplying operation, we define

$$\widetilde{R}_{C_2} \triangleq \Theta_{C-C_{1,2}} R_{C_1} \tag{14}$$

The remaining strategies are similar to the decoding of previous component code. This implies that the receiver must detect and decode the component codes in serial way. One should note that in using serial detection method, if an error occurs in detecting and decoding a lower order component code, such error will propagate to higher order component code in the detection process, which cannot be ignored [6]. However, this technique increases the data rate transmission and improves the diversity order gradually.

2.3 Summary of MGSTCs Method

As previously stated, the MGSTCs method decodes all received component codes in serial way. This implies that errors propagate in this method of detection. For example, if the received second component code detected and decoded with an error, then such error will propagate to all the remaining higher order received component codes. Figure 2 summarizes MGSTCs decoding process.

3 Proposed Decoding Method

A detection and decoding technique called parallel decoding algorithm was developed and applied to MIMO-MGSTCs system and the results were compared with the standard serial decoding algorithm of the same system. The proposed parallel decoding algorithm prevents error propagation. In addition, it increases the speed of signal processing at the receiver by extracting received modulated symbols independently and simultaneously [14].

3.1 Performance Analysis of MIMO – MGSTCs Using the Proposed New Decoding Technique

Under the assumption of channel state information at the receiver (CSIR) only, we will follow the same procedure for transmitting component codes at the transmitter that was used in standard MIMO – MGSTCs. Recall from (2), the transmit signal matrix is

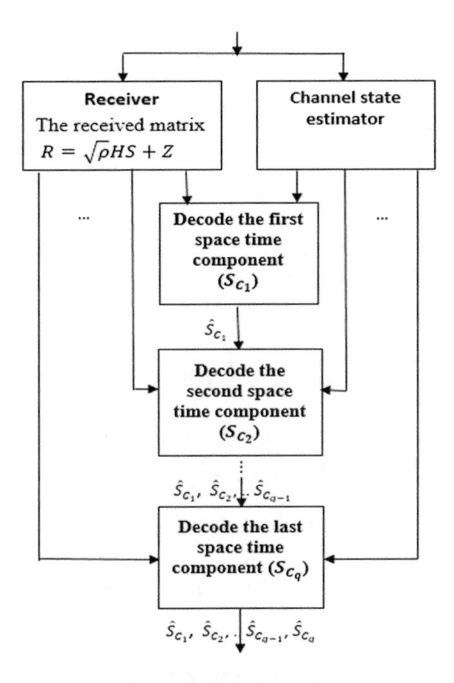

Fig. 2. Multi-group space time coding MGSTCs for standard serial detection and decoding.

$$S \triangleq \begin{bmatrix} s_1(1) & \ldots & s_1(p) \\ \vdots & \ddots & \vdots \\ s_{n_1}(1) & \ldots & s_{n_1}(p) \\ \vdots & \ddots & \vdots \\ s_{n_1+n_2}(1) & \ldots & s_{n_1+n_2}(p) \\ \vdots & \ddots & \vdots \\ s_{N_t}(1) & \ldots & s_{N_t}(p) \end{bmatrix} = \begin{bmatrix} S_{C_1} \\ S_{C_2} \\ \vdots \\ S_{C_q} \end{bmatrix} \tag{15}$$

Like before, the channel matrix $H(N_r \times N_t)$ is decomposed into components as (4) shows. However, the proposed new decoding algorithm allows that the detection and decoding of component codes to be done in a parallel manner instead of using a serial method. This implies that the receiver will be able to detect and decode all received component codes independently and simultaneously. We should note that using the proposed model, if an error occurs in the detecting and decoding of one component code, such error will not affect other component codes, since the component codes are processed simultaneously and independently from one another. Also, with the new technique the speed of signal processing at the receiver will highly be increased.

The decoding of the first component code is the same as it is for the standard serial decoding of MIMO - MGSTC. The estimated received signal after inter-group interference suppression is shown in (9). During same time of detecting and decoding the first component code, the receiver can detect and decode all the other component codes through the following new algorithm. Detection of the second component code can be achieved through suppressing the interference that is associated with the first component code. This is done by pre-multiplying the received signal as the following equation shows:

$$\widetilde{RR}_{C_2} \triangleq \Theta_{C_1} R \tag{16}$$

Where Θ_{C_1} consists of the set of null space vector of H_{C_1}. That is, each row of Θ_{C_1} matrix is an orthonormal vector of the null space of H_{C_1} as the following equation shows

$$\Theta_{C_1} H_{C_1} = 0_{[(N_r - n_1)\times(n_1)]} \tag{17}$$

Where the dimension of H_{C_1} is $[N_r \times n_1]$, and the dimension of Θ_{C_1} can be obtained by using the rank plus nullity theorem as follows:

$$\dim[\Theta_{C_1}] + rank[H_{C_1}] = N_r \tag{18}$$

Where the $rank[H_{C_1}] = n_1$, and from (18) the dimension of the null space matrix of H_{C_1} is

$$\dim[\Theta_{C_1}] \geq N_r - n_1 \tag{19}$$

Therefore, Θ_{C_1} is dimensioned $[(N_r - n_1) \times N_r]$, and its rows are orthonormal, this implies (20).

$$\Theta_{C_1} \Theta_{C_1}^H = I_{N_r - n_1} \tag{20}$$

Assuming Rayleigh distribution and flat fading environment, the received matrix can be expressed as $R = \sqrt{\rho} HS + Z$, and the resulting signal from (16) is

$$\widetilde{RR}_{C_2} = \sqrt{\rho}\left[\, \Theta_{C_1} H_{C_2} \quad \Theta_{C_1} H_{C - C_{1,2}} \,\right] \begin{bmatrix} S_{C_2} \\ S_{C-C_{1,2}} \end{bmatrix} + \Theta_{C_1} Z \tag{21}$$

Equation (21) contains inter-group interference that is associated with transmitting component codes, $(S_{C_3} \ldots S_{C_q})$. These interferences can be suppressed by pre-multiplying the received matrix for the second time to obtain the estimated received matrix $\widetilde{R}_{C_2}$ as shown in (22), and to decode the second component code S_{C_2} as follows:

$$\widetilde{R}_{C_2} = N\left[\Theta_{C_1} H_{C - C_{1,2}}\right] \widetilde{RR}_{C_2} \tag{22}$$

Where $N\left[\Theta_{C_1} H_{C-C_{1,2}}\right]$ refers to the null space of $\Theta_{C_1} H_{C-C_{1,2}}$, and it consists of the set of null space vector of $\Theta_{C_1} H_{C-C_{1,2}}$. The dimension of $\left[\Theta_{C_1} H_{C-C_{1,2}}\right]$ is $[(N_r - n_1) \times (N_t - n_1 - n_2)]$. Therefore, the dimension of $N\left[\Theta_{C_1} H_{C-C_{1,2}}\right]$ is as follows:

$$\dim\left[N\left[\Theta_{C_1}H_{C-C_{1,2}}\right]\right] \geq N_r - N_t + n_2 \tag{23}$$

Thus, $N\left[\Theta_{C_1}H_{C-C_{1,2}}\right]$ is dimensioned $[(N_r - N_t + n_2) \times (N_r - n_1)]$.
Substituting (21) in (22), the estimated received signal $\widetilde{R}_{C_2}$ is

$$\widetilde{R}_{C_2} = \sqrt{\rho}H_2^{(eff)}S_{C_2} + Z_2 \tag{24}$$

Where $H_2^{(eff)}$ refers to the effective channel matrix that is associated with decoding S_{C_2} and it represents the following expression:

$$H_2^{(eff)} = A_{C_2}H_{C_2} \tag{25}$$

Where A_{C_2} refers to the overall per-multiplying matrix, which is denoted by

$$A_{C_2[(N_r-N_t+n_2)\times(N_r)]} = N\left[\Theta_{C_1}H_{C-C_{1,2}}\right]\Theta_{C_1} \tag{26}$$

The noise term that is associated with receiving S_{C_2} is $Z_2 = A_{C_2}Z$. Therefore, by combining (16), (22), and (26), the compact form of the overall pre-multiplying process can be expressed as shown in (27).

$$\widetilde{R}_{C_2} = A_{C_2}R \tag{27}$$

Equation (24) shows that the energy of $\widetilde{R}_{C_2}$ is only from S_{C_2} component code. The received S_{C_2} component code can be estimated by applying the maximum likelihood decoding to the (24), where $\widehat{S}_{C_2}$ refers to the estimated value of S_{C_2} in (28).

$$\widehat{S}_{C_2} = \arg\min_{\{S_{C_2}\}}\left\{\left\|\widetilde{R}_{C_2} - \sqrt{\rho}H_2^{(eff)}S_{C_2}\right\|_F^2\right\} \tag{28}$$

The remaining decoding algorithm is similar to decoding of the second component code. In general, the decoding of the ith component code S_{C_i}, with $2 \leq i \leq q - 1$ will be achieved through the following pre-multiplying:

$$\widetilde{R}_{C_i} = A_{C_i}R \tag{29}$$

Where $\widetilde{R}_{C_i}$ is to the estimated received matrix after the inter-group interference suppression, and A_{C_i} refers to the overall per-multiplying matrix, which can be denoted by

$$A_{C_i[(N_r-N_t+n_i)\times(N_r)]} = N\left[\Theta_{C_1+C_2+\ldots+C_{i-1}}H_{C-C_{1,2,\ldots,i}}\right]\Theta_{C_1+C_2+\ldots+C_{i-1}} \tag{30}$$

Where $\Theta_{C_1+C_2+\ldots+C_{i-1}}$ refers to the null space of $[H_{C_1+C_2+\ldots+C_{i-1}}]$, and where the expression of the channel matrix is as follows:

$$H_{C_1+C_2+\ldots+C_{i-1}} = [\mathrm{H}_{C_1} \quad \mathrm{H}_{C_2} \qquad \mathrm{H}_{C_{i-1}}]_{\left[(\mathrm{N_r})\times\left(\sum_{j=1}^{i-1}\mathrm{n_j}\right)\right]} \tag{31}$$

Also, $N\left[\Theta_{C_1+C_2+\ldots+C_{i-1}} H_{C-C_{1,2,\ldots,i}}\right]$ refers to the null space of $\left[\Theta_{C_1+C_2+\ldots+C_{i-1}} H_{C-C_{1,2,\ldots,i}}\right]$ with the dimension $\left[\left(N_r-\left(\sum_{j=1}^{i-1} n_j\right)\right) \times \left(N_t-\left(\sum_{j=1}^{i} n_j\right)\right)\right]$. Thus, the dimension of $N\left[\Theta_{C_1+C_2+\ldots+C_{i-1}} H_{C-C_{1,2,\ldots,i}}\right]$ is

$$\dim\left[N\left[\Theta_{C_1+C_2+\ldots+C_{i-1}} H_{C-C_{1,2,\ldots,i}}\right]\right] \geq N_r - N_t + n_i \tag{32}$$

Therefore, $N\left[\Theta_{C_1+C_2+\ldots+C_{i-1}} H_{C-C_{1,2,\ldots,i}}\right]$ is dimensioned as

$$\left[(N_r - N_t + n_i) \times \left(N_r - \left(\sum_{j=1}^{i-1} n_j\right)\right)\right] \tag{33}$$

The estimated received matrix $\widetilde{R}_{C_i}$ can be expressed as

$$\widetilde{R}_{C_i} = \sqrt{\rho} H_i^{(eff)} S_{C_i} + Z_i \tag{34}$$

Where $H_i^{(eff)}$ refers to the effective channel matrix that is associated with decoding S_{C_i}, which is given by following expression.

$$H_i^{(eff)} = A_{C_i} H_{C_i} \tag{35}$$

The noise term that is associated with receiving the component code S_{C_i} is $Z_i = A_{C_i} Z$.

At the receiver, the S_{C_i} component code can be decoded by applying the maximum likelihood decoding algorithm to (34). The result is shown in (36), where $\widehat{S}_{C_i}$ refers to the estimate of S_{C_i}.

$$\widehat{S}_{C_i} = \arg\min_{\{S_{C_i}\}} \left\{ \left\| \widetilde{R}_{C_i} - \sqrt{\rho} H_i^{(eff)} S_{C_i} \right\|_F^2 \right\} \tag{36}$$

Decoding S_{C_q} component code is obtained using similar approach. The estimated received matrix $\widetilde{R}_{C_q}$ can be obtained from the following equation:

$$\widetilde{R}_{C_q} = A_{C_q} R \tag{37}$$

Where A_{C_q} refers to the overall pre-multiplying matrix, which represent the following:

$$A_{C_q\left[(N_r-N_t+n_q)\times(N_r)\right]} = \Theta_{C_1+C_2+\ldots+C_{q-1}} \tag{38}$$

Where $\Theta_{C_1+C_2+\ldots+C_{q-1}}$ refers to the null space of $\left[H_{C_1+C_2+\ldots+C_{q-1}}\right]$, and where this last matrix is

$$\left[H_{C_1} \quad H_{C_2} \quad \ldots \quad H_{C_{q-1}}\right]_{\left[(N_r)\times\left(\sum_{j=1}^{q-1} n_j\right)\right]=\left[(N_r)\times(N_t-n_q)\right]} \tag{39}$$

Thus, the estimated received matrix $\widetilde{R}_{C_q}$ can be expressed as

$$\widetilde{R}_{C_q} = \sqrt{\rho} H_q^{(eff)} S_{C_q} + Z_q \tag{40}$$

Where $H_q^{(eff)}$ refers to the effective channel matrix that is associated with decoding S_{C_q} and represents the following expression.

$$H_q^{(eff)} = A_{C_q} H_{C_q} \tag{41}$$

The noise term that is associated with receiving S_{C_q} is $Z_q = A_{C_q} Z$. At the receiver, the S_{C_q} component code can be decoded by applying the maximum likelihood decoding to (40). The result is shown in (42), where $\widehat{S}_{C_q}$ refers to the estimate of S_{C_q}.

$$\widehat{S}_{C_q} = \arg\min_{\{S_{C_q}\}} \left\{ \left\| \widetilde{R}_{C_q} - \sqrt{\rho} H_q^{(eff)} S_{C_q} \right\|_F^2 \right\} \tag{42}$$

As shown in (6), (27), (29) and (37), the proposed receiver can detect and decode all received component codes simultaneously and independently.

3.2 Summary of the New Decoding Method

Assuming the receiver has the channel state information, the new decoding method enables the receiver to detect and decode all received component codes independently and simultaneously as shown in Fig. 3. It is important to note that this algorithm has independent bit error probability for different component codes. In addition, the new algorithm reduces the overall computational time for detecting and decoding component codes. That is because of parallel processing, the new technique increases the speed of the overall signal processing and thus the data rate at the receiver.

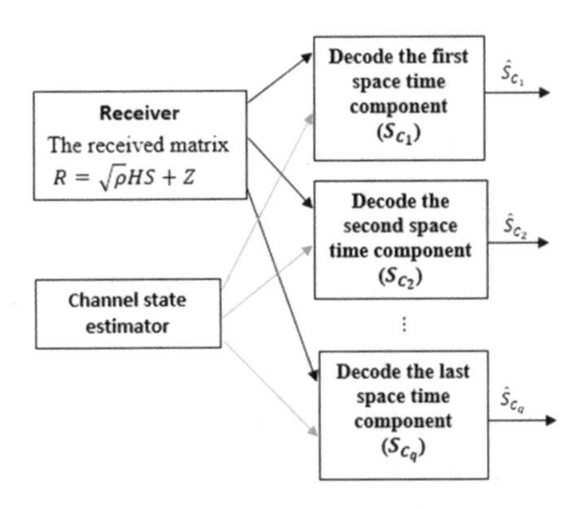

Fig. 3. Multi-group space time coding MGSTCs parallel decoding technique.

4 Simulation Results

This section presents performance analysis of systems that use the parallel decoding algorithm at the receiver and MIMO-MGSTCs method at the transmitter, and compares the results with the corresponding performance of similar systems that use serial decoding algorithm at the receiver. The minimum number of the transmit antennas that is required to implement MGSTCs method is four. Thus, this section focuses on $4 \times N_r$ MIMO configurations. At the transmitter, a 2×2 MIMO using Alamouti code will be used for each component code as shown in Fig. 4. This implies that the transmitter transmits four modulated symbols at two symbol periods.

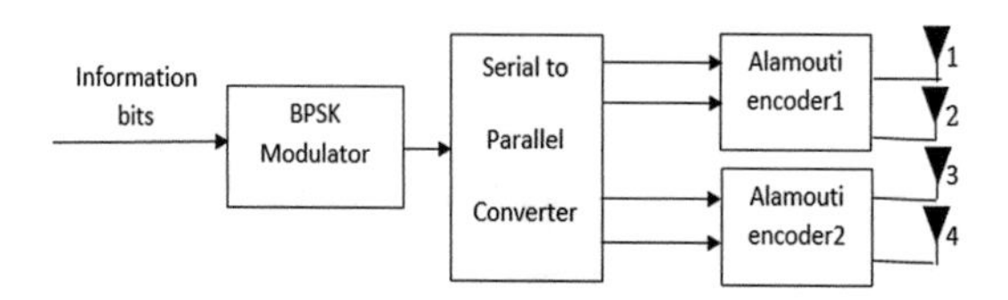

Fig. 4. MIMO – MGSTCs transmitter using $N_t = 4$.

Under assumption of $N_r = 4$, Fig. 5 shows the performance results in terms of the bit error probability [15, 16] versus the E_b/N_0 in dB for BPSK modulation in Rayleigh fading channels for both serial and parallel decoding algorithms [17, 18].

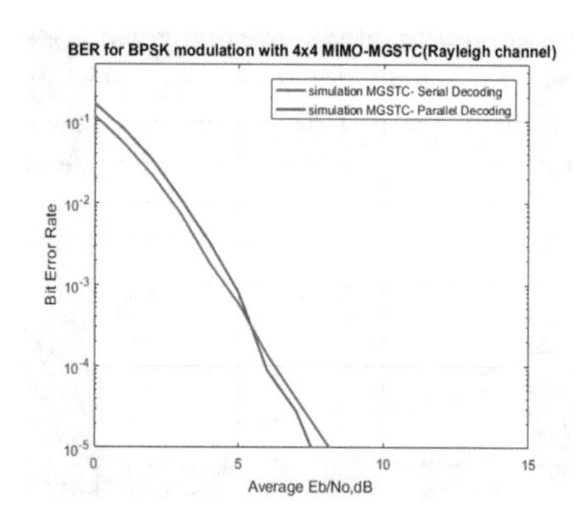

Fig. 5. The bit error probability of a 4×4 MIMO – MGSTCs method for both serial and parallel decoding techniques.

The results in this figure are based on computer simulations in which 1000000 information bits are used. This means that 250000 iterations are used.

As Fig. 5 shows, there is a small improvement in the bit error probability for $(E_b/N_0) \geq 6dB$ when using parallel decoding method compared to standard serial decoding algorithm. In addition, the new technique provides two more advantages.

Precisely, the parallel decoding technique prevents the error propagation and increases the speed of signal detection and decoding at the receiver due to decoding all received symbols independently and simultaneously. Figure 6 shows the bit error probability performances for a 4×5 MIMO configuration using both serial and parallel decoding for the same MGSTCs transmitter shown in Fig. 4. The results indicate that both systems have almost the same bit error probability.

According to the characteristic of MGSTCs method, as $N_r \geq N_t - n_1 + 1$, the number of the receive antennas can be $N_r = 3$. Figure 7 shows the BER of the parallel and serial decoding techniques using BPSK modulation in a 4×3 MIMO – MGSTC system. The results of Figs. 5, 6 and 7 shows that the parallel decoding method performs better than the serial decoding method when the signal to noise ratio is larger than 10 dB.

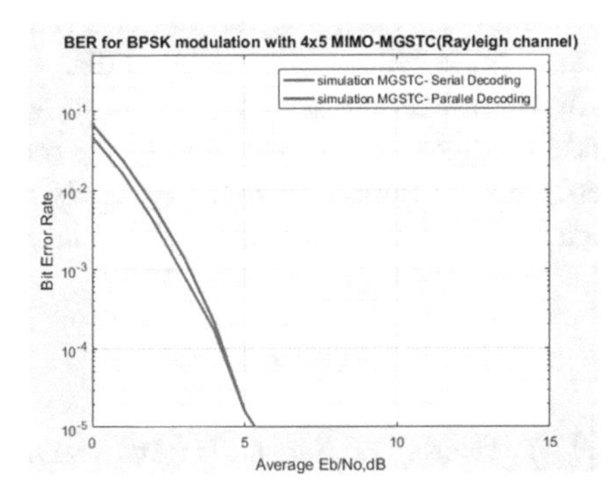

Fig. 6. The bit error probability of a 4×5 MIMO – MGSTCs method for both serial and parallel decoding techniques.

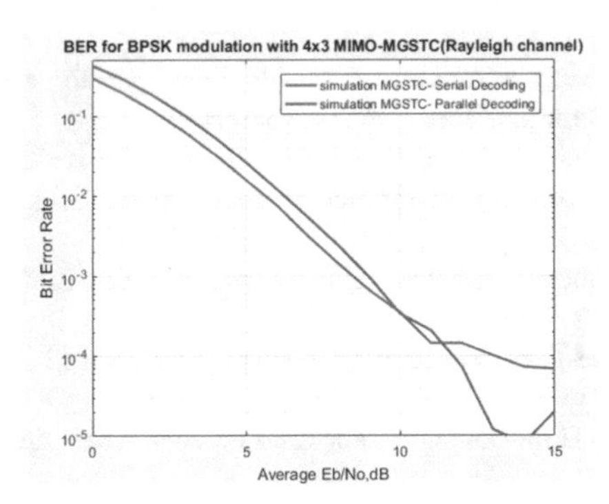

Fig. 7. The bit error probability of a 4×3 MIMO – MGSTCs method for both serial and parallel decoding techniques.

We should note that all $4 \times N_r$ MIMO – MGSTCs configurations achieve the same channel capacity due to transmitting four modulated symbols at two symbol periods.

5 Conclusion

In this paper, we focused on the study and investigation of MIMO - MGSTCs technique, and finding how to enhance the overall performance of this system. A new detection and decoding technique called parallel decoding algorithm was developed and applied to this system. The performance of the proposed algorithm compared with standard serial decoding technique. The performance results show the improvements provided by the new parallel decoding are as follows. It was shown that the new technique reduces the overall signal detection and decoding time due to parallel processing method. Therefore, it increases the speed of signal processing at the receiver. In addition, the new technique prevents error propagation by extracting all received modulated symbols independently. Also, the results show using BPSK modulation, when compared with the serial decoding methods, the proposed parallel decoding technique provides almost similar BER for signal to noise ratios smaller than 10 dB, but better BER for signal to noise ratios higher than 10 dB.

In the future, we will generalize the proposed model to apply it to other type of MIMO technologies, and investigate the performance of the proposed algorithm when symbol decoding is performed in parallel manner by applying a variety of MIMO detection techniques such as ML, Sphere Decoding (SD) and others.

References

1. Toshio, M., Tomoyuki, O., Hitoshi, Y., Narumi, U.: The overview of the 4th generation mobile communication system, pp. 1551–1555. IEEE (2005)
2. Seshadri, N., Sundberg, C.E.W., Weerackody, V.: Advanced technologies for modulation, error correction, channel equalization, and diversity. ATT Techn. J. (1993)
3. Wittneben, A.: Base station modulation diversity for digital simulcast. In: 41st IEEE Vehicular Technology Conference, Gateway to the Future Technology in Motion, pp. 848–853, May 1991
4. Alamouti, S.M.: A simple transmit diversity technique for wireless communications. IEEE J. Sel. Areas Commun. **16**(8), 1451–1458 (1998)
5. Driessen, P.F., Foschini, G.J.: On the capacity formula for multiple input-multiple output wireless channels: a geometric interpretation. In: 1999 IEEE International Conference on Communications, ICC 1999, vol. 3, pp. 1603–1607 (1999)
6. Hampton, J.R.: Introduction to MIMO Communications. Cambridge University Press, New York (2014)
7. Gregory, D.D.: Space-Time Wireless Channels. Prentice Hall PTR, NJ 07458: Pearson Education. Inc. (2003)
8. Jafarkhani, H.: Space-Time Coding: Theory and Practice. Cambridge University Press, New York (2010)
9. Rappaport, T.S.: Wireless Communications: Principles and Practice, 2nd edn. Prentice Hall, Upper Saddle River (2002)
10. Tarokh, V., Naguib, A., Seshadri, N., Calderbank, A.R.: Combined array processing and space-time coding. IEEE Trans. Inf. Theor. **45**(4), 1121–1128 (1999)
11. Yong, C.S., Jaekwon, K., Won, Y.Y., Chung, K.G.: MIMO-OFDM Wireless Communications with MATLAB. John Wiley & Sons (Asia) Pte., Singapore (2010)

12. Karim, H.: Introduction to Cellular Wireless Communication. Montezuma Publishing, San Diego (2015)
13. Karim, H.: Information Theory and Coding. Montezuma Publishing, San Diego (2014)
14. Wurod, Q.M.: Performance analysis of a new decoding technique for MIMO and MIMO – OFDM communication system, MS thesis, Fall 2016 California State University, Fullerton
15. Proakis, J., Salehi, M.: Digital Communications, 5th edn. McGraw Hill Science/Engineering/Math (2007)
16. Craig, J.W.: A new, simple and exact result for calculating the probability of error for two-dimensional signal constellations. In: Military Communications Conference, MILCOM 1991, Conference Record, Military Communications in a Changing World, vol. 2, pp. 571–575. IEEE, November 1991
17. Sklar, B.: Digital Communications Fundamentals and Applications, 2nd edn. Prentice Hall PTR, Upper Saddle River (2001)
18. Lathi, B.P., Ding, Z.: Modern Digital and Analog Communication Systems, 4th edn. Oxford University Press Inc., Oxford (2009)

A Novel Distributed Multi-access Platform for Broadband Triple-Play Service Delivery

Azrin Bin Aris(✉) and Mohd Kamil Abd Rahman

Faculty of Applied Science, Universiti Teknologi MARA, Shah Alam, Malaysia
azrin.aris@tm.com.my, drkamil@salam.uitm.edu.my

Abstract. Over the past years, Triple-play service through (Digital Subscribers Line (DSL) has gained attention. From a Central-Office base, DSL Access Multiplexer (DSLAM) has evolved into a remote-base that supports triple-play and its development has introduced new problems and challenges. This paper will discuss the problems encountered by ISPs as well as proposing a novel approach in solving these problems and challenges and providing a platform that can support multiple access technologies. A simulation using DSL access technology to analyse the performance of the proposed platform and a prototype has been built as a proof-of-concept.

Keywords: ATM · Digital Subscribers Line · FTTH · xDSL · Access network

1 Introduction

In the past few years, Digital Subscribers Line (DSL) services have expanded and more DSL Access Multiplexer (DSLAMs) have been installed to provide DSL services. Traditional DSLAMs are meant to be installed at the central office (telecom's building that houses the telephone exchange circuits) to provide DSL services [1]. However, there is a limitation to the technology; it only has a range of 5 km. Hence, Remote DSLAM architecture has been introduced. It is a fully featured DSLAM in a smaller box installed at the pedestal, outside the plant or DLC (Digital Loop Carrier) Cabinet [2]. Although the remote DSLAM has better coverage, it also instigates new problems.

A significant problem in implementing Remote DSLAMs is its high cost. Firstly, the deployment cost is higher as compared to Central base DSLAMs. This is due to the extra enclosure for outdoor protection. Remote DSLAM is also liable to vandalism and theft. The remote DSLAM unit itself is expensive due to the switching unit is built in the remote DSLAM box; thus, increasing the ISPs capital expenditure (CAPEX). Moreover, high operational expenditure (OPEX) is incurred as high-power consumption is needed to perform high speed switching. On top of that, the operation cost is also higher as truck roll is now part of the maintenance cost.

Excessive heat that is generated by a high frequency component in an enclosure is another issue especially in tropical countries. Without proper dissipation, the generated heat will cause the temperature of the system in the enclosure to increase which will eventually cause system failure.

K. Arai et al. (Eds.): FICC 2018, AISC 886, pp. 26–43, 2019.
https://doi.org/10.1007/978-3-030-03402-3_3

Thus, a new DSLAM architecture is necessary as an alternative to the existing setting should be deliberated to solve these problems. The DSLAM should be positioned in a controlled and secured environment and this will inevitably add up to the cost of deployment. To ensure the broadband service deployment is cheap enough to be mass deployed, a new DSLAM architecture is necessary to meet the increasing demand for broadband services.

Remote DSLAM and current CO-base DSLAM use ATM infrastructure between the CPE and the DSLAM. Using DMT (Discrete Multi-Tone) via the copper pair, data packets are converted into ATM cells at the CPE and transferred to the DSLAM [3]. The cells are then switched at the ATM network before reconverted to IP.

As Ethernet infrastructure is cheaper compared to ATM, IP-DSLAM is introduced to take advantage of it. However, with the usage of IP network, Quality of Service (QoS) has become an issue. Current IP-QoS is user based QoS [4]. What if different quality of service for different services is required by a user?

A new way to implement QoS and a new distributed architecture will be introduced in this paper in order to address the issues that have been mentioned. The proposal will also enable the integration of other access technology with the system, therefore making it a multi access platform providing the subscribers broadband services; thus, the introduction of the name Distributed Multi-Access Platform (D-MAP).

In this paper, an ideal architecture is deployed since there is no chipset/hardware in the market that is available for multi QoS control. However, to emulate the proposal, a prototype that uses ATM link and the existing chipset has been developed. The development of a multi-QoS chipset will potentially drive the next-generation of QoS controller.

2 Problem Statements

Telecommunication companies all over the world are continuously striving to enhance their infrastructure to provide better broadband services to the expectation of their end user. However, the demand for a higher bandwidth increase comes together with the complexity of the devices/boxes installed at the corner, which in the end will inevitably increase the implementation cost.

The most popular broadband implementation is the xDSL-based infrastructure as it uses the existing copper. This will ensure that the copper investment is not wasted and at the same time able to decrease the broadband deployment cost significantly. However, as xDSL infrastructure becomes more complicated (due to the requirement to deliver broadband at a higher bandwidth), the implementation cost is no longer cost-effective.

DSL Access Multiplexer has evolved from a central-office type to remotely distributed type [5]. This distributed architecture requires a full-featured DSLAM to be installed at the corner. This scaled-down version of DSLAM has all the features necessary of a Central DSLAM but with smaller number of ports which simply means that the cost of a remote DSLAM is almost as high as a Central DSLAM with the same number of ports.

In a typical remote DSLAM architecture, there will be at least the following components [6]:

(1) The Line Card: a single line card typically has from 16-32 DSL ports.
(2) The WAN Card (with all the switching components).
(3) The back-plane in which the entire card is interconnected.
(4) The Chassis where the entire component is housed.

From our study, the WAN card is the most expensive compared to the others. Any cost reduction in developing the WAN card will significantly reduce the cost of implementation.

The current architectural design of remote DSLAM suffers from another issue – heat. Switching unit, optical devices are some of the high frequency components in the DSLAM that generates heat. The ambient temperature within the remote DSLAM increases due to the heat generated from these components, which eventually leads to excessive heating. The functionality of the remote DSLAM might be an issue during the summer or in countries with tropical climate, as it might be unable to function properly if the temperature is higher than operating temperature of the remote DSLAM.

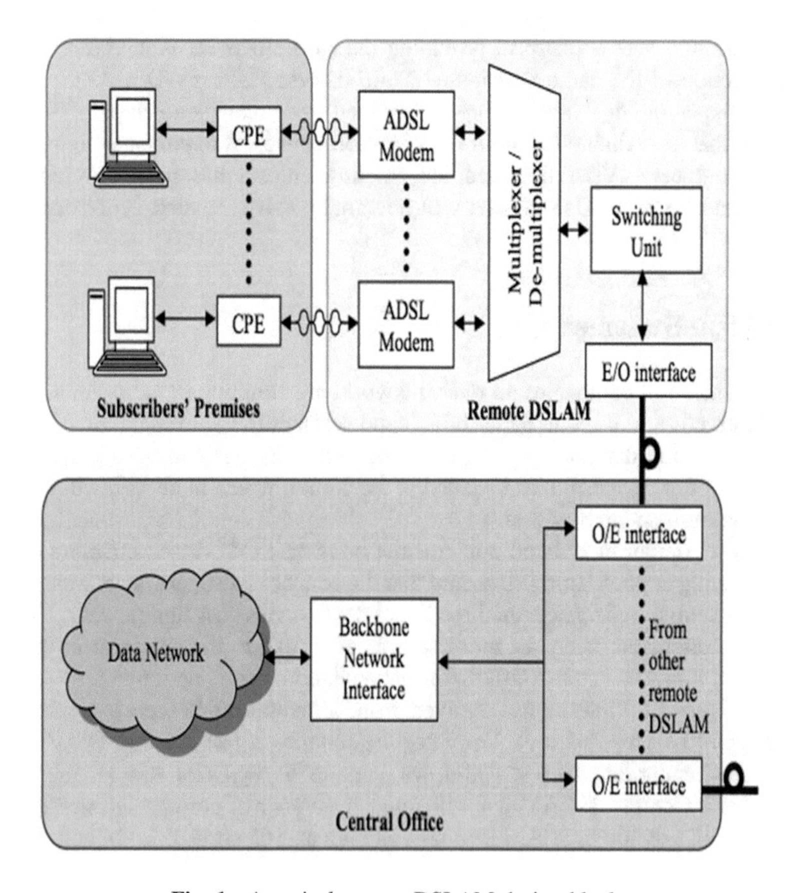

Fig. 1. A typical remote DSLAM design block.

Another major issue is the QoS when more users use the infrastructure. The current architecture implements QoS at the network core. However, traffic bottleneck has little effect at the core network because high capacity bandwidth equipment is installed everywhere.

Presently, efforts to reduce traffic bottleneck are mostly concentrated at the core network. A new location for QoS implementation has to be determined as bottleneck at the core network is no longer a significant problem. Looking at the existing remote DSLAM, specifically the network deployment topology (Fig. 1), it is apparent that traffic bottleneck will occur between the central office DSLAM aggregator and the remote DSLAM box. Hence, the implementation of QoS for packets going into the congested segment – i.e. packet going from Central to CPE and packet from CPE to Central, will resolve the issue.

Existing user-based QoS is no suitable as new types of services are introduced. The existing QoS is only able to provide a coarse control over the traffic and are meant to guarantee Service Level Agreement (SLA).

However, SLA does not identify service quality such as voice or video. Hence, there is a high probability for degradation of service performance or quality when two or more service in the same SLA section compete for resources as there are no differentiation between services.

3 Assumptions

Following the explanation on the earlier section, the proposed new architecture will be developed based on the subsequent assumptions.

3.1 At the Central Office

Due to high cost and liability to extreme environmental changes, complex network infrastructure should be placed in the Central Office. This is important due to the fact that in the Central Office, it is under controlled environment and the security is guaranteed.

The core bandwidth capacity has tremendously increased due to the evolution of today's core network technology. Hence, it is assumed that the core has virtually limitless bandwidth relative to the edge network.

3.2 At the Remote Installation

Within a certain reach, it is possible to get to 4 to 8 Mbps bit rate between the remote unit in the DLC and the subscribers CPE. Therefore, bottleneck is not a problem between the remote unit and the CPE.

The installation of the remote box will be on the outside the Central Office in which the environment is hostile due to extreme temperature changes, heat, acid rain and subject to theft, vandalism, etc.

Management and dimension constraints will be a challenge in the implementation of the remote box. For management constraint, approval from the local authority will be needed for the infrastructure and the installation of equipment. Dimensional constraint will involve the appearance and colour of the box for public appearance as well as the limitation of space for installation.

It is assumed that the bandwidth of outside plants (Digital Line Carrier – DLC units) is limited despite being optical fibre-ready. Devices with higher bandwidth dispels more heat hence contributes to the increase of temperature of the entire remote unit, which in turn increases the probability of system failure.

3.3 Other Assumptions

As bottleneck is moved to the subscribers' edge and is no longer situated at the core network, QoS implementation at the core is inefficient and management of a large subscriber base would be almost impossible.

As we are not constrained to only DSL technology, other access technology such as BFWA Access Points (Broadband Fixed Wireless Access), WiMAX, BPL (Broadband over Power Line) and 3G/4G micro base stations should be used to reach out to the subscribers.

The proposed architecture will address all three problems mentioned; the cost, heat and QoS and the design is based on the assumptions that were discussed earlier. The flexibility and scalability of the new proposed architecture will not only be able to accommodate the rapid changing of xDSL technology but other access technology as well. This new architecture will be named Distributed Multi Access Platform: D-MAP.

4 Technology Consideration

The major technology consideration involved in this new architecture is to select a suitable transfer mode between different segments of the network infrastructure and the QoS in between.

4.1 Transfer Mode

Transfer mode refers to the techniques used to transmit, switch and multiplex information [7]. It is how the information is actually being packed, sent and received throughout the network. There are basically two major types of transfer mode; circuit switch and packet switch.

Circuit switch is mostly used in telecommunications to carry voice. A point-to-point connection is established during call-setup phase and the information will only use that particular channel to communicate between the two ends. Since the channel is reserved, a good quality of information transfer can be obtained. However, since the bandwidth is only being used during information transfer, under-utilization will occur.

On the other hand, packet switching solves this issue by having the information "chopped" into small pieces call packets. In order for a packet to be sent correctly to its destination, an additional info - the destination address is added to the header of the packet. Packet is sent to its destination in a connectionless manner and there is no resource reservation as compared to circuit switch network. This transfer method is ideal for data communication where the data pattern is bursty in nature. Since the packet is processed and forwarded when it arrives, the reliability of a packet switched network is based on best effort. During high traffic condition, the quality might deteriorate rapidly. This transfer mode is commonly use in TCP/IP network.

Between these two extremities, another transfer mode called virtual circuit switching provides best for the both modes. It breaks the information like the packet switch but transfer the packet in a logical circuit. In this manner, the resource is reserved and efficiency is increase by chopping the data smaller and at a fixed size for deterministic performance. This however creates a larger header to payload ratio as compared to the other two. Having a higher header to payload ratio simply means lower data transfer efficiency. However, the deterministic performance outweighs the low efficiency when it comes to streaming data types. This transfer mode is the fundamental technique used in ATM infrastructure.

4.2 Quality of Service (QoS)

Quality of Service (QoS) refers to resource control mechanisms. QoS is the capability to provide different priority to different users, applications, data flows or to guarantee a certain level of performance to a data flow.

In an IP network, it is typically a best-effort type of infrastructure. All packets are indistinguishable and are given the same priority. There is no resource reservation in any classes of services. To implement QoS in this network, there has to be a mechanism to distinguish the packets and treat them differently according the class of service. Two main QoS mechanisms available for the IP network are the Integrated Service (IntServ) and Differentiated Services (DiffServ) [8]. IntServ is typically implemented at the user-network or network-network interfaces. It is a *flow-based* mechanism where reservation of resources is based on individual flow requirement.

On the other hand, DiffServ is typically implemented by the network where it employs a *class-based* differentiation mechanism. Traffic is shaped according to its type of service rather than individual flow.

ATM network is a connection-oriented packet network. And because of this, it is easier to implement QoS. ATM also implements a fixed length of packet with the size of 53 bytes/octets called an *ATM Cell*. The fixed length cell makes the network more predictable and dependable. The cells are transferred using virtual circuit. QoS is implemented in an ATM network by specifying some performance parameter for each virtual connection (VCs) known as the *Service Parameters*.

5 Technology Selection

Basically, there are three main segments in a typical access network, namely:

(1) Last mile access
(2) Edge network
(3) Core network

D-MAP is a device that connects the two segments that are the edge network and the last mile access. The upstream connection of D-MAP is connected to a node in the edge network elements and not directly to the core network. For that, the core network will not be discussed.

Today's core network is migrating to IP based network as it provides higher bandwidth at a lower cost as compared to other network infrastructure for example ATM core networks. With that in mind, to have a connection based on other than Ethernet will risk the device to be accepted in the market. To make D-MAP a good contender in the market, connection to the edge network is set to be using IP based network.

As for the last mile, there is still a huge network of copper connection from POTS services. And current technology that utilises this copper network to provide broadband is DSL. DSL mainly uses ATM cells to transfer the packet across. To reduce the complexity of the remote box and avoid from having to do Segmentation and Reassembly (SAR), connection to the central unit will be using SONET/SDH type with ATM cell transfer mode. With this, the packets can be sent straight to the central unit with minimal processing.

6 Basic Block Diagram

Distributed architecture is made used in the proposed architecture. The Remote D-MAP unit and the Central D-MAP unit are two significant parts in the architecture. Figure 1 shows a typical remote DSLAM design block.

The switching unit will generate high amount of heat to process all the traffic; with an average of 192 users per remote DSLAM and each subscriber with a maximum rate of 20 Mbps. The key difference is to move the switching unit to the central office instead of placing it in the remote D-MAP.

The proposed D-MAP architecture is shown in Fig. 2. The remote D-MAP unit is now a simple device that only aggregates the traffic from the subscriber, as the switching unit has been moved to the central office. This will cause the temperature of the remote D-MAP unit to significantly decrease to a safe operating temperature. The simplicity in design lowers the CAPEX of service providers as the production of the box is now cheaper.

Another major difference is that the central D-MAP unit aggregates traffic from multiple remote D-MAP units while the normal remote DSLAM is subtended from a CO-based DSLAM. This means that the central D-MAP unit can aggregate traffic from DSL remote D-MAP and any other access technology.

By only having the essential component at the remote along with all complex high-speed components in the central office equipment, simplicity is achieved.

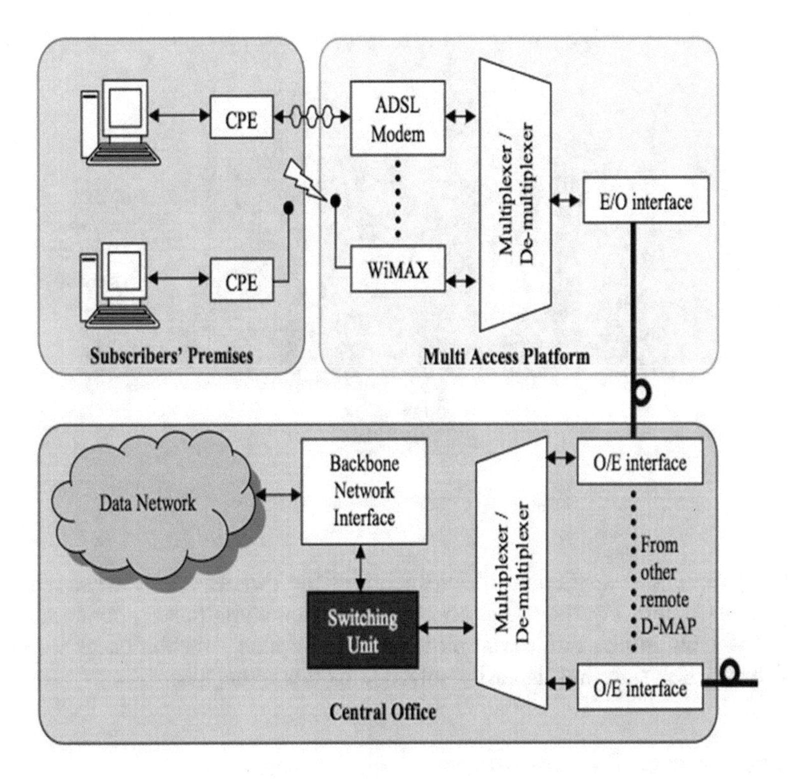

Fig. 2. DMAP architecture.

7 QoS Design Architecture

A new QoS architecture is implemented by D-MAP QoS, which is Distributed Multi-service QoS (DM-QoS). DM-QoS will provide a service based QoS; rather than connection, the QoS will be based on service. Connection based QoS will no longer be efficient as the number of subscribers increase.

With a multi-service QoS, services are now sorted into several categories. Each category has its own transmission rule and streaming services would have different transmission rule from data services. With this difference, fulfilment of different service requirements is possible.

D-MAP is distributed in two different locations whereas normal QoS implementation is located in a single location. The QoS is applied only to the packet going into the fibre-link between the central and remote D-MAP unit.

The block diagram of the DM-QoS architecture is shown in Fig. 3. The two locations of DM-QoS implementation - the remote and central D-MAP - is represented in two parts in the diagram. The two lines connecting the locations represent the subscriber's perspective of the upstream and downstream connection.

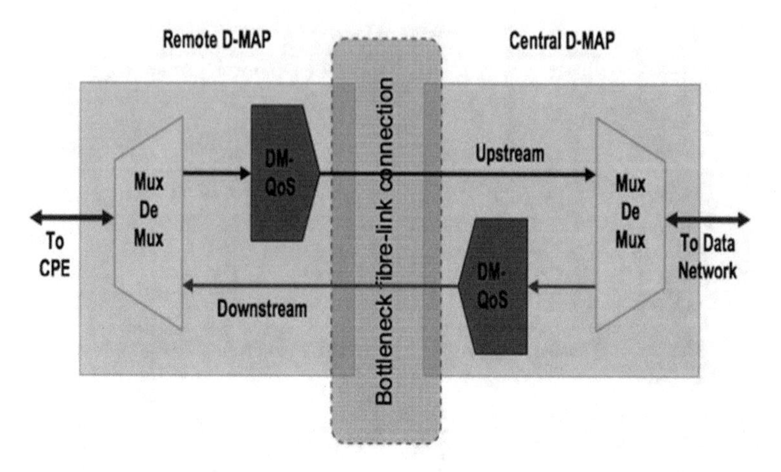

Fig. 3. DM-QoS block architecture.

The QoS is only applied to the upstream at the remote and downstream at the central D-MAP unit. For the traffic going into the limited/bottleneck connection – the OC-3/STM-1 fibre-link - traffic control/throttling is applied. The bottleneck issue at the limited bandwidth fibre-link is easily solved with this arrangement.

8 Performance Evaluation

The major performance measure includes three aspects – throughput, delay and loss, each of which is calculated for the total traffic and each class of traffic respectively [9]. In this paper, we will only discuss Unicast traffic from the GigE link (Internet) at the Central D-MAP to the CPE and analyses the packet's delay (between the central and the remote D-MAPs).

8.1 Simulation and Network Modelling

OpNet is being used as the simulation tool. One network model and four node models have been developed to simulate different parts in the D-MAP system. The network model, for a fully-connected D-MAP system is shown in Fig. 4. It contains one internet node, one central node, 32 remote nodes (all being connected to the central node), and 6144 CPE nodes (192 CPEs being connected to one remote node). Note that the number of active remote D-MAPs and CPEs can be adjusted.

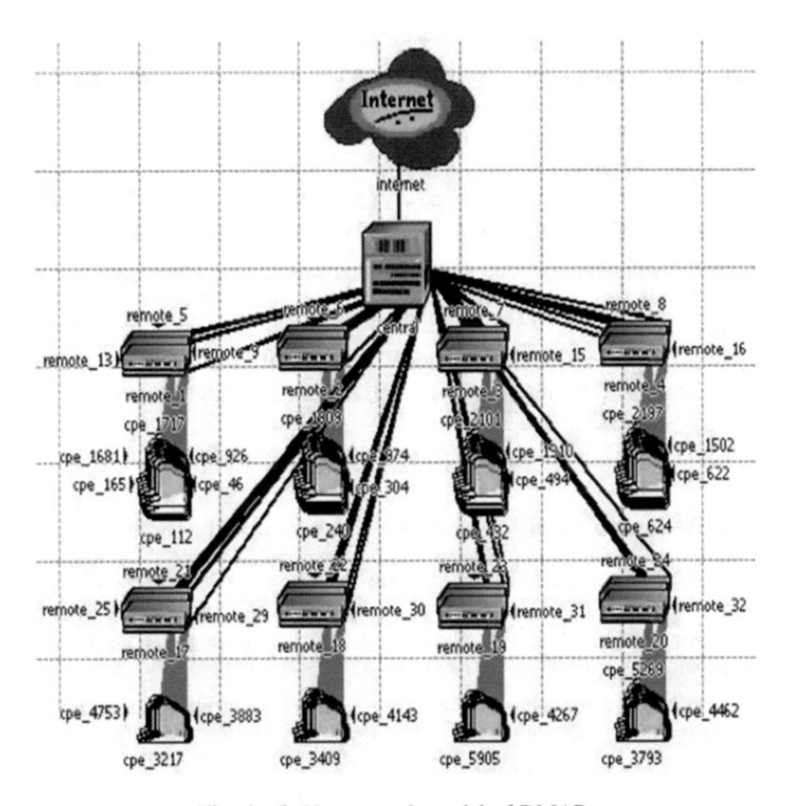

Fig. 4. OpNet network model of DMAP.

The Internet source, i.e. *UniFromInternet* acts as the packet source to the CPE and also as packet sink from the CPE. The Central D-MAP distributes the packets from internet to active Remote D-MAPS and also aggregates packets coming from them. The Remote D-MAP receives the packets from the Central unit and distributes them the CPE. It also aggregates and forwards the packets from the CPE to upper layer. Finally, the CPE sinks the packers generated from the Internet node and also generates packets for the upstream packet.

The OpNet is configured using the following parameters:

- 32 active remote D-MAPs.
- 48 active CPEs per remote D-MAP.
- Propagation delay: 100us on the GigE link (20 km), 10us on the OC-3 link (2 km), 1us on the ADSL line (200 m).
- Packet size (from the measured Internet traffic by CAIDA [10]: 46% of the packets are 40 bytes long, 18% are 552 bytes long, 18% are 576 bytes long, and 18% are 1500 bytes long.

- Packet arrival: ON-OFF mode, the ON/OFF period follows exponential distributions and the packet inter-arrival during the ON period follows an exponential distribution too.
- Up-stream per CPE: 300 Kbps.
- Down-stream/Up-stream ratio: 8:1 Mbps.
- Class of services: 3 classes (Class#0 - Video, Class#1 - Voice & Class#2 - Data).
- Percentage of traffic class: Class#0(highest priority)-40%, Class#1–30%, Class#2–30%.
- Number of buffers: 3 per D-MAP direction.
- Buffer size: 16 Mbytes.
- Scheduling scheme: FIFO, leaky-bucket/with priority for multi-class services.
- Multicast support: split multicast channel, not used in simulations.
- GigE rate: 4 Gbps.
- OC-3 rate: 135.631 Mbps (payload rate for ATM over OC-3).

8.2 Delay Performance with Respect to ADSL Down/Up Ratio

The objective of this simulation is to find out the delay performance with changing of ADLS Down/Up ratio. The Down/Up ratio is the ratio of Downstream over Upstream of an ADSL connection. Typically, the ratio is 8:1 simply means 8 Mbps downstream over 1 Mbps upstream.

In this simulation, four ratios are used to carry out the simulation starting with 2:1, 4:1, 6:1 and 8:1. These are the normal ratios that are being used by telecommunication service providers around the world.

Figure 5(a)–(d) depicts the delay performance of the unicast traffic from the GIGE link to the CPE when the up-stream is fixed at 300 Kbps. The Down/Up-stream ratios used are 2:1, 4:1, 6:1, and 8:1. There is no packet loss for all scenarios. Packet Burst (ON-OFF) model was used for the simulations, so the central delay becomes very large at a heavy load. It may exhibit different performance under different traffic patterns.

Figure 5(a) shows the delay at the GIGE link of the Central D-MAP. When the ratio is 8:1, the GIGE link has a load of 3.7 Gbps/4 Gbps. Due to the heavy load, the GIGE link generates 0.19 ms central delay (not shown due to out of scale) for all traffic. The heavy load is simply caused by all 48 CPEs at each of the 32 Remote D-MAP is using the bandwidth at the ratio of 8 Mbps Upstream and 1 Mbps downstream.

ATM network is a connection-oriented packet network. And because of this, it is easier to implement QoS. ATM also implements a fixed length of packet with the size of 53 bytes/octets called an *ATM Cell*. The fixed length cell makes the network more predictable and dependable. The cells are transferred using virtual circuit. QoS is implemented in an ATM network by specifying some performance parameter for each virtual connection (VCs) known as the *Service Parameters*.

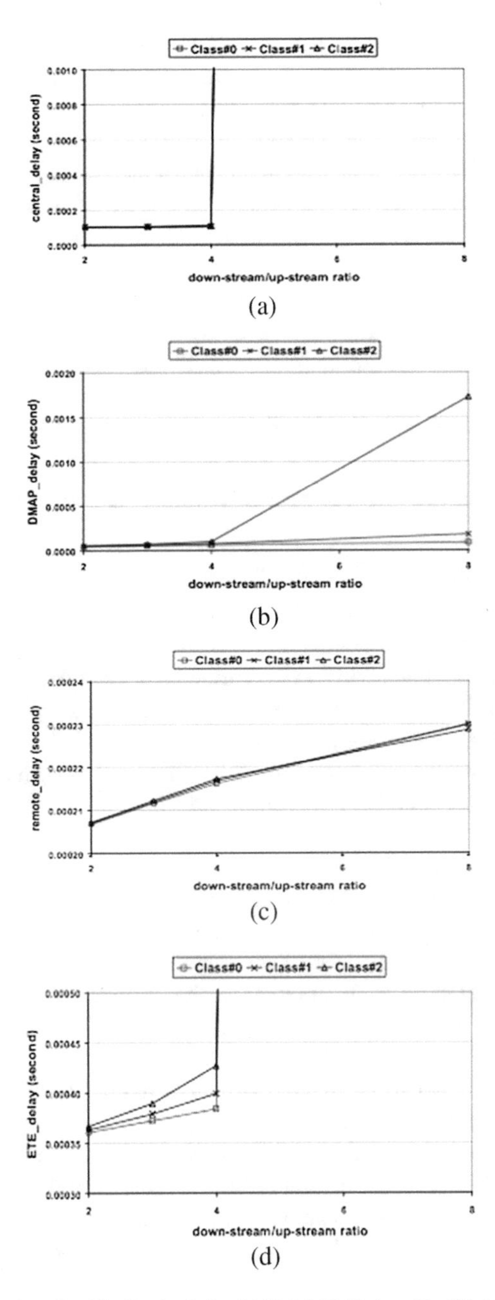

Fig. 5. (a) Central Delay (Dn/Up Ratio 8:1); (b) D-MAP Delay (Dn/Up Ratio 8:1); (c) Remote D-MAP Delay (Dn/Up Ratio 8:1); (d) D-MAP End-to-End Delay (Dn/Up Ratio 8:1).

Figure 5(b) shows the delay at each of the OC-3/STM-1 link to the remote D-MAP. At 8:1 ratio, the OC-3 links has a load of 115 Mbps/135 Mbps. The OC-3 link generates 1.7 ms delay for Class#2 traffic after prioritized scheduling. From the graph, there is a significant difference between Class#2 and the other two classes (Class#0 and Class#1).

The new QoS architecture proposed in this paper has been able to provide an almost linear delay on the Video and Voice packets as compared to the Data packets. This linearity provides the ISP the ability to provision their services in a more deterministic fashion.

Figure 5(c) shows the delay at the remote D-MAP. The constant increase of delay is as a result of the remote only acts as an aggregator and nothing is done on the downstream. The QoS is only done for the upstream packets (packets going into the fibre connection).

Figure 5(d) shows the End-To-End delay of the whole D-MAP platform.

8.3 Delay Performance with Respect to Number of Active Remote D-MAP

The objective of this simulation is to find out the delay performance with the changing number of active remote D-MAPs. In D-MAP architecture, a single Access Data Aggregator (ADA) card in the Central unit can be connected to up to 32 Remote units. In a Central unit, there can be up to 8 ADA cards. With each ADA card connected to 32 Remote units, and with each Remote unit can be connected to up to 192 CPE, a total of 6144 subscribers' connection can be achieved.

In this simulation a fix number of 48 CPEs is used for each Remote unit. Delay performances of different numbers of Remote units (i.e. 8, 16, 24 and 32) are simulated with the Down/Up ratio is set to 4:1.

Figure 6(a)–(d) depicts the delay performance of the unicast traffic from the Internet (UniFromInternet) when the number of active remote D-MAPs is changing from 8 to 32. When the number of active remote D-MAPs is 8, the total UniFromInternet traffic is 922 Mbps. Thus, the GIGE links is not a bottleneck. The bursty property generated by the Internet is transferred to the central D-MAP, generates a large delay (0.22 s, not shown) at the central D-MAP (D-MAP Delay) for Class#2 due to the congestion at the OC-3 link (115 Mbps/136 Mbps).

On the other hand, when the number of active remote D-MAPs becomes 32, the total traffic generated at the Internet is 3.67 Gbps. Then the GIGE link becomes the bottleneck that generates 0.19 s central delay (not shown). At the same time, the traffic is smoothed (streamlined) for the central D-MAP. Thus, even though the OC-3 link has the same high load, the D-MAP delay becomes smaller than the previous case. Summing up, the ETE delay exhibits different trends for three classes of packets.

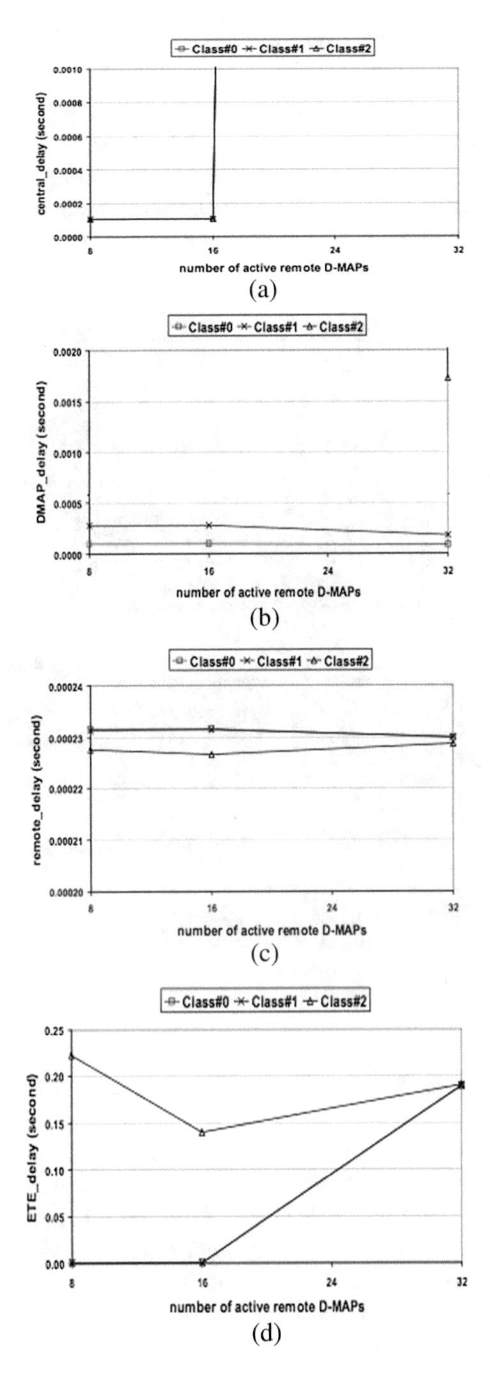

Fig. 6. (a) Central Delay w.r.t active nodes; (b) D-MAP Delay w.r.t active nodes; (c) Remote D-MAP Delay w.r.t active nodes; (d) End-to End D-MAP Delay w.r.t active nodes.

9 Current Prototype Development

To prove the effectiveness of the proposed, we have developed a prototype which uses ADSL2/2+ at the AN (Access Network) module, hence making D-MAP act as a distributed DSLAM system – a first of its kind. For the prototype's components, existing chipsets are used (Fig. 7). The components/chipsets that are used to develop the central D-MAP prototype is shown in Fig. 8.

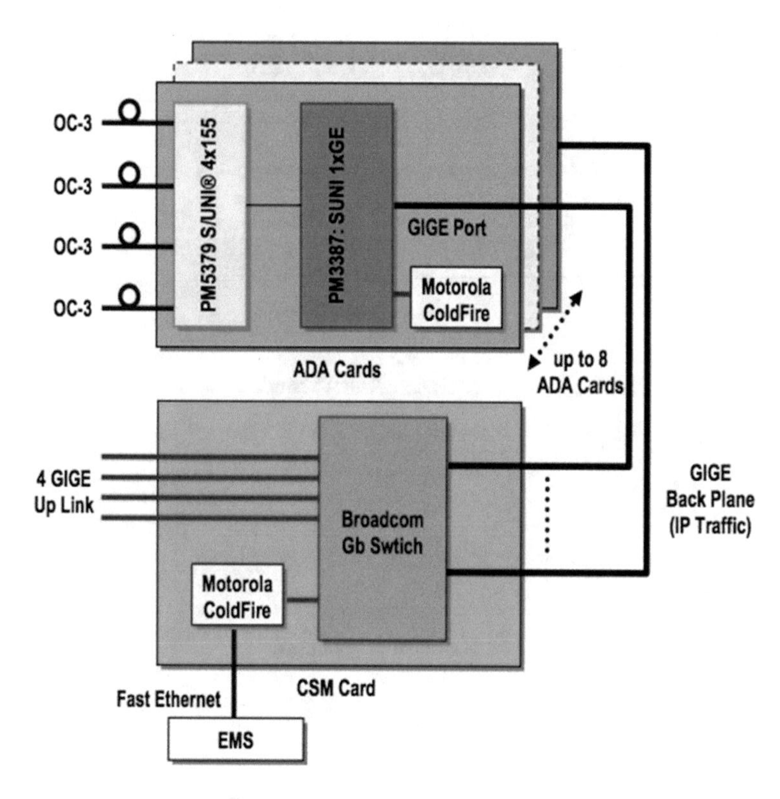

Fig. 7. D-MAP component block diagram.

Figure 9 illustrates the Remote D-MAP unit as an ADSL/2/2+ aggregator. Conexant G24 – 24 ports ADSL/2/2+ central office modem chipset is used for the prototype's line card, along with a Motorola ColdFire microcontroller to control the line card and PMC-Sierra DUPLEX for the high serial back plane interface (Fig. 10).

For the utopia connection, the WAN card uses PMC-Sierra VORTEX from the line cards. PMC-Sierra APEX is used for traffic management. However, it does not support Multi-Service QoS. In fact, from the author's own knowledge, traffic management chipset that supports Multi-Service QoS has yet to be developed. For OC-3/STM-1 physical link back to the central D-MAP unit, PMS Sierra ATLAS is used. To control the WAN card as well as management of the line cards, a Motorola ColdFire microcontroller is used.

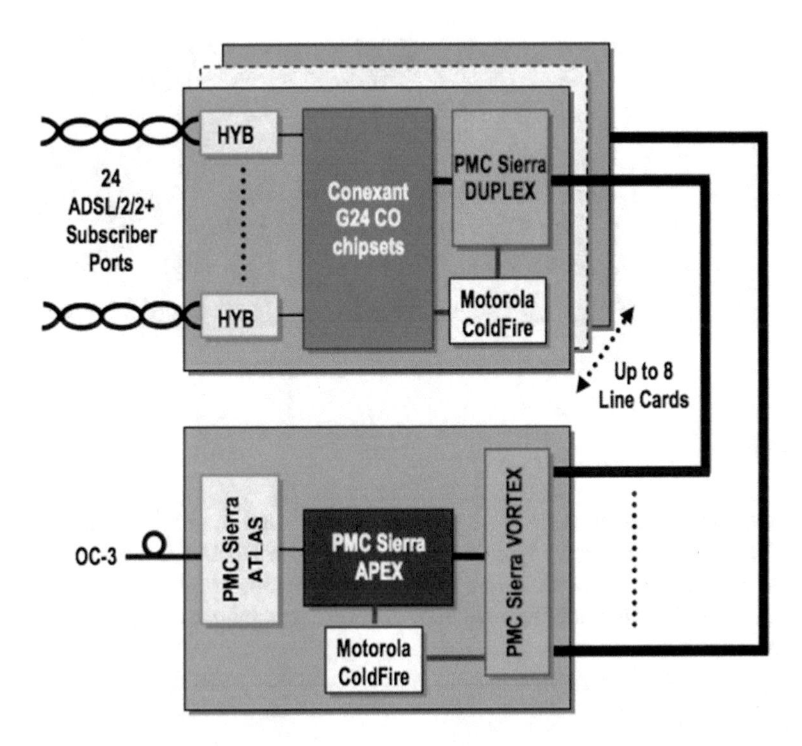

Fig. 8. D-MAP line-card component block diagram.

Fig. 9. Remote D-MAP PCBA.

Fig. 10. D-MAP Line Card PCBA.

10 Future Roadmap

The emergencies of generic network processors which are in general cheaper than the protocol specific chipset have ignited new excitement in D-MAP future version. These network processors are configurable and can be program to add new features to D-MAP. A new D-MAP architecture that is more cost-effective is currently being designed.

11 Conclusion

A D-MAP architecture has been outlined in this paper and its aim is to overcome issues faced by the current DSLAM implementation such as QoS, heat and cost.

An OpNet Simulation was done to simulate the performance of D-MAP architecture. Performance evaluation on the packet delay shows that D-MAP platform is able to perform sufficiently under the predefined scenarios. To prove the idea of the distributed architecture, a prototype is currently being implemented. Based from the results of the simulation, this has been concluded that designing a chipset that is capable of supporting multi-service QoS will be possible.

References

1. Kwok, T.C.: Residential broadband architecture over ADSL and G.Lite (G.992.2): PPP over ATM. IEEE Comm. Mag. **37**(5), 84–89 (1999)
2. Aris, A., Ramli, S., Yeap, T., Dimyati, K.: A novel distributed multi-access platform. In: TENCON 2006 (2006)
3. Azzam, A., Niel Ransom, M.: Broadband Access Technology. McGraw-Hill, New York (1999)
4. Wright, S., Anschutz, T.: QoS requirement in DSL network. In: GLOBECOM 2003, pp. 4049–4053 (2003)
5. Sauer, C., Gries, M., Sonntag, S.: Modular reference implementation of an IP-DSLAM. In: ISCC (2005)

6. van Wyk, J.H., Linde, L.P.: Comparison of theoretical models with practice for ADSL. In: IEEE AFRICON 2004, pp. 1023–1026 (2004)
7. Kasera, S.: ATM Networks – Concept and Protocols. McGraw-Hill, New York (2007)
8. Park, K.I.: QoS in Packet Networks. Springer, New York (2005)
9. Katzela, I.: Modelling and Simulating Communication Networks – A Hands-On Approach Using OPNET. Prentice Hall, Upper Saddle River (1999)
10. Fomenkov, M., Keys, K., Moore, D., Claffy, K.: Longitudinal study on Internet traffic in 1998–2003. In: Winter International Symposium on Information and Communication Technologies (WISICT) (2004)

Cognitive Channel Decoding

Nataša Živić(✉) and Christoph Ruland

Institute for Data Communications Systems, University of Siegen,
57076 Siegen, Germany
{natasa.zivic, christoph.ruland}@uni-siegen.de

Abstract. Improved authentication of images is presented, with additional improved quality of the image in sense of enhanced error correction. The algorithm is based on the idea on synergic effect of channel coding and cryptographic mechanisms used for authentication, e.g. Message Authentication Codes. The algorithm is not limited on image processing, but used for image authentication as a study case. Simulation results show the measure of gained image quality.

Keywords: Image authentication · Region of Interest
Joint Channel Coding and Cryptography · Message Authentication Codes
Coding gain

1 Introduction

This paper presents an algorithm for cognitive channel decoding, with a particular case of turbo decoding. The algorithm uses the synergy of cryptography and channel coding. The result of this approach, i.e. of the so called Joint Channel Coding and Cryptography algorithm, is additional improvement of robustness and BER characteristic of a telecommunication system.

Joint Channel Coding and Cryptography algorithm relies on the Soft Input Decryption method [1, 2] where the values on the reliability of decoded bits, i.e. the L-values [3, 4] are used. L-values are produced by the Soft Input Soft Output (SISO) decoder as the result of decoding which provides more information than "classical" decoder with hard output. They are brought to a block for authentication where some cryptographic mechanisms are implemented. In the authentication block L-values are used for the correction of its input, i.e. of the decoded bits from (SISO) channel decoder. Additionally, there is a feedback from authentication block to channel decoder which helps the decoder to achieve a stronger reduction of BER in an iterative process.

Section 2 of this paper recapitulates the Joint Channel Coding and Cryptography. Section 3 introduces the algorithm cognitive channel decoding. The results of performed simulations are presented in Sect. 4 and the conclusion is in Sect. 5.

K. Arai et al. (Eds.): FICC 2018, AISC 886, pp. 44–52, 2019.
https://doi.org/10.1007/978-3-030-03402-3_4

2 Joint Channel Coding and Cryptography

A standard communication system which uses SISO channel decoding gives a real number at the output of the demodulator (soft decision) which is then used by channel decoder as its "soft input". The result of decoding is also a real number (soft output), which is called reliability or L-value. The most reliable bits have maximal L-values, theoretically ∞, while the least reliable bits have L-values equal to 0 corresponding to the bits which have the same probability (0.5) of "1" and "0".

In a case of a cryptographic mechanism like e.g. Message Authentication Code (MAC), an additional code redundancy is added (Fig. 1), so such a communications system can be observed as the system with code concatenation [4].

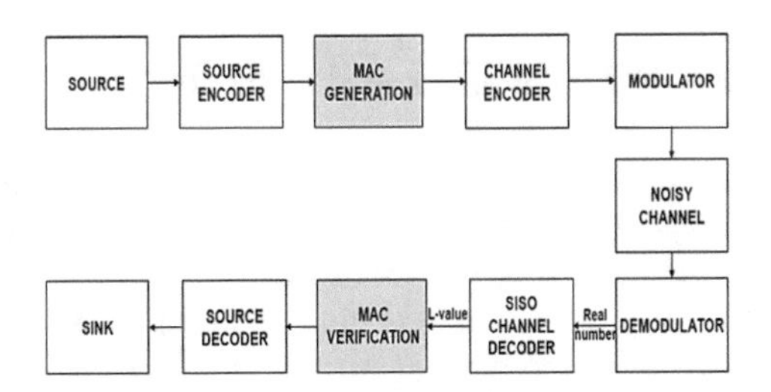

Fig. 1. Communications system with code concatenation.

The channel decoder's soft output is processed by the combination of Soft Input Decryption (SID) [1] algorithm and the cryptographic mechanisms based on the cryptographic check values added to the "payload data" – e.g. digital signature or MAC.

The received data block (i.e. an image or a message protected by its MAC value) will be successfully verified when the received check value MAC' matches the check value MAC'' calculated from the received data block M'(Fig. 2).

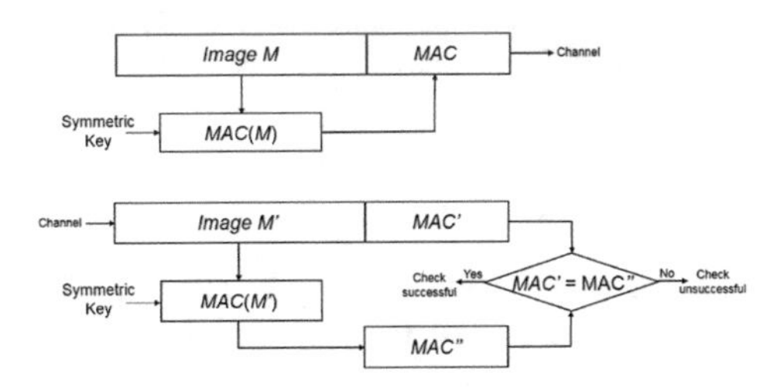

Fig. 2. Verification of message authentication codes.

The additional processing (introduced by SID) runs when the verification is not successful. In this case "a few bits" which have the smallest |L|-values are being "flipped" (i.e. inverted) through many iterations, where in each iteration a different combination of these bits is chosen for inversion. The bits with the lowest soft, i.e. absolute L-values are the "most unreliable bits" and therefore, through the inversion, they are tried to be corrected in a number of iterations. After each iteration, the verification is performed. Whole process continues after each unsuccessful verification, i.e. it enters the next iteration with a new combination of bits (among the set of bits with the lowest |L|-values) which are chosen to be inverted.

The end of the process occurs after the first successful verification, or when the dedicated resources (e.g. number of trials or memory capacity) are consumed. Different strategies can be used to choose the next candidate (i.e. the group of bits with the smallest |L|-values) for verification.

This approach, which assumes the inversion of the "most suspicious" bits (recognized as the bits with minimal reliability values), is not new – first time it has been implemented within the Chase decoding algorithm [3] as a generalized version of the GMD algorithm [4].

3 Algorithm for Cognitive Channel Decoding

In this chapter an algorithm for cognitive channel decoding is introduced. Soft Input Decryption used for the partly correction of information uses a feedback with corrected bit values needed for cognitive process iterations.

Let us observe a bit sequence, which is divided into two (or more) interleaved bit sequences: the bit sequence **a**, the bit sequence **b**, etc. (Fig. 3):

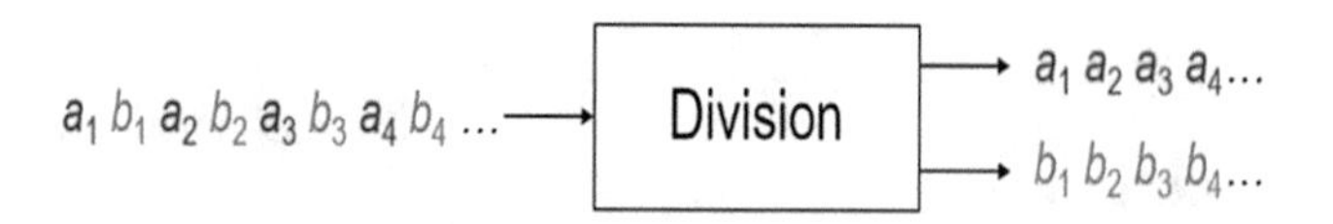

Fig. 3. Division of one into 2 (or more) bit sequences.

In the first step of the algorithm, Soft Input Decryption tries to decrypt (and, if necessary, in the same time to correct) the bit sequence **a** which contains a message and its MAC value for protection. If SID is successful, the obtained bit sequence **a** can be used as the feedback information for the improved correction of the bit sequence **b** (Fig. 4). The feedback information consists of new L-values (as soft input into SISO channel decoder) which are set as follows:

- Bit sequence a is known, so the corresponding L-values are initialized to $\pm\infty$ (the sign depends on a bit value, e.g. $+\infty$ for a bit "1" and $-\infty$ for "0"\\]).
- Bit sequence b is not known, so the corresponding L-values are initialized to 0.

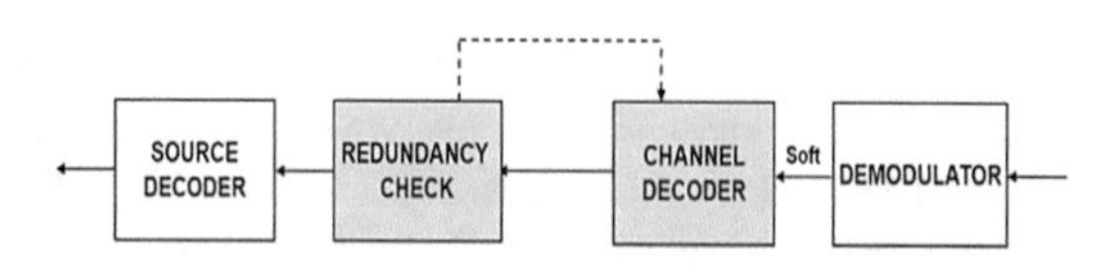

Fig. 4. Bit sequence separation into interleaved bit sequences.

The above described algorithm includes cognitive behavior considering channel decoding. One of the crucial cognitive properties is the ability to learn from the earlier experience in the decision process, and to apply the acquired knowledge with the continuation of the process. According to this definition, iterative channel decoding process learns from the previous successful iterative round, i.e., the successful authentication of each specific landmark. In case of the algorithm explained in Sect. 3, there are three iterative rounds of the proposed algorithm:

(1) Correcting iterations of channel decoding and authentication of the "Left eye" landmark.
Note: Success of this round, i.e., successful authentication is used as knowledge for the next, i.e., second round.
(2) Correcting iterations of channel decoding and successful authentication of the "Mouth" landmark.
Note: Success of this round, i.e., successful authentication is used as knowledge for the next, i.e., third round.
(3) Correcting iterations of channel decoding and successful authentication of the "Right eye" landmark.
Note: Success of this round, i.e., successful authentication could be further used as knowledge for the next, i.e., the fourth round, if it would be included in the algorithm.

Generally, the number of iterations when the process learns can be arbitrary and it depends on how many landmarks are applied.

The learning process is based on learning of parts of a decoding Trellis diagram, which are known after each successful channel decoding: knowing that the decoded bit has a value of "0" or "1" specifies certain paths of a transition between decoding states. In this way, the number of possibilities for the decoding of the remaining (previously not corrected) bits from the next round (or next rounds) is reduced and the probability of a successful channel decoding is increased.

The learning algorithm can be easily explained on the following example of channel decoding: let us observe the state diagram of a 1/2 convolutional code used as one of the two convolutional codes in a 1/3 Turbo code as shown in Fig. 5.

Let us observe an example two cases of the used landmarks:

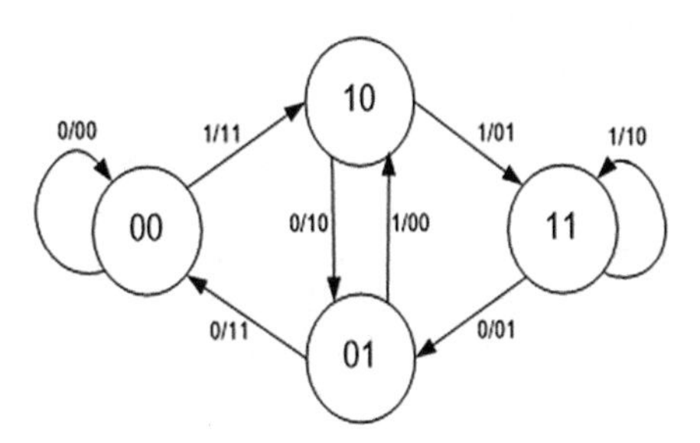

Fig. 5. State diagram of the used convolutional codes.

Case 1: Two landmarks are used: e.g. one eye protected with its authentication tag (block **a**) and a mouth protected with its authentication tag (block **b**), whereby:

$$\text{length}\,(\text{block } \mathbf{a}) = \text{length}\,(\text{block } \mathbf{b}) \tag{1}$$

$$a = a_1 b_1 a_2 b_2 a_3 b_3 \ldots \tag{2}$$

Case 2: Three landmarks are used: e.g., one eye protected with its authentication tag (block **a**), a mouth protected with its authentication tag (block **b**) and another eye protected with its authentication tag (block **c**), whereby:

$$\text{length}\,(\text{block } \mathbf{a}) = \text{length}\,(\text{block } \mathbf{b}) = \text{length}\,(\text{block } \mathbf{c}) \tag{3}$$

$$a = a_1 b_1 c_1 a_2 b_2 c_2 \ldots \tag{4}$$

Note: The equal block lengths are used for simplicity of the explanation; generally, any block lengths are supported by the proposed algorithm.

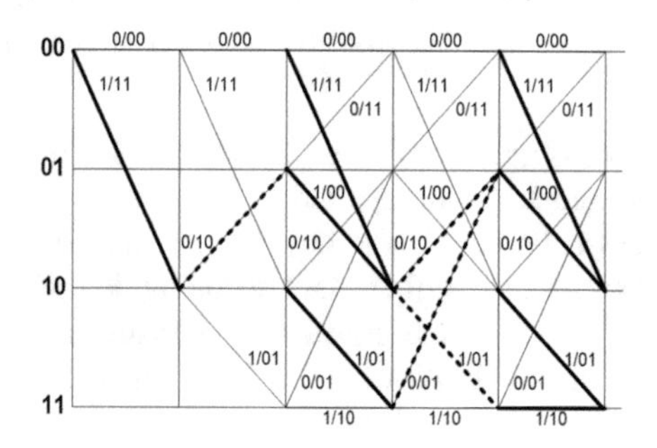

Fig. 6. Trellis diagram of case 1.

Feedback in Case 1 ("Fig. 6"):
Consider the following input bit sequences:

- Input of the channel encoder: 1 0 1 0 1 ($a_1 = 1$, $b_1 = 0$, $a_2 = 1$, $b_2 = 0$, $a_3 = 1$);

If Soft Input Decryption is successful, bits a_1, a_2 and a_3 are declared as authentic and their L-values are initialized as:

$$L(a_1) = L(a_2) = L(a_3) = -\infty \tag{5}$$

How to find the values of remaining bits b_1 and b_2?

Since the first bit $a_1 = 1$, the transition "1/11" in the trellis is only possible (bolded). After this, possible transitions are "0/10" and "1/01". Also knowing that bit $a_2 = 1$, there are three possible transitions in Trellis which accord to a_3 (also bolded): "1/11", "1/00" and "1/01". It is obvious that only transition "0/10" (according to bit b_1; dashed) can connect the first bolded transition on the left with one of the three possible bolded transitions assigned to bit a_2. This way only one path exists if $a_1 = 1$ and $a_2 = 1$, i.e. the path in which $b_1 = 0$. The achieved benefit is a reduced possibility of wrong decisions. For example, knowing only that $a_1 = 1$, from four possible paths in trellis after bits a_1, b_1 and a_2, the choice is narrowed on two paths. Knowing also that $a_2 = 1$, the choice is further narrowed on only one possible path.

In the same way, when $a_2 = 1$ and $a_3 = 1$ (i.e. when third and fifth transitions are known), three paths are possible – the paths which contain one of transitions according to bit b_2: "0/10", "1/01" or "0/01" (dashed). The consequence is that in Trellis, instead of four, exist three possible transitions that correspond to bit b_2, and this number is further reduced on two transitions ("0/10" and "1/01") after knowing the first three transitions assigned to bits a_1, b_1 and a_2 (see previous paragraph). Finally, since only two possible paths left (through the whole Trellis), the calculation of metrics of these paths gives a more reliable result, so the probability that the winner path is really correct is increased.

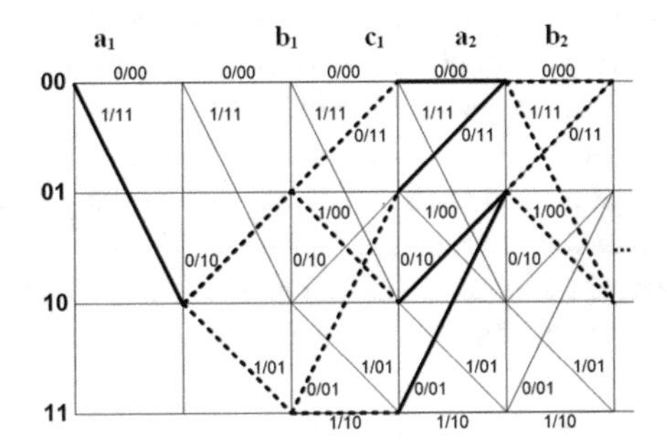

Fig. 7. Trellis diagram of case 2.

Feedback in Case 2 ("Fig. 7"):

- The known bits are $a_1 = 1$, $a_2 = 0$.

When SID successfully decrypts a_1 and a_2, these bits are declared as authentic and with the L-values initialized as:

$$L(a_1) = L(a_2) = -\infty \tag{6}$$

How to find the values of remaining bits b_1 and c_1?

Knowing that $a_1 = 1$, the transition "1/11" (on the left side; bolded) is only possible in the 1st stage of Trellis. After this, there are two possible branches – transitions "0/10" and "1/01" (dashed) according to bit b_1. Since bit a_2 is known as well, i.e. $a_2 = 0$, this bit value might be the cause of transitions 0/00", "0/11", "0/10" or "0/01". From "Fig. 7" it is clear that between the 1st and the 4th stage of Trellis exist only four paths (assigned to bits b_1 and c_1) which can connect those stages. These paths are: "0/10—0/11", "0/10—1/00", "1/01—0/01" and "1/01—1/10". Unlike the case when bit a_1 is unknown (with eight possible paths between 1st and 4th stage of Trellis), it is obvious that the knowledge of bit a_1 reduces the probability of finding the correct solution by half (from 1/8 to 1/4).

Continuing the process of decoding, and knowing the values of bits a_2 and a_3, between 4th and 7th stage of Trellis will be eight possible paths, instead of sixteen (in case when a_2 and a_3 wouldn't be known). Evidently, the probability of correct decoding increases twice (from 1/16 to 1/8).

The decoding process continues in the same way with finding new possible paths through the Trellis. Due to the narrowed choice of possible paths between every two consecutive (and known) bits a_i and a_{i+1}, the possibility of finding the correct path increases.

The soft output values of corrected **a** bits ($L(a_1)$, …, $L(a_{m1+n1})$) are set to $+\infty$ or to $-\infty$ (depending on the value of each **a** bit), and as such they are used as a feedback information to channel decoder, helping in better decoding of blocks **b** and **c**.

4 Simulation Results

Simulations are performed using a turbo channel encoder of rate 1/3 (performing 10 iterations) consisting of two channel (7,5) encoders. The channel model is based on an Additive White Gaussian Noise (AWGN). Maximum Aposteriori Probability (MAP) [5] algorithm is used for a SISO decoding of two individual SISO channel decoders. The Lena image is used as usual in image processing. The image resolution is set to 128 × 128 pixels and the length of the MAC tag to 160 bits. Soft Input Decryption with maximum 2^8 iterations (using the 8 lowest absolute L-values) was performed. The Region of Interest (RoI) is statically identified at both transmitter and receiver side.

Results showing improvement of the quality of the received and corrected Lena image are shown in Fig. 8 in case of SNR of 2 dB.

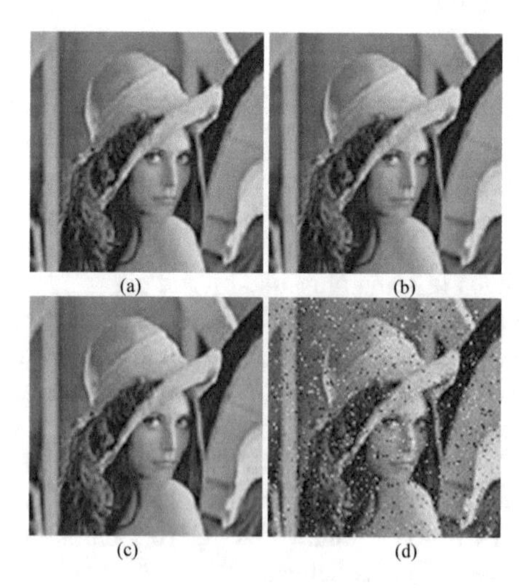

Fig. 8. Lena image: (a) Original, (b) with 1/2 of known bits, (c) with 1/4 of known bits, (d) with 1/10 of known bits.

The proposed algorithm enhances the quality of the image: firstly, image authentication is significantly enhanced buy Soft Input Decryption and, secondly, the overall coding rate is improved.

As expected, the amount of remaining errors is higher in case of a lower number of bits known at the receiver, so that the worse performances are shown in case that only every tenth bit is known at the receiver (Fig. 9). Vice versa, the best results are reached in case of the highest number of known bits at the receiver, in this case when every second bit is known.

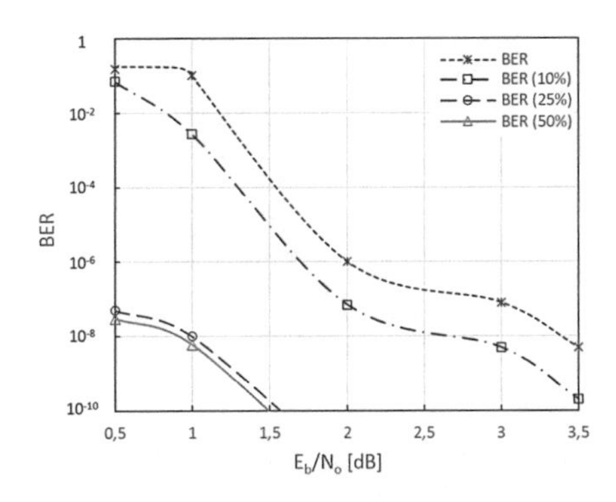

Fig. 9. Bit Error Rate in case of: (a) no known bits used for a correction; (b) 1/2 of known bits used for correction; (c) 1/4 of known bits used for correction; (d) 1/10 of known bits used for correction.

5 Conclusion

The paper presents an algorithm for cognitive channel decoding based on known bits as a result of Soft Input Decryption. SID performs flipping of low-reliable bits of a marked RoI protected by MAC, followed by the next turbo channel coding with new soft values, i.e. now known bits.

Simulation results show the significant coding gain after performance of the introduced algorithm. Nevertheless, the best results are reached in the cases of the biggest amount of known bits at the receiver, which are then used for the further correction. The algorithm is robust to transmission errors and applicable to all forms of messages, and especially interesting for applications dealing with very noisy transmission channels (satellite, mobile and wireless communications in general).

References

1. Ruland, C., Živić, N.: Soft input decryption. In: 4th Turbocode Conference, 6th Source and Channel Code Conference, VDE/IEEE, Munich, Germany, 3–7 April 2006
2. Ruland, C., Živić, N.: Feedbaack in joint channel coding and cryptography. In: 7th Source and Channel Code Conference, VDE/IEEE, Ulm, Germany, 14–16 January 2008

3. Chase, D.: A class of algorithms for decoding block codes with channel measurement information. IEEE Trans. Inf. Theory **IT-18**, 170–182 (1972)
4. Forney Jr., G.D.: Generalized minimum distance decoding. IEEE Trans. Inf. Theory **IT-12**, 125–131 (1966)
5. Bahl, L., Cocke, J., Jelinek, F., Raviv, J.: Optimal decoding of linear codes for minimizing symbol error rate. IEEE Trans. Inf. Theory **IT-20**, 284–287 (1974)

Scenario-Based Functional Safety for Automated Driving on the Example of Valet Parking

Valerij Schönemann[1(✉)], Hermann Winner[1], Thomas Glock[2], Stefan Otten[2], Eric Sax[2], Bert Boeddeker[3], Geert Verhaeg[4], Fabrizio Tronci[5], and Gustavo G. Padilla[6]

[1] Institute of Automotive Engineering, Technische Universität Darmstadt (TUD), Darmstadt, Germany
schoenemann@fzd.tu-darmstadt.de

[2] FZI Research Center for Information Technology, Karlsruhe, Germany
glock@fzi.de

[3] Research and Engineering Center, DENSO AUTOMOTIVE Deutschland GmbH, Eching, Germany
b.boeddeker@denso-auto.de

[4] TNO, The Hague, Netherlands

[5] Magneti Marelli, Corbetta, Italy

[6] Hella Aglaia Mobile Vision, Berlin, Germany

Abstract. New safety challenges have to be targeted due to the development of fully automated vehicles in the upcoming future. However, designing safe vehicle automation systems is essential. This work presents a scenario-based methodology for functional safety analysis according to the ISO 26262 using the example of automated valet parking (AVP). The vehicle automation system is decomposed into functional scenarios that can occur during operation. Potential malfunctions are identified for each scenario within a hazard analysis and risk assessment (HARA). Elaborated safety goals for automated valet parking are presented.

Keywords: Valet parking · Functional safety · ISO 26262 · Automated driving

1 Introduction

Modern vehicles are becoming more and more part of the Internet of Things (IoT) and are gradually transformed into a cyber-physical system (CPS). The European Union (EU) project ENABLE-S3 focuses on the testing and validation of autonomous CPS. The use case valet parking provides an autonomous parking procedure in interaction

The project ENABLE-S3 is partially funded by the German Federal Ministry of Education and Research and the ECSEL Joint Undertaking. This Joint Undertaking receives support from the European Union's HORIZON 2020 research and innovation program. The authors would like to thank all sponsors and partners within ENABLE-S3 for their support of our work.

K. Arai et al. (Eds.): FICC 2018, AISC 886, pp. 53–64, 2019.
https://doi.org/10.1007/978-3-030-03402-3_5

with the parking area management system (PAM). The consideration of functional safety as well as their testing and validation is an important aspect. During the automated procedure, it is required that no person in the car park is harmed by the self-driving vehicle. For this purpose, correct functionalities of the sensor system as well as the interaction between automated vehicle and PAM have to be ensured.

Current challenges lie in the interaction of subsystems and their distributed functions in various domains (automotive, industrial automation, etc.), which have to be targeted. At the same time, today's standards for functional safety have to be taken into account if such domain-specific subsystems are designed and developed. The ISO 26262 [1] as well as other standards from other domains are derived from the IEC 61508 [2]. A major challenge is to bring together involved domains and their corresponding derived norms. Furthermore, the valet parking system is considered as a hybrid system in which automated and manually driven vehicles are able to enter the parking garage. A safety concept is required in order to prevent hazards for persons in the car park. Sensor data of the interacting entities have to be combined. Distributed functionalities of the PAM and the vehicle have to be considered separately in the functional safety analysis with their corresponding standards applied for their domains. An overview of the valet parking system is given in Fig. 1. Thereby, the distributed system is defined as a set of independent subsystems with shared responsibilities that appear to its users as a single system.

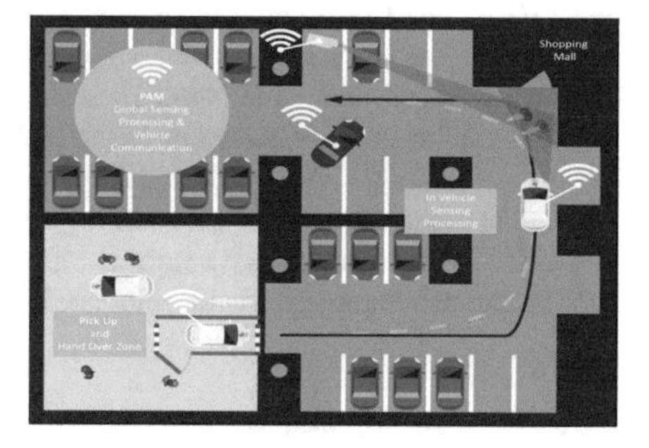

Fig. 1. Valet parking system in which automated parking is realized cooperatively between the parking area management system and the vehicle. Manually driven and automated vehicles as well as pedestrians are present in the parking garage.

The automated vehicle operates without the driver's presence and can be classified as level 5 of SAE International's taxonomy of driving automation [3] or level 4 according to the definition of the National Highway Traffic Safety Administration (NHTSA). Such a self-driving system requires safety mechanisms to transfer the system into a safe state at any point in time. A safe state is a full stop at a safe location. A safe location is a place in which the vehicle is no hazard for other participants in the parking garage. Possible hazards have to be identified in order to establish safety mechanisms to prevent any harm for other participants. The ISO 26262 is an international standard for functional safety of E/E systems in road vehicles. The standard provides a systematic approach to assess hazards and elaborate safety requirements in order to identify these risks and reduce them to acceptable levels. We applied the principle of ISO 26262 on the valet parking system. A detailed hazard and risk analysis was elaborated. Identified hazards were assessed and safety goals were defined in order to develop a safety concept for AVP.

This work is structured as follows: Sect. 2 presents the related work of valet parking concerning functional safety. Section 3 illustrates a general methodology for functional safety analysis, which can be adjusted to other use cases in the automated driving field. Thereafter, the illustrated methodology is applied using the example of automated valet parking. Section 4 shows the decomposition of the valet parking procedure into functional scenarios. Section 5 summarizes the results of the HARA and presents elaborated safety goals to develop a safety concept for the valet parking system. Finally, Sect. 6 provides a brief outlook for developing a safety architecture for AVP.

2 Related Work

2.1 Automated Valet Parking

In the course of accomplishing automated functions in modern vehicles, several challenges regarding the implementation of AVP systems exist. Different system versions and prototypical realizations of AVP systems are developed in order to solve these challenges.

Bosch and Daimler recently started a prototype system for automated parking in a car park [4]. The parking task is divided between vehicles and intelligent infrastructure. Localization and planning of the self-driving vehicle are covered by the infrastructure consisting of cameras, floor sensors, WiFi communication, and a management system. The vehicle only has to perform simple maneuvers that can be handled by state-of-the-art parking assistance systems.

The authors in [5] presented an integrated self-sufficient autonomous navigation system that cooperates with an intelligent parking facility without demanding the facility to be heavily adapted. For providing and using a valet parking system, no additional sensors are required to be installed into the parking facility and no other constructional changes are necessary.

In this work, a combination of both approaches is used for the functional safety analysis. Thereby, an automated vehicle interacts with the infrastructure of the parking garage. The parking garage can be entered by manually driven and automated vehicles. The vehicle not only uses its own sensors, it combines its data with the received information from the PAM.

2.2 Functional Safety

One of the obstacles for the release of autonomous driving is the issue of safety [6]. Up to now, neither a standard nor a methodology for elaborating a functional safety concept for automated vehicles exist. The safety concept is an important aspect for testing and validating automated driving functions e.g. with the help of model-based scenarios [7].

Chitnis et al. [8] cover the aspect of redundancy usage in order to avoid a single point of failure harming the overall functionality. Redundancy via multiple sensors and distributed hardware architectures is suggested to avoid hazardous events for humans. The authors listed pros and cons for sensors to show the importance of multi-sensor fusion. Since there is no sensor suitable for all driving scenarios, sensor fusion is elementary for automated driving.

Dijke et al. [9] describe investigations on certification processes of automated transport systems in various countries. A Failure Modes, Effects, and Criticality Analysis (FMECA) is used for the safety analysis. The certification program for an automated transport system consists of a combination of functional tests and evaluations and a series of FMECA. The functional tests are verified against the system specifications. This attempt only focuses on automated transport systems without any interaction with other subsystems from different domains.

Stolte et al. [10] elaborated a functional safety analysis for an automated unmanned vehicle according to ISO 26262. Together with experts from industrial members, a complete process of HARA is passed through consisting of item definition, identification of malfunctions, hazards, hazardous scenarios, and safety goals. Furthermore, the authors derived safety goals and functional requirements according to ISO 26262 for actuation systems of automated vehicles [11]. However, this approach addresses automated vehicles with a fallback solution consisting of a driver. A driverless vehicle with interaction of different subsystems is not targeted.

Reschka et al. [12] investigate safety concepts for autonomous driving without driver monitoring. Such a self-driving system requires safety mechanisms to transfer the system into a safe state. For an AVP system, the authors propose to utilize a remote operator. This kind of external mechanism ensures that a human observer located in the car park is able to execute an emergency stop of the driverless vehicle. This requires a secure and reliable communication channel between vehicle and remote control station. The study presents high-level safety mechanisms to handle hazards for AVP systems. More specific safety requirements are missing.

The authors in [12] also investigated safety concepts in other domains. In comparison to road vehicles, safety mechanisms for railways are integrated into the infrastructure. Control centers and monitoring components prevent a train to enter a track if it is occupied already. Since only lateral mechanical guidance is possible, the number of operations is limited and thus the complexity compared to a fully automated vehicle in an urban scenario is low. A methodology to determine functional and technical safety requirements, which are required for the test and validation, is not illustrated.

Model-based approaches for the application on functional safety analysis already exist. The development of a safety concept for distributed systems is a major challenge regarding the description, analysis, and planning of processes. The extension of model-based architecture description languages supports the illustration of processes, organizational structures, and resource assignment as indicated in [13]. The model-based approach only addresses an abstract process of ISO 26262. A process for automated driving is not part of that research study.

In summary, several investigations analyze functional safety according to ISO 26262 for automated vehicles. However, none of these consider a functional safety concept for distributed subsystems. A methodology for identifying functional and technical requirements of automated vehicles that cooperate with the infrastructure is not yet targeted.

3 Methodology

Developing a safety concept for an automated driving system by considering the overall system is complex and too extensive. During the concept phase, the ISO 26262 suggests an item definition followed by a HARA. The item definition describes the functionality, interfaces, and environmental conditions of the item. However, with regard to automated driving systems a broad range of parameters exists concerning the system's behavior and the environment. If the overall system is considered, the number of occurring situations might be unlimited. An abstraction of the system's behavior is necessary to limit the excessive number of existent parameters. The authors therefore propose a modified methodology. According to Ulbrich et al. [14], scenarios can be used to give a functional description of the system. The authors suggest decomposing the system's functional behavior into functional scenarios, as indicated in Fig. 2. Since the overall system is split into a manageable number of functional scenarios, the complexity is reduced and a situational analysis in the HARA can be performed for each scenario more specifically. As a result, a more complete set of safety goals can be elaborated which in turn leads to a more extensive safety concept. The methodology is applied to the use case automated valet parking to determine a safety concept. The scenarios are identified within the system development phase and serve as an input for scenario-based testing in the verification and validation process.

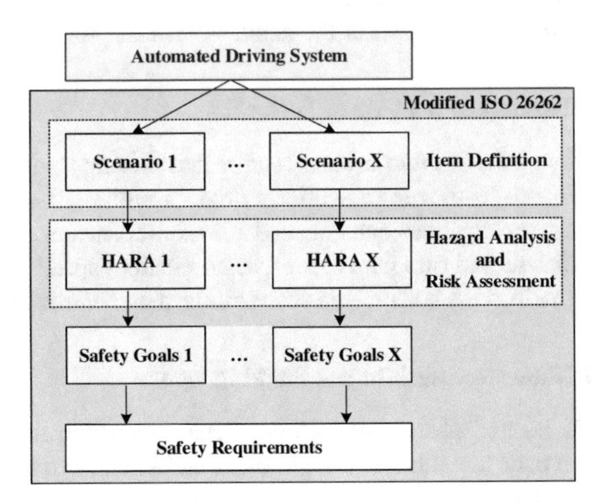

Fig. 2. Scenario-based functional safety analysis for automated driving. The system's functional behavior is decomposed into scenarios and for each scenario a hazard analysis and risk assessment is performed to determine corresponding safety goals.

4 Item Definition

Before the AVP system is decomposed into functional scenarios, some specifications are mentioned that are assumed for the AVP system when applying the ISO 26262:

(1) Automated valet parking is implemented as a distributed system by the parking management system and the automated vehicle with shared responsibilities.
(2) Instructions for handing over and requesting back the automated vehicle to/ from the PAM are ordered via a terminal (human-machine interface, HMI).
(3) Manually and automatically operated vehicles shall be able to park within the parking garage.
(4) Pedestrians, animals, obstacles, etc. sojourn in the car park.
(5) Drivers and passengers are not permitted to stay inside the automated vehicle during AVP.
(6) Parking construction prevents dangers caused by running engines.

These specifications serve as an input to break down the system's functional behavior into scenarios. In the following, the functional scenarios are illustrated.

4.1 Vehicle Handover to Parking Area Management System

Every automated valet parking procedure starts with the arrival of the vehicle at the handover zone of the parking garage. The driver and other passengers get off the vehicle. The driver or a passenger hands over the responsibility for the vehicle to the automated valet parking system by using an HMI to instruct a parking request. The system checks whether all requirements for automated valet parking are met. If the request is accepted, several possibilities exist: the PAM sends a predefined trajectory to the vehicle, the vehicle calculates a trajectory by itself, or the trajectory is determined in a cooperative mode. The PAM may transmit a static map of the car park to the vehicle. No passengers shall be inside the vehicle or located in the handover zone if AVP is activated.

4.2 Automated Driving to a Point of Interest

This scenario mainly addresses two aspects: automated driving to an assigned parking spot and back to the exit handover zone after a driver's request. The environment can be perceived through radar, lidar, camera, and ultrasonic sensors. The system determines an energy-efficient and safe path as well as an efficient speed profile to the point of interest. The scenario does not include the parking maneuver itself.

4.3 Automated Maneuvering into the Parking Space

Once the vehicle is located nearby the parking space, a parking maneuver has to be executed in order to park the vehicle. Longitudinal and lateral driving actions have to be planned until the ego vehicle is properly placed and the parking brake is set. Beside a precise positioning system, an analysis of the parking spot is crucial to decide whether the parking space is appropriate for parking, in case this was not checked by the PAM infrastructure.

4.4 Automated Leaving of the Parking Space

If the driver decides to continue his journey, a handback request is sent to the AVP system. Thereafter, the vehicle is triggered to leave the parking space. The ego vehicle either determines the trajectory by itself or loads the trajectory received for leaving the parking spot. If no obstacles are located in the area required for maneuvering out of the parking spot, longitudinal and lateral driving actions can be performed.

4.5 Vehicle Handover to Driver

After the driver requests his vehicle back, the ego vehicle drives to the exit of the parking garage. The HMI confirms the successful arrival at the pick-up location. If the vehicle is located in the pick-up zone in standstill and automated valet parking is deactivated, the driver can enter the vehicle in order to continue his journey.

4.6 Aborting the Valet Parking Procedure

This scenario characterizes the abort of the automated valet parking service while driving inside the car park. The abort is instructed via the HMI in order to get back to the vehicle or to cancel a previously ordered pick-up command to extend the parking process. The system shall determine a safe trajectory back to the handover zone or reinitiate the valet parking procedure in case of an abort of the pick-up request.

5 Hazard Analysis and Risk Assessment

The objective of the HARA is the identification of potential malfunctions to determine related safety goals. However, the authors propose a systematical methodology to identify hazards according to the divide and rule policy. Thereby, the overall AVP service is broken down into scenarios as mentioned in Sect. 3. For each functional scenario, a hazard analysis and risk assessment can be performed with the techniques suggested in the ISO 26262 such as brainstorming, checklists, quality history, FMEA, and field studies. Safety goals for automated valet parking are presented in Table 1 and HARA results are presented in Table 2 (Appendix). For clarity reasons, each safety goal is described by only one hazard. Elaborated safety goals inherit the hazard's Automotive Safety Integrity Level (ASIL) with ASIL D representing the highest and quality management (QM) the lowest safety risk. Thereby, the ASIL determination is a function f of severity S, exposure E and controllability C, whereby

$$S = \{S1, S2, S3\} \text{ with } S_i \in S \text{ and } i \leq 3 \wedge i \in \mathbb{N}, \tag{1}$$

$$E = \{E1, E2, E3, E4\} \text{ with } E_j \in E \text{ and } j \leq 4 \wedge j \in \mathbb{N}, \tag{2}$$

$$C = \{C1, C2, C3\} \text{ with } C_k \in C \text{ and } k \leq 3 \wedge k \in \mathbb{N}. \tag{3}$$

We conclude:

$$\text{ASIL} = \{\text{QM}, \text{A}, \text{B}, \text{C}, \text{D}\} \tag{4}$$

$$\Sigma_{SEC} = i + j + k = 10 \Rightarrow \text{ASIL D} \tag{5}$$

$$\Sigma_{SEC} = 9 \Rightarrow \text{ASIL C}$$

$$\Sigma_{SEC} = 8 \Rightarrow \text{ASIL B}$$

$$\Sigma_{SEC} = 7 \Rightarrow \text{ASIL A}$$

$$\Sigma_{SEC} \leq 6 \Rightarrow \text{QM}.$$

Table 1. Safety goals for automated valet parking

ID	Safety Goal	ASIL
SG01	Unintended activation of the valet parking function outside of the PAM-controlled parking area shall be prevented	D
SG02	The integrity of the communication between the PAM and the vehicle shall be ensured	D
SG03	The system shall prevent a collision between automated vehicles and persons	C
SG04	The vehicle shall not start moving during embarkment and disembarkment	C
SG05	The system shall prevent collisions with other vehicles	B
SG06	The system shall notify a human supervisor in case of a collision or fire	B
SG07	The system shall ensure that the vehicle stays within the (statically defined) drivable area during AVP	B
SG08	The valet parking function shall be disabled if people are inside the vehicle	A
SG09	The system shall prevent collision of automated vehicles with objects	A

The German Association of the Automotive Industry (VDA) released a situational catalogue for the classification of the exposure parameter [15]. A recommendation in both time and frequency domain is given for parking in a parking garage. For both cases, the exposure classification E4 is recommended. For a combinatorial analysis of the event parking and the given hazards (e.g. event parking in combination with unclear handover status) statistics are still missing. Therefore, the recommendation is accepted and most scenarios are classified conservatively as E4. Since a fully automated vehicle without driver presence is not controllable, the classification C3 in terms of controllability is preferred.

Safety goal SG01 accounts for an unintended activation of the valet parking function outside of the parking garage area such as highways or urban environments. This safety goal receives the ASIL D since high velocities can be present (S3). Additionally, no safety mechanisms can be expected beforehand and therefore it is assumed that the AVP service is always available (E4).

Safety goal SG02 considers the case that incorrect data such as maps or trajectories are transmitted to the automated vehicle by the PAM. Incorrect speed profiles may cause fatal injuries (S3). Data exchange is expected to last during AVP (E4). Additionally, from the controllability prospective the automated vehicle is assumed to be uncontrollable (C3). The authors therefore propose ASIL D.

Safety goal SG03 and SG05 describe a collision with persons and vehicles. The major difference between a direct collision with a person and a person in a vehicle is the protective body. The latter situation is less severe. However, both situations receive severity class S2. Encounters between vehicle and persons are expected during every AVP process (E4). In the context of uncontrollability for fully automated driving, risks are estimated to be ASIL C and ASIL B.

Safety goal SG04 prevents the vehicle from starting the valet parking function if passengers are getting in or out. The situation occurs two times within a valet parking procedure and as a result exposure is set to E4. In terms of severity the situation is assumed to be similar to a collision with a person (S2) which finally results in ASIL C.

Safety goal SG06 considers an emergency call after a collision inside the parking garage. In case of a severe or life-threatening injury (S2) every second counts. Exposure is determined similarly to SG03 (E4). It is expected that pedestrians not involved in the collision are still able to set up an emergency call and controllability is classified as C2. The setting yields to ASIL B.

Safety goal SG07 ensures that the vehicle does not leave its specified operating area during AVP (E4) since the automated vehicle might crash into persons (S3). However, other objects such as barriers will prevent the automated vehicle from leaving its drivable area (C1) and ASIL B is expected.

Safety Goal SG08 disables the valet parking function if persons are still inside the automated vehicle. The mechanism prevents the possibility to get off the vehicle during AVP. Severity and exposure is assumed to be similar as in SG04 (S2, E4). Since no safety mechanisms are implemented beforehand, passengers are not able to override or push the emergency off switch. However, passengers are able to stay in the vehicle (C1). The risk is classified to ASIL A.

Safety Goal SG09 indicates a collision with an object which in return injures persons. The kinetic energy of such an object is expected to be low (S1) and the combination of dangerous objects and moving persons is unlikely (E3). Probably less than 90% of persons are able to evade (C3) which results in ASIL A.

6 Conclusion and Outlook

The ISO 26262 has revealed challenges for automated driving in terms of functional safety. A modified ISO 26262 approach is proposed for automated driving. Specifications were defined to clarify boundary conditions. The functional description of the item was illustrated in scenarios to reduce overall complexity for the hazard analysis and risk assessment. For each scenario, a HARA was performed to determine potential hazards. The methodology was applied to automated valet parking. The authors identified an unintended activation of the valet parking function outside of the PAM-controlled parking area and incorrect data transmission between PAM and vehicle as most dangerous (ASIL D). In future work, safety goals will be further refined into safety requirements and assigned to functional blocks of the valet parking system architecture, which in turn raises the question of an optimal distribution of functionalities between automated vehicle and PAM with respect to functional safety. Based on the safety requirements, a safety concept will be developed in the future.

Appendix

Table 2 illustrates an extract of the developed HARA in the project ENABLE-S3 for the use case valet parking.

Table 2. Hazard analysis and risk assessment for automated valet parking

ID	Scenario	Failure Mode	Hazard	Specific Situation	Hazardous events & consequences	S	Rationale	E	Rationale	C	Rationale	ASIL	Safety Goal
1	A	Missed status	Unclear handover status	Persons getting into the vehicle just before it starts moving	Collision with person	S2	Severe or life-threatening crushing	E4	Getting in/ out happens every AVP	C3	Automated vehicle is uncontrollable	C	SG04
2	A	Incorrect data transmission by PAM	Incorrect data received by vehicle	Incorrect map/ path is loaded	Collision with person	S3	Incorrect speed profiles	E4	Communication between PAM and vehicle throughout every AVP	C3	Automated vehicle is uncontrollable	D	SG02
3	B	Missed detection	Collision with object	Automated vehicle crashes into object which in turn collides with persons	Medium structural damages or flying/ falling objects	S1	The vehicle's medium speed causes lower kinetic energy	E3	Limited combination of dangerous objects for moving persons	C3	Automated vehicle is uncontrollable; Probably less than 90% persons are able to evade	A	SG09
4	B	Missed detection	Collision with other vehicle	Automated vehicle crashes into other vehicle	The system does not detect the vehicle	S1	Person protected in vehicle	E4	Encounters with moving vehicles every drive	C3	Limited space for evading maneuvers; Automated system cannot be controlled	B	SG05
5	B	Missed detection	Collision with person	Person runs into moving vehicle	The AVP system does not detect the Person	S2	Collision at medium speed causes severe injuries	E4	Encounters with moving persons every drive	C3	Missed detection during fully automated driving	C	SG03
6	B	Missed emergency call	No emergency call after collision	Collision occurred	Unexpected continuation of AVP without emergency call	S2	Emergency call required if collision occurs; every second counts	E4	Encounters with moving persons every drive	C2	Other persons can still set up an emergency call	B	SG06
7	B	Unintended leaving of the drivable area	Vehicle falls off the brim and crashes into people	Automated driving close to the brim, people stand below the brim	Collision with person	S3	Fatal injury	E4	Staying in drivable area has to be always ensured during AVP	C1	Other objects prevent vehicle from leaving drivable area	B	SG07
8	F	Missed passenger	Undetected passenger in vehicle during AVP	Passenger tries to get off during AVP	Collision with person	S2	Collision at medium speed causes severe injuries	E4	Passengers almost every drive expected	C1	Passengers are able to stay in the vehicle	A	SG08
9	–	Unintended activation of AVP	Unexpected vehicle behavior	Driving on highway	Unexpected vehicle behavior on highway	S3	Life-threatening injuries due to accidents at high speeds	E4	AVP is available during normal vehicle use	C3	Assumption: it is not possible to override	D	SG01

References

1. ISO: ISO 26262: Road vehicles - Functional Safety. International Organization for Standardization, Geneva, Switzerland, International Standard (2011)
2. International Electrotechnical Commission: Functional Safety of Electrical/Electronic/Programmable Electronic Safety Related Systems. IEC 61508 (2000)
3. SAE: Taxonomy and Definitions for Terms Related to Driving Automation Systems for On-Road Motor Vehicles. Society of Automotive
4. Nordbruch, S., Nicodemus, R., Quast, G., Schweiger, R.: Automated valet parking. In: 7. TÜV Tagung Fahrerassistenz, München. TÜV Gruppe Süd, München (2015)
5. Klemm, S., Essinger, M., Oberländer, J., René Zofka, M., Kuhnt, F., Weber, M., Kohlhaas, R., Kohs, A., Roennau, A., Schamm, T., Zöllner, J.M.: Autonomous multi-story navigation for valet parking. In: IEEE International Conference on Intelligent Transportation Systems (2016)
6. Wachenfeld, W., Winner, H.: The release of autonomous vehicles. In: Maurer, M., Gerdes, J.C., Lenz, B., Winner, H. (eds.) Autonomous Driving: Technical, Legal and Social Aspects, pp. 425–449. Springer, Heidelberg (2016)
7. Bach, J., Otten, S., Sax, E.: A model-based scenario specification method to support development and test of automated driving functions. In: IEEE Intelligent Vehicles Symposium (2016)
8. Chitnis, K., Mody, M., Swami, P., Sivaraj, R., Ghone, C., Biju, M.G., Narayanan, B., Dutt, Y., Dubey, A.: Enabling functional safety ASIL compliance for autonomous driving software systems. Electron. Imaging **19**, 35–40 (2017)
9. Van Dijke, J., Van Schijndel, M., Nashashibi, F., De La Fortelle, A.: Certification of automated transport systems. Transportation Research Arena - Europe, Athènes, Greece, April 2012
10. Stolte, T., Bagschik, G., Reschka, A., Maurer, M.: Hazard analysis and risk assessment for automated unmanned protective vehicle. arXiv preprint arXiv:1704.06140 (2017)
11. Stolte, T., Bagschik, G., Maurer, M.: Safety goals and functional safety requiremnets for actuation systems of automated vehicles. In: IEEE 19th International Conference on Intelligent Transportation Systems (ITSC) (2016)
12. Reschka, A.: Safety concept for autonomous vehicles. In: Maurer, M., Gerdes, J.C., Lenz, B., Winner, H. (eds.) Autonomous Driving: Technical, Legal and Social Aspects, pp. 473–496. Springer, Heidelberg (2016)
13. Adler, N., Otten, S., Schwär, M., Müller-Glaser, K.D.: Managing functional safety processes for automotive E/e architectures in integrated model-based development environments. SAE Int. J. Passeng. Cars Electron. Electr. Syst. **7**(1), 103–114 (2014)
14. Ulbrich, S., Menzel, T., Reschka, A., Schuldt, F., Maurer, M.: Defining and substantiating the terms scene, situation, and scenario for automated driving. In: 2015 IEEE 18th International Conference on Intelligent Transportation Systems - (ITSC 2015), pp. 982–988 (2015)
15. German Association of the Automotive Industry (Verband der Automobilindustrie e.V.): VDA 702 E-Parameter according ISO 26262-3. VDA-Recommendations (2015)

Fair Comparison of DSS Codes

Natasa Paunkoska, Weiler A. Finamore(✉), and Ninoslav Marina

University of Information Systems and Technology St. Paul the Apostol (UIST), Ohrid, Macedonia
finamore@ieee.org

Abstract. The use of a distributed storage system (DSS), in a network with a large number of interconnected nodes, can increase significantly the storage efficiency. The main issues in this area are the reconstruction process or obtaining the entire original message out of the DSS and the repair process or recovering the lost stored data of a failed node. Finding an adequate code that will manage successfully both processes, to have an efficient system, is a challenge. Many codes for data distribution are in play. Finding a good code (good encoding procedure), one which achieves better performance, require choosing ways to make a fair comparison among codes. In this paper, we are proposing a way to compare two DSS codes by examining the system parameters. We conclude by comparing three different class of codes: Repetition, Reed-Solomon and Regenerating codes and deciding which one is the most efficient.

Keywords: Distributed storage system (DSS)
Reconstruction process · Repair process · Efficient DSS system

1 Introduction

Although distributed storage (DSS) systems have benefited largely from the past research in the area of coding for communication channels with erasures, there are several key aspects of a networked storage system that are not addressed in the classical coding theory. These nuances offer new challenges and exciting opportunities to system designers.

The storage of the digital information on a file $\mathcal{F}$ (a string with, say, $|\mathcal{F}|$ bits) in a system with physically separated storage places can have its reliability increased by spreading redundant information over the system. The immediate solution to the reliable distributed storage of digital information has been to keep copies of the file $\mathcal{F}$ in n nodes of the system. This solution makes use of what is well known in coding theory as a repetition code – a kind of code which falls in the category known as Maximum Distance Separable (MDS) codes [1–3].

If one node in the system fails there is no loss of information since there are copies still kept on the other nodes. When a node, say node i, fails a Repair Node (node i') is programmed to download information from any surviving nodes that kept the copy of the lost data and then the full content of the failed node i is

K. Arai et al. (Eds.): FICC 2018, AISC 886, pp. 65–80, 2019.
https://doi.org/10.1007/978-3-030-03402-3_6

recovered and will be stored in the i'-th node which then replaces the failed node. The reliability is thus increased by increasing the total amount of data stored by the system to $n|\mathcal{F}|$.

Further research showed that the increased amount of storage can be reduced by using coding for erasure channels [4–6]. These codes work as if they take the entire message $\mathcal{F}$, of size say $|\mathcal{F}|$, parses in blocks of size B which are then processed by an encoder which generates what is called a code-matrix (of size $k \times n$). A price to be paid, as it will be shown, is the bandwidth increase. Erasure codes require really high bandwidth during the node recovery (or the *repair process*). The repair process is activated whenever a node in the system, which stores α symbols-per-node, fail. A new node (a newcomer) that has to replace the node under failure contacts a set of d active (or alive) nodes, out of the n storage nodes in the network, and download β symbols ($\beta \leq \alpha$) from each of them. The data downloaded from the d nodes requires that a *bandwidth* of $\gamma = d\beta$ symbols be used to transmit data such that the recovery of the lost data, which was stored in the failed node, is possible. Another process taking place in a DSS is the *reconstruction process*, which is activated whenever the user (data collector) wants to retrieve the information he has stored in the system. In this case, the data collector contacts any set of $k \leq d$ nodes. Downloads all α symbols stored on them, and through a proper processing retrieve the entire message $\mathcal{F}$—of size say $|\mathcal{F}| = LB$. Even the use of plain erasure codes that reduce the system overload is not so efficient as far as the repair bandwidth (amount of information downloaded during the repair process) is concerned.

A seminal paper [7] revealed the existence of codes that performs better than MDS codes in general and better then, in particular, repetition and plain erasure codes. It should be considered that a better code is the one that minimizes the *download bandwidth* (the total amount of data required to be downloaded by the Repair-Node to setup the new node which will replace the failing node) while keeping the *total storage* (the total amount of data stored in the network) to a fixed value or vice-versa.

As we said the sophisticated solution are the codes introduced by Dimakis in [7,8], called regenerating codes. These codes are characterized by the parameters $[n, k, d]$ and secondary parameters $[\alpha, \beta, B]$. The authors established in their work that the parameters of a regenerating code that aim to store a file of size $|\mathcal{F}|$ bytes reliably and efficiently must satisfy the following condition

$$B \leq \sum_{i=0}^{k-1} \min\left\{\alpha, (d-i)\,\beta\right\}. \tag{1}$$

For these codes, if the bandwidth γ is constrained to be minimum, a trade-off between the repair bandwidth γ and the storage per node, α, which was previously studied in [4,5], can be set with two extreme points being achieved. The obtained extreme points are referred to as the MBR (*Minimum Bandwidth Regenerating*) codes and the MSR (*Minimum Storage Regenerating*) codes depending on which parameter is being reduced, the bandwidth or the storage.

Delving deeper in the subject one can see that many codes have been proposed as adequate and efficient to be used in the DSSs. One of these are the product-matrix codes [9]. It is claimed that these codes have good performances regarding the bandwidth and the storage. How to fairly compared such codes is a question that remains though. Therefore, the aim of our paper (our contribution) is to find a new way to compare disparaged codes (with parameters of diverse values). On more concrete grounds, what lead us to the current work was the goal to compare two class of codes the "regular DSS product-matrix codes" and the "shortened DSS product-matrix codes" [10].

The paper is organized as follows: In Sect. 2 is described the functioning of the three codes later used in the comparison procedure. Section 3 elaborates which is the current way of comparison and proposes the new way. Section 4 gives the results from the comparison process using the chosen codes. Section 5 includes the measurement of the capacity and Sect. 6 concludes the paper.

2 Overview of DSS Codes

In this section, we are elaborating the functioning and the efficiency of three diverse DSS codes: Regenerating Codes, Repetition and Reed-Solomon based DSS code.

2.1 Efficient DSS Codes (Dimakis Result)

To show the existence of efficient DS-codes, a simple and clever example (a constructive proof) was presented by Dimakis [8]. Its performance as compared to the performance of repetition codes can be shown to be better. The examples given next compare the performance of three schemes: the first one is based on Dimakis DS-code, the second a DS-code based on repetition codes and a third one based on RS-DSS-codes.

Example 1: For the code $\mathcal{C}[k,n,d]$ of this example, the chosen parameters were $k = 2$, $n = 4$, $d = 3$. Let us consider that a file $\mathcal{F}$, is a string of size $|\mathcal{F}|$ bits, that has been arranged as a vector $\underline{u} = (u_1, \ldots, u_B)$ of B q-ary symbols. which is, furthermore, arranged as a Matrix M. Consider for the current example that

$$M = \begin{pmatrix} u_1 & u_2 \\ u_3 & u_4 \end{pmatrix}. \tag{2}$$

In this case $B = 4$ q-ary symbols are taken. Let M have been somehow mapped into the code-matrix

$$\begin{aligned} \underline{C} &= \begin{pmatrix} c_{1,1} & c_{1,2} \\ c_{2,1} & c_{2,2} \\ c_{3,1} & c_{3,2} \\ c_{4,1} & c_{4,2} \end{pmatrix} \\ &= \begin{pmatrix} u_1 & u_2 \\ u_3 & u_4 \\ u_1 + u_2 + u_3 + u_4 & u_1 + 2u_2 + u_3 + 2u_4 \\ u_1 + 2u_2 + 3u_3 + u_4 & 3u_1 + 2u_2 + 3u_3 + 3u_4 \end{pmatrix} \end{aligned} \tag{3}$$

with dimensions α by n ($\alpha = 2$ and $n = 4$ in this case). If we let the number of distinct blocks (vectors) $\underline{u} = (u_1\ u_2\ u_3\ u_4)$ in $\mathcal{F}$ be L then $|\mathcal{F}| = LB\log_2 q$. In this example, since $|\mathcal{F}| = 4$ symbols, we have thus $L = 1$.

With such an encoding the Data-Collector at any terminal is able to reconstruct the original information by downloading data from any $k = 2$ nodes (two rows of the code-matrix $\underline{C}$). The solution of the system with four equations and four unknowns, so obtained, will lead to the vector $\underline{u} = (u_1\ u_2\ u_3\ u_4)$, the information to be recovered. Four q-ary symbols would have been downloaded in such a case, i.e., the downloaded data will have the same size as that of the original information.

Let us say, for instance, that node number $i = 4$ fails. To the new node, say $i' = 5$ (the repair-node), the following data—the parity-information $\underline{P}$ of dimension $d \times \beta$—would have to be sent.

$$\begin{aligned} \underline{P} = \begin{pmatrix} p_1 \\ p_2 \\ p_3 \end{pmatrix} &= \begin{pmatrix} c_{1,1} + 2c_{1,2} \\ 2c_{2,1} + c_{2,2} \\ 3c_{3,1} + c_{3,2} \end{pmatrix} \\ &= \begin{pmatrix} u_1 + 2u_2 \\ 2u_3 + u_4 \\ 4u_1 + 5u_2 + 4u_3 + 5u_4 \end{pmatrix} \end{aligned} \tag{4}$$

Notice that $\underline{P}$ is constructed using the vectors stored on $d = 3$ nodes (nodes 1, 2 and 3) and also $\beta = 1$. Upon receiving the parity symbols the repair-node regenerates the lost vector by properly combining the parity vector components. For the current example the regenerated vector is

$$(c_{5,1}\ c_{5,2}) = (5u_1 + 7u_2 + 8u_3 + 7u_4,\ 6u_1 + 9u_2 + 6u_3 + 6u_4),$$

which was obtained by multiplying $\underline{P}$ and

$$\begin{pmatrix} 1 & 2 \\ 2 & 1 \\ 1 & 1 \end{pmatrix}. \tag{5}$$

The repaired information now stored in the network is

$$\underline{C} = \begin{pmatrix} u_1 & u_2 \\ u_3 & u_4 \\ u_1 + u_2 + u_3 + u_4 & u_1 + 2u_2 + u_3 + 2u_4 \\ 5u_1 + 7u_2 + 8u_3 + 7u_4 & 6u_1 + 9u_2 + 6u_3 + 6u_4 \end{pmatrix}. \tag{6}$$

A few observations are needed at this point. First of all it should be noticed that the replacement-node does not necessarily holds the same information as before. And also that the linear combination coefficients which constitute the new equations have to be stored by the node. The coefficients in this technique are randomly generated.

- The bandwidth of the scheme by Dimakis [8], current example, is $\gamma = \beta d = 3$ symbols (if not explicit we will use the word symbols to refer to q-ary symbols), and the storage-per-node is $\alpha = 2$ symbols. If we consider a file $\mathcal{F}$ of size $|\mathcal{F}|$ bits (or $|\mathcal{F}|/\log_2 q$ symbols), that is to be stored into $n = 4$ nodes, and which has been parsed into L blocks of 4 symbols ($B = 4$) then we get (noticing that $|\mathcal{F}|/\log_2 q = BL$)

$$D_{\gamma}^{Dim} = \beta d L = \frac{\beta d}{B} \times \frac{|\mathcal{F}|}{\log_2 q} \text{ symbols} \tag{7}$$

$$= \frac{3}{4} \times |\mathcal{F}| \text{ bits;} \tag{8}$$

$$S_{\alpha}^{Dim} = \alpha n L = \frac{\alpha n}{B} \times \frac{|\mathcal{F}|}{\log_2 q} \text{ symbols} \tag{9}$$

$$= 2\,|\mathcal{F}| \text{ bits.} \tag{10}$$

2.2 MDS Code (Repetition Code)

Let the original file (the information to be stored) be represented by $\underline{u} = (u_1\ u_2\ \dots\ u_{LB})$, a sequence of symbols with length LB. The distributed information, kept by the storage system, represented by the matrix $\underline{C} = (\underline{C}^{[1]}; \dots; \underline{C}^{[n]})$ in which $\underline{C}^{[n]} = \dots = \underline{C}^{[1]} = \underline{M} = \underline{u}$ can be seen as generated by an encoder which multiply an $1 \times n$ matrix with all elements equal to one's, by the input vector $\underline{u}$, i.e.

$$\underline{C} = (1 \dots 1)\,\underline{M}. \tag{11}$$

Example 2: To set a simple example let $k = 1$, $n = 3$, $d = 1$ and $B = 1$, with $M = u_i$. We then get

$$\underline{C}^{[i]} = (1\,1\,1)\,M = \begin{pmatrix} u_i \\ u_i \\ u_i \end{pmatrix} \tag{12}$$

- The scheme bandwidth, for the current example, has an amount of downloaded data to repair a failed node $\gamma = \beta d = k = 1$ symbols and the storage-per-node is $\alpha = k = 1$ symbols. If we consider a file of size $|\mathcal{F}| = LB \log_2 q$ bits, that has been parsed into L blocks of q-ary symbols each block with size $B = 1$, we get

$$D_{\gamma}^{\text{rep}} = \beta d L = \frac{\beta d}{B} \times \frac{|\mathcal{F}|}{\log_2 q} \text{ symbols}$$

$$= \frac{\beta d}{B} \times |\mathcal{F}| \text{ bits;} \tag{13}$$

$$S_{\alpha}^{\text{rep}} = \alpha n L = \frac{\alpha n}{B} \times \frac{|\mathcal{F}|}{\log_2 q} \text{ symbols}$$

$$= \frac{\alpha n}{B} \times |\mathcal{F}| \text{ bits.} \tag{14}$$

and

$$(D_\gamma^{\text{rep}},\ S_\alpha^{\text{rep}}) = (|\mathcal{F}|,\ 3\,|\mathcal{F}|) \text{ in bits.} \tag{15}$$

One can easily see that the downloaded amount of data is 4/3 times larger when the repetition code is used instead of the DSS code. And this proves the existence of a better code.

2.3 MDS Code (RS Code)

The coding for DSS using Reed-Solomon (RS) codes, which are MDS codes that fit well in this role of *repair/regenerating* codes was suggested as a better alternative to repetition codes. Each input block is a vector of k q-ary symbols as well as the elements of the code-matrix belong to a Finite Field $\mathbb{F}_q$ (with all operations appropriately done in this field). To store a file of size $|\mathcal{F}^{RS}| = B^{RS}L^{RS}$ the file is parsed into L^{RS} vectors of size $k = B^{RS}$ which are encoded into a vector of size αn. To *reconstruct* the original file, k vectors of size $\beta = \alpha$ are downloaded from k distinct nodes and, after proper decoding, the file is obtained. To *repair* one node, k vectors of size α, selected among the $d = n - 1$ remaining nodes, are downloaded and, after proper decoding and encoding, the content of the failed node can be regenerating—the repair-node is thus ready to replace the failed node. Let us examine an example.

Example 3: A RS based DS-code $C[k, n, d]$ with parameters $[k, n, d]$ is obtained by taking $\underline{u} = (u_1, u_2, \ldots, u_k)$ and by using a RS-code generating matrix.

$$G = \begin{pmatrix} g_{1,1} & g_{1,2} & \cdots & g_{1,n} \\ g_{2,1} & g_{2,2} & \cdots & g_{2,n} \\ \vdots & \vdots & \ddots & \vdots \\ g_{k,1} & g_{k,2} & \cdots & g_{k,n} \end{pmatrix}. \tag{16}$$

Every block $\underline{u}$ will give rise to a RS-codeword $\underline{w} = (w_1, \ldots, w_j \ldots, w_n)$. With $B = k$, $\beta = \alpha = 1$ and $d = k$ we have the DS-code-matrix

$$\underline{C} = \begin{pmatrix} w_1 \\ \vdots \\ w_\ell \\ \vdots \\ w_n \end{pmatrix}.$$

- The scheme bandwidth, for this example, is $\gamma^{RS} = \beta d = k$. The storage-per-node is also $\alpha = 1$ symbols. For a file of size $|\mathcal{F}| = B^{RS}L^{RS} = kL^{RS}$, the total number of downloaded data is $D_\gamma^{RS} = \beta d L^{RS}$ and the total number of stored data is $S_\alpha^{RS} = \alpha n L^{RS}$. With both measured in bits, we get

$$\begin{aligned} D_\gamma^{\mathrm{RS}} = \beta d L^{RS} &= \frac{\beta d}{k} \times |\mathcal{F}| \text{ symbols} \\ &= |\mathcal{F}| \log_2(q) \text{ bits;} \end{aligned} \tag{17}$$

$$\begin{aligned} S_\alpha^{\mathrm{RS}} = \alpha n L^{RS} &= \frac{\alpha n}{k} \times |\mathcal{F}| \text{ symbols} \\ &= \frac{n}{k} |\mathcal{F}| \log_2(q) \text{ bits.} \end{aligned} \tag{18}$$

and

$$\left(D_\gamma^{\mathrm{RS}},\, S_\alpha^{\mathrm{RS}}\right) = \left(|\mathcal{F}| \log_2(q),\, \frac{7}{3} |\mathcal{F}| \log_2(q)\right) \text{ in bits.} \tag{19}$$

Let us compare the codes in Examples 1, 2 and 3. The RS-DSS-code to be considered has parameters $(n, k, d, \alpha, \beta, B, q) = (7, 3, 3, 1, 1, 3, 8)$ and thus we need $|\mathcal{F}| = 36$. We thus have, from (8), (13) and (17), and (10), (14) and (18)

$$\begin{aligned} \left(D_\gamma^{\mathrm{Dim}},\, S_\alpha^{\mathrm{Dim}}\right) &= (0.75,\, 2)|\mathcal{F}| \\ \left(D_\gamma^{\mathrm{rep}},\, S_\alpha^{\mathrm{rep}}\right) &= (1.00,\, 3)|\mathcal{F}| \\ \left(D_\gamma^{\mathrm{RS}},\, S_\alpha^{\mathrm{RS}}\right) &= (3.00,\, 7)|\mathcal{F}| \end{aligned}$$

The plot of the points D_γ versus S_α is shown in Fig. 1 for a file of size $|\mathcal{F}| = 108$, q-ary symbols (with $q = 8$) and an RS-code with parameters $(n, k) = (7, 3)$ was considered. It is easy to see, in this case, that of all the three schemes the Dimakis example DSS-code is a better choice.

3 General DSS Code and Comparison Procedure

Coding for DSS in a general scenario can be thought of as a processing that parses the input-string $\mathcal{F}$, of size $|\mathcal{F}|$, into L info-words $\underline{u}^{[\ell]}$ of size B and encode each info-word by arranging its components into a matrix $M^{[\ell]}$ of size $n \times k$, $(k < n)$ which is then multiplied by a generating-matrix Ψ. When one node of the system fails the repair-process is triggered by a new node that contacts a set of $d < n$ alive nodes (nodes under normal operation conditions) in the network and download β symbols $(\beta \leq \alpha)$ from each node. The downloaded data from the d nodes—which consumes a bandwidth of $\gamma = \beta d$ symbols—is used to recover the lost data that was stored in the failed node. Besides the repair process, another process that runs in a DSS is the *reconstruction process*. It is initiated whenever an authorized user wants to retrieve the information $\mathcal{F}$ stored in the system—in such a case the user contacts any set of $k \leq d$ nodes and downloads all α symbols stored on it and after due processing the entire message, of size say $|\mathcal{F}| = LB \log_2 q$ bits, is obtained.

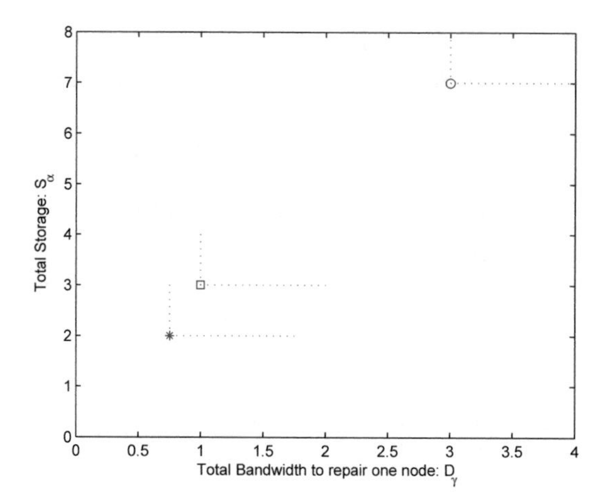

Fig. 1. Points displaying the Total repair bandwidth versus Total storage for $|\mathcal{F}| = 108$ bits and, Dimakis-DS-code ($n = 4, k = 2, d = 3, \alpha = 2, \beta = 1, B = 4, q = 8, L = 9$), marked by a star, Repetition-DS-code ($n = 3, k = 3, d = 1, \alpha = 1, \beta = 1, B = 1, q = 8, L = 36$), marked by a square, and, RS DSS-code ($n = 7, k = 3, d = 3, \alpha = 1, \beta = 1, B = 3, q = 8, L = 12$), marked by a circle.

The data from each info-word $\underline{u}$ (we will concentrate our discussion to the processing of one block and will thus drop the index ℓ) is used to build a matrix $\underline{M}$ which is subsequently mapped into a code-matrix $\underline{C}$. We can thus write

$$\underline{C} = \begin{pmatrix} \underline{C}_1 \\ \vdots \\ \underline{C}_i \\ \vdots \\ \underline{C}_n \end{pmatrix} = \begin{pmatrix} \psi_{1,1} & \dots & \psi_{1,n} \\ \vdots & \ddots & \vdots \\ \psi_{k,1} & \dots & \psi_{k,n} \end{pmatrix} \underline{M} \tag{20}$$

keeping in mind that each row of $\underline{C}$ is stored in each one of the n nodes in the network.

To reconstruct the original message the data-collector fetch the content stored in k (out of n) nodes or, in other words the $k \times \alpha$ matrix

$$\underline{C} = \begin{pmatrix} \underline{C}_{i_1} \\ \vdots \\ \underline{C}_{i_\ell} \\ \vdots \\ \underline{C}_{i_k} \end{pmatrix} . \tag{21}$$

It is of course required that $n > k$ and $n - 1 \geq d \geq k$. The repair of one failing node is achieved by fetching a $d \times \beta$, $(\beta \leq \alpha)$, parity-matrix

$$\underline{P} = \begin{pmatrix} \underline{P}_{j_1} \\ \vdots \\ \underline{P}_{j_\ell} \\ \vdots \\ \underline{P}_{j_d} \end{pmatrix}. \tag{22}$$

derived from $\underline{C}$.

Dimakis et al. [8] have shown that the parameters (k, d, α, β) of a q-ary regenerating code that aim to store one info-word of size B reliably must satisfy the following condition

$$B \leq \sum_{i=0}^{k-1} \min\left\{\alpha, (d-i)\,\beta\right\}. \tag{23}$$

Minimizing both the download-bandwidth and the amount of storage are conflicting goals and a tradeoff between the repair bandwidth, $\gamma = d\beta$, and the storage per node, α, has been established in [5]. In our paper, we are establishing a similar trade-off curve between the Total Download-bandwidth, defined as $D_\gamma = x\beta d = \gamma$, where x is the number of failed nodes in the system, and the Total Storage, $S_\alpha = \alpha n$. Here, we are assuming that $x = 1$, i.e., only one node fails at a time, so the optimum codes have points

$$(D_\gamma, S_\alpha) = (\beta d, \alpha n)$$

lying on this tradeoff curve. No code exists that are under this curve. The region above it is the feasible region.

The trade-off function relating S_α and D_γ is described by the theorem stated next.

Theorem 1 (Trade-off curve [8]*):* The trade-off function α versus γ is parametrically given by the set of points α and γ related by

$$\alpha = \begin{cases} \frac{B}{k} & \gamma \in [f_0, +\infty] \\ \frac{B - g_i\gamma}{k-i} & \gamma \in [f_i, +f_{i-1}] \end{cases} \tag{24}$$

in which $1 < i \leq k$ and

$$f_i \triangleq 2B \frac{d}{i(2k-i-1) + 2k(d-k+1)}, \tag{25}$$

$$g_i \triangleq \frac{i(2d-2k+i+1)}{2d}. \tag{26}$$

Two notable points of the trade-off function are obtained for $\gamma = \gamma_{min} = f_{k-1}$ and $\alpha = \alpha_{min} = B/k$. These two points, designated by the terminology minimum-storage parameters (MSR) and minimum-bandwidth parameter (MBR), are given by

$$(\gamma_{MSR}, \alpha_{MSR}) = \left(\frac{B}{k}, \frac{Bd}{k(d-k+1)}\right) \tag{27}$$

$$(\gamma_{MBR}, \alpha_{MBR}) = \left(\frac{2Bd}{2kd-k^2+k}, \frac{2Bd}{2kd-k^2+k}\right) \tag{28}$$

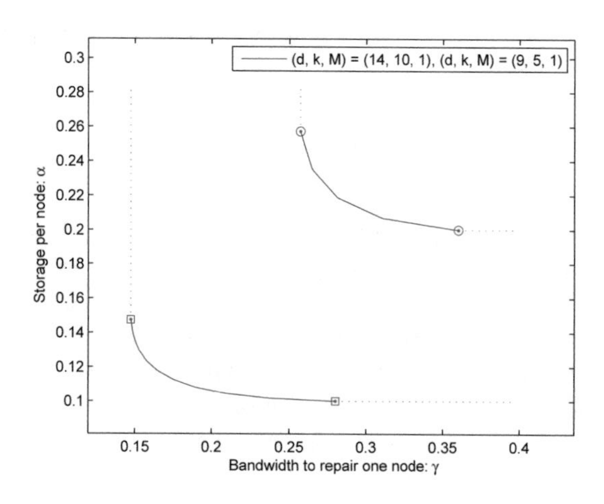

Fig. 2. Plot displaying the trade-off curve (Storage-per-node versus Total-repair-bandwidth) for parameters $(n, k) = (10, 5)$ (marked with a circle) and $(n, k) = (15, 10)$ (marked with a square). For both cases $B = 1$ (in Dimakis paper [8] B is notated as $\mathcal{M}$).

By Dimakis paper [8] to compare two codes applied in DSS they take into account the parameters: storage per node and repair bandwidth of one node. This kind of comparison is not the most adequate one because does not depict the real relation between the codes. In Fig. 2 is given an example of such comparison between two codes.

In our paper, we are showing that more sense for comparison of two codes (A and B—to examine both the graph of Code A (S_α^{A} versus D_γ^{A}), and, the graph of Code B (S_α^{B} versus D_γ^{B})) makes when we have as parameters total storage vs. total downloaded data. These plots take ino account the size of the file, namely, $|\mathcal{F}| = LB \log_2 q$ bits and we should ask which code wold do better when dealing with a file with same size in all cases.

Next we examine the following graphs shown in Fig. 3. Thus, we can claim that the code with parameters [15, 14, 10, 10, 1] (red curve) is better (they were compared under the same conditions).

4 Advantage of Plotting Total-Downloaded-Data Versus Total-Stored-Data

Equations (29) and (30) explicit the two quantities Total-downloaded-data (T_γ) and Total-Stored-data (T_α).

$$T_\gamma = \beta d L \log_2 q, \tag{29}$$

$$T_\alpha = \alpha n L \log_2 q. \tag{30}$$

It should be noticed that the new proposed trade-off curve aims at comparing two (or more) codes, for instance Code A and Code B under fair conditions. In every case (Code A and Code B), each curve $T_\alpha^A \times T_\alpha^B$ and $T_\alpha^A \times T_\alpha^B$ describes the total amount of stored data versus the total amount of downloaded data requirements for distributed storage of a given file of size. $|\mathcal{F}| = BL \log_2 q$ bits.

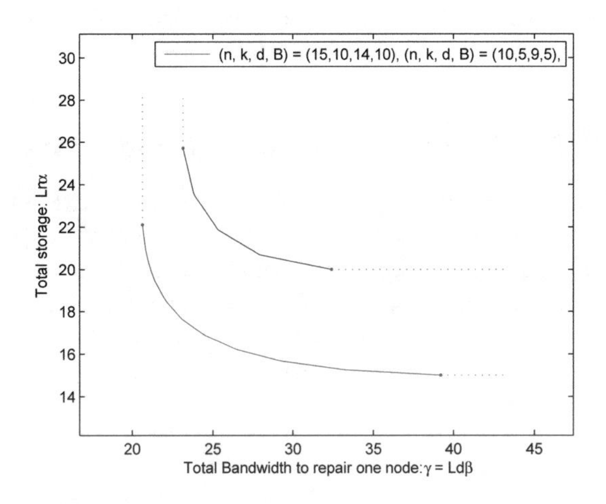

Fig. 3. Plot displaying the Total-Storage ($Ln\alpha$) versus Total-Bandwidth ($Ld\beta$) trade-off curve of codes with parameters $[n, k, d, B, L] = [10, 5, 9, 5, 2]$ (blue curve) and $[15, 10, 14, 10, 1]$ (red curve)—it is being assumed that the all codes use a field with the same q (otherwise this parameter should be taken into account too).

In Fig. 4 are shown the trade-off curve for each of the following three sets of the parameters (n, k):

- $(4, 2)$, (Dimakis code)
- $(3, 1)$, (Repetition code) and
- $(7, 3)$ (RS code)

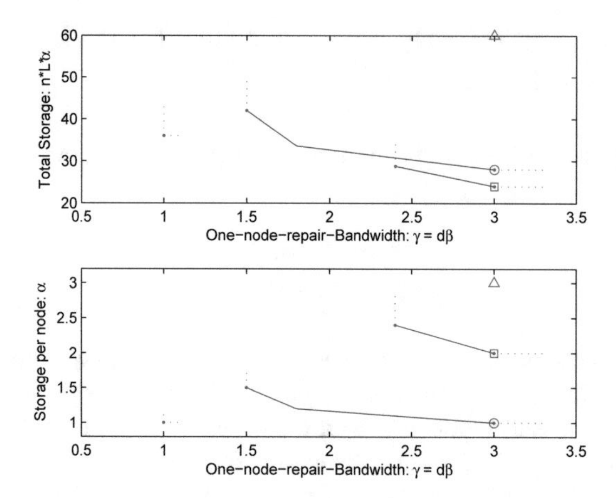

Fig. 4. Points displaying $T_\alpha \times T_\beta$ for Dimakis-DS-code (marked with a square), RS DSS-code (marked with a circle), and a Repetition-DS-codeRS DSS-code (marked with a triangle). The repetition code collapses to a single point.

4.1 Trade-Off for Changing Values of L (n and k Fixed)

Let us examine the trade-off plot when the size of the file $|\mathcal{F}| = L * B * \log_2 q$ is kept fixed but the values of L are changed. Under this which we consider a fair conditions as we can see from Fig. 5 the trade-off curve relating the Total Storage ($T_\alpha = nL\alpha$) and the Total Downloaded Data ($T_\gamma = \gamma L$) are all the same. On the other hand, the trade-off curves relating the Storage-per-Node (α) and Per-Node-Downloaded Data (γ) are distinct curves as it can be seen in Fig. 6.

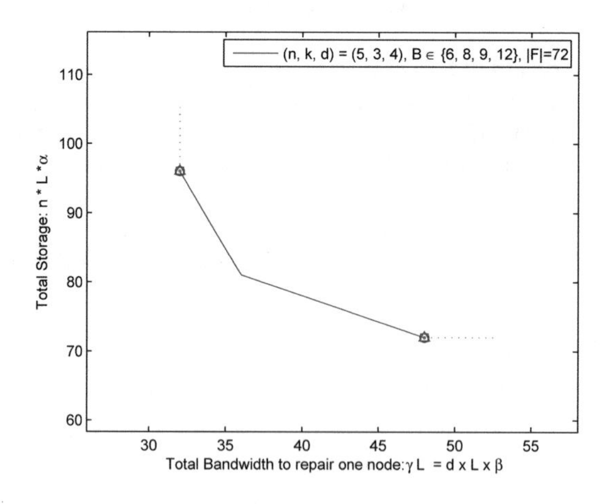

Fig. 5. Plot displaying the trade-off curves for several values of B. Since the file size does not change, all plots corresponds to the same line.

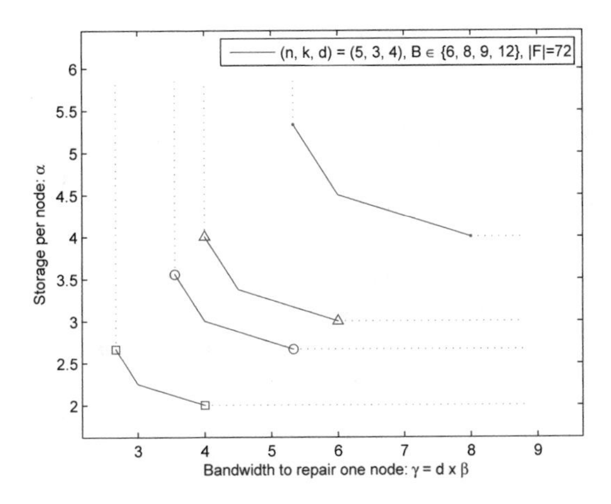

Fig. 6. Plot displaying the trade-off curves for several values of B. A distinct plot for each value of B.

4.2 Trade-Off Curve — Changing Values of d (n and k Fixed)

It has been pointed out by Dimakis [8] that a DSS-code with larger d is a better code. This statement remains true—when comparing codes with parameters $n, k, d, \alpha, \beta, B, L$, for fixed values of n and k—despite the trade-off curves under examination. Let us observe the current situation by examining Fig. 7. The bottom figure corresponds to the tradeoff curve $\alpha \times \gamma$ (since $L = 1$). In all cases the file size is $|\mathcal{F}| = LB = n = 15$. As it can be seen the relative positions of all cases remains the same (except for a scale change). With this much evidence we can say that the fair way to compare distinct codes is to fix the value of the file size and plot for the i code with parameters $n_i, k_i, d_i, \alpha_i, \beta_i, L_i, B_i$ the trade-off curve. It should be noticed that the file size $|\mathcal{F}| = L_i B_i$.

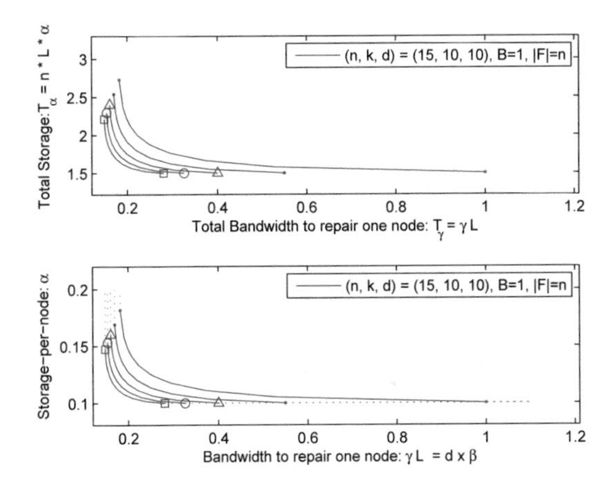

Fig. 7. Plot displaying the trade-off curves for several values of B.

5 Capacity of DSS Codes

The minimum cut is the cut between the source S and the data-collector (DC) in which the total sum of the edge capacities is smaller. Let us use the notation C_{mincut} to represent this sum. From Lema 2 in Dimakis paper [8], it is known the condition which guarantee the repair of a node (by downloading data from any set of d nodes) is to have the data-to-be-stored $\underline{u}$ with a number of symbols B larger then C_{mincut}. For a given a set of parameters $[n, k, d, \alpha, \beta, B, L]$ the engineering target would be to store the largest amount of symbol B as possible. But, since the relation $B \leq C_{\text{mincut}}$ has to be obeyed (according Lema 2) the capacity C of this scenario is defined as

$$C \triangleq \sum_{i=0}^{\min(d,k)} \min\{(d-i)\beta, \alpha\}. \tag{31}$$

Table 1. Maximum values of B for several values of (α, β)

n	k	d	α	β	$B = C$
5	3	4	2	1	6
5	3	4	3	1	8
5	3	4	4	1	9
5	3	4	2	2	6
5	3	4	3	2	9
5	3	4	4	2	12
2	1	1	1	1	1

T_α	T_γ
$120 = 5 \times 12 \times 2$	$48 = 4 \times 12 \times 1$
$135 = 5 \times 9 \times 3$	$36 = 4 \times 9 \times 1$
$160 = 5 \times 8 \times 4$	$32 = 4 \times 8 \times 1$
$120 = 5 \times 12 \times 2$	$96 = 4 \times 12 \times 2$
$120 = 5 \times 8 \times 3$	$64 = 4 \times 8 \times 2$
$120 = 5 \times 6 \times 4$	$48 = 4 \times 6 \times 2$
$144 = 2 \times 72 \times 1$	$72 = 1 \times 72 \times 1$

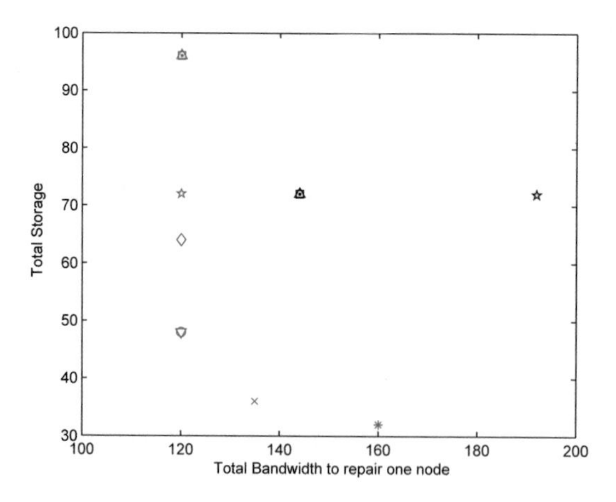

Fig. 8. Total Storage versus Total Downloaded data for several parameters set. $[n, k, d]$ as displayed on Table 1. The pair $[\alpha, \beta] = [2, 1]$ rendered a code with $[T_\alpha, T_\beta] = [120, 48]$ identified by a circle. The pair $[\alpha, \beta] = [4, 2]$ also rendered a code with $[T_\alpha, T_\beta] = [120, 48]$ (identified by a down-triangle). The other pairs $[\alpha, \beta]$ are: $[3, 1]$ - circle,$[4, 1]$ - x, $[2, 2]$ - $*$, $[3, 2]$ - square, $[1, 1]$ - lower triangle.

We can thus conclude that the capacity C is the maximum number of symbols that the sender is allowed to send (the maximum flow) to the DSS when the parameters set $[n, k, d, \alpha, \beta]$ has been selected (and thus the requirement $B \leq C$). Of course, if a scenario is characterized by the set $[n, k, d, \alpha, \beta]$, a file of size $\mathcal{F}$ will have to be parsed into L pieces such that $\mathcal{F} = LB \leq LC$.

Example 4: Let us consider a code to be built with parameters $[n, k, d] = [5, 3, 4]$. For this case, using (31) we have the maximum values of B for several values of (α, β) presented in Table 1. The table on the right shows the Total Storage (T_α) and the Total Downloaded data (T_γ) for each set of parameters when a file of size $\mathcal{F} = LB = 72$ is to be stored. Also shown (last line of the table) the value of C for a repetition code—just for comparison.

By examining Fig. 8 one can easily see that the best choice if Total Storage is of concern is the set of parameters $[n, k, d, \alpha, \beta, B] = [5, 3, 4, 2, 1, 6]$ or $[n, k, d, \alpha, \beta, B] = [5, 3, 4, 4, 2, 12]$ both corresponding to the point $(T_\alpha, T_\gamma) = (120, 48)$. To minimize the Total Download one should select $[n, k, d, \alpha, \beta, B] = [5, 3, 4, 4, 1, 9]$ which corresponds to the point $(T_\alpha, T_\gamma) = (160, 32)$.

Example 5: Let us consider now two codes with the following parameters:

$$(n, k, d) = (15, 10, 14),$$
$$(n, k, d) = (10, 5, 9).$$

Table 2. Maximum values of B for several values of (α, β)

n	k	d	α	β	$B = C$	L	T_α	T_β
15	10	14	1	1	10	2	180	168
15	10	14	2	1	20	6	180	84
15	10	14	3	1	30	4	180	56
15	10	14	4	1	40	3	180	42
15	10	14	2	2	20	6	180	168
15	10	14	3	2	30	4	180	112
15	10	14	4	2	40	3	180	84
15	10	14	3	3	30	4	180	168
15	10	14	4	3	40	3	180	126
10	5	9	1	1	5	24	240	216
10	5	9	2	1	10	12	240	108
10	5	9	3	1	15	8	240	72
10	5	9	4	1	20	6	240	54
10	5	9	2	2	10	12	240	216
10	5	9	3	2	15	8	240	144
10	5	9	4	2	20	6	240	108
10	5	9	3	3	15	8	240	216
10	5	9	4	3	20	6	240	162

For these cases, using (31) we have the maximum values of B for several values of (α, β) presented in Table 2. The table on shows the capacity, the total storage and the total data downloaded for each set of parameters. In the first code we can see that for maximum capacity and minimum Total Download the best choice is $[n, k, d, \alpha, \beta, B] = [15, 10, 14, 4, 1, 40]$ and for the second one is $[n, k, d, \alpha, \beta, B] = [10, 5, 9, 4, 1, 20]$.

6 Conclusion

Distributed storage of data is an important issue nowadays. Finding the best way to do it is also a big challenge. As we already mentioned there are many studies dealing with this problem each suggesting a cleverly distributed storage codes that can work properly with distributed storage networks. In this paper, we have proposed a fair comparison way which allows establishing which code gives better performances. Hence, the workings of three different DSS codes (Repetition, Reed-Solomon and Regenerating Codes) have been investigated and their performance compared.

References

1. Moon, T.K.: Error Correction Coding: Mathematical Methods and Algorithms, 1 edn. Wiley-Interscience (2005)
2. Weatherspoon, H., Kubiatowicz, J.: Erasure coding vs. replication: a quantitative comparison. In: Proceedings of 1st International Workshop Peer-to-Peer System (IPTPS), pp. 328–338 (2007)
3. Memorandum of understanding for the implementation of the COST Action "European Cooperation for Statistics of Network Data Science" (2015)
4. Rashmi, K.V., Shah, N.B., Kumar, P.V.: Regenerating codes for errors and erasures in distributed storage. In: Proceedings of IEEE International Symposium on Information Theory (ISIT) (2012)
5. Dimakis, A.G., Prabhakaran, V., Ramchandran, K.: Decentralized erasure code for distributed networked storage. IEEE/ACM Trans. Netw. **14**, 2809 (2006)
6. Rawat, A.S., Koyluoglu, O.O., Silberstein, N., Vishwanath, S.: Optimal locally repairable and secure codes for distributed storage system. IEEE Trans. Inf. Theory **60**, 212–236 (2013)
7. Dimakis, A.G., Godfrey, P.B., Wainright, M., Ramchadran, K.: Network coding for distributed storage systems. In: Proceedings of the 26th IEEE International Conference on Computer Communications, Anchorage, AK, pp. 2000–2008, May 2007
8. Dimakis, A.G., Godfrey, P.B., Wu, Y., Wainright, M.J., Ramchandran, K.: Network coding for distributed storage systems. IEEE Trans. Inf. Theory **57**(8), 5227–5239 (2011)
9. Rashmi, K.V., Shah, N.B., Kumar, P.V.: Optimal exact-regenerating codes for distributed storage at the MSR and MBR points via a product-Matrix Construction. IEEE Trans. Inf. Theory **57**(8), 5227–5239 (2011)
10. Paunkoska, N., Finamore, W., Karamachoski, J., Puncheva, M., Marina, N.: Improving DSS efficiency with shortened MSR codes. In: ICUMT (2016)

Comparative Analysis of Digital Circuits Using 16 nm FinFET and HKMG PTM Models

Satish Masthenahally Nachappa[1(✉)], A. S. Jeevitha[2], and K. S. Vasundara Patel[2]

[1] Department of Electronics and Communication, Middle East College, Muscat, Oman
satish@mec.edu.om

[2] Department of ECE, BMS College of Engineering, Bengaluru, India
jeevithagowda12@gmail.com, vasu.ece@bmsce.ac.in

Abstract. Low Power VLSI design has become the most important challenge of present chip designs. Advances in chip fabrication have made possible to design chips at high integration and fast performance. Reducing power consumption and increasing noise margin have become two major concerns in every stage of SRAM designs. In this paper the 6T and 8T SRAM cells are constructed using High-K Metal Gate and FinFET for low power embedded memory applications. These SRAM cells' performance are analyzed and compared in terms of basic parameters, such as power consumption and static noise margin (SNM).

Keywords: SRAM · High-K Metal Gate · FinFET · Low power Leakage current · Static noise margin

1 Introduction

As the technology node scales beyond 50 nm, there is need for inventive methodologies to ignore the hurdles due to the basic physics that bounds conventional MOSFETs. Design of digital circuits finds the prominent applications in storing data in the recent days. Since SRAMs are faster than DRAM, as it is faster to access SRAMs finds more applications in Cache memory. The simplest static random access memory cell consists of 6 transistors.

New demanding technologies have started emerging under the experiment and research to overcome the limitations of conventional MOSFETs. On an account of scaling philosophy, the conventional device bulk silicon (Si) technology airs the explosion of the chip due to higher leakage powers, whereas the upcoming devices like Silicon On Insulator, High K Metal Gate as shown in Fig. 1, FinFET, are emerged guaranteeing the low power resolution for the IC implementation. This proposed work shows the design of low power memory cells and also optimization of its performance is derived from the basic level of design.

K. Arai et al. (Eds.): FICC 2018, AISC 886, pp. 81–90, 2019.
https://doi.org/10.1007/978-3-030-03402-3_7

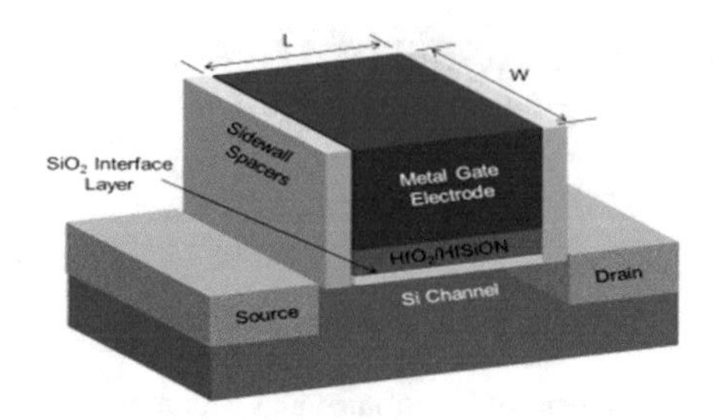

Fig. 1. Schematic diagram of High-K and Metal Gate MOSFET.

1.1 Static Random Access Memory

Static RAM or SRAM is a type of semi-conductor memory [7] which comprises of flip-flop or the bi-stable latching circuit to store data bits (0 or 1). Static Random Access Memory (SRAM) demonstrates data stability, but this is volatile in nature i.e., data is in time lost when the circuit or memory is powered off. The word static distinguishes SRAM by DRAM which needs to be refreshed at particular intervals. Static RAM is more expensive and faster than Dynamic RAM, Static RAM is usually in use for CPU cache memory while the DRAM finds its applications in the main memory of computers [9]. The power consumption of static random access memory (SRAM) varies extensively, which depends on how frequently the memory is accessed.

1.2 High-K Metal Gate

To tolerate the scaling of the device beyond 45 nm, manufacturers of semiconductor device have initiated that the High K and Metal Gate (HKMG) stacks within the MOSFET which used in the digital CMOS, which creates basis for the logic circuits inside the SoC (System on Chip) and microprocessors and used in cell phones, tablets, computers, etc. As well, architecture of memory such as DRAM also drifted to High K dielectrics. An insulator class with the metal oxide which has a relative dielectric constant which is greater than or equal to 9 and also which involves metals which belongs to the group 3–5, Al and lanthanides forms a High-K dielectrics. Equation (1) describes the dielectric constant K, where Dielectric permittivity = εd and free space permittivity = ε0

$$K = \epsilon_d / \epsilon_o \tag{1}$$

Some of the examples of High K dielectrics are Lu_2O_3, Ta_2O_5, La_2O_3, Sc_2O_3, Nb_2O_5, Y_2O_3, ZrO_2, TiO_2, $HfZrO_4$, HfO_2, Al_2O_3 and the mixtures of these.

1.3 FinFET

Scaling of traditional planar MOSFETs is been facing issues such as variations in device parameters, significant DIBL, leakage and subthreshold swing degradation. In order to solve these problems, three-dimensional (3-D) device structures as shown in

Fig. 2 possibly would be the solution and have been worked on. FinFETs are built on SOI wafers or bulk silicon. FinFETs among the 3-D devices are very promising competitor for the future nanoscale technology and memory applications with high-density. In recent days, FinFET technology has been seeing a huge increase in the implementation within the ICs.

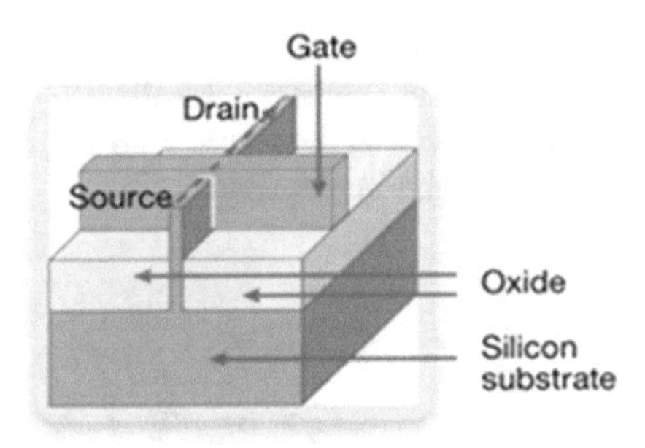

Fig. 2. Structure of FinFET

FinFET is primarily a multi-gate FET (Field Effect Transistor) which had been scaled further of MOSFET. The FinFET is a transistor design, first developed by Chenming Hu and colleagues at the University of California at Berkeley. These device structures have all the properties similar to a typical transistor, and also have few benefits on the CMOS. MOSFET has some practical issues like short channel effects, DIBL, power dissipation, performance degradation. Hence to avoid the difficulties faced by conventional MOSFETs, FinFET structure came into existence to build the transistors further efficient.

Different modes of operation of FinFET

- Shorted-gate mode [SG]: Two gates that are joined together as shown in Fig. 3(a).
- Independent gate mode [IG]: The gate is driven by independent signals as shown in Fig. 3(b).
- Low power mode [LP]: In this mode of operation one gate is connected to reverse bias, this reduces leakage current.

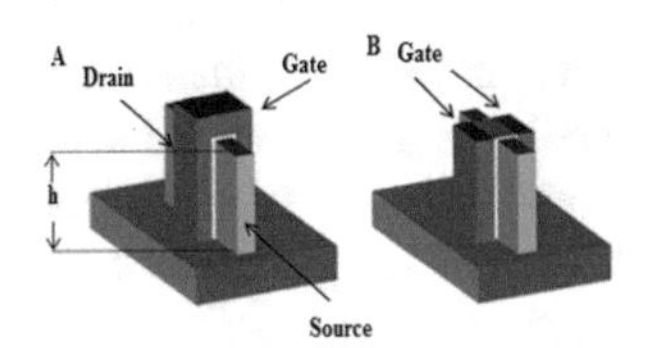

Fig. 3. (a) Shorted gate (b) Independent gate.

1.4 Conventional 6T SRAM Cell

The six transistors SRAM (static random access memory) cell configuration is implemented by connecting two inverters back to back as shown in Fig. 4. The four transistors at the center are modeled to create two cross coupled inverters [8]. The transistors are designed very tiny to save the area of chip. A ‘0’ (low) input on the inverter1 will generate a ‘1’ (high) value on the inverter2 which is because of the feedback structure, which amplifies the ‘0’ value on inverter2. Similarly, the first inverter that is having a ‘1’ (high) input value will results a ‘0’ input value on inverter2, so the ‘0’ (low) input value is given as feedback onto the first inverter. Hence, its logical value i.e., 0/1 is stored by the inverters.

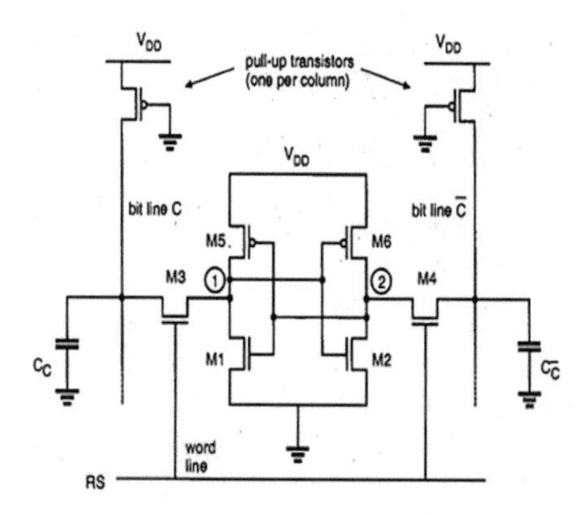

Fig. 4. CMOS SRAM cell.

1.5 Predictive Technology Models (PTM) for Multi-gate Transistors

A new generation of PTM for multi-gate transistor and metal-gate transistors, specifically FinFET and High-K Metal Gate for 16 nm technology nodes is developed. These model parameters are developed using BSIM model. It is a surface-potential-based skimmed model which can model double-gate, tri-gate and gate-all-around different multi gate models. The new PTMs for 16 nm multi-gate transistors have been designed in two specific application versions, high performance (HP) and low-standby power (LSTP).

This retains the standard framework of BSIM4 and BSIM-SOI models to incorporate the effect of 3D structure and QME (quantum mechanical effects) on device characteristics and short channel effects. The model for real-device effects velocity saturation, series resistance, mobility degradation, parasitic capacitance, etc. This allows ease in efficient extraction of model parameters. FinFET device behavior is most sensitive to the technology specifications, primary parameters and physical parameters.

1.6 Static Noise Margin (SNM)

SNM is measure of how strong the system is against the noise. More the SNM better will be the system. It is defined as the maximum quantity of noise voltage which can be introduced at the two inverters output, so the data is retained by the cell. By using a butterfly curve method, SNM can be calculated. In Fig. 5, we can see SNM which is graphically represented for the bit cells which is holding data. The resulting two-lobed curve is called as "butterfly curve" used to determine SNM. The largest side of the square diagonal length that can be embedded inside the lobes of the butterfly curve is known as SNM. From the plot shown in the figure, it causes the Inverter1 to move downwards and VTC of Inverter2 to move to the right. When both the NM value moved, then the curves will meet at only two points.

(1) SNM calculation

The SNM calculation with respect to the plot in Fig. 5 is explained here.

Butterfly curve = Maximum side of the square.

SNM = Maximum side of the Square = Maximum length of the diagonal of Square/$\sqrt{2}$.

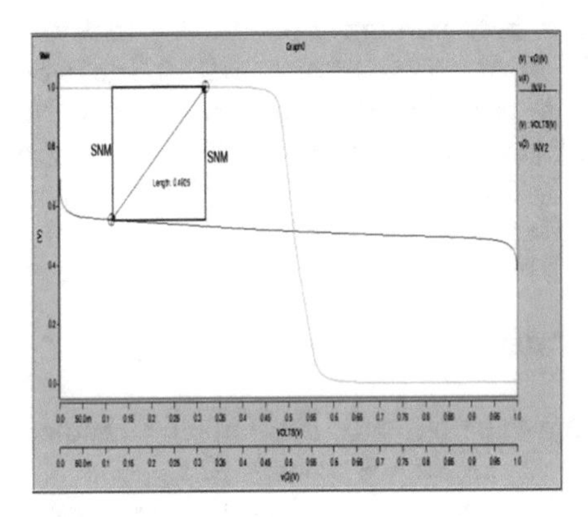

Fig. 5. Measuring SNM using butterfly curve.

2 Single Ended 8T SRAM Cell

An eight transistor, single ended S-RAM design as shown in Fig. 6 which by improving the read SNM, improves the data stability and also Power dissipation is reduced [1]. As per proposed design, the circuit which is used for read purpose is transmission gate. The inversion of the read word-line signal (RWL) is the additional signal, RWLB [4]. It regulates the M7 transistor of the transmission gate. Though the RWLB and RWL are asserted and when the transmission gate is ON, the data stored node gets linked with RBL. Hence the stored data at point Q is read through or moved to RBL.

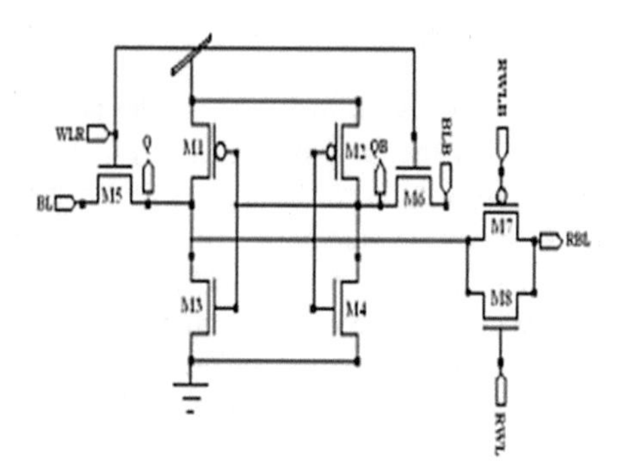

Fig. 6. Single-ended 8T SRAM cell.

The major advantage of the proposed circuit is that this is not essential to arrange a precharge circuit and the sense amplifier as essential in the prior six transistor Static RAM cell, since the saved data/value is straight away moved through the transmission gate [6]. The charging/discharging power of RBL is used up only after RBL is altered. [2] Accordingly, there will be zero power dissipation on RBL if an impending data is similar to earlier state. This proposal reduces the bit-line power in two cases, they are the successive 1's and successive 0's.

The simulations have been performed on HSPICE tool version [10] Z-2007 using 16 nm technology with input voltage of range 0.9 V. The two parameters total power dissipation and SNM are used for the simulation to verify SRAM performance. The performance of FinFET [11], HKMG and MOSFET are compared for each parameter [5]. The SNM of SRAM cell is demonstrated as the maximum noise magnitude that does not disturb the stored bit of the SRAM cell.

3 Simulation Results

HSPICE simulations are done using 16 nm FinFET technology. The 6T and the 8T cells are compared for several SRAM metrics. The widths of all the transistors were chosen to obtain optimal curves with least distortions in read-write and hold modes for the HSPICE simulations. To get the read failure analysis, a ramp voltage is given at one of the bit or $\overline{bit}$ lines and the output is read at the other and vice versa. On overlaying both the curves, a similar graph is obtained as to the one in Fig. 5, which is termed as the butterfly curve. SNM [3] is the side of the prime square that fits in the butterfly. By decreasing Vdd step by step, SNM can be tabulated. When the two curves flip, read failure is said to have occurred as shown in Fig. 7.

The SNM [3] was presented for both 6T SRAM and 8T SRAM as indicated by Tables 1 and 2 and it was found that read stability of the 8T SRAM is improved as compared to 6T SRAM.

Table 1. SNM for 6T and 8T SRAM cell using LP-HKMG

VDD (V)	SNM 6T SRAM (V)	SNM 8T SRAM (V)
0.9	0.3	0.2
0.8	0.2	0.16
0.7	0.17	0.14
0.6	0.17	0.12
0.5	0.14	0.09
0.4	0.1	0.06
0.3	0.07	0.05
0.2	Read failure	0.012
0.1		0.0037

Table 2. SNM for 6T and 8T SRAM cell using FINFET.

VDD (V)	SNM 6T SRAM (V)	SNM 8T SRAM (V)
0.85	0.29	0.38
0.7	0.26	0.28
0.6	0.24	0.23
0.5	0.21	0.18
0.4	0.19	0.14
0.3	0.14	0.1
0.2	0.07	0.07
0.1	Read failure	0.02

The below figures (Figs. 7, 8, 9 and 10) depict the read SNM and the read failure for the 6T and 8T SRAM cells respectively. The read-failure occurs when the two curves flips as shown in Figs. 8 and 10 for both LP-HKMG and LP-MultiGate PTM models.

SNM is tabulated for both 6T SRAM and 8T SRAM which is shown in Tables 1 and 2 and it is seen that read stability for both LP-HKMG and LP-MultiGate PTM models is improved compared to 6T SRAM cell.

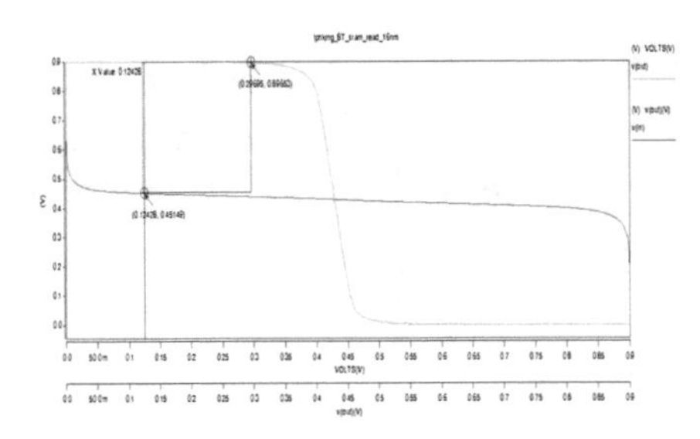

Fig. 7. LP HKMG 8T SRAM cell butterfly curve.

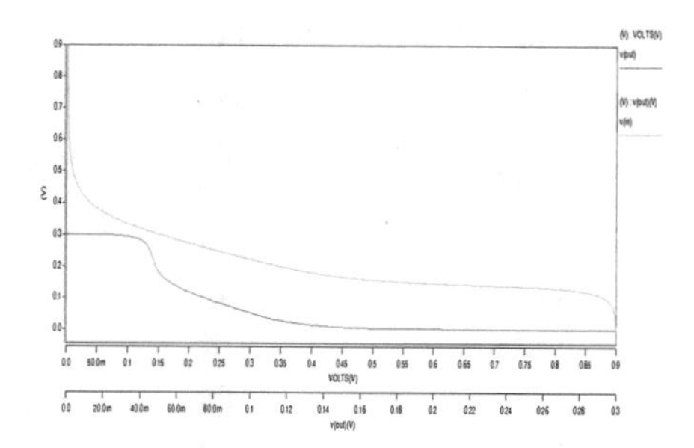

Fig. 8. Read failure in LP HKMG 8T SRAM cell.

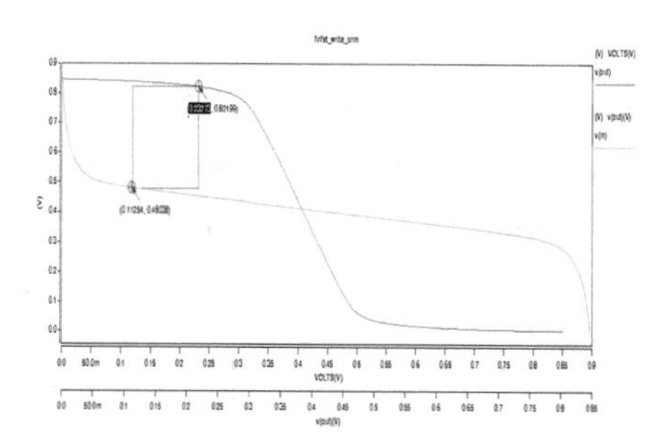

Fig. 9. FinFET 8T SRAM cell butterfly curve.

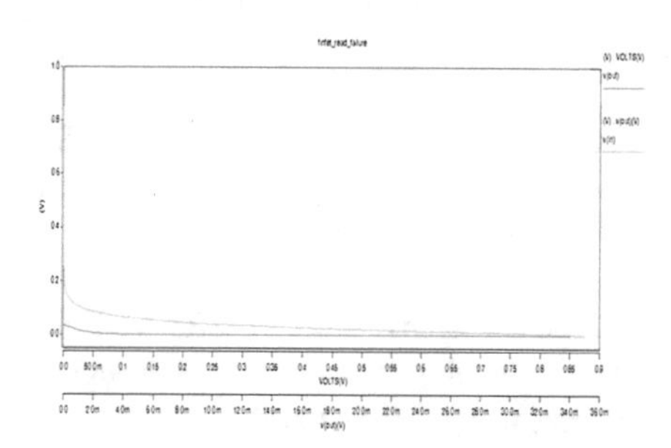

Fig. 10. Read failure in FinFET 8T SRAM cell.

Table 3. Delay and power dissipation for 6T SRAM cell.

Model	Write delay	Read delay	Power dissipated
CMOS	30.293 ps	3.884 ps	4.45753 mw
LP HKMG	39.45 ps	37.158 ps	46.5259 nw
FinFET	132.31 ps	3.125 ns	149.2712 nw

Table 4. Delay and power dissipation for 8T SRAM cell

Model	Write delay	Read delay	Power dissipated
CMOS	5.9663 ns	10.077 ps	4.45753 mw
LP HKMG	4.0052 ns	22.767 ps	46.5259 nw
FinFET	6.0061 ns	1.344 ps	149.2712 nw

Tables 3 and 4 show the simulation results for delay and power dissipation of 6T SRAM and 8T SRAM, respectively. The reduction in dissipated power and read-failure probability has been achieved better by using both models when compared to CMOS technology. Butterfly curve and read-failure representation for both models are displayed graphically.

4 Conclusion

In this paper 6T SRAM and 8T SRAM cells are designed and analyzed using 16 nm Low Power High-K Metal Gate and Low Power Multi-gate (FinFET) PTM models and their read-write operations were performed successfully. The reduction in access time and read-failure probability has been achieved better by using both models when compared to CMOS technology. Butterfly curve and read-failure representation for both models are displayed graphically.

References

1. Gopal, M., Prasad, D.S., Raj, B.: 8T SRAM cell design for dynamic and leakage power reduction. Int. J. Comput. Appl. **71**(9), 0975–8887 (2013)
2. Rahman, N., Singh, B.P.: Design and verification of low power SRAM using 8T SRAM cell approach. Int. J. Comput. Appl. **67**(18), 0975–8887 (2013). Department of electronics and communication FET-MITS (Deemed university), Lakshmangarh, India
3. Pal, P.K., Kaushik, B.K., Dasgupta, S.: Design metrics improvement for SRAMs using symmetric dual-k spacer (SymD-k) FinFETs. IEEE Trans. Electron Devices **61**(4), 1123 (2014)
4. Kushwah, C.B., Vishvakarma, S.K.: A sub-threshold eight transistor (8T) SRAM cell design for stability improvement. Nanoscale Devices and VLSI/ULSI Circuit and System Design Lab, Electrical Engineering, Indian Institute of Technology, Indore, M.P., India
5. Rahman, N., Singh, B.P.: Design of low power SRAM memory using 8T SRAM cell. Int. J. Recent Technol. Eng. **2**(1), 123 (2013). ISSN: 2277–3878
6. Pasandi, G., Fakhraie, S.M.: An 8T low-voltage and low-leakage half-selection disturb-free SRAM using bulk-CMOS and FinFETs. IEEE Trans. Electron Devices **61**(7), 235 (2014)
7. Kang, S.M., Leblebici, Y.: CMOS Digital Integrated Circuits Analysis and Design. Tata McGraw-Hill Education, New Delhi (2003)

8. Yeo, K.S., Roy, K.: Low – Voltage, Low – Power VLSI Subsystems. McGraw-Hill, New York (2005)
9. Baker, R.J.: CMOS: Circuit Design, Layout, and Simulation, vol. 1, 3rd edn. Wiley, New York (2010)
10. HSPICE LAB Manual – VSD Centre, by M. S. Ramaiah School of Advanced Studies
11. HSPICE® Simulation and Analysis User Guide, by Synopsis

Possibilities to Improve Online Mental Health Treatment: Recommendations for Future Research and Developments

Dennis Becker(✉)

Institute of Information Systems, Leuphana University, Lüneburg, Germany
dbecker@leuphana.de

Abstract. Online mental health treatment has the potential to meet the increasing demand for mental health treatment. But low adherence to the treatment remains a problem that endangers treatment outcomes and their cost-effectiveness. This literature review compares predictors of adherence and outcome for clinical and online treatment of mental disorders to identify ways to improve the efficacy of online treatment and increase clients' adherence. Personalization of treatment and client improvement tracking appears to provide the most potential to improve clients' outcome and increase the cost-effectiveness of online treatment. Overall, it was noticed that decision support tools to improve online treatment are commonly not utilized and that their influence on treatment is unknown. However, integration of statistical methods into online treatment and research of their influence on the client has begun. Decision support systems derived from predictors of adherence might be required for personalization of online treatments and to improve outcome and cost-effectiveness to ease the burden of mental disorders.

Keywords: Online treatment · e-mental-health · Outcome prediction

1 Introduction

Mental health disorders such as anxiety, depression, or substance abuse are an increasing problem in our society [1]. In Europe, one in four people have reported that they suffered from at least one mental disorder during their lifetime [2]. A similar situation is found in the USA where approximately 26% of the population suffer from at least one mental disorder [3]. According to the World Health Organization, the gap between the need for treatment of mental disorders and the accessibility of treatment is increasing [4]. As a result, approximately 35% to 50% mentally ill people receive no treatment due to long waiting lists and scarcity of therapists [5,6]. Mental disorders, besides having a high prevalence, create a struggle for the individuals, their families, and a high economic cost for society [7]. The costs arise from the direct treatment of depression but also due to the indirect costs from the loss of productivity or the workplace [8]. The lost

K. Arai et al. (Eds.): FICC 2018, AISC 886, pp. 91–112, 2019.
https://doi.org/10.1007/978-3-030-03402-3_8

earnings in America are estimated to be $193.2 billion per year [9]. The global costs of mental health were estimated at $2.4 trillion, in 2010, and is predicted to increase to $8.5 trillion by 2030 [10].

Online treatment has gained an increasing importance for the delivery of mental health interventions [11,12]. It can help in closing the gap between the demand and available treatment spots [13] and might be more cost-effective [14]. Although online treatment requires fewer intervention costs [15], studies regarding their cost-effectiveness report mixed results [16,17]. Typically, online treatments utilize computerized cognitive behavioral therapy in form of brief therapy that has been proven to reduce symptoms [18,19]. Brief therapies are typically shorter than traditional therapy. They are a focused application of therapeutic techniques to specifically targeted a symptom or behavior [20]. In general, computerized treatment applications encompass screening and treatment functionality. Using a computer or mobile device, these treatments can be processed and screening questionnaires can be filled in. During the treatment, the improvement of each client is supervised by a therapist. There also exist self-guided treatment programs that do not provide supervision by a therapist and the clients work at their own pace [21].

However, online treatments struggle with high drop-out rates [22]. A review study revealed that the drop-out rate of internet-based treatment programs for psychological disorders ranges from 2 to 83% and with a weighted average of 31% [23]. The identification of clients prone to drop-out is crucial because the participants who leave the treatment early are unlikely to recover and the failed treatment can negatively affect their attitude towards subsequent therapies [24]. In clinical treatment, therapists actively work with clients during therapy sessions where they can apply their knowledge to adjust the current treatment accordingly and keep the clients engaged into treatment. Online treatment, however, lacks face-to-face interaction and poses difficulties for therapists to provide clients with the same level of guidance.

Fortunately, a variety of data is collected during the online treatment process, which is already digitally available. To improve the efficacy of online treatment, this data can automatically be evaluated to aid therapists with their work. Predictions of the expected treatment outcome would allow to assign the client to the most beneficial treatment. Closely linked to an early prediction of outcome is the clients' adherence to the treatment. Research in clinical treatment has shown that data analyses using statistical models, can be used to identify trends and forecast events. These statistical models can be used to develop prognostic models or decision support tools for therapists. In clinical medicine, the use of such prognostic models to estimate, for example, clients' outcome or risk of developing a certain illness, is shown to be beneficial [25]. Similarly, in online treatment, the information provided by these tools would allow therapists to allocate their time more efficiently. They could focus on clients that show low adherence or little improvement and are prone to drop-out.

Because the clinical treatment of mental disorders is researched for a longer period of time than online treatment, this literature review compares the clinical treatment process to their online counterpart. This comparison allows to summarize the current state of the art of online treatment and potentially aids in identifying new research opportunities and beneficial methods to be integrated into online treatment. To provide a comprehensive view of the treatment process, this review covers adjacent fields such as diagnosis support and relapse preventions. The focus, however, lies on predictors of adherence to provide suggestions for further research directions and practical recommendations for online treatment improvement. This review could also be useful as a guide for online treatment developers to identify predictive methods to be incorporated into their treatment program. Development and integration of decision support tools, derived from predictors of adherence, could be a promising step towards treatment personalization and increase their cost-effectiveness. A review of the current online treatment process was necessary to gain a clearer idea of where improvements can be made and if aspects of clinical treatment were missed.

Section 2 describes the literature review process and introduces the different treatment phases that are used to structure this review, while Sect. 3 compares the predictors of treatment adherence to online and clinical mental health treatment. Section 4 discusses the presented literature in order to identify most promising predictors of adherence and point towards promising research directs. The limitations of this review are stated in Sect. 5, and a summary and conclusion is provided in Sect. 6.

2 Methods

In order to structure the literature search and to provide a better overview of the results, the online treatment process is split into three phases: screening, treatment, and relapse prevention. These three phases and their occurrence in a typical online treatment are shortly explained in the following. In the first phase, the symptoms of a potential client are assessed using standardized screening questionnaires. After assessing the symptom severity a decision has to be made if the applicant is suited for the treatment. For example, a candidate that is suffering from immediate suicidal thoughts is not suited for online treatment. Otherwise, people of any severity level can be treated and possibly benefit from online treatment. After access to an online program has been granted, interventions are provided to the new client. These interventions are based on psychological programs so that the clients can recognize their problems and develop solving strategies [26,27]. This can be considered as the second phase of treatment. During this period, the client can either improve or drop-out of treatment [28]. Besides the interventions, additional questionnaires can be inquired to track symptom development. The third phase begins after the active treatment, where clients are facing a high chance of deteriorating symptoms and a severe change of relapse. Especially, mental disorders such as depression are highly recurrent [29]. Therefore, it is essential to monitor, supervise, and provide additional interventions to counter relapse if necessary.

The literature search was conducted for all three phases of treatment to identify papers that research outcome or adherence in either clinical or online treatment. For the literature search, databases such as PsycINFO, Scopus, and PubMed were consulted for peer-reviewed sources. The keyword based search consisted of (but were not restricted to) the following terms: "adherence", "dropout", "mobile treatment", "outcome prediction", "aftercare", and "relapse", each in combination with the keyword "mental health".

Sources that are included regarding the clinical treatment had to research the influence of treatment factors on the outcome or adherence. Similarly, the papers that are included for online treatment could either use a smartphone or the Internet regardless if they were supervised or unsupervised. To broaden the focus, papers that discuss prediction of symptom development, diagnosis, or aim to increase adherence are included. Sources that do not relate to mental health or discuss medication treatment are excluded from the review. Additionally, meta-analyses and systematic reviews were sought for each relevant area of investigation to provide further inside and to retrieve additional papers. This review is not a systematic review and not based on a single search. Multiple searches were used to explore and refine factors for treatment outcome and adherence in both treatment types. Furthermore, the specific number of articles retrieved and excluded was not tracked.

3 Results

The results are separated according to the defined treatment phases. The clinical and online treatment are compared and identified predictors for treatment development are pointed out. A summary of the predictors and possibilities for improvement are found in Table 1.

3.1 Screening Phase

In the screening and diagnosis phase, little difference between clinical treatment and online treatment could be found. In clinical treatment, the screening phase consists of questionnaires and interviews to diagnosis the clients [30,31]. Similarly, the screening phase in online treatments consist of questionnaires to assess the clients and depending on the treatment program it can include telephone calls. This suggests that tools that are applicable for clinical treatment can also be used for the online counterpart such as automatic questionnaire evaluation or diagnosis support. Although an application of these techniques appears to be straight forward, research papers on such support were not found.

However, for online treatment, the lack of personal interaction already begins in this early stage, but text analysis could be a potential diagnoses support tool to help close this gap. Free text written by a client from processing a task can be helpful in identifying mental disorders such as schizophrenia, mania [32], and attention deficit disorders [33]. Text analysis can be also used to identify genuine suicide letters [34]. Therefore it might also be used to infer suicidal risk during the screening process.

Other data that is assessed in the screening phase, are the clients' demographic data. Research from the field of clinical psychiatry has examined the connection between the demographics and the personality profile of a client and the outcome of their psychological treatment. Demographic features such as age, sex, duration and severity of symptoms appear to have a small effect on treatment outcome [35,36]. Also, socio-demographic characteristics such as lower socioeconomic status, lower household income, lower level of education, and marital status are associated with earlier therapy termination [37,38]. Personality profile traits such as intelligence, social assets, and experience with psychotherapy could be identified [39]. The negative personality traits rejection, aggression, passivity, and conflict are found significantly related to the extent of which symptoms improve [40]. Also, a higher motivation for the treatment indicates a better treatment outcome [41], and self-reported socio-demographics could be used to identify clients at risk for treatment resistance [42].

For online treatments, similar results are found for the prediction of outcome and adherence based on demographic features. In terms of demographic factors, it is shown that male gender, lower educational level, and co-morbid anxiety symptoms increase drop-out risk significantly in self-guided web-based interventions [43]. Conversely, higher age, higher education level, and motivation are associated with online treatment adherence [44].

Smartphones are another tool that can facilitate the screening phase. Smartphones provide new possibilities for the detection and treatment of mental disorders. They are the daily companions of many people, and are equipped with a multitude of sensors including accelerometers, location estimation via GPS or WiFi ID, light sensors, and microphones. In addition, social features such as call logs, text logs, and contact list data. The built-in sensors allow a user's physical context, social setting, phone usage and GPS location to be estimated. Thus, smartphones provide the advantages that a user's behavior can be tracked without any effort from the user. For diagnosis support, the history of self-reported mood measures can be used to forecast depressive episode [45]. Furthermore, smart phones can be used to detect a change in sleep pattern, which might help to predict symptoms of depression [46]. In the following treatment phases, the benefits of smartphones will be further explored.

3.2 Treatment Phase

After the clients are diagnosed, they are assigned to a suitable treatment. During this treatment phase, a variety of information about the client is inquired that can be utilized for the prediction of outcome and adherence. The collected data during the therapy provides more insight into the outcome of the clients' psychotherapeutic treatment than solely the demographic features and personality profile. First, the in-session experience, which encompasses a variety of measures such as the early improvement, perception of the treatment seasons, and the therapeutic alliance. Perception of the intervention and treatment, in general, can offer the possibility of client improvement tracking, which will be presented second. Improvement tracking can also be accomplished using daily measures such as diaries or measures of mood, which emphasizes the use of the smartphone. Lastly, additional factors for adherence improvement are mentioned.

(1) In-Session Experience: Ratings of the in-session experience and involvement [47] can be an indicator of clinical treatment outcome. Measurements of early change and symptom improvement have been shown to be a favorable indicator in clinical treatment to predict the course of therapy. Especially in cognitive behavior therapy, the early response to the treatment can predict the outcome of psychological treatment. Early responding clients are more likely to achieve remission than nonresponsive clients. Although a small number of early nonresponsive clients can improve during the clinical treatment, they are more likely for treatment failure or early drop-out [48]. Similarly, in online treatment, an early response to the treatment can lead to improved outcomes and increased adherence [49].

Moreover, the perception of the latest treatment session is a valuable measure of client adherence in clinical treatment. Immediately after a face-to-face treatment session, a questionnaire is used for basic measures of the therapy processes from the clients' vantage point. The measure of clients' self-esteem, mastery, and clarification are a strong predictor for a successful completion of the treatment program [50]. For online treatment, the use of such questionnaires that measure the experience of the client after the treatment were not found. A review study that points out the lack of use of this method analysis 21 studies and only 1 study collected the clients' experience [51].

However, in contrast to clinical treatment, online treatment provides metrics derived from system use, which are similar to in-session experience. The clients' use behavior with the intervention platform can be utilized to help prevent early drop-out and estimate expected improvement. For example, the log data provides an insight into adherence and improvement during the treatment process. The number of logins, actions per login, completed modules, and time spent are metrics that can identify differences between adherers and nonadherers [52] and system use can indicate symptom improvement [53]. These predictors can also be used in conjunction with demographic features [54]. Furthermore, the activities of the user during the first week of treatment can predict the user's adherence to the remaining interventions [55], and a correlation between user login frequency and remaining symptoms after the treatment is shown for depression [56]. D'Alfonso et al. (2017) describe the use of clients' posts in their social therapy platform to suggest further content. Typically these suggestion are given by so called moderators. However, the automated system can provide suggestions around the clock. Initially, the content suggestions were based on the clients' use history of the platform, but text analysis enables to refer the clients better content regarding their needs. This is the only paper found in the literature review that decries the use of statistical methods during the treatment. In addition, it is envisioned to include a chatbot into the system that could either provide the user help in choosing new content or provide basic conversations.

Another in-session experience that is predictive of the therapeutic success in clinical treatment is therapeutic alliance [57,58]. The therapeutic alliance is defined as the collaborative and affective bond between therapist and client that is formed during the face-to-face meetings. This bond is an essential element of the therapeutic process, and aids in improving the clients' condition and is considered to be the most important component of care by the clients [59]. In clinical treatment, the therapeutic relationship is measured using questionnaires. Therapeutic alliance measured from the clients and therapist perspective shows correlation to the therapy outcome [60]. Where client ratings of the alliance and in-session experience appear to be the better predictor of outcome than the ratings of the therapists [61,62]. The Internet based communication, which many online treatments provide, can establish a personal relationship that is comparable to face to face treatment [63,64]. A healthy relationship with the therapist via e-mail and chat reduces the drop-out rate and is beneficial for client improvement during the treatment [65]. A measurement of the working alliance three weeks into the treatment correlated significantly with the outcome, while the expected working alliance measured pre-treatment had no significance [66]. Therefore, the therapeutic relationship could be recorded periodically to predict the course of the treatment and help to optimize treatment decisions.

(2) Clients' improvement tracking: Questionnaires for tracking the clients' improvement during the treatment are widely used in clinical treatment and have been found to be particularly useful in examining the efficacy of psychotherapy [67]. The tracking of the clients' symptoms and development during the treatment and informing the supervising therapist about nonresponders and clients prone to drop-out improves the adherence of the susceptible clients and their treatment outcome [68]. Outcome questionnaires are widely considered valid indicators of progress in treatment, which provides a measure of the weekly change. A variety of outcome questionnaires exist [69], such as the COMPASS Tracing Assessment System [70], Outcome Questionnaire-45 (OQ-45) [71], Clinical Outcomes in Routine Evaluation-10 (CORE-10) [72]. Besides these generic outcome questionnaires, disorder-specific questionnaires exist, that are typically better suited for treatment prediction [73]. The three major concepts that are measured with outcome questionnaires are: individual symptomatic, interpersonal relationship, and social and work performance. These measures allow to compare the recovery rate among different clients. Based on the recovery curves of previous clients the expected recovery curve can be calculated. This allows to build an early warning system that identifies clients that deviate dramatically from this curve and to informs the therapist about the current recovery rate [67]. Continues monitoring and an early warning system appears useful for the identification of clients, which do respond to the therapy [74]. Tracking of the clients' symptoms provides the possibility to initiate measures to prevent clients from leaving the therapy and choosing different interventions for them. There is also evidence that the overall therapy outcome can be improved if feedback regarding the clients is provided to the therapist [75,76]. Although these outcome questionnaires for improvement tracking are particularly useful in clinical treatment, their use in an online treatment was not found.

More fine grained measures to track the clients' development are daily diaries [77], activities or behavior [78], and ecological momentary assessment (EMA) [79]. EMA enables clients to regularly and actively report on their symptoms and behavior a few times a day. It is used to sample the clients' current experiences while they are engaged in their typical daily routine and captures momentary ratings of experiences, and the assessment of moods, thoughts, symptoms, or behaviors [79,80]. In contrast to clinical treatment, online treatment provides the possibility of automatically analyzing such data. It has been shown that the clients' mood can be inferred based on free text provided in an activity diary using text mining [81]. In clinical treatment, EMA traditionally has been assessed using pen and paper. However, nowadays smartphones can simplify this process in clinical and online treatment. Currently, research aims towards estimation of the clients' current condition without required interaction with their phone. This could replace the use of daily EMA measures, which would reduce the workload with the treatment which can increase adherence [82]. However, it appears that estimation of the current mood of a client based on previous self-rated measures is challenging [83–85] and that more context-relevant information might be required to infer more reliable information [86]. These context-related information could be assessed from the smartphones' sensors. The accelerometer can be used to estimate the current activity [87], the microphone to estimate the current stress level [88,89], and GPS to estimate the current location. The locations that clients visit and movement patterns have already been linked to symptoms of depression [90]. These context-related information could further be used to provide clients with situational interventions regarding their needs and symptoms [91].

(3) Additional factors for adherence improvement: Some possibilities that contribute to improve adherence, but are not yet directly linked to the prediction of adherence, are mentioned. Smartphones can be used to deliver reminder via text messages. Previous studies in the field of depression [92], smoking cessation [93], and weight loss [94], have demonstrated that regular reminders can increase the treatment adherence. Therefore, it is an opportunity to increase adherence. However, to use of mobile reminder is not limited to online treatment, it can increase the adherence in clinical treatment as well [95,96]. Also, the application design and features of personalization [97,98] can contribute to the adherence as well as elements of gamification [99,100]. Gamification aims at increasing the intrinsic motivation to engage with the treatment by using elements found in games and present the content of an intervention in a simplified form that is easier to grasp.

3.3 Relapse Prevention Phase

Relapse is a common problem after successful treatment of mental disorders [29,101], and the need for regular face-to-face meetings or additional interventions can still remain [3,102]. In clinical treatment, the aftercare can consist of follow-up screenings and additional interventions to reduce the risk of recurrence and relapse [3,102]. The required amount of aftercare is usually estimated at the end of the active treatment and can depend on the initial strength of the depressive symptoms. In the case of depression, clients with severe depression are more prone to relapse and require more aftercare to prevent relapse or suicide [103]. The required aftercare is estimated from the sociodemographic features such as employment and family. These features mostly indicate the social support of the client, which helps to prevent relapse [104,105]. The remaining symptom severity is another factor that influences the estimated aftercare, where the use of an illness specific aftercare questionnaire to capture the clients' condition is advantageous. In the case of depression these illness specific questionnaires capture the remaining symptoms of depression, anxiety, social and cognitive vulnerability [106], and in the case of substance dependence craving, early aftercare attendance added, and family history of addiction are relevant [107–109] for the estimation of the required aftercare.

Web-based aftercare programs for relapse prevention also exist and are similar to regular online treatment consisting of interventions, interaction with the therapist using messages, and regular self-ratings of symptoms [110–112]. Therefore the same methods as in the intervention phase are applicable in an aftercare program. Also, the methods found for clinical relapse prevention might be applicable. However, web based relapse prevention typically cannot respond to individual urgent clinical needs. Mobile treatment programs are developed and offer a promising platform for relapse prevention [113,114]. Context aware information appears especially useful for clients recovering from alcohol abuse [115] or eating disorders [116]. Context aware information can be used to provide clients with interventions that are near a place where they traditionally obtained alcohol [117]. Additionally, client self-reported measures can be used to predict possible relapse in advance [118]. For relapse in depression, it has been shown that monitoring of social and sleep behaviors can potentially be useful to predict the onset of future depressive episodes [119].

4 Discussion

In the previous section, predictors of treatment outcome and adherence in clinical and online treatment were explored. This section will discuss opportunities for further research directions and suggestions for decision support tool for inclusion in online treatment programs. To provide a better overview, the discussion is also structured according to the treatment phases.

4.1 Screening Phase

While reviewing the screening phase, it was noticed that the procedures used for diagnosis in online treatment are close to the one in clinical treatment. The diagnosis is based on standardized questionnaires and interviews. It certainly is difficult to integrate new methods such as text analysis, which are not widely accepted and standardized, into the diagnosis process. There might be additional

Table 1. Overview of possibilities for treatment improvement

Treatment phase	Clinical treatment	Online treatment
Screening		
Text analysis		Identify illness [32,33,120]
Self-reported measures	Treatment resistance [42]	Forecast future depressive episodes [45]
Demographic features	Treatment outcome [35–38]	Treatment outcome [43,44]
Personality traits	Treatment outcome [39,40]	
Motivation	Treatment adherence [41]	Treatment adherence [44]
Unobtrusive measures		Future depressive episodes [46]
Treatment		
Early response	Treatment outcome [48]	Treatment outcome [49]
Client symptom tracking	Treatment outcome [67,68,75,76]	
In-session experience	Treatment outcome [50]	
Intervention recommendation		Text analysis of latest posts [121]
Therapeutic alliance	Treatment outcome [60–62]	Treatment adherence [65] and outcome [66]
Log data evaluation		Treatment adherence [52,55] and outcome [53,56]
EMA assessment	Questionnaire based	Unobtrusive EMA [83–86]
Client diaries		Inferring clients' mood [81]
Client movement tracking		Depressive symptoms [90]
Treatment reminder	Increase in adherence [95,96]	Increase in adherence [92–94]
Gamification		Contributes to adherence [99,100]
Relapse prevention		
Level of aftercare	Questionnaire based [103]	
Client movement tracking		Interventions triggered [117]
Client self-reported measures		Relapse prediction [118,119]

benefit in identifying clients that are unsuited for the treatment as well as tracking clients that are at risk of developing a severe mental illness. Offering clients, which are unsuited for a specific treatment type, another kind of treatment and providing clients with early treatment that exhibit mood and sleep changes, which indicate depressive episodes, could contribute to the cost-effectiveness of treatment.

Although demographic features can indicate treatment outcome, they should be considered a rather weak predictor. The same applies to personality traits and motivation. However, no literature regarding personality traits and adherence to online treatment could be found. Only a study protocol that describes to measure the clients' personality profile in order to measure the influence on adherence was found [122]. The connection between one's personality profile and their adherence to online treatment certainly is an interesting research question, but the predictive performance might be expected to be minor. It is, however, likely that results of this will emerge in the future.

4.2 Treatment Phase

During the treatment phase, the most research on predictors of adherence and outcome is found. The client engages with the treatment platform, which allows to capture usage, intervention, and questionnaire data. First, the use of in-session data will be discussed, second tracking of client improvement, and last smartphone related enhancements.

(1) In-session experience: When estimating the clients' improvement and adherence, the in-session experience has been shown a valuable measure for clinical treatment. In online treatment, however, the experience the client has while processing the intervention is not captured. While the time the client spends with each part of the intervention can be measured unobtrusively, the perceived usefulness and difficulty could be assessed afterward with a questionnaire. Also, more traditional questionnaires could be useful, however, it should be considered that this increases the workload of each intervention which might increase dropout risks. Nonetheless, this could be a beneficial tool to measure the clients' satisfaction with the intervention and indicate their adherence and outcome. Where in face-to-face treatment the therapist would automatically adjust the treatment, this measure could be useful in online treatment to personalize the treatment and provide clients with suited interventions. Whenever clients access the treatment platform, their actions are tracked and saved into a log file. For this log file data, basic measures have been analyzed such as login frequency, program feature use, and time spend with interventions. Although program use can predict adherence and outcome, more complex measures have not been researched yet. For example, the path a client takes in the system or interventions can be compared to usage patterns of other users. This could allow to group clients according to their usage pattern.

(2) Clients' improvement tracking: Tracking of clients' improvement, which can be accomplished in many ways, has been shown to be a reliable predictor for outcome and adherence. While the client processes tasks in the intervention, for example, to state his problems and develop solving strategies, data is collected that is still un-utilized. This intervention data could be valuable for adherence and outcome prediction. The degree and granularity to which the client processes the task as well as text analysis could be considered. Text analysis can be used to estimate the client's sentiment or mood and change of these over the consequent interventions could be used to measure the treatment progress and if the client is improving as expected. Text analysis could further be helpful in deriving a measure of therapeutic alliance, which has been shown to be predictive. Text messages exchanged with the therapist can be analyzed for changes in word usage, emotions, or just the word count. This could potentially indicate the strength of the therapeutic bond which would allow to conclude on the adherence. The same applies to the analysis of clients' diaries. Analysis of these might be used to measure well-being, sentiment, or conducted activities. These measures could be a useful tool for the therapist to supervise the status and improvement of many clients. It would not be possible for the therapist, besides

moral concerns, to read all diaries to infer their current state and estimate their improvement. EMA data is a simplified measure, that allows to be displayed as a graph over time. Although a graph of mood development would be easy to sight, it would be less time consuming for the therapist when a statistical analysis provides the therapist with a notification about clients that are likely to diverge from the treatment. A well-researched tool for improvement tracking and outcome prediction in clinical treatment are outcome questionnaires. However, these type of questionnaires were not found in online treatment. Similarly to the previously mentioned tracking possibilities, they could be used for therapist notification.

Possibilities for its application and outcome tracking in online treatment are unexhausted and yet to be researched. Furthermore, it is unknown what the perception of therapists of potential supports tool is if they would consider them as useful, reliable, and use them. Also, the influence of these tool on the client has not been researched yet. These tools can encompass the recommendation of further interventions or provide the client with an outcome prediction and improvements curve in a self-help setting to allow the client to be their own therapist. A therapist or even client support tools based on improvement tracking appear to be the most promising aspect to be implemented into an online treatment to improve outcomes and increase their cost-effectiveness because such tools have been shown to improve outcomes in clinical treatment.

(3) Additional factors for adherence improvement: Smartphones are a valuable tool in online and regular treatment. Although unobtrusive EMA estimation appears very useful for client tracking, it is still difficult to estimate reliably. Context-related measures such as activity based on accelerometer or stress estimation using the microphone could provide more reliable estimates. However, this would increase the amount of data to be collected and analyzed considerably and is expected to be a difficult task. After all, these measures could also be unreliable and unobtrusive EMA estimates might still remain inaccurate. This would be an undertaking for future research, especially because context-related information could be highly beneficial for applications in relapse prevention. Although reminder via text message can increase adherence to treatment, many opportunities for research still exist. The potential negative effect on the motivation, when considered annoying or disruptive, is uncharted. Therefore, the optimal amount or time for a reminder could be considered as well as the effect of compliance or right to refuse. Also, personalization of messages might contribute to an increase in adherence, which is currently researched [123]. Gamification of intervention content is increasing and contributes to the adherence. But how to derive reliable predictors for outcome and adherence might be a promising and new area of research.

4.3 Relapse Prevention Phase

The relapse preventions programs that were found during the review process are similar to online treatment programs. This suggests that most of the tools applicable in the treatment phase are also viable in relapse prevention. It was noticed that in particular tracking of the client using the smartphone is beneficial in aftercare. After the treatment, clients are prone to fall back into old behavioral patterns. A change in daily behavior, sleep, or movement pattern can be a reliable measure to predict future relapse. Especially, context-related measures exhibit potential to prevent relapse in addiction. It expected that researcher and developers will make more use of the possibilities that smartphones provide and discover their usefulness for a variate of other disorders. However, contextual information appears difficult to estimate reliably. Furthermore, indicators of a particular mental illness and how to derive beneficial decisions from them has still to be researched.

5 Limitations

The aim of this review was to summarize the current state of the art and identify topics of future research and illustrate predictors to develop tools that can be used to increase cost-effectiveness when implemented into online treatment. However, this study has some limitations. First of all, it is not intended to give a comprehensive insight into all aspects of adherence improvement in online treatment. The focus was set on studies that do statistical analysis of the collected data, although some adjacent areas have been mentioned. This choice was based on the growing amount of research published based on statistical analysis of collected online treatment data and upcoming research of integrating such methods into the treatment process. Based on the search criteria, there might be some works that have not been considered for this review because the specified keywords have not been specified in the keywords or used in the paper abstract. In contrast to other reviews, no specific study type or outcome was focused but rather the methodological aspect. This allows to present a broader range of predictors to estimate further research in improvement for online mental health applications. But even if a variety of predictors and possibilities have been considered, some aspects might be missed.

6 Conclusion and Future Work

Decision support tools derived from adherence and outcome predictors offer exciting new opportunities to improve online based treatment and their perception. This review lays down a first summary of the current state of the art and presented new opportunities. In summary, tracking of clients' improvement appears to be a valuable tool to be implemented for the improvement of cost-effectiveness in online treatments. Also, unobtrusive EMA assessment using context-related measures suggests potential, but difficult to implement and still has to be realized. Context-related measures might also find more attention in the field of relapse prevention and their research is expected to increase in the future.

It is recommended that online treatment developer familiarize themselves with the literature in the field of outcome and adherence prediction in both clinical and online treatment before they develop and include potential tools. This review might help developers to get a starting point to engage with this topic. Moreover, the inclusion of such tools requires a multidisciplinary team of all stakeholders. The needs of therapists, clients, programmers, and finical stakeholders have to be considered because they might have contradicting needs. It might also be difficult to tailor these tools to the needs of therapists and clients and still deliver a simple and interactive design. Moreover, it is unknown how such tools could influence the treatment and cost-effectiveness. Although integration of decision support appears to be an effective way to personalize and improve online treatment, all these developments require experimental validation and are a topic of research to come. Overall, it was noticed that inclusion of statistical methods and their effect on the treatment has begun. Since mental disorder are predicted to increase, use of statistical models can improve online treatment to meet the rising demand in the future.

Regarding future work on this topic, it would be beneficial to focus on the treatment decision support part because this analysis suggests that decision support appears to provide the highest benefit to current online treatment. Focusing on the methodical possibilities to predict treatment adherence and outcome would provide further insight into this topic. Furthermore, we aim at developing a common framework that helps to categorize different treatment support tools as well as the collected data. Such a framework would provide therapist, software developers, and financial stakeholders a common ground to discuss decision support tools in the context of online treatment.

References

1. Lépine, J.P., Briley, M.: The increasing burden of depression. Neuropsychiatr. Dis. Treat. **7**(Suppl 1), 3–7 (2011)
2. Alonso, J., Angermeyer, M.C., Bernert, S., Bruffaerts, R., Brugha, T.S., Bryson, H., de Girolamo, G., Graaf, R., Demyttenaere, K., Gasquet, I., Haro, J.M., Katz, S.J., Kessler, R.C., Kovess, V., Lépine, J.P., Ormel, J., Polidori, G., Russo, L.J., Vilagut, G., Almansa, J., Arbabzadeh-Bouchez, S., Autonell, J., Bernal, M., Buist-Bouwman, M.A., Codony, M., Domingo-Salvany, A., Ferrer, M., Joo, S.S., Martínez-Alonso, M., Matschinger, H., Mazzi, F., Morgan, Z., Morosini, P., Palacín, C., Romera, B., Taub, N., Vollebergh, W.A.M.: Prevalence of mental disorders in Europe: results from the European Study of the Epidemiology of Mental Disorders (ESEMeD) project. Acta Psychiatr. Scandinavica. Suppl. **109**(420), 21–27 (2004)
3. Kessler, R., Chiu, W.: Prevalence, severity, and comorbidity of twelve-month DSM-IV disorders in the national comorbidity survey replication (NCS- R). Arch. Gen. Psychiatry**62**(6), 617–627 (2005)
4. WHO. 2014 Mental health atlas, WHO, p. 72 (2014)
5. The Executive Board. Global burden of mental disorders and the need for a comprehensive , coordinated response from health and social sectors at the country level, World Health, pp. 6–9 (2012)
6. Cameron, P.A., Thompson, D.R.: Changing the health-care workforce. Int. J. Nurs. Pract. **11**(1), 1–4 (2005)

7. Gustavsson, A., Svensson, M., Jacobi, F., Allgulander, C., Alonso, J., Beghi, E., Dodel, R., Ekman, M., Faravelli, C., Fratiglioni, L., Gannon, B., Jones, D.H., Jennum, P., Jordanova, A., Jönsson, L., Karampampa, K., Knapp, M., Kobelt, G., Kurth, T., Lieb, R., Linde, M., Ljungcrantz, C., Maercker, A., Melin, B., Moscarelli, M., Musayev, A., Norwood, F., Preisig, M., Pugliatti, M., Rehm, J., Salvador-Carulla, L., Schlehofer, B., Simon, R., Steinhausen, H.C., Stovner, L.J., Vallat, J.M., den Bergh, P.V., van Os, J., Vos, P., Xu, W., Wittchen, H.U., Jönsson, B., Olesen, J.: Cost of disorders of the brain in Europe 2010. Eur. Neuropsychopharmacol. **21**(10), 718–779 (2011)
8. Kessler, R.C., Heeringa, S., Lakoma, M.D., Petukhova, M., Rupp, A.E., Schoenbaum, M., Wang, P.S., Zaslavsky, A.M.: Individual and societal effects of mental disorders on earnings in the United States: results from the national comorbidity survey replication. Am. J. Psychiatry **165**(6), 703–11 (2008)
9. Insel, T.: Assessing the economic costs of serious mental illness. Am. J. Psychiat. **165**(6), 663–665 (2008)
10. Bloom, D.E., Cafiero, E., Jané-Llopis, E., Abrahams-Gessel, S., Reddy Bloom, L., Fathima, S., Feigl, A.B., Gaziano, T., Hamandi, A., Mowafi, M., O'Farrell, D., Ozaltin, E., Pandya, A., Prettner, K., Rosenberg, L., Seligman, B., Stein, A.Z., Weinstein, C., Weiss, J.: The global economic burden of noncommunicable diseases. World Economic Forum, pp. 1–46 (2011)
11. Christensen, H., Griffiths, K.M.: The prevention of depression using the internet. Med. J. Aust. **177**(7), S122–S125 (2002)
12. Cuijpers, P., Van Straten, A., Andersson, G.: Internet-administered cognitive behavior therapy for health problems: a systematic review. J. Behav. Med. **31**, 169–177 (2008). no. 0160-7715 (Print)
13. Kohn, R., Saxena, S., Levav, I., Saraceno, B.: Thee treatment gap in mental health care health. Bull. World Health Organ. **82**(11), 858–866 (2004)
14. Tate, D.F., Finkelstein, E.A., Khavjou, O., Gustafson, A.: Cost effectiveness of internet interventions: Review and recommendations. Ann. Behav. Med. **38**(1), 40–45 (2009)
15. Hedman, E., Andersson, E., Ljótsson, B., Andersson, G., Rück, C., Lindefors, N.: Cost-effectiveness of Internet-based cognitive behavior therapy vs. cognitive behavioral group therapy for social anxiety disorder: results from a randomized controlled trial. Behav. Res. Ther. **49**(11), 729–736 (2011)
16. Van Beugen, S., Ferwerda, M., Hoeve, D., Rovers, M.M., Spillekom-Van Koulil, S., Van Middendorp, H., Evers, A.W.M.: Internet-based cognitive behavioral therapy for patients with chronic somatic conditions: a meta-analytic review. J. Med. Internet Res. **16**(3), 1–15 (2014)
17. Donker, T., Blankers, M., Hedman, E., Ljótsson, B., Petrie, K., Christensen, H.: Economic evaluations of Internet interventions for mental health: a systematic review. Psychol. Med. **45**, 1–20 (2015)
18. Furmark, T., Carlbring, P., Hedman, E., Sonnenstein, A., Clevberger, P., Bohman, B., Eriksson, A., Hållén, A., Frykman, M., Holmström, A., Sparthan, E., Tillfors, M., Ihrfelt, E.N., Spak, M., Eriksson, A., Ekselius, L., Andersson, G.: Guided and unguided self-help for social anxiety disorder: randomised controlled trial. Br. J. Psychiatry **195**(5), 440–447 (2009)
19. McCrone, P., Knapp, M., Proudfoot, J., Ryden, C., Cavanagh, K., Shapiro, D.A., Ilson, S., Gray, J.A., Goldberg, D., Mann, A., Marks, I., Everitt, B., Tylee, A.: Cost-effectiveness of computerised cognitive-behavioural therapy for anxiety and depression in primary care: randomised controlled trial. Br. J. Psychiatry **185**, 55–62 (2004)

20. SAMHSA: TIP 34: Brief Interventions and Brief Therapies for Substance Abuse, Brief Interventions and Brief Therapies For Substance Abuse, pp. 105–121 (2012)
21. Van Straten, A., Cuijpers, P., Smits, N.: Effectiveness of a web-based self-help intervention for symptoms of depression, anxiety, and stress: randomized controlled trial. J. Med. Internet Res. **10**(1) (2008)
22. Van Ballegooijen, W., Cuijpers, P., Van Straten, A., Karyotaki, E., Andersson, G., Smit, J.H., Riper, H.: Adherence to internet-based and face-to-face cognitive behavioural therapy for depression: a meta-analysis. PLoS ONE **9**(7), e100674 (2014)
23. Melville, K.M., Casey, L.M., Kavanagh, D.J.: Dropout from Internet-based treatment for psychological disorders. Br. J. Clin. Psychol./Br. Psychol. Soc. **49**(4), 455–71 (2010)
24. White, K.S., Allen, L.B., Barlow, D.H., Gorman, J.M., Shear, M.K., Woods, S.W.: Attrition in a multicenter clinical trial for panic disorder. J. Nerv. Ment. Dis. **198**(9), 665–671 (2010)
25. Vogenberg, F.R.: Predictive and prognostic models: implications for healthcare decision-making in a modern recession. Am. Health Drug Benefits **2**(6), 218 (2009)
26. Cuijpers, P., van Straten, A., Warmerdam, L.: Problem solving therapies for depression: a meta-analysis. Eur. Psychiatry **22**(1), 9–15 (2007)
27. van Straten, A., Cuijpers, P., Smits, N.: Effectiveness of a web-based self-help intervention for symptoms of depression, anxiety, and stress: randomized controlled trial. J. Med. Internet Res. **10**(1), e7 (2008)
28. Boettcher, J., Rozental, A., Andersson, G., Carlbring, P.: Side effects in internet-based interventions for social anxiety disorder. Internet Interv. **1**(1), 3–11 (2014)
29. Burcusa, S.L., Iacono, W.G.: Risk for recurrence in depression. Clin. Psychol. Rev. **27**(8), 959–985 (2007)
30. Michael, F.G.L., Vergare, J., Binder, R.L., Cook, I.A., Galanter, M.: Practice guideline for the psychiatric evaluation of adults (2006)
31. Bhugra, D., Bhui, K.: Cross-cultural psychiatric assessment. Adv. Psychiatr. Treat. **3**(2), 103–110 (1997)
32. Song, I., Diederich, J.: Speech analysis for mental health assessment using support vector machines. In: Mental Health Informatics (2014)
33. Abussa, M., Diederich, J., Al-ajmi, A., Language, N., Group, M.L.: Web mining and mental health. In: IAWTIC 2004 Proceedings. International Conference on Intelligent Agents, Web Technologies and Internet Commerce, pp. 12–14 (2004)
34. Pestian, J., Nasrallah, H.: Suicide note classification using natural language processing: a content analysis. Biomed. Inform. Insights **3**, 19–28 (2010)
35. Conte, H.R., Plutchik, R., Picard, S., Karasu, T.B., Vaccaro, D.: Self-report measures as predictors of psychotherapy outcome. Compr. Psychiatry **29**(4), 355–360 (1988)
36. Steketee, G., Shapiro, L.J.: Predicting behavioral treatment outcome for agoraphobia and obsessive compulsive disorder. Clin. Psychol. Rev. **15**(4), 317–346 (1995)
37. Keijsers, G.P., Kampman, M., Hoogduin, C.A.L.: Dropout prediction in cognitive behavior therapy for panic disorder. Behav. Ther. **32**(4), 739–749 (2001)
38. Meulenbeek, P., Seeger, K., ten Klooster, P.M.: Dropout prediction in a public mental health intervention for sub-threshold and mild panic disorder. Cogn. Behav. Ther. **8**, e5 (2015)
39. Luborsky, L., Chandler, M., Auerbach, A.H., Cohen, J.: Factors influencing the outcome of psychotherapy: a review of quantitative research. Psychol. Bull. **75**(3), 145–185 (1971)

40. Conte, H.R., Plutchik, R., Picard, S., Karasu, T.B.: Can personality traits predict psychotherapy outcome? Compr. Psychiatry **32**(1), 66–72 (1991)
41. Keijsers, G.P., Hoogduin, C.A., Schaap, C.P.: Predictors of treatment outcome in the behavioural treatment of obsessive-compulsive disorder. Br. J. Psychiatry **165**(6), 781–786 (1994)
42. Perlis, R.H.: A clinical risk stratification tool for predicting treatment resistance in major depressive disorder. Biol. Psychiatry **74**, 7–14 (2013)
43. Karyotaki, E., Kleiboer, A., Smit, F., Turner, D., Pastor, A., Andersson, G., Berger, T., Botella, C., Breton, J., Carlbring, P., Christensen, H., De Graaf, E., Griffiths, K., Donker, T., Farrer, L., Huibers, M., Lenndin, J., Mackinnon, A., Meyer, B., Moritz, S., Riper, H., Spek, V., Vernmark, K., Cuijpers, P.: Predictors of treatment dropout in self-guided web-based interventions for depression: an 'individual patient data' meta-analysis. Psychol. Med. **45**(13), 2717–2726 (2015)
44. Alfonsson, S., Olsson, E., Hursti, T.: Motivation and treatment credibility predicts dropout, treatment adherence, and clinical outcomes in an internet-based cognitive behavioral relaxation program: a randomized controlled trial. J. Med. Internet Res. **18**(3), e52 (2016)
45. Suhara, Y., Xu, Y., Pentland, A.S.: DeepMood: forecasting depressed mood based on self-reported histories via recurrent neural networks. In: WWW, pp. 715–724 (2017)
46. Demirci, K., Akgönül, M., Akpinar, A.: Relationship of smartphone use severity with sleep quality, depression, and anxiety in university students. J. Behav. Addict. **4**(2), 85–92 (2015)
47. Gomes-Schwartz, B.: Effective ingredients in psychotherapy: prediction of outcome from process variables. J. Consult. Clin. Psychol. **46**, 1023–1035 (1978)
48. Van, H.L., Schoevers, R.A., Kool, S., Hendriksen, M., Peen, J., Dekker, J.: Does early response predict outcome in psychotherapy and combined therapy for major depression? J. Affect. Disord. **105**(1-3), 261–265 (2008)
49. Lutz, W., Arndt, A., Rubel, J., Berger, T., Schröder, J., Späth, C., Meyer, B., Greiner, W., Gräfe, V., Hautzinger, M., Fuhr, K., Rose, M., Nolte, S., Löwe, B., Hohagen, F., Klein, J.P., Moritz, S.: Defining and predicting patterns of early response in a web-based intervention for depression. J. Med. Internet Res. **19**(6), e206 (2017)
50. Kegel, A.F., Flückiger, C.: Predicting psychotherapy dropouts: a multilevel approach. Clin. Psychol. Psychother.**22**(5), 377–386 (2015)
51. Feather, J.S., Howson, M., Ritchie, L., Carter, P.D., Parry, D.T., Koziol-McLain, J.: Evaluation methods for assessing users' psychological experiences of web-based psychosocial interventions: a systematic review. J. Med. Internet Res. **18**(6), e181 (2016)
52. Proudfoot, J., Clarke, J., Birch, M.-R., Whitton, A.E., Parker, G., Manicavasagar, V., Harrison, V., Christensen, H., Hadzi-Pavlovic, D.: Impact of a mobile phone and web program on symptom and functional outcomes for people with mild-to-moderate depression, anxiety and stress: a randomised controlled trial. BMC Psychiatry **13**(1), 312 (2013)
53. Whitton, A.E., Proudfoot, J., Clarke, J., Birch, M.R., Parker, G., Manicavasagar, V., Hadzi-pavlovic, D.: Breaking open the black box : isolating the most potent features of a web and mobile phone-based intervention for depression , anxiety , and stress. JMIR Ment. Health **2**, 1–13 (2015)
54. Kelders, S.M., Bohlmeijer, E.T, Van Gemert-Pijnen, J.E.W.C.: Participants, usage, and use patterns of a web-based intervention for the prevention of depression within a randomized controlled trial. J. Med. Internet Res. **15**(8), e172 (2013)

55. Brindal, E., Freyne, J., Saunders, I., Berkovsky, S., Smith, G., Noakes, M.: Features predicting weight loss in overweight or obese participants in a web-based intervention: randomized trial. J. Med. Internet Res. **14**(6), e173 (2012)
56. Van Gemert-Pijnen, J.E.W.C., Kelders, S.M., Bohlmeijer, E.T.: Understanding the usage of content in a mental health intervention for depression: an analysis of log data. J. Med. Internet Res.**16**(1), e27 (2014)
57. Luborsky, L.: Therapeutic alliances as predictors of psychotherapy outcomes: factors explaining the predictive success. In: The Working Alliance: Theory, Research, and Practice (1994)
58. Ardito, R.B., Rabellino, D.: Therapeutic alliance and outcome of psychotherapy: historical excursus, measurements, and prospects for research. Front. Psychol. **2**, 1–11 (2011)
59. Johansson, H., Eklund, M.: Patients' opinion on what constitutes good psychiatric care. Scand. J. Caring Sci. **17**(4), 339–346 (2003)
60. Martin, D.J., Garske, J.P., Davis, M.K.: Relation of the therapeutic alliance with outcome and other variables: a meta-analytic review. J. Consult. Clin. Psychol. **68**(3), 438–450 (2000)
61. Horvath, A.O., Luborsky, L.: The role of the therapeutic alliance in psychotherapy. J. Couns. Clin. Psychol. **61**(4), 561–573 (1993)
62. Bachelor, A.: Clients' and therapists' views of the therapeutic alliance: similarities, differences and relationship to therapy outcome. Clin. Psychol. Psychother. **20**(2), 118–135 (2013)
63. Knaevelsrud, C., Maercker, A.: Internet-based treatment for PTSD reduces distress and facilitates the development of a strong therapeutic alliance: a randomized controlled clinical trial. BMC Psychiatry **7**, 13 (2007)
64. Reynolds, D.J., Stiles, W.B., Bailer, A.J., Hughes, M.R.: Impact of exchanges and client-therapist alliance in online-text psychotherapy. Cyberpsychology Behav. Soc. Netw. **16**(5), 370–7 (2013)
65. White, M., Stinson, J.N., Lingley-Pottie, P., McGrath, P.J., Gill, N., Vijenthira, A.: Exploring therapeutic alliance with an internet-based self-management program with brief telephone support for youth with arthritis: a pilot study. Telemed. J. e-Health **18**(4), 271–6 (2012). The official journal of the American Telemedicine Association
66. Bergman Nordgren, L., Carlbring, P., Linna, E., Andersson, G.: Role of the working alliance on treatment outcome in tailored internet-based cognitive behavioural therapy for anxiety disorders: randomized controlled pilot trial. JMIR Res. Protoc. **2**(1), e4 (2013)
67. Lueger, R.J.: Using feedback on patient progress to predict the outcome of psychotherapy. J. Clin. Psychol. **54**(3), 383–393 (1998)
68. Lambert, M.: Prevention of Treatment Failure: The Use of Measuring, Monitoring, and Feedback in Clinical Practice. American Psychological Association, Washington D.C. (2010)
69. Knaup, C., Koesters, M., Schoefer, D., Becker, T., Puschner, B.: Effect of feedback of treatment outcome in specialist mental healthcare: meta-analysis. Br. J. Psychiatry J. Ment. Sci. **195**(1), 15–22 (2009)
70. Lueger, R.J.: The Integra/COMPASS tracking assessment system. Integr. Sci. Pract. **2**(2), 20–23 (2012)
71. Lambert, M.J.: The outcome questionnaire-45. Integr. Sci. Pract. **2**(1), 24–27 (2012)
72. Evans, C.: The CORE-OM (Clinical Outcomes in Routine Evaluation) and its derivatives. Integr. Sci. Pract. **2**(2), 12–15 (2000)

73. Schibbye, P., Ghaderi, A., Ljótsson, B., Hedman, E., Lindefors, N., Rück, C., Kaldo, V.: Using early change to predict outcome in cognitive behaviour therapy: Exploring timeframe, calculation method, and differences of disorder-specific versus general measures. PLoS ONE **9**(6), e100614 (2014)
74. Finch, A.: Psychotherapy Quality Control: The Statistical Generation of Recovery Curves for Integration Into an Early Warning System. Brigham Young University, Department of Clinical Psychology (2000)
75. Lambert, M.J., Whipple, J.L., Smart, D.W., Vermeersch, D.A., Hawkins, E.J., Al, L.E.T.: The effects of providing therapists with feedback on patient progress during psychotherapy: are outcomes enhanced? Psychother. Res. **11**(1), 49–68 (2001)
76. Lambert, M.J., Whipple, J.L., Hawkins, E.J., Vermeersch, D.A., Nielsen, S.L., Smart, D.W.: Is it time for clinicians to routinely track patient outcome? A meta-analysis. Clin. Psychol. Sci. Pract. **10**(3), 288–301 (2003)
77. Bolger, N., DeLongis, A., Kessler, R.C., Schilling, E.A.: Effects of daily stress on negative mood. J. Pers. Soc. Psychol. **57**(5), 808–818 (1989)
78. Jacelon, C.S., Imperio, K.: Participant diaries as a source of data in research with older adults. Qual. Health Res. **15**(7), 991–7 (2005)
79. Wichers, M., Simons, C.J.P., Kramer, I.M.A., Hartmann, J.A., Lothmann, C., Myin-Germeys, I., van Bemmel, A.L., Peeters, F., Delespaul, P., van Os, J.: Momentary assessment technology as a tool to help patients with depression help themselves. Acta Psychiatr. Scand. **124**(4), 262–272 (2011)
80. Shiffman, S., Stone, A.A., Hufford, M.R.: Ecological momentary assessment. Annu. Rev. Clin. Psychol. **4**(5), 1–32 (2008)
81. Bremer, V., Becker, D., Funk, B., Lehr, D.: Predicting the individual mood level based on diary data. In: Proceedings of the Twenty-Fifth Conference on Information Systems (ECIS 2017) (2017)
82. Postel, M.G., De Haan, H.A., Ter Huurne, E.D., Becker, E.S., De Jong, C.A.: Effectiveness of a web-based intervention for problem drinkers and reasons for dropout: randomized controlled trial. J. Med. Internet Res.**12**(4), e68 (2010)
83. Asselbergs, J., Ruwaard, J., Ejdys, M., Schrader, N., Sijbrandij, M., Riper, H.: Mobile phone-based unobtrusive ecological momentary assessment of day-to-day mood: an explorative study. J. Med. Internet Res. **18**(3), e72 (2016)
84. Becker, D., Bremer, V., Funk, B., Asselbergs, J., Riper, H., Ruwaard, J.: How to predict mood? Delving into features of smartphone-based data. In: Twenty-second Americas Conference on Information Systems, pp. 1–10 (2016)
85. van Breda, W., Pastor, J., Hoogendoorn, M., Ruwaard, J., Asselbergs, J., Riper, H.: Exploring and comparing machine learning approaches for predicting mood over time. Smart Innovation, Systems and Technologies, vol. 60, pp. 37–47. Springer Science and Business Media Deutschland GmbH (2016)
86. Ma, Y., Xu, B., Bai, Y., Sun, G., Zhu, R.: Daily mood assessment based on mobile phone sensing. In: Proceedings - BSN 2012: 9th International Workshop on Wearable and Implantable Body Sensor Networks, pp. 142–147, May 2012
87. Kwapisz, J.R., Weiss, G.M., Moore, S.A.: Activity recognition using cell phone accelerometers. ACM SIGKDD Explor. Newsl. **12**(2), 74 (2011)
88. Lu, H., Frauendorfer, D., Rabbi, M., Mast, M.S., Chittaranjan, G.T., Campbell, A.T., Gatica-Perez,D., Choudhury, T.: StressSense: detecting stress in unconstrained acoustic environments using smartphones. In: Proceedings of the 2012 ACM Conference on Ubiquitous Computing - UbiComp 2012, p. 351 (2012)

89. Chang, K.H., Fisher, D., Canny, J.: AMMON: a speech analysis library for analyzing affect, stress, and mental health on mobile phones. In: Proceedings of the 2011 PhoneSense Conference (2011)
90. Saeb, S., Zhang, M., Kwasny, M.M., Karr, C.J., Kording, K., Mohr, D.C.: The relationship between clinical, momentary, and sensor-based assessment of depression. In: International Conference on Pervasive Computing Technologies for Healthcare : Proceedings of International Conference on Pervasive Computing Technologies for Healthcare, vol. 2015, pp. 7–10 (2015)
91. Burns, M.N., Begale, M., Duffecy, J., Gergle, D., Karr, C.J., Giangrande, E., Mohr, D.C.: Harnessing context sensing to develop a mobile intervention for depression. J. Med. Internet Res. **13**(3), e55 (2011)
92. Gill, S., Contreras, O., Munoz, R.F., Leykin, Y.: Participant retention in an automated online monthly depression rescreening program: Patterns and predictors. Internet Interv. **1**(1), 20–25 (2014)
93. Rodgers, A.: Do u smoke after txt? Results of a randomised trial of smoking cessation using mobile phone text messaging. Tob. Control. **14**(4), 255–261 (2005)
94. Patrick, K., Raab, F., Adams, M.A., Dillon, L., Zabinski, M., Rock, C.L., Griswold, W.G., Norman, G.J.: A text message-based intervention for weight loss: randomized controlled trial. J. Med. Internet Res. **11**(1), 1–9 (2009)
95. Gurol-Urganci, I., de Jongh, T., Vodopivec-Jamsek, V., Atun, R., Car, J.: Mobile phone messaging reminders for attendance at healthcare appointments. Cochrane Database Syst. Rev. no. 12, p. CD007458 (2013). (Review) SUMMARY OF FINDINGS FOR THE MAIN COMPARISON
96. Kannisto, K.A., Koivunen, M.H., Välimäki, M.A.: Use of mobile phone text message reminders in health care services: a narrative literature review. J. Med. Internet Res. **16**(10), e222 (2014)
97. Ludden, G.D., Van Rompay, T.J., Kelders, S.M., Van Gemert-Pijnen, J.E.: How to increase reach and adherence of web-based interventions: a design research viewpoint. J. Med. Internet Res. **17**(7), e172 (2015)
98. Kelders, S.M., Kok, R.N., Ossebaard, H.C., Van Gemert-Pijnen, J.E.W.C.: Persuasive system design does matter: a systematic review of adherence to web-based interventions. J. Med. Internet Res.**14**(6), 1–24 (2012)
99. Looyestyn, J., Kernot, J., Boshoff, K., Ryan, J., Edney, S., Maher, C.: Does gamification increase engagement with online programs? A systematic review. PLoS ONE **12**(3), 1–19 (2017)
100. Brown, M., O'Neill, N., van Woerden, H., Eslambolchilar, P., Jones, M., John, A.: Gamification and adherence to web-based mental health interventions: a systematic review. JMIR Ment. Health **3**(3), e39 (2016)
101. McKay, J.R.: Studies of factors in relapse to alcohol, drug and nicotine use: a critical review of methodologies and findings. J. Stud. Alcohol **60**(4), 566–576 (1999)
102. Vittengl, J.R., Clark, L.A., Dunn, T.W., Jarrett, R.B.: Reducing relapse and recurrence in unipolar depression: a comparative meta-analysis of cognitive–behavioral therapy's effects. J. Consult. Clin. Psychol. **75**, 475–488 (2009)
103. Kessing, L.V.: Severity of depressive episodes according to ICD-I0: prediction of risk of relapse and suicide. Br. J. Psychiatry **184**(2), 153–156 (2004)
104. Segal, Z.V., Kennedy, S., Gemar, M., Hood, K., Pedersen, R., Buis, T.: Cognitive reactivity to sad mood provocation and the prediction of depressive relapse. Arch. Gen. Psychiatry **63**(7), 749–755 (2006)

105. Pedersen, M.U., Hesse, M.: A simple risk scoring system for prediction of relapse after inpatient alcohol treatment. Am. J. Addict. **18**(6), 488–493 (2009). American Academy of Psychiatrists in Alcoholism and Addictions
106. Van Voorhees, B.W., Paunesku, D., Gollan, J., Reinecke, M., Basu, A.: Predicting future risk of depressive episode in adolescents: the chicago adolescent depression risk assessment (CADRA). Ann. Fam. Med. **6**(6), 503–512 (2008)
107. Ito, J.R., Donovan, D.M.: Predicting drinking outcome: demography, chronicity, coping, and aftercare. Addict. Behav. **15**(6), 553–559 (1990)
108. Farren, C.K., McElroy, S.: Predictive factors for relapse after an integrated inpatient treatment programme for unipolar depressed and bipolar alcoholics. Alcohol Alcohol. **45**(6), 527–533 (2010)
109. Farren, C.K., Snee, L., Daly, P., McElroy, S.: Prognostic factors of 2-year outcomes of patients with comorbid bipolar disorder or depression with alcohol dependence: importance of early abstinence. Alcohol Alcohol. **48**(1), 93–98 (2013)
110. Barnes, C., Harvey, R., Mitchell, P., Smith, M., Wilhelm, K.: Evaluation of an online relapse prevention program for bipolar disorder: an overview of the aims and methodology of a randomized controlled trial, pp. 215–224 (2007)
111. Holländare, F., Anthony, S.A., Randestad, M., Tillfors, M., Carlbring, P., Andersson, G., Engström, I.: Two-year outcome of internet-based relapse prevention for partially remitted depression. Behav. Res. Ther. **51**(11), 719–722 (2013)
112. Lobban, F., Dodd, A.L., Dagnan, D., Diggle, P.J., Griffiths, M., Hollingsworth, B., Knowles, D., Long, R., Mallinson, S., Morriss, R.M., Parker, R., Sawczuk, A.P., Jones, S.: Feasibility and acceptability of web-based enhanced relapse prevention for bipolar disorder (ERPonline): trial protocol. Contemp. Clin. Trials **41**, 100–109 (2015)
113. Lord, S., Moore, S.K., Ramsey, A., Dinauer, S., Johnson, K.: Implementation of a substance use recovery support mobile phone app in community settings: qualitative study of clinician and staff perspectives of facilitators and barriers. JMIR Ment. Health **3**(2), e24 (2016)
114. Kok, G., Bockting, C., Burger, H., Smit, F., Riper, H.: Mobile cognitive therapy: adherence and acceptability of an online intervention in remitted recurrently depressed patients. Internet Interv. **1**(2), 65–73 (2014)
115. Beckjord, E., Shiffman, S.: Background for real-time monitoring and intervention related to alcohol use. Alcohol Res. Curr. Rev. **36**(1), 9–18 (2014)
116. Juarascio, A.S., Manasse, S.M., Goldstein, S.P., Forman, E.M., Butryn, M.L.: Review of smartphone applications for the treatment of eating disorders. Eur. Eat. Disord. Rev. **23**(1), 1–11 (2015)
117. Gustafson, D.H., Shaw, B.R., Isham, A., Baker, T., Boyle, M.G., Levy, M.: Explicating an evidence-based, theoretically informed, mobile technology-based system to improve outcomes for people in recovery for alcohol dependence. Subst. Use Misuse **46**(1), 96–111 (2011)
118. Chih, M.Y., Patton, T., McTavish, F.M., Isham, A.J., Judkins-Fisher, C.L., Atwood, A.K., Gustafson, D.H.: Predictive modeling of addiction lapses in a mobile health application. J. Subst. Abus. Treat. **46**(1), 29–35 (2014)
119. Doryab, A., Min, J.K., Wiese, J., Zimmerman, J., Hong, J.I.: Detection of behavior change in people with depression. In: AAAI Workshops Workshops at the Twenty-Eighth AAAI Conference on Artificial Intelligence, pp. 12–16 (2014)
120. Diederich, J.: Ex-ray: text classification and the assessment of mental health. In: Eighth Australian Document Computing Symposium ADCS 2002–2003 (2003)

121. D'Alfonso, S., Santesteban-Echarri, O., Rice, S., Wadley, G., Lederman, R., Miles, C., Gleeson, J., Alvarez-Jimenez, M.: Artificial intelligence-assisted online social therapy for youth mental health. Front. Psychol. **8**, 1–13 (2017)
122. Graham, A.L., Cha, S., Papandonatos, G.D., Cobb, N.K., Mushro, A., Fang, Y., Niaura, R.S., Abrams, D.B.: Improving adherence to web-based cessation programs: a randomized controlled trial study protocol. Trials **14**(48), 1–15 (2013)
123. Graham, A.L., Jacobs, M.A., Cohn, A.M., Cha, S., Abroms, L.C., Papandonatos, G.D., Whittaker, R.: Optimising text messaging to improve adherence to web-based smoking cessation treatment: a randomised control trial protocol. BMJ Open **6**(3), e010687 (2016)

Multi-sensors 3D Fusion in the Presence of Sensor Biases

Cong Dan Pham(✉), Bao Ngoc Bui Tang, Quang Bang Nguyen, and Su Le Tran

C4I, Viettel Research and Development Institute, Viettel Group, Hanoi, Vietnam
{danpc,ngocbtb,bangnq,sult2}@viettel.com.vn

Abstract. In this paper, we study the problem fusion data from multi-sensors in presence of biases. We discuss both approaches, measurement data level, and track sensor level. A previous algorithm for local track fusion using a pseudo-equation with Jacobian matrix is presented for 2D sensors. In 2D case, this algorithm worked well and gave an equivalent performance and higher computational efficiency comparing with exact Kalman filter method. We extend the algorithm to 3D sensors to know how it works in this case. It is not totally straightforward when the computation of Jacobian matrix in 3D is very complex. We give the computation true Jacobian matrix for pseudo-equation using MATLAB and also give a simpler approximation Jacobian matrix. This helps to improve computational efficiency. The simulation compares the performance of the methods: Local track fusion and measurement fusion in the other cases as two sensors, four sensors, track fusion using Jacobian matrix, approximate Jacobian matrix.

Keywords: Local track · Measurement fusion · Track fusion
Biases sensors · Jacobian matrix

1 Introduction

In a multistatic-multi-target environment, where each radar processes its own observations and sends the resulting tracks to a data fusion center, the first step is to determine whether or not two or more tracks, coming from different radar systems with different accuracies, represent the same target (track-to-track association). The next step is to combine the radar tracks when it is determined that they indeed represent the same target (track fusion). Both problems arise when several radars carry out surveillance over a common volume (overlapping sensor coverage).

In this paper, we study the fusion process from multi-sensor in presence of systematic errors (registration sensors). The potential advantages of fusing information from disparate sensor systems used to achieve better surveillance has been recognised in [2,4,5]. The fusion process relies on the accurate registration of sensors which is regarded as a process to eliminate the effects caused

K. Arai et al. (Eds.): FICC 2018, AISC 886, pp. 113–131, 2019.
https://doi.org/10.1007/978-3-030-03402-3_9

by the sensor biases. If uncorrected, sensor biases would lead to large tracking errors and cause ghost tracks. Consequently, the fusion performance cannot be guaranteed to be optimal. There are two approaches to solve this problem. The classical approach is to use measurement fusion augmented state Kalman filter (ASKF) and improved versions, see [14, p. 81] [7–9,12,13,16]. The problem with this approach is that the implementation of this ASKF can be computationally infeasible. Then the fusion process at track level is applied for distributed local system, see [10,11,13,15]. In [17], a proposed algorithm gives equivalent performance as exact method in [10] and has higher computational efficiency. We extend previous algorithms (ASKF and the algorithm in [17]) from 2D to 3D. Our contribution in this paper is to give a computation for true Jacobian matrix using MATLAB and a approximate Jacobian matrix to improve the computational efficiency. Then we compare the performance of these algorithm for two sensors and multi-sensor.

A general motion model used in discrete extended Kalman filter for target tracking is as follows:

$$\begin{aligned} X_k &= FX_{k-1} + GW_{k-1} \\ Z_k &= h(X_k) + V_k \end{aligned} \tag{1}$$

where, X_k is the state vector, F is the state transition matrix and G is the process noise gain matrix. The process noise W_k and the measurement noise V_k are zero-mean, mutually independent, white, Gaussian with covariance Q and R, respectively. Z_k is the measurement vector at time k and $h(X_k)$ is a nonlinear function of the states computed at time k (see [6]). Through this paper, we consider the motion model is 2-DOF kinematic model (this consist the position and velocity).

1.1 3D Track Fusion in Presence of Biases

In [17], they propose an algorithm to sensor registration and track fusion in 2D sensor. We develop the model and the algorithm to 3D sensor.

We formulate the model:
Suppose that all m sensors are located on surface ($z_{om} = 0$), $X_{om} = (x_{om}, y_{om}, z_{om})$ are Cartesian coordinates. The local Cartesian coordinate systems (LCCS) have the rotation angles (α_m, β_m) respectively the axis rotation in direction $\vec{z} = (0, 0, 1)$ and $\vec{t} = (\frac{1}{\sqrt{2}}, -\frac{1}{\sqrt{2}}, 0)$ with respect to the global Cartesian coordinate system (GCCS).

Since there exist systematic biases and random errors, the original measurement $\{r_k^m, \theta_k^m, \varphi_k^m\}$ from sensor m at time k can be modeled as

$$\begin{aligned} r_k^m &= \overline{r}_k^m + \Delta r_k^m + \nu_{k,r}^m \\ \theta_k^m &= \overline{\theta}_k^m + \Delta \theta_k^m + \nu_{k,\theta}^m \\ \varphi_k^m &= \overline{\varphi}_k^m + \Delta \varphi_k^m + \nu_{k,\varphi}^m \end{aligned} \tag{2}$$

where $\overline{r}_k^m, \overline{\theta}_k^m$ and $\overline{\varphi}_k^m$ denote the real range, azimuth and elevation about the observed target, $\Delta r_k^m, \Delta\theta_k^m$ and $\Delta\varphi_k^m$ are the systematic biases, $\nu_{k,r}^m, \nu_{k,\theta}^m$ and $\nu_{k,\varphi}^m$ are random errors.

Furthermore, it is assumed that the biases invariant across the overall tracking time, that is

$$\Delta r_k^m = \Delta r^m, \Delta\theta_k^m = \Delta\theta^m, \Delta\varphi_k^m = \Delta\varphi^m.$$

In this way, the original measurement $\{r_k^m, \theta_k^m, \varphi_k^m\}$ from sensor m can be rewritten as

$$\begin{aligned} r_k^m &= \overline{r}_k^m + \Delta r^m + \nu_{k,r}^m \\ \theta_k^m &= \overline{\theta}_k^m + \Delta\theta^m + \nu_{k,\theta}^m \\ \varphi_k^m &= \overline{\varphi}_k^m + \Delta\varphi^m + \nu_{k,\varphi}^m. \end{aligned} \tag{3}$$

The random measurement noise $\tilde{V}_k^m = \left[\nu_{k,r}^m \; \nu_{k,\theta}^m \; \nu_{k,\varphi}^m\right]^T$ is modeled as independent white Gaussian noises in the range, azimuth and elevation with zero-mean and variances of $(\sigma_r^m)^2$, $(\sigma_\theta^m)^2, (\sigma_\varphi^m)^2$. The bias vector of two sensors is denoted by $b = [(b^1)^T, (b^2)^T]^T$, where $b^m = [\Delta r^m, \Delta\theta^m, \Delta\varphi^m]^T$.

In the next section, we represent the measurement method.

1.2 Measurement Fusion

(1) Measurement data fusion: Firstly, we give a overview on measurement fusion. Suppose that the system has M sensors, for some positive integer M. The measurement model of m-th sensor

$$Z_k^i = h(X_k) + V_k^i.$$

In this approach, all sensors sent data together to the fusion center

$$Z = [Z_1, Z_2, .., Z_M]^T, H = [H_1, H_2, ..., H_M]^T$$

The covariance matrix of noise measurement

$$R = \begin{bmatrix} R_1 & 0 & 0\,0 & 0 \\ 0 & R_2 & 0\,0 & 0 \\ 0 & 0 & \cdot\; 0 & 0 \\ 0 & 0 & 0\;\cdot & 0 \\ 0 & 0 & 0\,0 & R_M \end{bmatrix}.$$

And now we apply one time the Kalman Filter algorithm for all measurements.

$$\begin{aligned} \tilde{X}_{k+1}^f &= F\hat{X}_k^f \\ \tilde{P}_{k+1}^f &= F\hat{P}_k^f F^T + GQG^T. \end{aligned} \tag{4}$$

State and covariance measurement update

$$
\begin{aligned}
K_{k+1}^{f} &= \tilde{P}_{k+1}^{f} H^{T} [H \tilde{P}_{k+1}^{f} H^{T} + R]^{-1} \\
\hat{X}_{k+1}^{f} &= \tilde{X}_{k+1}^{f} + K_{k+1}^{f} [Z_{k+1} - H \tilde{X}_{k+1}^{f}] \\
\hat{P}_{k+1}^{f} &= [I - K_{k+1}^{f} H] \tilde{P}_{k+1}^{f}.
\end{aligned} \tag{5}
$$

(2) Data fusion in the case of bias sensors: We apply the method for our case

$$
\begin{aligned}
X_k &= F X_{k-1} + G W_{k-1} \\
Z_k &= h(X_k) + b + V_k.
\end{aligned} \tag{6}
$$

Model:
Suppose that the radar system has M sensors with m is the sensor index

$$
\begin{aligned}
X_{k+1}^{x} &= F^{x} X_k^{x} + G^{x} W_k^{x} \\
b_{k+1}^{m} &= b_k^{m} + 0.W_k^{b}.
\end{aligned} \tag{7}
$$

Measurement model

$$
Z_k = h(X_k) + b + V_k.
$$

Algorithm:
Input: The initial state X_0^x, b_0.
Output: Estimate of state and bias $(\hat{X}^x, \hat{b})$.
More precisely,

$$
b = \begin{bmatrix} b^1 \\ b^2 \\ \vdots \\ b^M \end{bmatrix}, \; X^x = \begin{bmatrix} x \\ \dot{x} \\ y \\ \dot{y} \\ z \\ \dot{z} \end{bmatrix}, \; b^m = \begin{bmatrix} \Delta r^m \\ \Delta \theta^m \\ \Delta \varphi^m \end{bmatrix}, \tag{8}
$$

$$
\begin{aligned}
& F_T = \begin{bmatrix} 1 & T \\ 0 & 1 \end{bmatrix}, \; F^x = diag(F_T, F_T, F_T), \; G_T = \begin{bmatrix} T^2/2 \\ T \end{bmatrix}, \\
& G^x = diag(G_T, G_T, G_T).
\end{aligned} \tag{9}
$$

Hence, couple the position state vector and bias vector to obtain coupled vector

$$
X = \begin{bmatrix} X^x \\ b \end{bmatrix}, \; F = diag[F^x, I_{3M}], \; G = \begin{bmatrix} G^x \\ O_{3M} \end{bmatrix}. \tag{10}
$$

Measurement model

$$
Z_k = h(X_k) + b + V_k.
$$

where

$$Z_k^m = \begin{bmatrix} r_k^m \\ \theta_k^m \\ \varphi_k^m \end{bmatrix}, \; Z = \begin{bmatrix} Z^1 \\ Z^2 \\ \vdots \\ Z^M \end{bmatrix}, \; V_k^m = \begin{bmatrix} v_{k,r}^m \\ v_{k,\theta}^m \\ v_{k,\varphi}^m \end{bmatrix}, \; V_k = \begin{bmatrix} V_k^1 \\ V_k^2 \\ \vdots \\ V_k^M \end{bmatrix}. \tag{11}$$

Next, we define the measurement function: $h(x) = [h^1(x),\, h^2(x), ..., h^{3M}(x)]^T$. Let the function to compute azimuth angle

$$a = \text{invert}(x, y) = \begin{cases} \arctan\left(\frac{y}{x}\right) & \text{if } x > 0, y > 0, \\ \pi + \arctan\left(\frac{y}{x}\right) & \text{if } x < 0, \\ 2\pi + \arctan\left(\frac{y}{x}\right) & \text{if } x > 0, y < 0. \end{cases}$$

We define the function $Xp = \text{convertcp}(X, \alpha, \beta, X_o)$ convert global Carterian coordinates to local polar coordinate when the location of local coordinate system is (X_o, α, β)

$$\begin{aligned} &Xl := R_{\alpha,\beta}^{-1}(X - X_o) \\ &Xp(1) = \sqrt{(Xl)^2}, Xp(2) = \text{invert}(Xl(1), Xl(2)) \\ &Xp(3) = \arctan\left(\frac{Xl(3)}{\sqrt{Xl(1)^2 + Xl(2)^2}}\right) \end{aligned} \tag{12}$$

where $R_{\alpha,\beta}$ is transform matrix as in (13). Hence, for $m = 1 : M$ we get

$$\begin{aligned} &h(3m-2) = \\ &\text{convertcp}([x(1), x(3), x(5)]^T, \alpha(m), \beta(i), X_{om})(1) + x_{3m+4}, \\ &h(3m-1) \\ &= \text{convertcp}([x(1), x(3), x(5)]^T, \alpha(m), \beta(i), X_{om})(2) \\ &+ x_{3m+5}, \\ &h(3m) = \\ &\text{convertcp}([x(1), x(3), x(5)]^T, \alpha(m), \beta(i), X_{om})(3) + x_{3m+6}. \end{aligned}$$

Next, using Extended Kalman Filter
The state function $f(x) = Fx$.
Predict:
Predicted state estimate

$$\hat{\boldsymbol{x}}_{k|k-1} = f(\hat{\boldsymbol{x}}_{k-1|k-1}, \boldsymbol{u}_{k-1}).$$

Predicted covariance estimate

$$\boldsymbol{P}_{k|k-1} = \boldsymbol{F}_{k-1}\boldsymbol{P}_{k-1|k-1}\boldsymbol{F}_{k-1}^{\top} + G\boldsymbol{Q}_{k-1}G^T.$$

Update:
Innovation or measurement residual

$$\tilde{\boldsymbol{y}}_k = \boldsymbol{z}_k - h(\hat{\boldsymbol{x}}_{k|k-1}).$$

Jacobian matrix

$$\boldsymbol{H}_k = \left.\frac{\partial h}{\partial x}\right|_{\hat{x}_{k|k-1}}.$$

Innovation (or residual) covariance

$$\boldsymbol{S}_k = \boldsymbol{H}_k \boldsymbol{P}_{k|k-1} \boldsymbol{H}_k^\top + \boldsymbol{R}_k.$$

Near-optimal Kalman gain

$$\boldsymbol{K}_k = \boldsymbol{P}_{k|k-1} \boldsymbol{H}_k^\top \boldsymbol{S}_k^{-1}.$$

Updated state estimate

$$\hat{\boldsymbol{x}}_{k|k} = \hat{\boldsymbol{x}}_{k|k-1} + \boldsymbol{K}_k \tilde{\boldsymbol{y}}_k.$$

Updated covariance estimate

$$\boldsymbol{P}_{k|k} = (\boldsymbol{I} - \boldsymbol{K}_k \boldsymbol{H}_k) \boldsymbol{P}_{k|k-1}$$

where the state transition and observation matrices are defined to be the following Jacobians:

$$\boldsymbol{F}_{k-1} = \left.\frac{\partial f}{\partial \boldsymbol{x}}\right|_{\hat{\boldsymbol{x}}_{k-1|k-1}, \boldsymbol{u}_{k-1}} \quad \boldsymbol{H}_k = \left.\frac{\partial h}{\partial \boldsymbol{x}}\right|_{\hat{\boldsymbol{x}}_{k|k-1}}.$$

Finally, we obtain State estimate: $\hat{X}^x = \hat{x}(1:6)$. Bias estimate of $m-$th sensor: $\hat{b} = \hat{x}((3i+4):(3i+6))$.

Now, we study the second method: local track fusion.

1.3 Track Fusion

(1) Notations: Suppose that the target Ta is detect by both two sensors (M = 2). Firstly, we introduce some notations for the sake of clarity.

$\overline{X}_k^m = \left[\overline{x}_k^m \; \overline{v}_{x,k}^m \; \overline{y}_k^m \; \overline{v}_{y,k}^m \; \overline{z}_k^m \; \overline{v}_{z,k}^m\right]^T$: real state in LCCS.

Conversion from equivalent measurement $\begin{bmatrix} x_k^m \\ y_k^m \\ z_k^m \end{bmatrix}$ in LCCS to the equivalent measurement in GCCS

$$\begin{bmatrix} \breve{x}_k^m \\ \breve{y}_k^m \\ \breve{z}_k^m \end{bmatrix} = R_{\alpha,\beta} \begin{bmatrix} x_k^m \\ y_k^m \\ z_k^m \end{bmatrix} + \begin{bmatrix} x_{om} \\ y_{om} \\ z_{om} \end{bmatrix}$$

where

$$R_{\alpha,\beta} = R_\alpha R_\beta = \begin{bmatrix} \cos(\alpha) & -\sin(\alpha) & 0 \\ \sin(\alpha) & \cos(\alpha) & 0 \\ 0 & 0 & 1 \end{bmatrix}$$

$$\begin{bmatrix} \frac{1+\cos(\beta)}{2} & -\frac{1-\cos(\beta)}{2} & -\frac{\sin(\beta)}{\sqrt{2}} \\ -\frac{1-\cos(\beta)}{2} & \frac{1+\cos(\beta)}{2} & -\frac{\sin(\beta)}{\sqrt{2}} \\ \frac{\sin(\beta)}{\sqrt{2}} & \frac{\sin(\beta)}{\sqrt{2}} & \cos(\beta) \end{bmatrix}. \tag{13}$$

Here we use the notation (A) (A is some formula) that denotes an element of matrix. Conversion from real state $\begin{bmatrix} \breve{\overline{x}}_k^m \\ \breve{\overline{y}}_k^m \\ \breve{\overline{z}}_k^m \end{bmatrix}$ in LCCS to real state $\begin{bmatrix} \overline{x}_k^m \\ \overline{y}_k^m \\ \overline{z}_k^m \end{bmatrix}$ in LCCS

$$\begin{bmatrix} \breve{\overline{x}}_k^m \\ \breve{\overline{y}}_k^m \\ \breve{\overline{z}}_k^m \end{bmatrix} = R_{\alpha,\beta}^{-1} \begin{bmatrix} \breve{\overline{x}}_k^m - x_{om} \\ \breve{\overline{y}}_k^m - y_{om} \\ \breve{\overline{z}}_k^m - z_{om} \end{bmatrix} \text{ where } R_{\alpha,\beta}^{-1} = R_{\alpha,\beta}^T.$$

(2) The pseudo-measurement equation and Jacobian matrix: Conversion from real state $\begin{bmatrix} \overline{x}_k^m \\ \overline{y}_k^m \\ \overline{z}_k^m \end{bmatrix}$ in LCCS to equivalent measurement $\begin{bmatrix} x_k^m \\ y_k^m \\ z_k^m \end{bmatrix}$ in LCCS

$$\begin{bmatrix} x_k^m \\ y_k^m \\ z_k^m \end{bmatrix} = \begin{bmatrix} (\overline{r}_k^m + \Delta r^m)\cos(\overline{\theta}_k^m + \Delta\theta^m)\cos(\overline{\varphi}_k^m + \Delta\varphi^m) \\ (\overline{r}_k^m + \Delta r^m)\sin(\overline{\theta}_k^m + \Delta\theta^m)\cos(\overline{\varphi}_k^m + \Delta\varphi^m) \\ (\overline{r}_k^m + \Delta r^m)\sin(\overline{\varphi}_k^m + \Delta\varphi^m) \end{bmatrix}$$

Note that the equivalent measurement $\breve{X}_k^m$ in GCCS can be written by

$$\breve{X}_k^m = f_m(\breve{\overline{X}}_k^m, b^m) = \begin{bmatrix} \breve{x}_k^m \\ \breve{v}_{x,k}^m \\ \breve{y}_k^m \\ \breve{v}_{y,k}^m \\ \breve{z}_k^m \\ \breve{v}_{z,k}^m \end{bmatrix}. \tag{14}$$

It is similar as in [17] we have the following theorem:

Theorem 1: Any local estimate $\hat{X}_k^m$ in GCCS can be approximated by the following linear form:

$$\hat{X}_k^m = \breve{\overline{X}}_k^m + J_{b^m} \begin{bmatrix} \Delta r^m \\ \Delta\theta^m \\ \Delta\varphi^m \end{bmatrix} + e_k^m \tag{15}$$

where J_{b^m} is the Jacobian matrix defined by

$$J_{b^m} = \frac{\partial f_m}{\partial b^m}\Big|_{(\hat{X}_k^m, [0\ 0\ 0]^T)}.$$

This matrix is approximated by a simpler matrix J that is expressed as following proof.

Proof. The position component of $\breve{X}_k^m$ can be computed by

$$\begin{bmatrix} \breve{x}_k^m \\ \breve{y}_k^m \\ \breve{z}_k^m \end{bmatrix} = R_{\alpha,\beta} \begin{bmatrix} x_k^m \\ y_k^m \\ z_k^m \end{bmatrix} + \begin{bmatrix} x_{om} \\ y_{om} \\ z_{om} \end{bmatrix}. \tag{16}$$

Moreover,

$$\sqrt{(\overline{x}_k^m)^2 + (\overline{y}_k^m)^2 + (\overline{z}_k^m)^2}$$
$$= \sqrt{\left(\breve{\overline{x}}_k^m - x_{om}\right)^2 + \left(\breve{\overline{y}}_k^m - y_{om}\right)^2 + \left(\breve{\overline{z}}_k^m - z_{om}\right)^2}.$$

We have

$$\begin{aligned} \overline{x}_k^m &= r_{11}(\breve{\overline{x}}_k^m - x_{om}) + r_{21}(\breve{\overline{y}}_k^m - y_{om}) + r_{31}(\breve{\overline{z}}_k^m - z_{om}) \\ \overline{y}_k^m &= r_{12}(\breve{\overline{x}}_k^m - x_{om}) + r_{22}(\breve{\overline{y}}_k^m - y_{om}) + r_{32}(\breve{\overline{z}}_k^m - z_{om}) \\ \overline{z}_k^m &= r_{13}(\breve{\overline{x}}_k^m - x_{om}) + r_{23}(\breve{\overline{y}}_k^m - y_{om}) + r_{33}(\breve{\overline{z}}_k^m - z_{om}). \end{aligned} \tag{17}$$

$$X_k^m = \lambda BA\overline{X}_k^m = \lambda C\overline{X}_k^m \tag{18}$$

where

$$\lambda = 1 + \frac{\Delta r^m}{\sqrt{\left(\breve{\overline{x}}_k^m - x_{om}\right)^2 + \left(\breve{\overline{y}}_k^m - y_{om}\right)^2 + \left(\breve{\overline{z}}_k^m - z_{om}\right)^2}}.$$

The matrices rotations have respectively the axis rotations $z = (0, 0, 1)$ and $u = (\cos(\overline{\theta}_k^m + \Delta\theta^m), -\sin(\overline{\theta}_k^m + \Delta\theta^m), 0)$ as follows:

$$A = \begin{bmatrix} \cos(\Delta\theta^m) & -\sin(\Delta\theta^m) & 0 \\ \sin(\Delta\theta^m) & \cos(\Delta\theta^m) & 0 \\ 0 & 0 & 1 \end{bmatrix} \tag{19}$$

$$B = \begin{bmatrix} \begin{bmatrix} \cos(\Delta\varphi^m) \\ +\sin^2(\overline{\theta}_k^m + \Delta\theta^m) \\ (1 - \cos(\Delta\varphi^m)) \end{bmatrix} & \begin{bmatrix} -\sin(\overline{\theta}_k^m + \Delta\theta^m) \\ \cos(\overline{\theta}_k^m + \Delta\theta^m) \\ (1 - \cos(\Delta\varphi^m)) \end{bmatrix} & \begin{matrix} -\cos(\overline{\theta}_k^m + \Delta\theta^m)\sin(\Delta\varphi^m) \\ -\sin(\overline{\theta}_k^m + \Delta\theta^m)\sin(\Delta\varphi^m) \\ \cos(\Delta\varphi^m) \end{matrix} \\ \begin{bmatrix} -\cos(\overline{\theta}_k^m + \Delta\theta^m) \\ \sin(\overline{\theta}_k^m + \Delta\theta^m) \\ (1 - \cos(\Delta\varphi^m)) \end{bmatrix} & \begin{bmatrix} \cos(\Delta\varphi^m) \\ +\cos^2(\overline{\theta}_k^m + \Delta\theta^m) \\ (1 - \cos(\Delta\varphi^m)) \end{bmatrix} & \\ \cos(\overline{\theta}_k^m + \Delta\theta^m)\sin(\Delta\varphi^m) & \sin(\overline{\theta}_k^m + \Delta\theta^m)\sin(\Delta\varphi^m) & \end{bmatrix} \tag{20}$$

The element of the matrix C as follows (C=BA):

$$\begin{aligned}
C_{11} &= \cos(\Delta\theta^m)\cos(\Delta\varphi^m) + \sin^2(\overline{\theta}_k^m + \Delta\theta^m)\cos(\Delta\theta^m) \\
&\quad (1-\cos(\Delta\varphi^m)) - \cos(\overline{\theta}_k^m + \Delta\theta^m)\sin(\overline{\theta}_k^m + \Delta\theta^m) \\
&\quad \sin(\Delta\theta^m)(1-\cos(\Delta\varphi^m)) \\
C_{12} &= -\sin(\Delta\theta^m)\cos(\Delta\varphi^m) - \sin^2(\overline{\theta}_k^m + \Delta\theta^m)\sin(\Delta\theta^m) \\
&\quad (1-\cos(\Delta\varphi^m)) - \cos(\overline{\theta}_k^m + \Delta\theta^m)\sin(\overline{\theta}_k^m + \Delta\theta^m) \\
&\quad \cos(\Delta\theta^m)(1-\cos(\Delta\varphi^m)) \\
C_{13} &= -\cos(\overline{\theta}_k^m + \Delta\theta^m)\sin(\Delta\varphi^m) \\
C_{21} &= -\cos(\overline{\theta}_k^m + \Delta\theta^m)\sin(\overline{\theta}_k^m + \Delta\theta^m)\cos(\Delta\theta^m) \\
&\quad (1-\cos(\Delta\varphi^m)) + \sin(\Delta\theta^m)\cos(\Delta\varphi^m) \\
&\quad + \cos^2(\overline{\theta}_k^m + \Delta\theta^m)\sin(\Delta\theta^m)(1-\cos(\Delta\varphi^m)) \\
C_{22} &= \cos(\overline{\theta}_k^m + \Delta\theta^m)\sin(\overline{\theta}_k^m + \Delta\theta^m)\sin(\Delta\theta^m) \\
&\quad (1-\cos(\Delta\varphi^m)) + \cos(\Delta\theta^m)\cos(\Delta\varphi^m) \\
&\quad + \cos^2(\overline{\theta}_k^m + \Delta\theta^m)\cos(\Delta\theta^m)(1-\cos(\Delta\varphi^m)) \\
C_{23} &= -\sin(\overline{\theta}_k^m + \Delta\theta^m)\sin(\Delta\varphi^m) \\
C_{31} &= \cos(\overline{\theta}_k^m + \Delta\theta^m)\cos(\Delta\theta^m)\sin(\Delta\varphi^m) \\
&\quad + \sin(\overline{\theta}_k^m + \Delta\theta^m)\sin(\Delta\theta^m)\sin(\Delta\varphi^m) \\
C_{32} &= -\cos(\overline{\theta}_k^m + \Delta\theta^m)\sin(\Delta\theta^m)\sin(\Delta\varphi^m) \\
&\quad + \sin(\overline{\theta}_k^m + \Delta\theta^m)\cos(\Delta\theta^m)\sin(\Delta\varphi^m) \\
C_{33} &= \cos(\Delta\varphi^m).
\end{aligned} \tag{21}$$

Combining the above equations we get

$$\begin{aligned}
\breve{X}_k^m &= R_{\alpha,\beta}X_k^m + X_{om} \\
&= R_{\alpha,\beta}\lambda C\overline{X}_k^m + X_{om} \\
&= R_{\alpha,\beta}\lambda C R_{\alpha,\beta}^{-1}(\breve{\overline{X}}_k^m - X_o) + X_{om}.
\end{aligned} \tag{22}$$

Hence, using MATLAB for symbolic we can compute true Jacobian matrix J_{b^m}. However, it is very complex formular so we do not write here. In simulation, we still execute algorithm using true Jacobian matrix (denote by track fusion ML).

When the local systems with the plan Oxy is approximate parallel to the ground (the angles β_m are approximate to 0), then we get the approximation as follows:

$$\begin{aligned}
\breve{X}_k^m &\approx \lambda C R_{\alpha,\beta}R_{\alpha,\beta}^{-1}(\breve{\overline{X}}_k^m - X_o) + X_{om} \\
&= \lambda C(\breve{\overline{X}}_k^m - X_o) + X_o.
\end{aligned} \tag{23}$$

More precisely,

$$
\begin{aligned}
\breve{x}_k^m &= \lambda[C_{11}(\breve{\bar{x}}_k^m - x_{om}) + C_{12}(\breve{\bar{y}}_k^m - y_{om}) \\
&\quad + C_{13}(\breve{\bar{z}}_k^m - z_{om})] + x_{om} \\
\breve{y}_k^m &= \lambda[C_{21}(\breve{\bar{x}}_k^m - x_{om}) + C_{22}(\breve{\bar{y}}_k^m - y_{om}) \\
&\quad + C_{23}(\breve{\bar{z}}_k^m - z_{om})] + y_{om} \\
\breve{z}_k^m &= \lambda[C_{31}(\breve{\bar{x}}_k^m - x_{om}) + C_{32}(\breve{\bar{y}}_k^m - y_{om}) \\
&\quad + C_{33}(\breve{\bar{z}}_k^m - z_{om})] + z_{om}.
\end{aligned}
\tag{24}
$$

For the velocity component

$$
\begin{aligned}
\breve{\bar{v}}_{x,k}^m &= \frac{\partial \breve{x}_k^m}{\partial \breve{\bar{x}}_k^m}\breve{\bar{v}}_{x,k}^m + \frac{\partial \breve{x}_k^m}{\partial \breve{\bar{y}}_k^m}\breve{\bar{v}}_{y,k}^m + \frac{\partial \breve{x}_k^m}{\partial \breve{\bar{z}}_k^m}\breve{\bar{v}}_{z,k}^m \\
\breve{\bar{v}}_{y,k}^m &= \frac{\partial \breve{y}_k^m}{\partial \breve{\bar{x}}_k^m}\breve{\bar{v}}_{x,k}^m + \frac{\partial \breve{y}_k^m}{\partial \breve{\bar{y}}_k^m}\breve{\bar{v}}_{y,k}^m + \frac{\partial \breve{y}_k^m}{\partial \breve{\bar{z}}_k^m}\breve{\bar{v}}_{z,k}^m \\
\breve{\bar{v}}_{x,k}^m &= \frac{\partial \breve{z}_k^m}{\partial \breve{\bar{x}}_k^m}\breve{\bar{v}}_{x,k}^m + \frac{\partial \breve{z}_k^m}{\partial \breve{\bar{y}}_k^m}\breve{\bar{v}}_{y,k}^m + \frac{\partial \breve{z}_k^m}{\partial \breve{\bar{z}}_k^m}\breve{\bar{v}}_{z,k}^m.
\end{aligned}
$$

We compute

$$
J \approx J_{b^m} = \frac{\partial f_m}{\partial b^m}|_{(\hat{X}_k^m, [0\ 0\ 0]^T)}. \tag{25}
$$

From (24) we get

$$
\begin{aligned}
J_{11} &= \frac{\partial \breve{x}_k^m}{\partial \Delta r^m}|_{(\hat{X}_k^m, [0\ 0\ 0]^T)} \\
&= \left(\frac{\partial \lambda}{\partial \Delta r^m}[C_{11}(\breve{\bar{x}}_k^m - x_{om}) + C_{12}(\breve{\bar{y}}_k^m - y_{om}) \right. \\
&\quad \left. + C_{13}(\breve{\bar{z}}_k^m - z_{om})]\right)|_{(\hat{X}_k^m, [0\ 0\ 0]^T)} \\
&= \frac{\hat{x}_k^m - x_{om}}{\sqrt{(\hat{x}_k^m - x_{om})^2 + (\hat{y}_k^m - y_{om})^2 + (\hat{z}_k^m - z_{om})^2}} \\
&\quad (\text{because } C_{11} = 1, C_{12} = 0, C_{13} = 0)
\end{aligned}
\tag{26}
$$

$$
\begin{aligned}
J_{12} &= \frac{\partial \breve{x}_k^m}{\partial \Delta \theta^m}|_{(\hat{X}_k^m, [0\ 0\ 0]^T)} \\
&= \left(\frac{\partial C_{12}}{\partial \Delta \theta^m}(\hat{y}_k^m - y_{om})\right)|_{(\hat{X}_k^m, [0\ 0\ 0]^T)} \\
&= -(\hat{y}_k^m - y_{om}) \\
&\quad (\text{because } \lambda = 1, \frac{\partial C_{11}}{\partial \Delta \theta^m} = 0, \frac{\partial C_{13}}{\partial \Delta \theta^m} = 0)
\end{aligned}
\tag{27}
$$

$$
\begin{aligned}
J_{13} &= \frac{\partial \breve{x}_k^m}{\partial \Delta\varphi^m}|_{(\hat{X}_k^m,\left[0\ 0\ 0\right]^T)} \\
&= \left(\frac{\partial C_{13}}{\partial \Delta\varphi^m}(\hat{z}_k^m - z_{om})\right)|_{(\hat{X}_k^m,\left[0\ 0\ 0\right]^T)} \\
&= -\frac{\overline{x}_k^m(\hat{z}_k^m - z_{om})}{\sqrt{(\overline{x}_k^m)^2 + (\overline{y}_k^m)^2}}|_{(\hat{X}_k^m)} \\
&\text{(because } \lambda = 1, \frac{\partial C_{11}}{\partial \Delta\varphi^m} = 0, \frac{\partial C_{12}}{\partial \Delta\varphi^m} = 0 \text{ and note that} \\
&\overline{X}_k^m = R_{\alpha\beta}(\hat{X}_k^m - X_{om}))
\end{aligned} \tag{28}
$$

$$
\begin{aligned}
J_{31} &= \frac{\partial \breve{y}_k^m}{\partial \Delta r^m}|_{(\hat{X}_k^m,\left[0\ 0\ 0\right]^T)} \\
&= \left(\frac{\partial \lambda}{\partial \Delta r^m}[C_{21}(\breve{\overline{x}}_k^m - x_{om}) + C_{22}(\breve{\overline{y}}_k^m - y_{om})\right. \\
&\left. + C_{23}(\breve{\overline{z}}_k^m - z_{om})]\right)|_{(\hat{X}_k^m,\left[0\ 0\ 0\right]^T)} \\
&= \frac{\hat{y}_k^m - y_{om}}{\sqrt{(\hat{x}_k^m - x_{om})^2 + (\hat{y}_k^m - y_{om})^2 + (\hat{z}_k^m - z_{om})^2}} \\
&\text{(because } C_{21} = 0, C_{22} = 1, C_{23} = 0);
\end{aligned} \tag{29}
$$

$$
\begin{aligned}
J_{32} &= \frac{\partial \breve{y}_k^m}{\partial \Delta\theta^m}|_{(\hat{X}_k^m,\left[0\ 0\ 0\right]^T)} \\
&= \left(\frac{\partial C_{21}}{\partial \Delta\theta^m}(\hat{x}_k^m - x_{om})\right)|_{(\hat{X}_k^m,\left[0\ 0\ 0\right]^T)} \\
&= (\hat{x}_k^m - x_{om}) \\
&\text{(because } \lambda = 1, \frac{\partial C_{22}}{\partial \Delta\theta^m} = 0, \frac{\partial C_{23}}{\partial \Delta\theta^m} = 0);
\end{aligned} \tag{30}
$$

$$
\begin{aligned}
J_{33} &= \frac{\partial \breve{y}_k^m}{\partial \Delta\varphi^m}|_{(\hat{X}_k^m,\left[0\ 0\ 0\right]^T)} \\
&= \left(\frac{\partial C_{23}}{\partial \Delta\varphi^m}(\hat{z}_k^m)\right)|_{(\hat{X}_k^m,\left[0\ 0\ 0\right]^T)} \\
&= -\frac{\overline{y}_k^m(\hat{z}_k^m - z_{om})}{\sqrt{(\overline{x}_k^m)^2 + (\overline{y}_k^m)^2}}|_{(\hat{X}_k^m)} \\
&\text{(because } \lambda = 1, \frac{\partial C_{21}}{\partial \Delta\varphi^m} = 0, \frac{\partial C_{22}}{\partial \Delta\varphi^m} = 0);
\end{aligned} \tag{31}
$$

$$
\begin{aligned}
J_{51} &= \frac{\partial \breve{z}_k^m}{\partial \Delta r^m}\Big|_{(\hat{X}_k^m, [0\;0\;0]^T)} \\
&= \Bigg(\frac{\partial \lambda}{\partial \Delta r^m} [C_{31}(\breve{\overline{x}}_k^m - x_{om}) + C_{32}(\breve{\overline{y}}_k^m - y_{om}) \\
&\quad + C_{33}(\breve{\overline{z}}_k^m - z_{om})] \Bigg)\Big|_{(\hat{X}_k^m, [0\;0\;0]^T)} \\
&= \frac{\hat{z}_k^m - z_{om}}{\sqrt{(\hat{x}_k^m - x_{om})^2 + (\hat{y}_k^m - y_{om})^2 + (\hat{z}_k^m - z_{om})^2}} \\
&\quad \text{(because } C_{31} = 0, C_{32} = 0, C_{33} = 1) \qquad (32)
\end{aligned}
$$

$$
\begin{aligned}
J_{52} &= \frac{\partial \breve{z}_k^m}{\partial \Delta \theta^m}\Big|_{(\hat{X}_k^m, [0\;0\;0]^T)} \\
&= 0 \\
&\quad \text{(because } \lambda = 1, \frac{\partial C_{31}}{\partial \Delta \theta^m} = 0, \frac{\partial C_{32}}{\partial \Delta \theta^m} = 0, \frac{\partial C_{33}}{\partial \Delta \theta^m} = 0) \qquad (33)
\end{aligned}
$$

$$
\begin{aligned}
J_{53} &= \frac{\partial \breve{z}_k^m}{\partial \Delta \varphi^m}\Big|_{(\hat{X}_k^m, [0\;0\;0]^T)} \\
&= \Bigg(\frac{\partial C_{31}}{\partial \Delta \varphi^m} (\hat{x}_k^m - x_{om}) \\
&\quad + \frac{\partial C_{32}}{\partial \Delta \varphi^m} (\hat{y}_k^m - y_{om}) \Bigg)\Big|_{(\hat{X}_k^m, [0\;0\;0]^T)} \\
&= \left(\frac{\overline{x}_k^m (\hat{x}_k^m - x_{om})}{\sqrt{(\overline{x}_k^m)^2 + (\overline{y}_k^m)^2}} + \frac{\overline{y}_k^m (\hat{y}_k^m - y_{om})}{\sqrt{(\overline{x}_k^m)^2 + (\overline{y}_k^m)^2}} \right)\Big|_{(\hat{X}_k^m)} \\
&= \frac{\overline{x}_k^m (\hat{x}_k^m - x_{om}) + \overline{y}_k^m (\hat{y}_k^m - y_{om})}{\sqrt{(\overline{x}_k^m)^2 + (\overline{y}_k^m)^2}}\Big|_{(\hat{X}_k^m)} \\
&\quad \text{(because } \lambda = 1, \frac{\partial C_{33}}{\partial \Delta \varphi^m} = 0). \qquad (34)
\end{aligned}
$$

We continue

$$
\begin{aligned}
J_{21} &= \frac{\partial \breve{v}_{x,k}^m}{\partial \Delta r^m}\Big|_{(\hat{X}_k^m, [0\;0\;0]^T)} \\
&= \frac{\partial J_{11}}{\partial \hat{x}_k^m} \hat{v}_{x,k}^m + \frac{\partial J_{11}}{\partial \hat{y}_k^m} \hat{v}_{y,k}^m + \frac{\partial J_{11}}{\partial \hat{z}_k^m} \hat{v}_{z,k}^m \\
&= \frac{\begin{array}{c}[(\hat{y}_k^m - y_{om})^2 + (\hat{z}_k^m - z_{om})^2] \hat{v}_{x,k}^m \\ -\ (\hat{x}_k^m - x_{om})(\hat{y}_k^m - y_{om}) \hat{v}_{y.k}^m \\ -\ (\hat{x}_k^m - x_{om})(\hat{z}_k^m - z_{om}) \hat{v}_{z,k}^m\end{array}}{\left[(\hat{x}_k^m - x_{om})^2 + (\hat{y}_k^m - y_{om})^2 + (\hat{z}_k^m - z_{om})^2 \right]^{3/2}}
\end{aligned}
$$

$$J_{22} = \frac{\partial \breve{v}^m_{x,k}}{\partial \Delta\theta^m}|_{(\hat{X}^m_k,[0\ 0\ 0]^T)} \tag{35}$$
$$= \frac{\partial J_{12}}{\partial \hat{x}^m_k}\hat{v}^m_{x,k} + \frac{\partial J_{12}}{\partial \hat{y}^m_k}\hat{v}^m_{y,k} + \frac{\partial J_{12}}{\partial \hat{z}^m_k}\hat{v}^m_{z,k} = -\hat{v}^m_{y,k}$$

$$J_{23} = \frac{\partial \breve{v}^m_{x,k}}{\partial \Delta\varphi^m}|_{(\hat{X}^m_k,[0\ 0\ 0]^T)} \tag{36}$$
$$= \frac{\partial J_{13}}{\partial \hat{x}^m_k}\hat{v}^m_{x,k} + \frac{\partial J_{13}}{\partial \hat{y}^m_k}\hat{v}^m_{y,k} + \frac{\partial J_{13}}{\partial \hat{z}^m_k}\hat{v}^m_{z,k}$$
$$= \frac{\begin{aligned}&[-r_{11}(\overline{y}^m_k)^2(\hat{z}^m_k - z_{om}) + r_{12}\overline{x}^m_k\overline{y}^m_k(\hat{z}^m_k - z_{om})]\hat{v}^m_{x,k}\\&+ [-r_{21}(\overline{y}^m_k)^2(\hat{z}^m_k - z_{om}) + r_{22}\overline{x}^m_k\overline{y}^m_k(\hat{z}^m_k - z_{om})]\hat{v}^m_{y,k}\\&+ [-(\overline{x}^m_k)^3 - r_{31}(\overline{y}^m_k)^2(\hat{z}^m_k - z_{om}) - \overline{x}^m_k(\overline{y}^m_k)^2\\&+ r_{32}\overline{x}^m_k\overline{y}^m_k(\hat{z}^m_k - z_{om})]\hat{v}^m_{z,k}\end{aligned}}{[(\overline{x}^m_k)^2 + (\overline{y}^m_k)^2]^{3/2}}$$

$$J_{41} = \frac{\partial \breve{v}^m_{y,k}}{\partial \Delta r^m}|_{(\hat{X}^m_k,[0\ 0\ 0]^T)}$$
$$= \frac{\partial J_{31}}{\partial \hat{x}^m_k}\hat{v}^m_{x,k} + \frac{\partial J_{31}}{\partial \hat{y}^m_k}\hat{v}^m_{y,k} + \frac{\partial J_{31}}{\partial \hat{z}^m_k}\hat{v}^m_{z,k}$$
$$= \frac{\begin{aligned}&- (\hat{x}^m_k - x_{om})(\hat{y}^m_k - y_{om})]\hat{v}^m_{x,k}\\&+ [(\hat{x}^m_k - x_{om})^2 + (\hat{z}^m_k - z_{om})^2]\hat{v}^m_{y.k}\\&- (\hat{x}^m_k - x_{om})(\hat{z}^m_k - z_{om})\hat{v}^m_{z,k}\end{aligned}}{\left[(\hat{x}^m_k - x_{om})^2 + (\hat{y}^m_k - y_{om})^2 + (\hat{z}^m_k - z_{om})^2\right]^{3/2}} \tag{37}$$

$$J_{42} = \frac{\partial \breve{v}^m_{y,k}}{\partial \Delta\theta^m}|_{(\hat{X}^m_k,[0\ 0\ 0]^T)}$$
$$= \frac{\partial J_{32}}{\partial \hat{x}^m_k}\hat{v}^m_{x,k} + \frac{\partial J_{32}}{\partial \hat{y}^m_k}\hat{v}^m_{y,k} + \frac{\partial J_{32}}{\partial \hat{z}^m_k}\hat{v}^m_{z,k} = \hat{v}^m_{x,k}$$

$$J_{43} = \frac{\partial \breve{v}^m_{y,k}}{\partial \Delta\varphi^m}|_{(\hat{X}^m_k,[0\ 0\ 0]^T)}$$
$$= \frac{\partial J_{33}}{\partial \hat{x}^m_k}\hat{v}^m_{x,k} + \frac{\partial J_{33}}{\partial \hat{y}^m_k}\hat{v}^m_{y,k} + \frac{\partial J_{33}}{\partial \hat{z}^m_k}\hat{v}^m_{z,k}$$
$$= \frac{\begin{aligned}&[- r_{12}(\overline{x}^m_k)^2(\hat{z}^m_k - z_{om}) + r_{11}\overline{x}^m_k\overline{y}^m_k(\hat{z}^m_k - z_{om})]\hat{v}^m_{x,k}\\&+ [-r_{22}(\overline{x}^m_k)^2(\hat{z}^m_k - z_{om}) + r_{21}\overline{x}^m_k\overline{y}^m_k(\hat{z}^m_k - z_{om})]\hat{v}^m_{y,k}\\&+ [-(\overline{y}^m_k)^3 - r_{32}(\overline{x}^m_k)^2\hat{z}^m_k - \overline{y}^m_k(\overline{x}^m_k)^2\\&+ r_{31}\overline{x}^m_k\overline{y}^m_k(\hat{z}^m_k - z_{om})]\hat{v}^m_{z,k}\end{aligned}}{[(\overline{x}^m_k)^2 + (\overline{y}^m_k)^2]^{3/2}}$$

$$J_{61} = \frac{\partial \breve{v}^m_{z,k}}{\partial \Delta r^m}|_{(\hat{X}^m_k,[0\ 0\ 0]^T)}$$

$$
\begin{aligned}
&= \frac{\partial J_{51}}{\partial \hat{x}_k^m}\hat{v}_{x,k}^m + \frac{\partial J_{51}}{\partial \hat{y}_k^m}\hat{v}_{y,k}^m + \frac{\partial J_{51}}{\partial \hat{z}_k^m}\hat{v}_{z,k}^m \\
&= \frac{\begin{array}{l} -(\hat{x}_k^m - x_{om})(\hat{z}_k^m - z_{om})\hat{v}_{x,k}^m \\ -(\hat{y}_k^m - y_{om})(\hat{z}_k^m - z_{om})\hat{v}_{y.k}^m \\ +[(\hat{x}_k^m - x_{om})^2 + (\hat{z}_k^m - z_{om})^2]\hat{v}_{z,k}^m \end{array}}{\left[(\hat{x}_k^m - x_{om})^2 + (\hat{y}_k^m - y_{om})^2 + (\hat{z}_k^m - z_{om})^2\right]^{3/2}} \\
J_{62} &= \frac{\partial \breve{v}_{z,k}^m}{\partial \Delta\theta^m}\Big|_{(\hat{X}_k^m, \left[0\ 0\ 0\right]^T)} \\
&= \frac{\partial J_{52}}{\partial \hat{x}_k^m}\hat{v}_{x,k}^m + \frac{\partial J_{52}}{\partial \hat{y}_k^m}\hat{v}_{y,k}^m + \frac{\partial J_{52}}{\partial \hat{z}_k^m}\hat{v}_{z,k}^m = 0 \\
J_{63} &= \frac{\partial \breve{v}_{z,k}^m}{\partial \Delta\varphi^m}\Big|_{(\hat{X}_k^m, \left[0\ 0\ 0\right]^T)} \\
&= \frac{\partial J_{53}}{\partial \hat{x}_k^m}\hat{v}_{x,k}^m + \frac{\partial J_{53}}{\partial \hat{y}_k^m}\hat{v}_{y,k}^m + \frac{\partial J_{53}}{\partial \hat{z}_k^m}\hat{v}_{z,k}^m \\
&= \frac{\begin{array}{l} [[r_{11}(\hat{x}_k^m - x_{om}) + \overline{x}_k^m + r_{12}(\hat{y}_k^m - y_{om})][(\overline{x}_k^m)^2 + (\overline{y}_k^m)^2] \\ -[\overline{x}_k^m(\hat{x}_k^m - x_{om}) + \overline{y}_k^m(\hat{y}_k^m - y_{om})][r_{11}\overline{x}_k^m + r_{12}\overline{y}_k^m]]\, \hat{v}_{x,k}^m \\ +[[r_{21}(\hat{x}_k^m - x_{om}) + r_{22}(\hat{y}_k^m - y_{om}) + \overline{x}_k^m][(\overline{x}_k^m)^2 + (\overline{y}_k^m)^2] \\ -[\overline{x}_k^m(\hat{x}_k^m - x_{om}) + \overline{y}_k^m(\hat{y}_k^m - y_{om})][r_{21}\overline{x}_k^m + r_{22}\overline{y}_k^m]]\, \hat{v}_{y,k}^m \\ +[[r_{31}(\hat{x}_k^m - x_{om}) + r_{32}(\hat{y}_k^m - y_{om})][(\overline{x}_k^m)^2 + (\overline{y}_k^m)^2] \\ -[\overline{x}_k^m(\hat{x}_k^m - x_{om}) + \overline{y}_k^m(\hat{y}_k^m - y_{om})][r_{31}\overline{x}_k^m + r_{32}\overline{y}_k^m]]\, \hat{v}_{z,k}^m \end{array}}{[(\overline{x}_k^m)^2 + (\overline{y}_k^m)^2]^{3/2}}.
\end{aligned} \tag{38}
$$

Similarly as in [17] we have the following approximation:

$$
\begin{aligned}
f_m(\breve{X}_k^m, b^m) = f_m(\hat{X}_k^m, \left[0\ 0\ 0\right]^T) &+ \frac{\partial f_m}{\partial \breve{X}_k^m}\Big|_{(\hat{X}_k^m, \left[0\ 0\ 0\right]^T)} \cdot [\breve{X}_k^m - \hat{X}_k^m] \\
&+ \frac{\partial f_m}{\partial b^m}\Big|_{(\hat{X}_k^m, \left[0\ 0\ 0\right]^T)} \cdot \begin{bmatrix} \Delta r^m \\ \Delta\theta^m \\ \Delta\varphi^m \end{bmatrix}.
\end{aligned} \tag{39}
$$

From the approximate equation $\hat{X}_k^m = f_m(\breve{X}_k^m, b^m) + e_k^m$ and $f_m(\hat{X}_k^m, \left[0\ 0\ 0\right]^T) = \hat{X}_k^m$, we have

$$
\hat{X}_k^m = \breve{X}_k^m + J_{b^m} \begin{bmatrix} \Delta r^m \\ \Delta\theta^m \\ \Delta\varphi^m \end{bmatrix} + e_k^m.
$$

Hence, combining with the approximation that J_{b^m} can be approximated by J, we finish the proof of Theorem 1.

(3) Bias estimation, correction and fusion of local estimates: Using cross-covariance in [1], the Bar-Shalom and Campo fusion formular [3], similarly as in

[17] to obtain bias estimation and accurate fusion. We hope to continuous the work on track fusion by Jacobian matrix for more than two sensors as in [15] for exact method.

1.4 Simulation

The numerical simulation using MATLAB with inputs as follows:
Simulation

$N = 200$ number of Monte Carlo runs, $M = 4$ number of sensors, $K = 200$ total dynamic steps, $T = 10$ time resampling, $xo = [20; 400; 30; 325]$, $yo = [50; 100; 40; 121]$, $zo = [0; 0; 0; 0]$ location of sensors, $\alpha = [0.242; 0.375; 0.124; 0.243]$ $\beta = [0.013; 0.01; 0.01; 0.2]$ angles of local system, $q = \sqrt{0.0002}$ standard of process, $r = 1$ parameter of measurement noise, $s = [35; 0.2128; 135; 0.1786; 5; 0]$ initial state, $b1 = [-1; -0.0042; -0.0001]$, $b2 = [1.2; 0.0035; 0.0002]$, $b3 = [1.6; 0.0022; 0.0007]$, $b4 = [-1.5; -0.0052; -0.0005]$ bias sensors system, $rr = 0.01, r\theta = 0.001, r\varphi = 0.0001$ standards of measurement.

Denote: The measurement fusion with 4 sensors by MF4, the measurement fusion with 2 sensors by MF2, track fusion using approximate Jacobian matrix by Track fusion, track fusion using true Jacobian matrix by Track fusion ML. The algorithms fusion from local tracks is treated for 2 sensors. For 2 sensors, the algorithms Track fusion and Track fusion ML give equivalent accuracy estimation and they are little better than measurement fusion (MF2). The measurement fusion with 4 sensors (MF4) gives the best estimation. In Fig. 3, for the elevation angle, the track fusion ML gives a most inaccurate estimation, and the measurement fusion gives a better estimation. For range bias and azimuth bias, the track fusion gives a better estimation (see Figs. 1 and 2). Figure 4 shows the real motion of the target and two local tracks of two sensors. The root means square errors of local tracks are simulated as in Fig. 5. The accuracy of esti-

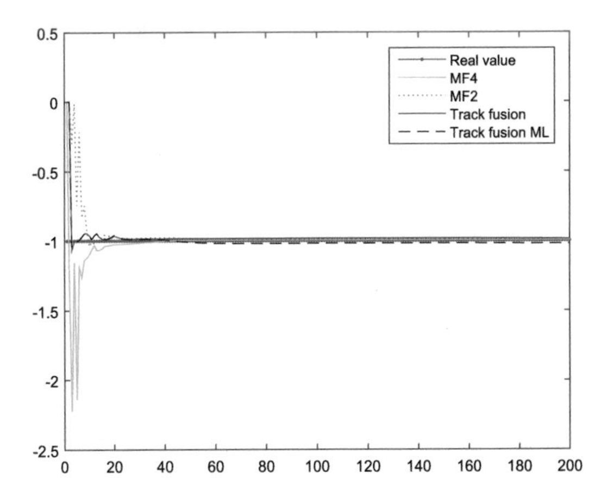

Fig. 1. Range bias estimation of sensor 1.

mation of the methods is measured by root mean square errors in position and velocity (see Figs. 6, 7 and 8).

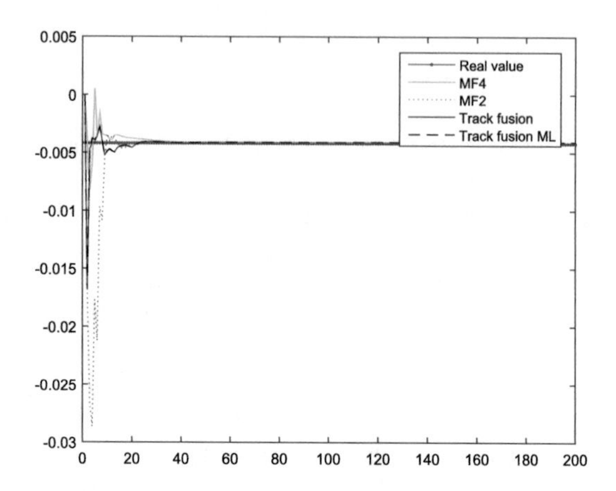

Fig. 2. Azimuth bias estimation of sensor 1.

Table 1. Executing time of algorithms

	Total time	Self time
MF4	0.507	0.003
MF2	0.215	0.003
Track fusion	0.277	0.000
Track fusion ML	0.434	0.000
Bias-ignorant fusion	0.212	0.000

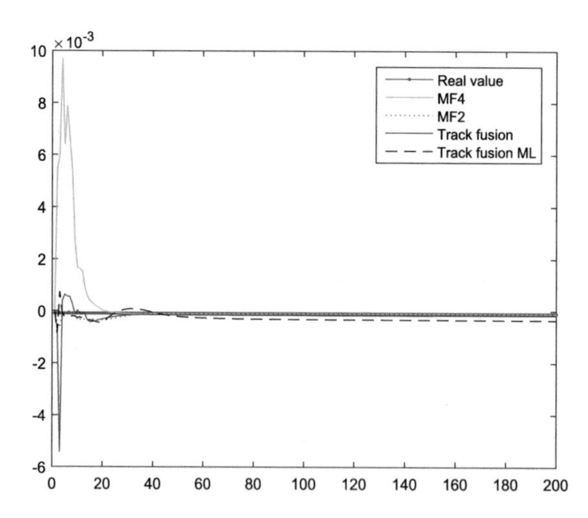

Fig. 3. Elevation bias estimation of sensor 1.

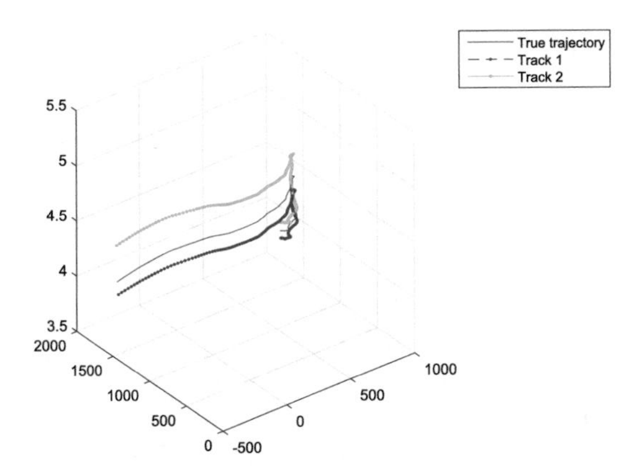

Fig. 4. Real and estimated track.

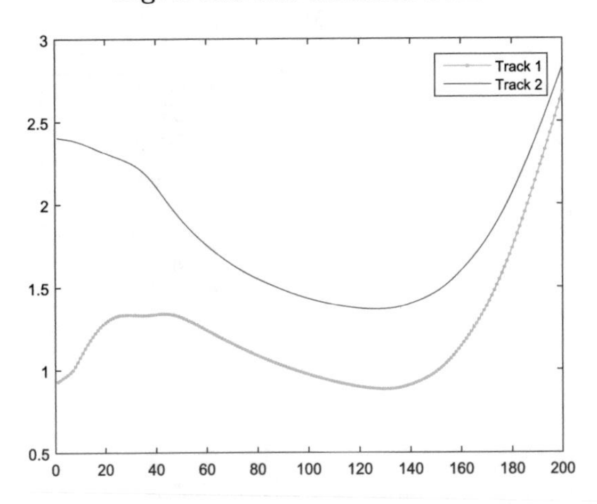

Fig. 5. RMSE local track.

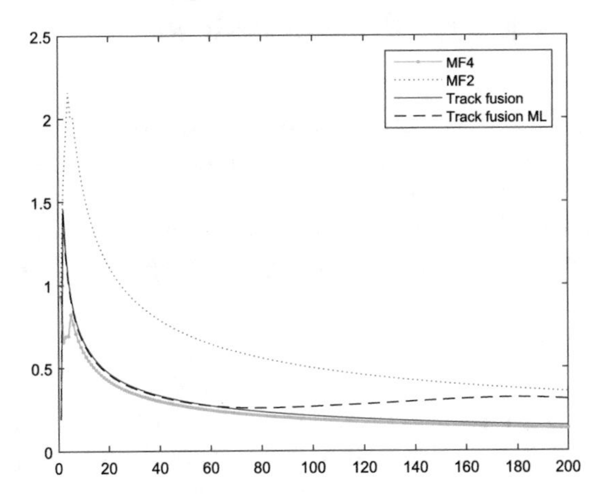

Fig. 6. RMSE position.

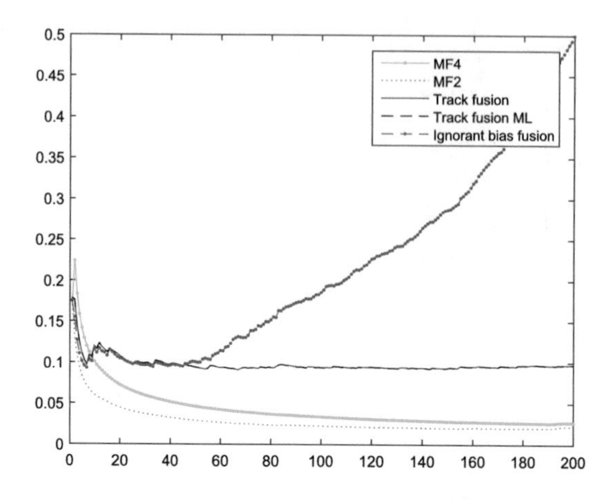

Fig. 7. RMSE velocity.

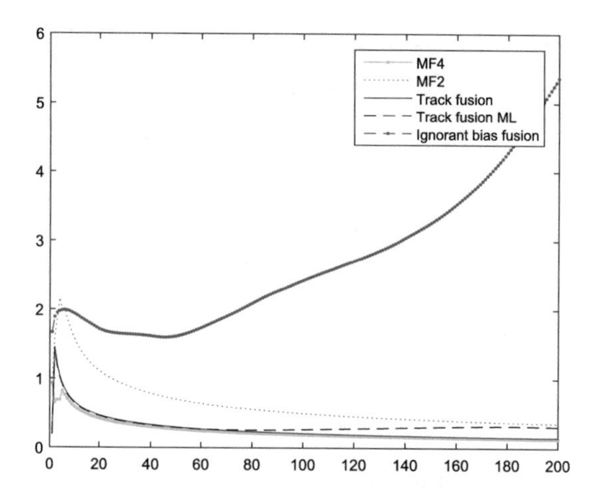

Fig. 8. RMSE position with ignorant bias.

The executing time of the algorithms is summarized in Table 1 (in s). It is evaluated on the computer configuration as follows: CPU: Intel Xeon(R) CPU E3-1245 v5@3.5 GHz 3.5 GHz; RAM 9.28 GB; Operating system Window 64bit. The executing time of Track fusion is more better than Track fusion ML because the approximation Jacobian matrix give higher computational efficiency.

2 Conclusion

The local track fusion for the 3D case is a well-done process and gives a better estimation than the measurement approach within the same number of radars (M = 2). When the number of sensors increases (M = 4) the estimation is better. For distributed sensor system, the local tracks fusion is preferred instead of the measurement fusion by the limitation of bandwidth. However, the exact Jacobian matrix is quite delicate, we need the help of MATLAB in computing it. The more complicated the matrix is the slower the program becomes. We give an approximation matrix for higher computational efficiency and it gives an equivalent accuracy estimation.

References

1. Bar-Shalom, Y.: On the track-to-track correlation problem. IEEE Trans. Autom. Control. **26**(2), 571–572 (1981)
2. Bar-Shalom, Y., Blair, W.D.: Multitarget-Multisensor Tracking: Applications and Advances, vol. III. Artech House Inc., Norwood (2000)
3. Bar-Shalom, Y., Campo, L.: The effect of the common process noise on the two-sensor fused-track covariance. IEEE Trans. Aerosp. Electron. Syst. **6**, 803–805 (1986)
4. Bar-Shalom, Y., Li, X.R.: Multitarget-Multisensor Tracking: Principles and Techniques, vol. 19. YBs, London (1995)
5. Bar-Shalom, Y., Li, X.R., Kirubarajan, T.: Estimation with Applications to Tracking and Navigation: Theory Algorithms and Software. Wiley, Hoboken (2004)
6. El-Badawy, E., Abd-ElShahid, T., Hafez, A.: A real time 3d multi target data fusion for multistatic radar network tracking. Session 1P0, p. 331 (2014)
7. Friedland, B.: Treatment of bias in recursive filtering. IEEE Trans. Autom. Control. **14**(4), 359–367 (1969)
8. Ignagni, M.: An alternate derivation and extension of friendland's two-stage kalman estimator. IEEE Trans. Autom. Control. **26**(3), 746–750 (1981)
9. Ignagni, M.: Optimal and suboptimal separate-bias kalman estimators for a stochastic bias. IEEE Trans. Autom. Control. **45**(3), 547–551 (2000)
10. Lin, X., Bar-Shalom, Y., Kirubarajan, T.: Exact multisensor dynamic bias estimation with local tracks. IEEE Trans. Aerosp. Electron. Syst. **40**(2), 576–590 (2004)
11. Lin, X., Bar-Shalom, Y., Kirubarajan, T.: Multisensor multitarget bias estimation for general asynchronous sensors. IEEE Trans. Aerosp. Electron. Syst. **41**(3), 899–921 (2005)
12. Nabaa, N., Bishop, R.: Solution to a multisensor tracking problem with sensor registration errors. IEEE Trans. Aerosp. Electron. Syst. **35**(1), 354–363 (1999)
13. Okello, N.N., Challa, S.: Joint sensor registration and track-to-track fusion for distributed trackers. IEEE Trans. Aerosp. Electron. Syst. **40**(3), 808–823 (2004)
14. Raol, J.R.:. Multi-sensor Data Fusion with MATLAB®. CRC Press, Boca Raton (2009)
15. Taghavi, E., Tharmarasa, R., Kirubarajan, T., Bar-Shalom, Y., Mcdonald, M.: A practical bias estimation algorithm for multisensor-multitarget tracking. IEEE Trans. Aerosp. Electron. Syst. **52**(1), 2–19 (2016)
16. van Doorn, B., Blom, H.: Systematic error estimation in multisensor fusion systems. In: Optical Engineering and Photonics in Aerospace Sensing, pp. 450–461. International Society for Optics and Photonics (1993)
17. Zhu, H., Chen, S.: Track fusion in the presence of sensor biases. IET Signal Process. **8**(9), 958–967 (2014)

3D Circular Embedded Antenna Mounted on Coaxial Feeding for Future Wideband Applications

Iram Nadeem[1], Dong-You Choi[1(✉)], Sun-kuk Noh[2], Jiwan Ghimire[1], and Ho-Gyun Yu[1]

[1] Department of Information and Communications Engineering, Chosun University, Gwangju, Republic of Korea
engineer.iram@gmail.com, ghimirejiwan2010@gmail.com, dychoi@chosun.ac.kr, 1030ghrbs@naver.com

[2] Department of Electronics, Chosun College of Science & Technology, Gwangju, Republic of Korea
nsk7078@cst.ac.kr

Abstract. This paper presents the simulation analysis of three dimensional circular embedded patch antenna mounted on coaxial cable for wideband applications. The proposed antenna design is constructed from taking a basic square substrate with ground at lower side which is fed by a cylindrical strip coaxial cable feed line satisfying the 50 Ω requirement. Two vertically symmetrical cylinder of equal diameter with perpendicular to each other are mounted on feed line; which acts as a radiator. High Frequency Structure Simulator (HFSS) is used to analyze the critical parameters and simulation results on the reflection coefficient/return loss, VSWR (voltage standing wave ratio), H-plane, E-plane, gain and radiation pattern of the proposed antenna at various frequencies. The antenna has ability to achieve the frequency range of 3.3–30 GHz along with 160%, 9.09:1 fractional bandwidth and bandwidth ratio (BWR), respectively.

Keywords: 3D antenna · Bandwidth ratio (BWR) · Coaxial cable High peak gain

1 Introduction

Wide band systems are growing dramatically and upon this demand there is great requirement of antennas for filling the requirements. Moreover many communication systems use circularly polarized or omnidirectional antennas having low profile. Several designs have been appeared in literature.

The monopolar patch antenna [1] is 0.09 λ tall and has wide bandwidth. A super wideband monopolar patch antenna [2] is similar to previous one but with a reasonable increased bandwidth. A wideband bi-cone antenna design is presented in [3]. In [4, 5], a sleeve monopole antennas are used for different communication system applications is presented. Similarly, a wideband low profile (STHMA) shorted top hat monocone antenna [6] is λ/14.7 tall and have 3:1 bandwidth ratio. All these design use shorting

K. Arai et al. (Eds.): FICC 2018, AISC 886, pp. 132–142, 2019.
https://doi.org/10.1007/978-3-030-03402-3_10

pins to achieve desired requirements. Previously, planner antenna designs [7, 8] like bow tie, dipole, Vivaldi, monopole and microstrip patch antennas which are two dimensional; ruled over communication industry for decades. But now the 3D printing has opened new methods for researcher to fabricate more complex shaped three dimensional antennas [9]. In [10], a 3D UWB diversity antenna consisting of eight ports monopoles perpendicular to each other is presented. A variety of 3D printed microwave antennas are offered in [11] including narrow band, wideband, multiband and reconfigurable designs methodologies. These antennas are folded E-patch, single layer patch, a bilateral Vivaldi, Lego-like assembled and Spartan logo antennas. In research article [12] planar vertical cross strips monocone antenna geometry is presented where aluminum is used as antenna radiator instead of copper, which results in wide impedance bandwidth and average efficiency 90%. Additionally a microstrip meander line 3D antenna [13], miniaturized tunable conical helix antenna [14] is described to support the literature.

In this paper, a three dimensional (3D) printed antenna is proposed as this technology provides significant flexibility in the designs that combine the assembly of metal layers and dielectric material properties to achieve required performance characteristics such as radiation pattern and maximum gain. The final structure achieves a maximum peak gain of 15 dB throughout the frequency ranging from 3.3 to 30 GHz. The proposed structure is compact in size and highly unidirectional towards the normal of the xy-plane.

2 Design of the 3D Structure

Figure 1 shows the geometry of the proposed antenna having size of $25 \times 25 \times 20\ \text{mm}^3$. The substrate material is Roger RT/duroid 5880 (tm) having the relative permittivity of 2.2 and dielectric loss tangent of 0.0009, respectively.

$$L_{Sub} = W_{Sub} = \frac{c}{2F}\left(\frac{\varepsilon_r + 1}{2}\right)^{-0.5} \tag{1}$$

Where, L_{Sub} and W_{Sub} is the length and width of antenna substrate, respectively, are equal to the 1/2 of the effective wavelength at F, the lowest initial operating frequency of antenna. C is the speed of light in free space and ε_r is the dielectric constant of substrate.

Spurious radiation that occurs by simple feeding can be reduced by coaxial probe feeding; so it is smeared to the proposed design structure. In this case due to high Q leakage value of the resonator, the coaxial probing illustrates a wide resonance bandwidth. Also with the increment in the cylindrical strip feeding length, the inductance along the line of the probe becomes more effective, which results wide bandwidth.

By reducing the length of the cylindrical strip feed line inductance effects can be reduced. The lower microwave frequency value range highly depends on overall radiating structure dimensions. So, the operating frequency exhibits important role in antenna miniaturization. To reach high bandwidth and miniaturization effectively, 3D

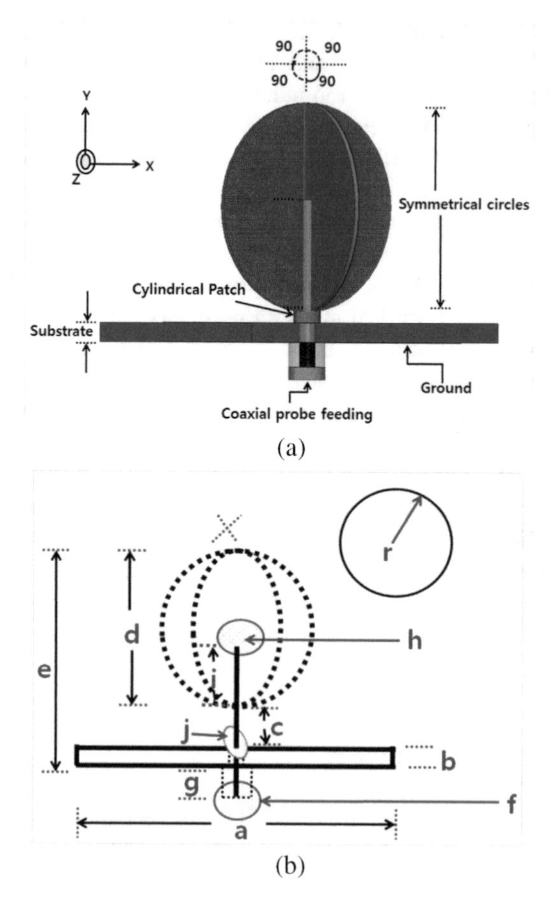

Fig. 1. (a) 3D geometry of the proposed antenna, (b) Structure along with dimensions.

embedded techniques are used. Figure 1 demonstrates the geometry and structure of the proposed antenna. The proposed antenna is etched at the center of a square-shaped substrate, where ground patch is on down side. This makes easy for antenna to resonate at lower frequency. The antenna is composed of substrate, coaxial probe fed, cylindrical patch and two symmetrical circles (main patch). A simplified process to construct the antenna structure in 2D is tried and shown in Fig. 1(a). A cylindrical patch of dimension "j" is directly connected with the coaxial probe. The main patch consists of two symmetrical circles "d" which are perpendicular to each other; embedded with cylindrical patch using 3D technology. To achieve low inductance value, the height of the cylindrical patch (i + c) is minimized. The structure has overall height of 20 mm (from the ground plane until main patch) and the probe feed height is 3 mm. Figure 1 (b) depicts the cross section view of the proposed structure. The proportion of coupling among two symmetrical circles has been used to tune the resonance frequency of the proposed structure.

To achieve higher bandwidth, gain and radiation efficiency both circles are stacked at perpendicular to each other position which is helpful to achieve high directivity normal to the antenna ground plane. With change in the size "d" of the circular patch (main patch) and cylindrical patch fed "j" the bandwidth and antenna radiation pattern changed steadily.

The circular patch is actively couple with the cylindrical patch fed; hence the reactance decreases at the feeding point with the reasonable increment in the input impedance resistance. Through parametric analysis, it can be concluded that "j" plays an important role in tuning the proposed antenna resonance frequency. Figure 2 depicts the surface current and vector E – field distribution. From Fig. 2(a), it can be observed that the direction of the current flows is heading from feeding point to the edges of the symmetrically stacked patch circles. Hence, direct connection between the patches is essential for the proposed antenna geometry. Moreover, the direction of current flow through the ground plane confirms the mechanism of ground in the proposed design antenna. Figure 2(b) shows the vector E-field where the field is stronger over the feeding point of the stacked patches; hence higher directivity can be seen in + z direction.

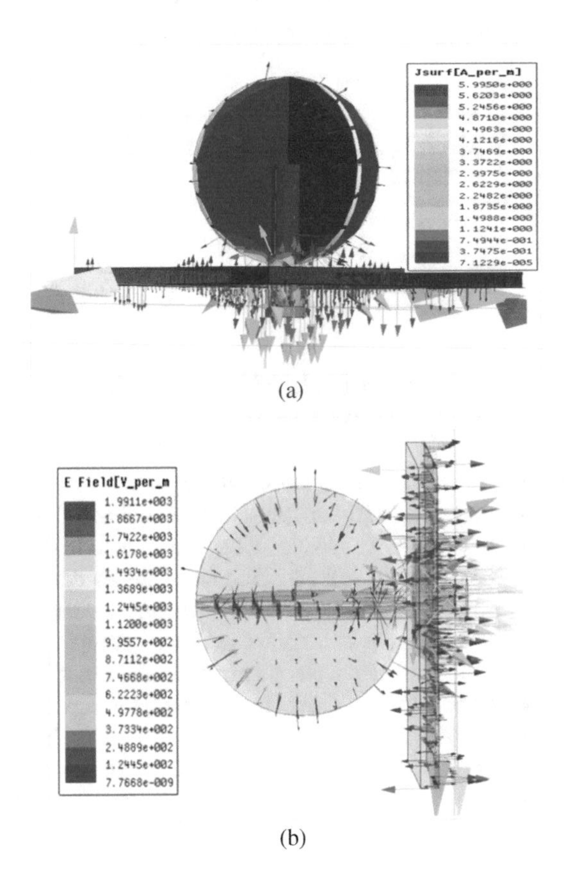

Fig. 2. (a) Surface current density, (b) E field distribution of the proposed antenna.

At the edges of the circular patches, amount of power is almost identical. However, due to the comparatively strong E-field at the lower portion shows glimpse of directional polarization in the proposed structure.

3 Results and Discussion

The optimal proposed designs dimension are listed in Table 1 are simulated on by using the HFSSv12 (High Frequency Simulator Software version 12). Figure 3 shows the simulated results of the proposed antenna to verify the wideband characteristics in term of reflection coefficient, axial ratio and impedance bandwidth of 26.7 GHz has been achieved, which can be confirm from Fig. 3(a) S11 ≤ -10 dB starting from 3.3 GHz (lower band) to 30 GHz (higher frequency value). The circular behavior of antenna can be observed from its axial ratio value, which should be axial value $\leq$ 3 dB is represented in Fig. 3(b). In this case, the antenna shows circular polarization properties at different frequency values but most prominent one is from 18.5 GHz to 20.3 GHz. VSWR (voltage standing wave ratio) represents the transmitted as well as the received power to the antenna. Figure 3(c) demonstrates that the VSWR $\leq$ 2 performances satisfy for the bandwidth of 3.3 to 30 GHz with fractional bandwidth of 160%.

Table 1. Proposed 3D antenna's parameters

Lengths	mm	Lengths	mm	Diameters	mm
a	25	e	20	f	1.6
b	1.6	g	3	j	0.7
c	0.5	i	9.5	h	1.5
d	9	-	-	r	9

Figure 4 depicts the simulation results of peak gain and radiation efficiency for whole range of operating frequencies achieved.

The gain increases starting from 1.5 dB at 3.3 GHz to a maximum of 15.5 dB at 29 GHz. The simulated gain has linear relation with input frequency band. Whereas the radiation efficiency decreases to 65% sharply at some higher values of frequency but has maximum assessment of 98%. Hence it can be concluded that the antenna is radiating effectively outside without greater amount of losses due to surface waves.

In order to verify capability of the proposed antenna to operate in as wideband antenna, it is necessary to achieve a constant or lower group delay values. This is because the shape of the transmitted electrical pulse must not be distorted by the antenna while transmission. Figure 5 shows the simulated group delay is non-uniform but maximum variation is below 1 ns for whole operating bandwidth which is acceptable. There is a group delay peak of value 1 ns at 28 GHz. If group delay variation is more than 1 ns it means the phases are no longer linear in the far field region and chances of pulse distortion is more.

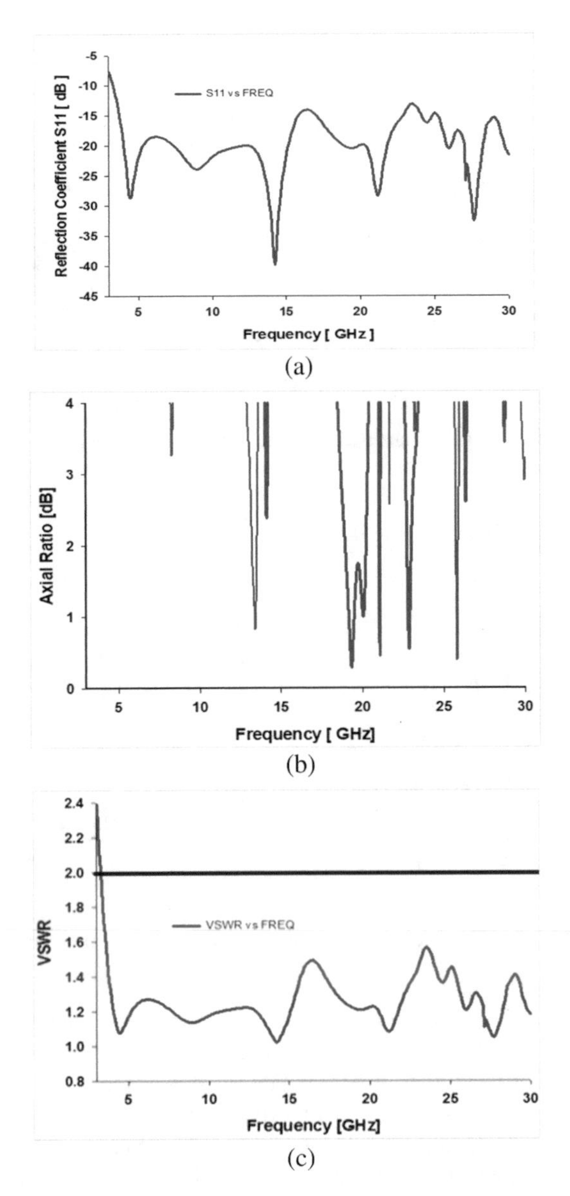

Fig. 3. Simulated (a) Reflection Coefficient, (b) 3 – dB axial ratio, (c) VSWR of the proposed geometry.

The 2D simulated radiation pattern of the proposed antenna at resonance frequencies of 5, 10, 15, 20, 25, 30 GHz with respect to total power in azimuthal plane and elevation plane are shown in Figs. 6 and 7, respectively. 2D patterns plot demonstrate the electric field components separately for $\varnothing = 0, 45$ and $90°$. Graph plot shows that the radiation pattern is symmetrical butterfly like for whole wideband. These features would allow the proposed antenna to be easily integrated into large platform installations.

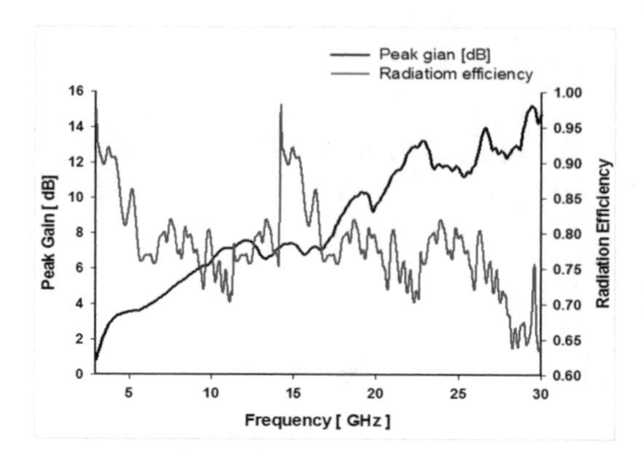

Fig. 4. Simulated peak gain and radiation efficiency of the proposed antenna.

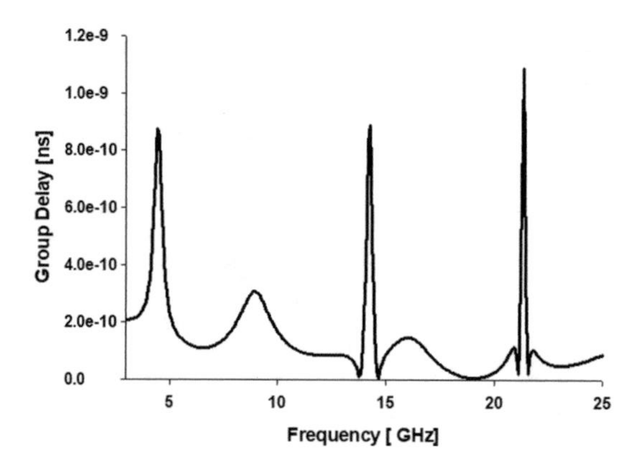

Fig. 5. Simulated group delay (ns) of the proposed antenna.

Table 2 show the reasonable comparison of antenna dimensions, reflection coefficient and fractional bandwidth of the proposed design with previous published work. In our case fractional bandwidth is 160 which is good compare to many other design reported except in [10]. Main reason of difference between reference [10] and the proposed design is antenna dimensions, as the size of our antenna is many times shorter than that of reference paper. However the antenna in reference [6] has length 5 times larger than the proposed antenna design length, while the antenna in [7] has small dimensions along with low value of bandwidth. So, it is concluded that dimension value difference is very obvious and can be clearly observed from pointed table.

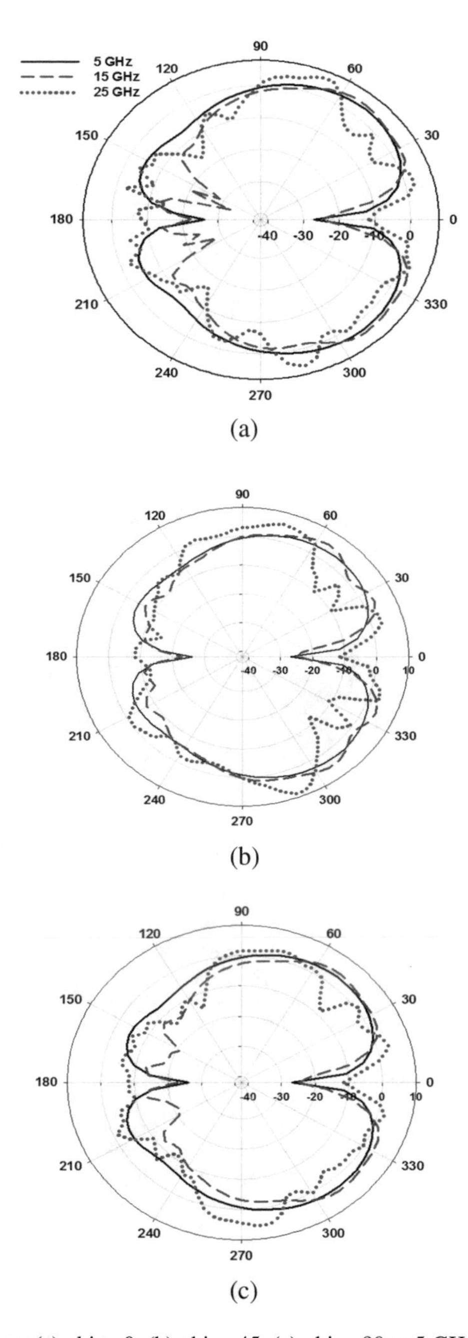

(a)

(b)

(c)

Fig. 6. Radiation pattern (a) phi = 0, (b) phi = 45, (c) phi = 90 at 5 GHz, 15 GHz and 25 GHz.

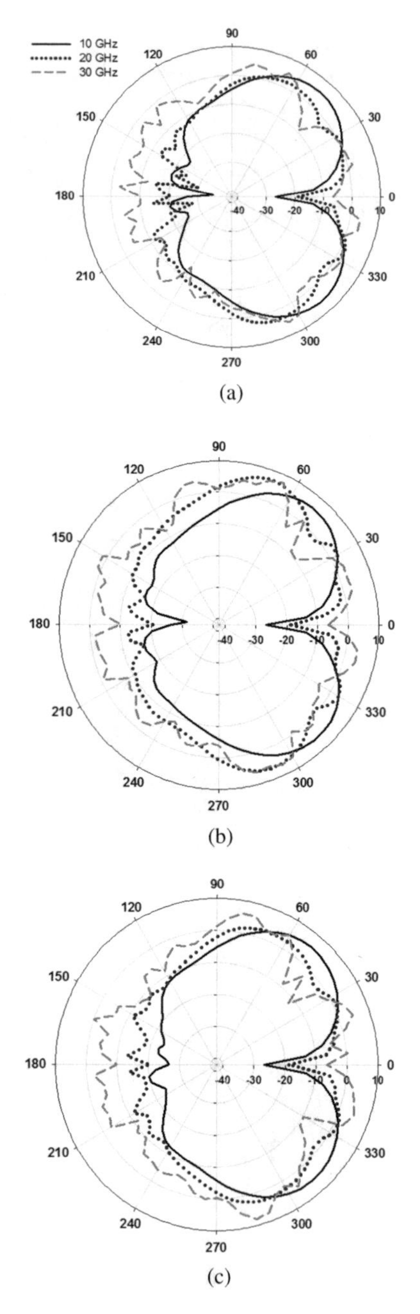

Fig. 7. Radiation pattern (a) phi = 0, (b) phi = 45, (c) phi = 90 at 10 GHz, 20 GHz and 30 GHz.

Table 2. Dimensions, reflection coefficient and fractional BW (%)

Ref.	Size (mm^3)	Bandwidth	Fractional BW (%)
[6]	126 × 25.4 × 29.3	800 MHz–2.5 GHz	103
[7]	17 × 17 × 15	4.8 GHz–29.8 GHz	144
[10]	57 × 80 × 112	550 MHz–18 GHz	188
This	25 × 25 × 20	3.3 GHz–30 GHz	160

4 Conclusion

In this paper, a 3D antenna is presented that utilizes perpendicular stacked patch with embedded techniques for antenna dimension miniaturization. The antenna is compact in size with dimensions of 25 × 25 × 20 mm^3. The proposed antenna has maximum directivity through the whole operating frequency band while maintaining the small size. A cylindrical patch is used between the main and coaxial probe as feeding line for the antenna, which controls the bandwidth and corresponding gain values. The radiation characteristics of the proposed antenna are appropriate for highly directive applications. The proposed antenna has maximum gain of 15.5 dB in the achieved operating bandwidth 3.3–30 GHz. With the further developments in 3D printing techniques, the front end RF (Radio Frequencies) chips can be embedded inside the antenna elements which will make future wireless communication systems compact as well as low-cost.

Acknowledgement. This research was supported by the MSIP (Ministry of Science, ICT and Future Planning), Korea, under the Global IT Talent Information and Communication Technology (IITP-2017-0-01658) supervised by the IITP (Institute for information and Communication Technology promotion) and partially by the Human Resource Training Program for Regional Innovation and Creativity through the Ministry of Education and National Research Foundation of Korea (2015H1C1A1035855).

References

1. Lau, K.-L., Li, P., Luk, K.-M.: A monopolar patch antenna with very wide impedance bandwidth. IEEE Trans. Antennas Propag. **53**, 1004–1010 (2005)
2. Lau, K.L., Kong, K.C., Luk, K.M.: Super-wideband monopolar patch antenna. Electron. Lett. **44**, 716–718 (2008)
3. Zuo, S.L., Yin, Y.Z., Zhang, Z.Y., Song, K.: Enhanced bandwidth of low-profile sleeve monopole antenna for indoor base station application. Electron. Lett. **46**, 1587–1588 (2010)
4. Zhang, Z.-Y., Fu, G., Gong, S.-X., Zuo, S.-L., Lu, Q.-Y.: Sleeve monopole antenna for DVB-H applications. Electron. Lett. **46**, 879–880 (2010)
5. Amert, A.K., Whites, K.W.: Miniaturization of the biconical antenna for ultrawideband applications. IEEE Trans. Antennas Propag. **57**, 3728–3735 (2009)
6. Aten, D.W., Haupt, R.L.: A wideband, low profile, shorted top hat monocone antenna. IEEE Trans. Antennas Propag. **60**, 4485–4491 (2012)

7. Nadeem, I., Kim, Y.J.: Design of ultra-wide band antenna with compact MCR structure for BW enhancement. J. Korean Inst. Commun. Sci. **42**, 798–805 (2017)
8. Nadeem, I., Shrestha, S., Kim, S.W., Han, S.J., Choi, D.Y.: Design of slotted UWB antenna with modified ground plane for bio medical applications. In: Proceedings of the International Conference on Electronics, Information and Communication, Thailand, pp. 621–624 (2017)
9. Canelón, E.E.L., López, A.G., Chandra, R., Johansson, A.J.: 3D printed miniaturized UWB antenna for wireless body area network. In: IEEE 8th European Conference on Antennas and Propagation (EuCAP), pp. 3090–3093 (2014)
10. Palaniswamy, S.K., Selvam, Y.P., Gulam Nabi Alsath, M., Kanagasabai, M., Kingsly, S., Subbaraj, S.: 3-D eight-port ultrawideband antenna array for diversity applications. IEEE Antennas Wirel. Propag. Lett. **16**, 569–572 (2017)
11. Ghazali, M.I.M., Gutierrez, E., Myers, J.C., Kaur, A., Wright, B., Chahal, P.: Affordable 3D printed microwave antennas. In: IEEE 65th Electronic Components and Technology Conference (ECTC), pp. 240–246 (2015)
12. Mazhar, W., Tarar, M.-A., Tahir, F.A., Gulistan, W.: A low profile cross strip 3D monocone antenna for UWB applications. Progr. Electromagn. Res. **46**, 51–61 (2014)
13. Chen, S.L., Lin, K.H., Mittra, R.: Miniature and near-3D omnidirectional radiation pattern RFID tag antenna design. Electron. Lett. **45**, 923–924 (2009)
14. Zhu, S., Ghazaany, T.S., Abd-Alhaeed, R.A., Jones, S.M., Noras, J., Suggett, T., Marker, S.: Miniaturized tunable conical helix antenna. In: IEEE Radio and Wireless Symposium (RWS), pp. 100–102 (2014)

Robustness of Intelligent Vehicular Rerouting Towards Non-ideal Communication Delay

Christian Backfrieder[1](✉), Manuel Lindorfer[1], Christoph F. Mecklenbräuker[2], and Gerald Ostermayer[1]

[1] Research Group Networks and Mobility, FH Upper Austria, Hagenberg, Austria
christian.backfrieder@fh-hagenberg.at, manuel.lindorfer@fh-hagenberg.at, gerald.ostermayer@fh-hagenberg.at

[2] Institute of Telecommunications, TU Wien, Vienna, Austria
cfm@nt.tuwien.ac.at

Abstract. One of the main goals of Intelligent Transport Systems (ITSs) is to optimize traffic flow for the sake of saving fuel, decreasing travel time and/or reducing congestion. In order to achieve this goal, most of the numerous approaches from literature require some kind of information exchange between vehicles and the environment. Vehicles on the one hand need to provide data containing predicates, such as current velocity, position or route destination. On the other hand, a router needs a functional communication infrastructure to contribute route guidance to vehicles which are affected by traffic jams. However, variable delay or complete message loss can influence the rerouting performance significantly, since either route advices could fail to reach their recipient, or the supposed knowledge of the road conditions could be outdated. The delay requirements of various routers may be divergent, and therefore we propose two delay models which are independent of the underlying communication standard. Furthermore, this paper evaluates the existing PCMA* routing algorithm concerning its performance with varying delays and message loss probabilities by applying the introduced delay models in microscopic traffic simulations. We define constraints of both the delay and message loss probability which are required to achieve certain improvements ensuing from intelligent rerouting. The results further reveal a high robustness of the algorithm with regard to delays and message loss probabilities, which expresses itself by similarly low achieved average vehicle travel times for a large amount of the investigated simulation setups.

Keywords: Vehicular communication · Vehicle-to-everything (V2X) Communication delay model · Vehicle routing · Traffic simulation

1 Introduction

Traffic congestion is a big problem all over the world and especially in areas of high population density. Therefore, high values of effort are invested on working

K. Arai et al. (Eds.): FICC 2018, AISC 886, pp. 143–164, 2019.
https://doi.org/10.1007/978-3-030-03402-3_11

towards a reduction of the negative effects of traffic jams on both the environment and drivers. Vehicles are getting more intelligent rapidly due to various sensors and assistant systems. In addition, advanced communication capabilities help to not solely monitor the environment, but also make use of information from other vehicles or traffic coordination systems. Thus, vehicles increasingly transform from autonomous to cooperative systems, by communicating with each other and with Road Side Units (RSUs) along the road or via cellular networks. Applications of various communication types are manifold, from reducing crashes and increasing safety to route optimization and traffic guidance, summarized as ITSs.

A widely investigated research area deals with centralized traffic coordination [1–8]. Independent of the applied algorithm to improve traffic flow and reduce congestion, the routing entity requires information of the current conditions on the road, the vehicles positions, their destinations and other constraints regarding the route itself or a vehicles' capabilities. Based on that information, it can analyze the situation and decides which actions to be taken to improve the situation. In the end, the proposed decisions for optimization need to be transferred to the affected vehicles. Various approaches from literature work in this or a similar way, and therefore require the availability of any communication system, be it Vehicle-to-Vehicle (V2V), Vehicle-to-Infrastructure (V2I) or a combination of them. The requirements of routing approaches further differ in delay and latency constraints, transmission capacities or number of supported users. Currently available communication standards such as 3rd Generation Partnership Project (3GPP) Release 14, which introduces Long Term Evolution (LTE) support for Vehicle-to-everything (V2X) services [9] or the Wi-Fi based Institute of Electrical and Electronics Engineers (IEEE) 802.11p [10] imply different limitations regarding driving velocity, capacity, latency and other relevant properties.

Standardization organizations naturally tend to improve existing communication standards and develop new ones, in the interest of fulfilling today's and tomorrow's requirements and which have even shorter delays, latencies and higher capacities than their predecessors. However, those standards in case of LTE or IEEE 802.11p can solely define the maximum technically possible achievable transmission speed, latency, allowed driving velocities of communication partners and other crucial characteristics. The average transmission speed is rather dependent on the environment, number of active subscribers, geographical properties, movement speed and capabilities of the subscriber's device. However, while there are approaches to somehow guarantee the latency for 802.11p [11,12] or LTE [13,14], the question is whether the effort is worth the benefit resulting from it. Some approaches may have more restrictive requirements on the underlying communication network, while others may be more robust in respect of delays or transmission errors. In other words, the performance of intelligent cooperative vehicle routing algorithms can also depend on the quality of the information exchange by V2X communication. Ideal information exchange with no delays and perfect communication conditions does not exist in reality. In fact, the conditions can vary substantially depending on the infrastructure and environment [15] and therefore could significantly reduce the functioning ability of an intelligent traffic routing.

This paper addresses this topic from the communication environment point of view. The border between achieved improvements that satisfy the operator of such an intelligent traffic optimization system to a bad performance is not hard, but to a greater degree follows a smooth behavior - depending on the communication quality, as our investigation shows. We define abstract delay and loss models for two different message types, which are independent of any communication standard. Further, we analyze the cooperative routing algorithm Predictive Congestion Minimization in combination with an A*-based Router (PCMA*) which was initially proposed in [16] for its robustness against delays and data loss. Additionally, the traffic density is considered. The result is a precise quantification of improvements concerning travel time but also fuel consumption due to intelligent cooperative routing depending on the features and characteristics as well as applied parameters of the used communication system.

The remainder of this paper is structured as follows. In the next section, related work regarding routing algorithms and communication protocols are presented. Section 3 introduces the abstract delay and loss models for the communication. In Sect. 4, the system model, used scenarios and simulation setup is shown. Section 5 presents and discusses the simulation results. Finally, we conclude this paper in Sect. 6 and give an outlook of potential future work.

2 Related Work

Dynamic route optimization can significantly reduce travel time and result in fuel savings by providing an accurate and situation-dependent calculation of the shortest or fastest route to a defined destination, of course on top of reducing the possibility of getting lost [17]. While the early research on this topic focused on unidirectional information transfer to vehicles (e.g. from satellites) which then need to interpret the information and choose whether it is relevant [18], the majority of recently published articles requires bidirectional communication.

Firstly, many of the approaches from literature require an existing, fully functional information exchange mechanism. While assuming an error- and delay-free communication as given, some investigations one the one hand focus more on traffic predictions [5,8,19]. On the other hand, different solutions for the vehicle routing optimization problem itself are proposed, which also do not analyze lossy or delayed information exchange. Examples are the savings algorithm [20], seep algorithm [21], approaches applying ant colony optimization [22,23], tabu search [24] or particle swarm optimization [25–27]. All of those different algorithms solve the Vehicle Routing Problem (VRP), but however presuppose an ideal, fully functioning information exchange.

Secondly, another group of approaches that can be classified together are those which investigate existing communication standards for Vehicular Ad-hoc Networks (VANETs), i.e. V2V or V2I communication with RSUs or cellular communication. Araniti et al. [28] and Hameed et al. [15] evaluate the performance of LTE and IEEE 802.11p, compare those and point out technical details of latencies, ranges, used frequencies and potential applications. Investigations on the Medium Access Control (MAC) of VANETs [29,30] and message routing protocols [31–33] are also subject to intense research. A very recent paper from Sondi et al. [34] evaluates the performance regarding loss rate and delay of the 802.11p standard. A different area of research focuses on the distribution of safety messages within VANETs by simulation of vehicle's behavior and communication [35,36].

The investigation of intelligent vehicular routing in combination with non-ideal communication environments is a much less investigated research field. Elbery et al. [37] propose a so-called eco-routing approach. The authors define the applied communication standard to be 802.11p and optimize the resulting fuel consumption by the help of communication through VANETs. The results show how much fuel can be saved by applying the presented approach for a single scenario.

However, depending on the applied routing algorithm, the full capabilities of a communication system regarding delays, latencies and transmission speed might not be necessary in order to achieve the desired reductions in average travel time, fuel consumption or other performance indicators [37]. The requirements on the information exchange for messages of the routing might differ from one approach to another, and the results could reveal a worse performance, but still be satisfiable while assuming higher delays or lost data due to communication errors. As a consequence, freed capacities of communication networks can be assigned to other services. Alternatively, cheaper equipment or a lower network coverage is conceivable, if the goals for improvements due to intelligent routing are still met. Both mentioned implications are advantageous and of great benefit for network operators or service vendors, since support of multiple services (beyond route optimization) within the same information system could possibly not be a scalable way, if both systems require more than the available capacity [38].

This paper focuses on the investigation of cooperative routing with non-ideal communication capabilities. It studies the effects of delayed or lossy message distribution on travel time and fuel consumption, with varying delays and message error probabilities. Instead of assuming defined properties of the network and determining how well a routing algorithm can perform by applying them, the problem is redefined in a different light. The proposed models are independent from any existing (or future) cellular or Dedicated Short Range Communication (DSRC) standard, but define abstract communication constraints concerning delays and message loss probabilities. The results show the performance of the predictive router PCMA* [16] depending on varying communication constraints. To the best of our knowledge, none of the previous research efforts covers this issue and presents such an approach.

3 Communication Delay Models

The time for transmission of a message from its origin to the destination is one of the most important performance characteristics of a data network [39]. In almost the same manner, delay is an important criteria for information propagation for PCMA*. Basically, we separate the message exchange into the following two types:

- On-Demand Messages (ODMsgs): We define ODMsgs to be messages which are not transmitted in regular intervals, but only when needed. Those are comparable to Decentralized Environmental Notification Message (DENM) as defined by the European Telecommunications Standards Institute (ETSI) in [40]. This asynchronous notifications are primarily used to transmit new route information to a vehicle. Furthermore, the vehicle somehow needs to confirm that first it has received the message and second it was able to apply the provided route, replace the currently active one and that it follows the new route with immediate effect. In addition, vehicles initially send their chosen route and driving destination when they start their trip.
- Periodic Messages (PMsgs): This type of message is used to report current information and is transmitted in defined intervals. The pendant of the ETSI specification [40] are Common Awareness Messages (CAMs). The vehicles periodically send updates of their current status containing velocity and current position, which is required by the router to obtain the current conditions on the road and make use of this information to calculate optimal routes.

As delay is a major criterion for PCMA* in order to increase traffic efficiency, the models for message delay must be able to provide the required times until the message is received successfully. Depending on the type of message, also complete loss of messages is modeled. These models comprise the whole protocol stack and reduce the output to either a certain time delay or a flag indicating whether or not a message was lost, without modeling detailed channel characteristics, Automatic Repeat Request (ARQ) or Forward Error Correction (FEC) protocols or comparable mechanisms.

3.1 Geometric Delay Model

For ODMsgs, the delay is modeled as a geometrically distributed random variable. It is assumed that potential unsuccessful transmissions are repeated until the full message can be interpreted by the receiver (e.g. by an expression of an ARQ mechanism). The distribution itself is described by a single parameter λ in the range $0 \leq \lambda \leq 1$ which represents the probability of success. The influence of a varying parameter λ on the Probability Density Function (PDF) is shown in Fig. 1.

The next message delay Δt is determined by calculation of the next sample of the distribution f, which defines the number of failures until the message is received successfully. Therefor, a second parameter is introduced, which defines the Roundtrip Time (RTT). This RTT is multiplied by the number of failures resulting from a realization of the random variable. In order to represent a potential fixed time delay which should be applied to each message transfer, an offset t_o is added. Therewith, an overall delay of zero, which would be the result of the distribution in many cases can be avoided, since this situation is impossible to achieve under real conditions according to (1).

$$\Delta t = t_o + f * \text{RTT} \tag{1}$$

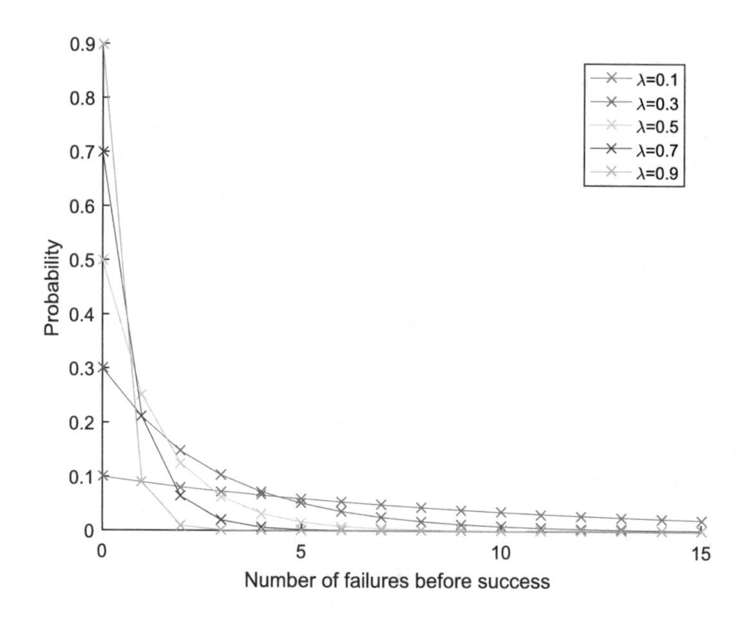

Fig. 1. PDF of geometric distribution with varying parameter λ.

The model also allows specification of the parameters t_o and RTT as normally distributed statistical variables with parameters mean m_{OD} and standard deviation σ_{OD}.

3.2 Uniform-Loss Model

The communication of PMsgs is modeled differently than for ODMsgs. It is assumed that the information transfer is successful with a certain probability p_{success} and fails with $p_{fail} = 1 - p_{\mathrm{success}}$. As the content of all PMsgs has the same payload size, the distribution does not depend on the message size. If a single message transfer is decided to be successful by the model, a certain delay can be added. This delay is either zero or can also specified by a Gaussian normal distribution with a mean value m_{PM} in milliseconds and standard deviation σ_{PM} of the delay distribution, analogous to the offset t_o of ODMsgs. This model is applied to messages containing vehicle data including the current position and speed of the vehicle. A consequence of failing message transfers is obviously outdated information on the receiver side. Any error correction is not considered explicitly, but is included in p_{fail}. Hence, the effects of powerful error correction mechanisms of the Modulation and Coding Scheme (MCS) are to be considered by setting lower probabilities of failure. This further means a message is either assumed to be fully except errors, or lost completely. Partly received information is not possible, nor is it possible to get wrong information when using this channel model.

4 Environment and System Model

For the carried out simulations, the microscopic traffic simulator *TraffSim* [41] is used. In addition to the functionality described in [41], the simulator was extended to support configurable communication models for message exchange, such as the ones which are proposed in Sect. 3. As longitudinal model we use the widely accepted Intelligent Driver Model (IDM) by Kesting et al. [42], the model Minimizing Overall Braking Induced by Lane Changes (MOBIL) [43] is applied to model lane-change behaviour on multi-lane roads and highway ramps and exits, in connection with the cooperative behavior extension Cooperative Longitudinal and Lane-Change Behavior Extension (CLLxT) [44]. All intersections within the presented scenarios are unregulated using the intersection control as described in [45] in order to avoid influence of complex intersection regulation algorithms. The fuel consumption is calculated by a physics based model, as described in [46].

4.1 PCMA* Routing Algorithm

PCMA* is a comprehensive traffic optimization algorithm which uses both present and future road conditions for traffic optimization. It continuously evaluates the current conditions on the road in order to be able to predict potential future bottlenecks and counteract timely. The algorithm achieves good performance compared to other algorithms from literature. For a detailed introduction of PCMA* it is referred to [16]. In order to keep the current digital representation of the road conditions up to date and to inform vehicles about route updates, information exchange is required. The proposed delay models from Sect. 3 are applied for this purpose.

4.2 Evaluation Environments

Analogous to the previous investigations of PCMA* in [16], we use an artificial and a real world scenario for the evaluations within this paper, as shown in Figs. 2 and 3, respectively. Scenario 1 is a generated, pattern-oriented scenario [47]. Scenario 2 represents an extract of the city of Linz in Austria, extracted from OpenStreetMap (OSM). The scale bar is valid for both scenarios.

The start and end locations for vehicles are immutable for all simulations and defined a priori. Vehicles start within the green area and have their route destination within the blue ellipse. The network is empty before simulation start, and the simulation ends as soon as the last vehicle reached its destination. We further assume that if a vehicle receives a new route information by the router which is valid and possible to apply, the driver does not ignore it but rather uses this new route immediately.

4.3 Simulator Parametrization

(1) Communication models: For PMsgs, the probability whether a message is transmitted successfully or lost completely is varied between 0 and 1. However,

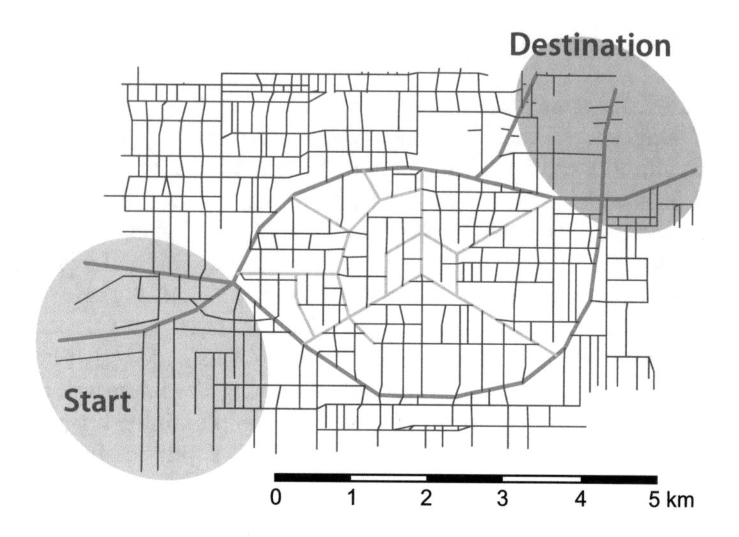

Fig. 2. Scenario 1: Artificial road network.

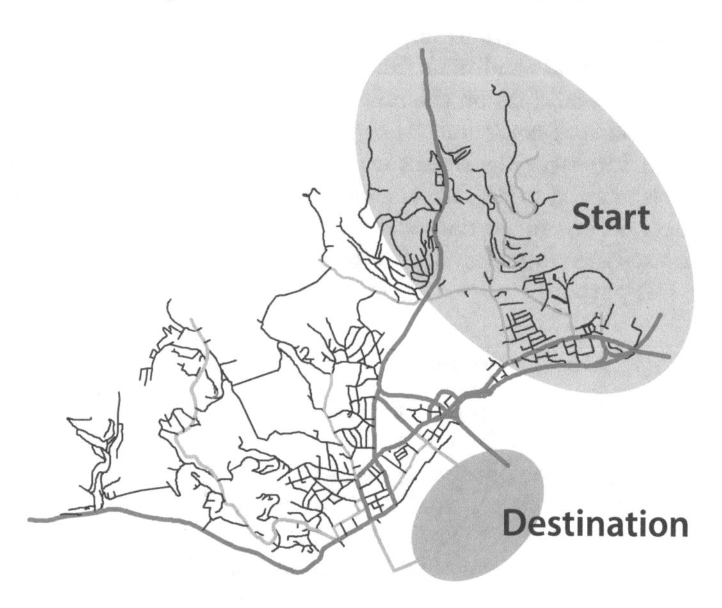

Fig. 3. Scenario 2: City of linz.

as the results look quite similar for high values of p_{success}, the lower range is pointed out supplementally in order to get a better view onto it. Each message is considered independently. The regular update interval of periodic messages to 30 seconds, what conforms to the transmission interval if p_{success} equals 1.

For the geometric distribution which is used for ODMsgs, the distribution parameter λ is varied in the interval $[0, 1]$. Each realization of this random variable is multiplied by a defined roundtrip time (RTT), which is specified in milliseconds. The RTT is set to values between 0 (which is equal to a λ-independent delay of zero) and 200.000. In concrete terms, the values $\{0, 500, 1000, 2000, 5000, 10000, 30000, 60000, 200000\}$ were set up. Although the probability to come upon some of these delays in reality may be very low, we intentionally also included very high delays in this set to investigate how the system behaves in these extreme cases. The RTT distribution parameters m_{OD} and σ_{OD} of the model are set to zero.

(2) Vehicle density: The grade of improvement of traffic flow regarding average travel time and fuel consumption the rerouter is able to achieve depends on the vehicle density [16]. Therefore, we varied the vehicle density by variation of the arrival interval of vehicles from 500 ms to 3500 ms time between the start of the two consecutive vehicles, while the origin and destinations of the vehicles are kept unchanged.

The arrival interval can be mapped to an average amount of vehicles for a single simulation. Table 1 shows the average total number of vehicles for selected densities and both scenarios for the simulation duration. Naturally, the duration until a simulation is finished (i.e. the last vehicle reached its destination) strongly depends on the optimization efficiency in addition to the density as well. Thus, we chose a 'good' situation with excellent communication conditions and $p_{\mathrm{success}} = \lambda = 1$, and a 'bad' one with $p_{\mathrm{success}} = 0$ and $\lambda = 0.005$ to point out the differences. Furthermore, one representative configuration with arrival time 1700 ms was selected for the purpose of depicting the number of vehicles over time for both scenarios and situations in Fig. 4. The red and blue curves indicate the different scenarios 1 and 2, respectively, while the line style represent 'good' (solid line) and 'bad' (dash-dot line) situations.

Table 1. Average number of vehicles for both scenarios and various setups

Arrival interval	Scenario 1		Scenario 2	
	good	*bad*	*good*	*bad*
500	1001,3	1292,9	1077,7	1259,2
1100	644,5	1036,5	686,7	913,2
1700	392,9	860,6	337,4	508,1
2300	272,3	675,5	193,2	271,3
2900	206,6	597,2	140,6	140,8
3500	158,8	567,8	113,6	116,7

4.4 Explanation of Arising Situations

For a better understanding of the effects of the differently parametrized communication models as explained in Sect. 4.3, we would like to point out how the internal knowledge of the PCMA* algorithm is influenced. The following potential problematic situations and conceivable effects can be identified:

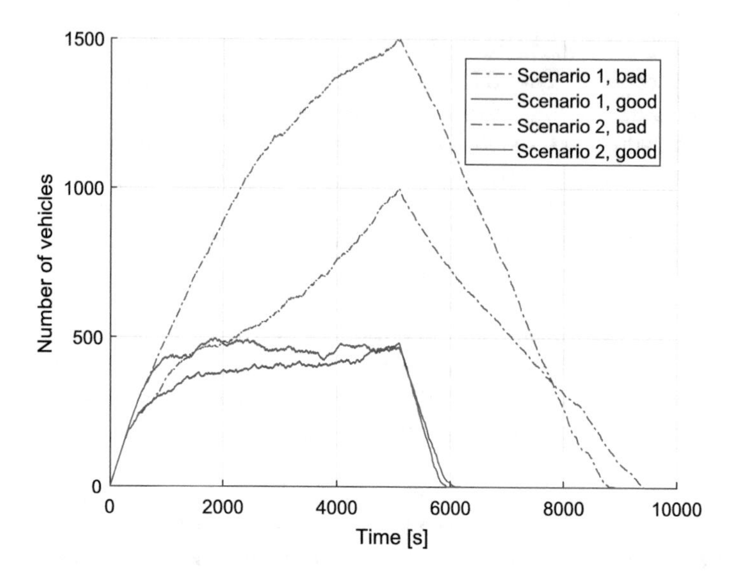

Fig. 4. Number of vehicles over time for both scenarios, inter-arrival interval of 1700 ms and in each case a 'good' and a 'bad' situation.

(1) Success probability for PMsgs: As a consequence of a low p_{success} for periodic messages, the router's knowledge about possible velocities on certain road segments becomes inaccurate. This issue may not be serious enough to severely distort the known road network status as soon as the traffic density is high enough that at least one functional communication per road segment is possible in a reasonable time interval. If the success probability becomes very low, the router weighs its graph wrongly and uses this wrong basis for its routing decision. This has on the one hand strong influence on the reactive component of PCMA*, since congestion is recognized on the basis of the current load factor of an intersection. On the other hand, also the congestion prediction mechanism has a hard time, since the predicted times when a certain vehicle passes an intersection also depend on the currently detected average velocity of all passed nodes, and those time predictions are essential for congestion predictions.

(2) Delay time for ODMsgs: The impact of delay time for aperiodic messages is threefold. The first influenced communication protocol is the initial registration of a vehicle on the system. A too late reception of this message shrinks the possibilities for the router for early reaction on possible traffic jams by spacious detours. Second, the assignment of new routes can be delayed, which could invalidate the route for the specific vehicle. And third, the confirmation message of vehicles that a new route is applied is delayed. This results in a discrepancy between the real assigned routes and those the router thinks that are assigned, which further has negative effects on the prediction mechanism due to a wrong calculation of the occupancy of intersections.

5 Simulation Results

5.1 Success Rate of PMsgs

Within this subsection, the success probability p_{success} of periodic messages was analyzed and set up to be from 0 to 1, in steps of 0.1. As described in Sect. 4.4, low values for p_{success} lead to inaccurate and outdated knowledge base of the current road conditions for the router. Figures 5 and 6 (scenarios 1 and 2, respectively) show the average travel time per vehicle over the success probability. As the shape of the surface basically behaves similar for the majority of investigated delay times ($0.1 \leq \lambda \leq 0.9$), we decided to choose medium values for the geometric model of $\lambda = 0.5$ and $RTT = 1000$.

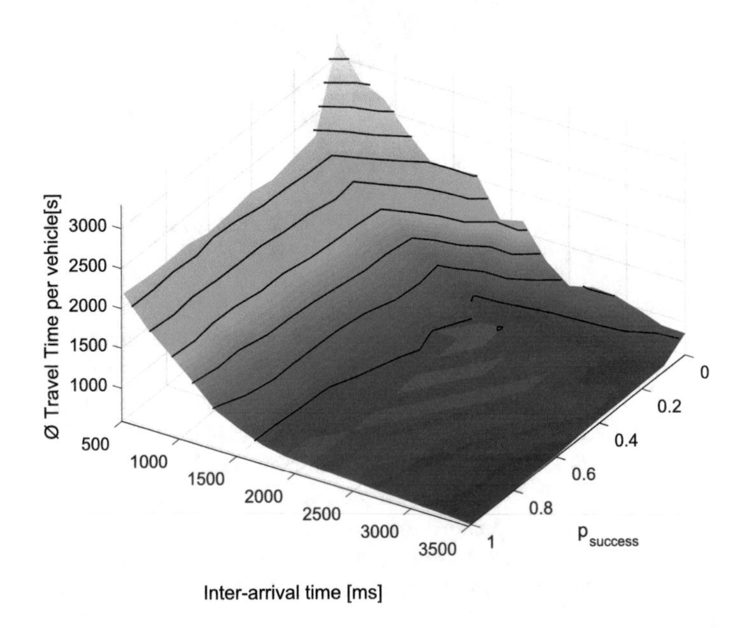

Fig. 5. Average travel times for scenario 1 with varying p_{success}, $\lambda = 0.5$ and $RTT = 1000$.

In both Figs. 5 and 6, the travel time decreases slightly with higher success probability of the PMsgs. However, a huge portion of the gain is achieved already at success probabilities of 15 to 30 % in both scenarios. From then on, the improvements already attenuate until the minimum average travel times are unsurprisingly reached with $p_{\text{success}} = 1$, what is accented by the trend of the black contour lines. This fact emphasizes the strength of the algorithm in suboptimal environmental conditions. Obviously, the result values are negatively influenced with a higher traffic density, since congestion is much more severe and the potential for improvements shrinks.

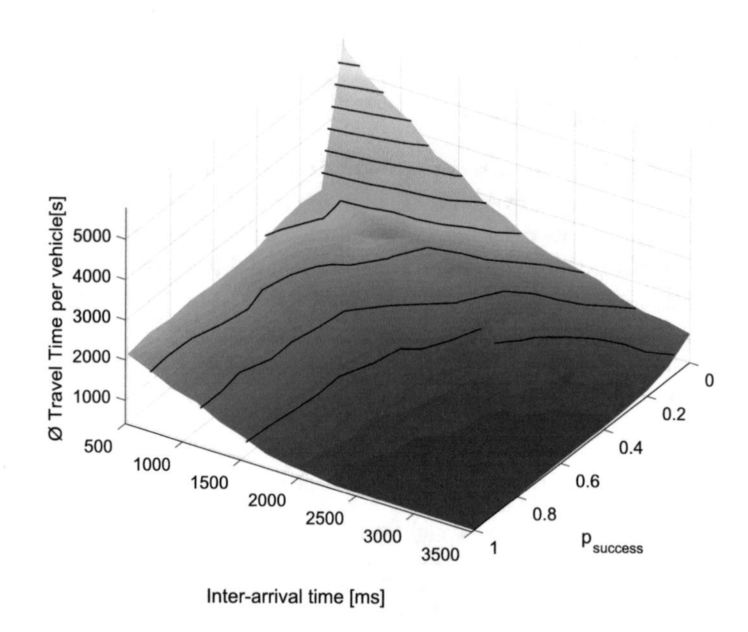

Fig. 6. Average travel times for scenario 2 with varying p_{success}, $\lambda = 0.5$ and $RTT = 1000$.

While the surfaces in Figs. 5 and 6 basically behave similarly, they are still not identical. Minor deviations become visible in the right part of the diagrams, with $0.1 \leq \lambda \leq 0.3$. In Scenario 1, the resulting average travel times are closer to the minimum much earlier, i.e. with low success rates. In other words, in this scenario the router can deal with sparse information better than in Scenario 2. We ascribe this effect to the higher amount of possibilities for alternative routes in scenario 2. So while there are no more route possibilities for a certain situation in Scenario 1, there might still be some in scenario 2. Therefore, a better information quality and density due to higher success probabilities of PMsgs does still lead to better results for $p_{\text{success}} = 0.4$ compared to a value of 0.2.

In contrast, the results are similar for the entire set of success rates (except from 0) for very low values of λ and, as a consequence therefrom, high delays for ODMsgs (average delay time >200 s). In such cases, only a miniscule minority of route messages reach their destination timely and are therefore not applicable any more. This explains the independence of the success rate, which is pointed out in Fig. 7 and the close proximity of the travel time curves therein. However, a success rate of zero, which means no knowledge of the router about the road conditions at all still adds up in worse results. The still existing minority of ODMsgs that reach their destination (despite the low λ) is useless for the case $p_{\text{success}} = 0$, and explains the considerably higher results indicated by the green curve in Fig. 7.

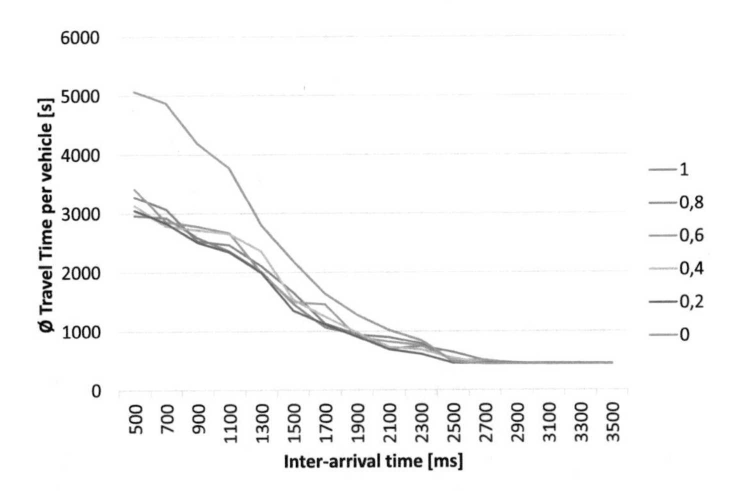

Fig. 7. Average travel times with different line colors for multiple values of p_{success} for scenario 2 with very low $\lambda = 0.005$ and $RTT = 1000$.

5.2 Roundtrip Time and Lambda of ODMsgs

Both parameters of the delay model for on-demand messages λ and RTT were adapted for the purpose of analyzing the influence of message delay length on achievable average travel time. Basically, the parameters have an effect on the distribution of the overall delay for an ODMsgs, as shown in Fig. 8. This figure shows the resulting average delay time per on-demand message for different setups of the geometric delay model. The delay naturally is independent of the vehicle density and equal for both scenarios. What also can be observed is that there exist parameter combinations which lead to equal average delays. Those pairs can easily be found by looking for columns which have similar values in y-dimension in Fig. 8.

The consequences of delayed reception for ODMsg is shown in Figs. 9 and 10 for Scenarios 1 and 2, respectively. In these simulations, a 100 % probability of success for PMsgs is assumed in order to exclude the influence of them and be able to ascribe the changes in achievable travel time solely to described message delay variations.

As expected, the average travel time per vehicle increases with longer average delay, which becomes visible in the rear peak of the diagram (low lambda, high roundtrip time) in both scenarios. Accordingly, the decrease in average travel time per vehicle flattens with increasing lambda and decreasing RTT, while reaching its minimum at $\lambda = 1$ and RTT$= 0$ (i.e. zero delay). For reasons of validation of the model, these extreme values were also added to the parameters. As expected zero roundtrip time lead to a pretty straight distribution of average travel time. The same results become apparent for high values of λ. However, slight irregularities are visible in Scenario 2 in Fig. 10 with RTTs of 60000 and 200000 and λ between 0.3 and 0.6: The achievable average travel time decreases although the message delay becomes higher. We ascribe this effect to the fact

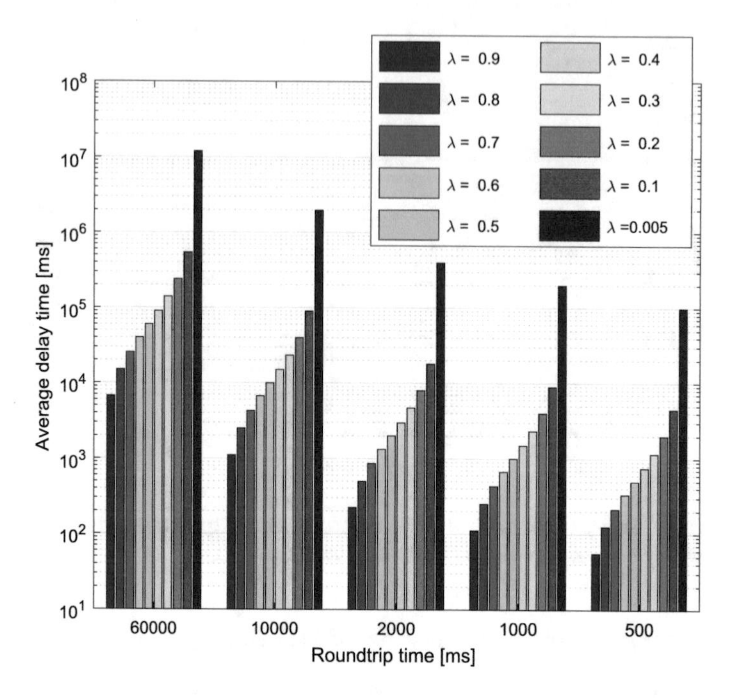

Fig. 8. Average message delay times with different geometric delay model parameters λ and RTT for Scenario 1 and inter-arrival interval of 1700 ms.

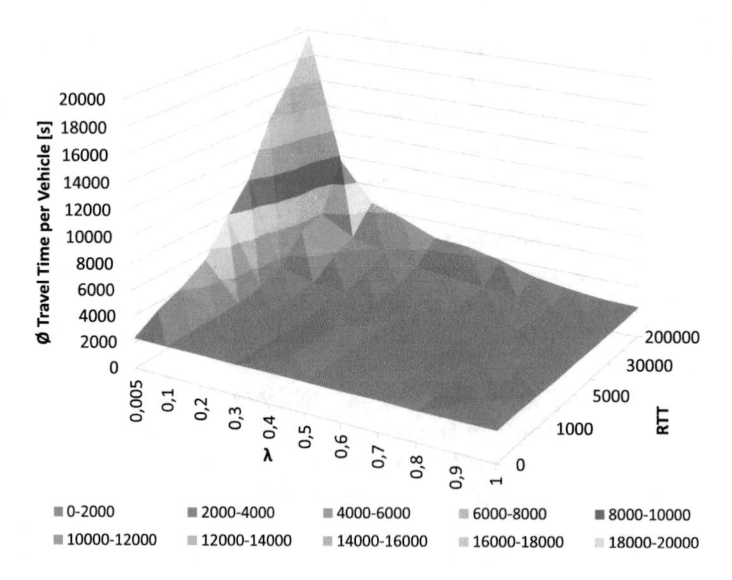

Fig. 9. Average travel times with different geometric delay model parameters λ and RTT for Scenario 1 and inter-arrival interval of 1700 ms.

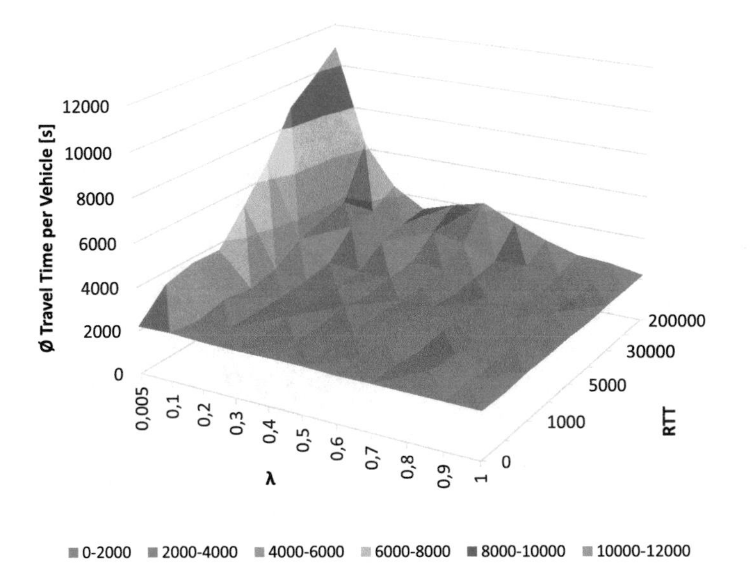

Fig. 10. Average travel times with different geometric delay model parameters λ and RTT for Scenario 2 and inter-arrival interval of 1700 ms.

that in these cases only very few messages arrive in a reasonable time, and the internal knowledge of the router where the vehicles like to go is scarce. Therefore, also routing decisions are made based on that outdated internal knowledge of the router. These routes lead to unnecessary detours and therefore increase the travel time with no gains elsewhere. This effect is becoming clearer when decreasing the vehicle density and setting the vehicle inter-arrival time to 3500 ms, as Fig. 11 illustrates. It should be noted that only the interesting range is extracted here. The peak disappears with low values for λ, since the probability of receiving wrong information is also very low then, and therefore the negative impact therefrom is mitigated.

5.3 Fuel Consumption and Travel Distance

The consumed fuel basically is strongly related to the travel time for all investigated scenarios. However, we picked out one scenario for visualization with medium success rate of 0.5 and medium value for λ of 0.5 as well. The RTT of the selected configuration is 1000 ms. As illustrated in Fig. 12, the fuel consumption also flattens at a medium density of about 2000 ms inter-arrival time, as does the average travel time. The travel distance does not change to the same extent as consumed fuel and time, but slightly decreases with lower vehicle density. The reason are the detours due to rerouting, which of course lead to longer distances since the alternative routes are longer than the original, shortest ones.

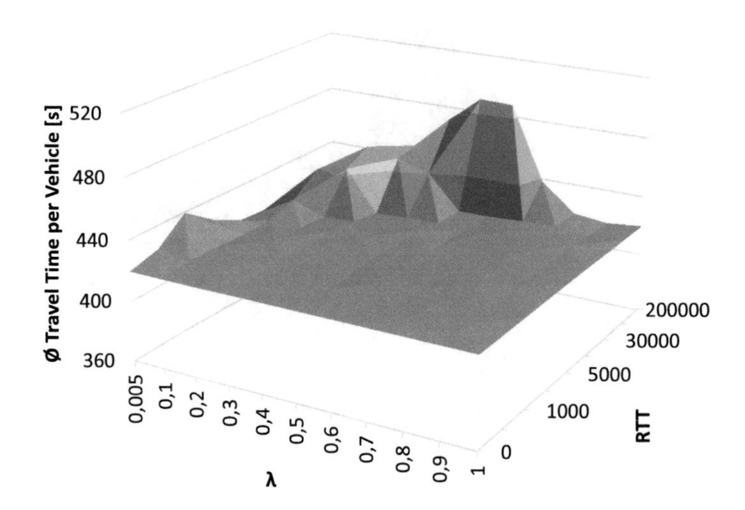

Fig. 11. Zoomed view of average travel times with different geometric delay model parameters λ and RTT for scenario 2 and inter-arrival interval of 3500 ms.

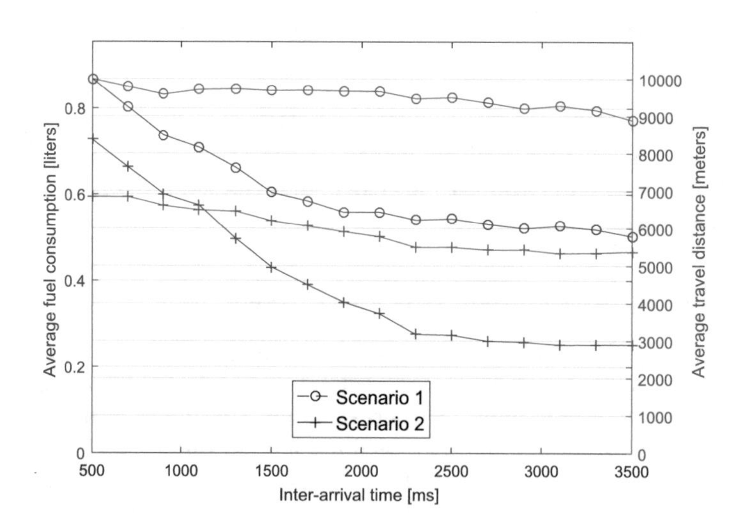

Fig. 12. Fuel consumption and travel distance for both scenarios and $p_{\text{success}} = 0.5$, $\lambda = 0.5$ and $RTT = 1000$ ms.

Table 2. Amount of transmitted data volume and messages for various densities

Inter-arrival interval	700 ms	1900 ms	3300 ms
Data volume [kB]	19.89	9.36	5.91
Transmitted messages	242328	108818	83773

5.4 Volume of Transmitted Data

Due to the relatively high regular update interval of 30 s, the amount of transmitted data is at a level that is basically manageable by most current communication networks. Table 2 presents the amount of transmitted messages (successful or not) during a complete simulation as well as the sum of transferred data for three selected vehicle inter-arrival time setups. The higher the density, the higher is also the required amount of messages due to a greater need of intervention of the router and hence more messages containing new routes. However, the data amount comprises only the payload itself, no overhead due to transmission protocols, error correction or similar mechanisms. We assume the following payloads:

- Periodic Messages: 3 bytes (position as pair of latitude and longitude, and current velocity, one byte each).
- On-Demand Messages: 3 bytes + 1 byte × length of route (position and velocity as for PMsgs, in addition to 1 byte for each road segment identifier of the containing route).

5.5 Discussion

Summarizing, it has been discovered that the algorithm is very tolerant regarding both transmission delays and failures. This robustness can be explained by several arguments:

1. Delayed transmission of ODMsg is not necessarily a problem in all situations. If it is still possible for targeted vehicle to apply the route because the road segment which would differ to the current route is not yet reached, the route is applied anyway. Figure 13 illustrates such a situation, where the vehicle continues its original route (thick, red) from position 1 to position 2, where it receives the route update. The delay does not matter in this case, since the routes differ not until at a later point.
2. Situations where the intended receivers are not able to make use of the transmitted routing messages because of too high delays are resolved automatically by the routing algorithm itself by the included method of vehicle selection [16]. If the predicted congestion is not eliminated by communication with the initially selected vehicles, the next potential vehicle out of the affected ones is chosen. This process is repeated and retried until the situation gets better, i.e. any of the vehicles which pass the congested node in the defined time interval finally manages to choose another route.

3. As the method of gaining knowledge about the current situation on the road is based on average speed, it is less critical if some of the vehicles' reports regarding their position and speed included in PMsgs is lost, since the speed of all vehicles on the very same road segment is likely to be similar.

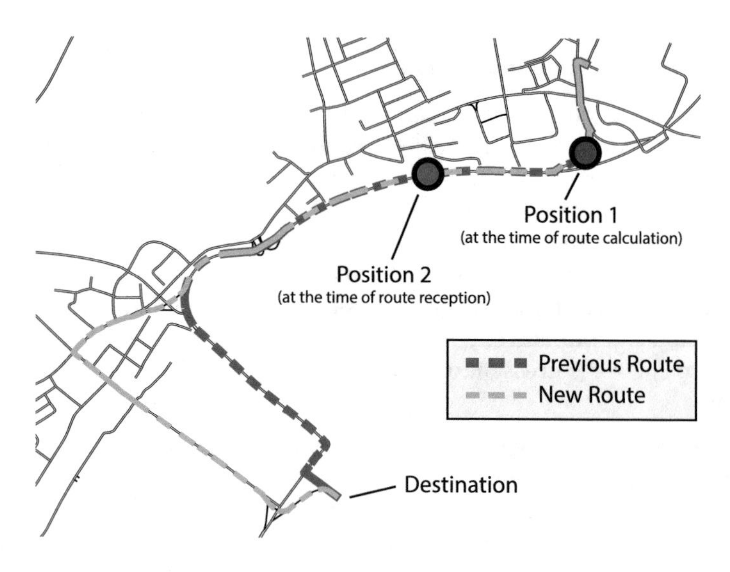

Fig. 13. Unproblematic delay between positions of route calculation and reception

6 Conclusion and Future Work

In this paper, we initially introduce two versatile delay models which are independent of any existing or future communication standard. Firstly, the abstraction of the message delay represents the latency by a geometrically distributed random variable, which is applied to messages where loss is not tolerated, i.e. which are retransmitted as long as they are correctly received. Secondly, a uniform distribution with a parameter defining the success probability is proposed, which is intended for messages which are transmitted in regular intervals, and are lost if transmission is unsuccessful.

In the second part, we apply those models to the existing routing algorithm PCMA* and evaluate the robustness against differently parametrized setups of the delay models regarding the average vehicle travel time. The simulations are executed with the microscopic traffic simulator [41] for two different scenarios. The results basically show that the algorithms performs very well for a large set of setups and even high delays and high loss probabilities. They comprise parameter variations for communication properties for both types of messages, that are status messages which are sent from vehicles to the router periodically as well as on-demand messages with routing information.

The presented results can help operators to decide how much communication capacity is required for a well-functioning intelligent rerouting mechanism. From another point of view, we defined the preconditions for the underlying communication standard which must be fulfilled to reach a defined or required improvement, while the rest of the capacity can used for other services. It may not be necessary to further invest in lower delays and latencies and an increase of the network capacity, if the current features of the communication network already fulfill the desired requirements.

Future work could contain investigations with local deviations of communication network parameters due to poor signal reception or environments with different characterizations, such as urban or rural areas. In addition, the influence of older vehicles which are not equipped with any communication possibility could be analyzed.

Acknowledgment. This project has been co-financed by the European Union using financial means of the European Regional Development Fund (EFRE). Further information to IWB/EFRE is available at www.efre.gv.at.

Europäische Union Investitionen in Wachstum & Beschäftigung. Österreich.

References

1. Liang, Z., Wakahara, Y.: Real-time urban traffic amount prediction models for dynamic route guidance systems. EURASIP J. Wirel. Commun. Netw. **2014**(1), 85 (2014)
2. Li, Y., Ren, W., Jin, D., Hui, P., Zeng, L., Wu, D.: Potential predictability of vehicular staying time for large-scale urban environment. IEEE Trans. Veh. Technol. **63**(1), 322–333 (2014)
3. Gramaglia, M., Calderon, M., Bernardos, C.: ABEONA monitored traffic: VANET-assisted cooperative traffic congestion forecasting. IEEE Veh. Technol. Mag. **9**(2), 50–57 (2014)
4. Chan, K.Y., Dillon, T., Chang, E.: An intelligent particle swarm optimization for short-term traffic flow forecasting using on-road sensor systems. IEEE Trans. Ind. Electron. **60**(10), 4714–4725 (2013)
5. Daraghmi, Y.-A., Yi, C.-W., Chiang, T.-C.: Negative binomial additive models for short-term traffic flow forecasting in urban areas. IEEE Trans. Intell. Transp. Syst. **15**(2), 784–793 (2014)
6. Nafi, N., Khan, R., Khan, J., Gregory, M.: A predictive road traffic management system based on vehicular ad-hoc network. In: 2014 Australasian Telecommunication Networks and Applications Conference (ATNAC), pp. 135–140, November 2014
7. Cao, Z., Jiang, S., Zhang, J., Guo, H.: A unified framework for vehicle rerouting and traffic light control to reduce traffic congestion. IEEE Trans. Intell. Transp. Syst. **18**(7), 1958–1973 (2016)

8. Kim, K., Kwon, M., Park, J., Eun, Y.: Dynamic vehicular route guidance using traffic prediction information. Mob. Inf. Syst. **2016**, e3727865 (2016)
9. 3GPP: Release 14, June 2017
10. Bilstrup, K.: A survey regarding wireless communication standards intended for a high-speed vehicle environment. Halmstad University (2007)
11. Zhang, B., Ma, M., Liu, C., Shu, Y.: Delay guaranteed MDP scheduling scheme for HCCA based on 802.11p protocol in V2R environments. Int. J. Commun. Syst. **30**, e3307 (2017)
12. Zafar, B.A., Ouni, S., Boulila, N., Saidane, L.: Communication delay guarantee for IEEE 802.11 p/wave Vehicle networks with RSU control. In: 2016 IEEE 13th International Conference on Networking, Sensing, and Control (ICNSC), pp. 1–7. IEEE (2016)
13. Madueño, G.C., Pratas, N.K., Stefanović, E., Popovski, P.: Massive M2M access with reliability guarantees in LTE systems. In: 2015 IEEE International Conference on Communications (ICC), pp. 2997–3002. IEEE (2015)
14. Chen, X., Zhang, Z., Yuen, C.: Resource allocation for cost minimization in limited feedback MU-MIMO systems with delay guarantee. IEEE Syst. J. **9**(4), 1229–1236 (2015)
15. Hameed Mir, Z., Filali, F.: LTE and IEEE 802.11p for vehicular networking: a performance evaluation. EURASIP J. Wirel. Commun. Netw. **2014**(1), 89 (2014). http://jwcn.eurasipjournals.com/content/2014/1/89
16. Backfrieder, C., Ostermayer, G., Mecklenbräuker, C.F.: Increased Traffic flow through node-based bottleneck prediction and V2X communication. IEEE Trans. Intell. Transp. Syst. **18**(2), 349–363 (2017)
17. Barth, M., Boriboonsomsin, K., Vu, A.: Environmentally-friendly navigation. In: 2007 IEEE Intelligent Transportation Systems Conference, ITSC 2007, pp. 684–689. IEEE (2007)
18. Videtich, M.: Receiving traffic update information and reroute information in a mobile vehicle. US Patent 7,062,379, 13 June 2006. https://www.google.com/patents/US7062379
19. Blokpoel, R., Vreeswijk, J., Leistner, D.: Micro-routing using accurate traffic predictions. IET Intell. Transp. Syst. **6**(4), 380–387 (2012)
20. Clarke, G., Wright, J.W.: Scheduling of vehicles from a central depot to a number of delivery points. Oper. Res. **12**(4), 568–581 (1964)
21. Gillett, B.E., Miller, L.R.: A heuristic algorithm for the vehicle-dispatch problem. Oper. Res. **22**(2), 340–349 (1974)
22. Abdulkader, M.M., Gajpal, Y., ElMekkawy, T.Y.: Hybridized ant colony algorithm for the multi compartment vehicle routing problem. Appl. Soft Comput. **37**, 196–203 (2015)
23. Reed, M., Yiannakou, A., Evering, R.: An ant colony algorithm for the multi-compartment vehicle routing problem. Appl. Soft Comput. **15**, 169–176 (2014)
24. Belhaiza, S., Hansen, P., Laporte, G.: A hybrid variable neighborhood tabu search heuristic for the vehicle routing problem with multiple time windows. Comput. Oper. Res. **52**, 269–281 (2014)
25. Khouadjia, M.R., Alba, E., Jourdan, L., Talbi, E.-G.: Multi-swarm optimization for dynamic combinatorial problems: a case study on dynamic vehicle routing problem. In: ANTS Conference, vol. 6234, pp. 227–238. Springer (2010)
26. Kaiwartya, O., Kumar, S., Lobiyal, D.K., Tiwari, P.K., Abdullah, A.H., Hassan, A.N.: Multiobjective dynamic vehicle routing problem and time seed based solution using particle swarm optimization. J. Sens. **2015**, 1–14 (2015)

27. Chen, A.-L., Yang, G.-K., Wu, Z.-M.: Hybrid discrete particle swarm optimization algorithm for capacitated vehicle routing problem. J. Zhejiang Univ. Sci. A **7**(4), 607–614 (2006)
28. Araniti, G., Campolo, C., Condoluci, M., Iera, A., Molinaro, A.: LTE for vehicular networking: a survey. IEEE Commun. Mag. **51**(5), 148–157 (2013)
29. Bilstrup, K., Uhlemann, E., Ström, E., Bilstrup, U.: On the ability of the 802.11 p MAC method and STDMA to support real-time vehicle-to-vehicle communication. EURASIP J. Wirel. Commun. Netw. **2009**(1), 902414 (2009). http://link.springer.com/article/10.1155/2009/902414
30. Bilstrup, K., Uhlemann, E., Strom, E.G., Bilstrup, U.: Evaluation of the IEEE 802.11 p MAC method for vehicle-to-vehicle communication. In: IEEE 68th Vehicular Technology Conference, VTC 2008-Fall, pp. 1–5. IEEE (2008). http://ieeexplore.ieee.org/abstract/document/4657278/
31. Jerbi, M., Senouci, S.-M., Rasheed, T., Ghamri-Doudane, Y.: Towards efficient geographic routing in urban vehicular networks. IEEE Trans. Veh. Technol. **58**(9), 5048–5059 (2009). http://ieeexplore.ieee.org/abstract/document/5061874/
32. Taleb, T., Ochi, M., Jamalipour, A., Kato, N., Nemoto, Y.: An efficient vehicle-heading based routing protocol for VANET networks. In: 2006 IEEE Wireless Communications and Networking Conference, WCNC 2006, vol. 4, pp. 2199–2204. IEEE (2006). http://ieeexplore.ieee.org/abstract/document/1696637/
33. Borsetti, D., Gozalvez, J.: Infrastructure-assisted geo-routing for cooperative vehicular networks. In: 2010 IEEE Vehicular networking conference (VNC), pp. 255–262. IEEE (2010). http://ieeexplore.ieee.org/abstract/document/5698271/
34. Sondi, P., Wahl, M., Rivoirard, L., Cohin, O.: Performance evaluation of 802.11 p-based ad hoc vehicle-to-vehicle communications for usual applications under realistic urban mobility. IInt. J. Adv. Comput. Sci. Appl. IJACSA **7**(5), p221–230 (2016)
35. Ucar, S., Ergen, S.C., Ozkasap, O.: Multihop-cluster-based IEEE 802.11 p and LTE hybrid architecture for VANET safety message dissemination. IEEE Trans. Veh. Technol. **65**(4), 2621–2636 (2016). http://ieeexplore.ieee.org/abstract/document/7081788/
36. Toor, Y., Muhlethaler, P., Laouiti, A.: Vehicle ad hoc networks: applications and related technical issues. IEEE Commun. Surv. Tutor. **10**(3) (2008). http://ieeexplore.ieee.org/abstract/document/4625806/
37. Elbery, A., Rakha, H., Elnainay, M., Drira, W., Filali, F.: Eco-routing using V2I communication: system evaluation. In: 2015 IEEE 18th International Conference on Intelligent Transportation Systems (ITSC), pp. 71–76. IEEE (2015). http://ieeexplore.ieee.org/abstract/document/7313112/
38. Santa, J., Pereñguez, F., Moragón, A., Skarmeta, A.F.: Experimental evaluation of CAM and DENM messaging services in vehicular communications. Transp. Res. Part C Emerg. Technol. **46**, 98–120 (2014). http://www.sciencedirect.com/science/article/pii/S0968090X14001193
39. Chan, K.: Future Communication Technology and Engineering: Proceedings of the 2014 International Conference on Future Communication Technology and Engineering (FCTE 2014), Shenzhen, China, 16–17 November 2014. CRC Press, Boca Raton (2015). Google-Books-ID: Oo7YCQAAQBAJ
40. T. ETSI, 102 637-3 (2010): Intelligent Transport Systems (ITS); Vehicular Communications; Basic Set of Applications; Part 3: Specifications of Decentralized Environmental Notification Basic Service. Technical Specification V1, vol. 1 (2010)

41. Backfrieder, C., Ostermayer, G., Mecklenbräuker, C.F.: TraffSim - a traffic simulator for investigations of congestion minimization through dynamic vehicle rerouting. ResearchGate **15**(4), 38–47 (2014)
42. Kesting, A., Treiber, M., Helbing, D.: Enhanced intelligent driver model to access the impact of driving strategies on traffic capacity. Philos. Trans. R. Soc. A Math. Phys. Eng. Sci. **368**(1928), 4585–4605 (2010). http://rsta.royalsocietypublishing.org/content/368/1928/4585
43. Kesting, A., Treiber, M., Helbing, D.: General lane-changing model MOBIL for car-following models. Transp. Res. Rec. J. Transp. Res. Board **1999**, 86–94 (2007). http://trrjournalonline.trb.org/doi/abs/10.3141/1999-10
44. Backfrieder, C., Ostermayer, G., Lindorfer, M., Mecklenbräuker, C.F.: Cooperative lane-change and longitudinal behaviour model extension for TraffSim. In: Alba, E., Chicano, F., Luque, G. (eds.) Smart Cities. Lecture Notes in Computer Science, pp. 52–62. Springer, Cham (2016). https://doi.org/10.1007/978-3-319-39595-1_6
45. Backfrieder, C., Ostermayer, G.: Modeling a continuous and accident-free intersection control for vehicular traffic in TraffSim. In: 2014 European Modelling Symposium, pp. 332–337, October 2014
46. Treiber, M., Kesting, A.: Fuel consumption and emissions. In: Treiber, M., Kesting, A. (eds.) Traffic Flow Dynamics. Springer, Heidelberg (2013)
47. Lindorfer, M., Backfrieder, C., Kieslich, C., Krösche, J., Ostermayer, G.: Environmental-sensitive generation of street networks for traffic simulations. In: 2013 European Modelling Symposium, pp. 457–462, November 2013

Ranking Based System to Reduce Free Riding Behavior in P2P Systems

Sanjeev K. Singh[1](✉), Chiranjeev Kumar[1], and Prem Nath[2]

[1] Department of Computer Science and Engineering, Indian Institute of Technology (ISM), Dhanbad, India
sanjeevsingh.erdr@gmail.com, k_chiranjeev@yahoo.co.uk

[2] Department of Computer Engineering, Mizoram University, Aizawl, India
pmnath26@gmail.com

Abstract. Peer-to-peer (P2P) network is a distributed system in which the autonomous peers participate at their will and resources are shared in distributed manner. The massive increase in Internet has given wing to the P2P applications and the search protocols are gainfully utilized in the resource discovery process for Internet related applications. Today, the research community has found interest in the P2P networks. There are several challenges in designing an efficient protocol for the P2P networks. These networks suffer from various problems, such as free riding behaviors, whitewashing, poor search scalability, fake/malicious content distribution, lack of a robust trust model, etc. Moreover, the peers can join and leave the network (churn rate) at any time which makes trust management and searching files more challenging task. We have proposed a ranking based P2P system which collects the statics of data shared by the participating peers and defines the rank of peers. The ranking of a peer increases/decreases in ratio of the data uploaded/downloaded by the peer. The proposed system promotes the peers to share their resources in order to increase their rank and reduces free riding behaviors. We have analytically analyzed the performance of proposed system and observed that proposed system performs better than existing reputation based system.

Keywords: P2P networks · Reputation · Trustworthiness · Social friends Free riding

1 Introduction

P2P networks provide an open environment where peers participate and share their resources. Effectiveness of the P2P networks relies on cooperation of the participating peers. But several studies revealed that most of the peers in P2P networks are free riders and misbehaving. The free riders consume resources without contributing. The misbehaving peers upload the malicious file which is spread from one peer to another. For example, in Gnutella 85% peers are free riders and 45% of files downloaded through the Kazaa file sharing application contain malicious code [1, 2].

After downloading the content, most of the peers are not graceful enough to share the downloaded content and such peers disappear after download. This situation leads toward unfairness in the P2P systems. In P2P systems, a free rider is a peer that uses

K. Arai et al. (Eds.): FICC 2018, AISC 886, pp. 165–177, 2019.
https://doi.org/10.1007/978-3-030-03402-3_12

P2P network services but doesn't contribute to the system or other peers at an acceptable level. Free riding behavior is unwanted for better P2P system performance. Many studies revealed that high degrees of free riding exist in P2P networks. So, it is necessary to have some mechanism to promote resource sharing in the P2P networks. We have proposed a rank based system which promotes the peers in P2P networks to share their resources and reduces the free riding behaviors.

Rest of the paper is organized as follows. The literature survey is given in Sect. 2. The proposed system has been discussed in detail in Sect. 3. The derivations of different costs in proposed and existing systems are provided in Sect. 4. In Sect. 5, analytical modeling and performance analysis of proposed and existing systems are furnished. The conclusion about the performance of the proposed system is given in Sect. 6 followed by references.

2 Related Work

In a free-riding environment, a small number of peers serve a large number of peers, which can lead to scalability and single-point-of- failure problems. It has been observed that structured P2P systems appear to be more vulnerable to free riding behaviors than unstructured and centralized ones. In a structured P2P system, a free rider can ignore the incoming queries and does not store the key-value pairs.

There are many schemes proposed to encourage the peers in P2P networks to cooperate like reputation based incentive methods [3–8], pricing-based schemes [9], and game theoretical method [10]. In a reputation system, a peer's reputation is built based on a collection of feedbacks from other peers. A Page Rank mechanism has been proposed in [11] to rate the web pages based on the popularity of web pages. Eigen-Trust [12] determines the global contribution of the peers based on the contribution (upload/download) of the peers and to whom contribution has been made. In tit-for-tat [13], the transactions between two peers are made according to a policy "Give and Take". There are several other mechanisms [14–19] proposed for reducing free riding behaviors in P2P networks.

Our proposed system is similar to the social network based reputation system proposed by Chen et al. [7]. This existing system [7] is a credit based system based on the two factors: social trust and reputation. The social trust is based on the number of social friends and reputation is based on credit received from non-social friends/partners. There is a new entity named TA (Trust Authority) proposed to look after the reputation credit. A peer with high social trust or reputation or both are considered as server to provide the services to the client. The rating is received from client as bad/good after receiving a complete file. The problem associated with this system [7] is consideration of the extent of free riding behaviors and churn rate of the designated servers. It has been reported in many studies that 70% of peers didn't share any files at all and 25% peers in the system provided 99% of all query hits in the P2P systems. It has been also observed that the existing system [7] considers the fact that entire file is downloaded from a single server (social friend) and rating is given by the client after receiving the complete file. But this is not the fact in the P2P systems. A file is divided into many pieces (like 512 bytes per piece) and different pieces may be downloaded from different servers. Further, the QoS (Quality of Service) is to be rated based on the amount of time involved in downloading the file along with the fact that whether downloaded file is malicious or not. The impact factor proposed in the existing reputation system [7] is derived from number of social friends and feedback received from the friends. Although the existing system [7] incorporates the quality (good/bad) of the downloaded file (malicious or not) but it does not incorporate the download time of the file.

We have proposed a ranking system based on incentive mechanism. We have considered two factors: reputation similar as proposed in [7] and trustworthiness based on amount of data upload/download with respect to churn rate. We have analytically analyzed and compared the existing system [7] and proposed system and found that the proposed system performs better than existing reputation system.

3 Proposed System Model

We have proposed a new P2P system model in which statics of the peers' activity (download/upload and number of social friends) in the P2P overlay has been collected and ranking of the peers has been decided accordingly. The ranking of the peers is distributed among participated peers in the overlay and each peer maintains a ranking table which comprises the ranks of the associated peers. The service request is satisfied based as per the ranking of the peers. Peers with higher rank get priority and peers having lower rank are encouraged to share their resources. A peer requests for a file to higher rank peer always. The proposed ranking system introduces fairness and restricts the free riding behaviors. Based on the rank of a peer, availability and reliability of the resources are also judged. We have used some parameters in performance evaluation which are given in Table 1.

We have proposed a ranking based system comparable to the reputation system suggested in [7]. The churn rate of the peers in P2P networks is a Poison process and plays crucial role in performance of the P2P systems. In downloading the important files or streaming the live video, peers with high churn rate can spoil the performance of the P2P systems. So, we have defined DCR, UCR, and trustworthiness for the same considering the total data uploaded/downloaded in t time by a peer i.

$$\mathrm{DCR_i} = \frac{\sum \mathrm{d_{ij}}}{\mathrm{CR}} \tag{1}$$

$$UCR_i = \frac{\sum u_{ij}}{CR} \tag{2}$$

Table 1. Parameters and nomenclature

Nomenclature	Description
u_{ij}	Data volume uploaded by Peer i to Peer j
d_{ij}	Data volume downloaded by Peer i from Peer j
α	Free parameter to tune the rank
β	Free parameter to fine tune resource sharing $0 \leq \beta \leq 1$
S	Total size of a file (in Mb)
ρ_i	Probability that a Peer sends request for a file and i number of Peers respond positively
N	Total no. of Peers in the P2P overlay
h_j^i	No. of hops between Peer i and Peer j
h	Average no. of hops between two Peers
C_{reg}	Cost of sending a request (no. of hops) and receiving confirmation
W_{avg}	Average bandwidth to download a file (Mbit/sec)
C_d	Cost of downloading file (in sec)
C_t	Cost of updating the rank/reputation Table
CR	No. of join/leave of a Peer per unit time
DCR	Download to Churn Ratio
UCR	Upload to Churn Ratio
λ_i	Trustworthiness of Peer i
t_{ina}	The average inactive time during which either the feeder peer is switch off or does not upload the data

$$\lambda_i = \frac{\sum u_{ij}}{\sum d_{ij}} = \frac{UCR_i}{DCR_i} \quad \textit{if} \sum d_{ij} \neq 0 \, \textit{and} \sum u_{ij} \neq 0 = \\ 0 \quad \textit{Otherwise} \tag{3}$$

We assume that peer i has F_i number of social friends out of N peers available in the P2P overlay. The ratio F_i/N is considered as the reputation of the peer i and rank of peer i is defined as follows:

$$R_i = \alpha\lambda_i + (1 - \alpha)\frac{F_i}{N} \tag{4}$$

Where, $0 \leq \alpha \leq 1$ is a tuning parameter.
Based on the value of R_i, the ranking of the peers has been defined in Table 2.

Table 2. Ranking for peers

Value of R_i	Rank	Remark
$\lambda_i > 1$	Top	Highly reliable
$0.5 < \lambda_i \leq 1$	Medium	Reliable
$0.1 < \lambda_i \leq 0.5$	Low	Satisfactory
$\lambda_i < 0.1$	Poor	Not reliable

The proposed system comprises a TA (Trust Authority) which is a designated server to store the ranking of the peers and all the peers can access it. Each peer is associated with its social friends as illustrated in Fig. 1. As shown in Fig. 1, peer P has three social friends namely A_1, A_2, and A_3. Peer A_1 has n social friends namely A_{11}, A_{12}, ..., A_{1n}. Peer A_2 has m social friends namely A_{21}, A_{22}, ..., A_{2m}. Peer A_3 has r social friends namely A_{31}, A_{32}, ..., A_{3r}. Peers Q_1, Q_2, ...Q_u are non-social friends including friends of peers A_1, A_2, and A_3.

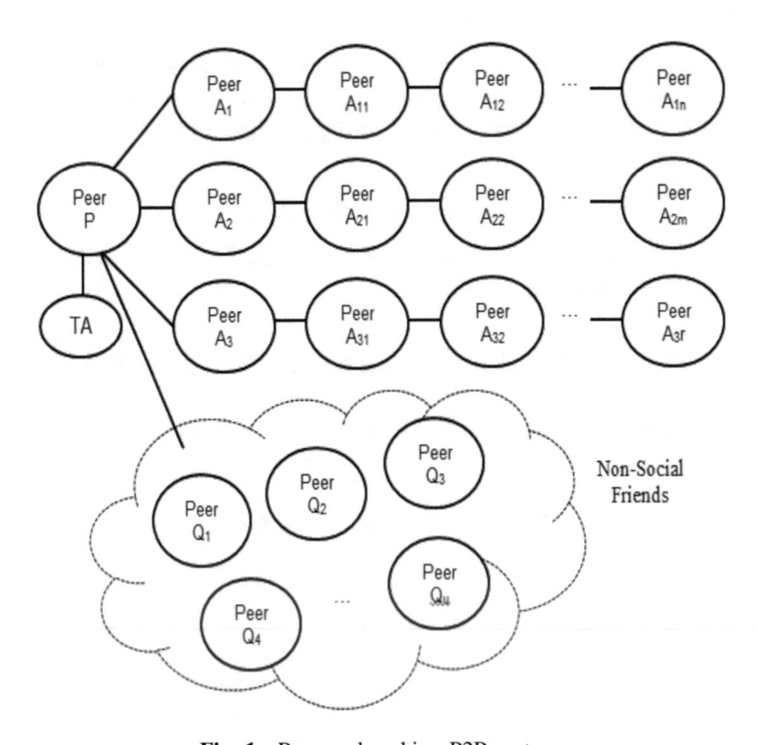

Fig. 1. Proposed ranking P2P system.

3.1 Procedure to Query Availability of File

(1) Peer P sends query to friend A_1 which has highest rank. If file is found then P downloads from A_1 else:

(a) *Peer A_1 suggests another peer A_{1j} from its social friends list.*

(b) *Peer P sends request to the peers A_{1j}, A_2, A_3 based on their rank. If desired file is found at some peer then P downloads it else P sends query to the TA.*

(c) *In case of equal rank of A_{1j}, A_2 or A_3, peer P select A_2 or A_3.*

4 Different Costs Expression

The different costs such as requesting peers to download a file, downloading the file, updating the ranking table, total cost, etc. are defined in this section. Suppose there are N peers in the P2P system and peer P sends request to download some file to i peers having equal rank. The probability ρ_i is defined as i/N. We consider t_{ina} (in sec) is the average inactive time between two consecutive churns of feeding peers.

4.1 Different Costs in Proposed System

Suppose there are m peers who are non-social friends/participants and peer k is selected among these and n = i + m. If no suitable peer is found to download the file, peer P sends request to other non-social friends/participants based on the response from TA and peer r is selected.

(a) *Cost of Requesting Peers for File Download (C_{req})*

The cost involved in sending the request and deciding the appropriate peers to download a file is calculated as follows.

$$C_{\mathrm{req}} = \rho_{\mathrm{i}}\left(2 \times \sum\nolimits_{\mathrm{j}=1}^{\mathrm{i}} h_j^P + 2\sum\nolimits_{i=1}^{m} h_i^P + h_k^P\right) + 2(1-\rho_{\mathrm{i}})^n\left(h_r^P + h_{TA}^P\right) \tag{5}$$

For simplification purpose, let $h_1^P = h_2^P = h_3^P == \ldots h_i^P = h_k^P = h_r^P = h_{TA}^P = h$ then

$$C_{\mathrm{req}} = \rho_{\mathrm{i}} h(2i + m + 1) + 4h(1-\rho_{\mathrm{i}})^n \tag{6}$$

(b) *Cost of Downloading a File (C_d)*

Suppose S is the size of a file and n is the number of data pieces of S. Suppose n data pieces S_1, S_2, S_3, …, S_n are downloaded from n different peers having upload bandwidth b_1, b_2, b_3, …, b_n. The cost of downloading a file depends on bandwidth. We consider the average time to download the data pieces S_1, S_2, S_3, …, S_n by n peers. There are average m peers inactive and CR is the churn rate during complete file download. So, cost of downloading a file is expressed as follows:

$$C_{\mathrm{d}} = \beta\sum\nolimits_{\mathrm{i}=1}^{\mathrm{n}} \frac{S_i}{\mathrm{nb_i}} = \frac{\beta}{\mathrm{n}}\left(\frac{S_1}{\mathrm{b}_1} + \frac{S_2}{\mathrm{b}_2} + \frac{S_3}{\mathrm{b}_3} + \ldots + \frac{S_n}{\mathrm{b_n}}\right) + (1-\beta)\frac{t_{ina}(CR-1)}{m} \tag{7}$$

Where β is a parameter to tune the effect of intermittent bandwidth and defined as:

$$\beta = \frac{\text{Average Upload Speed}}{\text{Max. Upload Speed}}$$

For simplification purpose, let $b_1 = b_2 = b_3 = \ldots = b_n = W_{avg}$ (Mbit/sec), and $S_1 = S_2 = S_3 = \ldots = S_n = S/n$. So, the download cost is expressed as follows.

$$C_d = \frac{\beta S}{nW_{avg}} + (1-\beta)\frac{t_{ina}(CR-1)}{m} \tag{8}$$

(c) *Cost of Updating Rank Table (C_t)*

The rank of a peer is updated by the TA. Each peer sends its feedback (λ_i) to the TA after every successful file download. If some social friends are inactive, a peer also intimates to the TA and TA updates the rank table. Let C_t (in sec) is the cost of updating the table by TA and rank table is under active maintenance.

$$C_{tc} = 2h_{TA}^P + C_t \tag{9}$$

$$C_{total} = C_{req} + C_d + C_t \tag{10}$$

$$\begin{aligned} &= \rho_i h(2i+m+1) + 4h(1-\rho_i)^n + \frac{\beta S}{nW_{avg}} \\ &\quad + (1-\beta)\frac{t_{ina}(CR-1)}{m} + 2h_{TA}^P + C_t \end{aligned} \tag{11}$$

4.2 Different Costs in Reputation System [7]

In this section, we have defined the different costs in existing reputation system [7]. The request for downloading a file is made to the social friends or non-social friends based on TA response. Only one peer (say k) is selected by needy peer P to download a file. Suppose there are i peers who are social friends and peer k is selected among these. In case of no social friend available for as a server, one peer r is selected from the non-social friends/participants.

(a) *Cost of Requesting Peers for File Download $\left(C_{req}^R\right)$*

The cost involved in sending the request and deciding the appropriate peer to download a file is expressed similarly as in (5) and (6).

$$\begin{aligned} C_{req}^R &= \rho_i\left(2\times\sum_{j=1}^{i} h_j^P + h_k^P\right) \\ &\quad + 2(1-\rho_i)^i\left(h_r^P + h_{TA}^P\right) \end{aligned} \tag{12}$$

For simplification purpose, let $h_1^P = h_2^P = h_3^P = \ldots h_i^P = h_k^P = h_r^P = h_{TA}^P = h$ then

$$C_{req}^R = \rho_i h(2i+1) + 4h(1-\rho_i)^i \tag{13}$$

(b) *Cost of Downloading a File* $\left(C_d^R\right)$

The average inactive time of the feeding peer per churn is t_{ina} (sec) and CR is the churn rate during complete file download. So, the download cost is expressed as follows.

$$C_d^R = \frac{\beta S}{W_{avg}} + (1 - \beta)(CR - 1) \times t_{ina} \tag{14}$$

c) *Cost of Updating Social Trust Table* $\left(C_t^R\right)$

Each peer after every successful file download sends its feedback (good/bad) to the TA and TA updates the social trust table. If feeding social friend is inactive, the beneficiary peer intimates to the TA and TA updates the social trust table. In case of any disagreement about the feedback, TA resolves the issue with the help of feedbacks received from other peers. Let C_t (in sec) is the cost of updating the social trust table by TA and table is under active maintenance.

$$C_t^R = 2h_{TA}^P + C_t \tag{15}$$

(d) *Total Cost* $\left(C_{total}^R\right)$

The total cost $\left(C_{total}^R\right)$ is expressed as below.

$$\begin{aligned} C_{total}^R &= C_{req}^R + C_d^R + C_t^R \\ &= \rho_i\left(2ih + h_k^P\right) + 4h(1 - \rho_i) + \frac{\beta S}{W_{avg}} \end{aligned} \tag{16}$$

$$+ (1 - \beta)(1 - CR)t_{ina} + 2h_{TA}^P + C_t \tag{17}$$

5 Evaluation and Analysis of Costs

In this section, we have analytically analyzed the different costs involved in proposed system and existing system. The effect of file size and number of social friends over the rank of a peer is illustrated in Figs. 2 and 3. Referring to (4), we have selected $\alpha = 0.5$, $F_i/N = 0.5$ and vary the u_{ij} from 0 to 10 Mb. The value of d_{ij} is kept as 1 Mb. As the value of u_{ij} increases, the rank of the feeding peer also increases as illustrated in Fig. 2.

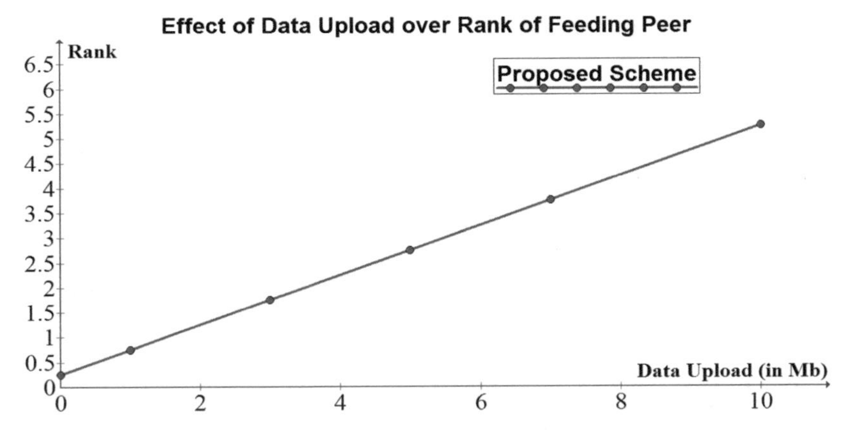

Fig. 2. Effect of data upload over rank of a peer.

Referring to Fig. 3, the total number of peers in P2P overlay (N) is considered as 500 and value of social friends (F_i) varies from 0 to 500. In both Figs. 2 and 3, the rank of a peer increases as upload data or number of social friends increases. It has been observed that the rank of a peer increases by 0.5 if peer uploads 1 Mb data whereas rank is increased by 0.1 if peer makes 50 new social friends. It means, the effect of data uploads over rank of a peer is more as compared to effect of number of social friends.

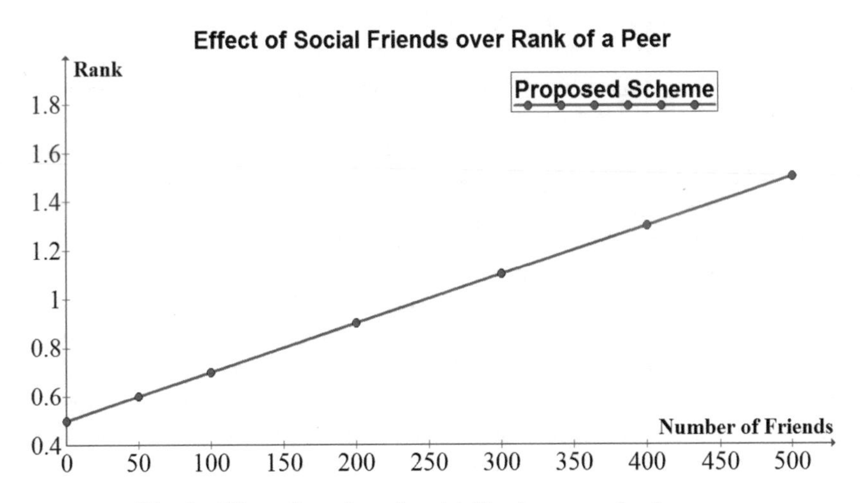

Fig. 3. Effect of number of social friends over rank of a peer.

The cost of downloading a file while varying the CR from 2 to 10 during complete file download is illustrated in Fig. 4. We have consider, W_{avg} = 1 Mb/s, $\beta = 0.5$, S = 1 Mb, n = 4, m = 2 (50% of n), and t_{ina} = 3600 s. The download cost is more in

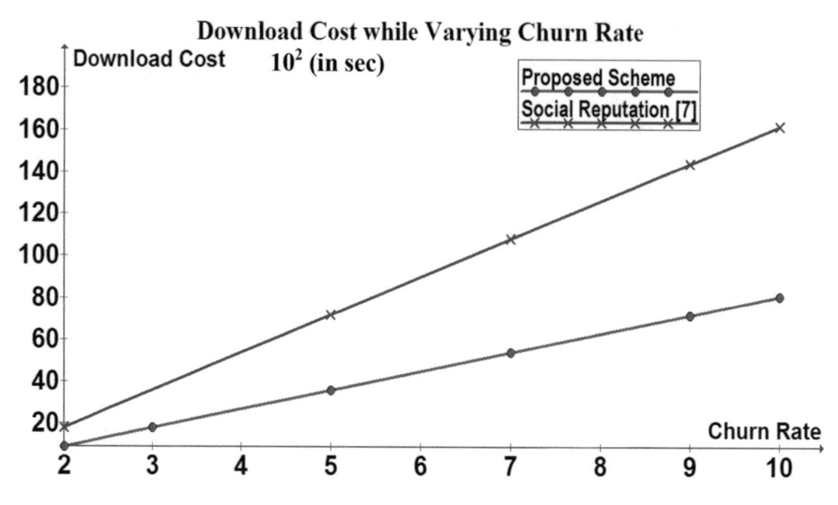

Fig. 4. Download cost varying the churn rate.

the existing reputation system [7] because there is only one feeder peer and its inactive time (t_{ina}) is 3600 s per churn. But in the proposed system, there are n = 4 and m = 2 feeding peers and inactive peers respectively. We consider 50% of n as inactive peers.

The cost of downloading a file while varying the inactive time (t_{ina}) from 60 to 3600 s during complete file download is illustrated in Fig. 5. We have consider, W_{avg} = 1 Mb/s, $\beta = 0.5$, S = 1 Mb, n = 4, m = 2 (50% of n), and CR = 2. The download cost increases in the proposed system and existing system as t_{ina} increases. The download cost is higher in the existing reputation system [7] because there is only one feeder peer and if feeder is inactive then download time is higher. But in the proposed system, there are n = 4 and m = 2 feeding peers and inactive peers respectively. So, as the inactive time (t_{ina}) increases, the download cost also increases but less than the existing reputation system [7].

The cost of downloading a file while varying the file size from 1 to 100 Mb is illustrated in Fig. 6. We have consider, W_{avg} = 1 Mb/s, $\beta = 1$, t_{ina} = 1200 s, n = 4, m = 2 (50% of n), and CR = 1. The download cost increases in the both proposed system and existing system as file size increases. The download cost is higher in the existing reputation system [7] because there is only one feeder peer. But in the proposed system, there are n = 2 and m = 0 feeding peers and inactive peers respectively. So, as the file size (S) increases, the download cost also increases but less than the existing reputation system [7] due to more feeding peers.

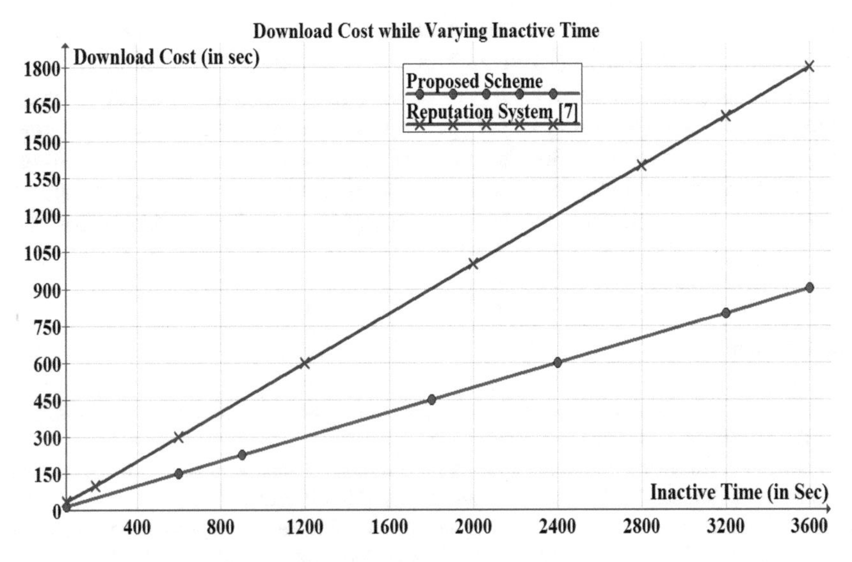

Fig. 5. Download cost varying the inactive time of feeding peers.

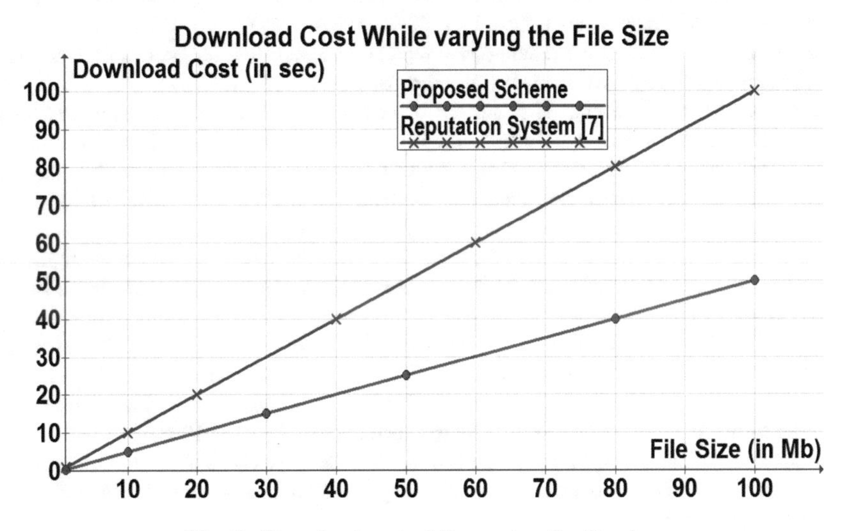

Fig. 6. Download cost while varying the file size.

6 Conclusion

We have observed that existing reputation system does not consider the churn rate and inactive time of the feeding peers in performance evaluation. We have considered two factors: reputation and trustworthiness based on data upload/download and churn rate. We have analytically analyzed and compared the existing reputation system and proposed system. It has been observed that the download cost in proposed system is less than the existing reputation system. The trustworthiness factor is useful in deciding the appropriate peer to download file in case of multiple feeders are available. The introduction of trustworthiness factor promotes the peers to share their resources in order to increase their rank. So, the proposed system introduces fairness among the participating peers and reduces free riding behaviors. We have not discussed the effect of interrupted bandwidths of the feeding peers which may be considered for further study.

References

1. Hughes, D., Coulson, G., Walkerdine, J.: Free riding on Gnutella revisited: the bell tolls? IEEE Distrib. Syst. Online 6(6) (2005)
2. Satsiou, A., Tassiulas, L.: Reputation-based resource allocation in P2P systems of rational users. IEEE Trans. Parallel Distrib. Syst. **21**(4), 466–479 (2010)
3. Zhou, R., Hwang, K.: Power trust: a robust and scalable reputation system for trusted peer-to-peer computing. IEEE Trans. Parallel Distrib. Syst. **18**(4), 460–473 (2007)
4. Zhang, Y., Fang, Y.: A fine-grained reputation system for reliable service selection in peer-to-peer networks. IEEE Trans. Parallel Distrib. Syst. **18**(8), 1134–1145 (2007)
5. Zhou, R., Hwang, K., Cai, M.: Gossip trust for fast reputation aggregation in peer-to-peer networks. IEEE Trans. Knowl. Data Eng. **20**(9), 1282–1295 (2008)
6. Xiong, L., Liu, L.: Peer trust: supporting reputation-based trust for peer-to-peer electronic communities. IEEE Trans. Knowl. Data Eng. **16**(7), 843–857 (2004)
7. Chen, K., Shen, H., Sapra, K., Liu, G.: A social network based reputation system for cooperative P2P file sharing. IEEE Trans. Parallel Distrib. Syst. **26**(8), 2140–2153 (2015)
8. Papaioannou, T.G., Stamoulis, G.D.: Reputation-based policies that provide the right incentives in peer-to-peer environments. Comput. Netw. **50**(4), 563–578 (2006). Special issue on management in peer-to-peer systems: trust, reputation, and security
9. Vishnumurthy, V., Chandrakumar, S., Sirer, E.G.: KARMA: a secure economic framework for peer-to-peer resource sharing. In: Proceedings of the Workshop on Economics of Peer-to-Peer Systems (2003)
10. Ma, R.T.B., Lee, S.C.M., Lui, J.C.S., Yau, D.K.Y.: Incentive and service differentiation in P2P networks: a game theoretic approach. IEEE/ACM Trans. Netw. **14**(5), 978–991 (2006)
11. Page, L., Brin, S., Motwani, R., Winograd, T.: The PageRank Citation Ranking: Bringing Order to the Web. Stanford Digital Library Technologies Project (1998)
12. Kamvar, S.D., Schlosser, M.T., Garcia-Molina, H.: The EigenTrust algorithm for reputation management in P2P networks. In: Proceedings of 12th International World Wide Web Conference (2003)
13. Lin, C.S., Cheng, Y.-C.: A barter-based incentive mechanism for peer-to-peer media streaming. In: Proceedings of IEEE 13th ISCE, pp. 871–875, May 2009
14. Junfeng, T., Lidan, Y., Juan, L., Zhongyu, L.: A distributed and monitoring-based mechanism for discouraging free riding in P2P network. In: 2009 Computation World: Future Computing, Service Computation, Cognitive, Adaptive, Content, Patterns, pp. 379–384. IEEE Computer Society (2009)
15. Li, Y., Liu, Y., Xu, K., Chen, W.: Analysis and balanced mechanism on free-rider in P2P network. In: Proceedings of Second International Conference on Computer Modeling and Simulation, pp. 462–466. IEEE Computer Society (2010)

16. Ge, T., Manoharan, S.: Mitigating free-riding on bittorrent networks. In: Proceedings of Fifth International Conference on Digital Telecommunications, pp. 52–56. IEEE Computer Society (2010)
17. Wang, C., Feng, J.: A study of mutual authentication for P2P trust management. In: Proceedings of Sixth International Conference on Intelligent Information Hiding and Multimedia Signal Processing, pp. 474–477. IEEE Computer Society (2010)
18. Liu, Q., Qin, F., Ge, L.: Modeling and analysis of free riding in peer-to-peer streaming systems. In: Proceedings of 6th International ICST Conference on Communications and Networking in China (CHINACOM), pp. 780–784. IEEE Computer Society (2011)
19. Sheshjavani, A.G., Akbari, B., Ghaeini, H.R.: A free-riding resiliency incentive mechanism for VoD streaming over hybrid CDN-P2P networks. In: Proceedings of 8th International Symposium on Telecommunications (IST 2016), pp. 771–776. IEEE Computer Society (2016)

Partition Based Product Term Retiming for Reliable Low Power Logic Structure

S. Jalaja[1(✉)] and A. M. Vijaya Prakash[2]

[1] Department of Electronics and Communication Engineering, Bangalore Institute of Technology, Research Scholar, Visvesvaraya Technological University, Bangalore, India
jalajabit@gmail.com

[2] Department of Electronics and Communication Engineering, Bangalore Institute of Technology, Bangalore, India
am_vprakash@yahoo.co.in

Abstract. Filtering is the one of the core element in any of the low power VLSI signal processing architecture. Increasing filter tap length will cause on the hardware complexity and lead to more power dissipation. The digital circuit requires some specialized algorithm to achieve high speed or low power consumption thereby to increase the chip performance. In this research work a modified retiming algorithm is proposed to reduce power dissipation by placing the Flip flops at the resultant multiplication of the output nodes. The proposed architecture equations are simplified in terms of sum of product term and distribute the weight of product terms by applying retiming method. In the proposed architecture design the flip flops are placed to fan out of partition multiplication to minimize the switching activity factor. In this paper, to prove the performance of chip sum-of-product term retiming is applied, this can be implemented for any digital circuit to reduce total power. The proposed algorithms have been implemented in cadence EDA tool and the results are proved by using finite-impulse response (FIR) and infinite-impulse response (IIR) filters. Using Digital Signal Processing (DSP) application, proposed algorithm is synthesized to ensure that power saving is achieved compared to existing method. The method proves Energy per sample (EPS) as much better than node-splitting and node-merging technique. Experimental results shows power dissipation is minimized compared to other FIR architecture. Complete design task is modeled using Data Flow Graph (DFG) to achieve more precise result.

Keywords: Sum-of-product term retiming · Low power logic
Partition node · Retimed data flow graph

1 Introduction

The hot spot in Very Large Scale Integration (VLSI) chip is excessive power dissipation, causing much parameter degradation. In challenging applications like

K. Arai et al. (Eds.): FICC 2018, AISC 886, pp. 178–189, 2019.
https://doi.org/10.1007/978-3-030-03402-3_13

Processing of Text, graphics, video decompression, hand writing recognition, speech recognition etc., the battery life plays a very important role. The lack of switching activity during the operation of the circuit, will cause for power dissipation due to unwanted signal transition like glitches, short circuit current etc. These are considered as the main causes in low power devices and an important component for source of power dissipation. In digitized world main challenging topic of research is the development of Internet of things (IoT) application. Here also power saving positions and sleep mode techniques contribute to a great deal in future, to develop IoT devices without intervention of user. To reduce dynamic power dissipation, Partition based Product term retiming is used in this paper. To achieve this factor the output Flip flops are shared to each product term by preserving its output result. There are many optimization methods adopted to enhance capability of VLSI chip. Lots of challenges are involved in terms of interconnect capacitance, node to node accuracy, total power dissipation, speed, portability of the device etc. Balancing delay in the entire circuit is another optimization technique. There are two ways of balancing delay and can be restructured as shown in Fig. 1. Any Boolean equation are represented in the form Flow Graph transformation for easy analysis. The graph comprises a set of gate delay and wire delay, balanced using Retiming concepts and is formulated as; $G = \langle G_d, W_d \rangle$.

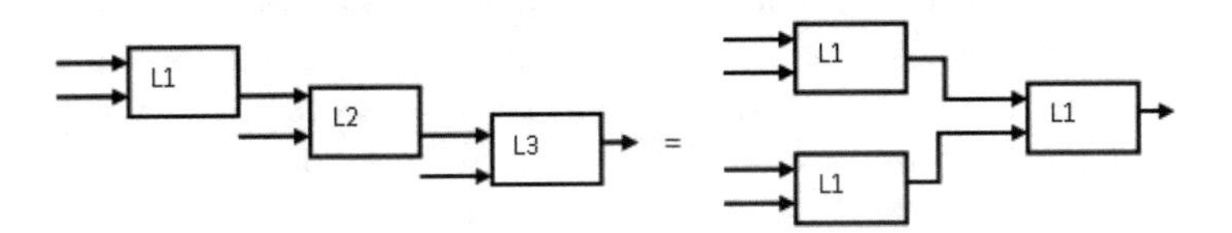

Fig. 1. Balanced tree structure.

2 Review of Literature

One of the powerful transformation techniques in digital system design is Retiming. The Retiming is a method in which Flip flops are moved from input node to output node without changing its functionality. In any logic circuit the amount of time required for switching is modified using retiming approach. The capacitance change due to fanout of FFs is further affects the power dissipation. The phase circuit retiming is done to preserve the testability of the circuit. There are many ways to improve the circuit performance with different objectives of Retiming. Some of commonly used VLSI signal processing Design Methodology are pipelining, parallel processing, retiming, folding and unrolling. etc. In paper [15] unrolling and retiming approaches for scheduling stream applications on embedded multicore processors performance result was analyzed. Similarly in paper [11] Retiming and Unfolding concepts are implemented for notch filter to achieve high frequency. Two more new techniques like functionally equivalent retiming and clock-rate pipelining techniques are implemented in paper [5].

Many researchers and industry are exploring the Retiming with different approach [10,14]. and it was Leiserson and Saxena [2,3] who introduced the Retiming concepts. The power consumption is caused by the switching activity that occurred in circuit is modified using retiming. The Flipflops are placed near the high switching activity node to reduce the power by applying retiming [6]. The fixed phase retiming algorithm [7] is used to reduce the power by retiming in two level circuits. For many DSP applications, same approach [12] is implemented to decrease the power dissipation. The author [16] deals with feasible retiming and optimal retiming to achieve the smallest iteration period. Retiming transformation is also implemented in scan architecture [8]. The transformation is applied either forward or backward direction. Here they used backward retiming to scan multiplexer. Usually in this technique Flip flops are moved from input to output node. The techniques implemented to analyze the critical path of the multiplexer are original flip-flop with normal data path; the scan path is appeared with shadow flip-flop and last flip-flop. The experimental results prove the higher speed and reduced hardware overhead compared to existing results. One of the researchers [14] presented the solution for concurrent implementation of software-hardware codesign systems. As a result they investigated the technique to implement synthesis for high level design, same method they adopted to reduce less number of multiplexer to improve the performance and also area overhead of the circuit. Register transfer level retiming algorithm is applied to minimize their clock period. To improve latency of system-level design, optimization is performed on FPGA based design specification [4]. Here Taylor Series expansion data is represented in the form of data flow graph. In this paper mainly author concentrated on critical path delay minimization.

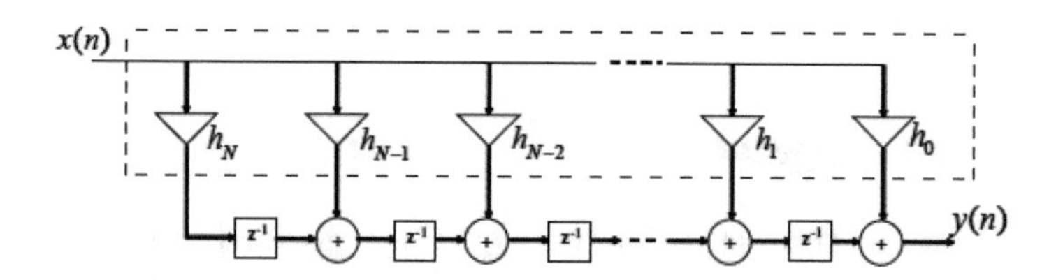

Fig. 2. Transform fir filter.

Logic equation or architecture of the design model is represented in the form of data flow graph (DFG) for easy analysis. Edges and nodes are the two main parameters are represented in the data flow graph. The set of operation is represented by every node (Circle) and edges show the connection between two vertexes. In this paper we used data flow graph representation for proposed algorithm to enhance the retiming concepts. For FIR and IIR filter the modified retiming algorithm is adopted to estimate the power consumption compared to

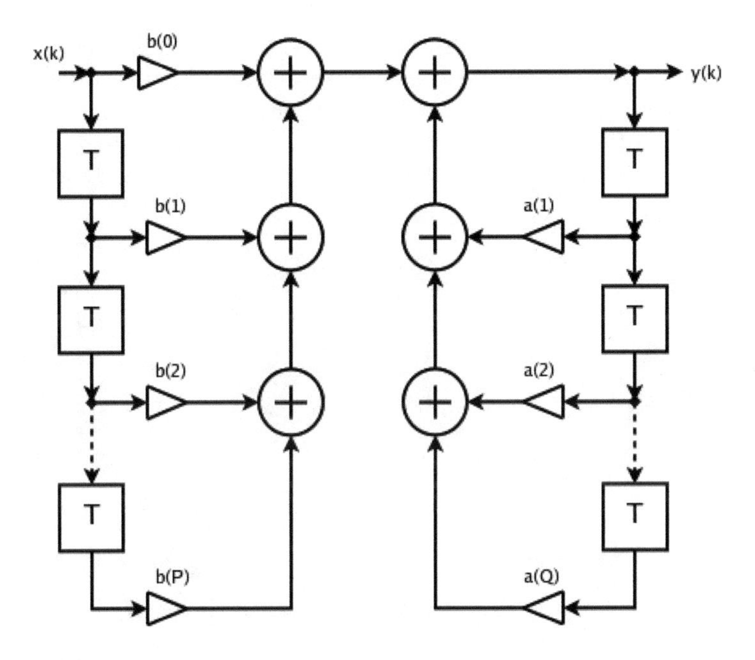

Fig. 3. IIR filter.

existing method [1,9]. There are many ways to realize the filter structure to improve the performance. The design model of FIR and IIR filter is shown in Figs. 2 and 3, respectively. Several different structures of FIR filter are proposed from many researchers that suits in numerous applications such as medical imaging, Bio-signal processing, removing interference, radar imaging etc. The use of digital filter in any real time application though is economical and but, computationally expensive. In general both the filters requires addition and multiplication, so frequency required to driving for adder is more than that of multiplier. The proposed work in each multiplication block is partitioned into number of sub-blocks and is retimed backward to reduce switching activity factor. Same method is implemented to get IIR and FIR filter output.

3 Product Term Retiming Algorithm

One of the key parameter of power dissipation is switching activity factor on every vertex of DFG. In any of the signal processing architectural design most commonly used blocks are multiplier, adder, comparator and subtractor, etc. There are many optimize algorithms are adopted in architecture design abstraction in present scenario. Simplify the architectural logic equation into a sum-of-product form, later apply the transformation technique to reduce power. The proposed modified retiming technique, Flip flops are inserted at cutset line to reschedule the computation. Output FFs are reschedule stage by stage at each high switching activity path to reduce the power dissipation. In this paper retiming algorithm is represented by the circuit graph derived from the actual circuit. The graph comprises combinational blocks, treated as node and its weights are distributed by path balancing technique at sum of product term to reduce power consumption. The outline of proposed algorithm is framed using success history on retiming.

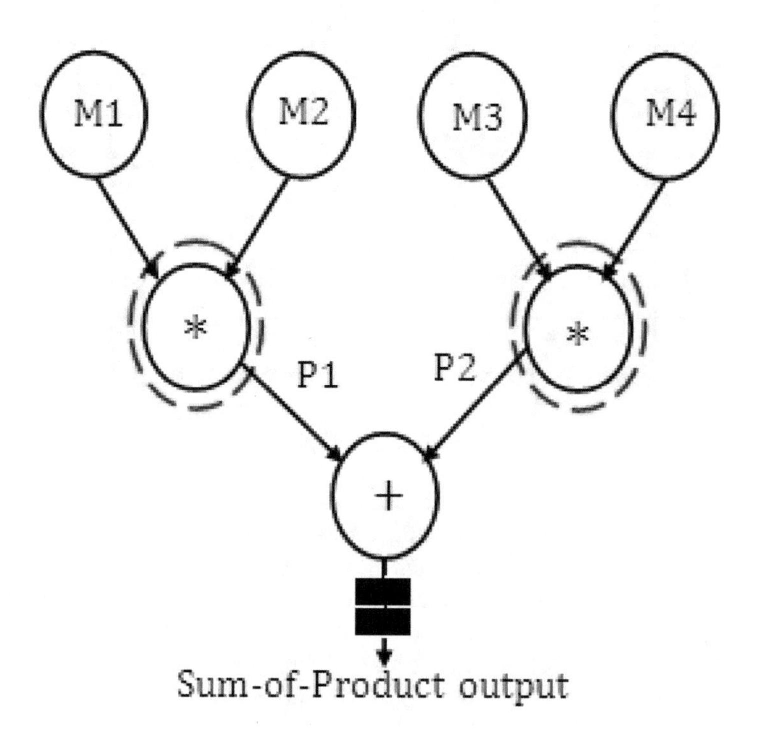

Fig. 4. Partition multiplication.

Sum-of-Product form multiplication output as shown in Fig. 4. In this scenario M1, M2, M3 and M4 represents multiplication inputs, P1 and P2 represents set of product term. Each product term is partitioned into three clusters G_{HL}, G_{LH} and G_{LH+HL}. For each cluster backward retiming performed, individual P1/P2 fanout loading can be split in different way to avoid the longest multiplication computation. Each path that ends with FFs with different delay (D) combination is as shown in Fig. 5. Decomposing a resultant multiplication became a solution to low power dissipation that is analyzed using synthesize result. Above method is used recursively in both IIR and FIR filters to minimize the switching factor; it leads to dynamic power reduction.

Pseudo code describes register allocation to each sum of product term output edge and same registers are shifted to each partition multiplication. Each multiplication block is partitions into three group of subsets with lower to higher

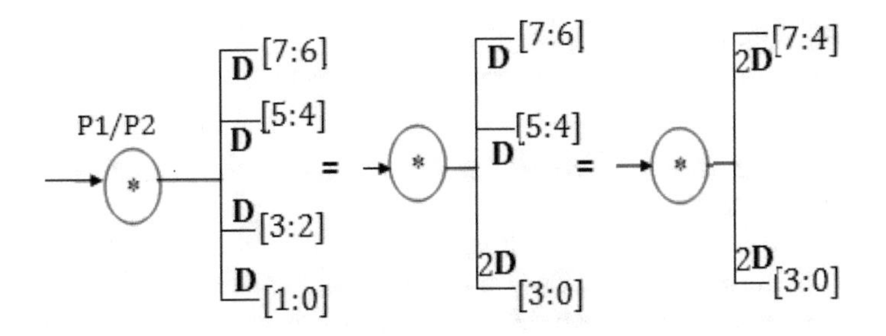

Fig. 5. Decomposition of multiplication delay.

nibble bit, higher to lower nibble bit and summation of both lower to higher nibble bit, higher to lower nibble bit. Each sum outer lag FFs are mapped to each partition product term. The pseudo code for proposed technique is as shown below.

Pseudo code for Sum-of-Product term retiming to determine switching activity factor

---

Build a conventional DFG for digital circuit(G)
Measure the switching activity factor for G
M_o = Number of FF's
$M_o \Leftarrow M_o + 1$
For each sum-of-product output edge(v) = [sum$_r$]
Map[sum$_r$] $\Leftarrow r_{M_o}$
split[productterm]$\Leftarrow P_{LH}, P_{HL}, P_{LH+HL}$ P_o =split[productterm] do
{
foreach(P_o)
$P_o$$\Leftarrow$move[sum$_r$] }
Rebuild retimed DFG (G$_r$) after shifting FF's at

High Switching activity factor power dissipation is measured for Gr.

---

4 Low Power Sum of Product Term Realization

Power management is one of the significant research challenge in most of the hand held devices available in the market. Minimization of the power is done with different knobs of the design. Power dissipation can be avoided by partitioning the multiplication block into multiple parts and retimed the path by shifting the FF's. Introduce the flops at the output stage and shift towards the each product term to save power. During shifting the flip flops are move towards the split multiplication block to reduce the power consumption. For example consider the multiplication of 8 bit number is retimed with set of different bit combination like [7:4] & [3:0] or [7:6][5:4][3:0]. Then following two methods are implemented to improve the performance level.

(1) Shift the FFs to sum-of-product terms.
(2) Partition the each multiplier into different set, & shift the FFs at cutset line.

Both the techniques aim to reduce overall dynamic power during the each switching transition.

As a proof, the proposed algorithm, is first implemented for full adder, then to the carry save adder, to save power consumption. Conventional XOR based full adder sum is having single gate delay and carry out generation having multiple gate delay, but the same full adder is implemented using Sum-of-Product form, then carry out is having single gate delay. Same concept is adopted to propose a product term retiming algorithm. The full adder carry out data flow graph is shown in Fig. 6, to represent how we can minimize the clock cycle after retiming. The product term retiming critical path of the circuit is passed through only one gate, so that clock speed is increased. Sample time period with proposed technique decreases, compared to fixed retiming [9].

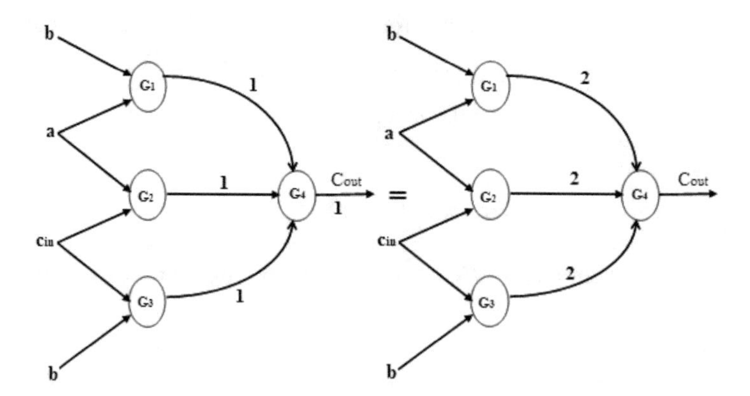

Fig. 6. Sum-of-product term realization for full adder carry out.

In this paper adder block is replaced by the combination of carry save adder and carry select adder to save power. The addition of large number result is obtained from modified carry save adder with carry select adder is represented in Fig. 7. Dotted line represents cutset line to implement the pipelining and output FFs are transferred backward towards the cutset line. The proposed Algorithm is incorporated to carry save adder to reduce the transition activity. The complexity for higher order bit addition is substantially reduced by applying retiming technique.

5 FIR and IIR Filter Retiming Model

In DSP-architecture the complexity of the circuit is reduced by applying various optimization techniques to get better performance in silicon chip. The sum-of-product retiming is proposed to reduce the transition density in FIR and IIR

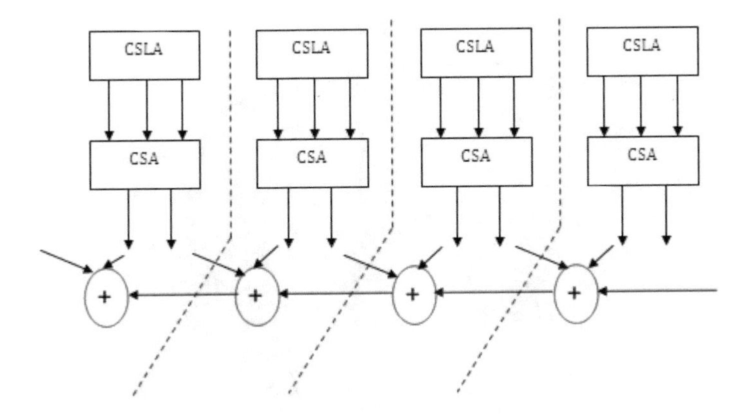

Fig. 7. Modified carry save adder using cutset.

filters. The Product term cutset FIR filter restructure is shown in Figs. 8, 9 and 10, shows graphical representation of the FIR and IIR filter, respectively. The proposed modified adder and multiplication blocks are used in both the filter to increase the performance. To realize the FIR and IIR filters block Backward retiming is applied to each sum-of-product term.

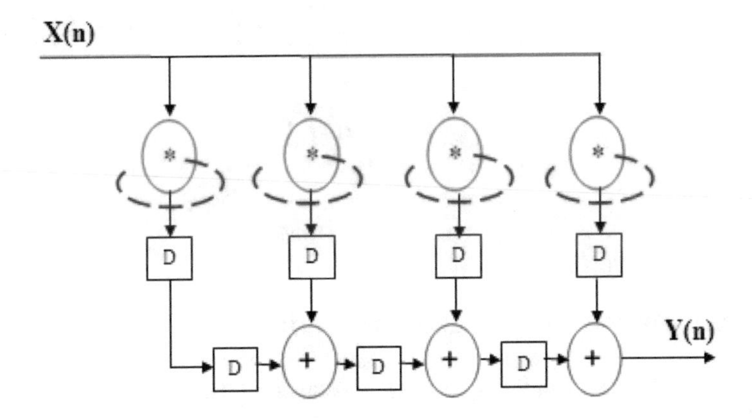

Fig. 8. Product term cutset FIR filter.

The FIR filter block is reconfigured with different architecture [1]. Author of this paper the multiple constant multiplications of horizontal and vertical sub expression techniques are used to implement FIR filter. For 8 bit block size with different tap length is synthesized and it is compared with the conventional method. In this paper we applied the proposed method to enhance the performance level [1,9].

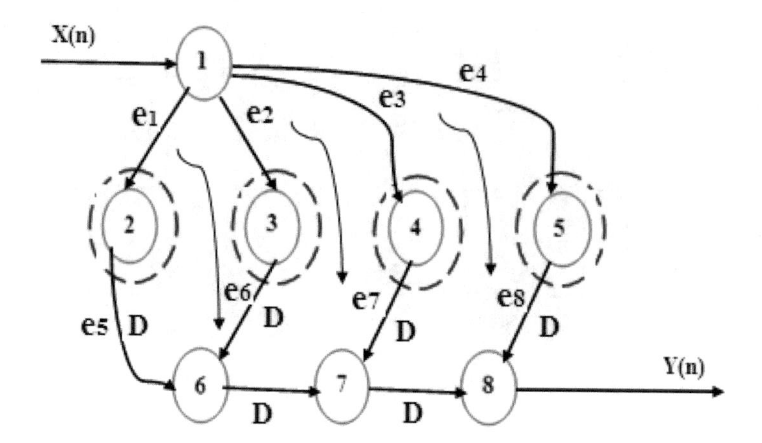

Fig. 9. DFG for FIR filter.

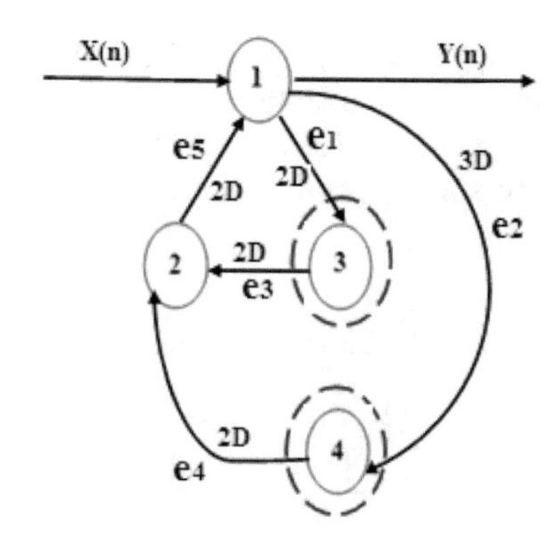

Fig. 10. DFG for IIR filter.

6 Performance Level Measurement

Performance of the circuit is usually measured interms of low power and computational speed. There are various design methodology are available to reduce power dissipation. As thumb rule says the switching activity factor is directly proportional to dynamic power. Therefore generally power dissipation is expressed by $P = \alpha CV^2 f$. In this paper switching activity factor is controlled by applying the sum of product partition retiming method. The delay distribution starts from output node and recursively passed to all partition nodes. The power consumption of high switching activity node is reduced by embedding the FFs in several ways. For above mentioned design the functional simulation is verified using EDA tool.

The IIR and FIR filter topology is synthesized using register transfer logic (RTL) compiler from cadence. Tables 1 and 2 shows synthesis result for 16,32

Table 1. Fir filter comparison results for Minimum Clock Period (MCP) and Area-Delay Product (ADP)

Design	Filter length	MCP (ns)	ADP ($u\mu^2us$)
Prmod Kumar [9]	16	2.3	111739.4
	32	2.3	223910.2
	64	2.3	448251.9
Proposed method	16	1.034	68372.216
	32	1.034	182463.77
	64	1.034	365097.128

Table 2. Fir filter comparison results for area and power dissipation

Design	Filter length	MCP (ns)	Area ($u\mu^2$)	Power (mW)	EPS (ns*mW)
Park [17]	16	0.95	56325	24.6	23.37
	32	1.10	112446	47.3	52.03
	64	1.28	218351	85.6	109.568
B.K.Mohnaty [1]	16	1.4	232089	100.9	141.26
	32	1.4	476503	186.4	260.96
	64	1.4	957186	366.4	512.96
Proposed method	16	1.034	66124	1.022	1.0567
	32	1.034	176464	2.766	2.86
	64	1.034	353092	5.58	5.77

and 64 taplength of FIR filter block using 45 nm technology. Synthesis results of IIR filter shows area and MCP in Table 3. The proposed algorithm is synthesized for different filters without applying constrained to analyze the area, power and delay. The switching power and internal power are used to calculate dynamic power. The sum-of-product term retiming shows reduction in switching power and the zero transition states are neglected to reduce power consumption. Many architecture [1,9,13] are proposed for FIR and IIR filters but problem solution proves more realistic outcome, compared to Energy per Sample(EPS) and ADP. It reports approximately 55% and 20% reduction MCP [9] and ADP [9] respec-

Table 3. IIR filter MCP and area synthesis result

Filter length	MCP [1] (ns)	Area [1] ($u\mu^2$)	MCP (ns)	Area ($u\mu^2$)
16	2.27	6010.6	1.08	1664

tively. The performance of the design is synthesize using RTL complier, results shows that 90% power reduction [1,17]. The comparison results are represented in the bar chart and it is shown in Fig. 11.

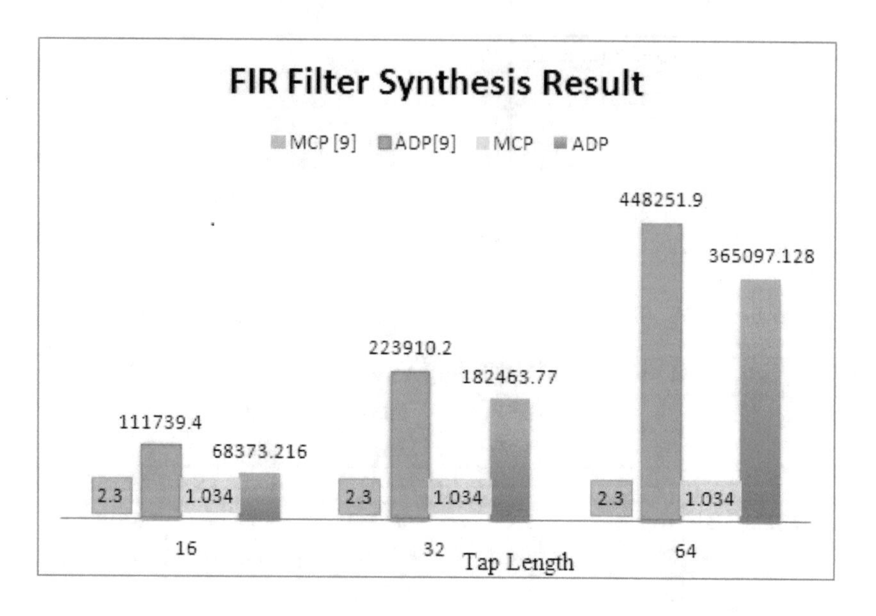

Fig. 11. Bar chart analysis.

7 Conclusion

The contribution of this paper is to analyze the switching activity factor by applying sum-of-product term retiming. Proposed algorithm is explicitly eliminating the high switching activity by retiming each partition net. Multiplier transition density is reduced by placing the register at each partition node. The modified Retimed logic structure increases the computation efficiency without modifying the functionality of the design. The synthesis result shows minimum power consumption compared to node splitting and node merging architecture. The transition delay is minimized by applying proposed method. The investigated pseudo code prove the satisfactory result than the fixed retiming method. Retimed filter structure is significantly verified with different practical filter length to measure computation time. Three parameters are extracted to verify the performance level, so it may fit for rapidly growing application.

References

1. Kumar, B., Meher, M.P.K.: A high-performance FIR filter architecture for fixed and reconfigurable applications. IEEE Trans. Very Large Scale Integr. (VLSI) Syst. (2015)
2. Leiserson, C., Rose, F., Saxe, J.B.: Optimizing synchronous circuitry by retiming. In: Proceeding of the Caltech Conference on VLSI, 3rd edn. (1983)
3. Leiserson, C., Rose, F., Saxe, J.B.: Retiming synchronous circuitry. Algorithmica (1991)
4. Gomez-Prado, D., Ciesielski, M., Tessier, R.: FPGA latency optimization using system-level transformations and DFG restructuring. In: DATE Conference (2013)
5. Venkataramani, G., Gu, Y.: System-level retiming and pipelining. In: IEEE 2nd Annual International Symposium on Field Programmable Custom Computing Machines (2014)
6. Monteiro, J., Devadas, S., Ghosh, A.: Retiming sequential circuits for low power. In: Proceedings of the IEEE/ACM International Conference on Computer-Aided Design (1993)
7. Lalgudi, K.N., Papaefthymiou, M.: Fixed phase retiming for low power. In: Proceedings of the International Symposium of Low Power Electronic Devices (1996)
8. Sinanoglu, O., Agrawal, V.D.: Retiming scan circuit to eliminate timing penalty. In: Proceedings of IEEE International Conference (2012)
9. Meher, P.K.: On efficient retiming of fixed-point circuits. Trans. Very Large Scale Integr. (VLSI) Syst. (2016)
10. Nanda, R.: DSP architecture optimization in MATLAB/Simulink environment, M.S. thesis, University of California, Los Angeles (2008)
11. Samanta, S., Chakraborty, M.: FPGA based implementation of high speed tunable Notch Filter using pipelining and unfloding. In: IEEE Conference (2014)
12. Simon, S., Schimpfle, C.V., Wroblewski, M., Nossek, J.A.: Retiming of latches for power reduction of DSP design. In: Proceedings of the IEEE International Symposium on Circuits and Systems (1997)
13. Elloumi, Y., Akil, M., Bedoui, M.H.: Execution time optimization using delayed multidimensional retiming. In: IEEE/ACM 16th International Symposium on Distributed Simulation and Real Time Applications (2012)
14. Hara-Azumi, Y., Matsuba, T., Tomiyama, H., Honda, S., Takada, H.: Selective resource sharing with RT-level retiming for clock enhancement in high-level synthesis. In: 2012 IEEE 14th International Conference on High Performance Computing and Communications (2014)
15. Che, W., Chatha, K.S.: Unrolling and retiming of stream applications onto embedded multicore processors. In: DAC (2012)
16. Zhu, X.-Y., Basten, T., Geilen, M., Stuijk, S.: Efficient retiming of multirate DSP algorithms. IEEE Trans. Comput.-Aided Des. Integr. Circuits Syst. **31**(6) (2012)
17. Park, S.Y., Meher, P.K.: Efficient FPGA and ASIC realizations of a DA-based reconfigurable FIR digital filter. IEEE Trans. Circuits Syst., II, Exp. Briefs **61**(7) (2014)

Energy Efficient Cognitive Radio Network Using High Altitude Platform Station

Shital Joshi[1(✉)] and Umar Albalawi[2]

[1] Department of Computer Science and Engineering, Oakland University, Rochester, MI 48309, USA
ShitalJoshi@oakland.edu
[2] Department of Computer Science, University of Tabuk, Tabuk, Kingdom of Saudi Arabia
UAlbalawi@ut.edu.sa

Abstract. To meet the demand of future communication, it is utmost important to efficiently utilize various communication resources. Cognitive radio network is one of the most promising area for future communication, which aims to increase the frequency utilization. With the cell size growing smaller with each generation, the number of base stations and the time to develop the required infrastructure will increase exponentially. As a "last mile" solution, this paper considers a high altitude platform station (HAPS) to implement the cognitive radio network where the optimization problem is formulated to maximize the energy efficiency (EE) of the overall network. Two primary objectives are addressed: increase the EE of the network and decrease the interference to the primary users. The non-convex optimization problem is transformed to semi-definite problem. The results obtained are promising for HAPS as a future radio platform.

Keywords: Energy efficiency · Cognitive radio network
High altitude platform station (HAPS) · Convex optimization
Semidefinite relaxation

1 Introduction

Over a past decade, wireless communication has undergone tremendous improvement in its technology making it possible to provide large variety of services like high quality voice and video services including live multi-party audio/video conferencing, online interactive games, high data rate file download and so on. However, the improvement in the underlying radio frequency (RF) hardware has not been able to meet the demand of higher quality services. Similarly, bandwidth posses another big issue to meet this accruing demand. Over these course of time, various alternative approaches like cognitive radio networks (CRN) [1], massive multiple-input-multiple-output (MIMO) [2], femto/pico cells, beamforming [3], high altitude platform station (HAPS) [4], visible light communication [5] have been proposed, all aimed to efficiently utilize the available network

K. Arai et al. (Eds.): FICC 2018, AISC 886, pp. 190–202, 2019.
https://doi.org/10.1007/978-3-030-03402-3_14

resources and exploits all possible gaps so that to provide new dimension to future communication.

CRN has been recently gaining lot of research interest due to its ability to utilize the available frequency spectrum in a much efficient way. At present, the allocated spectrum has not been used efficiently by the specified application. There are significant amount of spectrum that are either underutilized or partially utilized most of the time [6]. CRN makes use of these spectrum, which have been originally licensed to primary users (PUs), to provide service to the unlicensed users (i.e. secondary users, SUs) in such a way that the primary users are not affected. So the interference that SUs produce must always be within the tolerable acceptable limit for the PUs.

Due to increase demand of high-capacity wireless services, future cellular terrestrial network is confined to just few meters of cell radius, called femto cells and pico cells. Thus future cellular network employs base stations at every corner of the road and even at house. Due to inherent delays in satellite systems, they cannot be employed for delay constraints applications like voice and video communications. So HAP is as a "last mile" solution for future communication, which takes the advantage of both terrestrial as well as satellite system. It is an aerial platform flying over 70,000 feet above the Earth's surface to achieve maximum footprint per HAP, minimize wind effect and ensure no blockage of air plane traffic (all classes of commercial air planes fly under 55,000 feet). HAPS can serve as a multipurpose platform for future world. The HAPS payload can provide services like broadband wireless access, navigation, monitoring, remote sensing, etc. They are also extremely suitable to provide immediate communication facility in disaster scenario where the terrestrial network has been severely damaged. Similarly, they can also be employed to connect two remote location which otherwise would require long relays of number of base station, which would be extremely cost inefficient solution.

This paper considers HAPS based CRN for future communication with the aim to maximize the energy efficiency of the network. The concept of improving energy efficiency has been extensively studied over past few years. In [7], authors have shown that the energy efficiency can be improved with the heterogeneous cell with relays serving at the pico cells and base station with massive MIMO serving at the macro cell. In [8], cooperation between primary transmitter (TX) and secondary BS is compared with the case when there is no cooperation between them and it has been shown that the signal-to-interference-plus-noise-ratio (SINR) requirements of all users SUs can be met when there is cooperation between BS and TX. In [9], it has been shown that there is energy efficiency improvement with the heterogeneous network (HetNet) deployment compared to that of the single cell scenario. In [10], low-altitude and high mobile unmanned aerial vehicles (UAVs) have been considered for providing wireless communication. Whilst both HetNets and massive MIMO concepts have been in cognitive radio networks (CRNs) [11,12] separately, the unique contribution in this research work is the employment of both massive MIMO and HetNets in the CRN. However the area of cognitive radio in HAPS is a relatively new domain where not much research have been done till date. As far as author's knowledge is concerned, this is one of the first attempt on this area.

In order to address the critical design aspect for HAPS based CRN, this paper consider a formulation of optimization problem with an objective to maximize the energy efficiency of the overall CRN while satisfying the QoS requirement at each SUs as well as limiting the interference caused from SUs to PUs to within the tolerable limit. The optimization problem also need to satisfy the per antennas power constraint as well the total power constraints. The remaining of the paper is organized as follows: Sect. 2 presents the system description which includes the system model and the HAPS channel model. Optimization problem formulation is presented in Sect. 3, followed by results in Sect. 4. Section 5 presents the conclusion and future work.

2 System Description

In this section, the system model and the channel model are presented.

2.1 System Model

An underlay CRN for downlink transmission is considered as shown in Fig. 1. A HAPS provides coverage to all the secondary users in presence of the primary base station (PBS). Since the coverage area of a HAPS is much larger than that of the PBS, obviously it serves more users than the PBS. The PBS serves the primary users (PUs) within its own coverage area. The Secondary users (SUs) can be located within the coverage area of the HAPs. Since the HAPS system is using the unlicensed spectrum which is primarily meant for PUs, the signals from the HAPS to its SUs can become source of interference to the PUs. As PUs are equipped with normal handheld terminals with no interference suppression techniques implemented, it becomes the responsibility of the HAPS to allocated power wisely to its SUs, so that the interference to the PUs are within its tolerable limit. Thus the HAPS is subjected to interference constraint, while the PBS is not subjected to any interference constraints, in terms of SUs. However, with the PBS scenario, one PU signal can be interference to other PUs, thus it is also subjected to its own interference constraint, which is not similar to that of the case of the HAPS. Both stations (i.e. HAPS and PBS) are subjected to power constraints. Since HAPS is an aerial platform, at an altitude of 17–21 km, it can provide coverage to large area and large number of users. The diameter of the HAPS footprint can be computed using the formula [13]. One of the most important aspect in the HAPS system is that there is no external power supply that can be given to it. So all its operation has to be done by the battery energy or via solar energy. So, it must use its power efficiently as it is not directly connected to any external power source. All the communication has to be performed using the battery located within its payload. So, energy efficiency aspect in such a network becomes very crucial.

$$d = 2R\left(cos^{-1}\left(\frac{R}{R+h}Cos\theta\right) - \theta\right) \tag{1}$$

where R is the Earth's radius (6378 km), θ is the minimum elevation angle and h is the altitude. Even though the HAPS can provide the coverage of up to 960 km in diameter [14], for this paper, a footprint of diameter 152 km is considered which yield an elevation angle of $\theta = 15\,^\circ$.

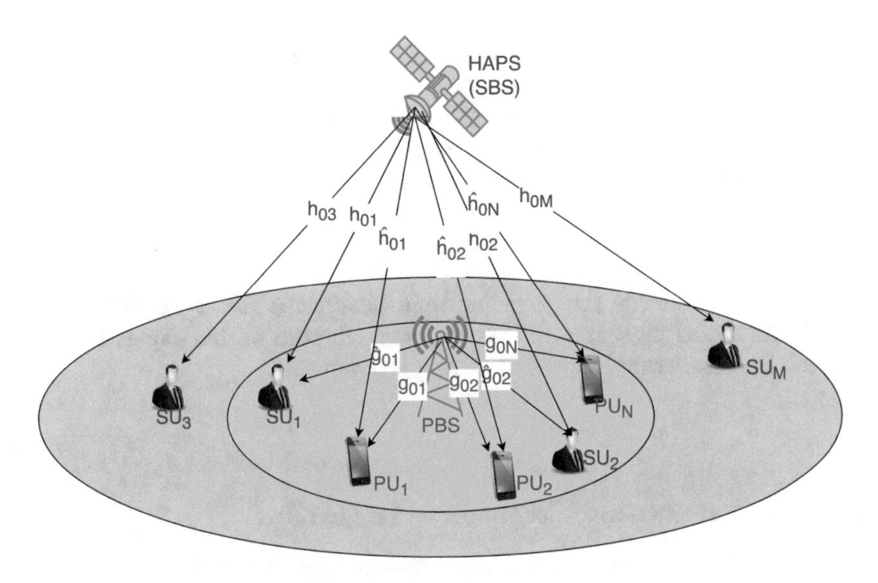

Fig. 1. System model for underlay CRN with a secondary cell covering a primary cell.

Each of the users are equipped with a single antenna while the SBS is assumed to have omnidirectional antenna (i.e. $K_{SB} = 1$) and the HAPS consists of K_{HAPS} number of antennas. This paper assumes that there is no coordination between the HAPS and the PBS hence all the schemes have to be independently applied to the HAPS. Hence the amount of power that needs to be allocated to each SU depends on the solution of the optimization problem. As shown in Fig. 1, the terms $\mathbf{h}_{0i} \in C^{K_{HAPS} \times 1}$ and $\mathbf{g}_{0j} \in C^{K_{PBS} \times 1}$ are used to represent the signal channel from HAP to i^{th} SU and PBS to j^{th} PU respectively where $i \leq M$, $j \leq N$ and $M \gg N$. Similarly, the terms $\hat{\mathbf{h}}_{0j} \in C^{K_{HAPS} \times 1}$ and $\hat{\mathbf{g}}_{0i} \in C^{K_{PBS} \times 1}$ are used to represent the interference channel from HAP to j^{th} PU and PBS to i^{th} SU, respectively.

The received signal power at the ith SU is given by:

$$y_i = \mathbf{h}_{i0}^* \mathbf{w}_i s_i + \sum_{\substack{p=1 \\ p \neq i}}^{N_{SU}} \mathbf{h}_{i0}^* \mathbf{w}_p s_p + \hat{\mathbf{g}}_{i0}^* x_p + n_i \qquad (2)$$

Where, N_{SU} represents the number of SUs in the HAPS system, vector $\mathbf{h}_{i0}$ represents the channel gain (i.e. spatial characteristic of the channel including shadowing effect, fading and the path loss) from the HAPS to the ith SU, the superscript * represents the complex conjugate, $\mathbf{w}_i$ represents the beamforming vector, s_i represents modulated signal intended for the ith SU from the HAPS, vector $\hat{\mathbf{g}}_{i0}$ represents the channel gain from the PBS to the ith SU and n_i represents the noise terms.

In (2), the first term represents the actual intended signal for the ith SU while the second term represents the interference from all other SUs in the system. The third term represents the interference due to the PBS. It should be noted that the third term may or may not be present for SUs. It depends on the location of SUs in the overall system. If the SUs are located within the coverage zone of PBS then it will receive the signal from the PBS as well. For the PUs, all the signal from HAP is an interference.

2.2 Channel Model

In order to account for the path loss and shadowing effect for HAPS based channel, the model presented in [15] has been adopted in this paper. Hence the channel loss considering large scale fading is given as:

$$L = \begin{cases} L_{FSL} + \zeta_{LOS}, & \text{for LOS.} \\ L_{FSL} + L_s + \zeta_{NLOS}, & \text{for NLOS.} \end{cases} \tag{3}$$

where L_{FSL} is the free space loss in dB and is given by:

$$L_{FSL} = 20log_{10}(d_{\text{km}}) + 20log_{10}(f_{\text{GHz}}) + 92.4 \tag{4}$$

where d_{km} is the propagation distance between transmitter and receiver, f_{GHz} is the carrier frequency used in the transmission (measure in GHz), L_s represents the elevation angle dependent random shadowing.

3 Problem Formulation

The objective of the optimization is to maximize the energy efficiency of the overall CRN while satisfying the Quality of service (QoS) requirement of the SUs and restricting the interference to PUs within the tolerable limit. The QoS requirement for i^{th} SU can be specified in terms of information rate (bits/Hz) and hence can be related to signal-to-interference-plus-noise ratio (SINR) as:

$$\begin{aligned} log_2(1 + SINR_i) &\geq \gamma_t \\ \text{or,} \quad SINR_i &\geq 2^{\gamma_i} - 1 = \hat{\gamma}_i \end{aligned} \tag{5}$$

where γ_t is the required QoS for i^{th} SU in the system. It is assumed that all the SUs requires same QoS requirement i.e. $\hat{\gamma}_1 = \hat{\gamma}_2 = ... = \hat{\gamma}_{N_{SU}} = \hat{\gamma}$.

From (2), the SINR for i^{th} SU can also be defined as:

$$SINR_i = \frac{\mid \mathbf{h}_{i0}^* \mathbf{w}_i \mid^2}{\sum\limits_{\substack{p=1 \\ p \neq i}}^{N_{SU}} \mid \mathbf{h}_{i0}^* \mathbf{w}_p \mid^2 + \mid \hat{\mathbf{g}}_{i0}^* \mid^2 + \sigma_i^2} \geq \hat{\gamma} \tag{6}$$

Similarly, the interference for the ith PU is given by:

$$\zeta_i = \sum_{j=1}^{N_{SU}} \mid \hat{\mathbf{h}}_{i0}^* \mathbf{w}_j \mid^2 \ \leq \zeta \tag{7}$$

where ζ is the interference limit required by the primary network (or by the PUs).

Since power consumption can be a critical design challenge in the HAPS system, it is very important to limit the maximum transmission power. With each antenna having its own power amplifier, it is a common practice to impose per-antennas power constraint on HAPS, which can be expressed as:

$$\sum_{i=1}^{N_{SU}} \mathbf{w}_i^* \mathbf{Q}_i \mathbf{w}_i \leq q_i \tag{8}$$

where $Q_i \in C^{K_{HAPS} \times K_{HAPS}}$ represents the weighting matrix which defines per-antenna power constraint, q_i represents the per-antenna power limit. The matrix $\mathbf{Q}_i$ is a square matrix with 1 in the diagonal and 0 elsewhere [16].

The total power consumption per subcarrier is given by [17]:

$$\begin{aligned} P_{tot} &= P_{dynamic} + P_{static} \\ P_{tot} &= \left(\rho \sum_{i=1}^{N_{SU}} \|\mathbf{w}_i\|^2\right) + \left(\frac{\eta_0}{C} K_{HAPS}\right) \end{aligned} \tag{9}$$

where ρ is the power amplifier inefficiency, $\|\mathbf{w}_i\|^2$ refers norm-square of $\mathbf{w}_i$, η_0 is the power dissipation that occurs in antenna circuits like filters, mixers, converters and baseband processing and C represents the total number of subcarriers.

Since the objective is to maximize the energy efficiency (EE) of the overall network, the normalized EE has been defined in terms of the bandwidth as [7]:

$$EE = N_{SU}\left(\frac{log_2(1+\hat{\gamma})}{P_{tot}}\right) \tag{10}$$

Hence the final optimization problem with all the constraints would be:

$$\begin{aligned} \max_{\forall \mathbf{w}}.\quad & EE = \frac{N_{SU} \cdot log_2(1+\hat{\gamma})}{\rho \sum_{i=1}^{N_{SU}} \|\mathbf{w}_i\|^2 + P_{static}} \\ \text{subject to:}\quad & log_2(1+SINR_i) \geq \gamma \quad \forall i \\ & \sum_{i=1}^{N_{SU}} \mid \hat{\mathbf{h}}_{j0}^* \mathbf{w}_i \mid^2 \ \leq \zeta \quad \forall j \\ & \sum_{i=1}^{N_{SU}} \mathbf{w}_i^* \mathbf{Q}_i \mathbf{w}_i \leq q_i \end{aligned} \tag{11}$$

The optimization problem given in (11) is a non-convex problem and is NP-hard in general. However, there are approaches which can be employed to solves such problem and one such approach is using semidefinite relaxation. Defining the matrix $\mathbf{W}_i = \mathbf{w}_i\mathbf{w}_i^*$ and considering channel to have a random fading with known second-order statistics, it is possible to transform (11) into a form, given in (12) which is convex and easier to solve.

$$\begin{aligned} &\underset{\forall \mathbf{w}}{\text{minimize}} \quad \psi \\ \text{s.t.:} \quad &\psi N_{SU} \cdot log_2(1+\hat{\gamma}) \geq \rho \sum_{i=1}^{N_{SU}} Tr(\mathbf{W}_i) + P_{static} \\ &\mathbf{h}_{i0}^* \left(\left(\frac{1}{\hat{\gamma}+1} \right) \mathbf{W}_i - \sum_{p=1}^{N_{SU}} \mathbf{W}_p \right) \mathbf{h}_{i0} \geq | \, \hat{\mathbf{g}}_{i0} \, |^2 + \sigma^2 \quad \forall i \\ &\zeta_i = \sum_{j=1}^{N_{SU}} | \, \hat{\mathbf{h}}_{i0}^* \mathbf{w}_j \, |^2 \leq \zeta \\ &\sum_{i=1}^{N_{SU}} \mathbf{w}_i^* \mathbf{Q}_i \mathbf{w}_i \leq q_i \\ &\mathbf{W}_i \succeq 0 \\ &\mathbf{W}_i = \mathbf{W}_i^* \end{aligned} \tag{12}$$

where, $||\mathbf{w}_i||^2 = Tr(\mathbf{W}_i)$, Tr(.) is a trace of a matrix (.), matrix $\mathbf{W}_i$ is a correlation matrix which has to be Hermitian and positive semi-definite which are ensured by the last two constraints in (12) and ψ is the new objective function which is given as:

$$\psi = \frac{P_{tot}}{log_2(1+\hat{\gamma})} \tag{13}$$

4 Simulation Results

All simulations were performed using MATLAB R2016a on Dell PowerEdge C6320 server with two 2.4 GHz Intel Xeon E5-2680 v4 fourteen-core processors on Linux distro. The results were obtained after averaging the instances of data obtained from 50 iterations. To solve the optimization problem given in (12), CVX (a package for specifying and solving convex programs) has been used [18].

The simulation results have been obtained for the scenario developed in Fig. 1. SUs and PUs are randomly distributed within their respective coverage zone. The coverage area for the PBS, the number of PUs within PBS coverage range, SU density within the HAPS coverage area and the number of SUs within the coverage are of PBS are shown in Table 1. Table 2 shows the various channel and system parameters used in the simulation.

For the channel, L_s has been modeled as a normal distributed random variable while ζ_{LOS} and ζ_{NLOS} follow log-normal distribution with zero mean. It is assumed that the different data signals s_i are uncorrelated and have normalized power i.e. $E[|s_i|^2] = 1$. Noise is modeled as a circularly symmetric complex Gaussian with zero mean and variance σ^2.

Figure 2 shows the EE curve versus HAPS transmission power for various SNR levels at SUs and interference constraints at PUs. With the increases in

Table 1. Coverage area and user distribution

Description	Values
Number of SUs with HAPS coverage	10–70
PBS coverage radius (m)	150–300
Number of PUs in PBS coverage area	5
Minimal separable distance between PUs and PBS	5 m
Number of SUs within PBS coverage areas	3

Table 2. Channel parameters in the numerical results

Parameters	Values
Number of antennas at HAPS	100
QoS constraint of SU	2–3 bits/sec/Hz
Interference Temperature(IT)	0.1 mW
Carrier frequency	2 GHz
Number of subcarriers	10–80
Total bandwidth	10 MHz
Subcarrier bandwidth	15 kHz
Packet length	200 bits
Data rate	100 Kbps
Modulation	QPSK
Standard deviation of log-normal shadowing (within PBS)	7 dB
Path and penetration loss at distance d (km) (within PBS)	$148.1 + 37.6 \times log_{10}(d)$ dB
Standard deviation for local variability	4 dB (for LOS)
(for HAPS)	10 dB (for NLOS)
Noise variance	–127 dBm
Noise figure	5 dB
Efficiency of power amplifiers	0.3
Static power (Fixed)	15 W (43.9794 dBm)
Maximum total power	50 W (46.02 dBm)

transmission power at HAPS, the energy efficiency of the network decreases sharply up to a point, after which it remains almost constant irrespective of an increase in the transmission power. It is due to the fact at that higher transmission power, interference limit at PUs restricts HAPS from allocating more power to SUs. Hence the curve almost saturate once this limit is reached. The figure also shows that there is no significant variations in EE at various SNR value for the same interference limit. Since EE has been defined as the ratio of SINR and the total power, the net increase of one will be cancelled out by the net decrease in another parameter. Hence the overall effect is almost constant.

Figure 3 shows the EE versus interference level at PUs at various transmission power. When the interference level is high, the EE is high due to low transmission

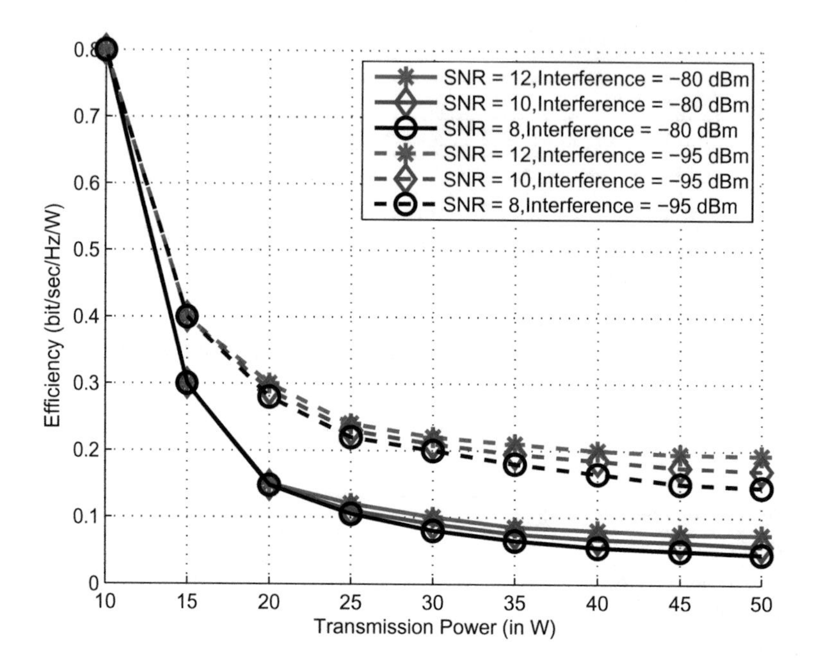

Fig. 2. Energy efficiency versus Transmission power for various Interference levels.

power as shown by red curve as compared to the case with high transmission power (shown by black curve). As the interference limit for SUs decreases, the EE saturates at a point. This saturation point varies for different transmission power. Higher the transmission power, lower will be the saturation point and lower the transmission power, higher will be this point.

In Fig. 4, the number of SUs in the system is varied and the EE of the HAPS system is evaluated at a transmission power of 30 W and SNR value of 10 dB. The results show that the EE increases with the increase in the number of SUs and this increase is pronounced at more strict interference level. This is due to the fact that as the number of SUs increases in the system, the channel occupancy increases for the given transmission power, which is further restricted by the interference limit. Hence a high interference limits the transmission power and the network becomes more energy efficient.

In Figs. 5 and 6, the energy efficiency is evaluated for variable number of channels available to serves the SUs. In Fig. 5, EE is evaluated at an interference limit of −95 dBm while in Fig. 6, EE is evaluated at an interference limit of −80 dBm. When the available number of channels are less than the number of SUs in the system, the EE increases more due to more channel utilization. This difference is more visible with more strict interference limit.

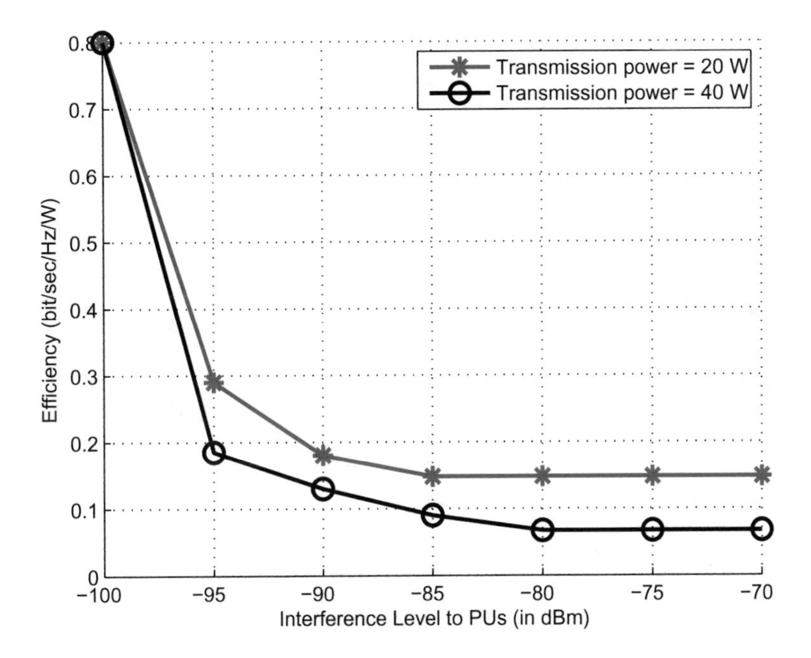

Fig. 3. Energy efficiency versus Interference level at various Transmission powers.

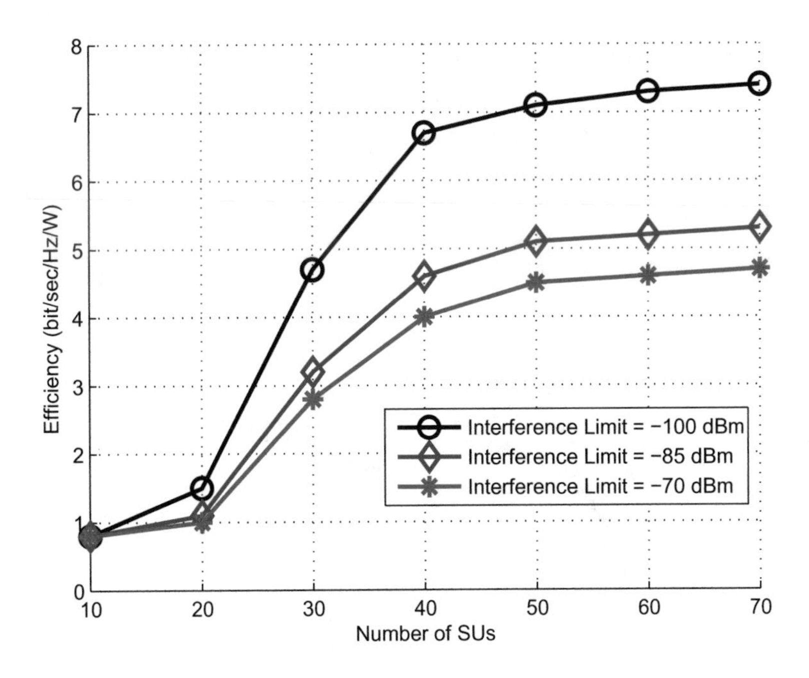

Fig. 4. Energy efficiency versus Number of SUs for various Interference levels. Transmission power = 30 W and SNR = 10 dB.

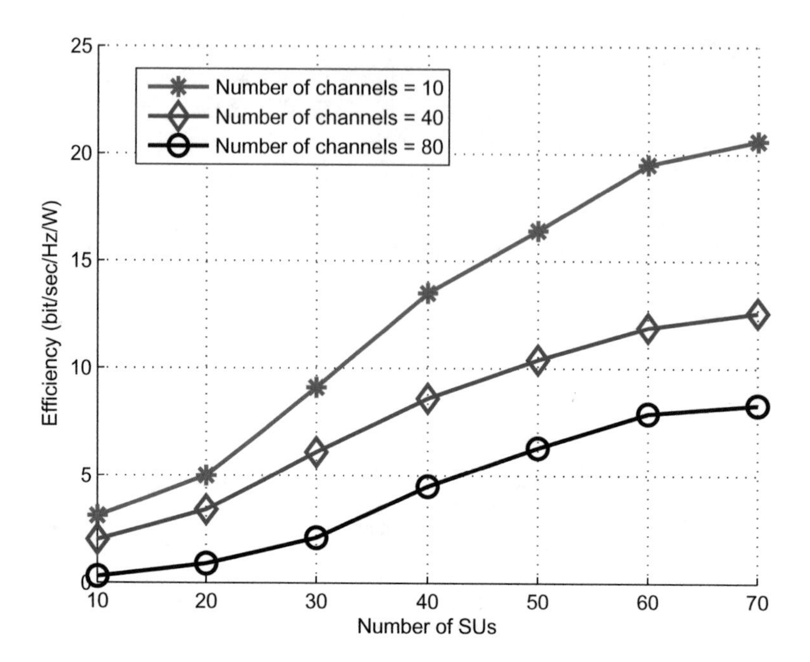

Fig. 5. Energy efficiency versus Number of SUs for different numbers of channels. Interference level = −95 dBm; Power = 30 W and SNR = 10 dB.

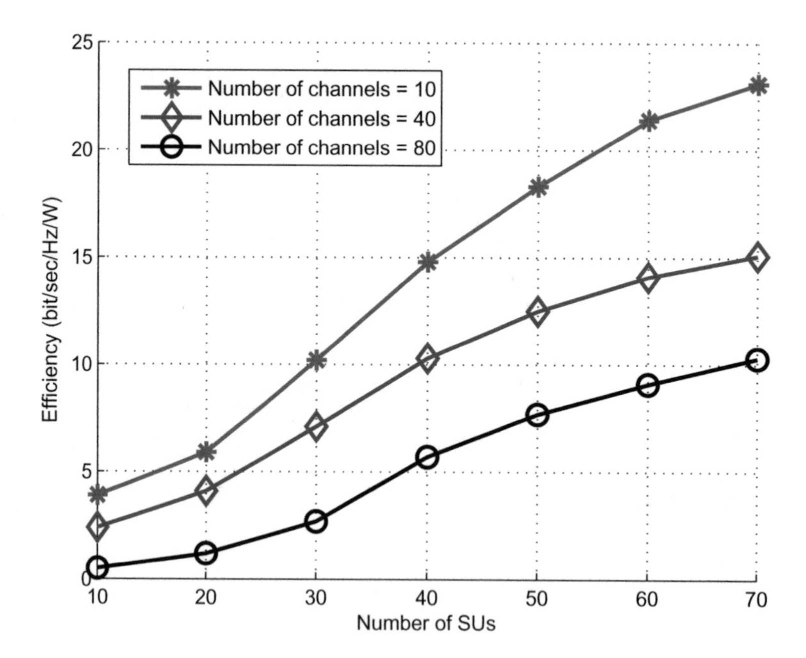

Fig. 6. Energy efficiency versus Number of SUs for different numbers of channels. Interference level = −80 dBm; Power = 30 W and SNR = 10 dB.

5 Conclusion and Future Work

An energy efficient HAPS based cognitive radio network has been presented in this paper. The results showed that the EE is better at low transmit power, at strict interference level and when the number of SUs increase. The results proved that HAPS are better for CR type environment, as it has larger coverage area and hence can improve the efficiency of the network, utilizing more channels. When the number of channels are less than the number of users, it may not be possible to simultaneously serve all SUs. This can results in inherent delays. Similarly, the coverage area of the HAPS is very large and in real scenario it will be covering multiple cells. The network performance of the cells which are nearer to the HAPS (i.e. elevation angle is higher) will be significantly different from the cell which are farther away from the HAPS. These key aspects will be considered in a future work where a more practical multi-cell, heterogeneous network and service type scenarios will be implemented, with an objective to improve the energy efficiency of the overall HAPS system.

References

1. Mitola, J., Maguire, G.Q.: Cognitive radio: making software radios more personal. IEEE Pers. Commun. **6**(4), 13–18 (1999)
2. Hoydis, J., Brink, S.T., Debbah, M.: Massive MIMO in the UL/DL of cellular networks: how many antennas do we need? IEEE J. Sel. Areas Commun. **31**(2), 160–171 (2013)
3. Bengtsson, M., Ottersten, B.: Optimal and Suboptimal Transmit Beamforming. CRC Press, Boca Raton (2001)
4. Djuknic, G.M., Freidenfelds, J., Okunev, Y.: Establishing wireless communications services via high-altitude aeronautical platforms: a concept whose time has come? IEEE Commun. Mag. **35**(9), 128–135 (1997)
5. Rajagopal, S., Roberts, R.D., Lim, S.K.: IEEE 802.15.7 visible light communication: modulation schemes and dimming support. IEEE Commun. Mag. **50**(3), 72–82 (2012)
6. Report of the spectrum efficiency working group. Spectrum Policy Task Force, November 2002
7. Barnes, S.D., Joshi, S., Maharaj, B.T., Alfa, A.S.: Massive MIMO and Femto Cells for Energy Efficient Cognitive Radio Networks, pp. 511–522. Springer, Cham (2015)
8. Islam, M.H., Liang, Y.C., Hoang, A.T.: Joint beamforming and power control in the downlink of cognitive radio networks. In: Wireless Communications and Networking Conference, vol. 9, no. 11, pp. 21–26. IEEE, Kowloon (2007)
9. Saker, L., Elayoubi, S.E., Chahed, T., Gati, A.: Energy efficiency and capacity of heterogeneous network deployment in LTE advanced. In: European Wireless Conference, Poznan, no. 11, pp. 1–7 (2012)
10. Zhang, J., Zeng, Y., Zhang, R.: Spectrum and energy efficiency maximization in UAV-enabled mobile relaying. In: IEEE International Conference on Communications (ICC), pp. 1–6, May 2017
11. Yan, Z., Zhou, W., Chen, S., Liu, H.: Modeling and analysis of two-tier hetnets with cognitive small cells. IEEE Access **5**, 2904–2912 (2017)

12. Cui, M., Hu, B.J., Li, X., Chen, H., Hu, S., Wang, Y.: Energy-efficient power control algorithms in massive mimo cognitive radio networks. IEEE Access **5**, 1164–1177 (2017)
13. Aragón-Zavala, A., Cuevas-Ruíz, J.L., Delgado-Penín, J.A.: High-Altitude Platforms for Wireless Communications. Wiley (2008)
14. Feasibility of high altitude platform station (HAPS) in the fixed and mobile services in the frequency bands above 3 GHz allocated exclusively for terrestrial radiocommunication. International Telecommunication Union (ITU), 2000, iTU-R Resolution 734
15. Holis, J., Pechac, P.: Elevation dependent shadowing model for mobile communications via high altitude platforms in built-up areas. IEEE Trans. Antennas Propag. **56**(4), 1078–1084 (2008)
16. Wang, F., Wang, X., Zhu, Y.: Transmit beamforming for multiuser downlink with per-antenna power constraints. In: IEEE International Conference on Communications (ICC), pp. 4692–4697, June 2014
17. Bjornson, E., Kountouris, M., Debbah, M.: Massive MIMO and small cells: improving energy efficiency by optimal soft-cell coordination. In: 20th International Conference on Telecommunications, Casablanca, no. 11, pp. 1–5, May 2013
18. Grant, M., Boyd, S.: CVX: Matlab software for disciplined convex programming, version 2.1, March 2014. http://cvxr.com/cvx

Evaluation of Parameters Affecting the Performance of Real Time Streaming on Real Time Communication Library in Named Data Networking

Yudi A. Phanama(✉), F. Astha Ekadiyanto, and Riri Fitri Sari

Department of Electrical Engineering, Universitas Indonesia,
Kampus Baru UI Depok, Depok 16424, Indonesia
{yudi.andrean,astha.ekadiyanto,riri}@ui.ac.id

Abstract. Named Data Networking (NDN) shifts the current Internet protocol's networking paradigm from addresses of machines or 'where' to get data into the needed data or 'what' data to get directly through naming data packets. NDN architecture has been proven with various applications, including multi-party real-time video-conferencing, which had been implemented in NDN-RTC library (Real Time Communication library in NDN). This work presents the analysis and evaluation of parameters that have impact on NDN-RTC's mechanism in fetching real-time streaming data using Interest queuing. Assessments on NDN-RTC are also performed through several runs on different network conditions, tuning NDN-RTC's multiplier factor in the Interest Expression Control Module, and running NDN-RTC on multiple clients simultaneously with different network conditions. The results shows that there is no linear relationship between the increases of network delay in influencing the quality of playout on application layer. NDN-RTC also suffers from the early uncertainty of application-level round-trip-time (RTT), which is the result of NDN-RTC's implicit RTT averaging, for around the first 3 s. A more explicit estimation mechanism is suggested for future works. On varying the multiplier factor, the factor of 0.25 is found to be the best for the network delay of 100 and 200 ms, and 0.25 and 0.75 for the 300 ms network delay. The findings open the discussion of developing a more adaptive strategy. NDN-RTC's implicit fetching mechanism and its lack of synchronization methods also leads to the out-of-sync state of multiple clients, with up to 4 s playout difference in this work. Further study and development can be done for NDN-RTC's strategy to control *Interest* expression, and for a new inter-consumer playback synchronization method.

Keywords: Named Data Networking · Information centric networking
Real Time Communication

1 Introduction

Internet has grown rapidly over time, moving from a total global traffic of 100 GB per day in 1992, 100 GB per second in 2002, and 20.235 GB per second in 2015, forecasting to 61.386 GB per second for 2020, which means a total IP traffic of 2.3

K. Arai et al. (Eds.): FICC 2018, AISC 886, pp. 203–220, 2019.
https://doi.org/10.1007/978-3-030-03402-3_15

zettabytes in that year. Video content streaming over internet has gained a massive popularity among users, contributing 82% of all IP traffic by 2020 [1]. This leads to a challenge for the current host-oriented IP architecture to provide stable and reliable network for video communication. IP possesses a prone design to provide efficient network communication for the future internet. While improving IP connection capacity is an option, it requires the addition of hardware infrastructure, demanding increased installation and running costs.

Content Centric Networking [2] or Named Data Networking (NDN) [3] emerged as a potential solution for IP's host-oriented architecture. NDN moves the network pattern of host-centric communication into data-centric communication, shifting the job of the network to connect and find hosts, into fetching data directly from the network with its own forwarding mechanism [4]. NDN names every data packet using certain naming scheme, letting consumers send Interest packets to fetch data packets using the name of the Data. NDN also caches the data in every network device to reduce redundant data delivery through links that had been used to transport the same data before. This offers a potential solution to provide future effective and reliable network.

NDN has been proven through the implementation of various application designs, such as messaging service [5], VANET [6], mHealth [7], Internet of Things [8], video streaming [9], and the recent real-time conferencing implementation, NDN-RTC [10]. The works has proved not only that NDN is capable to run as a basis of future application of services, but also has the potential for effective network delivery through its network. Howbeit, the current implementation of real-time video streaming (NDN-RTC) shows an inferior performance compared to the current implementation in IP, and possesses several left-out problems, such as inter-consumer synchronization, which will be discussed later in Sect. 2.

The aim of this work is to evaluate NDN-RTC's mechanism in fetching real-time streaming data through assessments with varying network parameters, tuning NDN-RTC's bursting and withholding multiplier factor in the Interest Expression Control module, and running NDN-RTC instance on multiple clients simultaneously with different network conditions.

The rest of the paper is organized as follows: Sect. 2 discusses related works engaging video streaming topics on NDN/CCN. Section 3 describes the evaluation method. Section 4 presents the results and analysis on the evaluation. Section 5 provides conclusions of the paper.

2 Backgrounds and Related Works

Several related work, such as Audio Conferencing Tool (ACT) over NDN [11] presents a design of a distributed ACT service demonstrating NDN's design flexibility in delivering real-time data for conference and shows its capability of user discovery. However, some technical issues regarding user experience shows the occurrence of bad quality of echo cancellation, due to the chosen library it was built on.

Video streaming implementations, which includes pre-recorded, YouTube-like streaming and live streaming, such as a recent work in [12] explores the implementation of MPEG-DASH (Dynamic Adaptive Streaming over HTTP) on NDN to achieve dynamic adaptive video streaming on NDN, and adopted an adaptation algorithm to determine the best bitrate for the consumer. The work does not explain the aspect of scalability through the experiment. It discussed the pre-recorded video stream instead of live video streaming. NLB (NDN Live Video Broadcasting) [13] presents a cross-layer mechanism for video broadcasting over WLAN connection. They used WLAN's built-in mechanism to deliver same data packet to multiple clients simultaneously. Their implementation shows a promising scalability aspect over WLAN. However, the design is constrained into using specific kind of medium, creating an application limited to the medium. An Interest Aggregation approach in [14] discussed the energy consumption and network performance in NDN. They managed to improve the network performance and energy consumption on embedded devices compared to legacy NDN. However, the work covers transport layer mechanism; which is out of the scope of work and objective discussed in this paper; to build user-driven application running independently from the constraint of another layer.

Work in [15] demonstrates the ability of NDN to deliver live media streaming using somewhat direct mechanism to split video data into chunks. Their results show a better performance compared to HTTP streaming, but lack of the discussion on multi-party conferencing usage. The iSER (information centric network-based Service Edge Routers) presented in [16] delivers a transport layer solution for video conferencing service. Despite showing the ability to serve video conferencing, the design shifts the work of conference service back to the network, which requires the router to be the service anchor for the service to be able to run. This leads the service into consumer-router relationship, making the consumer depends on certain type of routers, and adding implementation's complexity to the existing NDN architecture. The most recent works are NRTS (Content Name-based Real-time Streaming) [17] and NMRTS (Content Name-based Mobile Real-time Streaming) [18], which use symbolic interests to enable per-content streaming, working on the stream level of the content. This leads to a rate control scheme, avoiding Interest bursts coming from pipeline-based implementations. The symbolic interest extends the use of Interest Lifetime, changing how routers act towards the component. Such mechanism shifts the application work into the network, leaving consumers with less control over their interest rate, giving a kind of pseudo-congestion control driven by routers. This also differs from the objective of this paper which is working on receiver-driven application of real-time conferencing.

A recent work from NDN Project [19] delivers an implementation library, NDN-RTC [20]. They continued the prior work on [9, 21], targeting to achieve low-delay audio-video conference over NDN capable of delivering realtime data, producing a set of library for realtime conferencing uses. NDN-RTC is built on top of WebRTC library to use the audio-video processing capabilities and then porting it into NDN architecture usage. They presented an implementation capable of hosting multiparty realtime conferencing tested on NDN Testbed network across several nodes. They show that NDN can deliver scalable RTC usage leveraging NDN's intrinsic multicast to many users. However, the lack of study and evaluation against the working mechanism left some open questions on its performance in high delay and jittery network conditions, tuning multiplier factor in the library, and its performance in multiple users with different network conditions. This motivates our study to evaluate NDN-RTC.

2.1 Real Time Communication Library

As a realtime streaming library, NDN-RTC aims to fetch realtime data from producers of data stream in the network. NDN-RTC uses an Interest Expression Control mechanism to yield continuous stream of data. A consumer issues multiple concurrent Interests as in-flight Interests, Interests which was sent but not answered. The library leverages Interest Pipeline to do this. The pipeline is characterized by the size, λp, which represents the number of Interests that the pipeline must issue to keep the size λp of in-flight Interests. The consumer controls λp by adjusting it with Interest bursting and withholding, as shown in Fig. 1.

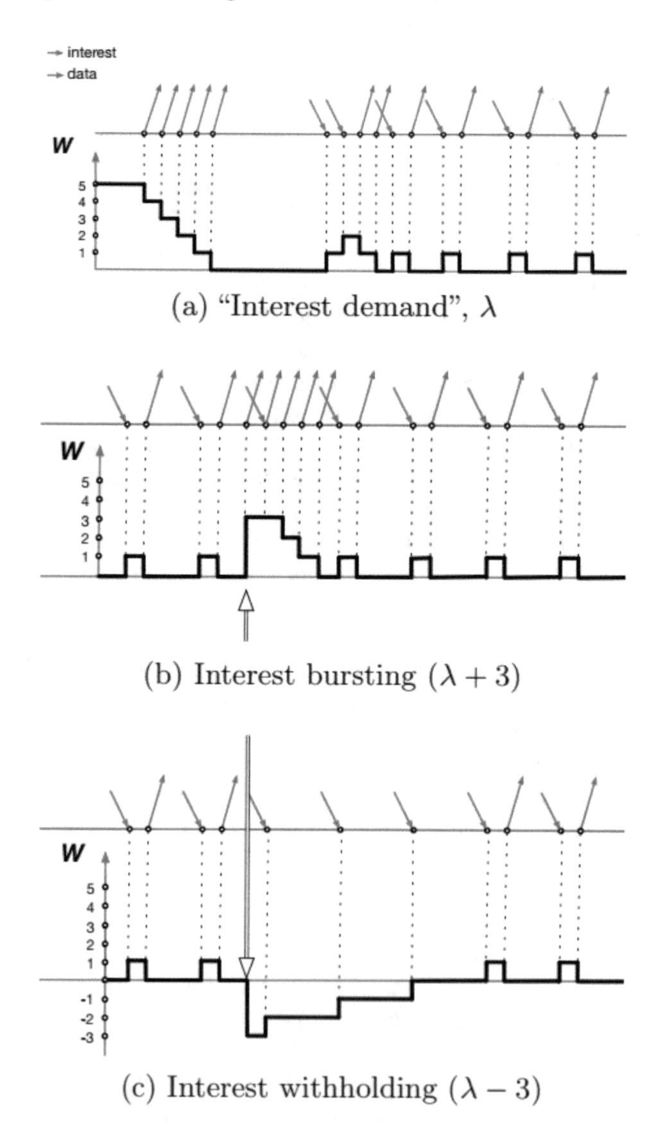

Fig. 1. Interest expression patterns [10].

RTT in NDN is called Data Retrieval Delay (DRD) for generalization purpose, which is the consumer-observed time between an Interest expression and the responding Data packet receiving. To this point, the DRD is a coarse measurement of RTT from consumer's point of view. On the producer's side, the delay between Interest's arrival and the availability of the responding data is called data generation delay, dgen, which adds to the real network DRD to form the DRD measured by consumer, effective DRD:

$$\mathrm{DRD}' = \mathrm{DRD} + \mathrm{dgen} \tag{1}$$

Consumers fetch sequential data names from the producers, thus enabling the use of producer's sample rate and using NDN's packet selectors at bootstrapping to know the latest sequence number. It will then issue Interests based on the information afterwards, faster than the producer's sampling rate. When the data arrive at the same rate as the producer's sampling rate, then the consumer will know that the data is realtime data. NDN-RTC assumes realtime data as packets arrive no faster than the producer's sampling rate, and old data as the data arriving at quicker rate. The rate of data arrival is defined as Inter-arrival delay, or Darr, used to monitor the freshness of received data.

Consumers perform running average on the effective DRDs measured throughout the run of a session, DRDest = DRD′ - dgen, to determine interest demand λd, the minimum number of in-flight interests required to keep received data real-time. The consumer will then adjust the pipeline size, λp, to match the interest demand, using bursting and withholding technique previously discussed. The interest demand is defined as:

$$\lambda_d = \left\lceil \frac{DRD_{est}}{T} \right\rceil \tag{2}$$

Where, T is the sampling rate provided by the producers in the data packets.

During bootstrapping, consumers aim to fetch the latest data and chase the producer's latest data production. To do so, consumers will initialize λp using the value of λd and then adjusts it accordingly after waiting for some period to detect the impact of pipeline adjusting to the network. The consumer adjusts the interest pipeline with these equations:

$$\text{Interest bursting}: \ \lambda_p^n = \lambda_p^{n-1} + \lambda_p^{n-1} \times K \tag{3}$$

$$\text{Interest withholding}: \lambda_p^n = (\lambda_p^{n-1} - \lambda d) \times K \tag{4}$$

Where, K is a constant multiplier factor to adjust λ_p over time, defaulting to 0.5. Until the period finished, no more interest pipeline adjustment will be done.

3 Evaluation Setup

Four scenarios with different setups are created to evaluate NDN-RTC, using a headless client as the implementation of NDN-RTC library. The clients are implemented in a Docker container [22], running on the same host machine. NetEm [23], a network emulator capable of creating different network parameters on an interface was used to emulate network conditions such as delay and jitter. The scenarios are described in the following section. For each scenario, playout frame number as a metric to show the quality of the user's playout over time, interarrival delay, DRD estimation, jitter buffer size, and the lambda value of the corresponding consumer are collected. In Scenario IV only the playout frame number was collected. The general parameters for each scenario is shown in Table 1.

Table 1. Generic parameters for each scenario

Parameter	Value
Video resolution	720 p
Jitter buffer size	150 ms
Frame rate	30 fps
Video codec	VP8
Audio codec	G722
Interest lifetime	2000 ms

3.1 Scenario I

The purpose of the scenario is to evaluate the effect of different network delays to NDN-RTC consumers. The scenario contains a basic topology with 1 consumer, 1 producer, and 2 routers. The second router is given varying value of 50, 100, 150, 200, 250, and 300 ms network delays for each run, without jitter. The topology for this scenario is show in Fig. 2.

3.2 Scenario II

This scenario aims to evaluate the effect of network jitter to NDN-RTC consumers. This scenario uses the same topology as scenario I, with a router being given varying jitter on each run, with the same network delay. This scenario runs with varying network jitter value of 10, 20, 30, 40, 50, 100, and 150 ms. The run with 150 ms of network jitter will be given 200 ms network delay, while the other runs with 100 ms network delay. The topology for this scenario is shown in Fig. 2.

3.3 Scenario III

Scenario III explores the tuning of NDN-RTC's Interest Expression Control strategy by changing the bursting and withholding multiplier factor (default value 0.5). The purpose is to evaluate the aggressiveness of bursting and withholding of Interests to the performance of NDN-RTC in different network conditions. The topology used is a

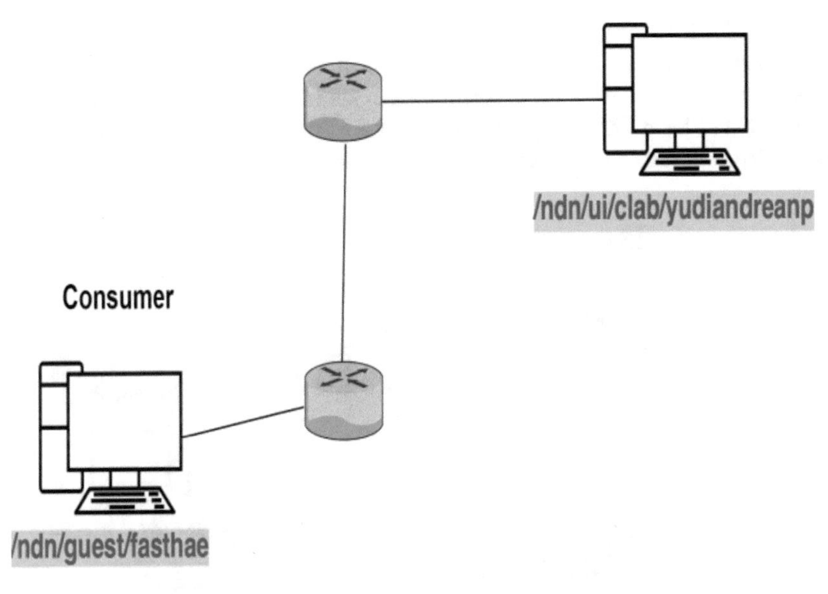

Fig. 2. Network topology for Scenario I and II.

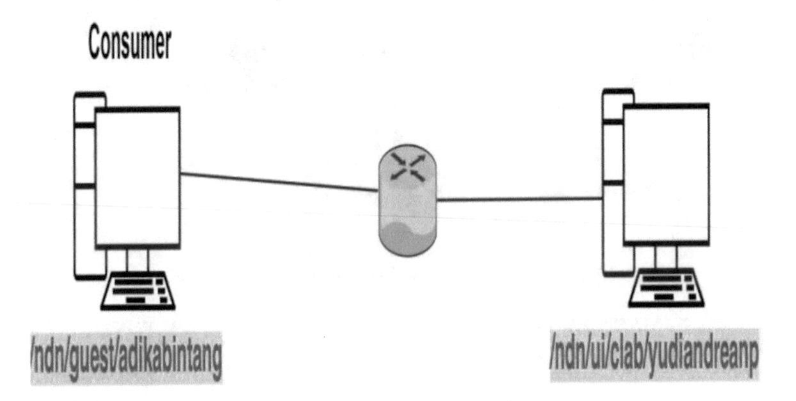

Fig. 3. Topology for Scenario III.

basic 1 consumer, 1 producer, and a router without any jitter. We will tune the multiplier factor with 0.25, 0.75, and 1.00, along with the default value 0.5. The topology for the scenario is shown in Fig. 3.

3.4 Scenario IV

In this scenario, the scalability of NDN-RTC was evaluated. The aim is to run NDN-RTC clients with many consumers consuming one producer in NDN network. The consumers are given different network delays on their interfaces with the value of 10,

20, 30, 40, 50, 100, 150, 200, 250, 300 ms for client #1-10 consecutively. One thing to note is that NDN-RTC provides no inter-consumer playback synchronization. This scenario evaluates NDN-RTC's performance under conditions in which the users potentially will have different playback because of different network conditions on all clients. The topology for the scenario is shown in Fig. 4.

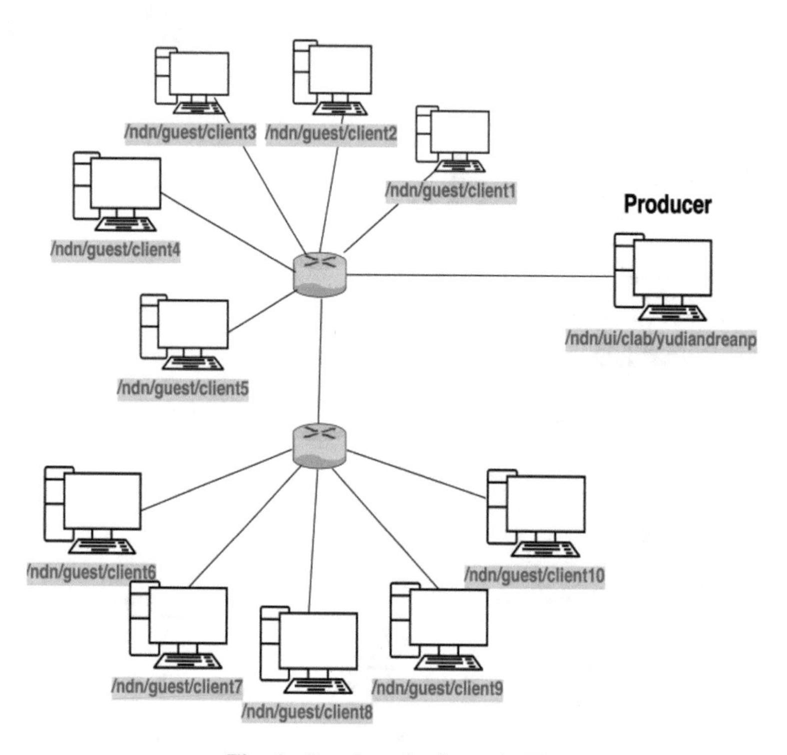

Fig. 4. Topology for Scenario IV.

4 Results and Analysis

This section presents the test results and analysis, and evaluates NDN-RTC's performance based on our setup. The overall quality of a consumer is represented using a playout graph. A smooth playout graph represents good playout quality, and a jagged playout graph represents bad playout quality.

4.1 Scenario I

In this scenario, the latency value for a consumer consuming the stream of a producer is adjusted to different values. Figure 5 shows the Consumer's playout with 50 ms and 100 ms latency. The playout differs, as the 100 ms run suffers a jagged playout at the beginning of the run. This is an anomaly for the 100 ms run, as the consumer cannot play the latest frame as the impact of the consumer unable to catch up with the latest data. This is shown by the jitter buffer, in Fig. 6. The jitter buffer's content fluctuates until the 37th second of the run. The prime suspect for this is anomaly the NDN-RTC's Interest Expression Control mechanism, as the Interest pipeline control is unable to adjust the right level of Interests needed to be sent for the network latency condition.

The Interest sending rate is shown in Fig. 7. The figure shows that the Interest Pipeline size keeps the level at 4 until the 34th second, which affects the Interest sending rate in fetching data, ultimately getting the latest data to catch up with the playout, which started to play frames at the 38th second.

There scenario also reveales that there is no linear relation between latency and the overall playout quality, as shown in Fig. 8. The consumer with 200 ms latency suffers a jagged playout, compared to a smooth playout of the consumer with 300 ms latency. This shows that latency gives a significant impact on the performance, with no linear relation of latency-quality. The Interest Expression Control's decision depends on the DRD, which changed with network latency. It is also shown that the Interest rate needs to change according to the DRD, and the right level of pipeline size is needed for different latencies. The current NDN-RTC's Interest control is too rigid to adapt with network latency, as previously shown in Figs. 5, 6 and 7, on the case with 100 ms latency.

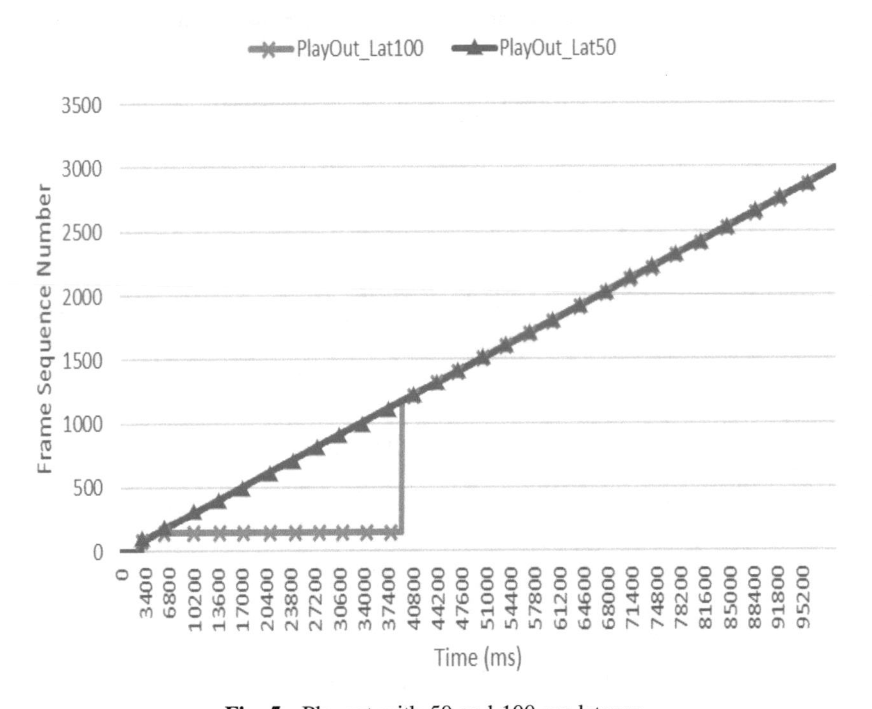

Fig. 5. Playout with 50 and 100 ms latency.

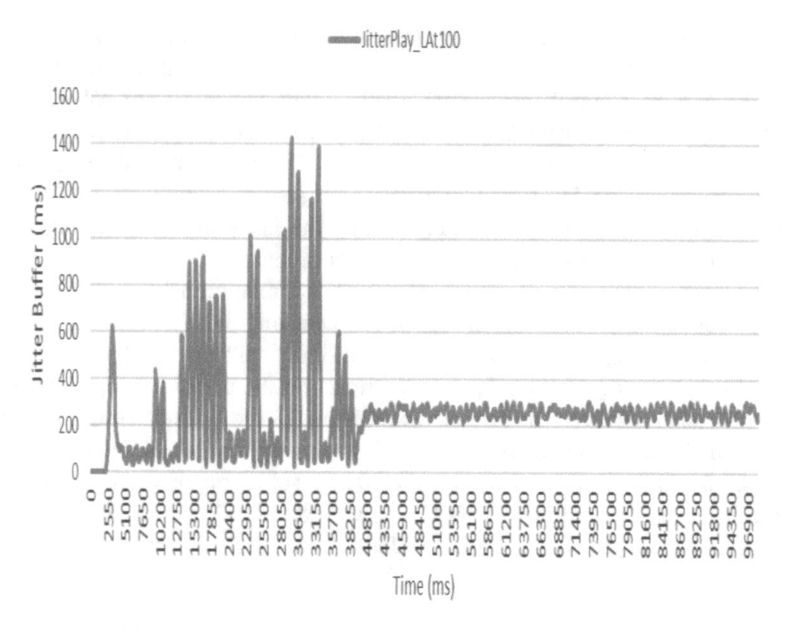

Fig. 6. Jitter Buffer content with 100 ms network latency.

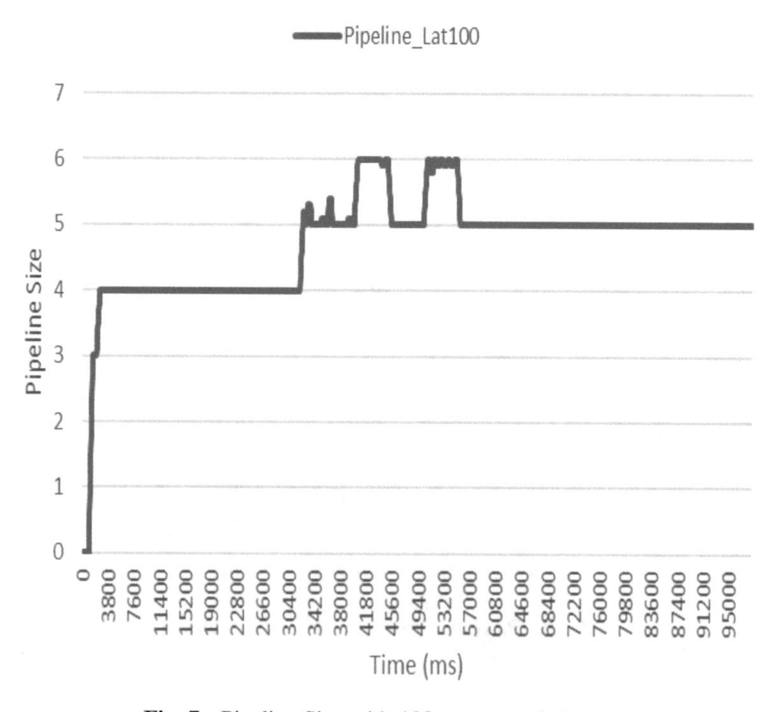

Fig. 7. Pipeline Size with 100 ms network latency.

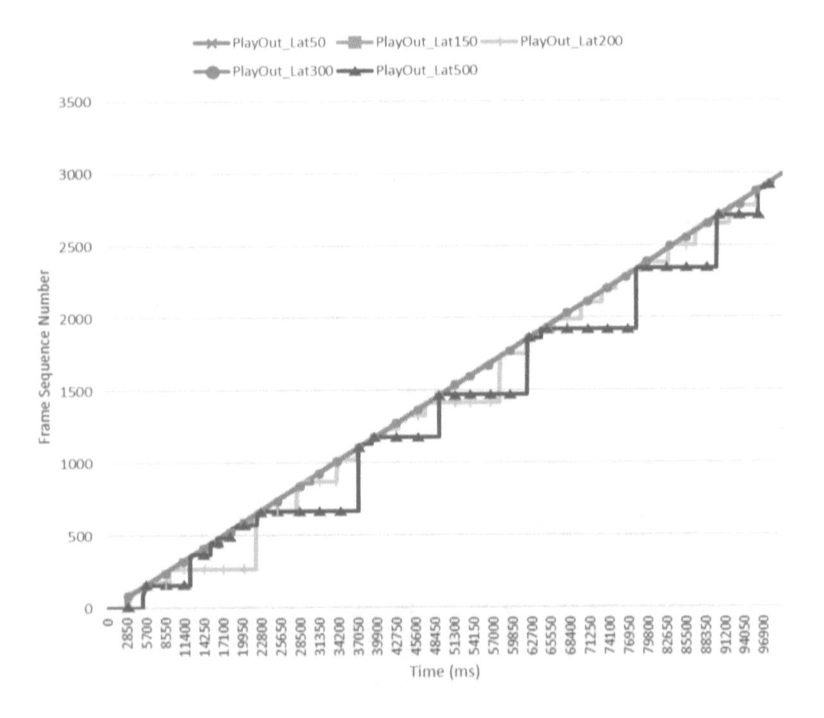

Fig. 8. Playout at 50, 150, 200, 300, and 500 ms latency.

4.2 Scenario II

In this scenario the network buffer was adjusted to introduce fluctuating latency within the network, in order to observe the impact to the overall performance.

Figure 9 shows the consumer's playout with no jitter and 10 ms jitter, at a base latency of 100 ms. That the anomaly in the previous discussion does not happen when the latency fluctuates, compared to the stable latency of 100 ms. The playout still suffers a jagged frame playout, but still smoother than the 0 jitter counterpart. This shows that the Interest Expression Control needs to adapt to the network condition. Evaluation in Scenario III will show how the Interest Expression Control handles this condition by changing the bursting-withholding multiplier factor.

Figure 10 shows the playout with 20, 30, 40, 50, 100, and 150 ms jitter with 100 ms base latency, and 200 ms base latency for the 150 ms jitter. The run shows that jitter does not affect the performance significantly. The playouts with 100 ms base latency and varying jiter run relatively smooth, compared to the one with 200 ms base latency and 150 ms jitter, which suffers a jagged playout. Changing the base latency impacts the performance, while the changing of jitter does not significantly give impact to the performance. Moreover, in Fig. 11, NDN-RTC has a somewhat slow DRD estimation averaging in a jittery network, with the DRD starting to stabilize on at least the 31st second, on the run with 20 ms jitter. This might impact the overall performance, as NDN-RTC relies on DRD estimation value to adjust its pipeline size. A more

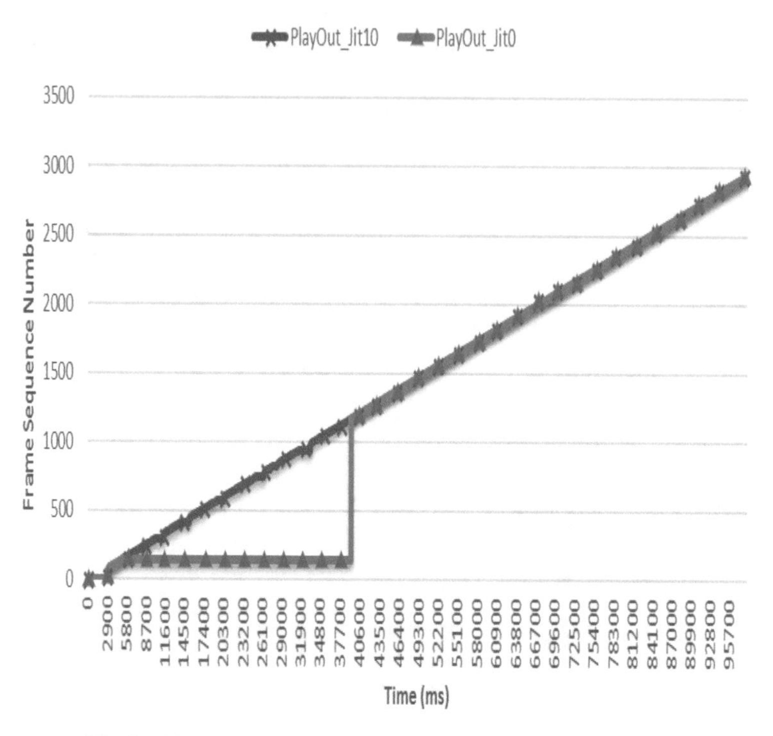

Fig. 9. Playout with 0 and 10 ms jitter at 100 ms base latency.

explicit approach such as direct synchronization with the producer can be used on future development to mitigate the problem.

4.3 Scenario III

In this scenario, the performance evaluation result on changing the value of Interest Expression Control's multiplier factor is evaluated. The values are set to 0.25, 0.5, 0.75, and 1.00, consecutively and focusing it on three latency conditions: 100, 200, and 300 ms without network jitter. Figure 12 shows the result for 100 ms network latency. The figure shows that applying 0.25 multiplier factor mitigates the problem at 100 ms network latency. The smooth playout can be compared with all the other multiplier factors. Factor 1.00 surprisingly shows no playout, because it's too aggressive bursting and withholding.

Figure 13 shows the results at 200 ms network latency. Multiplier factor of 0.25 yields smoother playout compared to default, but still suffer from a jagged playout. All other factors also produced the same result. Further detailed fine-grained study on the multiplier factor between 0–1.00 needs to be done to find the best multiplier factor for the network condition.

Figure 14 shows the pipeline changes on different multiplier factors. The larger the factor, the more aggressive is the bursting and withholding, jumping with larger Interest gap.

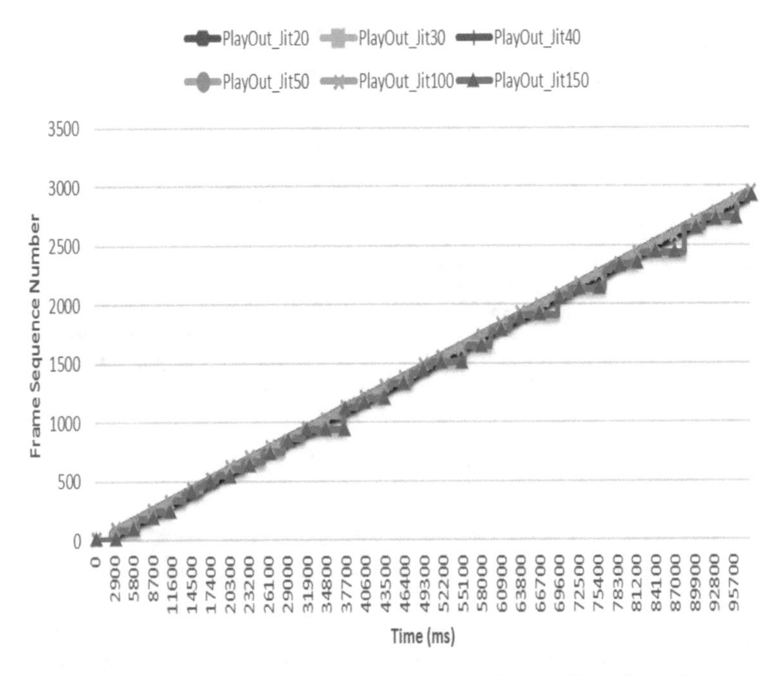

Fig. 10. Playout with 20, 30, 30, 50, 100, and 150 ms jitter with 100 ms base latency and 200 ms latency for 150 ms jitter.

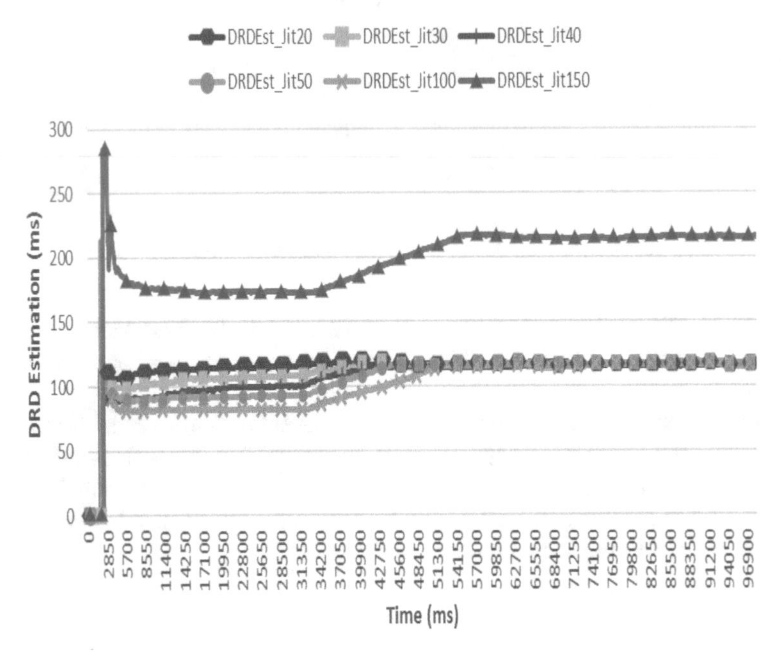

Fig. 11. DRD Estimation at 20, 30, 30, 50, 100, and 150 ms jitter.

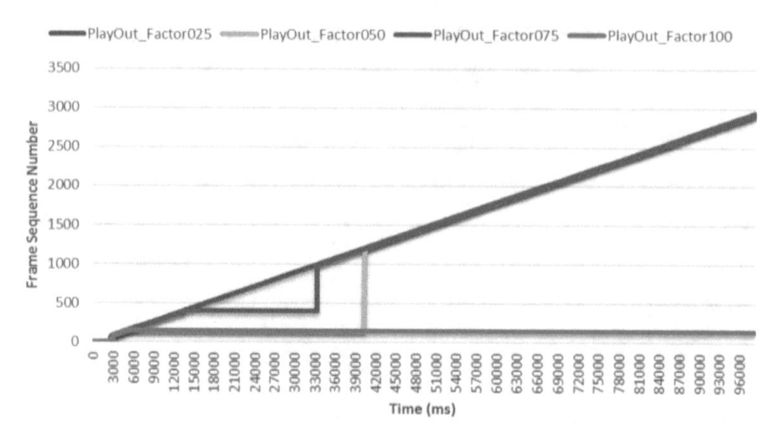

Fig. 12. Playout at 100 ms latency with different multiplier factors.

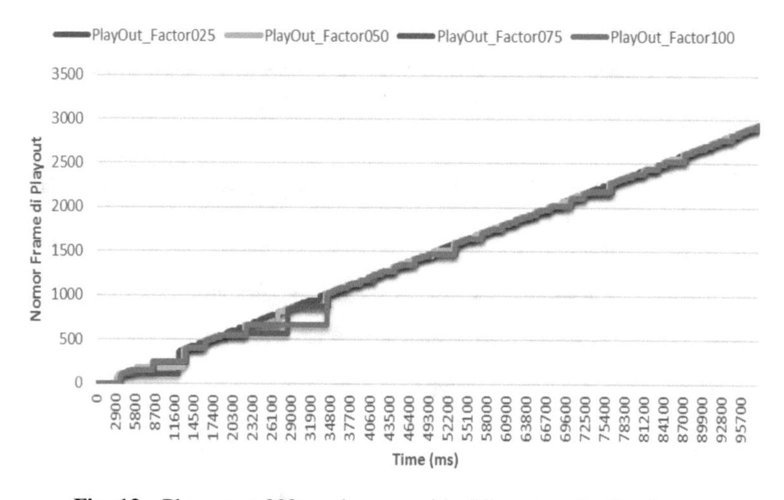

Fig. 13. Playout at 200 ms latency with different multiplier factors.

Figure 15 shows that with 300 ms network latency, the best factor for Interest Expression Control are 0.25 and 0.75. Both values yield to best performance with smooth playout.

Results of the experiment show how certain multiplier factor can be best for some network latencies. A further experiment is recommended on a fine-grained factor between 0–1.00. The knowledge can be used to develop a more adaptive Interest Expression Control strategy based on the known factor for the suitable network conditions, compared to the static approach of NDN-RTC.

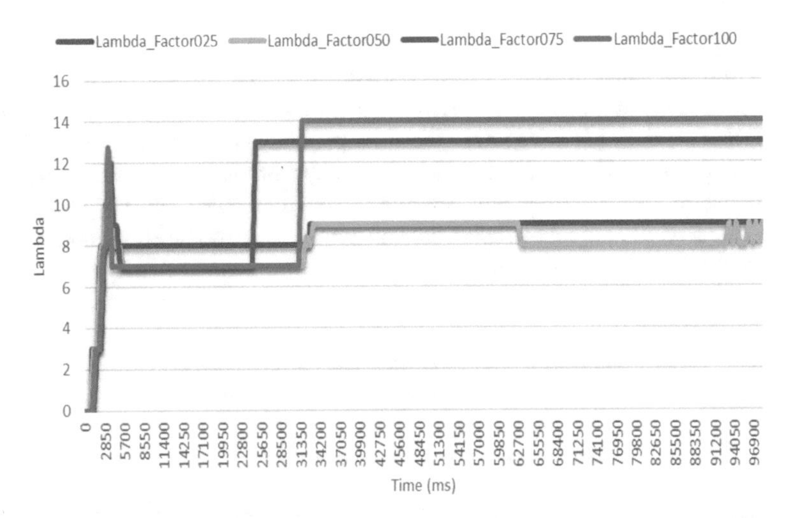

Fig. 14. Pipeline size with different multiplier factors.

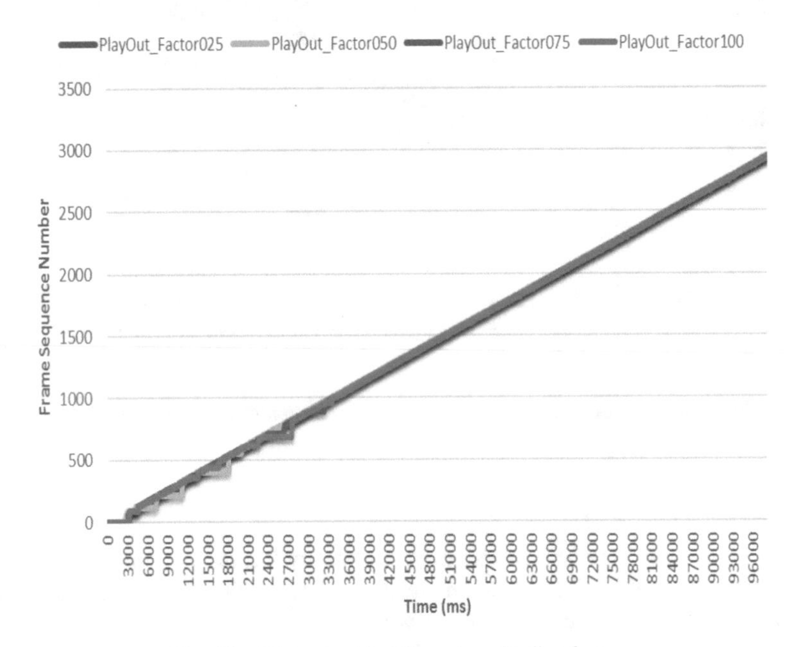

Fig. 15. Playout with different multiplier factors.

4.4 Scenario IV

This scenario runs a scalability test of NDN-RTC, with 10 consumers consuming the stream of 1 producer. Each consumer will have different network latency and jitter. The latency is ranging from 10 ms to 500 ms, and the jitter are from 10 ms to 30 ms. The result for the scenario shown in Fig. 16, exhibits different playouts at some points of time with jagged playout at the clients with higher latencies. This shows that clients can have different playouts on the run of NDN-RTC session, as the cause of different network conditions in the clients, which means inter-consumer playout state is out-of-sync. NDN-RTC does not have inter-consumer playout synchronization mechanism to mitigate the problem. For further development, a synchronization mechanism for real-time application such as streaming real-time data from NDN can be done, by performing direct communication between consumers and producer to maintain synchronization state between consumers.

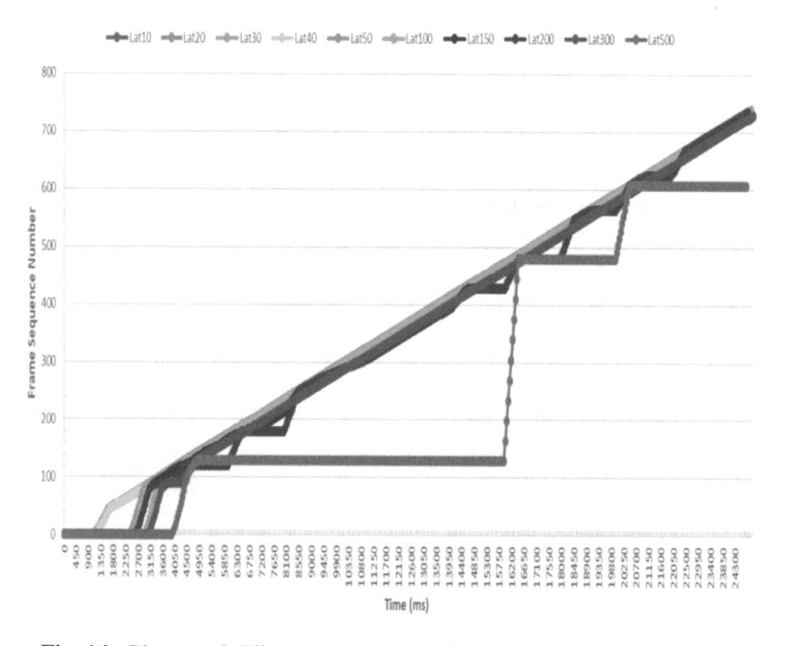

Fig. 16. Playout of different consumers with different network conditions.

5 Conclusion

Experiments' results have shown that network latency has the most significant impact to the performance of realtime streaming with NDN-RTC. The increase of latency does not correspond to the decrease of playout quality, suggesting different Interest Pipelining response to different network latency without linear relation to the quality. This paper also shows that NDN-RTC has a slow DRD averaging estimation, reaching stable state sometime after the beginning of the run, especially on network with high jitter. Tuning NDN-RTC's Interest Expression Control's bursting and withholding mechanism by changing its multiplier factor, yields in better values compared to the default value, which opens the opportunity to implement a more adaptive approach of changing strategies according to network conditions. NDN-RTC also lacks of inter-consumer playout synchronization. Future works on synchronization mechanism to improve NDN-RTC can be done, such as direct coordination between consumers and producer to maintain the sync-state of user playouts on a real-time manner.

References

1. Cisco, Zettabyte Era: Trends and Analysis, Cisco White Paper, July 2016
2. Jacobson, V., Smetters, D.K., Thornton, J.D., Plass, M.F., Briggs, N.H., Braynard, R.L.: Networking named content. In: Proceedings of the 5th International Conference on Emerging Networking Experiments and Technologies, pp. 1–12. ACM (2009)
3. Zhang, L., Afanasyev, A., Burke, J., Jacobson, V., Crowley, P., Papadopoulos, C., Wang, L., Zhang, B.: Named data networking. ACM SIGCOMM Comput. Commun. Rev. **44**(3), 66–73 (2014)
4. Yi, C., Afanasyev, A., Moiseenko, I., Wang, L., Zhang, B., Zhang, L.: A case for stateful forwarding plane. Comput. Commun. **36**(7), 779–791 (2013)
5. Zhu, Z., Afanasyev, A.: Let's chronosync: decentralized dataset state synchronization in named data networking. In: 2013 21st IEEE International Conference on Network Protocols (ICNP), pp. 1–10. IEEE (2013)
6. Grassi, G., Pesavento, D., Pau, G., Vuyyuru, R., Wakikawa, R., Zhang, L.: VANET via named data networking. In: 2014 IEEE Conference on Computer Communications Workshops (INFOCOM WKSHPS), pp. 410–415. IEEE (2014)
7. Zhang, H., Wang, Z., Scherb, C., Marxer, C., Burke, J., Zhang, L.: Sharing mHealth data via named data networking. In: Proceedings of the 2016 Conference on 3rd ACM Conference on Information-Centric Networking, pp. 142–147. ACM (2016)
8. Shang, W., Bannis, A., Liang, T., Wang, Z., Yu, Y., Afanasyev, A., Thompson, J., Burke, J., Zhang, B., Zhang, L.:. Named data networking of things. In: 2016 IEEE First International Conference on Internet-of-Things Design and Implementation (IoTDI), pp. 117–128. IEEE (2016)
9. Wang, L., Moiseenko, I., Zhang, L.: Ndnlive and ndntube: Live and prerecorded video streaming over ndn. NDN, Technical report 0031 (2015)
10. Gusev, P., Burke, J.: NDN-RTC: real-time videoconferencing over named data networking. In: Proceedings of the 2nd International Conference on Information-Centric Networking, pp. 117–126. ACM (2015)
11. Zhu, Z., Wang, S., Yang, X., Jacobson, V., Zhang, L.: ACT: audio conference tool over named data networking. In: Proceedings of the ACM SIGCOMM Workshop on Information-Centric Networking, pp. 68–73. ACM (2011)
12. Shao, Y., Tan, X., Wu, X.: Dynamic adaptive streaming in named data networking. In: 2016 35th Chinese Control Conference (CCC), pp. 6855–6860. IEEE (2016)
13. Li, M., Pei, D., Zhang, X., Zhang, B., Xu, K.: NDN live video broadcasting over wireless LAN. In: 2015 24th International Conference on Computer Communication and Networks (ICCCN), pp. 1–7. IEEE (2015)
14. Ishizu, Y., Kanai, K., Katto, J., Nakazato, H., Hirose, M.: Energy-efficient video streaming over named data networking using interest aggregation and playout buffer control. In: 2015 IEEE International Conference on Data Science and Data Intensive Systems, pp. 318–324. IEEE (2015)
15. Xu, H., Chen, Z., Chen, R., Cao, J.: Live streaming with content centric networking. In: 2012 Third International Conference on Networking and Distributed Computing, pp. 1–5. IEEE (2012)

16. Jangam, A., Ravindran, R., Chakraborti, A., Wan, X., Wang, G.: Realtime multi-party video conferencing service over information centric network. In: 2015 IEEE International Conference on Multimedia & Expo Workshops (ICMEW), pp. 1–6. IEEE (2015)
17. Matsuzono, K., Asaeda, H.: NRTS: content name-based real-time streaming. In: 2016 13th IEEE Annual Consumer Communications & Networking Conference (CCNC), pp. 537–543. IEEE (2016)
18. Matsuzono, Kazuhisa, Asaeda, Hitoshi: NMRTS: content name-based mobile real-time streaming. IEEE Commun. Mag. **54**(8), 92–98 (2016)
19. Zhang, L., Estrin, D., Burke, J., Jacobson, V., Thornton, J.D., Smetters, D.K., Zhang, B., et al.: Named data networking (NDN) project. Relatório Técnico NDN-0001, Xerox Palo Alto Research Center-PARC (2010)
20. Gusev, P., Wang, Z., Burke, J., Zhang, L., Yoneda, T., Ohnishi, R., Muramoto, E.: Real-time streaming data delivery over named data networking. IEICE Trans. Commun. **99**(5), 974–991 (2016)
21. Kulinski, D., Burke, J.: NDNVideo: random-access live and pre-recorded streaming using NDN. University of California, Los Angeles, Technical report NDN–0007, pp. 1–17 (2012)
22. Docker Inc. What is Docker? Docker Website. https://www.docker.com/what-docker. Accessed 1 June 2017
23. Canonical Ltd. NetEm – Network Emulator, Ubuntu Manpage Repository. http://manpages.ubuntu.com/manpages/trusty/man8/tc-netem.8.html. Accessed 2 June 2017

Strong Degrees in Single Valued Neutrosophic Graphs

Said Broumi[1(✉)], Florentin Smarandache[2], Assia Bakali[3], Seema Mehra[4], Mohamed Talea[1], and Manjeet Singh[5]

[1] Laboratory of Information Processing, Faculty of Science Ben M'Sik, University Hassan II, 7955 Sidi Othman, Casablanca, Morocco
broumisaid78@gmail.com, taleamohamed@yahoo.fr
[2] Department of Mathematics, University of New Mexico, 705 Gurley Avenue, Gallup, NM 87301, USA
smarandache@gmail.com, smarand@unm.edu
[3] Ecole Royale Navale, Boulevard Sour Jdid, 16303 Casablanca, Morocco
assiabakali@yahoo.fr
[4] Department of Mathematics, Maharshi Dayanand University, Rohtak, India
Sberwal2007@gmail.com, mehra.seema@yahoo.co.in
[5] Department of Mathematics, K.L.P. College, Rewari, Rohtak, India
manjeetmaths@gmail.com

Abstract. The concept of Single Valued Neutrosophic Graphs (SVNGs) generalizes fuzzy graphs and intuitionistic fuzzy graphs. The purpose of this research paper is to define different types of strong degrees in SVNGs and introduce novel concepts, such as the vertex membership of truth-values, vertex membership of indeterminate-values and vertex membership of false-values, which are sequence of SVNG with proof and numerical illustrations.

Keywords: Single valued neutrosophic graph (SVNG) · Neutrosophic set Sequence · Strong degree

1 Introduction

In [1, 3] Smarandache explored the notion of sets which are neutrosophic in nature (NS abbreviated) as a powerful tool which extends the analysis of crisp sets, ambiguous sets such as fuzzy sets and intuitionistic fuzzy sets which are vague in nature [2–6]. This idea deals ambiguous, incomplete and indeterminate information, which exist in real time. The concept of Neutrosophic sets associate to each element of the set a membership degree, $T_A(x)$, an indeterminate degree $I_A(x)$, and a false degree $F_A(x)$, in which each degree of membership is a standard real system or non-standard subset of the real valued nonstandard unit $]^{-}0, 1^{+}[$. Smaranadache [1], [2] and Wang [7] defined the idea of sets in single valued neutrosophic logic termed as single valued neutrosophic sets (SVNS), an occurrence of NS, to accord with real application. In [8], the readers can found a rich literature on SVNS.

In more recent times, combining the concepts of NSs, interval based neutrosophic sets (IVNSs) and bipolar neutrosophic sets sets in the field of graph theory, *Broumi*

K. Arai et al. (Eds.): FICC 2018, AISC 886, pp. 221–238, 2019.
https://doi.org/10.1007/978-3-030-03402-3_16

et al. defined various neutrosophic graphs including Single valued neutrosophic graphs (SVNGs for short) [9, 11, 14], interval valued neutrosophic graphs [13, 18, 20], bipolar neutrosophic graphs [10, 12], all these graphs are studied deeply. Later on, the same authors presented some papers for determining the shortest path problem on a some network having single valued neutrosophic edges length [17, 32], interval valued neutrosophic edge length [32], bipolar neutrosophic edge length [21], trapezoidal neutrosophic numbers [15], SV-trapezoidal neutrosophic numbers [16], triangular fuzzy neutrosophic [19]. Our approach of neutrosophic graphs are different from that of Akram et al. [26–28] since while Akram considers, for the neutrosophic environment (<=, <=, >=) we do (<=, >=, >=) which is better, since while T is a positive quality, I, F are considered negative qualities. Akram et al. include "I" as a positive quality together with "T". So our paper improves Akram et al.'s papers. After that, several authors are focused on the study of SVNGs and many extensions of SVNGs have been developed. Hamidi and Borumand Saeid [25] defined the notion of accessible-SVNGs and apply it social networks. In [24], Mehra and Manjeet defined the idea of SVN signed graphs. Hassan et al. [30] proposed some kinds of bipolar neutrosophic graphs. Naz et al. [23] studied some basic operations on SVNGs and introduced vertex degree of these operations for SVNGs and furnished an application of directed single valued neutrosophic graphs (SVNDG) in travel time. Ashraf et al. [22] defined new classes of SVNGs and studied some of its important properties. They solved a multi-attribute decision making problem using a SVNDG. Mullai [31] analyzed the concept of spanning tree problem in bipolar neutrosophic conditions and gave a numerical example, motivated by the Karunambigai work [29]. The concept of strong degree of intuitionistic fuzzy graphs is extended to strong degree of SVNGs.

This paper has been organized in five sections. In Sect. 2, we firstly review some preliminary notions related to that will be used in the paper. In Sect. 3, different strong degree of SVNGs are proposed and studied with proof and example. In Sect. 4, the concepts of vertex truth-membership, vertex indeterminate-membership, and vertex false-membership are discussed. Lastly, in Sect. 5 some conclusion and directions for future work are initiated.

2 Preliminaries and Definitions

Let us start with some fundamental definitions related to neutrosophic sets, SVNS and SVNGs.

Definition 2.1 [1]. Given the universal sets ζ. A neutrosophic set A ζ is expressed by three function as: truth membership function, an indeterminacy membership, and a falsity membership F_A, where T_A, I_A, $F_A : \zeta \to]^{-}0, 1^{+}[$. $\forall$ all $x \in \zeta$, $x = (x, T_A(x), I_A(x), F_A(x)) \in$ A is a neutrosophic element of A.

The neutrosophic set can be dictated in the following form:

$$\mathrm{A} = \{ < \mathrm{x}: T_A(x), I_A(x), F_A(x) > , \ \mathrm{x} \in \zeta \} \quad (1)$$

With the constraint

$$^{-}0 \leq \mathrm{T_A(x)} + \mathrm{I_A(x)} + \mathrm{F_A(x)} \leq 3^{+} \tag{2}$$

Definition 2.2 [7]. The universal sets ζ A single valued neutrosophic set A on ζ is represented by a truth-membership function T_A, an indeterminacy-membership function, and a falsity membership function F_A, where T_A, I_A, $F_a : \zeta \rightarrow [0, 1]$. For all $x \in \zeta$, $\mathrm{x} = \mathrm{x}, T_A(x), I_A(x), F_A(x)) \in \mathrm{A}$ is *a single valued neutrosophic element* of A.

The SVNS can be written in the following form:

$$\mathrm{A} = \{ < \mathrm{x}: T_A(x), I_A(x), F_A(x) > , \mathrm{x} \in \zeta \} \tag{3}$$

with the condition

$$0 \leq T_A(x) + I_A(x) + F_A(x) \leq 3 \tag{4}$$

Definition 2.3 [14]. A *SVN-graph* G is of the form G = (A, B) where A

1. $\mathrm{A} = \{v_1, v_2, \ldots v_n\}$ Such that the functions T_A: $\mathrm{A} \rightarrow [0, 1]$, $I_A : A \rightarrow [0, 1]$, F_A : $\mathrm{A} \rightarrow [0, 1]$ denote the *truth-membership function*, an *indeterminate-membership function* and *false-membership function* of the element $v_i \in \mathrm{A}$ respectively and

$$0 \leq t_A(v_i) + i_A(v_i) + f_A(v_i) \leq 3 \forall v_i \in A, \; i = 1, 2, \ldots, n$$

2. $\mathrm{B} = \{ (v_i, v_j); (v_i, v_j) \in A \times A \}$ and *the* function $T_B : \mathrm{B} \rightarrow [0, 1]$, $\mathrm{I_B} : \mathrm{B} \rightarrow [0, 1]$, $\mathrm{F_B} : \mathrm{B} \rightarrow [0, 1]$ are defined by

$$T_B(v_i, v_j) \leq \min (T_A(v_i), T_A(v_j)) \tag{5}$$

$$I_B(v_i, v_j) \geq \max (I_A(v_i), I_A(v_j)) \tag{6}$$

$$F_B(v_i, v_j) \geq \max (F_A(v_i), F_A(v_j)) \tag{7}$$

Where T_B, I_B, F_B denotes *the truth-membership function, indeterminacy membership function and falsity membership* function of the edge $(v_i, v_j) \in \mathrm{B}$ respectively where

$$0 \leq T_B(v_i, v_j) + I_B(v_i, v_j) + F_B(v_i, v_j) \leq 3 \tag{8}$$

$$\forall (v_i, v_j) \in B, \; \mathrm{i, j} \in \{1, 2, \ldots, \mathrm{n}\}$$

A is named the vertex sets of and B is the edge sets of G.

The following Fig. 1 represented a graphical representation of SVNG.

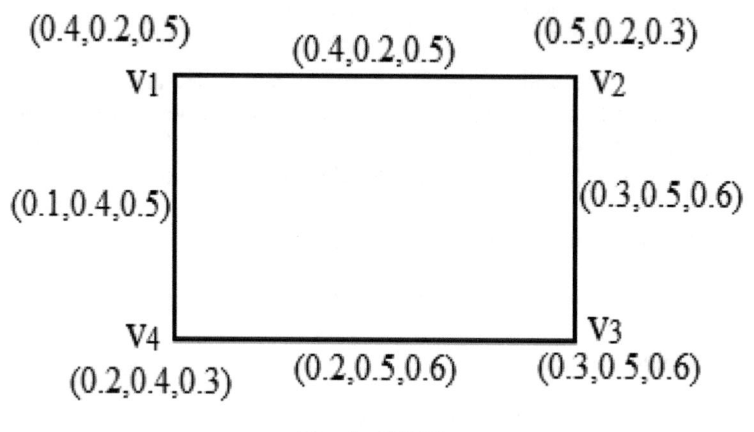

Fig. 1. SVNG.

3 Strong Degree in Single Valued Neutrosophic Graph

The following section introduces new concepts and proves their properties.

Definition 3.1. Given the SVN-graph G. The T-strong degree of a vertex is defined as $d_{s(T)}(v_i) = \sum_{e_{ij} \in E} T_{ij}$, e_{ij} are *strong degree incident* at v_i.

Definition 3.2. Given the *SVN-graph* G. The I-strong degree of a vertex is defined as $d_{s(I)}(v_i) = \sum_{e_{ij} \in E} I_{ij}$, e_{ij} are strong edges incident at v_i.

Definition 3.3. Given the *SVN-graph* G. The F-strong degree of a vertex is defined as $d_{s(F)}(v_i) = \sum_{e_{ij} \in E} F_{ij}$, e_{ij} are strong edges incident at v_i.

Definition 3.4. Given a *SVN-graph* G. The *strong degree* of a vertex is defined as follow $d_s(v_i) = \left[\sum_{e_{ij} \in E} T_{ij}, \sum_{e_{ij} \in E} I_{ij}, \sum_{e_{ij} \in E} F_{ij} \right]$, where e_{ij} are *strong edge* incident at v_i.

Definition 3.5. Given A SVNG. The *minimum strong degree* of G is defined as

$$\delta_s(\mathrm{G}) = (\delta_{s(T)}(\mathrm{G}), \delta_{s(I)}(\mathrm{G}), \delta_{s(F)}(\mathrm{G}),$$

where

$\delta_{s(\mathrm{T})}(\mathrm{G}) = \wedge\{d_{s(\mathrm{T})}(\mathrm{v}_i)/\mathrm{v}_i \in V\}$ is the *minimum* T-*strong degree* of G.
$\delta_{s(I)}(G) = \wedge\{d_{s(\mathrm{I})}(\mathrm{v}_i)/\mathrm{v}_i \in V\}$ is *the minimum* I-strong *degree* of G.
$\delta_{s(F)}(G) = \wedge\{d_{s(F)}(\mathrm{v}_i)/\mathrm{v}_i \in V\}$ is *the minimum* F-*strong degree* of G.

Definition 3.6. Given a *SVN-graph* G = (V, E). The maximum *strong degree* of G is defined as

$$\Delta_s(G) = (\Delta_{s(\mathrm{T})}(\mathrm{G}), \Delta_{s(\mathrm{I})}(\mathrm{G}), \Delta_{s(\mathrm{F})}(\mathrm{G})),$$

where

$\Delta_{s(T)}(G) = \vee\{d_{s(\mathrm{T})}(\mathrm{v}_i)/\mathrm{v}_i \in V\}$ is *the maximum* T-*strong degree* of G.
$\Delta_{s(I)}(G) = \vee\{d_{s(I)}(\mathrm{v}_i)/\mathrm{v}_i \in V\}$ is *the maximum* I-*strong degree* of G.
$\Delta_{s(\mathrm{F})}(\mathrm{G}) = \vee\{d_{s(\mathrm{F})}(\mathrm{v}_i)/\mathrm{v}_i \in V\}$ is the maximum F-*strong degree* of G.

Definition 3.7. Let G be a SVNG, the T-total strong degree, I-total strong degree, and F-total strong degree are defined degree of a vertex in G are defined

$$td_{s(\mathrm{T})}(v_i) = d_{s(\mathrm{T})}(v_i) + T_i$$

$$td_{s(F)}(v_i) = d_{s(F)}(v_i) + F_i$$

$$td_{s(\mathrm{I})}(v_i) = d_{s(\mathrm{I})}(v_i) + I_i$$

Definition 3.8. Let G be a SVNG, the *total strong degree of a vertex* $v_i \in V$ in G is defined as

$$td_s(v_i) = \left[td_{s(T)}(v_i), td_{s(I)}(v_i), td_{s(F)}(v_i)\right]$$

Definition 3.9. Given the *SVN-graph* G minimum *total strong degree* of G is defined as

$$\delta_{ts}(G) = (\delta_{ts(T)}(G), \delta_{ts(I)}(G), \delta_{ts(F)}(G)),$$

where

$\delta_{ts(T)}(G) = \wedge\{d_{ts(T)}(\mathrm{v}_i)/\mathrm{v}_i \in V\}$ is the minimum T-*total strong degree* of G.
$\delta_{ts(I)}(G) = \wedge\{d_{ts(\mathrm{I})}(\mathrm{v}_i)/\mathrm{v}_i \in V\}$ is the minimum I-*total strong degree* of G.
$\delta_{ts(F)}(G) = \wedge\{d_{ts(F)}(\mathrm{v}_i)/\mathrm{v}_i \in V\}$ is the minimum F-*total strong degree* of G.

Definition 3.10. Given the SVN-graph G = (V, E). The maximum total strong degree of G is defined as

$$\Delta_{ts}(G) = (\Delta_{ts(T)}(G), \Delta_{ts(I)}(\mathrm{G}), \Delta_{ts(F)}(G)),$$

where

$\Delta_{ts(T)}(\mathrm{G}) = \vee\{d_{ts(\mathrm{T})}(\mathrm{v}_i)/\mathrm{v}_i \in V\}$ is the *maximum* T-*total strong degree* of G.
$\Delta_{ts(I)}(G) = \vee\{d_{ts(I)}(\mathrm{v}_i)/\mathrm{v}_i \in V\}$ is the *maximum* I-*total strong degree* of G.
$\Delta_{ts(\mathrm{F})}(\mathrm{G}) = \vee\{d_{ts(\mathrm{F})}(\mathrm{v}_i)/\mathrm{v}_i \in V\}$ is the *maximum* F-*total strong degree* of G.

Definition 3.11. Given the *SVN-graph* G = (V, E). The T-strong size, the I-strong size, and the F-strong size of a SVNG are defined as

$$S_{s(T)}(G) = \sum_{v_i \neq v_j} T_{ij},\ S_{s(I)}(G) = \sum_{v_i \neq v_j} I_{ij},$$
$$S_{s(F)}(G) = \sum_{v_i \neq v_j} F_{ij}$$

Definition 3.12. Given the SVN-graph G = (V, E). The strong size of a SVNG is defined as

$$S_s(G) = \left[S_{s(\mathrm{T})}(\mathrm{G}), S_{s(\mathrm{I})}(\mathrm{G}), S_{s(F)}(\mathrm{G})\right]$$

Definition 3.13. Given the SVN-graph G = (V, E). The T-strong order of a SVNG is defined as
$O_{s(T)}(G) = \sum_{v_i \in V} T_i$ where v_i is the strong vertex in G.

Definition 3.14. Given the *SVN-graph* G. The I-strong order of a SVNG is defined as
$O_{s(\mathrm{I})}(G) = \sum_{v_i \in V} I_i$ where v_i is a *strong vertex* in G.

Definition 3.15. Given the *SVN-graph* G. The F-strong order of a SVNG is defined as
$O_{s(\mathrm{F})}(G) = \sum_{v_i \in V} F_i$ where v_i is the strong vertex in G.

Definition 3.16. Given the *SVN-graph G*. The strong order of a SVNG is defined as

$$O_s(G) = \left[O_{s(T)}(G), O_{s(I)}(G), O_{s(F)}(G)\right]$$

Definition 3.17. Let G be a SVNG. If $d_{s(T)}(v_i) = k_1$, $d_{s(I)}(v_i) = k_2$ and, $d_{s(F)}(v_i) = k_3$ for all $v_i \in V$, then the SVNG is called as (k_1, k_2, k_3) - strong constant SVNG (or) Strong constant SVNG of degree (k_1, k_2, k_3).

Definition 3.18. Let G be a SVNG. If $td_{s(T)}(v_i) = r_1$, $td_{s(I)}(v_i) = r_2$ and, $d_{s(\mathrm{F})}(v_i) = r_3$ for all $v_i \in V$, then the SVNG is called as (r_1, r_2, r_3) – totally strong constant SVNG (or) totally strong constant SVNG of degree (r_1, r_2, r_3).

Proposition 3.19. In a SVNG G

$$2\,S_{s(T)}(G) = \sum_{i=1}^{n} d_{s(T)}(v_i),\ 2S_{s(I)}(G) = \sum_{i=1}^{n} d_{s(I)}(v_i) \text{ and}$$
$$2S_{s(F)}(G) = \sum_{i=1}^{n} d_{s(F)}(v_i)$$

Proposition 3.20. In a connected SVNG,

(1) $d_{S(T)}(v_i) \leq d_{Ti}$, $d_{S(I)}(v_i) \leq d_{Ii}$ and $d_{S(F)}(v_i) \leq d_{Fi}$

(2) $td_{S(T)}(v_i) \leq td_{Ti}, td_{S(I)}(v_i) \leq td_{Ii}$ and
$td_{S(F)}(v_i) \leq td_{Fi}$.

Proposition 3.21. Let G be a SVNG where crisp graph G^* is an odd cycle. Hence G is a strong constant if f$< T_{ij}, I_{ij}, F_{ij} >$ is constant function for every $e_{ij} \in$ E.

Proposition 3.22. Let G be a SVNG where crisp graph G^* is an even cycle. Hence G is strong constant if f$< T_{ij}, I_{ij}, F_{ij} >$ t is constant function or substitute arcs have same true-membership, indeterminate-membership and false membership for every $e_{ij} \in$ E.

Remark 3.23. The above Propositions 3.21 and 3.22 hold for totally *strong constant* SVNG, if $< T_i, I_i, F_i >$ is a constant function.

Remark 3.24. A complete SVNG need not be a strong constant SVNG and *totally strong constant* SVNG.

Remark 3.25. A strong SVNG need not be a strong constant SVNG and totally strong constant SVNG.

Remark 3.26. For a strong vertex $v_i \in V$,

(1) $d_T(v_i) = d_{sT}(v_i)$, $d_I(v_i) = d_{sI}(v_i)$ and $d_F(v_i) = d_{sF}(v_i)$
(2) $td_T(v_i) = td_{sT}(v_i)$, $td_I(v_i) = \mathrm{t}\,d_{sI}(v_i)$ and $\mathrm{t}\,d_F(v_i) = td_{sF}(v_i)$

Theorem 3.27. Let G be a complete SVNG with $V = \{v_1, v_2, \ldots, v_n\}$ such that $T_1 \leq T_2 \leq T_3 \leq \ldots \leq T_n$, $I_1 \geq I_2 \geq I_3 \geq \ldots \geq I_n$ and $F_1 \geq F_2 \geq F_3 \geq \ldots \geq F_n$ Then

(1) T_{1j} is minimum edge truth-membership, I_{1j} is the maximum edge indeterminate-membership and F_{1j} is the maximum edge false-membership of e_{ij} emits from v_1 for all j = 2, 3, 4, ..., n.
(2) T_{in} is maximum edge truth membership, I_{in} is the minimum edge indeterminate-membership and F_{in} is the minimum edge falsity-membership of among all edges from emits from v_i to v_n for all $i = 1, 2, \ldots,$ n − 1.
(3) $td_T(v_1) = \delta_{td_T}$ (G) = n.T_1, $td_I(v_1) = \Delta_{td_I}(G)$ = n.I_1 and $td_F(v_1) = \Delta_{td_F}$ $(G) = n.F_1$.
(4) $td_T(v_n) = \Delta_{td_T}(G) = \sum_{i=1}^{n} T_i$, $td_I(v_n) = \delta_{td_I}(G) = \sum_{i=1}^{n} I_i$, *and* $td_F(v_n) = \delta_{td_F}$ $(G) = \sum_{i=1}^{n} F_i$.

Proof: Throughout the proof, suppose that $T_1 \leq T_2 \leq T_3 \leq \ldots \leq T_n$, $I_1 \geq I_2 \geq I_3 \geq \ldots \geq I_n$ and $F_1 \geq F_2 \geq F_3 \geq \ldots \geq F_n$.

(1) To prove that T_{1j} is minimum edge truth-membership, I_{1j} is the maximum edge indeterminate-membership and F_{1j} is the maximum edge false-membership of e_{ij} emits from v_1 ∀ j = 2, 3, ..., n. Assume the contrary i.e. e_{1l} is not an edge of minimum true membership, maximum indeterminate membership and maximum false membership emits from v_l. Also let e_{kl}, $2 \leq k \leq n, k \neq l$ be an edge with

minimum true-membership, maximum-indeterminate membership and maximum false-membership emits from v_k.

Being a complete SVNG,
$T_{1l} = min\{T_1, T_l\}$, $I_{1l} = max\{I_1, I_l\}$ and $F_{1l} max\{F_1, F_l\}$
Then $T_{kl} = min\{T_k, T_l\}$, $I_{kl} = max\{I_k, I_l\}$ and
$F_{kl} = \max\{F_k, F_l\}$
Since $T_{kl} < T_{1l} \Rightarrow \min\{T_k, T_l\} < \min\{T_1, T_l\}$ Thus either $T_k < T_1$ or $T_l < T_1$.
Also since $I_{kl} > I_{1l} \Rightarrow \max\{I_k, I_l\} > \max\{I_1, I_l\}$, so either $I_k > I_1$ or $I_l > I_1$.
Since *l*, $k \neq l$, this is contradiction to our vertex assumption that is the unique minimum true-membership, I_1 is the maximum vertex indeterminate-membership and F_1 is the maximum vertex false-membership.

Hence T_{1j} is minimum edge true-membership, I_{1j} is the maximum edge indeterminate-membership and F_{1j} is the maximum edge false-membership of e_{ij} emits from v_1 to v_j for all $j = 2, 3, 4, \ldots$, n.

(2) On the contrary, assume let e_{kn} is not an edge with *maximum true-membership, minimum indeterminate-membership and minimum false-membership* emits from v_k for $1 \leq k \leq n - 1$. On the other hand, let e_{kr} be an edge with maximum true-membership, *minimum indeterminate-membership* and *minimum false-membership* emits from v_r from $1 \leq r \leq n - 1$, $k \neq r$.

Then $T_{kr} > T_{kn} \Rightarrow min\{T_k, T_r\} > min\{T_k, T_n\} = T_k$, so $T_r > T_k$,
$I_{kr} < I_{kn} \Rightarrow max\{I_k, I_r\} < max\{I_k, I_n\} = I_k$, so $I_r < I_k$ and
Similarly $F_{kr} < F_{kn} \Rightarrow max\{F_k, F_r\} < max\{F_k, F_n\} = F_k$, $\Rightarrow F_r < F_k$.
So $T_{kr} = T_k = T_{kn}$, $I_{kr} = I_k = I_{kn}$ and $F_{kr} = F_k = F_{kn}$, which is a contradiction. Hence e_{kn} is an edge with *maximum* true-membership, minimum indeterminate-membership and minimum false-membership among all edges emits from v_k to v_n.

(3) Now

$$\begin{aligned} td_T(v_1) &= d_T(v_1) + T_1 \\ &= \sum\nolimits_{e_{ij} \in E} T_{1j} + T_1 = \sum\nolimits_{j=2}^{n} T_{1j} + T_1 \\ &= (n-1).T_1 + T_1 = nT_1 - T_1 + T_1 = nT_1, \end{aligned}$$

Similarly, we have for $td_I(v_1) = n.I_1$ and $td_F(v_n) = n.F_1$
Suppose that t $d_T(v_1) \neq \delta_{td_T}(G)$ and let v_k, $k \neq 1$ be a vertex in G with minimum T-*total degree.*
Then,

$$td_T(v_1) > td_T(v_k)$$

$$\Rightarrow \sum_{i=2}^{n} T_{1i} + T_1 > \sum_{k \neq 1, k \neq j} T_{kj} + T_k$$

$$\Rightarrow \sum_{i=2}^{n} T_1 \bigwedge T_i + T_1 > \sum_{k \neq 1, k \neq j} T_k \bigwedge T_j + T_k$$

Since $T_1 \bigwedge T_i = T_1$ for i $= 1, 2, 3, \ldots,$ n and for all other indices j, $T_k \bigwedge T_j > T_1$, it follow that

$$(\mathrm{n}-1) \cdot T_1 + T_1 > \sum\nolimits_{k \neq 1, k \neq j} T_k \bigwedge T_j + T_k > (\mathrm{n}-1).T_1 + T_1$$

Hence, $td_T(v_1) > td_1(v_1)$, *a contradiction.*

Therefore, $td_T(v_1) = \delta_{td_T}(G)$.

Suppose that $\mathrm{t}d_I(\mathrm{v}_1) \neq \Delta_{\mathrm{t}d_I}(\mathrm{G})$ and let v_k, $k \neq 1$ *be a* vertex in G with maximum I-*total degree.*

Then,

$$td_I(v_1) < \mathrm{t}d_I(v_k)$$

$$\Rightarrow \sum_{i=2}^{n} I_{1i} + I_1 < \sum_{k \neq 1, k \neq j} I_{kj} + I_k$$

$$\Rightarrow \sum_{i=2}^{n} I_1 \vee I_i + I_1 < \sum_{k \neq 1, k \neq j} I_k \vee I_j + I_k$$

Since $I_1 \vee I_i = I_1$ for i $= 1, 2, 3, \ldots,$ n and for all other indices j, $I_k \vee I_j < I_1$, it follow that

$$(n - 1.)I_1 + I_1 < \sum\nolimits_{k \neq 1, k \neq j} I_k \bigvee I_j + I_k < (n-1).I_1 + I_1$$

So that $td_I(v_1) < td_I(v_i)$ a contradiction.

Therefore, $td_I(v_1) = \Delta_{\mathrm{t}d_I}(\mathrm{G})$.

Also, Suppose that $\mathrm{t}d_F(\mathrm{v}_1) \neq \Delta_{\mathrm{t}d_F}$ and let v_k, $k \neq 1$ be a vertex in G with maximum F-*total degree.*

Then

$$\mathrm{t}d_F(v_1) < \mathrm{t}d_F(v_k)$$

$$\Rightarrow \sum_{i=2}^{n} F_{1i} + F_1 < \sum_{k \neq 1, k \neq j} F_{kj} + F_k$$

$$\Rightarrow \sum_{i=2}^{n} F_1 \vee F_i + F_1 < \sum_{k \neq 1, k \neq j} F_k \vee F_j + F_k$$

Since $F_1 \vee F_i = F_1$ for i = 1, 2, 3, …, n and for all other indices j, $F_k \vee F_j < F_1$, it follow that

$$(\mathrm{n}-1.)F_1 + F_1 < \sum_{k \neq 1, k \neq j} F_k \bigvee F_j + F_k < (n-1).F_1 + F_1$$

So that $td_F(v_1) < td_F(v_1)$ a contradiction.
Therefore, $td_F(v_1) = \Delta_{td_F}(\mathrm{G})$.
Hence,

$td_T(\mathrm{v}_1) = \delta_{td_T}(\mathrm{G}) = \mathrm{n}.\mathrm{T}_1,$
$td_I(\mathrm{v}_1) = \Delta_{td_I}(\mathrm{G}) = \mathrm{n}.\mathrm{I}_1$ and
$td_F(\mathrm{v}_1) = \Delta_{td_F}(\mathrm{G}) = \mathrm{n}.\mathrm{F}_1.$

(4) Since, $T_n > T_i$, $I_n < I_i$ and $F_n < F_i$, $i = 1, 2, \ldots, \mathrm{n}-1$ and G is complete $T_{ni} = T_n \wedge T_i = T_i$, $I_{ni} = I_n \vee I_i = I_i$ and $F_{ni} = F_n \vee F_i = F_i$.

Hence, $\mathrm{td}_T(v_n) = \sum_{i=1}^{n-1} T_{ni} + T_n = \sum_{i=1}^{n-1} (T_n \bigwedge T_i) + T_n = \sum_{i=1}^{n-1} T_i + T_n = \sum_{i=1}^{n} T_i,$
Similar to $td_T(v_n)$, we got for

$$td_I(v_n) = \sum_{i=1}^{n} I_i$$

Also $\mathrm{td}_F(v_n) = \sum_{i=1}^{n-1} F_{ni} + F_n$

$$= \sum_{i=1}^{n-1} (F_n \vee F_i) + F_n = \sum_{i=1}^{n-1} F_i + F_n$$
$$= \sum_{i=1}^{n} F_i.$$

Suppose that $\mathrm{td}_T(v_n) \neq \Delta_{td_T}(\mathrm{G})$. Let v_l, $l \leq 1 \leq \mathrm{n} - l$ $\mathrm{td}_T(v_n) < \mathrm{td}_T(v_l)$. In addition,

$$\mathrm{td}_T(v_l) = \left[\sum_{i=1}^{l-1} T_{il} + \sum_{i=l+1}^{n-1} T_{il} + T_{nl}\right] + T_l$$
$$\leq \left[\sum_{i=1}^{l-1} T_i + (\mathrm{n}-1)T_l + T_l\right] + T_l$$
$$\leq \sum_{i=1}^{n-1} T_i + T_l$$

$\leq \sum_{i=1}^{n} T_i = \mathrm{td}_T(v_n)$. Thus $td_T(v_n) \geq \mathrm{td}_T(v_l)$, contradiction. So, $\mathrm{td}_T(v_n) = \Delta_{td_T}(\mathrm{G}) = \sum_{i=1}^{n} T_i$.

Suppose that $\mathrm{td}_I(v_n) \neq \delta_{td_I}(\mathrm{G})$. Let v_l, $1 \leq 1 \leq \mathrm{n}-1$ be a vertex in G such that $\mathrm{td}_I(v_l) = \delta_{td_I}(\mathrm{G})$ and $\mathrm{td}_I(v_n) > \mathrm{td}_I v_l)$.

In addition,

$$\begin{aligned} td_I(v_l) &= \left[\sum\nolimits_{i=1}^{l-1} I_{il} + \sum\nolimits_{i=l+1}^{n-1} I_{il} + I_{nl}\right] + I_l \\ &\geq \left[\sum\nolimits_{i=1}^{l-1} I_i + (\mathrm{n}-1)I_l + I_l\right] + I_l \\ &\geq \sum\nolimits_{i=1}^{n-1} I_i + I_l \end{aligned}$$

$\geq \sum_{i=1}^{n} I_i = \mathrm{td}_I(v_n)$. Thus $\mathrm{td}_I(v_n) \leq \mathrm{td}_I(v_l)$, contradiction. So, $\mathrm{td}_I(v_n) = \delta_{\mathrm{td}_I}(\mathrm{G}) = \sum_{i=1}^{n} I_i$.

Also, suppose that $\mathrm{td}_F(v_n) \neq \delta_{\mathrm{td}_F}(\mathrm{G})$. Let v_l, $1 \leq 1 \leq \mathrm{n}-1$ be a vertex in G such that $td_F(v_l) = \delta_{td_F}(G)$ and $\mathrm{td}_F(v_n) > \mathrm{td}_F(v_l)$. In addition,

$$\begin{aligned} td_F(v_l) &= \left[\sum\nolimits_{i=1}^{l-1} F_{il} + \sum\nolimits_{i=l+1}^{n-1} F_{il} + F_{nl}\right] + F_l \\ &\geq \left[\sum\nolimits_{i=1}^{l-1} F_i + (\mathrm{n}-1)F_l + F_l\right] + F_l \\ &\geq \sum\nolimits_{i=1}^{n-1} F_i + F_l \end{aligned}$$

$\geq \sum_{i=1}^{n} F_i = \mathrm{td}_F(v_n)$. Thus $td_F(v_n) \leq td_F(v_l)$, contradiction. So, $td_F(v_n) = \delta_{\mathrm{td}_F}(\mathrm{G}) = \sum_{i=1}^{n} F_i$.

Hence the lemma is proved.

Remark 3.28. In a complete SVNG G,

(1) There exists at least one pair of vertices v_i and v_j such that $d_{T_i} = d_{T_j} = \Delta_T(\mathrm{G})$, $d_{I_i} = d_{I_i} = \delta_I(\mathrm{G})$ and $d_{F_i} = d_{F_i} = \delta_F(\mathrm{G})$,
(2) $\mathrm{td}_T(v_i) = O_T(G) = \Delta_{td_T}(G)$, $\mathrm{td}_I(v_i) = O_I(G) = \delta_{\mathrm{td}_I}(\mathrm{G})$ and $td_F(v_i) = O_F(\mathrm{G})$ $\delta_{\mathrm{td}_F}(\mathrm{G})$ for a vertex $v_i \in \mathrm{V}$,
(3) $\sum_{i=1}^{n} td_T(v_i) = 2S_T(\mathrm{G}) + O_T(G)$, $\sum_{i=1}^{n} td_I(v_i) = 2S_I(G) + O_I(G)$ *and* $\sum_{i=1}^{n} \mathrm{td}_F(v_i) = 2S_F(G) + O_F(G)$.

4 Vertex Membership of Truth-Values, Indterminate-Values, and False-Values Sequence of SVNG

In the present section, we will define the concept of vertex true-membership, vertex indeterminate-membership, and vertex false-membership sequences in SVNGs.

Definition 4.1. Given *a SVN-graphs* G with $|V| = \mathrm{n}$. The vertex truth membership sequence of G is proposed to be $\{x_i\}_{i=1}^{n}$ with $x_1 \leq x_2 \leq x_3 \leq \ldots \leq x_n$ where x_i,

$0 < x_i \leq 1$, is the truth membership value of the vertex v_i when vertices are arranged so that truth-membership values are non-decreasing.

Particularly, x_1 is smallest vertex truth-membership value and x_n is largest vertex truth-membership value in G.

Note 4.2. If vertex truth-membership sequence x_i is repeated more than once in G, say r $\neq$ 1 times, then it is denoted by x_i^r in the sequence.

Example 4.3. In Fig. 2 the vertex truth membership sequence of G is {0.1, 0.1, 0.3, 0.3, 0.4, 0.8} or $\{0.1^2, 0.3^2, 0.4, 0.8\}$.

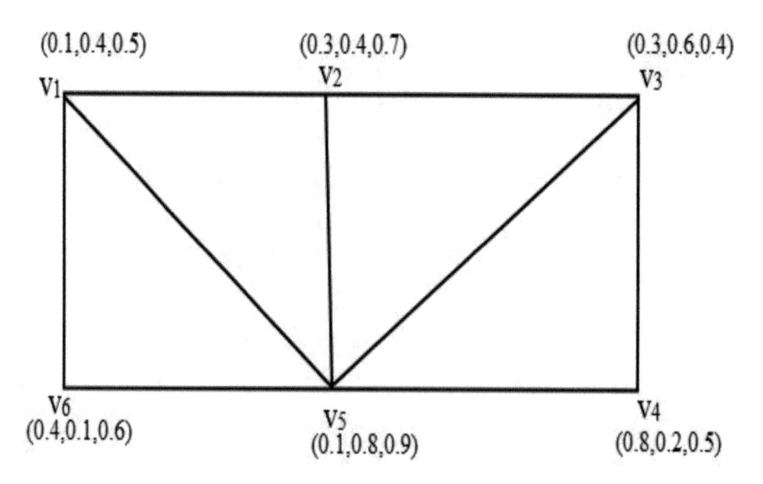

Fig. 2. Vertex truth membership sequence.

Definition 4.4. Let G be a SVNG with $|V| = $ n. The vertex indeterminate-membership sequence of G is proposed to be $\{y_i\}_{i=1}^n$ with $y_1 \leq y_2 \leq y_3 \leq \ldots \leq y_n$ where y_i, $0 < y_i \leq 1$, is the indeterminate-membership value of the vertex v_i, when vertices are arranged so that their indeterminate-membership values are non-increasing.

Particularly, y_1 is largest vertex indeterminate-membership value and y_n is smallest vertex indeterminate-membership value in G.

Note 4.5. If vertex indeterminate-membership sequence y_i is repeated more than once in G, say r $\neq$ 1 times, then it is denoted by y_i^r in the sequence.

Example 4.6. In Fig. 3 the vertex indeterminate- membership sequence of G is {0.7, 0.6, 0.6, 0.5, 0.4, 0.4} or $\{0.7, 0.6^2, 0.5, 0.4^2\}$.

Definition 4.7. Let G be a SVNG with $|V| = $ n. The vertex false-membership sequence of G is proposed to be $\{z_i\}_{i=1}^n$ with $z_1 \leq z_2 \leq z_3 \leq \ldots \leq z_n$ where z_i, $0 < z_i \leq 1$, is the falsity membership value of the vertex v_i when vertices are arranged s that their false-membership values are non-increasing. Particularly, z_1 is largest vertex falsity Y membership value and z_n is smallest vertex false-membership value in G.

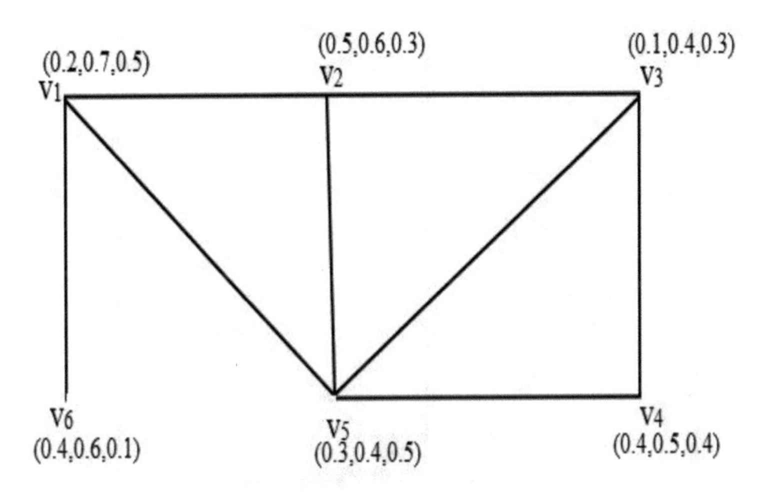

Fig. 3. Vertex indeterminacy membership sequence.

Note 4.8. If vertex false-membership sequence z_i is repeated more than once in G, say r $\neq$ 1 times, then it is denoted by z_i^r in the sequence.

Example 4.9. In Fig. 4 the vertex falsity membership sequence of G is {0.8, 0.8, 0.7, 0.6, 0.6, 0.5} or $\{0.8^2, 0.7, 0.6^2, 0.5\}$.

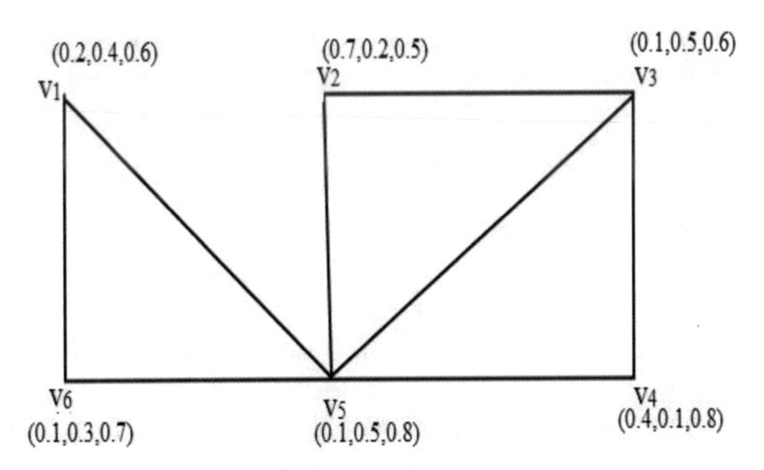

Fig. 4. Vertex falsity membership sequence.

Definition 4.10. If a SVNG with $|V| = \text{n}$ has vertex true-membership sequence $\{x_i\}_{i=1}^{n}$, vertex indeterminate-membership sequence $\{y_i\}_{i=1}^{n}$, and vertex false-membership sequence $\{z_i\}_{i=1}^{n}$ in same order, then it said to have vertex single valued neutrosophic sequence and denoted by $\{<x_i, y_i, z_i>\}_{i=1}^{n}$.

Example 4.11. In Fig. 5 the vertex true-membership, vertex indeterminate-membership, and vertex false-membership sequence of G is {<.4, .4, .5>, <.2, .3, .5>, <.1, .2, .6>, <.5, .4,.8>, <.4, .5,.4>, <.3, .1,.7>}.

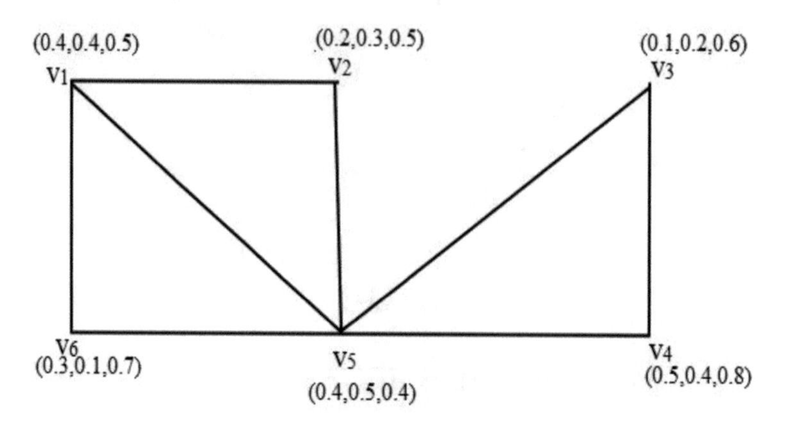

Fig. 5. Vertex SVN-sequence

In the following, we discuss some properties of vertex true-membership, indeterminate-membership, and false-membership sequences of complete SVNGs:

Theorem 4.12. Given a complete SVNG with $|V| = \text{n}$. Hence

(1) If the vertex truth-membership sequence of G having the form $\{x_1^{n-1}, x_2\}$, vertex indeterminate-membership sequence having the form $\{y_1^{n-1}, y_2\}$ and vertex false-membership sequence of G having the form $\{z_1^{n-1}, z_2\}$, then

a. $\delta_{td_T}(\text{G}) = \text{n}.T_1$ and $\Delta_{td_T}(\text{G}) = \sum_{i=1}^{n} T_i$

b. $\Delta_{td_I}(\text{G}) = \text{n}.I_1$ and $\delta_{td_I}(\text{G}) = \sum_{i=1}^{n} I_i$

c. $\Delta_{td_F}(\text{G}) = \text{n}.F_1$ and $\delta_{td_F}(\text{G}) = \sum_{i=1}^{n} F_i$

(2) If vertex truth-membership sequence of G having the form $\{x_1^{r_1}, x_2^{n-r_1}\}$, vertex indeterminate-membership of G is of the form $\{y_1^{r_1}, y_2^{n-r_1}\}$ and vertex false-membership sequence having the form $\{\{z_1^{r_1}, z_2^{n-r_1}\}$ with $0 < r_1 \leq \text{n} - 2$, then there exists exactly r_1 vertices with minimum T-*total degree* , $\delta_{td_T}(G)$ maximum I-*total degree* $\Delta_{td_I}(G)$ and maximum F-*total degree* Δ_{td_F} and exactly $(\text{n} - r_1)$ vertices with maximum T-*total degree* $\Delta_{td_T}(\text{G})$, minimum I-*total degree* $\delta_{td_I}(\text{G})$ and minimum F-*total degree* $\delta_{td_F}(\text{G})$.

(3) If the vertex truth-membership sequence of G having the form $\{x_1^{r_1}, x_2^{r_2}, x_3^{r_3}, \ldots, x_k^{r_k}\}$, vertex indeterminate-membership sequence of G having the form $\{y_1^{r_1}, y_2^{r_2}, y_3^{r_3}, \ldots, y_k^{r_k}\}$ and vertex false-membership sequence of G is of the

form $\{z_1^{r_1}, z_2^{r_2}, z_3^{r_3}, \ldots, z_k^{r_k}\}$ with $r_k > 1$ and k > 2, thus there exists exactly r_1 vertices with minimum T-*total degree*, $\delta_{td_T}(G)$ maximum I-*total degree* Δ_{td_I} and maximum F-*total degree* Δ_{td_F}. Also, there exists exactly r_k vertices with maximum T-*total degree* Δ_{td_T}(G), minimum I-*total degree* δ_{td_I}(G) and minimum F-*total degree* δ_{td_F}(G).

Proof. The proof of (1) and (2) are obvious. 3 Let $v_i^{(j)}$ be the set of vertices in G, for $j = 1, 2, 3, \ldots, r_i$, $1 \leq i \leq k$. Then by the Theorem 3.27

$$td_T\left(v_1^{(j)}\right) = \delta_{td_T}(G) = \text{n}.T_1 = \text{n}.x_1,$$
$$td_I\left(v_1^{(j)}\right) = \Delta_{td_I}(G) = \text{n} \cdot I_1 = \text{n}.y_1 \text{ and}$$
$$td_F\left(v_1^{(j)}\right) = \delta_{td_F}(G) = \text{n}.F_1 = \text{n}.z_1,, \text{ for, } j = 1, 2, \ldots, r_1.$$

Since $\text{T}\left(v_i^{(j)}, v_{i+1}^{(l)}\right) = \text{T}\left(v_i^{(j)}\right) > x_1 \forall 2 \leq i \leq k, j = 1, 2, \ldots, r_i, \text{l} = 1, 2, 3, \ldots, r_{i+1}$, no vertex with truth membership more than x_1 can have degree δ_{td_T}(G).

$\text{I}\left(v_i^{(j)}, v_{i+1}^{(l)}\right) = \text{I}(v_i^{(j)}) < y_1$ for $2 \leq i \leq k$, $j = 1, 2, .., r_i$, $l = 1, 2, \ldots, r_{i+1}$, no vertex with indeterminate- membership less than y_1 can have degree Δ_{td_I}(G).

And $\text{F}\left(v_i^{(j)}, v_{i+1}^{(l)}\right) = \text{F}\left(v_i^{(j)}\right) < z_1 \Big) \forall 2 \leq i \leq k, j = 1, 2, \ldots, r_i, l = 1, 2, \ldots, r_{i+1}$, no vertex with false-membership less than z_1 can have degree Δ_{td_F}(G).

Thus, there exist exactly r_1 vertices with degree δ_{td_T}(G), $\Delta_{td_I}(G)$, Δ_{td_F}(G).

To prove $td_T\left(v_k^{(t)}\right) = \Delta_{td_T}(G)$,

$$td_I\left(v_k^{(t)}\right) = \delta_{td_I}(\text{G})\text{and}$$
$$td_F\left(v_k^{(t)}\right) = \delta_{td_F}(G), \qquad \forall t = 1, 2, \ldots, r_k$$

Since, T $(v_k^{(t)})$ is maximum vertex truth-membership,

$$T\left(v_k^{(t)}, v_k^{(j)}\right) = x_k, t \neq j, \forall t, j = 1, 2, \ldots, r_k$$

$$T\left(v_k^{(t)}, v_i^{(j)}\right) = \min\left\{T\left(v_k^{(t)}\right), T\left(v_i^{(j)}\right)\right\} = T\left(v_i^{(j)}\right) \qquad \text{for} \qquad t = 1, 2, \ldots, r_k,$$

$j = 1, 2, \ldots, r_i$, $i = 1, 2, \ldots, k - 1$

Thus for $t = 1, 2, \ldots, r_k$,

$$td_T\left(v_k^{(t)}\right) = \sum\nolimits_{i=1}^{k} \sum\nolimits_{j=1}^{r_i} T\left(v_i^{(j)}\right) + (r_k - 1)x_k$$
$$= \sum\nolimits_{i=1}^{n} T_i$$

$= \Delta_{td_T}$(G) by Theorem 3.27.

Now, if v_m is vertex such that $T_m = x_{k-1}$, then

$$\begin{aligned} td_T(v_m) &= \sum_{i=1}^{k-2}\sum_{j=1}^{r_i} T\left(v_m, v_i^{(j)}\right) + (r_{k-1} - 1 + r_k)x_{k-1} + T_m \\ &= \sum_{i=1}^{k-2}\sum_{j=1}^{r_i} T\left(v_i^{(j)}\right) + \sum_{j=1}^{r_{k-1}} T\left(v_{k-1}^{(j)}\right) + (r_k - 1)x_{k-1} + T_m \\ &< \sum_{i=1}^{k-2}\sum_{j=1}^{r_i} T\left(v_i^{(j)}\right) + \sum_{j=1}^{r_{k-1}} T\left(v_{k-1}^{(j)}\right) + (r_k - 1)x_k + T_m \\ &= \Delta_{td_T}(\mathrm{G}) \end{aligned}$$

Thus, there exist exactly r_k vertices with degree $\Delta_{td_T}(G)$.

To prove $\mathrm{td}_I\left(v_k^{(t)}\right) = \delta_{\mathrm{td}_I}(\mathrm{G})$, for $t = 1, 2, 3, \ldots, r_k$

Since $\mathrm{I}(v_k^{(t)})$ is minimum vertex indeterminate-membership,

$\mathrm{I}\left(v_k^{(t)}, v_k^{(j)}\right) = y_k, t \neq j,\ \forall t, j = 1, 2, \ldots, r_k.$

$I\left(v_k^{(t)}, v_i^{(j)}\right) = \max\left\{I\left(v_k^{(t)}\right), I\left(v_i^{(j)}\right)\right\} = I\left(v_i^{(j)}\right), \forall\, t = 1, 2, \ldots, r_k,\ \mathrm{j} = 1, 2, 3, \ldots, r_i$
$i = 1, 2, \ldots, k-1$

Thus, $\forall\ t = 1, 2, .., r_k$,

$td_I\left(v_k^{(t)}\right) = \sum_{i=1}^{n} I_i = \delta_{\mathrm{td}_I}(\mathrm{G})$ by Theorem 3.27 (the proof is similar to vertex truth-membership).

Now, if v_m is vertex such that $I_m = y_{k-1}$, then $td_I(v_m) = \delta_{\mathrm{td}_I}(\mathrm{G})$ (the proof is similar to vertex truth-membership).

So, there exist exactly r_k vertices with degree $\delta_{td_I}(G)$.

Similarly, it can be proved that t $d_F\left(v_k^{(t)}\right) = \delta_{\mathrm{td}_F}(\mathrm{G})$, for $\mathrm{t} = 1, 2, \ldots, r_k$.

Since $\mathrm{F}(v_k^{(t)})$ is minimum vertex false-membership,

$$F\left(v_k^{(t)}, v_k^{(j)}\right) = z_k, t \neq j,\ t, j = 1, 2, \ldots, r_k$$

$F\left(v_k^{(t)}, v_i^{(j)}\right) = \max\left\{F\left(v_k^{(t)}\right), F\left(v_i^{(j)}\right)\right\} = F\left(v_i^{(j)}\right)$ for $t = 1, 2, \ldots, r_k$, $j = 1, 2, \ldots, r_i,\ i = 1, 2, 3, ..k-1$.

Thus for $\mathrm{t} = 1, 2, 3, \ldots, r_k$,

$\mathrm{td}_F\left(v_k^{(t)}\right) = \sum_{i=1}^{n} F_i = \delta_{\mathrm{td}_F}(\mathrm{G})$ by Theorem 3.27.

Now, if v_m is vertex such that $F_m = z_{k-1}$, then

$$\mathrm{td}_F(v_m) = \delta_{td_F}(G)$$

So, there exist exactly r_k vertices with degree $\delta_{\mathrm{td}_F}(\mathrm{G})$.

5 Conclusion

In this paper, the idea of strong degree is imposed on the existing concepts of degrees in SVNGs. After that, we defined the vertex true-membership, vertex indeterminate-membership, and vertex false-membership sequence in SVNG with proofs *and* suitable examples. In the next research, the proposed concepts can be extended to labeling neutrosophic graph and also characterize the corresponding properties.

Acknowledgment. The authors would like to thank anonymous reviewers for the constructive suggestions that improved the quality of the paper.

References

1. Smarandache, F.: Neutrosophy. Neutrosophic Probability, Set, and Logic, ProQuest Information & Learning, Ann Arbor, 105 p. (1998)
2. Atanassov, K.: Intuitionistic fuzzy sets: theory and applications, Physica, New York (1999)
3. Smarandache, F.: A Unifying Field in Logic. Neutrosophy: Neutrosophic Probability, Set, Logic, 4th edn. American Research Press, Rehoboth (2005)
4. Zadeh, L.: Fuzzy logic and approximate reasoning. Synthese **30**(3–4), 407–428 (1975)
5. Zadeh, L.: Fuzzy sets. Inf. Control **8**, 338–353 (1965)
6. Atanassov, K.: Intuitionistic fuzzy sets. Fuzzy Sets Syst. **20**, 87–96 (1986)
7. Wang, H., Smarandache, F., Zhang, Y., Sunderraman, R.: Single valued neutrosophic sets. Multispace Multisrtuct. **4**, 410–413 (2010)
8. http://fs.gallup.unm.edu/NSS
9. Broumi, S., Talea, M., Bakali, A., Smarandache, F.: Single valued neutrosophic graphs. J. New Theor. **10**, 86–101 (2016)
10. Broumi, S., Talea, M., Bakali, A., Smarandache, F.: On bipolar single valued neutrosophic graphs. J. New Theor. **11**, 84–102 (2016)
11. Broumi, S., Bakali, A., Talea, M., Smarandache, F.: Isolated single valued neutrosophic graphs. Neutrosophic Sets Syst. **11**, 74–78 (2016)
12. Broumi, S., Smarandache, F., Talea, M., Bakali, A.: An introduction to bipolar single valued neutrosophic graph theory. Appl. Mech. Mater. **841**, 184–191 (2016)
13. Broumi, S., Smarandache, F., Talea, M., Bakali, A.: Decision-making method based on the interval valued neutrosophic graph. In: Future Technologie, pp. 44–50. IEEE (2016)
14. Broumi, S., Talea, M., Smarandache, F., Bakali, A.: Single valued neutrosophic graphs: degree, order and size. In: IEEE World Congress on Computational Intelligence, pp. 2444–2451 (2016)
15. Broumi, S., Bakali, A., Talea, M., Smarandache, F.: Shortest path problem under trapezoidal neutrosophic information. In: Computing Conference 2017, 18–20 July 2017, pp. 142–148 (2017)
16. Broumi, S., Bakali, A., Talea, M., Smarandache, F., Vladareanu, L.: Computation of shortest path problem in a network with SV-trapezoidal neutrosophic numbers. In: Proceedings of the 2016 International Conference on Advanced Mechatronic Systems, Melbourne, pp. 417–422 (2016)

17. Broumi, S., Bakali, A., Talea, M., Smarandache, F., Vladareanu, L.: Applying Dijkstra algorithm for solving neutrosophic shortest path problem. In: Proceedings of the 2016 International Conference on Advanced Mechatronic Systems, Melbourne, pp. 412–416 (2016)
18. Broumi, S., Talea, M., Bakali, A., Smarandache, F.: Interval valued interval valued neutrosophic graphs. In: Critical Review, XII, pp. 5–33 (2016)
19. Broumi, S., Bakali, A., Mohamed, T., Smarandache, F., Vladareanu, L.: Shortest path problem under triangular fuzzy neutrosophic information. In: 10th International Conference on Software, Knowledge, Information Management & Applications (SKIMA), pp. 169–174 (2016)
20. Broumi, S., Smarandache, F., Talea, M., Bakali, A.: Operations on interval valued neutrosophic graphs. In: Smarandache, F., SurpatiPramanik (eds.) New Trends in Neutrosophic Theory and Applications, pp. 231–254 (2016). ISBN 978-1-59973-498-9
21. Broumi, S., Bakali, A., Talea, M., Smarandache, F., Ali, M.: Shortest path problem under bipolar neutrosphic setting. Appl. Mech. Mater. **859**, 59–66 (2016)
22. Ashraf, S., Naz, S., Rashmanlou, H., Malik, M.A.: Regularity of graphs in single valued neutrosophic environment. J. Intell. Fuzzy Syst. **33**(1), 529–542 (2017)
23. Naz, S., Rashmanlou, H., Malik, M.A.: Operations on single valued neutrosophic graphs with application. J. Intell. & Fuzzy Syst. **32**(3), 2137–2151 (2017)
24. Mehra, S., Singh, M.: Single valued neutrosophic signed graphs. Int. J. Comput. Appl. **157** (9), 32–34 (2017)
25. Hamidi, M., Bourumand Saeid, A.: Accessible single valued neutrosophic graphs. J. Appl. Math. Comput. 21 (2017)
26. Akram, M., Shahzadi, G.: Operations on single-valued neutrosophic graphs. J. Uncertain Syst. **11**, 1–26 (2017)
27. Akram, M.: Single-valued neutrosophic planar graphs. Int. J. Algebra Stat. **5**(2), 157–167 (2016)
28. Akram, M., Shahzadi, S.: Neutrosophic soft graphs with application. J. Intell. Fuzzy Syst. 1–18 (2016). https://doi.org/10.3233/jifs-16090
29. Karunambigai, M.G., Buvaneswari, R.: Degrees in intuitionistic fuzzy graphs. Ann. Fuzzy Math. Inform. (2016)
30. Hassan, A., Malik, M.A., Broumi, S., Bakali, A., Talea, M., Smarandache, F.: Special types of bipolar single valued neutrosophic graphs. Ann. Fuzzy Math. Inform. **14**(1), 55–73 (2017)
31. Mullai, M., Broumi, S., Stephen, A.: Shortest path problem by minimal spanning tree algorithm using bipolar neutrosophic numbers. Int. J. Math. Trends Technol. **46**(2), 80–87 (2017)
32. Broumi, S., Bakali, A., Talea, M., Smarandache, F., Kishore Kumar, P.K.: Shortest path problem on single valued neutrosophic graphs. In: International Symposium on Networks, Computers and Communications (ISNCC) (2017)

A Pro-Active and Adaptive Mechanism for Fast Failure Recovery in SDN Data Centers

Renuga Kanagavelu(✉) and Yongqing Zhu

Data Center Technology Division, A*STAR Data Storage Institute, Singapore, Singapore
{Renuga_k,ZHU_yongqing}@dsi.a-star.edu.sg

Abstract. As modern data centers continue to grow in size and complexity to host different kinds of applications, it is required to have an efficient proactive failure management for Data Center reliability. Although Software-Defined Networking (SDN) and its implementation OpenFlow facilitate dynamic management and the configuration of Data center networks, network failure recovery in a timely manner remains great challenging. The centralized SDN controller is responsible for monitoring the entire network health status and maintain the end-to-end connectivity between the hosts. In the event of a link failure, the controller either computes a new backup path reactively on demand and creates flow table entries for the new backup path, or pro-actively computes the backup path a-priori and set up flow table rules for the pre-defined backup path. Switching to the predefined backup path locally results in faster recovery time compared to switching to the backup path that establish on demand. In this paper, we propose a proactive mechanism to provide fast recovery upon a link failure. With the proposed proactive approach, we compute the recovery (or backup) paths for the flows prior to failures and install appropriate rules in the forwarding tables at the switches in advance. Such recovery paths are adaptively updated based on the current load state of the network to improve resource efficiency and reduce congestion. By providing the backup forwarding rules in advance, upon a failure, the failed traffic is rerouted without interacting with the controller, thus ensuring fast recovery. We demonstrate the effectiveness of the proposed mechanisms using an experimental testbed with Openstack platform and simulated environment with Mininet.

Keywords: Software-defined network · Fast-failover · OpenFlow Recovery · Data center network

1 Introduction

Ever-increasing cloud data traffic drives immense interest in improving the cloud data center performance. Cloud data centers contain tens of thousands of servers to deliver huge computational power needed by the cloud applications. Inside the

K. Arai et al. (Eds.): FICC 2018, AISC 886, pp. 239–257, 2019.
https://doi.org/10.1007/978-3-030-03402-3_17

Data Center, these servers are connected by data center network topology. The selection of data center topology depends on the performance metrics. Failures are common in such a large scale Data Center network topology as they contain huge number of servers, switches and links. Once failure occurs, the time taken to detect failure, finding the alternate paths and rerouting the flows in the alternate paths results in unacceptably much longer recovery time. The longer recovery time significantly impacts the latency-sensitive real time applications like stock-trading, online banking, and may cause severe performance degradation. It is crucial to design the network topology to quickly recover from failures to provide high availability.

Software-defined networking (SDN) [1,2] is a special architecture in which the control plane is decoupled from data plane. In SDN, network devices focus on data plane functionalities and the control plane functionalities are managed by a centralized entity called Controller. The controller is responsible for monitoring the network status, routing and failure detection. OpenFlow [3] is a popular protocol used in SDN architecture to forward the packets in the data plane based on the rules in the centralized controller. It is flexible enough to develop routing/rerouting policies that are specifically tuned for particular needs. Apart from this, the control plane separation leads to fine-grained, adaptive traffic management solutions.

High availability can be achieved by having redundant links and switches present in the topology. Failure recovery can be done in a reactive or proactive way in SDN Data centers. In the reactive approach, a failure recovery algorithm at the SDN Controller chooses new alternate paths in the network topology upon a switch or link failure. Hence, when a link/switch breaks, the controller needs to reconfigure the network to restore or maintain end-to-end connectivity for all paths. The broken path restoration time includes the failure detection time, time to notify the controller, alternate path computation and rerouting time. As a result, controller-initiated reactive path restoration may take longer time to recover with lots of messages exchanged between controller and the switches in the network. In the proactive approach, the backup paths are installed a priori at the switchs forwarding table along with the primary paths. When a link failure is detected at the switch, it can use the backup path directly without the intervention of the controller. This reduces the failure recovery time.

Our work proposes an efficient proactive and adaptive fast recovery mechanism for SDN based data center networks. In the proposed mechanism, we compute a set of candidate link disjoint paths (with the working or primary path) between each source-destination pair prior to the failures. Those paths are considered as backup paths (or recovery paths) that can be used for data transmission upon failures with improved resource utilization. Appropriate rules in the forwarding tables and group tables are installed at the switches in advance. Such recovery paths are adaptively updated based on the current load state of the network to improve resource efficiency and reduce congestion. Upon link failure, the least-cost backup path among the candidate backup paths for a source-destination pair will be selected so as to achieve better performance.

By providing the backup forwarding rules in advance, upon a failure, failed traffic is rerouted without interacting with the controller, thus ensuring fast recovery.

We demonstrate the effectiveness of the proposed mechanisms using an experimental testbed with Openstack platform and simulated environment using Mininet.

The rest of the paper is organized as follows. In Sect. 2, we discuss the related work. In Sect. 3, we present the architecture of the proposed work. We present our proposed technique in Sect. 4. Section 5 introduces the evaluation methodology and experimental setup. We make concluding remarks in Sect. 6. We present the future work in Sect. 7.

2 Background and Related Work

2.1 Failure Detection and Recovery

In this section, the basic concepts and the realization of various failure detection mechanisms in SDN are described. In general, there are two ways to recover from failures. In a reactive restoration, upon the link/node failure, the new paths are computed on demand. On the other hand, in a proactive protection, the backup paths are pre-configured in advance enabling fast switch-over upon failure.

2.2 Controller-Driven Recovery

This is a reactive method in which the controller will be notified through OFPort-Down message when a link failure occurs. Upon receiving this notification, the controller will compute the new alternate path and install new entries in the effected switch via Flow-mod message. The reactive time taken for this method is longer as the following things happened upon link failure as shown in Fig. 1: (1) the failure has to be detected; (2) the SDN controller should be notified; (3) alternate path needs to be computed; and (4) switches to be notified about new entries. The time taken for the entire process cannot be tolerated in carrier-grade networks in which the failure recovery requirement is 50 ms [4].

2.3 Fast Failover Groups

The openflow's fast failover group (FFG) mechanism is used to recover the failures in a proactive way. The SDN Controller pre-computes working and backup paths and pre-inserted into all switches flow table and Group table in advance; Upon the link failure in the working path, the failure recovery involved the following steps as shown in Fig. 2: (1) upon the link failure, once the switch detects port down, the switch can select the backup path from its Group table; and (2) send packets to the backup path without consulting controller. Since controller is not involved, it takes faster time to recover from failures. The Group table entries are shown in Fig. 3, allows OpenFlow switches to do the recovery locally without consulting the controller. FFG connects many action buckets. Each bucket has

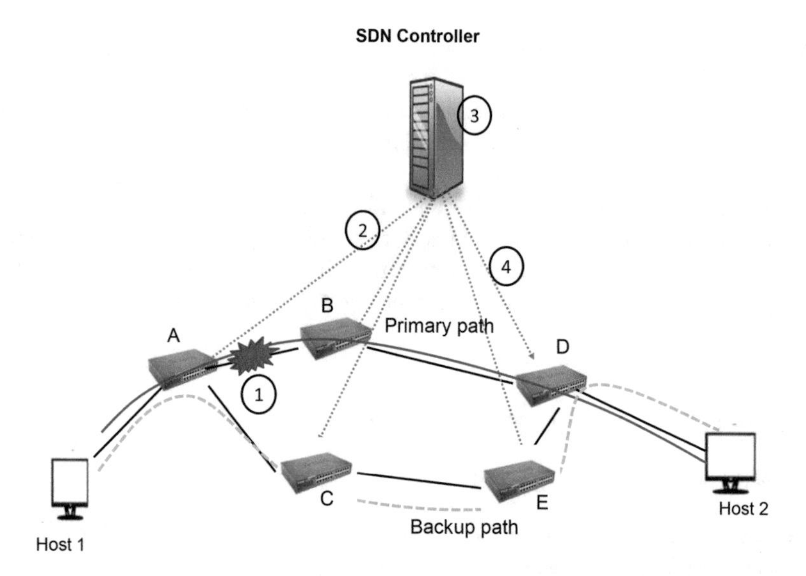

Fig. 1. Controller-driven reactive recovery.

Watch Port, actions and associated parameters. The watch port is to monitor the status of the indicated Port. The bucket will not be used if the associated port is down. A buckets action is to forward incoming packets on a defined output port. OpenVswitch beyond version 2.3 supports group chaining in which one group forwards to another group.

2.4 Bidirectional Forwarding Detection

Bidirectional Forwarding Detection (BFD) [5] is a protocol used to detect failures in a fast manner proposed by Juniper Networks in 2010. BFD sends periodic BFD control packets in both directions over the path connecting two end system. The system which receives this packet replies its status through echo message. If one of the two systems stops receiving BFD messages and is not responding within a time frame, the link is assumed to be failed.

2.5 Related Work

To meet carrier-grade networks recovery requirements (less than 50 ms), Sharma et al. [6] used the group table [3] in OpenFlow to meet the carrier grade requirement with the protection approach. They used per-path Bidirectional Forwarding Detection (BFD), to detect the failure between end hosts. Van et al. [7] used per-link BFD method for faster recovery.

In [8], Sgambelluri et al. proposed a segment-protection approach based on pre-installed backup paths with different priorities. They suggested the mechanism which react to the failures by auto rejecting flow entries of the failed

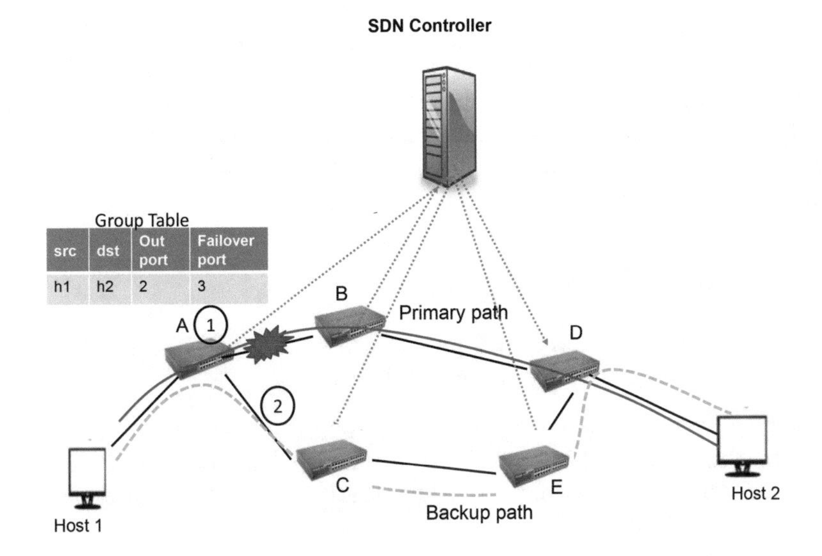

Fig. 2. Proactive recovery.

Group Table

Group ID	Group Type	Counter	Action Buckets
#1	Fast-failover	777	Outport X,Y

Flow Table

Switch Port	MAC src	MAC dst	Ether Type	VLAN ID	Src IP	Dst IP	Proto No.	TCP S Port	TCP D Port	Action
Port 1	*	*	*	*	1.2.2	*	*	*	*	Port 3
Port 1	00:FF …	*	0800	*	1.2.3 …	11.1...	*	*	*	Group #1

Fig. 3. Fast failover group.

interface. Lee et al. in [9], aim to reduce the load at the controller by proposing controller based monitoring and optimization scheme. Borokhovich et al. in [10] proposed graph search algorithms to randomly try new ports to reach traffic demands destination. Mohan et al. in [11], considered TCAM usage reduction while routing backup paths. They considered bandwidth reservation and backup sharing on the paths and do not consider congestion.

Ramos et al. [12] propose SlickFlow, in which the controller calculates a primary path with minimum latency and a disjoint backup path. Then they are encoded in the packet header along with an alternative bit to indicate that the path being used. Upon the link failure in the primary path, the switch forwards the packet through the backup path, changing the alternative bit to indicate to the subsequent switches, which one must be used. This method has the limitation on the number of backup path hops as the header is encoded.

Borokhovich et al. [10] use graph search problem to model the link failover. Modulo algorithm, Breath-First search (BFS) and Depth-first search (DFS) algorithms are applied. Modulo algorithm forwards the packets to the switch ports in a round-robin fashion until the packet reaches destination. Depth-first search algorithm forwards the packets to the next hop until the failed link and sends the packets to the parent to choose another port. BFS algorithm sends the packet immediately back to its parent, forwarding it only when all neighbors of its parent have been visited.

3 System Architecture

The functional components of our proposed framework is shown in Fig. 4. Our system architecture has the following modules:

- Monitor: This module is responsible for collecting statistics from the OpenFlow switches that will be used by the Routing Component to compute and compare the load on various links. These statistics are collected at fixed intervals (every 5 ms).

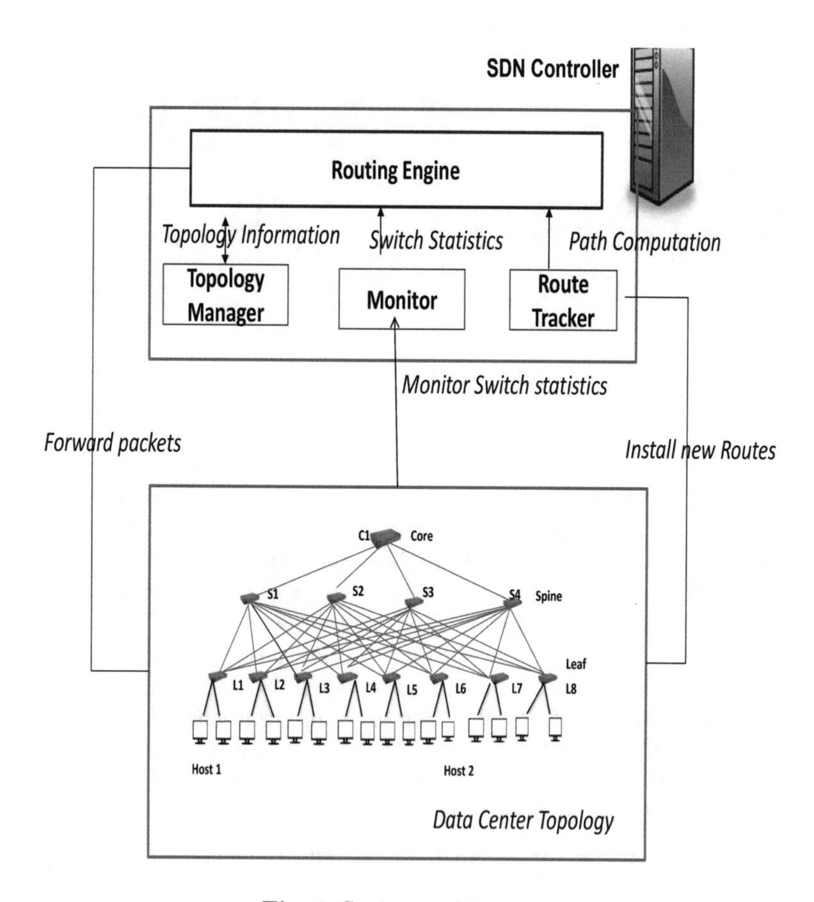

Fig. 4. System architecture.

- Topology Manager: This module keeps track of all the switches and hosts detected on the network.
- Routing Engine: The main functionality of this module is to route/re-route the flows along the least-loaded path. Based on the most recent load statistics collected by the Monitor module, it computes the possible candidate least loaded-disjoint paths between any pair of end hosts.
- Route tracker: This module is responsible for the installation of the primary routing paths and optimal backup paths chosen by the Routing Engine into the series of OpenFlow switches as flow entries.

The following sections describe the functionalities of host tracker, routing engine module and route tracker.

3.1 Host Tracker

The *Host Tracker* module maintains the list of all the hosts that are detected on the network. It maintains the information about the MAC address, IP address of the interface of the hosts, the OpenFlow switch id and the OpenFlow switch port number to which each of the hosts is connected. Table 1 shows an sample of the entries stored in this module.

Table 1. Host tracker entries

Switch ID	MAC Addr	IP	Port No.
$Switch_1$	$11 : AA : 23 : FE : 1C : 4A$	10.200.1.1	1
$Switch_2$	$22 : BB : 34 : EA : 2D : 9B$	10.200.1.2	2

3.2 Routing Engine Module

This section describes in detail the functionalities of the *Routing Engine.* The routing engine is responsible for computing the least-loaded shortest disjoint paths between the end hosts based on the statistics collected from the *Monitor* component. To illustrate the routing mechanism, we consider a Spine-and-Leaf Topology. Figure 5 shows a two-tier Clos network based spine-and-leaf architecture [13]. This Leaf-Spine architecture is designed to provide high bandwidth, low latency and scalability. This Clos architecture has two tiers namely spine layer (top-tier) and leaf layer (lower-tier). Each leaf layer (lower-tier) switch is connected to every spine layer (top layer) switch in a full-mesh topology. The servers are connected to leaf layer switches. The spines are not interconnected with each other. The leaf switches are not interconnected with each other. The spine layer is the backbone of the network and is responsible for interconnecting all leaf switches. Since each leaf layer switch is connected to every spine layer switch in a fabric, there are multiple redundant paths available for traffic between any pair of leaf switches.

Consider the case of routing between host 1 and host 2 in Fig. 5. Let S represent the set of candidate shortest paths s_1, s_2, s_3, s_4.

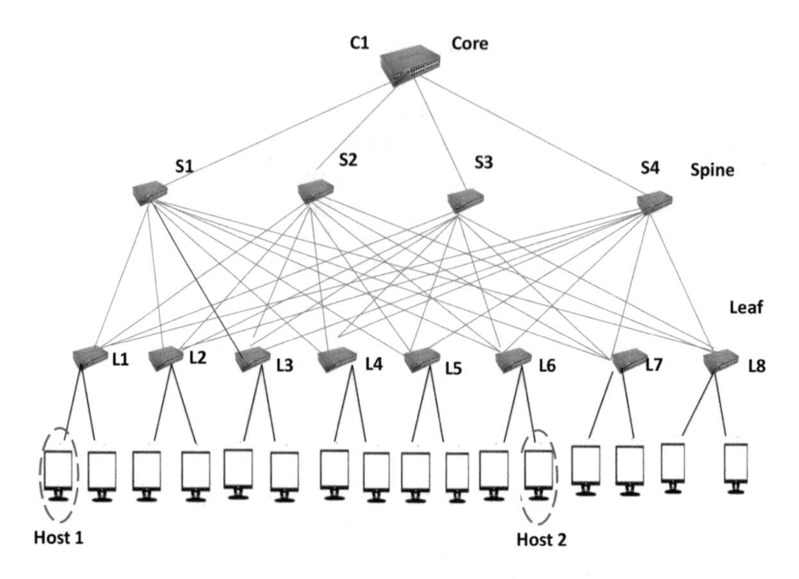

Fig. 5. Spine-and-leaf topology.

s_1 traverses $Host\ 1 - L_1 - S_1 - L_6 - Host\ 2$,
s_2 traverses $Host\ 1 - L_1 - S_2 - L_6 - Host\ 2$,
s_3 traverses $Host\ 1 - L_1 - S_3 - L_6 - Host\ 2$, and
s_4 traverses $Host\ 1 - L_1 - S_4 - L_6 - Host\ 2$.

Each $link_i$ has a weight W_i. The value of link load L_i is estimated from the change in byte count for all flow entries in a switch between the two most recent snapshots from the "Monitor" module. The controller calculates the path load for each of the candidate paths for a pair of ToRs as the maximum load on any of the links traversed by the path. Mathematically this is represented as path load pl^i_{mn} where m and n are edge switches and i is the path identifier. Let L_j be the load on link j; $P^i_{m,n}$ be the ith path from ToR_m to ToR_n. $L^i_{m,n}$ be the load on i^{th} path from ToR_m to ToR_n.

$$\begin{aligned} Let\ P^i_{m,n,j} &= 1\ if\ link\ j\ is\ on\ path\ P^i_{m,n};\\ &= 0\ otherwise; \end{aligned} \tag{1}$$

The path load is calculated as follows:

$$L^i_{m,n} = max_{j\ \epsilon\ network}(L_j * P^i_{m,n,j}) \tag{2}$$

After computing $L^i_{m,n}$ for all $s_i\ \epsilon$ S, controller chooses the path with minimum pathload $L^i_{m,n}$ as the primary path and the next least-loaded path as the backup path. The backup path entity is installed in the switch group table and the flow entries are inserted into the series of switches on the chosen primary path. In our example, suppose s_3 is a shortest path with minimum path load, it is considered as a primary path and flow entries are inserted into the switches $L_1 - S_3 - L_6$. Based on the fact that the network is not always stable and might congested, the controller computes path load on the backup paths between any two hosts periodically and update the flow tables and group tables accordingly.

3.3 Route Tracker

The *Route Tracker* is used to install the flow entries into the OpenFlow switches as chosen by the *Routing Engine* module. It maintains the information all the installed routes as well as their optimal backup routes shown in Table 2.

Table 2. Route tracker entries

Entry	S.MAC	D.MAC	Type	IP PROTO	Route
1	11:AA:23:FE:1C:4A	22:BB:22:EA:2D:9B	IP	TCP	$L_1 - S_2 - L_6$
2	22:BB:22:EA:2D:9B	11:AA:23:FE:1C:4A	IP	UDP	$L_2 - S_3 - L_4$

4 Proposed Proactive and Adaptive Failure Recovery Algorithm

The proposed *Proactive and Adaptive Failure Recovery Algorithm* is described in the following steps as shown in Fig. 6.

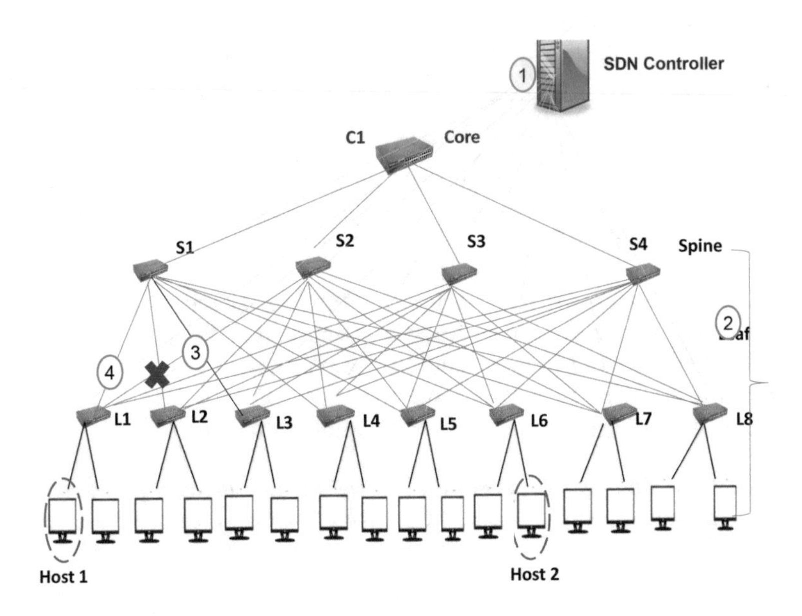

Fig. 6. Proactive and adaptive failure recovery algorithm illustration

(1) The controller Computes least-loaded shortest primary path and candidate backup paths between each source-destination pair and return the ordered list of backup paths according to their residual capacity.
(2) The primary path and the ordered list of backup paths are pre-inserted to all switches in advance.
(3) Upon the node/link failure, the switch will chose the highest priority backup path locally for forwarding.
(4) The controller will recompute the primary and reorder the list of the backup paths according to their residual capacity in a periodic manner.

The Pseudo code is given in Algorithm 1.

Algorithm 1. Proactive and Adaptive Failure Recovery

```
1: Input: Network Topology G(V,E) /* Here ToR_m and ToR_n are edge switches */
2: for each new flow F do
3:    Compute candidate disjoint paths between ToR_m and ToR_n switches
4:    for each candidate path do
5:       compute path load using equation (2)
6:    end for
7:    Find the primary path with minimum path load
8:    Choose the next least-loaded path as backup path
9:    Install flow table and group table entries for primary and backup path
10:   Add GROUP TYPE =fast failover
11:   add bucket with ID
12:   install backup path
13: end for
14: if node/link failure then
15:    Reroute the traffic locally to the backup as per the group table entry
16: end if
```

We now introduce the failure recovery mechanism using an example on Spine-and-Leaf network as shown in Fig. 7. Suppose that link L_1 - S_3 fails in the network as shown in Fig. 7.

The Host 1 is connected to Port 1 of low-tier leaf switch L_1.
Port 2 of leaf switch L_1 is connected to spine switch S_1.
Port 3 of leaf switch L_1 is connected to spine switch S_2.
Port 4 of leaf switch L_1 is connected to spine switch S_3.
Port 5 of leaf switch L_1 is connected to spine switch S_1.

Suppose that link L_1 - S_3 fails and leaf switch L_1 detects *"portdown"* event, leaf switch L_1 can locally fail over the affected flows to another backup path through Port 3 as installed in group table which greatly reduces the recovery time.

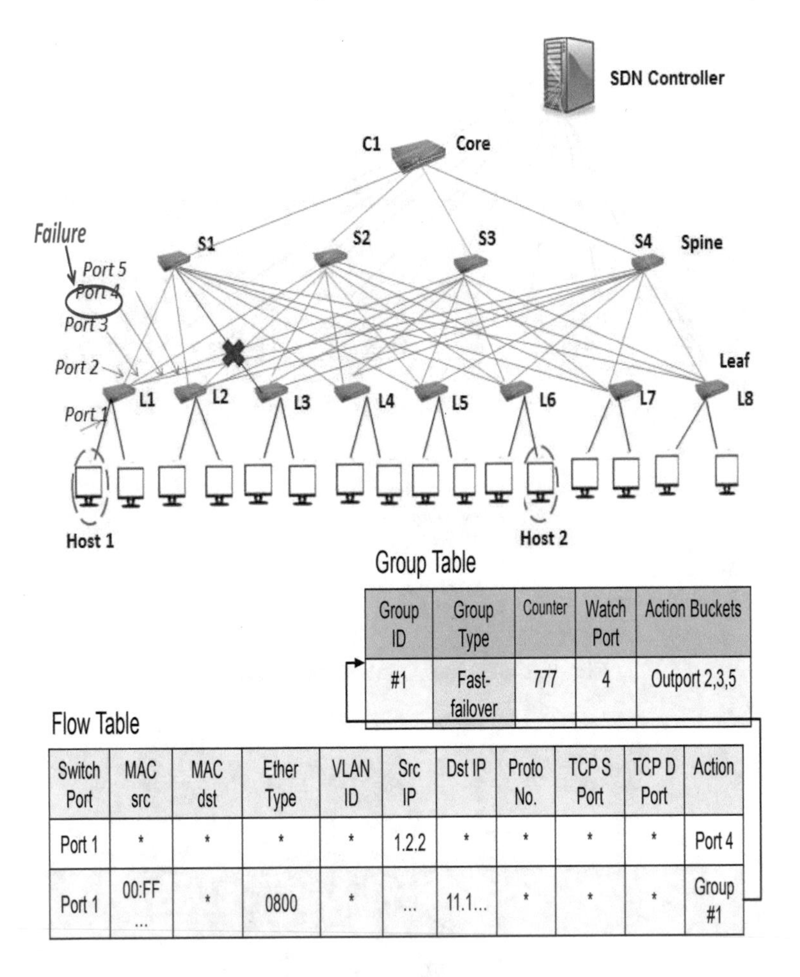

Group ID	Group Type	Counter	Watch Port	Action Buckets
#1	Fast-failover	777	4	Outport 2,3,5

Switch Port	MAC src	MAC dst	Ether Type	VLAN ID	Src IP	Dst IP	Proto No.	TCP S Port	TCP D Port	Action
Port 1	*	*	*	*	1.2.2	*	*	*	*	Port 4
Port 1	00:FF ...	*	0800	*	...	11.1...	*	*	*	Group #1

Fig. 7. Illustration of failure recovery.

5 Performance Study

We evaluate the performance of the proposed Proactive and adaptive failure recovery algorithm by using two test beds, one is Openstack environment and the other one is simulation environment. The Openstack environment is shown in Fig. 8. The physical test bed consists of three neutron nodes (network nodes), two hosts installed with OVS on compute nodes (Nova nodes) and are managed by Openstack Manager. We use Ryu [14] Controller as SDN controller. We use OpenVswitch version 2.5.0. and OpenFlow version 1.3, which supports OpenFlow group tables. The hardware configurations of our experiments are shown in Tables 3 and 4. For simulation, we use Mininet to create a two-tier spine-leaf topology. Each low-tier leaf switch acting as OpenVswitch(OVS) and two hosts are connected to each leaf switch as shown in Fig. 9.

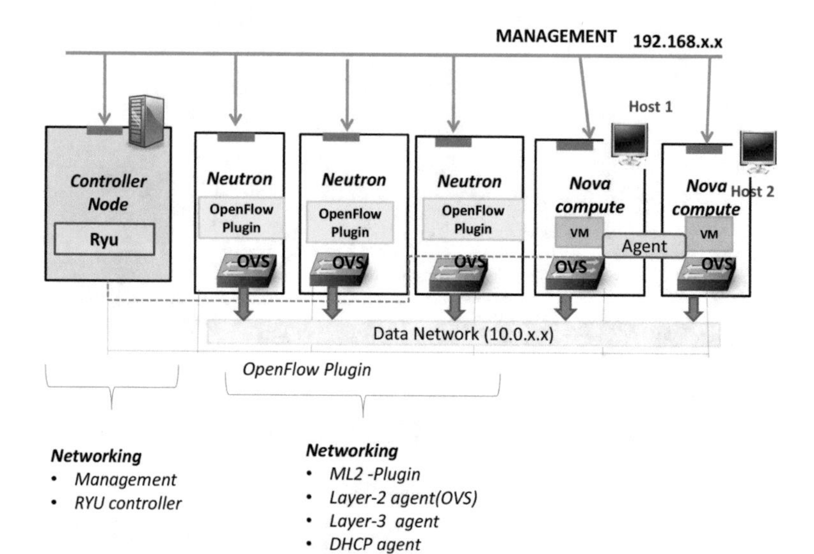

Fig. 8. Openstack testbed environment.

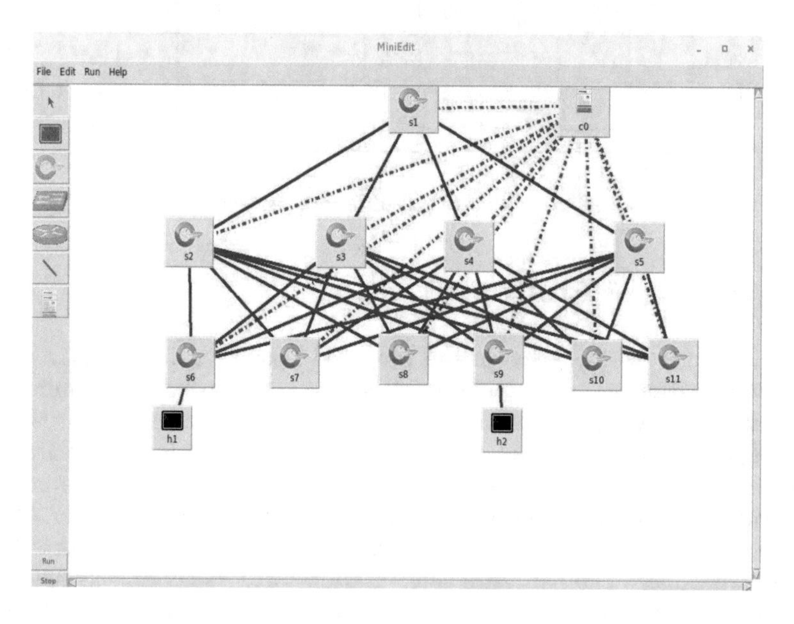

Fig. 9. Mininet testbed - spine-and-leaf topology.

Table 3. Openstack testbed configuration

	Controller node	Compute node	Network node
Operating System	CentOS Linux release 7.3.1611 (Core)	CentOS Linux release 7.3.1611 (Core)	CentOS Linux release 7.3.1611 (Core)
Memory	8.0 GB	8.0 GB	8.0 GB
Processor	Intel Xeon CPU	Intel Xeon CPU	Intel Xeon CPU
# of Processor	2 Cores	2 Cores	2 Cores

Table 4. Mininet testbed configuration

	Controller Node (Ryu)
Operating System	CentOS Linux release 7.3.1611 (Core)
Memory	8.0 GB
Processor	AMD Opteron (tm) Processor 6172
# of Processor	2 Cores

5.1 Bandwidth Measurement in Openstack Environment

In this experiment, we evaluate the effectiveness of our proactive and adaptive recovery mechanism by measuring the network bandwidth before and after link/node failure in the Openstack Environment. We compare the performance of our algorithm with reactive algorithm.

1) Proactive and Adaptive Failure Recovery Algorithm. Figure 10 shows the network bandwidth evaluation for our proactive and adaptive failure recovery algorithm in the Openstack Environment. We use Jperf [15] which is the most famous industry standard for network bandwidth evaluation. It is observed that the network bandwidth before link failure is 495 MBytes/s and network bandwidth after link failure is 473 MBytes/s. The results show that upon link failure only there is the minimal decrease in the throughput that shows the effectiveness of our algorithm.

2) Reactive Algorithm. Figure 11 shows the network bandwidth evaluation for reactive algorithm in the Openstack Environment. It is observed that the network bandwidth before link failure is 575 MBytes/s and network bandwidth after link failure is 379 MBytes/s. The results show that upon link failure, there is a significant decrease in the throughput.

5.2 Network Bandwidth Measurement in Mininet Environment

In this experiment, we evaluate the effectiveness of our proactive and adaptive recovery mechanism by measuring the network bandwidth before and after link/node failure in the Openstack Environment. We compare the performance of our algorithm with reactive algorithm.

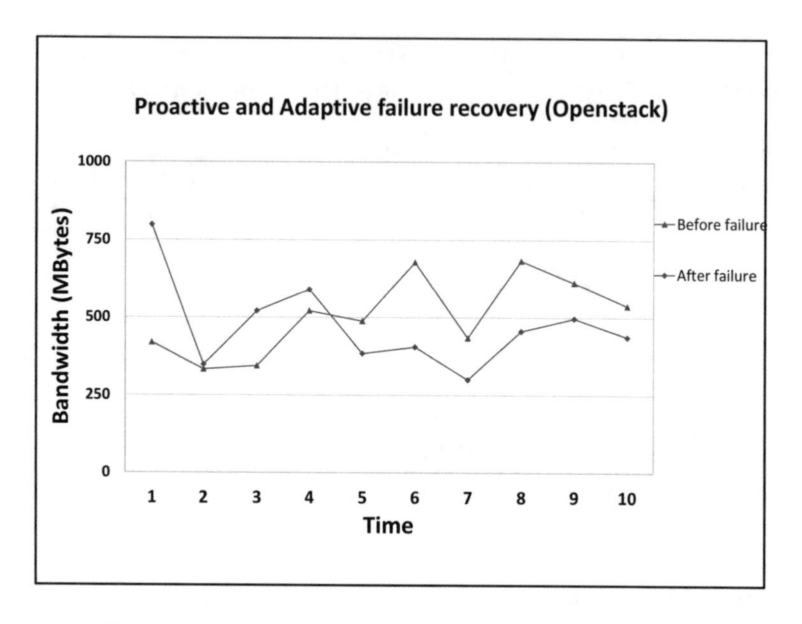

Fig. 10. Proposed proactive approach, Openstack testbed.

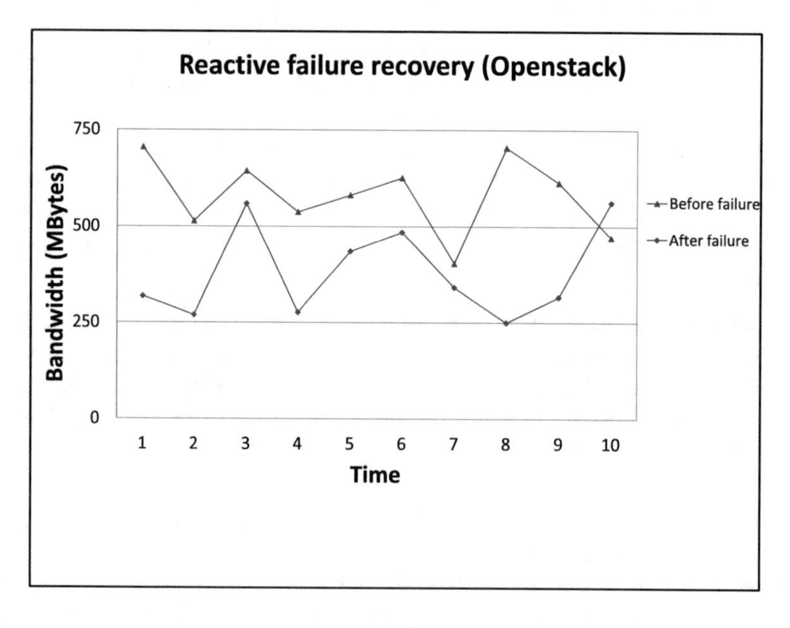

Fig. 11. Reactive approach, Openstack testbed.

1) Proactive and Adaptive Failure Recovery Algorithm. Figure 12 shows the network bandwidth evaluation for our proactive and adaptive failure recovery algorithm in the Mininet Environment. It is observed that the network bandwidth before link failure is 434 MBytes/s and network bandwidth after link failure is 424 MBytes/s. The results show that upon link failure only there is the marginal decrease in the throughput that shows the effectiveness of our algorithm.

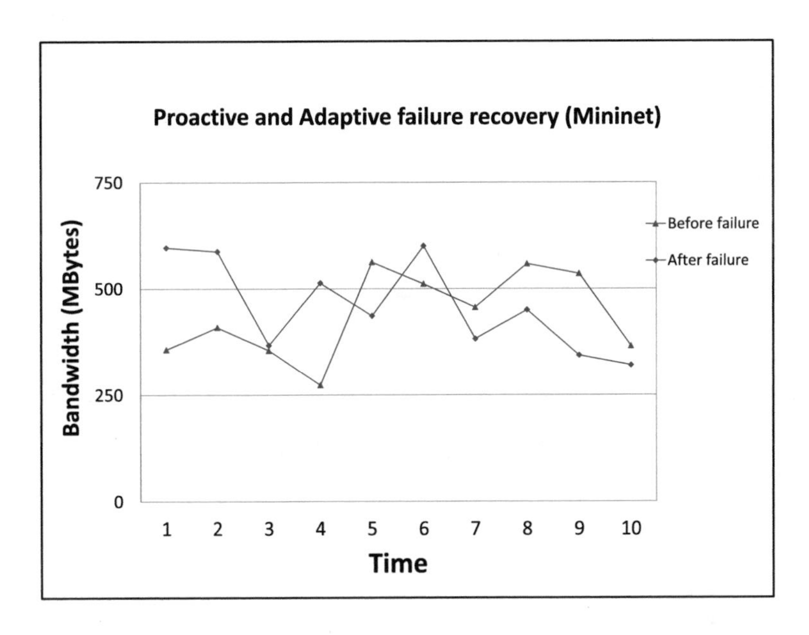

Fig. 12. Proposed proactive approach, Mininet testbed.

2) Reactive Algorithm. Figure 13 shows the network bandwidth evaluation for reactive algorithm in the Mininet Environment. It is observed that the network bandwidth before link failure is 499 MBytes/s and network bandwidth after link failure is 320 MBytes/s. The results show that upon link failure, there is a significant decrease in the throughput.

5.3 Recovery Time Measurement in Openstack Environment

In this experiment, we demonstrate the recovery time of the proactive and adaptive failure recovery mechanism in Openstack Environment. We compare it with the recovery time of the reactive algorithm. The recovery time is the time interrupted during the link failure and it is estimated as the time between the reception of the last packet before the link failure and the reception of the first packet after the link failure at the destination host. The link failure is repeated for 25 times to measure the average recovery time. Figure 14 shows the recovery time of our Proactive and reactive algorithms. It is observed that the average recovery

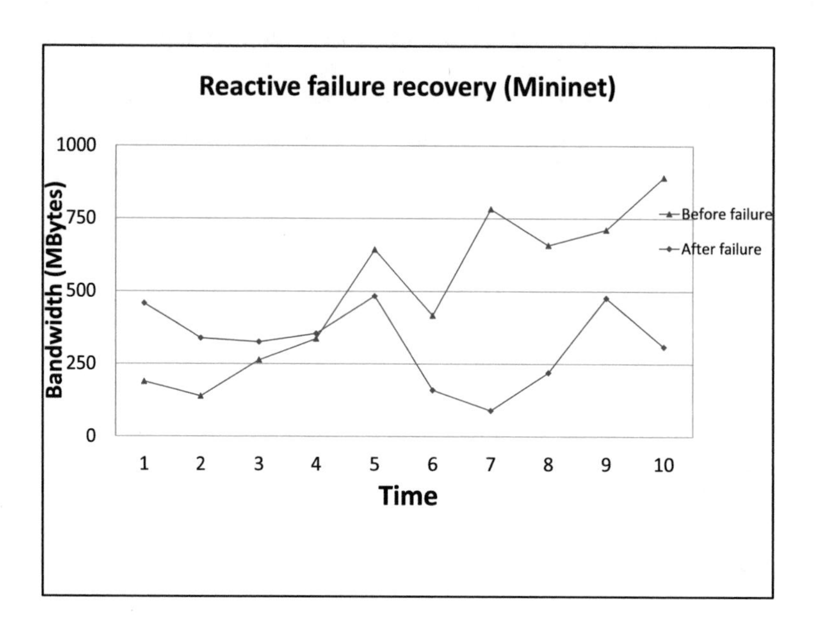

Fig. 13. Reactive approach, Mininet testbed.

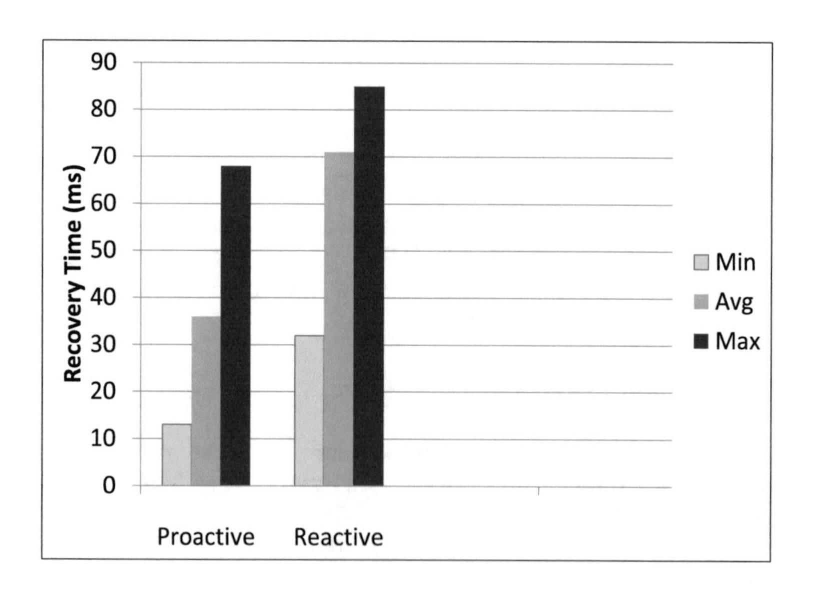

Fig. 14. Average recovery time in Openstack testbed.

time for the proactive mechanism is less than the reactive mechanism because the controller is not involved in the failure recovery whereas in the reactive mechanism the controller has to be involved in the computation of the on demand backup path upon the link failure.

5.4 Recovery Time Measurement in Mininet Environment

In this experiment, we demonstrate the recovery time of the proactive and adaptive failure recovery mechanism in Mininet Environment. We compare it with the recovery time of the reactive algorithm. The recovery time is the time interrupted time during the link failure and it is estimated as the time between the reception of the last packet before the link failure and the reception of the first packet after the link failure at the destination host. The link failure is repeated for 25 times to measure the average recovery time. Figure 15 shows the recovery time of our Proactive and reactive algorithms. It is observed that the average recovery time for the proactive mechanism is less than the reactive mechanism because the controller is not involved in the failure recovery.

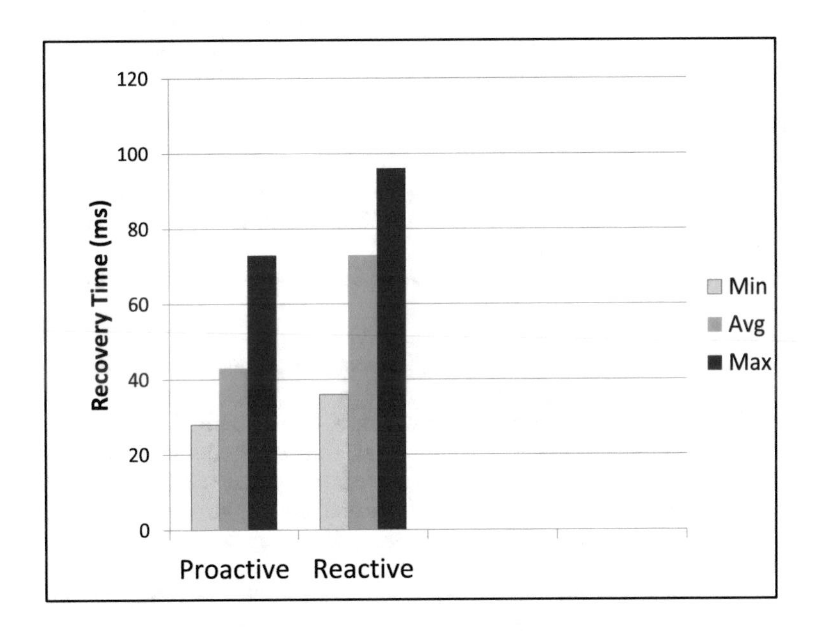

Fig. 15. Average recovery time in Mininet testbed.

6 Conclusion

In this paper, we proposed a proactive mechanism to provide fast recovery upon a link failure. With the proposed proactive approach, we compute the recovery (or backup) paths for the flows prior to failures and install appropriate rules in the forwarding tables at the switches in advance. Such recovery paths are adaptively updated based on the current load state of the network to improve resource efficiency and reduce congestion. By providing the backup forwarding rules in advance, upon a failure, the failed traffic is rerouted without interacting with the controller, thus ensuring fast recovery. We demonstrated the effectiveness of the proposed mechanisms using an experimental testbed with Openstack and simulation testbed with Mininet platform.

7 Future Work

We can further extend this work to consider a global and integrated control framework considering failure prediction and recovery in virtualized resources of network, computing server, and storage systems. Such an integrated approach has an advantage of improved resource usage, quick failure recovery and predictable performance for tenants. We can further extend this work to consider a global and integrated control framework considering the failure prediction and recovery in virtualized resources of network server (computing and storage) systems.such an integrated approach has an advantage of improved resource usage, quick failure recovery and predictable performance for tenants.

References

1. McKeown, N., Anderson, T., Balakrishnan, H., Parulkar, L. Peterson, G., Rexford, J., Shenker, S., Turner, J.: OpenFlow: enabling innovation in campus networks. In: Proceedings of SIGCOMM (2008)
2. McKeown, N.: How SDN will shape networking. Open Networking Summit 2011 (2011)
3. Thomas, F.H.: SDN, Openflow, and Open Vswitch: Pocket Primer
4. Jenkins, B., Brungard, D., Betts, M., Sprecher, N., Ueno, S.: MPLS-TP requirements, RFC 5654, IETF (2009)
5. Katz, D., Ward, D.: Bidirectional forwarding detection (BFD) (2010)
6. Sharma, S., Staessens, D., Colle, D., Pickavet, M., Demeester, P.: OpenFlow: meeting carrier-grade recovery requirements. Comput. Commun. **36**(6), 656–665 (2013)
7. Van Adrichem, N.L.M., Van Asten, B.J., Kuipers, F.A.: Fast recovery in software-defined networks. In: 2014 Third European Workshop on Software Defined Networks. IEEE (2014)
8. Sgambelluri, A., Giorgetti, A., Cugini, F., Paolucci, F., Castoldi, P.: OpenFlow-based segment protection in ethernet networks. IEEE/OSA J. Opt. Commun. Netw. **5**(9), 1066–1075 (2013)
9. Lee, S., Li, K.Y., Chan, K.-Y., Lai, G.-H., Chung, Y.-C.: Path layout planning and software based fast failure detection in survivable OpenFlow networks. In: 2014 10th International Conference on the Design of Reliable Communication Networks (DRCN), pp. 1–8 (2014)
10. Borokhovich, M., Schiff, L., Schmid, S.: Provable data plane connectivity with local fast failover: introducing openflow graph algorithms. In: Proceedings of the Third Workshop on Hot Topics in Software Defined Networking, HotSDN 2014, pp. 121–126. ACM (2014)

11. Mohan, P.M., Truong-Huu, T., Gurusamy, M.: TCAM-aware local rerouting for fast and efficient failure recovery in software defined networks. In: IEEE Global Communications Conference (GLOBECOM). IEEE (2015)
12. Ramos, R.M., Martinello, M., Esteve Rothenberg, C.: Slickflow: resilient source routing in data center networks unlocked by openflow. In: 2013 IEEE 38th Conference on Local Computer Networks (LCN), pp. 606–613. IEEE (2013)
13. Cisco Data Center Spine-and-Leaf Architecture: Design Overview White Paper
14. RYU SDN framework - ebook
15. Blum, R.: Network Performance Toolkit: Using Open Source Testing Tools

Flow Characteristic-Aware Cache Replacement Policy for Packet Processing Cache

Hayato Yamaki(✉)

Department of Computer and Network Engineering,
The University of Electro-Communications, Chofu, Tokyo 182-8585, Japan
yamaki@uec.ac.jp

Abstract. The increase in internet traffic amount becomes a serious problem for routers from the aspects of the packet processing throughput and the power consumption. Packet processing cache (PPC) is a promising approach to meet the requirements. PPC can reduce the number of accesses to ternary content addressable memory (TCAM), which accounts for a large percentage of the power consumption of a router, by storing the TCAM lookup results into a cache memory and reusing them. For PPC, the cache miss rate has significant impact on the throughput and the power consumption. Thus, reducing the number of cache misses is a main concern for PPC. In this study, we first analyze the elephant flows and mice flows in networks to reveal the packet behavior in PPC and propose a novel cache replacement policy based on the analysis. Hit dominance cache (HDC), proposed in this paper, gives high priority to the elephant flows and evicts the mice flows rapidly. Our simulation showed HDC can reduce the number of cache misses in PPC by up to 29.1% compared to conventional 4-way LRU PPC. In addition, we estimated the hardware cost of HDC by using Verilog-HDL and showed that it is comparable to those of 4-way LRU though HDC performs as if the cache was composed of 8-way set associative cache.

Keywords: Core router · Packet processing · Cache replacement policy

1 Introduction

In recent years, the amount of internet traffic has increased explosively and applications dominant in the traffic are utilizing various large stream data, such as video streams and IoT-related data. According to the reports of The Ministry in Japan [1, 2], the total download traffic generated by Japanese broadband service contractors increased by 52.2% compared to the last year. It is expected that the network traffic will increase approximately 190-fold in 2025 compared to 2006. With the increase of internet traffic, power consumption of internet routers will account for a large part of the total power consumption in the world [3, 4]. It is also estimated that the power consumption of network equipment in Japan will increase more than 10-fold in 2025 compared to 2006 [2]. Hence, both high-throughput and low-power consumption technologies are

K. Arai et al. (Eds.): FICC 2018, AISC 886, pp. 258–273, 2019.
https://doi.org/10.1007/978-3-030-03402-3_18

required for routers and especially for internet core routers, which process a huge number of packets in the center of the internet.

In a core router, from the viewpoint of both processing throughput and power consumption, table lookups in packet processing are known as bottlenecks [5]. When processing a packet, the router needs to retrieve some information from various tables such as the routing table, the address resolution protocol (ARP) table, the access control list (ACL), and the quality of service (QoS) table. Recent core routers configure these tables with ternary content addressable memory (TCAM), which is a memory specialized for retrieving data with low latency. TCAM can obtain the data within one cycle by simultaneously comparing between an input and all data in TCAM. However, concerns exist about the throughput of TCAM not being good enough for processing packets in future networks, such as 400-Gbps networks. This is because TCAM can process packets at a rate of approximately 200 M packets per second, which is equivalent to achieving a rate of 100 Gbps, in the case where the smallest packets come continuously. Moreover, the power consumption of TCAM is more than 16 times as large as that of static random access memory (SRAM) [6]. It was estimated that the power consumed by TCAM accounts for approximately 30% of a router's total power consumption because all packets need to access to multiple tables [7, 8]. For these reasons, the improvement of the table lookup function is an important issue for achieving high-throughput and low-power processing with regard to core routers.

There are some approaches to improve the table lookup in core routers. The fundamental approaches are to modifying the TCAM architecture. In [7], the power consumption of TCAM was reduced by dividing TCAM into multiple arrays and selectively accessing them with an index calculated from the five-tuple of a packet. Furthermore, a method of processing packets with parallelized TCAM was proposed to realize high-throughput table lookups [5]. However, it is difficult for the TCAM-based solution to achieve both 400-Gbps throughput and the power efficiency comparable to SRAM. An approach with less access to TCAM is needed for future core routers.

As another approach, a cache-based solution, called packet processing cache (PPC), has been proposed. It utilizes the temporal locality of internet traffic, which means that many packets with an identical five-tuple appear in short period [9]. TCAMs provide the same results for these packets. Based on this, PPC stores the results of TCAM lookups into a small cache memory and then tries to retrieve the data from the cache before accessing to TCAMs. If a packet hits the cache, PPC can process the packet without the access to TCAMs and thus achieve low-latency and low-power packet processing. Because cache misses cause the performance degradation of packet processing with the increase of the power consumption, reducing them is an important issue for PPC. In this study, we propose a novel cache replacement policy based on the flow characteristics to reduce the cache misses in PPC.

2 Packet Processing Cache

In packet processing, one or several fields of a packet header are used to access to tables. Thus, the tables produce the same output for multiple packets if they have an identical header. Based on this fact, PPC defines packets with an identical header as a flow. PPC simply represents a header with a five-tuple (i.e., source/destination IP address, source/destination port #, and protocol #), which is used for most table lookups in a router. PPC stores all data retrieved from the tables by a flow into a cache line. As a result, a subsequent packet can be processed with the cached data (i.e., without the accesses to TCAMs) if it belongs to the same flow. Because the latency and the power consumption of a cache are smaller than those of TCAMs, PPC can achieve high-throughput and low-power packet processing.

Figure 1 shows the outline of packet processing with PPC. A cache is typically accessed with a hash value calculated from a five-tuple. A cache tag consists of the 104-bit flow information of the five-tuple. A data field contains the result of multiple table lookups, as above mentioned. The typical size of a data field is 15 bytes, which includes 1 byte as an output interface number in the routing table, 12 bytes as a destination MAC address in the address resolution protocol (ARP) table, 1 byte as results of filtering in the access control list (ACL), and 1 byte as a priority number in the quality of service (QoS) table. Furthermore, PPC can store not only the results of table lookups but also the results of encapsulation, encryption, and packet inspection conducted by network intrusion detection systems. PPC can substitute many processing of routers for one cache access.

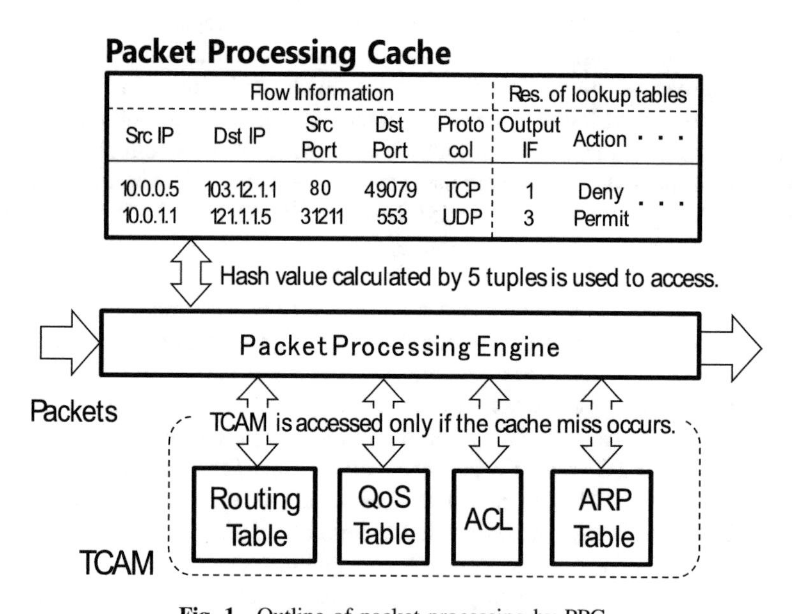

Fig. 1. Outline of packet processing by PPC.

In PPC, the cache miss rate is directly impacts on the packet processing throughput and the power consumption of routers. The throughput achieved by PPC that is represented as T_{PPC} can be calculated as:

$$T_{PPC} = \begin{cases} \dfrac{1}{l_{Cache}} \cdot 64 \text{ byte } (l_{Cache} > l_{TCAM} \cdot m) \\ \dfrac{1}{l_{TCAM} \cdot m} \cdot 64 \text{ byte } (l_{Cache} < l_{TCAM} \cdot m) \end{cases} \quad (1)$$

Here, l_{Cache} and l_{TCAM} represent the latency of the cache memory and TCAM, respectively; m represents the cache miss rate in PPC. 64 byte refers to the shortest length of a packet. Equation (1) denotes that the throughput of PPC is limited by the smaller effective throughput of the cache and TCAM. Likewise, the average dynamic power consumption of PPC, which is represented as E_{PPC}, can be calculated as:

$$E_{PPC} = E_{Cache} + E_{TCAM} \cdot n \cdot m. \quad (2)$$

Here, E_{Cache} and E_{TCAM} represent the dynamic power consumption of the cache memory and TCAM, respectively; n represents the number of TCAM accesses needed for processing a packet. We ignore the impact of the static power consumption of the cache and TCAM on the router because the dynamic power consumption of TCAM dominates the power consumption of a router [10]. These equations indicate that both the throughput and the power consumption of PPC mainly depend on the memory performance and the cache miss rate. Therefore, reducing both the cache size and the number of cache misses is important. Our previous work, mentioned in Sect. 3, shows that the cache hit rate of PPC reaches around 80% for many types of network traffic patterns even if a cache which has the same size as an L1 cache of a microprocessor is used. On the contrary, our result also shows that remaining cache misses are remarkable (about 20%); thus, PPC still has a chance to improve the cache-management algorithm.

One of the important elements for utilizing the cache space effectively is to consider the cache replacement policy. The cache replacement policy is a way how to decide an entry replaced on a set associative cache, which sets several entries on a cache line. Figure 2 shows simulation results of the cache miss rate in case of applying least recently used (LRU) and optimal page replacement algorithm (OPT) [11] to 4 way set associative PPC. Details of the simulator and network traces are explained in Sect. 5. LRU is an algorithm which replaces an entry referenced oldest. LRU is widely used because LRU achieves high cache hit rate in many cache systems. On the other hand, OPT is an algorithm which replaces an entry under an ideal situation that all subsequent data is known. Although OPT achieves the best performance, it is impossible to implement in many cache systems. The results of Fig. 2 show that the cache miss can be reduced by approximately 30% compared to LRU in any networks by applying the best cache replacement policy for PPC. In this study, we consider the more effective cache replacement policy than LRU and close the gap of the cache miss. The contribution of this paper is summarized as follows:

- This paper analyzes the cache miss causes in PPC and shows that elephant flows and mice flows make a great impact on the cache miss rate.
- This paper shows that LRU, which is used in many types of cache systems, is not suitable for PPC and proposes an advanced cache replacement policy for PPC.

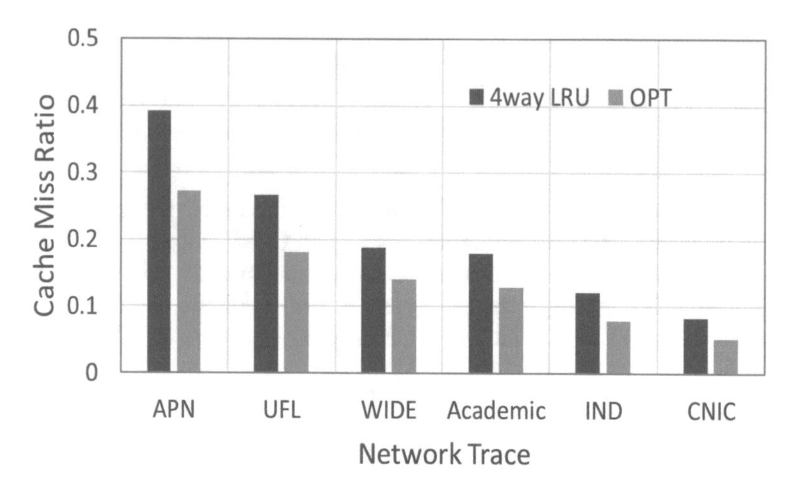

Fig. 2. Comparison of cache miss ratios between LRU and OPT.

- This paper reveals the performance of the cache replacement policies for various types of network patterns. Most previous studies showed the cache performance for only a few network traces; however, our simulation uses 15 types of network traces as workloads.

3 Related Work

There are some studies of reducing the number of cache misses in PPC. In this section, we introduce them and then discuss their problems.

Cache tag compression is one of the approaches to reducing the number of capacity misses in PPC. This approach increases the number of cache entries in exchange for reducing the entry size. Chang et al. proposed the digest cache to increase the number of cache entries [12]. In the digest cache, a 32-bit hash value calculated by a five-tuple was used as a cache tag instead of the 104-bit flow information. Likewise, Ata et al. proposed a tuned cache that used the source and destination IP addresses and the lower port number of a flow as a cache tag [13]. However, the digest cache needs additional hardware to prevent hash conflicts and the tuned cache is unable to conduct fine-grained processing because of the lack of the flow information.

Li et al. considered the suitable cache design of PPC from the three aspects; the hash function which is used to calculate the cache index, cache associativity, and the cache replacement policy [14]. Li et al. evaluated several types of cache designs and concluded that the 4-way LRU is the best from the balance between the cache miss rate and the implementation costs. In addition, they also evaluated the impact of hash function on the cache miss rate and concluded that the effect of the hash functions is small. In [14], only three policies, LRU, least frequently used (LFU), and round robin are considered as the cache replacement policies. However, other effective approaches were not considered. As mentioned later, LRU is not suitable for PPC.

Kim et al. indicated that LRU cannot achieve high performance in PPC because LRU focuses only the temporal locality and cannot utilize activities of networks [15]. In [15], the method which classifies entries to switching entries, which are entries hit at least once, or non-switching entries, which are entries never hit, and replaces the entry from non-switching entries. Furthermore, the authors proposed two line replacement policies called weighted priority LRU Scheme and L2A cache scheme. In weighted priority LRU scheme, the non-switching entries referenced again with a high probability are reserved by not replacing entries inserted within a threshold time. On the other hand, L2A cache scheme selects the replaced entry based on the sum of the last two packet timestamps. The authors showed that L2A cache scheme can reduce the number of cache misses especially in the case of small caches. However, concrete hardware architecture of proposed methods, such as the number of bits for storing the timestamp to the cache and the way for getting the time, were not considered and evaluated. The increase in the memory costs becomes a critical problem especially in PPC.

Our previous work also considered the cache replacement policies for reducing the number of conflict misses in PPC [16, 17]. We focused on some specific applications that create flows composed of a single packet, such as domain name system (DNS) and network attacks. Our policies can reduce the cache misses by 8% by not allowing PPC to store these flows into a cache because they never hit in the cache. However, our policies cannot prevent the useless entry insertion caused by various small factors because the policies prevent the useless entry insertion of only specific applications. Other approaches should be considered to reduce the remaining misses.

4 Cache Replacement Policy

As introduced in Sect. 3, there are many approaches to reduce the cache misses in PPC. In this paper, we focus on the cache replacement mechanism from the perspectives of little loss of the cache potential and the balances between the cache miss reduction and implementation costs. In this section, as a first step of improving the cache replacement mechanism, we analyze the characteristics of flows to reveal flow behavior in the cache. After that, we propose hit dominance cache (HDC) based on the flow characteristics to achieve low cache miss rate.

4.1 Analyzation of Flow Characteristics

In PPC, one of the reasons for increasing the cache miss of PPC is that flows composed of a small number of packets (it is known as mice flows) evict useful entries [17]. Although the mice flows have few chances to hit in PPC, they remain in the cache and pollute the cache entries. In our previous work [17], it was analyzed that mice flows which were composed of less than 10 packets occupy 99% of all flows in networks. Furthermore, around half of all flows were flows which were composed of one packet, while these one packet flows have no chance to hit in PPC. This analysis shows that mice flows greatly decrease the PPC performance due to the occupation of most entries in PPC, and thus it is important to evict the mice flows rapidly.

We also focus on elephant flows, which are composed of more than 100,000 packets, such as video streaming. For PPC, the elephant flows are quite important because they have many chances to hit in PPC and impacts on the cache miss rate though they are few in networks. This trend that the mice flows and elephant flows have strong influence in networks is known as the elephant and mice phenomenon [18]. First, we verify the impact of the elephant flows on the cache misses by simulating PPC and show some results in Fig. 3. This graph shows the occurrence of the cache miss or cache hit for an elephant flow. Contrary to the expect that the most packets in an elephant flow hit in the cache throughout, our simulation shows a large number of cache misses are occurred in elephant flows. It means that entries of the elephant flows are evicted and re-inserted repeatedly and leads to the cache pollution. From this reason, it is also important for PPC to remain the entries of elephant flows preferentially.

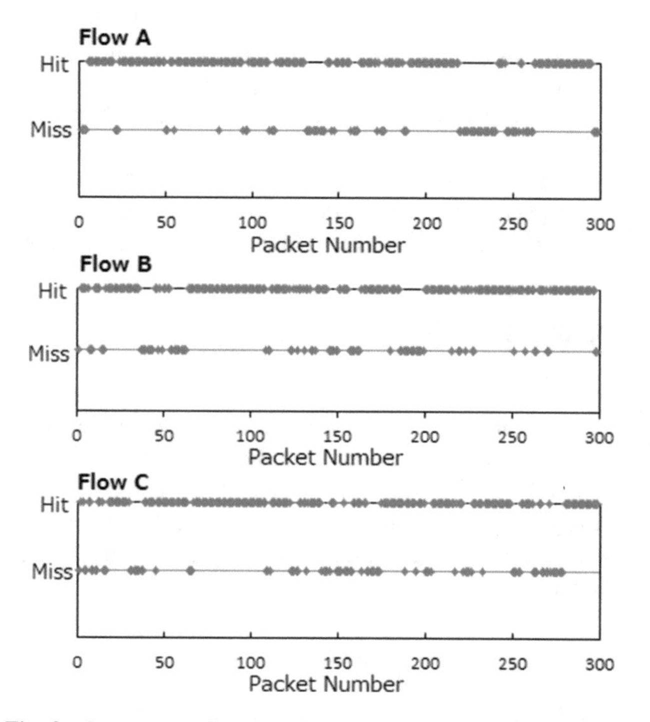

Fig. 3. Occurrence of cache miss or cache hit in an elephant flow.

4.2 Hit Dominance Cache

From the above discussion, we propose hit dominance cache (HDC), which retain elephant flows preferentially and evicts mice flows immediately. HDC gives different cache priorities to the elephant flows and mice flows by allocating different cache spaces to them.

Figure 4 shows how to insert and evict cache entries by HDC. HDC has two cache area: a primary area and hit area. Basically, entry insertion and eviction are performed in the primary area. Unlike LRU, a new entry is inserted to the least recently used (LRU) position in the cache line. This change enables to evict the mice flows at one replacement in the shortest. When the cache entry hits, the hit entry is shifted to the MRU position in each area like LRU. However, HDC shifts the most recently used (MRU) position entry in the primary area to the LRU position in the hit area when the entry hits threshold times. This behavior gives high cache priority to the elephant flows which has many chances to hit in PPC.

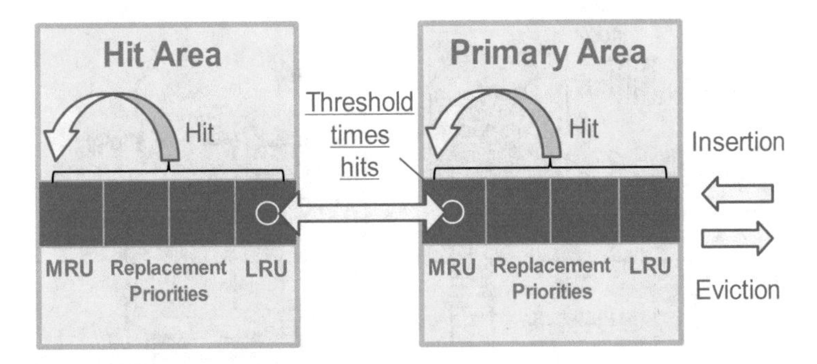

Fig. 4. Entry insertion and eviction by HDC.

5 Evaluation

This section shows the evaluation results of HDC. We measured the following two items with a few in-house PPC simulators to discuss the effectiveness of the proposed methods.

5.1 Cache Miss Reduction

To simulate PPC, we used an identical cycle-accurate PPC simulator written in C++ and various real-network traces. The processing flow of a PPC simulator is shown in Fig. 5. The simulator models the cycle-level behavior of PPC such as reading packets, extracting flow information, creating the cache index, referring to the cache, and referring to TCAM. We assume that the clock frequency is 1 GHz. A packet is read out from a network trace at the appropriate cycle designated by the timestamp, and then accesses to the cache with a hash value calculated from the five-tuple. The cache is L1-sized and has the latency of 0.5 ns as shown in Table 1. If a cache hit occurs, the processing of the packet is finished. On the other hand, if a cache miss occurs, the packet is forwarded to cache miss table (CMT). CMT manages flows being retrieved in TCAM for preventing cache misses caused by subsequent packets which belong to the flows managed in CMT, and the subsequent packets are sent to cache miss queue (CMQ). Packets in CMQ are processed after the processing of the precedent packet in TCAM is completed. The details of CMT and CMQ are described in [19]. If a CMT miss occurs, the packet is forwarded to a TCAM module and then accesses to TCAM. The processing of the packet in the TCAM module is completed after 5 ns, which is equivalent to the TCAM latency shown in Table 1. After that, the cache is updated with the result of the packet processing in the TCAM module. Various types of real-network traces shown in Table 2 were used as workloads. These traces were acquired from RIPE Network Coordination Centre [20] and Widely Integrated Distributed Environment (WIDE) [21]. In addition, an academic trace acquired from a core network in Japan was used as a 10 Gbps workload.

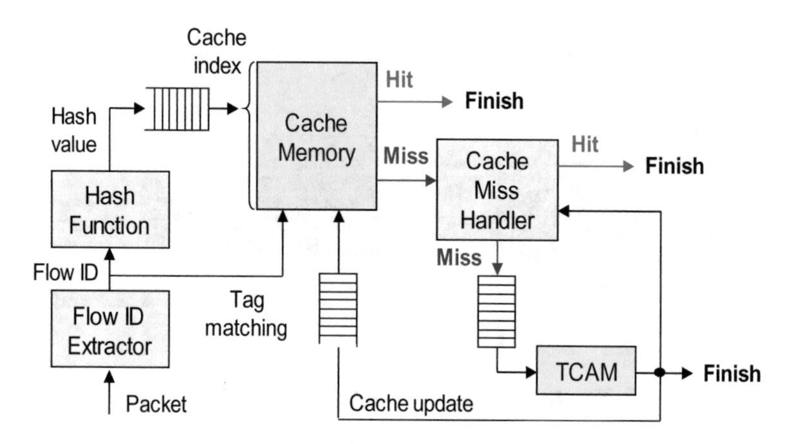

Fig. 5. Processing flow of PPC simulator.

First, the impacts of the threshold in HDC on the cache miss rate are evaluated. Figure 6 shows the difference of the cache miss rates among 4-way LRU and various types of HDC. The threshold is set to 1, 2, 4, 8, and 16. In Fig. 6, we showed the results of only three networks because the trends of all the results are mostly the same. Figure 7 shows the improvement ratios against 4-way LRU. These results indicate that it is the best to set the threshold to eight. Hereinafter, the threshold of HDC is set to eight.

Next, the cache performance of HDC is evaluated by comparing other policies. In this paper, we implemented four typical policies for comparison: 4-way LRU, 8-way LRU, OPT, and segmented LRU (SLRU) [22]. SLRU splits the cache space into probationary segment and protected segment and resembles HDC in the replacement scheme. Figure 8 shows the outline of entry insertion and eviction by SLRU. Unlike HDC, SLRU inserts a new entry into the MRU position of the probationary segment. In addition, when an entry is hit in the probationary segment, the entry is shifted to the MRU position in the protected segment. It means that SLRU is a kind of 8-way LRU whose entry insertion position is modified.

Figure 9 shows the cache miss ratios measured by the simulation with various patterns of network traffics. Figure 10 shows the cache miss improvement against 4-way LRU. The results indicate that HDC performs better than 4-way LRU and 8-way LRU in the most network traces and better than SLRU in more than half of all network

Table 1. Parameters of PPC Simulator

Parameter	Value
Number of cache entries	1,024 entries
Cache access latency	0.5 ns
TCAM access latency	5 ns

Table 2. Details of network traces

Trace	Captured date	Avg. # of packets [pps (packets per second)]
IND [20]	2003/1/6	15,540
BUF [20]	2003/1/18	8,380
TXG [20]	2004/3/26	12,475
APN [20]	2004/3/26	20,330
IPLS3 [20]	2004/6/1	116,778
BWY [20]	2004/10/7	17,922
COS [20]	2005/1/8	8,051
CNIC [20]	2005/3/17	28,440
MRA [20]	2005/3/21	41,372
UFL [19]	2005/3/21	50,769
FRG [20]	2006/1/10	32,955
PSC [20]	2006/2/20	26,912
PUR [20]	2006/2/20	42,515
Academic	2010/6/17	371,013
WIDE [21]	2017/4/12	58,776

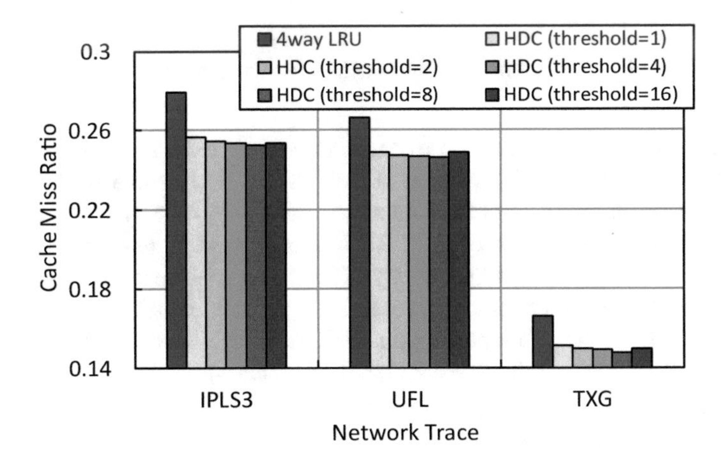

Fig. 6. Cache miss ratios of various types of HDC with three network traces.

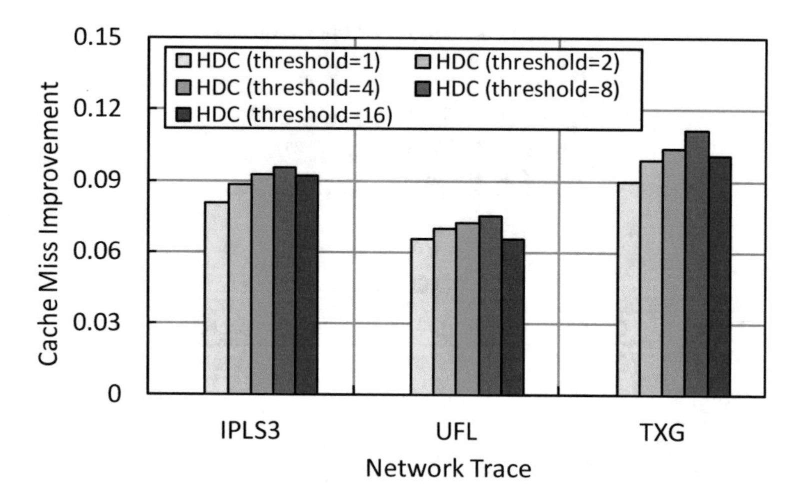

Fig. 7. Cache miss improvements with various thresholds of HDC in three network traces.

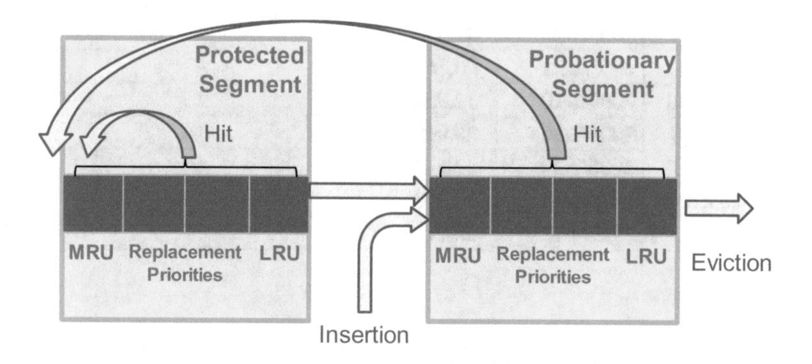

Fig. 8. Outline of entry insertion and eviction by SLRU.

traces. HDC can reduce the number of cache misses by up to 29.1% (10.0% on avg.) compared to conventional 4-way LRU. Likewise, SLRU can reduce the number of cache misses by up to 20.2% (11.1% on avg.) compared to 4-way LRU and performs well in all network traces. However, SLRU cannot implement with realistic hardware costs as described later. Meanwhile, 8-way LRU cannot achieve major improvement. It is because short flows stay in the cache at least eight times insertion of new entries and occupy the cache entries. Compared with OPT, HDC can attain 27.9% of the performance of OPT on average. This result shows HDC has still room for improvement.

5.2 Implementation Costs

This section evaluates the hardware implementation cost of HDC. For the cache replacement policy, not only the cache miss reduction but also the implementation

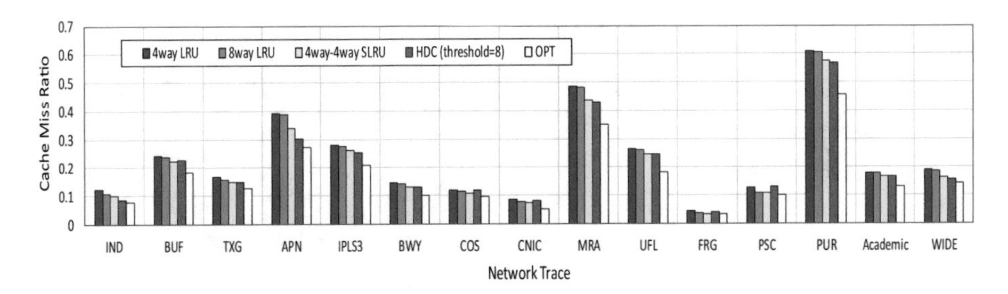

Fig. 9. Cache miss ratios of conventional replacement policies and HDC with various network traffics.

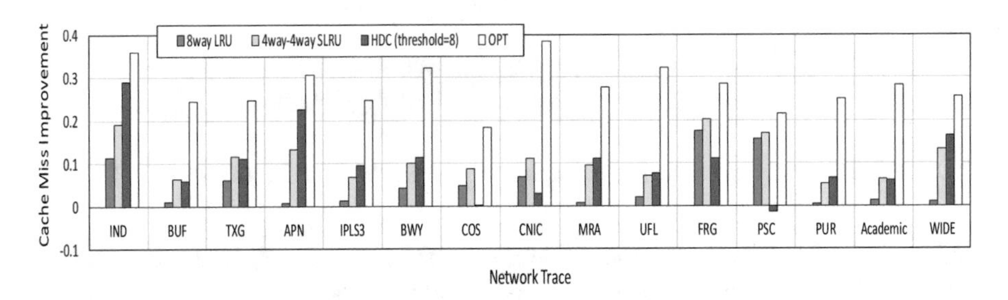

Fig. 10. Cache miss Improvement against 4-way LRU with various network traffics.

costs are important in practical use due to the large circuit areas. In general, hardware caches require two modules for implementation: the priority management memory, which stores the replacement priorities of cache entries in a cache line, and the priority updater, which updates the replacement priorities stored in the priority management memory.

To represent the cache priorities in a cache line, 4-way LRU uses 5 bits per cache line as the state (because there are 24 combinations of the cache priorities) instead of a string of the way numbers. On the other hand, in the case of 8-way LRU, 16 bits per cache line are required (because there are 40,320 possible combinations of the cache priorities). In addition, 8-way LRU becomes more complex than 4-way LRU because the priority updater of 8-way LRU must manage $40{,}320 \times 7 = 282{,}240$ state transitions while 4-way LRU manage $24 \times 3 = 72$ state transitions. As a result, it is not realistic to implement 8-way LRU on hardware. Similarly, SLRU cannot implement with realistic hardware costs.

Against above consideration, HDC can implement with the same hardware costs of 4-way LRU though it performs as the pseudo 8-way cache. Figure 11 shows the way to manage the cache replacement priorities by HDC. Hit counters in this figure counts the number of hits of each entry in the primary area to shift it to the hit area. In the case of HDC (threshold = 8), the size of the hit counter is 3 bits $\times$ 1,024 entries/2, namely, 1.5 K bits. When the hit counter reaches the threshold, the corresponding entry is swapped for the entry of LRU position in the hit area. At this time, the priority updater

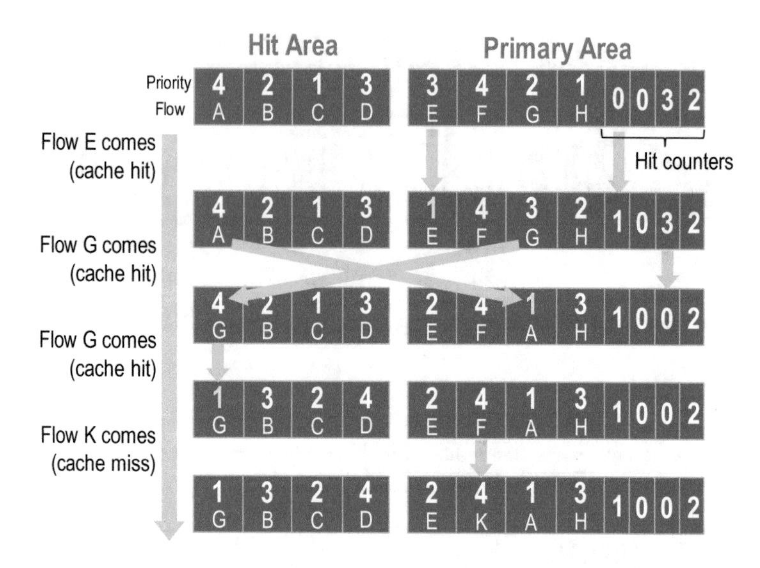

Fig. 11. The way to manage the replacement priorities by HDC (threshold = 4).

updates only the cache replacement priorities in the primary area. The cache replacement priorities in the hit area is not changed. In addition, when a new entry is inserted, HDC also does not need to update the replacement priorities in the primary area.

We evaluated the implementation costs of the cache replacement policies by implementing and synthesizing them. Tools used for evaluation is summarized in Table 3. Figure 12 shows the modules implemented in this evaluation. The priority updater updates the replacement priorities based on the cache hit signal and the hit way number and returns them to the priority management memory. The cache index of PPC is used for accessing the priority management memory. Table 4 shows the estimation results of the circuit areas and memory requirements. In this estimation, SLRU and 8-way LRU did not evaluate because the implementation costs of them is obviously oversize as mentioned above. The results showed that the circuit area of HDC is 72.6% of that of 4-way LRU. It is because HDC does not require the replacement priory updates when an entry is newly inserted. However, the memory requirement of HDC is 2.2 times larger than that of 4-way LRU. We consider this memory requirement difference is negligible because the size of the priority management memory is enough small compared to the cache memory. As a conclusion, the implementation cost of HDC is comparable to 4-way LRU.

Table 3. Simulation environment

Item	Tool name
Hardware description language	Verilog-HDL
Logic simulation	Cadence NC-Verilog LDV5.7
ASIC synthesis	Synopsys Design Compiler X-2005.09
Libraries for ASIC synthesis	Free PDK OSU Library (45 nm) [23]

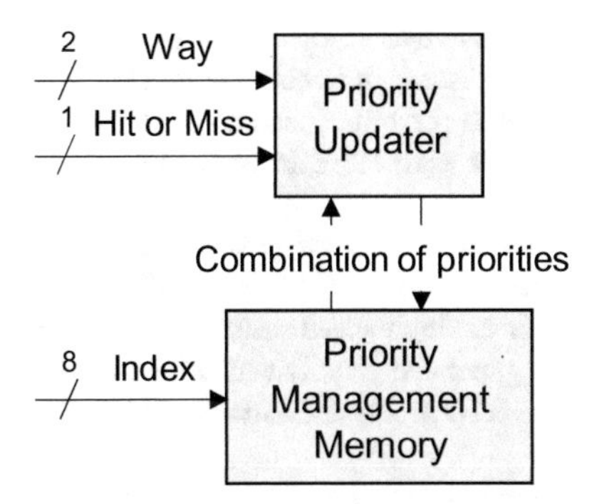

Fig. 12. Modules required to implement cache replacement policy with hardware.

Table 4. Estimation of circuit areas and memory requirements of 4-way LRU and HDC

	4-way LRU	HDC
Circuit area [μm^2]	110.66	80.332
Memory requirement [bit]	1,280	2,816

6 Conclusion

In this paper, to reduce the number of cache misses in PPC, a novel cache replacement policy was proposed. Least recently used (LRU), which is used empirically in many cache systems, cannot good performance in PPC due to the unique access patterns of internet traffics. For the effective replacement for PPC, it is required to retain elephant flows, which is composed a large number of packets, preferentially and evict mice flows, which is composed a few number of packets, immediately.

HDC splits the cache into two areas and allocates flows to the suitable area depending on the number of hits in the cache. HDC can evict the mice flows immediately and retain the elephant flows preferentially. The simulation results with various network traces showed that HDC can reduce the number of cache misses by up to 29.1% (10.0% on avg.) compared to 4-way LRU. Furthermore, we showed the implementation cost of HDC is comparable to that of 4-way LRU, conventional used in PPC.

References

1. The Ministry, "Tabulation and Estimation of Internet Traffic in Japan," (2016). http://www.soumu.go.jp/main_content/000462459.pdf. Accessed 16 Mar 2017
2. METI, "Green IT Initiative in Japan". http://www.meti.go.jp/english/policy/GreenITInitiativeInJapan.pdf. Accessed 16 Mar 2017
3. Fan, J., Hu, C., He, K., Jiang, J., Liuy, B.: Reducing power of traffic manager in routers via dynamic on/off-chip scheduling. In: 2012 Proceedings of IEEE INFOCOM, Orlando, FL, pp. 1925–1933 (2012)

4. Zheng, X., Wang, X.: Comparative study of power consumption of a NetFPGA-based forwarding node in publish–subscribe Internet routing. Comput. Commun. **44**, 36–43 (2014)
5. Gamage, S., Pasqual, A.: High performance parallel packet classification architecture with popular rule caching. In: 2012 18th IEEE International Conference on Networks (ICON), Singapore, pp. 52–57 (2012)
6. Agrawal, B., Sherwood, T.: Ternary CAM power and delay model: extensions and uses. IEEE Trans. Very Large Scale Integr. (VLSI) Syst. **16**(5), 554–564 (2008)
7. Nawa, M. et al.: Energy-efficient high-speed search engine using a multi-dimensional TCAM architecture with parallel pipelined subdivided structure. In: 2016 13th IEEE Annual Consumer Communications & Networking Conference (CCNC), Las Vegas, NV, pp. 309–314 (2016)
8. Hewlett-Packard Development Company: Energy Efficient Networking - Business white paper (2011). http://h17007.www1.hp.com/docs/mark/4AA3-3866ENW.pdf. Accessed 16 Mar 2017
9. Girish, C., Govindarajan, R.: Improving performance of digest caches in network processors. In: Proceedings of the 15th International Conference on High performance computing (HiPC 2008), India, pp. 6–17 (2008)
10. Agrawal, B., Sherwood, T.: Modeling TCAM power for next generation network devices. In: 2006 IEEE International Symposium on Performance Analysis of Systems and Software, pp. 120–129 (2006)
11. Belady, L.A.: A study of replacement algorithms for a virtual-storage computer. IBM Syst. J. **5**(2), 78–101 (1966)
12. Chang, F., Feng, W.C., Li, K.: Efficient packet classification with digest caches. In: Proceedings of Third Workshop Network Processors and Applications (NP-3) (2005)
13. Ata, S., Murata, M., Miyahara, H.: Efficient cache structures of IP routers to provide policy-based services. In: IEEE International Conference on Communications (ICC 2001), Helsinki, vol. 5, pp. 1561–1565 (2001)
14. Li, K., Chang, F., Berger, D., Wu-chang, F.: Architectures for packet classification caching. In: The 11th IEEE International Conference on Networks (ICON2003), Sydney, pp. 111–117 (2003)
15. Kim, N., Jean, S., Kim, J., Yoon, H.: Cache replacement schemes for data-driven label switching networks. In: 2001 IEEE Workshop on High Performance Switching and Routing, Dallas, TX, pp. 223–227 (2001)
16. Yamaki, H., Nishi, H.: An improved cache mechanism for a cache-based network processor. In: Proceedings of the International Conference on Parallel and Distributed Processing Techniques and Applications (PDPTA 2012), Las Vegas, NV, pp. 1–7 (2012)
17. Yamaki, H., Nishi, H.: Line replacement algorithm for l1-scale packet processing cache. In: Adjunct Proceedings of the 13th International Conference on Mobile and Ubiquitous Systems: Computing Networking and Services (MOBIQUITOUS), Hiroshima, Japan, pp. 12–17 (2016)
18. Mori, T., Uchida, M., Kawahara, R., Pan, J., Goto, S.: Identifying elephant flows through periodically sampled packets. In: Proceedings of the 4th ACM SIGCOMM Conference on Internet Measurement (IMC 2004), pp. 115–120. ACM, New York (2004)
19. Okuno, M., Nishi, H.: Network-processor acceleration-architecture using header-learning cache and cache-miss handler. In: The 8th World Multi-Conference on Systemics, Cybernetics and Informatics (SCI2004), pp. 108–113
20. RIPE Network Coordination Centre: Réseaux IP Européens Network Coordination Centre RIPE NCC. http://www.ripe.net/. Accessed 16 Mar 2017
21. WIDE MAWI WorkingGroup: MAWI Working Group Traffic Archive. http://mawi.wide.ad.jp/mawi/. Accessed 17 Aug 2016

22. Karedla, R., Love, J.S., Wherry, B.G.: Caching strategies to improve disk system performance. Computer **27**(3), 38–46 (1994)
23. North Carolina State University: FreePDK45:Contents. http://www.eda.ncsu.edu/wiki/FreePDK45:Contents. Accessed 1 Sept 2017

Forecasting Player Behavioral Data and Simulating In-Game Events

Anna Guitart(✉), Pei Pei Chen, Paul Bertens, and África Periáñez

Yokozuna Data unit, Silicon Studio, 1-21-3 Ebisu, Shibuya-ku, Tokyo, Japan
{anna.guitart,peipei.chen,paul.bertens,
africa.perianez}@siliconstudio.co.jp

Abstract. Understanding player behavior is fundamental in game data science. Video games evolve as players interact with the game, so being able to foresee player experience would help to ensure a successful game development. In particular, game developers need to evaluate beforehand the impact of in-game events. Simulation optimization of these events is crucial to increase player engagement and maximize monetization. We present an experimental analysis of several methods to forecast game-related variables, with two main aims: to obtain accurate predictions of in-app purchases and playtime in an operational production environment, and to perform simulations of in-game events in order to maximize sales and playtime. Our ultimate purpose is to take a step towards the data-driven development of games. The results suggest that even though the performance of traditional approaches, such as ARIMA is still better, the outcomes of state-of-the-art techniques like deep learning are promising. Deep learning comes up as a well-suited general model that could be used to forecast a variety of time series with different dynamic behaviors.

Keywords: Social games · Time series · Forecasting
Sequential analysis · Deep learning · ARIMA models
Gradient boosting

1 Introduction

In the last few years, we have witnessed a genuine paradigm change in the development of video games [1,2]. Nowadays, petabytes of player data are available, as every action, in-app purchase, guild conversation or in-game social interaction performed by the players is recorded. This provides data scientists and researchers with plenty of possibilities to construct sophisticated and reliable models to understand and predict player behavior and game dynamics. Game data are time-dependent observations, as players are constantly interacting with the game. Therefore, it is paramount to understand and model player actions taking into account this temporal dimension.

K. Arai et al. (Eds.): FICC 2018, AISC 886, pp. 274–293, 2019.
https://doi.org/10.1007/978-3-030-03402-3_19

In-game events are drivers of player engagement (Fig. 1). They influence player behavior due to their limited duration, strongly contributing to the in-game monetization (for instance, an event that offers a unique reward could serve to trigger in-app purchases). Those events break the monotony of the game, and thus it is essential to have a variety of types, such as battles, rewards, polls, etc. Anticipating the most adequate sequence of events and the right time to present them is a determinant factor to improve monetization and player engagement.

Fig. 1. Screenshot of an in-game gacha event in *Grand Sphere* developed by *Silicon Studio*. The left panel shows an item that players can purchase, which on opening reveals an in-game card (shown in the right panel), in this case having 3 stars. The card obtained is random, and cards with more stars are more valuable and also rarer. Different in-game events can modify the probability of getting cards with more stars or the types of cards that can be obtained.

Forecasting time series data is a challenge common to multiple domains, e.g. weather or stock market prediction. Time series research has received significant contributions to improve the accuracy and time horizon of the forecasts, and an assortment of statistical learning and machine learning techniques have been developed or adapted to perform robust time series forecasting [3,4].

Time series analysis focuses on studying the structure of the relationships between time-dependent features, aiming to find mathematical expressions to represent these relationships. Based on them, several outcomes can be forecast [5].

Stochastic simulation [6] optimization consists on finding a local optimum for a response function whose values are not known analytically but can be inferred. On the other hand, the analysis of *what-if* scenarios (or just simulation optimization) is the process of finding the best inputs among all possibilities that maximize an outcome [7].

The aim of this work is twofold: on the one hand, to accurately forecast time series of in-game sales and playtime; on the other, to simulate events in order to find the best combination of in-game events and the optimal time to publish them. To achieve these goals, we performed an experimental analysis utilizing several techniques such as ARIMA (dynamic regression), gradient boosting, generalized additive models and deep neural networks [8–11].

Pioneering studies on game data science in the field of video games, such as [12–15], concentrated in churn prediction. Other related articles that analyze temporal data in the game domain [16–20] focused on unsupervised clustering, not in supervised time series forecast. To the best of our knowledge this is the first work in which forecasts of time series of game data and simulations of in-game events are performed.

2 Model Description

2.1 Autoregressive Integrated Moving Average (ARIMA)

ARIMA was firstly introduced by Box and Jenkins [8] in 1976, in a book that had a tremendous impact on the forecasting community. Since then, this method has been applied in a large variety of fields, and it remains a robust model, used in multiple operating forecast systems [21]. ARIMA characterizes time series focusing on three aspects: (1) the autoregressive (AR) terms model the past information of the time series; (2) the integrated (I) terms model the differencing needed for the time series to become stationary, e.g. the trend of the time series; (3) the moving average (MA) terms control the past noise of the original time series.

Specifically, the AR part represents the time series as a function of p past observations,

$$y_t = \varphi_1 y_{t-1} + \cdots + \varphi_p y_{t-p}, \tag{1}$$

with $\varphi_1, \ldots, \varphi_p$ the AR coefficients and p the number of past observations needed to perform a forecast at the current time. The MA component, rather than focusing on past observations, uses a moving average of past error values in a regression model,

$$y_t = \varepsilon_t + \theta_1 \varepsilon_{t-1} + \cdots + \theta_q \varepsilon_{t-q}, \tag{2}$$

where q is the number of moving average lags, $\theta_1, \ldots, \theta_q$ are the MA coefficients and $\varepsilon_t, \ldots, \varepsilon_{t-q}$ the errors. Finally, a d parameter is used to model the order of differencing, i.e. the I term of ARIMA:

$$y_t^* = y_t - y_{t-1} - \cdots - y_{t-d}. \tag{3}$$

Taking $d = 1$ is normally enough in most cases [5]. If $d = 0$, the model is reduced to an ARMA(p, q) model.

The ARIMA model can only analyze stationary time series; however most of the stochastic processes exhibit non-stationary behavior. Only through the differencing operation, or by applying a previous additional transformation to the time series, e.g. a log or Box–Cox [22] transformation, we can convert the time series into stationary objects.

We can write the ARIMA model as

$$\varphi(L)(1-L)^d y_t = \theta(L)\varepsilon_t, \tag{4}$$

where L is the lag operator, i.e. $L\varphi_t = \varphi_{t-1}$. Therefore, we can express the ARIMA model in terms of the p, d and q parameters as

$$\left(1-\sum_{i=1}^{p}\varphi_i L^i\right)(1-L)^d y_t = \left(1+\sum_{j=1}^{q}\theta_j L^j\right)\varepsilon_t. \tag{5}$$

Box and Jenkins also generalized the model to recognize the seasonality pattern of the time series by using $y_t = \text{ARIMA}(p,d,q)(P,D,Q)_m$ [8]. The parameters P, D, Q represent the number of seasonal autoregressive, differencing and moving average terms. The order of the seasonality (e.g. weekly, monthly, etc.) is set by m. In order to determine the ARIMA parameters, the autocorrelation function (ACF) and partial autocorrelation function (PACF) are analyzed [8]. Once the parameters are fixed, the model is fitted by maximum likelihood and selected (among the different estimated models) based on the Akaike (AIC) [23] and Schwarz Bayesian (BIC) [24] information criteria.

Dynamic Regression: To include the serial dependency of external covariates, a multivariate extension of (5) is proposed, the so-called *dynamic regression* (DR). The relationship between the output, its lags and the variables is linear. DR takes the form [25]

$$\begin{aligned}\left(1-\sum_{i=1}^{p}\varphi_i L^i\right)(1-L)^d y_t \\ &= \sum_{r=1}^{k}\gamma_r\left(1-\sum_{l=1}^{p}\varphi_l L^l\right)(1-L)^d x_{rt} \\ &\quad + \left(1+\sum_{j=1}^{q}\theta_j L^j\right)\varepsilon_t. \end{aligned} \tag{6}$$

The first term represents the AR and I parts, the last corresponds to the MA component and the middle term is where the k external variables x_t are included. The γ_r are the corresponding coefficients of the covariates.

2.2 Gradient Boosting Models

Gradient boosting machines (GBMs) [26] are ensemble-based machine learning methods capable of solving classification and regression problems [9]. A GBM consists of an ensemble of weak learners (i.e. learners that perform only slightly better than random classifiers), commonly decision trees. Each weak learner is sequentially added to the ensemble, thus continuously improving its accuracy [27]. The approach taken by GBMs stems from the idea that boosting (using multiple weak models to create a strong model) can be seen as an optimization algorithm on some suitable loss function [28]. If this loss function is differentiable, the gradient can be calculated and gradient descent can be used for optimization [9]. This makes it possible to recursively add weak learners that follow the gradient, minimizing the loss function and reducing the error at each step. The construction of an ensemble to fit a desired function $f(x)$ is illustrated below (a more in-depth analysis can be found in [29]). For each weak learner in the ensemble, we sequentially do the following: First, we compute the negative gradient:

$$z_i = -\frac{\delta}{\delta f(x_i)}\mathcal{L}(y_i, f(x))|_{f(x_i)=\hat{f}(x_i)}, \tag{7}$$

where, $\mathcal{L}$ is the loss function and i the current weak model.

Then, we fit a regression model $g(x)$ predicting z_i from x_i and apply gradient descent, with step size given by

$$\rho(x) = \arg\min_{\rho} \sum_{i=1}^{N} \mathcal{L}(y_i, f(x_i)) + \rho g(x_i). \tag{8}$$

Finally, we update the estimate of $f(x)$ through

$$\hat{f}(x) \leftarrow \hat{f}(x) + \rho g(x). \tag{9}$$

GBMs can effectively capture complex non-linear dependencies [30] and several strategies to avoid overfitting can be applied, e.g. shrinkage [9] or early stopping [31].

2.3 Generalized Additive and Generalized Additive Mixed Models

The generalized additive models (GAMs) derived by [32] are a combination of generalized linear models (GLMs) and additive models (AMs). In this way, GAMs exhibit the properties of both, namely, the flexibility to adapt to any distribution of GLMs and the non-parametric nature of AMs.

The structure of a GAM is

$$g(E(y)) = \beta_0 + s_1(x_1) + \cdots + s_n(x_n), \tag{10}$$

where n is the number of predictors, $E(y)$ is the expected value, $g(\cdot)$ is a link function, and $s_i(\cdot)$ are smooth functions.

Even though the distribution of the response variable is the same as for GLMs, there is a generalization that allows GAMs to accommodate different kinds of responses, for example binary or continuous. GAMs assume that the means of the predictors x_i are related to an additive response y through a nonlinear link function g (such as the identity or logarithm function). The model distribution can be selected from e.g. a Gaussian or Poisson distribution [10].

As additive models, in contrast to parametric regression analysis (which assumes a linear relation between responses and predictors), GAMs serve to explore non-parametric relationships, as they make no assumptions about those relations. Instead of using the sum of the individual effects of the predictors as observations, GAMs employ the sum of their smooth functions, which are called *splines* [33] and include a parameter that controls the smoothness of the curve to prevent overfitting. With the link and smooth functions mentioned above, the GAM approach has the flexibility to interpret and regularize both linear and nonlinear effects [34].

However, GAMs do not assume correlations between observations (such as the time series temporal correlation in this study). As a consequence, when these are present, another approach would be better suited to perform the forecast. Generalized additive mixed models (GAMMs) [35], an additive extension of generalized linear mixed models (GLMMs) [36], constitute such an approach. GAMMs incorporate random effects u and covariate effects γ_j (which model the correlations between observations by taking the order of observations into account) into the previous GAM formulation. When estimating the jth observation, the structure of a GAMM is

$$g(E(y_j)) = \beta_0 + s_i(x_{ji}) + \gamma_j^\top u. \tag{11}$$

Smoothness selection can be automatized [37] and the estimation of the GAMM is conducted via a penalized likelihood approach.

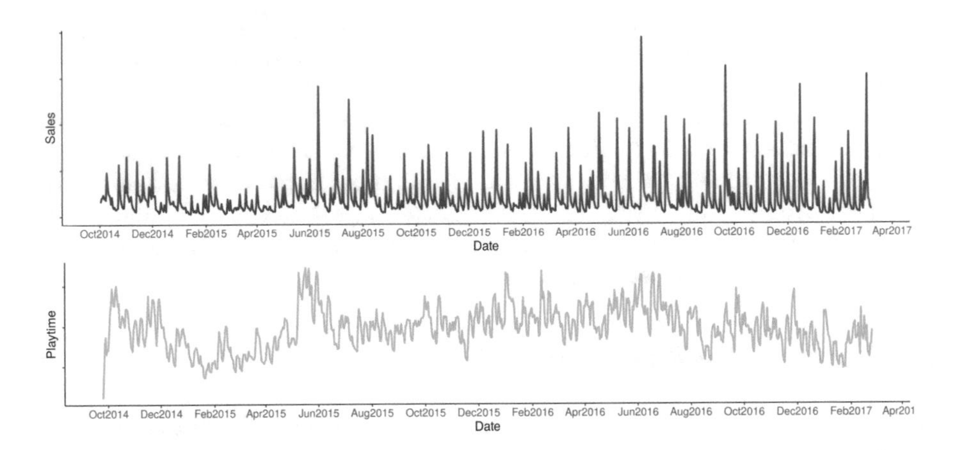

Fig. 2. Time series of the total daily sales (above) and total daily playtime (below) from *Age of Ishtaria* over a period of ~2.5 years. In this and other figures below, the quantitative sales and playtime values are not shown in the vertical axes for privacy reasons.

2.4 Deep Belief Networks

Deep neural networks (DNNs) [38] have been used with great success in recent years, achieving cutting-edge results in a wide range of fields [39]. This method has outperformed alternative models on almost all tasks from image classification [40] to speech recognition [41] and natural language modeling [39].

Deep belief networks (DBNs) [42] are an extension of DNNs where the units in the hidden layers are binary and stochastic. They are capable of learning a joint probability distribution between the input and the hidden layers and can be used as a generative model to draw samples from the learned distribution. A DBN uses a stack of restricted Boltzman machines (RBMs) [43], where each RBM is trained on the result of the previous layer as follows:

$$P(h_i^{(k)} = 1 | h^{(k+1)}) = \sigma\left(b_i^{(k)} + \sum_j W_{ij}^{(k)} h_j^{(k+1)} \right), \tag{12}$$

with $W^{(k)}$ being the weight matrix between the layers k and $k+1$ and σ being the logistic sigmoid.

For supervised learning, a DBN can be used as a way to pre-train a DNN [44]. Each layer of the RBM is first trained separately, and then the weights in all layers are fine-tuned through standard back-propagation. A more comprehensive overview of DBNs and DNNs can be found in [38,42].

There have been previous works on applying DNNs to time series forecasting problems (see e.g. [11,45]) and comparisons between ARIMA and artificial neural networks (ANNs) have also been performed [46]. However, when dealing with small datasets, standard DNNs can quickly overfit due to having too many parameters. Careful regularization is required to ensure that the model can be generalized to unseen data. Recent techniques like drop-out (which randomly masks some of the hidden units) [47] or stochastic DBNs can significantly help with this problem, as they constrain the number of parameters that can be modified. Additionally, using L_2 regularization on the weights [48] and ensuring proper selection of the number of hidden units can make DBNs an effective model even when faced with smaller datasets.

3 Forecast Performance Metrics

In order to validate the accuracy of forecasting models, several performance measures have been proposed in the literature [49]. The use of various metrics to analyze time series is a common practice. The ones selected in this study possess different properties, which is important to correctly assess the forecasting capabilities of the models from different perspectives (e.g. in terms of the magnitude or direction of the error). Each of the metrics summarized here is a function of the actual time series and forecast results. In all the equations below, n refers to the number of observations, f_i denotes the forecast value and a_i represents the actual value.

3.1 Root Mean Squared Logarithmic Error (RMSLE)

The RMSLE can be defined as

$$\mathrm{RMSLE} = \sqrt{\frac{1}{n}\sum_{i=1}^{n}\Big(\log(f_i+1) - \log(a_i+1)\Big)^2}. \tag{13}$$

Because the RMSLE uses a logarithmic scale, it is less sensitive to outliers than the standard root mean square error (RMSE). Additionally, the RMSE has the same tendency to underestimate and overestimate values, whereas the RMSLE penalizes more the underestimated predictions.

3.2 Mean Absolute Scaled Error (MASE)

The MASE is a scale independent metric, i.e. it can be used to compare the relative accuracy of several forecasting models applied to different datasets. To calculate this metric, errors are scaled with the in-sample mean absolute error (MAE). The expression for the MASE is

$$\mathrm{MASE} = \frac{1}{n}\sum_{i=1}^{n}\left(\frac{|f_i - a_i|}{\frac{1}{n-1}\sum_{i=2}^{n}|a_i - a_{i-1}|}\right). \tag{14}$$

3.3 Mean Absolute Percentage Error (MAPE)

The MAPE is estimated by

$$\mathrm{MAPE} = \frac{100}{n}\sum_{i=1}^{n}\left|\frac{f_i - a_i}{a_i}\right|. \tag{15}$$

Since it is a percentage error measure of the average absolute error, the MAPE is also scale-independent. However, if the time series have zero values, the MAPE yields undefined results because of a division by zero [49]. The MAPE is also biased towards underestimated values and does not penalize large errors.

4 Dataset

4.1 Data Source

The study presented in this article focuses on the analysis of two different daily time series, those of *playtime* and *sales*. The data was collected from two Japanese game titles developed by Silicon Studio: *Age of Ishtaria* (hereafter, *AoI*) and *Grand Sphere* (hereafter, *GS*). Playtime corresponds to the total amount of time spent in the game by all users, while sales represents the total amount of in-app purchases. For *AoI*, daily information was extracted from October 2014 until February 2017 (~2.5 years) and for *GS*, from July 2015 until

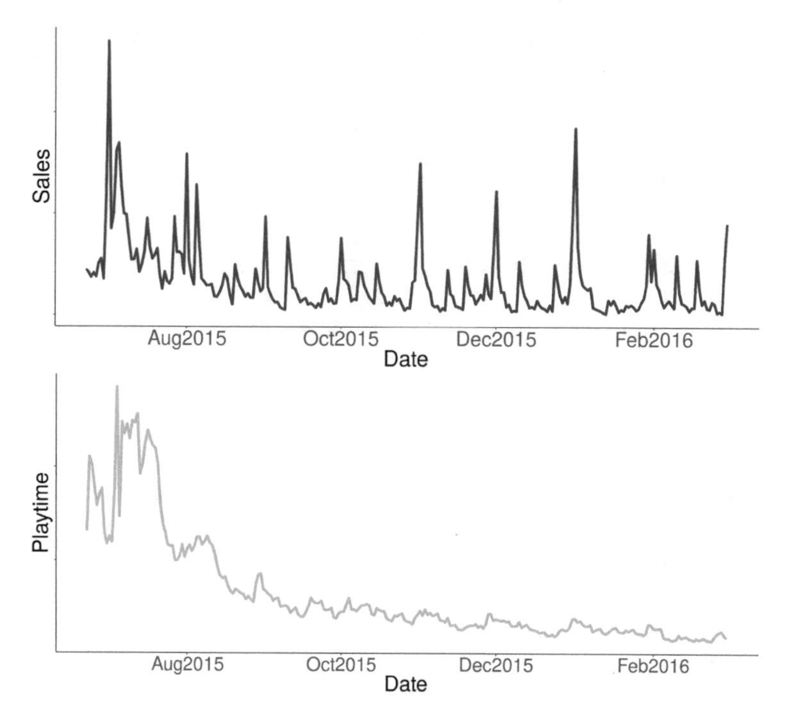

Fig. 3. *Grand Sphere* time series of total daily sales (above) and total daily playtime (below) over a period of ~9 months. Quantitative values are not shown in the vertical axes for privacy reasons.

March 2016 (~9 months). Additionally, data from all the game events, marketing and promotion campaigns within the collection period were also gathered to be used as external variables for the model. Figures 2 and 3 show the two daily time series for *AoI* and *GS*, respectively, which present clear differences concerning trends and seasonal behavior.

4.2 External Features

Proper identification and subsequent removal of outliers caused by unexplained phenomena can significantly improve the modeling of the time series [50]. To that end, anomaly detection using a deep autoencoder [51] was performed. This technique is capable of finding subtler anomalous behaviors than traditional methods, and shows that the outliers coincide with the external events of the game that take place on the same particular day. These events are derived from in-game information and included in the model as external variables. They can be:

- *Game Events*: Events such as raid battles, boss battles, etc. Each type of event is input separately.

- *Gacha*: A monetization technique used in many successful Japanese free-to-play games. It describes an event where players can spend money to randomly pull an item from a large pool (inspired by capsule toy vending machines).
- *Promotions*: Release of new cards and discounts to engage current users.
- *Marketing Campaigns*: Acquisition campaigns launched to obtain new users.

In the case of *gacha* and *promotions*, the corresponding *event scale* was also taken into account. This scale is used to quantify the outlier effect. It represents the influence or importance of the event in question and is related to the amount of money invested in it. Specifically, the event is assigned a value 1, 2, 3 or 4 to denote low, medium, high or super-high influence.

Other non-game-related features included in the model are:

- *National Holidays*: National Holidays in Japan [52].
- *Temperature*: Average daily temperature in Tokyo [53].

Moreover, the day of the week and month of the year were added as extra input for the GAMM and GBM, to make it possible for the model to learn the seasonality effects. For the ARIMA model these data do not need to be included as they are already inherently considered by the seasonal parameters.

5 Method

5.1 Data Preparation

To make the original time series stationary, a *Box–Cox* transformation [22] is applied to the sales data, and a *logarithm* transformation to the playtime data, for the DR, GBM and DBN models. The GAMM is the only technique that does not require a prior transformation of the data.

The categorical values for the external features (game events, *gacha*, promotions, marketing campaigns, national holidays, etc.) are included in the model as step functions (dummy variables). For each day in the training and forecasting data, the covariates are encoded with a vector that is either 0 or 1 depending on the absence or presence of the event on that date. However, for events with an *event scale*, the vector value matches the corresponding scale value instead.

Table 1. Tuned parameters for each model

Dataset		Models														
		GBM		ARIMA							DBN					
		max_depth	eta	p	d	q	P	D	Q	m	h	n	plr	tlr	k	b
Sales	Age of Ishtaria	100	0.20	2	1	1	1	1	1	7	2	50	0.0001	0.01	5	50
Sales	Grand Sphere	1	0.76	2	1	1	1	0	1	7	2	300	0.001	0.1	2	10
Playtime	Age of Ishtaria	1	0.66	2	1	2	1	1	1	7	2	50	0.0001	0.01	2	50
Playtime	Grand Sphere	1000	0.23	1	1	1	1	1	1	7	2	300	0.0001	0.1	2	50

Since promotions and marketing campaigns can have some delayed effects on the series, input is added to the days after the campaign release by means of a decay function. For marketing campaigns, one-week effects are considered, and their values are assumed to decrease linearly with time; on the other hand, for promotions, the decrease is dependent on the scale of the campaign.

5.2 Model Specification

(1) *DR:* ARIMA parameters are calculated through the ACF and PACF functions. The parameter tuning of the dynamic regression model is performed using cross-validation, based on the MAPE metric presented in Sect. 3. Table 1 shows the parameters used to fit the models.
(2) *GBM:* The implementation used for the GBM model is XGBoost [54], an efficient and scalable tree boosting model. Table 1 contains the optimal parameters found for XGBoost using cross-validation and grid search. The parameter *max_depth* refers to the maximum depth of a tree, and *eta* is the step size used to prevent overfitting. In the case of GBMs, tuning the model is computationally expensive and time-consuming as there are many parameters. However, the parameter search can be automatized and directly re-used for equivalent time series data from other game titles, which makes it more flexible.
(3) *GAMM*: As we have continuous data, we consider the *identity* link function with a Gaussian distribution. For the GAMM, weekly and monthly seasonalities are introduced as cyclic P-splines [55]. *Gacha* is added by applying a P-spline with 4-knots corresponding to the four values of the event scale. For the temperature variable, we employ a cyclic cubic regression spline [56] that estimates a periodic smooth regression function for seasonal patterns. For the other variables (holidays, events, day of the week and the month), the default spline corresponding to low-rank isotropic smoothers is used [57].
(4) *DBN:* The parameters obtained by grid search are shown in Table 1 (h: number of hidden layers, n: number of nodes per layer, plr: pre-train learning rate, tlr: fine-tuning learning rate, k: number of steps to perform Gibb sampling in the RBM, b: mini-batch size). Before training the model, training data are shuffled and 80% of them are randomly assigned to the training set and the other 20% to the validation set. The model is first trained with the training set, and then validated with the validation set, for every epoch. To avoid overfitting, early stopping [58] is applied and the fine-tuning iteration is terminated when the loss of the validation set stops decreasing for 20 consecutive epochs.

To perform the predictions, a period of thirty days is chosen in order to obtain monthly forecasts of sales and playtime. For the test evaluation, we use a rolling forecasting technique [59], taking steps of 7 days for each new forecast and with a minimum of 6 months of training data. For *AoI* the forecasts were performed weekly from Nov 2, 2015 until Jan 10, 2017. In the case of *GS*, they were carried out from Oct 5, 2015 until Feb 1, 2016. The average errors are then computed over the rolling prediction results, which serves to compare the RMSLE, MASE and MAPE values.

6 Results

Figure 4 shows an example of the predictions within a given period for each of the models. It illustrates that, while the performance of both DR and the GAMM is relatively similar, the GBM model has much more difficulty capturing the valleys of the series, while the DBN overestimates the peaks. The prediction errors for sales and playtime displayed in Table 2 also reflect this, showing that both of these models perform worse than the GAMM and DR in the case of *AoI*.

DR does provide better results for playtime forecasting than the GAM; still, to achieve a proper parameter selection, we need to check the autocorrelation and other measures to have more control over the fitting. This evaluation turns DR into a rigid model, which cannot adapt easily to time series data from other games, while the GAMM is more flexible and easier to tune. Once the smooth functions in the GAMM are selected, they automatically adjust to fit the distribution of the external variables. This way, the model can also learn to fit time series data from other games of the same nature.

Table 2 also shows the forecasting error results for *GS*. We can see that the pattern is similar to that for *AoI*. The performance of the GAMM and DR is approximately the same, and both methods outperform GBM on the sales forecasting. In this case, the DBN does perform significantly better than the GBM model, and also slightly better than the GAMM. Overall, however, taking into account all error measures for both games and dimensions (i.e. playtime and sales), the GAMM yields the most consistent results.

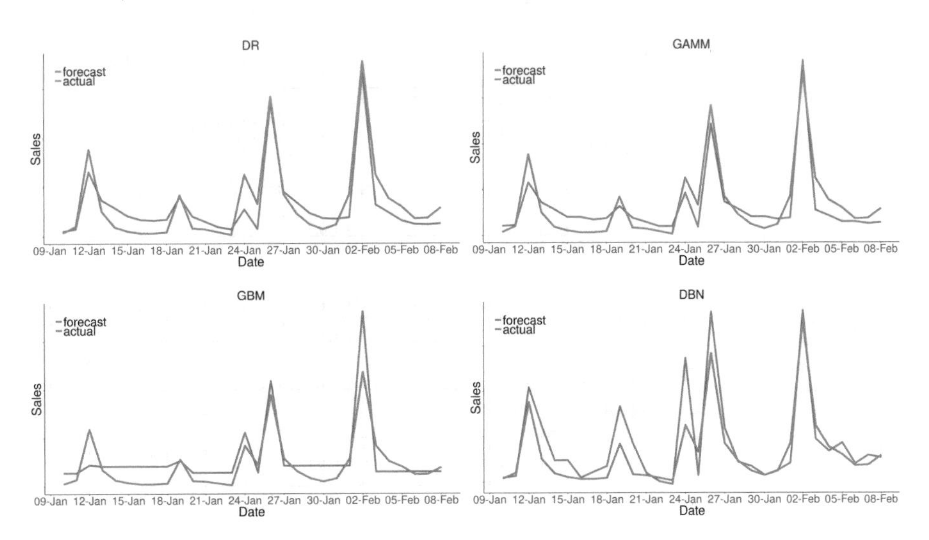

Fig. 4. Actual and forecast sales time series for *AoI*, for the period from Jan 10, 2017 until Feb 8, 2017. The different panels correspond to the forecast with DR (top left), GAMM (top right) gradient boosting (bottom left) and DBN (bottom right).

6.1 Model Validation

6.2 Forecasting Horizon

The forecasting horizon was evaluated for all the models, since the performance can significantly decrease as the number of days forecast increases [60]. Figure 5 depicts the resulting RMSLE, MASE and MAPE as a function of time, illustrating thus the predictability of the different models. The GAMM performs better than all the other models, staying much flatter for all error measures, which indicates that the forecast accuracy does not decrease much even when forecasting two or three months into the future. The GBMmodel also shows a steadier behavior, with a reasonably stable forecasting accuracy. However, this method has much higher initial and overall errors than the GAMM. The DBN has more difficulty keeping the predictions stable, showing divergent behavior as the number of days forecast increases. Finally, the DR results diverge rapidly as the forecast period becomes larger. For a short prediction range of just a few days, this technique performs better than the GAMM, but when the forecast horizon increases, the errors also become significantly larger.

Table 2. Error results for time series forecasting

Dataset	Model	RMSLE	MASE	MAPE
Sales Age of Ishtaria	DR	0.106	0.669	8.9
	GBM	0.118	0.727	10.2
	GAMM	0.106	0.672	9.1
	DBN	0.122	0.770	11.1
Sales Grand Sphere	DR	0.119	0.741	9.8
	GBM	0.164	0.814	14.8
	GAMM	0.142	0.874	11.9
	DBN	0.128	0.762	11.4
Playtime Age of Ishtaria	DR	0.243	0.921	18.7
	GBM	0.237	1.057	19.5
	GAMM	0.246	0.931	22.7
	DBN	0.357	1.057	31.5
Playtime Grand Sphere	DR	0.362	1.034	35.6
	GBM	0.559	1.000	73.5
	GAMM	0.495	1.011	56.2
	DBN	0.265	0.991	24.6

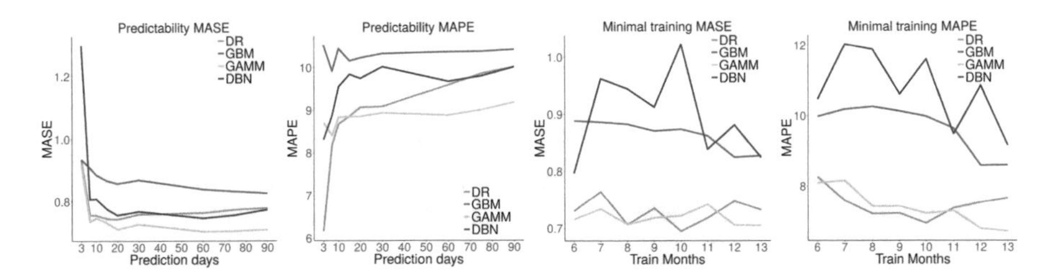

Fig. 5. MASE and MAPE of the forecasting horizon (left panels) and the minimal training period (right panels) for DR, GBM, GAMM and DBN. Results from AoI.

6.3 Minimal Training Set

We performed an error analysis of the model performance as a function of the training set size in order to evaluate the minimal training time required to obtain robust predictions.

Figure 5 shows a significant drop in the prediction errors after 12 months of training time for the GAMM, DBN and GBM. For the GAMM, the error consistently decreases with training time, while the DR performance is more unstable. The latter behavior can also be seen for the DBN and could be explained by the variable nature of the time series, which causes instability during training. In general, a 12-month training set should be sufficient to obtain the most accurate forecasts, but even after just 6 months of training data the errors are already relatively low, especially for DR and the GAMM.

7 Event Simulation

Simulation optimization (i.e. the analysis of *what-if* scenarios) is used to find the optimum input value that maximizes the response [6,7]. Using the time series forecasting models proposed in this work, a simulation was carried out to analyze the effect of future events on the total playtime and sales. The order of the upcoming events can be changed to evaluate how a different event planning impacts the forecasts.

7.1 Simulation Results and Analysis

Figure 7 shows an example of a simulation, with different event sequences being input into the models. In the case of DR, the sales for Sequences 1 and 2 were 37% higher and 25% lower, respectively, than for the predefined original event sequence (the sequence that had been planned). Thus, using a different sequence of events the total sales could have been increased by 37%. For the GAMM, Sequence 1 results in an amount of sales lower than that for DR (by 32%). Although all models are suitable to perform simulations, their forecasts present different levels of accuracy. As shown in the results section, the GAMM model provides the most accurate estimate of the predicted sales and therefore it can also be expected to produce the most precise simulation results.

In DR models, the response has a linear relationship with the features, which has the drawback of not capturing the non-linear behavior of real phenomena. However, this also makes simulations easier, as the interpretation of the parameters is straightforward compared to other strongly non-linear models, like DBNs.

7.2 Simulation Engine Tool

A common business problem faced by many game studios is how to plan future acquisition and in-game events so as to maximize sales and playtime [61]. While it is possible to manually investigate the success of past events, there are too many potentially correlated variables to be considered. The proposed forecasting models can do this automatically, and learn all the potential impacts of external variables on the future sales and playtime, providing a better estimate of the effect of future event sequences.

In order to provide a solution to the event-planning problem from a business perspective, a web-based system for time series simulation was developed. With this system, users can easily plan in-game events by means of an event planner user interface. After inputting the planned events, the daily sales and playtime for the next 30 days will be simulated and shown within a few seconds.

The system structure is shown in Fig. 6. A database stores the forecasting models, the model parameters, and the external variables mentioned in Sect. 4.2 (such as temperature and holidays). The parameters are tuned once for each game, and then models are trained once per month with the parameters. When the server receives a simulation request, it connects the database, the file system and the modeling script, so that the modeling script can obtain the data required for the simulation. After the simulation is finished, the server reads the output file and sends the results to the front-end for display.

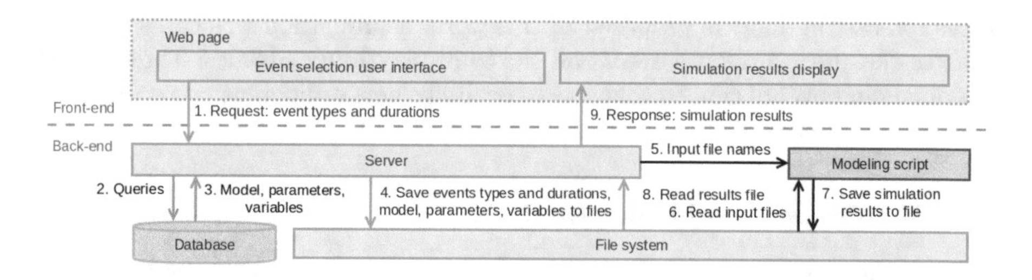

Fig. 6. Structure of the time series simulation system. The arrows show how the web page, server, database, file system and modeling script interact with each other. From the user interface, one can specify future game or marketing events for desired dates. Then, the server will take the previously trained model, perform daily predictions of the sales and playtime for the input events and return the simulation results to the user.

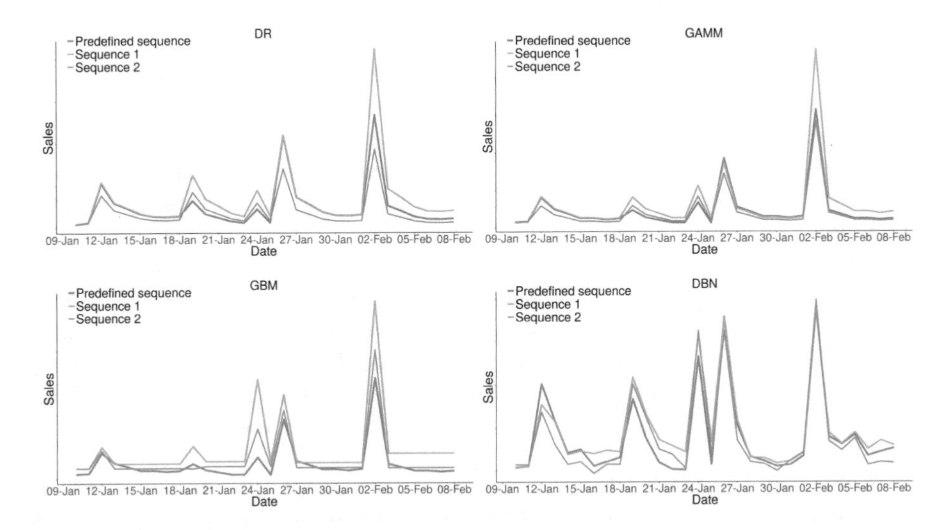

Fig. 7. Simulation results for the sales time series of *AoI*, forecasting 30 days from Jan 10, 2017. The forecast resulting from the predefined sequence of events (consisting of the planned events for that period of time) is compared with two other event sequences, named Sequence 1 and Sequence 2. The different panels correspond to the simulation with DR (top left), GAMM (top right), GBM (bottom left) and DBN (bottom right).

8 Summary and Conclusion

Overall, we found that the GAMM and ARIMA models are the most accurate for daily forecasting of sales and playtime. This result held for both of the evaluated games, *Age of Ishtaria* and *Grand Sphere*. However the GAMM has the advantage of requiring less manual tuning, which makes it more practical in a production environment. The gradient boosting model is less suitable for forecasting with these particular time series, as it showed difficulty capturing the peaks and valleys of the data.

Similarly, the DBN overestimated the peaks, which could be explained by the fact that the model has many parameters, while the dataset used was relatively small (less than 1000 observations). The GAMM and ARIMA models, on the other hand, have less parameters, avoiding altogether the overfitting issue.

Alternatively, when dealing with much larger datasets that present long-range dependencies, a long short-term memory (LSTM) [62] model could be used to properly capture and learn these dependencies and provide accurate future predictions. However, for small datasets with very short-range dependencies, as is the case for daily sales and playtime data, the input can still be fit by a standard DNN using a sliding window over the past days. This approach, though, still suffers from the overfitting problem due to the small number of data.

Nevertheless the DBN or DNN models still show room for improvement in time series forecasting, even when dealing with small datasets. ARIMA is a common forecasting model applied in a wide range of fields [63] and has the advantage of inherently containing parameters for time series forecasting. Potentially, we could incorporate such parameters into DNNs, and inspiration from the ARIMA model could be drawn to construct a similar model suitable for non-linear forecasts. However this approach is beyond the scope of this paper, as the aim here was to have an interpretable, well-established model capable of performing accurate predictions and simulations that can be applied to time series forecasting in games.

The GAMM allows for a generalizable model that can correctly capture the time series dynamics. It can be used not only to forecast both future playtime and sales, but also to simulate future game and marketing events. Instead of randomly deciding the event planning, we can employ a model-based approach that uses past information to automatically learn the interactions that are relevant for predicting event success (e.g. the weather, the day of the week and the national holidays).

We provided a solution that can be used operationally in a business setting to get real-time simulation results. Allowing game studios to accurately simulate future events can help them to optimize their planning of acquisition campaigns and in-game events, ultimately leading to an increase in the amount of user playing time and to an overall rise of the in-game monetization.

9 Software

All analysis except DBN was performed with R version 3.3.2 for Linux, using the following packages from CRAN: *forecast* 1.0 [64], *mcgv* 1.0 [65] and *xgboost* 2.38 [54]. DBN was performed with Python 2.7.12, using *Theano* [66].

Acknowledgments. We thank Sovannrith Lay for helping to gather the data and Javier Grande for his careful review of the manuscript.

References

1. El-Nasr, M.S., Drachen, A., Canossa, A.: Game Analytics. Sprint, New York (2013)
2. Yannakakis, G.N., Togelius, J.: Artificial Intelligence and Games. Springer (2017). http://gameaibook.org
3. De Gooijer, J.G., Hyndman, R.J.: 25 years of time series forecasting. Int. J. Forecast. **22**(3), 443–473 (2006)
4. Brockwell, P.J., Davis, R.A.: Introduction to Time Series and Forecasting. Springer, Heidelberg (2016)
5. Adhikari, R., Agrawal, R.: An introductory study on time series modeling and forecasting. arXiv preprint arXiv:1302.6613 (2013)
6. Asmussen, S., Glynn, P.W.: Stochastic Simulation: Algorithms and Analysis, vol. 57. Springer Science and Business Media, Heidelberg (2007)
7. Carson, Y., Maria, A.: Simulation optimization: methods and applications. In: Proceedings of the 29th Conference on Winter Simulation, pp. 118–126. IEEE Computer Society (1997)

8. Box, G.E., Jenkins, G.M.: Time series analysis: forecasting and control, revised ed. Holden-Day (1976)
9. Friedman, J.H.: Greedy function approximation: a gradient boosting machine. Ann. Stat. **29**, 1189–1232 (2001)
10. Hastie, T.J., Tibshirani, R.J.: Generalized Additive Models, vol. 43. CRC Press, Boca Raton (1990)
11. Busseti, E., Osband, I., Wong, S.: Deep learning for time series modeling. Technical report, Stanford University (2012)
12. Bauckhage, C., Kersting, K., Sifa, R., Thurau, C., Drachen, A., Canossa, A.: How players lose interest in playing a game: an empirical study based on distributions of total playing times. In: 2012 IEEE Conference on Computational Intelligence and Games (CIG), pp. 139–146. IEEE (2012)
13. Hadiji, F., Sifa, R., Drachen, A., Thurau, C., Kersting, K., Bauckhage, C.: Predicting player churn in the wild. In: 2014 IEEE Conference on Computational Intelligence and Games (CIG), pp. 1–8. IEEE (2014)
14. Periáñez, Á., Saas, A., Guitart, A., Magne, C.: Churn prediction in mobile social games: towards a complete assessment using survival ensembles. In: 2016 IEEE International Conference on Data Science and Advanced Analytics (DSAA), pp. 564–573. IEEE (2016)
15. Bertens, P., Guitart, A., Periáñez, Á.: Games and big data: a scalable multi-dimensional churn prediction model. In: Accepted in IEEE CIG (2017)
16. Bauckhage, C., Drachen, A., Sifa, R.: Clustering game behavior data. IEEE Trans. Comput. Intell. AI Games **7**(3), 266–278 (2015)
17. Drachen, A., Sifa, R., Bauckhage, C., Thurau, C.: Guns, swords and data: clustering of player behavior in computer games in the wild. In: 2012 IEEE Conference on Computational Intelligence and Games (CIG), pp. 163–170. IEEE (2012)
18. Drachen, A., Thurau, C., Sifa, R., Bauckhage, C.: A comparison of methods for player clustering via behavioral telemetry. arXiv preprint arXiv:1407.3950 (2014)
19. Sifa, R., Bauckhage, C., Drachen, A.: The playtime principle: large-scale cross-games interest modeling. In: 2014 IEEE Conference on Computational Intelligence and Games (CIG), pp. 1–8. IEEE (2014)
20. Saas, A., Guitart, A., Periáñez, Á.: Discovering playing patterns: time series clustering of free-to-play game data. In: 2016 IEEE Conference on Computational Intelligence and Games (CIG), pp. 1–8. IEEE (2016)
21. Lawrence, K.D., Geurts, M.D.: Advances in Business and Management Forecasting, vol. 4. Emerald Group Publishing, Bingley (2006)
22. Box, G.E., Cox, D.R.: An analysis of transformations. J. Roy. Statist. Soc. Ser. B (Methodol.) **26**, 211–252 (1964)
23. Akaike, H.: A new look at the statistical model identification. IEEE Trans. Autom. Control **19**(6), 716–723 (1974)
24. Schwarz, G.: Estimating the dimension of a model. Ann. Stat. **6**(2), 461–464 (1978)
25. Cragg, J.G.: Estimation and testing in time-series regression models with heteroscedastic disturbances. J. Econom. **20**(1), 135–157 (1982)
26. Dietterich, T.G.: Ensemble methods in machine learning. In: International Workshop on Multiple Classifier Systems, pp. 1–15. Springer (2000)
27. Mason, L., Baxter, J., Bartlett, P.L., Frean, M.R.: Boosting algorithms as gradient descent. In: NIPS, pp. 512–518 (1999)
28. Breiman, L.: "Arcing the edge," Technical Report 486, Statistics Department. University of California at Berkeley, Technical report (1997)
29. Ridgeway, G.: Generalized boosted models: a guide to the gbm package. Update **1**(1), 2007 (2007)

30. Natekin, A., Knoll, A.: Gradient boosting machines, a tutorial. Front. Neurorobot. **7**, 21 (2013)
31. Zhang, T., Yu, B.: Boosting with early stopping: convergence and consistency. Ann. Stat. **33**(4), 1538–1579 (2005)
32. Hastie, T., Tibshirani, R.: Generalized additive models: some applications. J. Am. Stat. Assoc. **82**(398), 371–386 (1987)
33. Maindonald, J.: Smoothing terms in GAM models (2010)
34. Larsen, K.: GAM: the predictive modeling silver bullet. Multithreaded. Stitch Fix, vol. 30 (2015)
35. Chen, C.: Generalized additive mixed models. In: Communications in Statistics-Theory and Methods, vol. 29, no. 5–6, pp. 1257–1271 (2000)
36. Breslow, N.E., Clayton, D.G.: Approximate inference in generalized linear mixed models. J. Am. Stat. Assoc. **88**(421), 9–25 (1993)
37. Wood, S.N.: Fast stable restricted maximum likelihood and marginal likelihood estimation of semiparametric generalized linear models. J. Roy. Stat. Soc.: Ser. B (Stat. Methodol.) **73**(1), 3–36 (2011)
38. Bengio, Y.: Learning deep architectures for AI. Found. Trends® Mach. Learn. **2**(1), 1–127 (2009)
39. Deng, L., Yu, D.: Deep learning: methods and applications. Found. Trends® Sig. Process. **7**(3–4), 197–387 (2014)
40. Krizhevsky, A., Sutskever, I., Hinton, G.E.: ImageNet classification with deep convolutional neural networks. In: Advances in Neural Information Processing Systems, pp. 1097–1105 (2012)
41. Graves, A., Mohamed, A.-R., Hinton, G.: Speech recognition with deep recurrent neural networks. In: 2013 IEEE International Conference on Acoustics, Speech and Signal Processing (ICASSP), pp. 6645–6649. IEEE (2013)
42. Hinton, G.E., Osindero, S., Teh, Y.-W.: A fast learning algorithm for deep belief nets. Neural Comput. **18**(7), 1527–1554 (2006)
43. Ackley, D.H., Hinton, G.E., Sejnowski, T.J.: A learning algorithm for Boltzmann machines. Cognit. Sci. **9**(1), 147–169 (1985)
44. Larochelle, H., Bengio, Y.: Classification using discriminative restricted Boltzmann machines. In: Proceedings of the 25th International Conference on Machine Learning, pp. 536–543. ACM (2008)
45. Längkvist, M., Karlsson, L., Loutfi, A.: A review of unsupervised feature learning and deep learning for time-series modeling. Pattern Recognit. Lett. **42**, 11–24 (2014)
46. Zhang, G., Patuwo, B.E., Hu, M.Y.: Forecasting with artificial neural networks: the state of the art. Int. J. Forecast. **14**(1), 35–62 (1998)
47. Srivastava, N., Hinton, G., Krizhevsky, A., Sutskever, I., Salakhutdinov, R.: Dropout: a simple way to prevent neural networks from overfitting. J. Mach. Learn. Res. **15**(1), 1929–1958 (2014)
48. Ng, A.Y.: Feature selection, l 1 vs. l 2 regularization, and rotational invariance. In: Proceedings of the Twenty-first International Conference on Machine Learning, p. 78. ACM (2004)
49. Hyndman, R.J., Koehler, A.B.: Another look at measures of forecast accuracy. Int. J. Forecast. **22**(4), 679–688 (2005)
50. Fox, A.J.: Outliers in time series. J. Roy. Stat. Soc. Ser. B (Methodol.) **11**, 350–363 (1972)
51. Sakurada, M., Yairi, T.: Anomaly detection using autoencoders with nonlinear dimensionality reduction. In: Proceedings of the MLSDA 2014 2nd Workshop on Machine Learning for Sensory Data Analysis, p. 4. ACM (2014)

52. Japan national holidays. https://www.timeanddate.com/holidays/japan/
53. Tokyo daily temperature 2014 to 2017. https://www.wunderground.com/
54. Chen, T., Guestrin, C.: Xgboost: a scalable tree boosting system. In: Proceedings of the 22Nd ACM SIGKDD International Conference on Knowledge Discovery and Data Mining, pp. 785–794. ACM (2016)
55. Eilers, P.H., Marx, B.D.: Flexible smoothing with B-splines and penalties. Stat. Sci. **11**, 89–102 (1996)
56. Wood, S.N.: Generalized Additive Models: An Introduction with R. CRC Press, Boca Raton (2017)
57. Wood, S.N.: Thin plate regression splines. J. R. Stat. Soc. Ser. B Stat. Methodol. **65**(1), 95–114 (2003)
58. Prechelt, L.: Early stopping-but when? In: Neural Networks: Tricks of the trade, pp. 55–69. Springer (1998)
59. Gilliland, M., Sglavo, U., Tashman, L.: Business Forecasting: Practical Problems and Solutions. Wiley, Hoboken (2016)
60. Makridakis, S., Hibon, M.: The M3-Competition: results, conclusions and implications. Int. J. Forecast. **16**(4), 451–476 (2000)
61. Julkunen, J.: Feature Spotlight: In-Game Events and Market Trends (2016). http://www.gamerefinery.com/in-game-events-market-trends/
62. Hochreiter, S., Schmidhuber, J.: Long short-term memory. Neural Comput. **9**(8), 1735–1780 (1997)
63. Dwyer, L., Gill, A., Seetaram, N.: Handbook of Research Methods in Tourism: Quantitative and Qualitative Approaches. Edward Elgar Publishing, Cheltenham (2012)
64. Khandakar, Y., Hyndman, R.J.: Automatic time series forecasting: the forecast Package for R (2008)
65. Wood, S.N.: MGCV: Mixed GAM computation vehicle with GCV/AIC/REML smoothness estimation (2012)
66. Theano Development Team: Theano: A Python framework for fast computation of mathematical expressions, arXiv e-prints, abs/1605.02688, http://arxiv.org/abs/1605.02688, May 2016

SQL and NoSQL Database Comparison

From Performance Perspective in Supporting Semi-structured Data

Ming-Li Emily Chang(✉) and Hui Na Chua

Department of Computing and Information Systems,
Sunway University, Subang Jaya, Selangor, Malaysia
14063796@imail.sunway.edu.my, huinac@sunway.edu.my

Abstract. In this digital era, social media web applications have churned out huge amount of unstructured data each day. These social media data may be processed into meaningful data through text analytics. With the rapid growth of the volume of unstructured data produced daily, NoSQL database is increasingly popular that it has become the chosen database to store data. However, little research is done on the comparison of SQL and NoSQL in terms of indexing, performance tuning, and amount of records supported. This paper aims to provide a thorough comparative evaluation of MongoDB and MySQL, a tool for SQL and NoSQL databases, respectively, in terms of their performance in populating and retrieving big data after performance tuning. The findings presented in this paper give a new insight from the aspect of how these databases support semi-structured social media data by considering the options of performance tuning. The methodology for this research consists of four performance measurements, namely, insert, select, update, and delete up to 1 million Twitter data stored, to evaluate SQL and NoSQL databases. Our result findings indicate that MongoDB does perform faster for all the four operations. However, there are more performance tuning options provided by MySQL for more flexible performance optimization.

Keywords: Database management systems · Semi-structured data model NoSQL · Big data · Twitter data streaming

1 Introduction

The rapid growth of technology has produced billions of data each day. According to Gartner [1], more than 70% of the world's data was "buried" or "lost" in unstructured formats like books, documents, and journals. The majority of data in the world can only be accessed through unstructured sources. These data should not continue to be buried. Instead, it should be harvested and integrated.

In this digital era, social media plays a huge role in the modern society. According to research, on average, a Malaysian spends a shocking number of 12 h on social media daily [2]. This leads to the high volume of semi-structured and unstructured data to be created every day in the Internet. These semi-structured data may be processed into meaningful data through text analytics. Among popular social media sites, Twitter has

K. Arai et al. (Eds.): FICC 2018, AISC 886, pp. 294–310, 2019.
https://doi.org/10.1007/978-3-030-03402-3_20

been able to establish itself as one of the top engagement platform. The micro-blogging site allows users to send and read short 140-character messages in real time.

Through Twitter Streaming API, developers are able to extract data from Twitter. These data are considered as semi-structured data because they are in JSON format. These social media data can be stored in NoSQL database or in SQL database. There are several researches have been carried out on the performance comparison between SQL database and NoSQL database, but there is a lack of research of comparing their performances to store and retrieve Twitter data with the consideration of performance tuning options. This paper presents a detailed evaluation of storing and retrieving semi-structured data in SQL and NoSQL after performance tuning. Throughout the research work of this paper, four measurements were taken to compare between SQL and NoSQL databases. These measurements can be found in Sect. 4.

2 Literature Review

2.1 Types of Data Models

(1) *Structured data*: Structured data refers to any set of data values conforming to a common schema or type [3]. When information is highly structured and predictable, it can be easily entered, stored, queried and analyzed into structured data, usually stored in relational database management systems.
(2) *Unstructured data*: Data that is not contained in a database or any type of data structure is known as unstructured data. It can range from textual unstructured data like documents, messages to non-textual unstructured data like media files, namely images and videos.
(3) *Semi-structured data*: The semi-structured data model is designed as a progression of the relational data model that supports a flexible structure for data representation [4]. Semi-structured data refers to data that may be irregular or incomplete. However, generally, it has some structure, but does not conform to a fixed schema. This makes it schema-less and self-describing, where data carries information about its own schema. For example, XML and JSON formats that are commonly used to support the social media data in internet applications such as Twitter and Facebook.

2.2 SQL Database

Structured Query Language (SQL) database refers to relational database, where SQL is a standard language for accessing relational database. SQL statements are able to carry out tasks such as insert, retrieve or update data from a database such as interactive queries for information from the database and for generating reports through the database.

SQL database is user-friendly in terms of creation and access. More importantly, it is easy to extend. A SQL database is a set of tables containing structured data, which is data fitted into predefined categories. Each table, or sometimes known as entity or relation, contains one or more attributes which is also known as data categories. Each row contains a unique instance of data for the attributes. Many organizations use SQL database to manage their data and it has been proven to be an effective approach.

One of the advantages of SQL database is its simple concept of two-dimensional table. It has good theoretical basis as it is also based on mathematical basis called relational algebra. It also promotes data independence where data descriptions are independent of the data itself. It provides improved security and integrity using high level queries language and better organization of data in the tables. Lastly, it is suitable for high level languages. Its simple structures enable the use of declarative languages.

However, SQL database does have its shortcomings. It fails to reflect the real-world object or entity. This is because the process of storing relation tables involves normalization which causes the proliferation of table. The result is that a simple application will produce many tables. This also increases the processing overheads as simple queries may require joining many tables. Moreover, it lacks sematic power. Normalization causes the tables to have less meaning thus limiting the way the relationship between the data is shown. It is also difficult to express specialist relationships.

For this project, social media obtained through the Twitter Streaming API is considered as a semi-structured data which can be stored into SQL database. This project will investigate the advantages and disadvantages of using SQL database to store semi-structured data. The following section considers potential SQL database management system that can be implemented.

Example of SQL Databases:

(1) *Microsoft SQL server*: Microsoft SQL Server is a SQL database management system developed by Microsoft. There are many editions of Microsoft SQL Server that is catered for a wide range of audiences and workloads, from small single-machine applications to large web applications with distributed users [5]. However, because of its robust technology, this system is not free to use, it comes with a price that is not within the budget of this project.
(2) *Oracle database*: Oracle database is one of the pioneers for database management systems. Oracle Database is produced and marketed by Oracle Corporation; it is an object-relational database management system. It has features that enable administrators and developers to adapt to changing business requirements timely.
(3) *Sybase*: Sybase is a SQL database management system that supports creating, updating and administrating a database. It uses SQL to access the database. The default system databases are the master database, model database, system procedure database (`sybsytemprocs`), two-phase commit transaction database (`sybsystemdb`) and temporary database (`tempdb`). While Sybase is able to load data quickly, the deployment of the system is rather complex and has lesser features than Oracle databases.
(4) *MySQL*: SQL database can be implemented in many ways such as MySQL, which is dubbed as the world's most prominent open source database. It is a database system that uses standard SQL and is considered very fast, reliable and easy to use. It runs on a server, ideal for small and large application. Moreover, MySQL is developed, distributed, and supported by Oracle Corporation. Due to its open source feature, there are many forks of MySQL. Lastly, there is a free edition of

MySQL that will be sufficient for this project. Therefore, MySQL is deemed more suitable for our research of this paper as compared to Microsoft SQL server.

2.3 NoSQL Database

SQL database is born in the era of mainframes and business applications. However, lately, there is a shift to the digital economy, an economy powered by the Internet and latest technology like the cloud, big data, social media and many more. Thus, there is the birth of Not Only SQL (NoSQL) database. It is a new approach to database where it caters for large sets of distributed data by having fewer restrictions as compared to SQL database.

It is implemented by many prestigious Internet companies like Google and Facebook as a result of the exponential growth of the Internet and the rise of web applications. It has become the first alternative to SQL databases, with scalability, availability, and fault tolerance being key deciding factors. NoSQL data model is able to address the issues that SQL data model is incapable of. For example, it can manage the large volume of rapidly changing structured, semi-structure and unstructured data. It also uses object-oriented programming that is flexible and user-friendly. Lastly, it uses inexpensive architecture that is geographically distributed.

There are four types of NoSQL databases, namely, key-value store, column store, document database and graph database. As the name suggest, key-value contains an indexed key and a value which able to store data in a schema-less way. Column store is designed for storing data tables as sections of columns of data. It offers very high performance and its architecture is highly scalable. Next, document database expands on the basic idea of key-value stores. Each document is assigned a unique key, which is used to retrieve the document-oriented information. The model is basically versioned documents that are collections of other key-value collections. The semi-structured documents are stored in formats like JSON. Document databases support nested value keys and allow more efficient querying. Lastly, the graph database is based on graph theory. It is a representation of the relations using graph and has interconnected elements, with an undetermined number of relations between them. It is scalable across multiple machines using a flexible graph model.

For this research, the semi-structured data is stored in JSON format. Therefore, document databases systems were considered.

Example of NoSQL Databases:

(1) *HBase*: Apache HBase, short for Hadoop database, is a column-oriented NoSQL database that runs on top of Hadoop as a big data store that is scalable and distributed. Hadoop is an open-source framework for storing data and running applications on clusters of commodity hardware. This means that HBase can leverage the distributed processing paradigm of the Hadoop Distributed File System (HDFS) and benefit from Hadoop's MapReduce programming model. This database adopts master-slave architecture where Master is the Name Node and Slaves are the Data Nodes. Moreover, this system is efficient in the 'read' mode. Although so, it does not support multiple 'writes' [6].

(2) *Cassandra*: Apache Cassandra is a NoSQL database that is a scalable and open source. It was initially developed at Facebook and then adopted into an Apache project. It provides availability consistently, linear scale performance, operational simplicity and easy data distribution across multiple data centres and cloud availability zones. It is column-base store where each block contains data from only one column. Cassandra offers a constant-time write regardless of data size. However, querying options for retrieving data are very limited and it focuses on availability over consistency.

(3) *MongoDB*: MongoDB is a scalable, high-performance, open source, document-oriented NoSQL database. It focuses on consistency. In the MongoDB replication model, a replica set will host the same dataset using a group of database nodes; a primary node will be defined and the others will be secondary nodes. Replica sets provide strict consistency because by default, a primary node is used for all read and write operations. MongoDB implements JSON-like documents with dynamic schemas, making the integration of data in certain types of applications easier and faster. Lastly, it supports many languages and application development platforms, including Python programming. Therefore, MongoDB is chosen to be used as the NoSQL database system.

2.4 Performance Tuning

Performance tuning is considered as a crucial requirement of running a mission-critical application to achieve and sustain high performance with database. Database performance tuning encompasses the steps users may take to maximize performance with the goal of optimizing the usage of system resources. The overall performance of the database is highly dependent on the tuning of certain database elements such as data model, indexing, query structure, system configuration and so on. Therefore, it is essential to do performance tuning before taking the measurements for evaluation so that both databases can perform in their best condition.

(1) *MySQL*: In addition to performance tuning like query processing and indexing, MySQL also have different storage engines. Storage engines are components of the MySQL that handle the SQL operation for different table types [7]. Each has its own strength and limitation, but the two commonly used engines are `InnoDB` and `MyISAM`. The default storage engine is `InnoDB`, it is the most-general-purpose storage engine that acts like an in-memory database. Meanwhile, `MyISAM` tables have small footprint, and is optimized for read-intensive environment, with little or no write operations. Table 1 below shows a comparison of the advantages and disadvantages between `InnoDB` and `MyISAM`.

(2) *MongoDB*: MongoDB performance tuning is similar to the usual query processing, indexing like SQL databases. Since MongoDB is considered schema-less, the way the documents are stored also affects the performance [8]. MongoDB also has features like logging and query explain which allows users to track their database performance.

Table 1. Comparison of InnoDB and myisam storage engine

Database engine	Advantages	Disadvantages
InnoDB	• Faster in heavy write operation tables because it utilizes row-level locking.	• No full-text indexing. • Consumes more RAM and other system resources.
MyISAM	• Simpler structure allows it to perform faster compared to InnoDB thus much less costs of server resources. • Especially good for read-intensive tables. • Full-text indexing. • Small footprint.	• No data integrity and checking, which increases overhead of the system administrators and developers. • Does not support transactions. • The whole table is locked for any insert or update which results it to perform slower than InnoDB during write operations.

2.5 Quantitative Performance Measurement

In this research, it is critical that experiments and the observations recorded are repeatable. A quantitative description of a relevant characteristic involves a numerical measurement. When a numerical value is given to a characteristic, it allows for direct comparison between observations made by different people or at different times.

Several researches on evaluating SQL and NoSQL databases have used quantitative measurements. We adopted these measurements as it is validated by literature as follows:

(1) *Study on performance evaluation of MySQL and MongoDB databases*: There was a research done [9] where the performance of the mentioned databases on application of hypermarket was evaluated. This study focused on the insertion and searching on the hypermarket databases for auditing purposes. The performance evaluation is measured by the time taken for a certain number of records to execute using MongoDB and MySQL. Each round, the number of record increases steadily from 100 to 25,000 and the execution time to run the operation is recorded in terms of millisecond.

The study concluded that there is no difference in the execution time taken for insertion and searching operations to complete for both MongoDB and MySQL databases when the number of records is small. However, when number of records increased, MongoDB showed significant reduction in the time taken for execution compared to MySQL. Thus, when the number of records is higher, MongoDB takes lesser time compared to MySQL. The authors have drawn a conclusion that MongoDB can be preferred for better performance when dealing with large amount of records with 25,000 being the maximum number of records used in this study.

(2) *Study on performance evaluation of Cassandra over MySQL for big data*: Research work in [6] uses web crawler to collect data from websites. The record is generated by a parser and inserted into MySQL and Cassandra data servers. This

study ran a workload executor at the "`Write`" Phase and "`Read`" Phase. The "Write" phase, known as load phase, working set is created from 100 to 1 million records. Then, during the "Read" phase, which is the retrieval phase, some queries will be generated to read data from clusters. The queries retrieve from small data to large data sets using simple "`SELECT`" statement to complex "`JOIN`" statements. As for the performance benchmarking, an open source platform called Datastax Corporation (2014) [10] for the Cassandra database was used. This study concluded that Cassandra performed better with "`WRITE`" operations as compared to MySQL, but it performed poorly for the "`READ`" operations. The authors proposed a configuration to maintain the balance of "`READ`" and "`WRITE`" operations for Cassandra.

(3) *Study on performance evaluation of SQL and MongoDB databases*: Research work in [11] used Microsoft SQL Server and MongoDB for their SQL and NoSQL databases management system. For their database, they adopted an actual simulation of real model of transaction registration used by big companies' database engines. Using Java programming language, they conducted `INSERT`, `UPDATE`, `DELETE` and `SELECT` operations on records with the scale of 1,000, 10,000, 100,000 and 1,000,000. Moreover, they ran three queries involving functions such as `COUNT()`, `SUM()` and `AVG()`. The authors showed the comparison of the time taken to run each of the operations as well as the queries. The study concluded that MongoDB provides good flexibility in database designing process and suggested the use of MongoDB for large databases or databases in which the structure is constantly changing. For those users that use a relatively stronger and more structured schema, MongoDB still presents better results in terms of performance and speed in many cases. However, MongoDB performed poorly when aggregate functions are done on non-key attributes.

2.6 Twitter and Its Streaming API

Twitter is a social networking micro-blogging platform that allows registered users to share quick messages in no more than 140 characters. It aims to produce the power to create and share ideas and information in an instant for users. When Twitter first started, many criticized its concept of limiting messages into 140 characters, much like Short Message Service (SMS). However, it is now a top player and is people's prime selection for expressing opinions, especially live events. It has become an extremely popular way to share content, communicate, and broadcast information. Thus, it became a powerful force in how modern society uses the Internet now, with 313 million monthly active users.

These 140-character messages are known as "tweets" and only registered users can post a tweet. However, users are free to switch between private and public profiles, whereby all tweets published by a public profile can be read by everyone, with or without an account. Meanwhile, all tweets of a private profile will only be available to the followers, users who are subscribed to the account. Users may send a follower request to a private user if they wish to gain access to read their content, however the private user has the right to accept or reject follower requests. At the same time, public users whose content is of interest of others will be followed accordingly.

Tweets of the users followed will appear on their own personalized Twitter Homepage, arranged in reverse chronological order, enabling them to read the latest update first. If they wish to forward a tweet to their own followers, they may "retweet" it onto their own timeline or "quote tweet" which allows users to add their comment on the tweet while forwarding it. Users may also favorite a tweet, where their favorite tweets can be viewed from their profile.

Another powerful feature of Twitter is its hashtags, it is a simple way to organize content around certain topic such as `#RioOlympics`, `#election2016` and so on. Whenever a major event takes place, Twitter will light up with tweets, thus generating a new "Trending Topic", whereby a hashtag is made popular by number of tweets. For example, the Oscars could be a trending topic recognized by this hashtag: `#oscars`. Users discussing on this topic would include the hashtag in their tweets to increase the tweets' popularity and be part of the greater conversation. Twitter is also used to share information in real-time.

Twitter provides developers low latency access to their global stream of millions of users and billions of Tweet data through Twitter Streaming API. API users are able to obtain a sample of tweets matching some preferred parameters. This service has been used by many researchers, companies and governmental institutions that want to extract knowledge in accordance with a diverse array of questions pertaining to social media.

There are three streaming endpoints offered by Twitter, namely, user streams, site streams and public streams. Firstly, user streams are single-user streams that contain approximately all the data corresponding with a single user's view of Twitter. Secondly, site streams are the multi-user version of user streams. Lastly, public streams are streams of public data flowing through Twitter. This is suitable for subscribing to specific users or topics, and for data mining [12] like text analytics. For Twitter Streaming API, the stream returns more than 1% of the all tweets being tweeted at that given moment without rate limit but multiple streams from one user is not allowed.

To acquire authorized access to Twitter's API, developers must be a user of Twitter and then proceed to their Developer website and obtain the API key, API secret, Access token and Access token secret. The data from the stream is in JSON (JavaScript Object Notation) format. This format is convenient for machines to parse as well as easy for humans to decipher.

The tweet not only contains the text of a message, but it also provides a huge variety of data such as date, unique Tweet ID, source, user data such as screen name, location, URL, biodata description, account verification, number of followers, number of favorite tweets, time zone, retweet count, favorite count etc. All these data may be obtained through the Twitter Streaming API.

JSON format has fixed syntax, rules and data types, so it is not unstructured data. It does not have high degree of organization like structured data. Hence, it is considered as semi-structured data which can be stored in SQL database and NoSQL database.

2.7 Python Programming Language

Python is a clear and powerful object-oriented programming language, which is comparable to Java, Ruby, Perl and many more. It is distinguished by its large and active scientific computing community. Adoption of Python for scientific computing in both industry applications and academic research has increased significantly since the early 2000s [13].

Python is suitable for conducting research and developing prototype, as well as building the production systems. Besides, it has an extensive library support, including file processing and connecting to databases such as MySQL and MongoDB.

3 Problem Statement

With the rapid growth of the volume of unstructured and semi-structured data produced daily, NoSQL database has been increasingly popular as the chosen database to store data. However, no comparison of SQL and NoSQL is done in terms of performance tuning and storing social media data as semi-structured data. In the past, there are researches done on using Cassandra to store database by using MySQL to store relational structured database [6]. There is also comparison done on MongoDB and SQL server for big e-commerce transactional structured data [11]. Another performance evaluation is done by comparing MongoDB and MySQL [9], however the dataset used was structured data. In addition, these researches did not include a detailed description of how the measurements were taken, for example, what queries are used.

While all these researches done are beneficial, there is still a gap that is yet to be filled, which is from the aspect of how different databases support semi-structured social media data. A huge number of social media data is generated every second and these data can be stored in both MongoDB and MySQL database. However, there is currently no research done (at the time of conducting this research) on addressing the performance of implementing each of the database for storing social media data and considered the factor of performance tuning.

Moreover, the comparison between the databases will only be unbiased if both databases are in their best conditions. In order to reach optimum performance, both databases should have performance tuning. Existing researches did not cover the part of performance tuning for SQL and MySQL. Particularly, MySQL is established for over twenty years, and has provided users range of features for different usage of database. For example, data warehouses use `MyISAM` engine for their tables while write-intensive tables use `InnoDB`. On the other hand, MongoDB provides little option to fine tune the database due to its nature as the recent developing and evolving technologies of database management system in supporting semi-structured big data.

This paper presents the research that aims to fill in the gap by providing a thorough evaluation of comparing MongoDB and MySQL, a tool for SQL and NoSQL databases respectively, in terms of their performance in populating and retrieving data using semi-structured data. The research questions that we seek to answer in this paper are:

(1) How different in terms of performance of SQL and NoSQL databases in supporting semi-structured data?
(2) How different in terms of complexity is the process of data populating, retrieval and storing comparing SQL and NoSQL databases implementation?

The research questions are answered based on MySQL and MongoDB databases and by comparing their performances in terms of time taken.

4 Research Methodology

4.1 Data Collection

For this research, a data collection program was written in Python using the Twitter Streaming API. This program ran consistently to store the social media data. All the records collected were temporarily stored as a text file (`.txt`). The targeted volume of data to be collected is 1,000, 000 entries. Previously, research work done in [11] had used this amount of records to conduct their research using big e-commerce data which is structured data. Moreover, a study [9] was conducted using MySQL and MongoDB as the database systems concluded that MongoDB performed better when dealing with large amount of records after using 25,000 number of records.

Therefore, the number of records increased 40 times, which is 1,000,000 records for this research. This number was chosen after much consideration due to the fact that although there are 500 million tweets generated per day [14], Twitter Streaming API is only capable of retrieving 1% at the given moment and does not allow multiple streams from one user. The data collection program ran automatically when the Windows 7 machine was powered up with stable internet connect. Within 12 weeks, a total of 1,000,000 tweets were successfully collected.

4.2 Data Transformation

The data collected contains plentiful tweeter-specific identifiable information. For example, a tweet consists of timestamp, user id, text, number of favorites and retweets, even the profile of the user such as how many followers, how many likes they gave, etc. are stored in the JSON object. However, for the purpose of our research, the focus is on testing the performance of the two databases, so it is necessary to filter some fields; after much consideration, only three attributes were chosen to be inserted into the databases, namely, created_at, id and text.

In MySQL, these data are stored inside a table called tweets, while in MongoDB, they are stored in a collection called tweets. A script is written to process all the tweets collected and output two files formatted for data population in MySQL and MongoDB respectively.

It is important to note that the original form of the tweets, which is JSON format is supported to be inserted into MongoDB directly, without any formatting. However, for an unbiased comparison for both databases, the schema was simplified into three comparable attributes.

4.3 Performance Tuning

Performance tuning is essential to optimize each database. In this case, the database must be able to handle read and write efficiently. Before taking performance measurements on both databases, performance tuning was done to reach its optimum performance. This was to ensure both databases were in their best state before comparing one with another. In terms of system configuration, both databases ran on the same exact machine, hence their software and hardware are identical.

(1) *MySQL*: The version of MySQL used for our investigation is 5.7. Prior to taking measurements, suitable configuration steps were taken. Firstly, the database engine used is very important as different engines supports different usage. For example, for write-intensive operations, `InnoDB` engine is recommended for a more optimum performance while `MyISAM` is suitable for data warehousing as it is catering for read-intensive tables.

Next, for `InnoDB`, it maintains a buffer pool, which is a storage area for caching data and indexes in memory. For this research, the buffer pool size was increased to 4 GB, which was half of the RAM available in the machine used for testing. The maximum size of one packet was also increased to 64 GB.

(2) *MongoDB*: The version of MongoDB used in this research is 3.4. In MongoDB, there is little performance tuning options available in querying and indexing. The database profiler collected fine grained data about MongoDB write operations, cursors, and database commands on a running `mongod` instance.

4.4 Measurements

The following measurements were adapted from methods performed by various researchers on the similar topic. This is because these methods were already scientifically validated as a benchmark to evaluate between different database systems. On top of that the database complexity and software complexity were measured for both SQL and NoSQL databases.

(1) *Measurement 1*: *Inputting data*: The measurement recorded the time required to insert the social media data into SQL and NoSQL databases using the `INSERT` operation. The observation consisted of number of records and the execution time (seconds). This method was adopted from research done in [6, 9, 11].

For this `INSERT` operation, the suitable MySQL storage engine would be `InnoDB` engine. A new table named tweets was created in `InnoDB` engine with three fields, `created_at` which stores dates, `id` which stores numerical identification for the tweets, and `text` for the content of the tweets. A script was written for each of the amount to be bulk inserted into the table. The script was then called in the MySQL console. The execution time was recorded in a table.

In MongoDB, a new collection called tweets was created. The structure of the collection was determined based on the first input. For MongoDB, a script was also written for each of the amount to be bulk inserted into the collection. Then, the scripts were run in `mongoimport`. The execution time was recorded in a Table

(2) *Measurement 2: Retrieving data:* This measurement recorded the time required to retrieve the social media data into SQL and NoSQL databases using the `SELECT` operation. This method was adopted from research done in [6, 9, 11].

In MySQL database, the database can be treated as a data warehouse as the data are only expected for read-only operations. Therefore, the same data was pumped into a new table that runs in `MYISAM` engine. Then, different number of rows was selected and the execution time was recorded.

In MongoDB, the command db.collection.find() is similar to the `SELECT` operation in MySQL. The query execution time was recorded in the unit of milliseconds inside the system profile, which can be retrieved through `db.system.profile.find()`.

For both databases, different number of rows was retrieved and their execution time was recorded. However, due to the fact that MongoDB records execution time in milliseconds, it returns 0 ms for the retrieval of 100 and 1000 rows. Due to the insignificant value returned, there were no comparison between SQL and MySQL retrieval for 100 and 1000 rows.

(3) *Measurement 3: Updating data:* This measurement recorded the time required to update the social media data from SQL and NoSQL databases using the `UPDATE` operation. The observation consisted of number of records and the execution time (milliseconds). This method is adopted from research work done in [15].

For UPDATE operation, to avoid table scanning, new tables were created in InnoDB engine with different table sizes, 100, 1000, 10 000, 100 000, 1 000 000 respectively. Then all the id in the tables were updated to id + 1 and the execution time will be recorded. Below (1) shows the SQL statement for UPDATE operation.

$$\text{update hundred set id} = \text{id} + 1; \tag{1}$$

However, in MongoDB, the data types of the fields were determined upon the first insertion. The id field was treated as a `NumberLong()` which is a floating point value. In MongoDB, to increase the value of id, an increment field (`$inc`) is needed to be included into the `update()` statement. By default, the `update()` statement only modifies the first row that matches with the condition. An additional `{"multi": true}` field was added in order to update all the documents.

(4) *Measurement 4: Removing data:* This measurement recorded the time required to remove the social media data from SQL and NoSQL databases using the `DELETE` operation. The observation consisted of number of records and the execution time (milliseconds). This method was adopted from a research done in [6, 9, 11].

For `DELETE` operation, the tables with different table sizes 100, 1000, 10 000, 100 000, 1 000 000, respectively were removed and the time taken was recorded.

5 Results

5.1 Measurement Results

The measurement results are presented in Figs. 1, 2, 3 and 4 as follows:

(1) *Data input*: Ratio difference between MySQL and MongoDB as the number of rows increased were 10:8, 10:3, 10:2, 10:7 and 10:5 respectively. Upon inserting 1 million records, MongoDB took approximately half the time compared to MySQL. Despite the increase of number of rows, MongoDB consistently performed faster than MySQL.

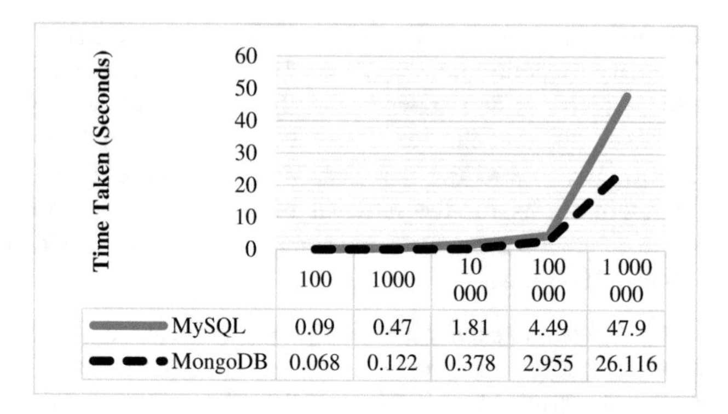

	100	1000	10 000	100 000	1 000 000
MySQL	0.09	0.47	1.81	4.49	47.9
MongoDB	0.068	0.122	0.378	2.955	26.116

Fig. 1. Line graph of data input for MySQL and MongoDB.

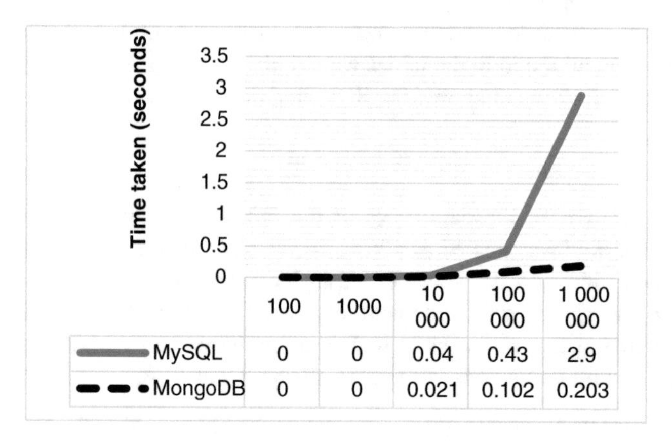

	100	1000	10 000	100 000	1 000 000
MySQL	0	0	0.04	0.43	2.9
MongoDB	0	0	0.021	0.102	0.203

Fig. 2. Line graph of data retrieval for MySQL and MongoDB.

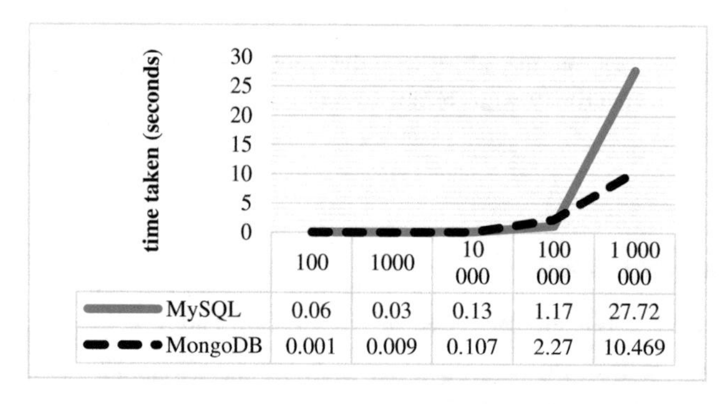

	100	1000	10 000	100 000	1 000 000
MySQL	0.06	0.03	0.13	1.17	27.72
MongoDB	0.001	0.009	0.107	2.27	10.469

Fig. 3. Line graph of data modification for MySQL and MongoDB.

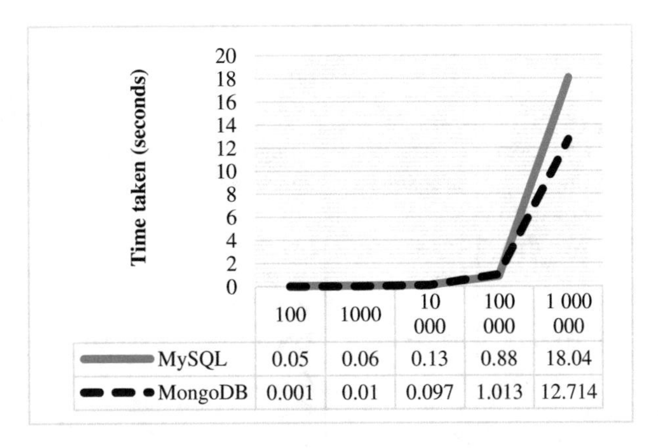

Fig. 4. Line graph of data removal for MySQL and MongoDB.

(2) *Data retrieval*: MongoDB stored the execution time in the unit of milliseconds. The retrieval time for 100 and 1000 entries returned 0 ms. Therefore, these two results will not be compared.

Ratio difference between MySQL and MongoDB as the number of rows increased were 10:5, 10:2, and 10:0.7 respectively. Despite with the use of `MyISAM` storage engine, MongoDB was significantly faster than MySQL. When retrieving 1 million records, MongoDB took only 0.203 s while MySQL which took 2.9 s.

(3) *Data modification*: Ratio difference between MySQL and MongoDB as the number of rows increased were 10:0.2, 10:3, 10:8, 10:19 and 10:4 respectively. At 100 000 records, MySQL performed better than MongoDB. However, when the number of records was increased to 1 million, MongoDB performed twice as efficient in terms of time taken when compared with MySQL.
(4) *Data removal*: Ratio difference between MySQL and MongoDB as the number of rows increased were 10:02, 10:2, 10:7, 10:12 and 10:7 respectively. Upon removing 100 000 records, MySQL performed 14.06% better than MongoDB. At 1,000,000 records, the ratio for data removal was the smallest as compared to previous operations.

5.2 Analysis with Existing Research

The following Table 2 shows the work done in existing literature comparing with our researches.

The existing literature covered unstructured data collected through web crawler, structured data from hypermarket, transaction registration data and user's profiles. Up till date, only the research work presented in this paper covers semi-structured data using social media data collected through Twitter by considering performance tuning options.

Previously, the existing literature did not discuss on performance tuning on the databases before measurements were taken. Our research presented in this paper covers

Table 2. Analysis with existing research

	[6]	[11]	[9]	[15]	This paper
SQL database	MySQL	SQL Server	MySQL	Oracle	MySQL
NoSQL database	Cassandra	MongoDB	MongoDB	MongoDB	MongoDB
Data model	Un-structured	Structured	Structured	Structured	Semi-structured
Data source	Web data	Transactional data	Hypermarket transactional data	User profiles data	Social media data
No. of records	100000	1000000	25000	1000000	1000000
Insert operation	✓	✓	✓	✓	✓
Select operation	✓	✓	✓	N/A	✓
Update operation	N/A	✓	N/A	✓	✓
Delete operation	N/A	N/A	N/A	✓	✓
Tuning performed	Resource (Disk) Utilisation in Cassandra	N/A	N/A	N/A	For MySQL, InnoDB engine is used for write-intensive operations and MyISAM engine for read-intensive operations.

performance tuning and switches storage engines of MySQL depending on the given operation situations for performance optimization.

6 Discussion

The social media data collected from Twitter is in a complex JSON schema format that contained many fields that needs to be further processed for the purpose of this research. Therefore, although MongoDB supports JSON format, the data collected still had to go through data transformation.

In terms of extracting the useful data and writing it into scripts to insert into MySQL and NoSQL databases, there is not much difference because the fields extracted are the same across both databases. Although the data schema is very much simplified, it gives an equal start for both databases. The data cleaning process consumes a larger amount of time than expected as there are many inconsistencies between acceptable formats for both databases. There is also the problem of merging all the data collected together and then separating them into the respective number of entries.

It is a challenging task in taking the measurement of execution times. For MySQL, there is not much of a problem as it returns the execution time in the unit of seconds with two decimal places. As for MongoDB, the profiling level of the database must be set in order to record the execution time of the query performed. The execution time does not display after each query. Instead, users have to run another statement to read the previous query profile, which contains the execution time. Moreover, the execution time is stored in the unit of milliseconds, which results in return value of 0 for queries that took less than 1 ms. To obtain a more accurate execution time, an external script is needed to run along with the Mongo console.

In terms on performance tuning, MySQL offers an extensive options and configurations to cater different situations such as populating, updating, removing large amount of data or with read-only operation data like a data warehouse.

Meanwhile MongoDB is limited to choices. While both database supports `explain()` features, for queries in MongoDB, the explain features does not provide an execution time, instead it provides information on the execution of other operations.

In terms of populating the data, MongoDB supports JSON objects because it supports documents, and is not limited to a rigid format. However, as mentioned, data transformation is still needed so that both databases contained the same data before taking measurements for comparison.

Overall, despite performance tuning in MySQL, MongoDB still performs faster than MySQL from small number of rows to up to 1 million rows. In the future, MySQL may be catching up with the latest Big Data trend by adding more features like in the past twenty years of development. For example, users may now get valuable insights from their data using MySQL with the Hadoop platform. On the other hand, the initial release of MongoDB was eight years ago, yet it has achieved very impressive results and has been serving renowned companies such as Expedia, Cicso, e-bay, etc. The amount of unstructured data produced grows exponentially as time goes on. In the coming years, both SQL and NoSQL databases will have to evolve to store and organize data more efficiently.

7 Conclusion

This paper presents our investigation of the differences of MySQL and MongoDB databases in performance from the aspects of execution time as well as the complexity in the process of data populating, retrieval and storing by comparing SQL and NoSQL databases implementation. Our findings indicate that MongoDB performs faster in inserting, selecting, updating, and deleting data, while MySQL provides more performance tuning features.

Compared to existing literature, the findings of our research work contribute to a new insight by examining a more thorough comparative evaluation of MongoDB and MySQL as a tool for SQL and NoSQL databases, respectively from the aspects of their response times in populating and retrieving data after performance tuning.

Our research has some limitations due to time constraints. The operations chosen to be measured are limited to four criteria. Moreover, indexing was not fully compared as the nature of the data is social media data, which is in text form. The database design for both databases is also simple, consisting only one table with three attributes.

For future work, investigations and experiments can be performed to compare different databases that support different data modelling in terms of indexing on more complex datasets, as well as considering other means of performance evaluation. For example, the resource allocation of each database and other aspects such as security, scalability, integrity and design complexity can be considered.

References

1. Pervasive Software Inc. 2003. Harvesting Unstructured Data, p. 2
2. Kaur, M.: Malaysians spend 12 hours daily on phone and online. [Online]. New Straits Times (2015). http://www.nst.com.my/news/2015/12/116437/malaysians-spend-12-hours-daily-phone-and-online. Accessed 20 Oct 2016
3. Arasu, A., Garcia-Molina, H.: Extracting structured data from web pages. In: SIGMOD 2003: Proceedings of the 2003 ACM SIGMOD International Conference on Management of Data (2003). Accessed 20 Aug 2017
4. Sucio, D.: Encyclopedia of Database Systems: Semi-Structured Data Model, p. 144. Springer, Washington
5. Zoulfaghari, R.: SQL server versions in distribution, parallelism and big data. Int. J. Comput. Appl. (0975–8887) **148**(14), 1 (2016)
6. Gupta, S., Narsimha.: Performance evaluation of NoSQL – cassandra over relational data store – MYSQL for bigdata. Int. J. Technol. 2015 **6**, 640 (2016)
7. Dubois, P., et al.: MySQL 5.0 Certification Study Guide, p. 541 (2006)
8. Chodorow, K., Dirolf, M.: MongoDB: The Definitive Guide, p. 7 (2010)
9. Damodaran, D., et al.: Performance Evaluation Of Mysql And MongoDB. Databases Int. J. Cybern. Inf. (IJCI) **5**(2) (2016)
10. Datastax Corporation. The Modern Online application for the Internet economy: 5 Key Requirements that Ensure Success. White paper by Datastax Corporation, Santa Clara, Calif (2014)
11. Moradi, M., Ghadiri, N.: Performance Evaluation of SQL and MongoDB Databases for Big E-Commerce Data, p. 2 (2015)
12. Twitter Developers. Streaming APIs | Twitter Developer Documentation (2016). https://dev.twitter.com/streaming/overview. Accessed 20 Oct 2016
13. McKinney, W.: Python for Data Analysis, p. 5. O'Reilly Media Inc. (2013)
14. Internet Live Stats. 2013. Twitter Usage Statistics. Internet Live Stats. http://www.internetlivestats.com/twitter-statistics/. Accessed 20 Oct 2016
15. Boicea, A., et al.: MongoDB vs Oracle – database comparison. In: Emerging Intelligent Data and Web Technologies (EIDWT) 2012 (2012)

The IBS as a Catalyst: Data Driven Insights

Manchuna Shanmuganathan(✉)

Researcher and Management Consultant, Toronto, Canada
s_manchuna@hotmail.com

Abstract. This paper examines the repositioned shared services paradigm to Integrated Business Services (IBS), as changing the operational and organizational set-up of shared services towards functional-based and process driven service delivery philosophy. The rationale behind combining services and data driven business models together to understand the phenomenon of big data in terms of, how data must be treated in relation to gathering, analysis and customers' behaviour patterns together with perceived value proposition to create better services via IBS models. A number of researchers' have identified IBS as an extension of business servitization which links service with technology to provide far greater value for its customers. Further, based on literature and IBS model, a case study was conducted for an Italian Bank such as UniCredit, to understand the extent to which they have evolved their business lines and service lines to provide multi-service with data driven insights, using the same resources, networks and customers.

Keywords: Integrated Business Services (IBS) · Business analytics
Reporting and master data

1 Introduction

Over the past two decades, shared services have been the prominent operating model for many business support services as it has established its successful delivery of sustainable cost and service improvements. However, the IBS model contributes more towards business value. Modern Systems are developed in aligned with the rationale and understanding of how to utilize those data to generate more value and profits at the same time [12]. The IBS is often referred to as multifunctional-shared services, which have now become globally accustomed. Some organizations view IBS simply as a fundamentally different way of thinking about all support services throughout an organization. Where a common leadership and governance structure that connect overall company objectives. There are many forms of IBS, most vital features has been identified as necessary to drive sustainable performance improvements, that an organization might seek are: multifunction, globalization, involvement of external service providers, service concept and service excellence. For example, multifunction indicates that where an organization might have started with single function or business process and may have significant integration across those functions later, due to services, product lines and integrated services management, such as global services. Globalization provides an organization to achieve maximum flexibility through a combination

K. Arai et al. (Eds.): FICC 2018, AISC 886, pp. 311–329, 2019.
https://doi.org/10.1007/978-3-030-03402-3_21

of globally positioned integrated service team together with external partners. Who deliver services locally, regionally and globally? Involvement of external service providers; the IBS operates as independent service provide, while it can also incorporate external service providers, such as payroll services in a responsible manner. Service concept enhances customer experience through a service-oriented culture with knowledge-based services along with adding-value. Service excellence implies that IBS needs to employ skill-set that are needed to change from a transactional to an advisory focus. The IBS require more cooperation from shared service centres. The move towards IBS indicates a fundamental shift in businesses, about how they think and manage these shared services and outsourcing. Thus, the IBS provides enormous improvements in performance and the implementation of shared services centres, whereas companies are now approaching services that have been impossible up to now, for example mergers and acquisitions and business analytical services. Figure 1 illustrates the evolution of shared services' organizational structures as follows:

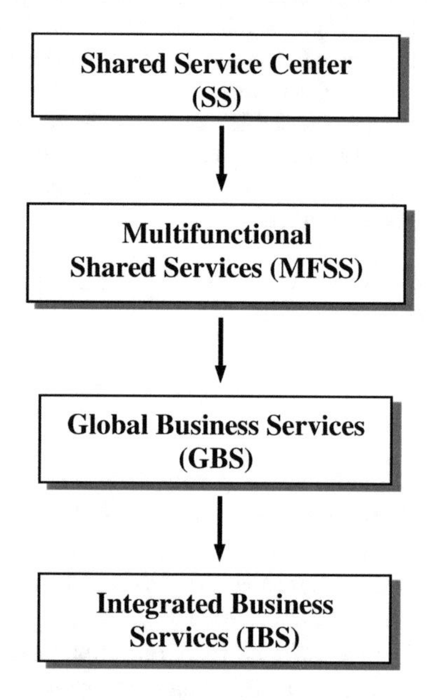

Fig. 1. Evolution of Shared Service.

The concept of enhanced reporting and business analytics for improving company insights in the context of IBS has gained advantage over big data. As a result, many businesses might invest significantly on data integration such as master data, reporting together with business analytics to design and improve their customer knowledge. In order to re-position products in market sectors, generate more profitability through sales and to improve resource management, ultimately to delivery better financial performances. Numerous companies have learned that by broadening their perspective on how business service models can yield value; have allowed them to compete in the market place efficiently by fulfilling multiple industrial responses. Therefore, those firms have converted into IBS model, where its service delivery has pushed the traditional shared services concept into a new era. As IBS integrate functions across organizations provides those pioneering firms to accomplish their business objective and goals at a first-rate with reliability and cost efficiency. As such, the IBS changed Shared Service Centres for the past.

2 Literature Review

The concept of data driven business model have been established a decade ago, however, there are still gaps in literatures. The existing business models have evolved significantly in recent years. Thus, the concept is now used in the context of e-business, strategy and innovations management. In [5], author first introduced the business model framework, where they have described the business model concept by functions it performs. Their business model explains the value proposition of an organization, as an example the value created for the user groups while trading with them through technology based solutions [6]. Further, they have also identified the market segment in which these user groups or customers use those technological solutions and for what purposes, while it generates specific revenue for an enterprise. In [11], it has identified a unique blend of the three streams as conceptual components for business model, as an example Value stream, Revenue stream and Logistical stream for e-commerce and Internet based businesses. In [9], author has proposed another business model based on strategic theory and related business model research. Where they have integrated different components from strategic theory such as customers and competitors (industry), by presenting generic strategy, activities, organization–value chain, resource base and production inputs (supply chain), market factors and sourcing including the process by which these business models evolved. Derived from those components they have proposed seven components framework as their business model, which consist of customers, competitors, commerce (what the company offers to its customers), activities and organization (structure), resources, supply of factors and production inputs along with the dynamics of business model to cover, cognitive and cultural constraints that managers have to cope up with over time.

In 2008 Harvard Business review article, [10], proposed a business model framework, which consist of four inter-connected components. As an example; customer value proposition, which consist of important intricacy or demand that need to be satisfied through commerce for target customers. Where profit formula defines the creation of value for the company that must have consistent revenue and margin models and cost structure together with relevant resources that are required to deliver value proposition. The most important key resources must be people, business partnership or brand royalty and technology. This normally excludes generic resources that every company has which do not create competitive advantage for the company. Further, [1] and [18] have proposed some of the frequently cited business model frameworks. Where [18] defines how a firm delivers value to its customers, and persuade those customers to pay for that value and converts payments into profits, based on three conceptual components as such; supplier specification value proposition, related appropriation mechanism and how to avoid imitators, in the context of business strategy and innovation.

An IBS model concept provides enhanced reporting and business analytics for number of companies. As these companies have to make significant investment in data integration on master data, reporting and analytics to design and improve their service deliveries, while achieving their objective through financial performances. The conception of Master Data Management (MDM) is not a new idea, but the notion of enhanced reporting and analytics for improved business insights have gained momentum. Today's business leaders have established the notion behind the advantages of leveraging shared services and outsourcing consistently across multiple functions and regions through IBS. The evolution of IBS was apparent through function-oriented structures, as these structures are now growing and replacing service-oriented structures. However, cost reductions and service improvements are still considered to be important factors for those companies and it is necessary to focus on improving business results from all perspective. These investments are usually made through at an independent function or business unit level rather than at an organization wide basis. Where, the IBS have the potential to change service delivery for those firms' customers as well. The IBS have capabilities to provide business value beyond the traditional cost and operational efficiencies related to share services, while supporting key business priorities. These have been identified under six different categories as follows:

- Strengthen Corporate Governance; an independent IBS organization operates as a business and makes its strategic decisions on how to provide services.
- Flexibility in Sourcing; where it uses multifunctional service models with the mix of service centres alongside strategic outsourcing partners.
- Globalization; provides a talented group of people and adoption of global technologies to support company expansions.
- Process Excellence; global process owners for process enhancements and securing enterprise compliance. Global end-to-end (supporting) process landscape with shared responsibility.
- Service Excellence; cutting-edge customer experience (service-based not function-based). Service simplification at all levels of the organization.
- Changes in Services; proactive development of new service products dependent on business needs. Moving from cost efficiency to value-add and knowledge based service offerings.

3 IBS Servitization: Organization

This section provides an examination of how an established master data, reporting and business analytics capabilities perform within a single IBS in an organization. Many companies have transformed their operations from multiple industries response to their industry challenges by expanding their perspectives on how these business models can yield value, through value proposition. Gradually more companies have adopted IBS model for end-to-end business services, which have taken the shared service concept to the next level. Where, the integration of IBS across functions must be capable of executing business objectives of these companies at an exceptional speed, reliability, and cost efficient manner. Indeed, the opportunities are plentiful with an IBS model as a catalyst for data driven insight with high performance. The IBS have even shifted its integrated services by providing more mature functions such as human resources, finance and information technology. As these service have been provided under traditional shared service models for a long-time. It has now become more transformed into whole and interdependent services intended to facilitate and deliver business strategies. Using this model how an enterprise can manage the four most common issues that they might have to face within these current capabilities, which have spread across functions and business units. These four topics have been categorized under following heading;

3.1 Data Structure and Quality Management

Once master data, reporting and analytics capabilities have been set-up as services within the IBS for an organization, it represents a single ownership of governance and management of data across that company. This paves the way towards one vision and mission to the entire company. Still with durable corporate governance structure, a firm can encounter higher risk of data quality, in particular, when functions and business units build their own master data, reporting and analytics capabilities. However, to eliminate data quality risk, after moving the data control to IBS, for different functions and business units an independent authority over data standards must be determined to avoid these situations. The IBS is designed to provide responsibility for maintaining data consistency across the enterprise while obtaining cross-functional agreements before making any major changes. In particular, when different business units have control over how they use information to report and analytics purposes, they must present data build on standardised formats and same time period. Meanwhile, this can avoid circumstances where business units might reflect in most favourable manner. At the same time, corporate decision-makers can make use of these reports and presentations to drive conclusions about their performances and future investments. As an example, a company might have two different ways of defining daily sales and outstanding (debtors). The inconsistency in definition and classification of daily sales must have an impact on sales, customers and financials. Eventually, this might undermined the confidence in data integrity itself. Management waste time on reconciling these differences and makes decisions based on inconsistent data sources. When, making important business decisions based on higher-order with multi-functional business analytics that might have distributed directly into managements' hands. A well-defined ownership can protect against any proliferation of master data elements from function to function as well as form business unit to business unit. An established IBS can process data with consistent and standardised output, which account for each party's requirements and needs. Indeed, the IBS provides consistency in data structure and centralizing quality management. Where, the proliferation of ad hoc reports and revisions by functions or business units can be eliminated. The shared data can bring key measures from multiple functions, which add great value to the business as it facilitates better decision-making.

3.2 Developing Data Driven Insight to Cross-Functional

The fundamental concepts behind a well-developed shared master data, reporting and analytics capabilities are to allow a single entity to have ownership of access to consistent date across functions and business units. Where this entity can become a cornerstone for the development and to provide an organization with the capability for common shared and standardized reports along with to derive genuine business insight and to take necessary action on it. When it comes to data driven insights, the entire model can be beneficial rather than sum of its parts. At the same time, functions and business units might realize the initial success in developing their own master data, reporting and analytics capabilities. The precise insight often comes when entities are equipped to swiftly, easily and confidently link the data and the insights from multiple functions or business units, as an example finance and sales division. Further, insights and value are often generated if an entity can adeptly layer in and manage third party data source as well. These consistent data must be used to generate reports focused on key functional measures, or combined with third-party information, as such macro-economic forecasts. This on the other hand can multiple value of all information exponentially. In a nutshell, it can accelerate the ability for the core business to perform a full range of reporting and analytics—from foundational, standard reporting to more advanced, predictive business analytics. As an example, a leading consumer product manufacturer lately utilized multi-functional analytics across finance, sales, marketing and supply chain to inform that an anticipated poor crop season in one sales region might drive down demand for migrant workers and this might have an impact on sales for that particular product. Where this is considered to be a macro-economic predictive business analytics capability that the organization may be able to forecast decrease in sales and adjust accordingly the inventory level, make relevant marketing decision to increase sales in other regions and or products. In order to reflect these revised forecast in the management reports.

3.3 Master Data, Reporting and Analytics Resources

The skill sets required for run master data, reporting and analytics capabilities vary from transactional data management to basic reporting and at the very end of the value spectrum predictive analytics, which facilitates better decision-making. Building a single mater data, reporting and analytics team that is designed to provide services to the entire organization may have the opportunity to consider new sourcing strategies while benefits to the company can be substantial. As an example, when considering the basic reporting function for low sensitive and low complexity might be adequate for low-cost regions. Instead, more advanced capabilities may perhaps consider sourcing in a differentiated manner. For a leading company in global communication and technology, some researchers have identified that potential for a net cost saving might be within the range of 35% to 45%. In fact, these saving can come from wage negotiations, workforce optimization and efficiency obtained through outsourcing core reporting services. As the requirement, for higher-value analytics becomes apparent then both the services company and businesses that have been served can benefit from by working together to build the business case for the right sourcing model. Where

Fig. 2 (Sourcing Model), explains the details of these an IBS internal and outsourced activities–complexity and value added, into four different headings.

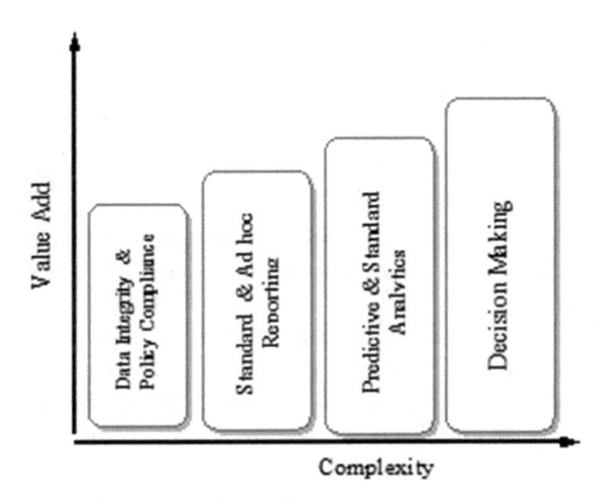

Fig. 2. IBS sourcing model.

(1) *Data Integrity and Policy Compliance*

- Data Integrity: provides data validation alongside data strategy, architecture, maintenance, data set-up and user accesses.
- Policy Compliance: related to service level approvals such as data and report governance, compliance reporting and service level reporting.

(2) *Ad hoc Reporting and Standard*

- Ad hoc Reporting: performs report and query requirement for real-time queries and iterative report developments.
- Standard Reporting: carry out the report requirements, for example: report design, maintenance, production and water marking.

(3) *Predictive and Standard Analytics*

- Predictive Analytics: are conducted on forecasting, optimization, simulation, model development and requirements, where it provides relationship identification, visualization, advanced statistical analysis and model execution/monitoring.
- Standard Analytics: identifies the root cause for conclusion and communication. Categorization, Data mining, Text analytics and basic statistical analysis.

(4) *Decision Making*

- Performance management
- Customer engagement

Both these activities are done through target monitoring.

3.4 Achieving Evolving Analytical Requirements – Company

The IBS model by its very nature, a customer focused one. Where moving the ownership to IBS capabilities, an enterprise can benefit from its structural support that it needs to scale and modify easily to improve the services to meet changing needs of the business. Consistency in data structure across entities, and the combination of governance together with a dedicated team whose skills-set range from transactional data management to higher-order of analysis can provide timely information and insight through IBS which leadership team desire to make decisions designed to keep them ahead of the competition. This facilitates businesses in today's fast-paced environment to move swiftly into new territories or new segments to stay ahead of their competitors. For example, an entertainment and media company needs to better understand consumer consumption patterns of newly released windows, such as DVD, on demand and theatrical, etc. The company have hypothesized that it is necessary to connect with consumption habits and behaviours across windows. Thereby, it might develop a holistic and effective social media strategy to better engage customer and influence consumption. Thus, make possible to increase the sales and improve margins. In order to address this situation company needs to structure data and business goal across business units quickly, such as theatrical, television and home entertainment. As IBS model, master data, reporting and analytics might have accelerated its ability to capitalize on a strong idea and generate a more optimized value chain for the company.

4 IBS: Management Performance

At the start, centralized and standardized service brought significant benefits to organizations. However, it often lacked shared responsibility to continuously improve these services. It always accompanied with increased cost due to shadow processes. The SSC allowed companies to use individual databases along with excel spreadsheets while IBS service management regulates relationship among the organization and its operations including internal and external partners, which enhance service delivery. Where, the IBS governance defines a structure that provides strategic guidance to entire organization. Governance includes formal process of interaction and decision-making with stakeholders who are involved in the IBS, in which it undertakes the formation and reinforcement of SSC norms and directions. One of main objective is to integrate IBS into overall organizational hierarchical structure and reporting lines. The formal IBS governance bodies which include advisory board, steering committee and user groups which are essential for the governance structure of the IBS. It is essential for governance bodies to promote strong liaison between shared service within the organization and its operating units.

In the context of globalization, to have significant business changes an organization must evaluate its scenarios more than one time or sometime multiple times to consider the factors that might impact the long-term benefits. There are multiple viewpoints for any given dimension, in order to gain consensus it takes time and effort. For many companies spreadsheet data model and graphics can be used to run one to three scenarios. Running sensitivity analysis, needs models to be capable of re-run several scenarios and then review them. Quite often, this requires more analyses along with further questions. At times, it can identify as each changes organization tries make, take them further away from the decision to act. If an organization can evaluate any given scenario and run sensitivity analysis in real time then it might help them to eliminate any concerns as it may arise. Where IBS predictive analytics provides much faster and easier way to conduct these analyses now, which allow them to have visual analytics for decision-making. For example, large retail companies are under constant pressure to improve their margins and overall performance. The key to improved performance, in many cases, is data – not just better data, but data organized and accessed in ways that make it easier and faster to make enterprise-wide decisions. These retail chains sometimes face particular problems in this regard. When a potential customer enters a store looking for a specific item and doesn't find it on the shelves, the store can experience a lost sale. Similarly, when shelves are stacked with merchandise that customers don't want, that can represent lost opportunities and wasted resources. These are challenging problems, but some retailers have found a promising approach to data management by evolving their shared services to create the IBS organizations. These IBS organizations, which are designed to deliver back, middle and front office services from areas including HR, IT, procurement, supply chain and logistics, and sales and marketing among others are typically accountable and collect a tremendous amount of data which can be aggregated and analyzed to provide useful insights to the enterprise.

5 Research Approach: Case Study

This case study was developed with an intension to explore following research questions:

- Can mainstream IBS model, techniques and technologies be successfully implemented in a banking sector environment?
- What are the advantages of applying IBS for UniCredit bank?

A case study approach was carried out because of the method's effectiveness in examining application in real-world scenarios, specifically as emerging research field of influences it have in shared services [7] and [19]. From the practical perspective, it provides an opportunity to contribute further into IBS application in a baking service sector such as UniCredit, a Milan based Italian bank. This example was important as IBS has gained momentum within banking industry and shared service has evolved into IBS, nowadays, which is a superior model to strive for. Where, the case study has provided evidence that may of use in understanding the concepts of IBS as a catalyst into data driven insights. Drivers for IBS such as increased global operations give an indication that enhanced connective and or increase in scope of higher value-added activities within existing operations. Further leverage lessons learned form the past indicates that companies are skipping single-function concept and pursuing with multi-function shared services while focusing on continuous improvement. It is apparent that insights gained from this case study may serve as a motivation to further research into this important area in different sectors of businesses and operational environments.

5.1 UniCredit: Sights of an IBS Model

A detailed exercise on opportunities related to business activities was conducted to gain an understanding of the case organisation's plans, strategy, method of client (Customers) engagement and organisational structures. The UniCredit bank's stakeholders and the IBS initiatives are:

- *Management*: Directors of the organisation who currently holds executive management responsibilities.
- Staff: Personnel involved in day-to-day service delivery processes and operational activities.
- *Consultants*: Specialist who provides domain expertise to the organisation (IBS – Solutions).

Importantly, the evolution of business' structure and the culture of process conscious, the organisation has seeing that as strategic imperative. There has been minimal culture or practice that have been embedded within the organisation, meaning there is more focus on establishing healthy attitudes and modes-of-operation rather than the need to change the status quo. Management desires to ensure that the business evolves with a strong focus on process in order to avoid the need to change dysfunctional systems and behaviours in the future to satisfy customer expectations in the global market. Following those developments, the UniCredit bank introduced IBS model in 2012 for its multi-services. In order to make information technology (IT) and its operations perform together within the bank's business divisions to improve customer expectations. As a matter of fact, integrating and consolidating sixteen of its existing service companies have created the transformation of multi-service model for UniCredit. Now it operates within eleven countries from United States to Singapore and manages a cost base of €2.5 billion with 11,000 staff [17]. The UniCredit bank have expanded its horizons through acquiring other banks initially in Italy, Central and Eastern Europe, since mid-1990 and later in Germany and Austria. As these banks naturally have their own service staff. Mission was to present UniCredit as one of Europe's biggest banks, while servicing globally through information and communication technology (ICT), with back and middle office operations, real-estate, security and procurement. First initiative began in 2000s, as to centralize its operations by country and IT service, where it has included real-estate services and procurement. Followed by its second initiative to create shared services centres (SSC) across national boards in order form global functions, in which have assured synergies and also have harmonized services across countries. The financial crisis of 2008 had played a major role in creating bank's IBS model, as it was critical to retain costs down and to respond rapidly to changing market conditions. At that time, Internet based customers' expected an ultra convenience and personalized service. For example, the bank has started a company responsible for infrastructure, development and maintenance of all ICT applications in all main countries. In addition, the bank formed another company for back-office activities in those countries. This enabled the bank to develop other banking products such as credit cards and specific customer loans. Indeed, ICT, back-office functions, and other support services were centralized to all locations together had the

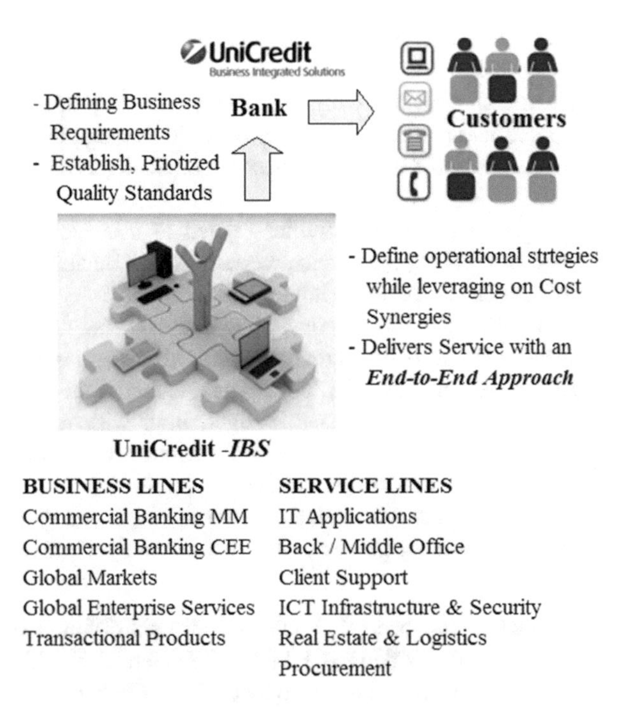

Fig. 3. UniCredit Bank's IBS model.

human resources and other resources to fulfill the task in hand. As such, Fig. 3 demonstrates UniCredit bank's IBS model.

Once UniCredit has built on those two initiatives the third and most important one relied on the transformation of six pillars such as flexibility, agility, innovation, transparency, cost efficiency and market readiness. The main objective was to create value for UniCredit group through reducing costs and further integrating service lines to better support the commercial lines, develop solutions that can benefit end customers. In order to accommodate these objectives bank have reorganized its Service Lines to facilitate, optimize and develop production factors that are responsible for planning overall capacity. For example there are four service lines; ICT, Operations, Client Support and Security along with two infrastructure service lines, namely, real estate and procurement. Appendix 1 explains these in greater detail. These service lines and infrastructure lines have been categorised under following headings:

– Information, Communication and Technology (ICT): This service line directly or indirectly manages all issues related to ICT applications, including development and maintenance of solutions, in accordance with Group guidelines and providing cutting-edge, economically sustainable infrastructure while ensuring Business support services.
– Security: Service handles all activities related to this service line, in compliance with Group's security rules together with local laws and regulations. Designed to prevent and manage any criminal and harmful events or business interruptions that

could damage the bank, its tangible and intangible assets or that could impact negatively on customers.
- Operations: Manages all activities with regard to operations processes and develops an innovative operational model and or process, while sharing the best practices and achieving economies of scale in order to maximize effectiveness and efficiency for the bank.
- Client Support: Administrates all activities related to user technical support and technical requests, monitoring the service requests of the Business Lines and other Service Lines of UniCredit Business Integrated Solutions.
- Real Estate: Offers technical and administrative real estate services to internal and external customers; it supports the parent company in managing and optimizing its proprietary real estate portfolio, for both strategic and non-strategic assets; it manages and optimizes the Real Estate budget, both with regard to costs and investments; and it encourages, supports and monitors real estate projects and transactions carried out by other legal entities.
- Procurement: Manages all activities related to management and optimization of the product and service purchasing process, while contributing effectively rationalizing costs.

As for business lines, it has been differentiated into six different sectors, such as Commercial Banking Mature Markets (CBMM), Commercial Banking Central and Eastern Europe (CBCEE), Transactional Products (TP), Global Markets (GM) and Global Enterprise Services (GES) using an end-to-end service model which provides for single point of reference that act as a fulcrum, which can improve relationship effectiveness and operating efficiency. These six business lines have been identified under following headings:

- Commercial Banking Mature Markets (CBMM): This business Line ensures the provision of end-to-end global services for Multifunctional and Core Banking, Credit along with Securities. It integrates and handles both application solutions and operations components. It is expected to act as a single point of reference for all customers within the Commercial Banking sector of the mature countries.
- Commercial Banking Central and Eastern Europe (CBCEE): This sector determines the specification of end-to-end solutions. It also provides information and communication technology (ICT) platforms to Central and Eastern European countries. Where, it acts as a single point of reference and provides direction to ICT services for Commercial Banking sector and ICT-Operations services for credit and debit card sections.
- Transactional Products: Ensures global supply of end-to-end services regarding Payments, Other Transactions and Cards. It integrates and handles both application solutions and Operations components, and is expected to act as a single point of reference for all customers in that Products category.
- Global Markets (GM): Set ups the global supply of end-to-end services for the Markets sectors and represents a single point of reference for all customers, focusing on the Market Utilities and Trading & Treasury production structures, which integrate and manage both application solutions and Operations components.

– Global Enterprise Services (GES): Ensures the global supply of end-to-end services related to Competence Line Planning, Finance & Administration, Human Resources, Legal and Compliance, and Audit. It offers ICT platforms to support the Risk Management Competence Line and the activities of other Governance functions – Appendix 1 (Organization Structure).

5.2 Advantages of Applying IBS for UniCredit Bank

There are many advantages while collaboration among other divisions with one person accountable for end-to-end service delivery was the core concept of the bank. However, it happened in many divisions, except for few other divisions according bank's CEO, Cederle, (2013). As it, takes time to achieve this goal because bank's structure is quite a complex one. The main objective was to define accountability more clearly and to simplify on work processes to build a more adaptive firm. This matrix structure defines accountability especially at all intersections of the bank. Indeed the IBS model was first introduced at UniCredit bank with an intention to make manager more accountable and to reduce shared services costs by 16 percent by 2015, while achieving faster and more flexible delivery of innovative products and services which improve the customer expectations and experience. As Fig. 4, demonstrates the reduction in the shared services costs from 2010 to 2016. In particular, after the introduction of the IBS in 2012 (as the base year), the reduction in shared service costs represents 16.04% in 2015, which illustrates that bank was able to achieve its objectives through IBS, with yearly cost decreases, starting from 2013 (€0.54 billion), 2014 (€0.71 billion) and 2015 (€1.43 billion), representing 6.06%, 8.02% and 16.04% respectively. These shared service costs reductions made by the on a yearly basis has improved its customer service by introducing new product lines and service, in order to keep pace with constantly evolving customer needs.

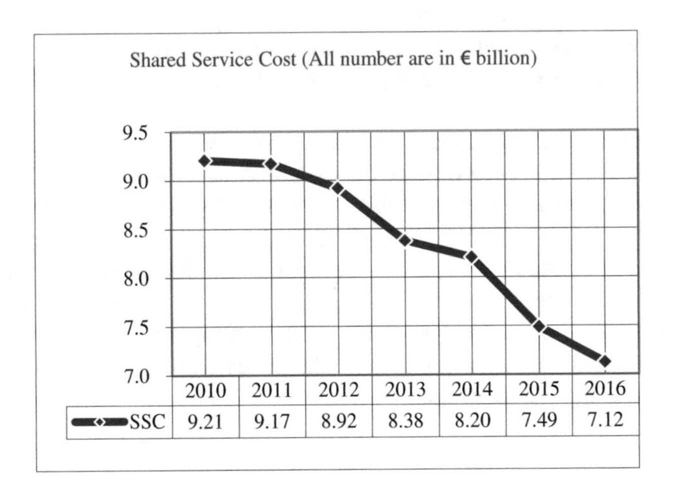

	2010	2011	2012	2013	2014	2015	2016
SSC	9.21	9.17	8.92	8.38	8.20	7.49	7.12

Fig. 4. Shared Service Cost (SSC)–(All numbers are in €, billion).

Figure 5 explains the administrative cost structure of UniCredit bank, which includes shared service cost (SSC) and other administrative costs (OAC). It is evidence that IBS model have provided the bank with relevant information and data to control the shared services cost and to achieve their set target by year 2015, which represents within the range of 16%. As the bank was able to maintain the same set target and have seen a further reduction in shared services costs by €0.36 billion in 2016 compare to year 2015. This indicates that UniCredit bank's IBS model have provided solutions to achieve their set targets within reasonable time period.

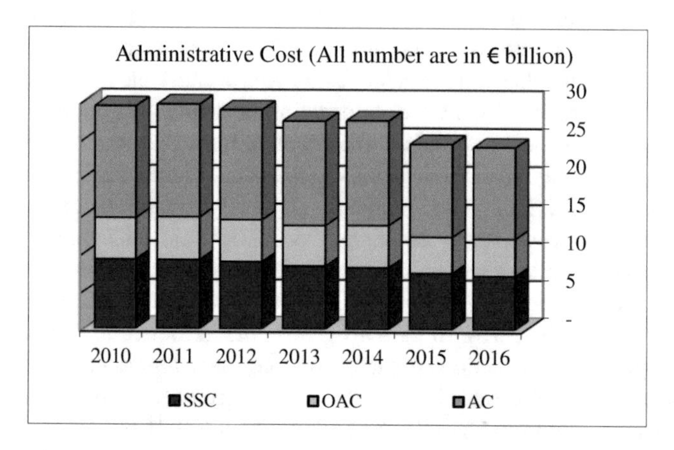

Fig. 5. Administrative Cost (AC)–(All numbers are in €, billion).

This new organizational model designed for UniCredit bank helps to foster better IBS and solutions. Where, operations have placed all service areas involved in a business-focused single point of accountability. In the past work across the operational processes were defined to improve customer experience. One person is responsible for both IT and operations; for managing services delivery, people, and costs; and for driving innovation in an end-to-end value chain. It is very different from the organization that bank had before, where IT and back office were separate legal entities. For an example; in order to illustrate how business and services lines work together as a single point of accountability, in particular, when developing a new type of credit-card, which typically involves, operations, application development, client support, IT infrastructure and IT security functions. In this new model the head of the cards unit within the global-products business line (Appendix I) is directly responsible for all those functions and acts as single point accountability for the UniCredit business division. This way of organizing the work allows the bank to be part of the business rather than just a support function.

The cultural challenges of shifting to integrated-business organization, was disruptive. In effect, the UniCredit bank depends on a group of people with different backgrounds and skill-sets, who are quite different. Based on the organization structure, how they organize their work, interact with each other varies. For example, IT staff must be using their IT skills for different divisions rather than within the IT department itself. As it, ranges from managerial styles and the complexity of the organization's matrix design for multi-services banking systems. In order to over come some of these difficulties, the bank has introduced a new service-manager role. That why, head of mortgage department becomes responsible for both, IT as well as mortgages within this complex organization structure, which includes back-office, which is quite different and involves managing a lot of staff and processes. In order to overcome this situation the bank established a set of governance functions to develop the kind of managers they needed for their multi-services. For example, the human-resources division spent significant amount of time and effort in supporting these managers to understand their new role and how to facilitate the transformation.

In addition, they have also provided training for top 300 managers to help them develop these new skill-sets that are required by the bank for IBS. They have also changed the agents in various locations, such as, employees who, in addition to doing their ordinary day-to-day jobs, help their colleagues and other staffs understand this new concepts and ways of working within the IBS banking systems. Its very difficulty, but benefits to the bank are in credible. Further, communication changes made across multi-country organization with different languages and cultures in many ways was made easier through digital technology, such as Internet, video conferencing through personal computers, and narrative videos about the concepts of transformation.

These transformation and strategic planning across the countries has strengthened UniCredit's position in the commercial banking sector through IBS in the European network. Where, it allowed them to better meet the needs of their customer expectations. Customer satisfaction is and will remain a key performance indicator and a benchmark against which performance are measured for management and the sustainability of the bank's results. Investment banking services will remain as a core activity but focused on supporting the commercial banking operations and always within very strict risk appetite. The diversification of the business lines and of bank's presence across the different European countries is an asset for the bank and the key competitive advantage over other banks in that region.

Figure 6 explains the correlation between the operating income (OI) and shared service costs.

Where, after an in-depth strategic review with the use of IBS model in year 2015 reveals that in order to optimize UniCredit's capital position, profitability, and ensuring continuous transformation of its operations, while maintaining flexibility to seize value creating opportunities, certain segments of bank has to be disposed for example, bank agrees to sell-off 32.8% of Poland bank–Bank Pekao in 2016 and balance 7.3% in 2017. It have resulted, a decrease in the operating income by €3.6 billion compare to year 2015 (Fig. 6).

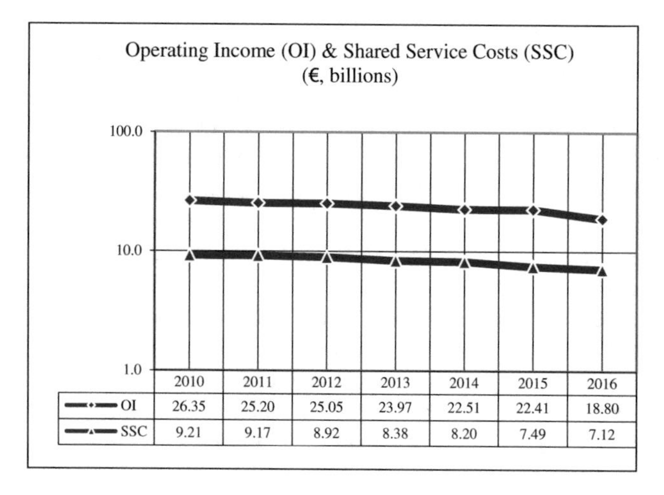

	2010	2011	2012	2013	2014	2015	2016
OI	26.35	25.20	25.05	23.97	22.51	22.41	18.80
SSC	9.21	9.17	8.92	8.38	8.20	7.49	7.12

Fig. 6. Operating Income & SSC (all numbers are in €, billion).

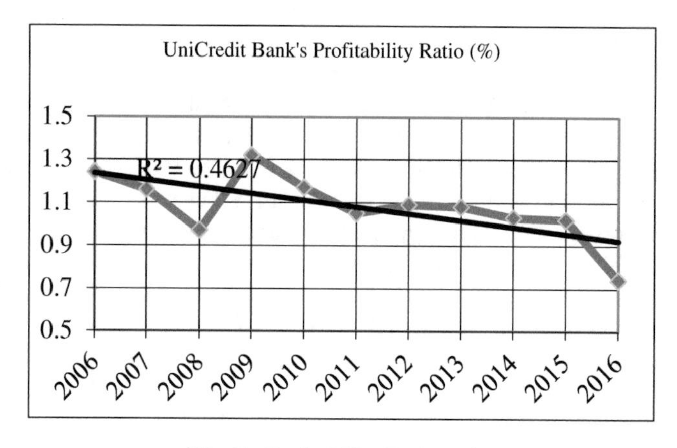

Fig. 7. Profitability Ratios (%).

Profitability ratio was calculated for the past ten years (as net profits divided by total assets), which indicates that year 2008 ratio has been dropped sharply to 0.97% from 1.16% (2007) due financial crisis of 2008. Following that the bank had its highest profitability ratio of 1.32%. However, in year 2016 it dropped again to 0.74%, as result of disposal of some division in Poland subsidiary (Fig. 7). Looking back at the transformation of IBS so far and events that have been taken placed at the UniCredit bank it is evidence that integrating of shared services companies was the best option for this bank in the direction that all managers and staffs must work together towards achieving the common goal. This pace was the recognition for the bank by lunching the IBS and informal networks that made these things happen.

6 Conclusion

As organizations continue to focus on developing data driven business insight, master data, reporting and analytics can be a prime target to be delivered as a service through the IBS organization. By embedding this capability within IBS organization and offering it as a service. Where entities can place data governance, quality and standards into the hands of an organization, which can continue to drive efficiencies, improve processes and to innovate their services to ensure consistency in data analytics. It can also provide a line of business and a pathway for master data, reporting and analytics resources that might be beneficial for both the organization and the individuals as well. More importantly, this might be able to advance organizations toward the quite often considered to be as necessary but elusive, the truth of one version, across the businesses and functions. In particular when combined with third party data, it might become the basis for accelerating insights which might help today's leaders to answer their analytics questions in an effort to take advantage and outperform their competitors.

When implementing an IBS model high level of integration with business partners are required. The focus should be to create an enterprise value and not only on cost efficiency, but with one shared vision, strategy, goals and performance objectives. An end-to-end view of the value chain made it possible to rationally leverage technology for process improvement, which has immediate impact on all levels of productivity in the UniCredit bank shared services. Where the bank has also benefited from the following improvement, for example:

- Having a single manager as the contact person on the service side have helped to provide all the necessary business data and information, which has increased the level of accountability on both sides for the bank.
- Redesigning end-to-end processes have lowered the bank's overheads (running cost) while leveraging global operations and developing economies of scale.
- Strengthening client focus by further improving customer experience and expectations, through product standardization and engaging them in more client-focused activities.
- Investments in IT that have supported the business transformation with greater digitalization. Where technological improvement of core systems, and ongoing infrastructure updates.

In the context of shared service concept, the shared services leadership research has emphasized its importance, while demonstrating that large global companies have discussed their transition from multi-functional shared service to GBS along with the recent move towards IBS. The IBS has given prominence to its shared services model based on an end-to-end approach. Where, it merely simplified customer expectations and experience, which considered being the next-step beyond the end-to-end process-based model.

Appendix 1 – Unicredit Bank's Matrix Organization Structure (IBS Model)

Governance Layer
Human Resources
Legal and Compliance
Finance, Planning & Administration
Foreign Branches Management
Chief Security Officer
Risk Management
Identity and Communication
Organization and Logistices
Mature Countries Legal Entities & Branches
Operational layer
Business Line
IT Governance
Germany Austria
Strategic Planning
Transformation Office
Commercial Banking, Mature Markets
Commercial Banking, Central & Eastern Europe
Transactional Products
Global Markets
Global Enterprise Services
Banking Service Lines
Multichannel and Core Banking
Credits
Securities
Payments
Other Transaction services
Cards
Trading and Treasury
Post-Trade Services
CRO & Support Platform
CFO
Human Resources
Information Communication Technology (ICT) Applications
Operations
Client Support
Security
Infrastructure Service Lines
Real Estate
Procurement

References

1. Al-Debei, M., Avison, D.: Developing a unified framework of the business model concept. Eur. J. Inf. Syst. **19**(3), 359–376 (2010)
2. Andersen, T.C.K., Bjerrum, T.C.B.: Service and Big Data Business Model in a World of Persuasive Technologies. Aarhus University Business of Social Science, Aarhus (2016)
3. Baines, T., Lightfoot, H.: Made to Serve: How Manufacturers can Compete Through Servitization and Product Service Systems. Wiley, Chichester (2013)
4. Brownlow, J., Zaki, M., Neely, A., Urmetzer, F.: Data and Analytics - Data-Driven Business Models: A Blueprint for Innovation. University Of Cambridge Institution for Manufacturing, Cambridge (2015)
5. Chesbrough, H., Rosenbloom, R.: The role of the business model in capturing value from innovation: evidence from Xerox Corporation's technology spin-off companies. Ind. Corp. Change **11**(3), 529–555 (2002)
6. Chesbrough, H.: Open Services Innovation Rethinking your Business to Grow and Compete in a New Era. Wiley, San Francisco (2013)
7. Dallas, I., Wynn, M.T.: Business Process Management in Small Business: A Case Study, vol. XXVI, 420 p. Springer, Heidelberg (2014)
8. Hou, J., Neely, A.: Barriers of servitization: results of a systematic literature review. In: Proceeding of the Spring Servitization Conference. University Of Cambridge, Cambridge (2013)
9. Hedman, J., Kalling, T.: The business model concept: theoretical underpinnings and empirical illustrations. Eur. J. Inf. Syst. **12**(1), 49–59 (2003)

10. Johnson, M.W., Christensen, C., Kagermann, H.: Reinventing your business model. Harv. Bus. Rev. **86**(12), 57–68 (2008)
11. Mahadevan, B.: Business models for internet-based E-commerce. Calif. Manag. Rev. **42**(4), 55–69 (2000)
12. Mcafee, A., Brynjolfsson, E.: Big data: the management revolution. Harv. Bus. Rev. (2012)
13. Lindgren, P., Jørgensen, R.: Towards a multi business model innovation model. J. Multi Bus. Mod. Innovat. Technol. **1**, 1–22 (2012)
14. Looy Van, B., Visnjic, I.: Successfully implementing a service business model In: A Manufacturing Firm. Cambridge Service Alliances, Cambridge Service Alliances, Institution for Manufacturing, Cambridge (2013)
15. Osterwalder, A.: The business model ontology. a proposition in design science research, these. Ecole des Hautes Etudes Commerciales de l'Université de Lausanne (2004)
16. Osterwalder, A., Pigneur, Y., Clark, T.: Business Model Generation. A Handbook for Visionaries, Game Changers, and Challengers. Wiley, Hoboken (2010)
17. Cederle, P.: Creating a business-integrated services company: An interview with UniCredit's Paolo Cederle," L. D'Aversa, A. Del Miglio, and R. Lancellotti (2013). (in press)
18. Teece, D.J.: Business models, business strategy and innovation. Long Range Plann. **43**(2–3), 172–194 (2010)
19. Yin, R.K.: Case Study Research: Design and Methods, 3rd edn. Sage Publications, London (2003)

Accuracy of Clustering Prediction of PAM and K-Modes Algorithms

Marc-Gregory Dixon[1(✉)], Stanimir Genov[2], Vasil Hnatyshin[2], and Umashanger Thayasivam[1]

[1] Department of Mathematics, Rowan University, Glassboro, NJ 08062, USA
dixonm7@students.rowan.edu, thayasivam@rowan.edu
[2] Department of Computer Science, Rowan University, Glassboro, NJ 08062, USA
genovs6@students.rowan.edu, hnatyshin@rowan.edu

Abstract. The concept of grouping (or clustering) data points with similar characteristics is of importance when working with the data that frequently appears in everyday life. Data scientists cluster the data that is numerical in nature based on the notion of distance, usually computed using Euclidean measure. However, there are many datasets that often consists of categorical values which require alternative methods for grouping the data. That is why clustering of categorical data employs methods that rely on similarity between the values rather than distance. This work focuses on studying the ability of different clustering algorithms and several definitions of similarity to organize categorical data into groups.

Keywords: Clustering · Partitioning around medoids · K-modes
Similarity functions

1 Introduction

Clustering the data set points with similar characteristics often relies on the notion of distance: i.e., the points with the small distance between them are placed into the same group or cluster. The real life data often consists of non-numerical or categorical values, for which the notion of distance, usually computed using Euclidean measure, is not defined. Categorical variables typically contain labels or names for the dataset instances. Categorical variables take on the values that belong to a finite set of logically independent discrete labels for which the notions of distance or any numerical operation are undefined. That is why clustering of categorical datasets employs methods that rely on similarity between the values rather than distance [12]. While both the notion of similarity and distance help to identify how closely the variables are related to one another, logically these definitions have opposite meaning. That is, short distance between variables corresponds to high similarity and long distance between variables corresponds to low similarity. In other words, similarity can be thought of as inverse of distance. That is why the notion of dissimilarity (inverse of similarity) is often used in the literature.

K. Arai et al. (Eds.): FICC 2018, AISC 886, pp. 330–345, 2019.
https://doi.org/10.1007/978-3-030-03402-3_22

The data clustering typically works as follows. First, the algorithm computes initial distances or dissimilarities between data points and cluster heads (determined using certain bootstrap algorithm). Next, using computed distances or dissimilarities, the algorithm organizes the data into clusters and computes the new cluster heads. The algorithm then re-computes the distances between the data points and the new cluster heads. The process continues until there is no more change in the cluster head assignment or cluster membership [17].

Two leading clustering algorithms for grouping categorical data are *Partitioning Around Medoids* (PAM) and K-modes. This project started as an extension of the previous study [8], which focused on performance of PAM algorithm, implemented using R programming language. Since then, the work evolved into a comparison of PAM and K-modes algorithm, the implementations of which were recently added to the R library. This study examines performance of two clustering algorithms and several definitions of similarity. It attempts to find trends and to determine which algorithm and which similarity functions are best suited to cluster categorical and mixed type datasets accurately and reliably.

The rest of the paper is organized as follows. Section 2 describes the PAM and K-modes algorithms in detail. Section 3 provides a brief overview of the datasets used to evaluate the algorithms. It is followed by Sect. 4 that describes evaluation methodology. Analysis of results is presented in Sect. 5. Finally, the paper concludes in Sect. 6.

2 Algorithm Overview

The K-modes algorithm partitions datasets into k clusters. The initial central points, or centroids, around which the remaining data points are grouped, are selected according to a certain bootstrap algorithm or randomly. For each cluster, the algorithm computes a centroid, not necessarily a data point, as a vector of the most common features or modes of all the data points in the given cluster. Next, the algorithm compares distances between computed centroid and every point in the dataset. These distances are determined by counting the number of the individual data point attributes that match those of the centroid. The distance function for K-modes algorithm is defined as follows:

$$d(X,Y) = \sum\nolimits_{j=1}^{m} \delta(x_j, y_j) \tag{1}$$

Where, $\delta(x_j, y_j) = \begin{cases} 0, x_j = y_j \\ 1, x_j \neq y_j \end{cases}$

The data points with the smallest distance to a central node are placed into that node's cluster. After that, the central nodes are recomputed based on new cluster membership. The algorithm iterates until it finds a stable clustering or reaches the maximum number of iterations.

The *Partitioning Around Medoids* (PAM) algorithm, which is also known as *K-medoids*, is similar to *K-modes*. However, rather than using modes (i.e., vector of most common characteristics), PAM initializes central nodes, or medoids, to randomly selected data points. This makes the algorithm less susceptible to outliers.

In our study, R *implementation* of K-modes algorithm employs a single internal similarity function for computing distance between data points and centroids [4]. On the other hand, PAM implementation can work with any similarity function. The R implementation of PAM accepts dissimilarity matrix computed for a given dataset using a certain similarity function, which allows researchers to conduct data clustering using many varying criteria [9]. In this study, we evaluated performance of PAM algorithm with such similarity functions as *Overlap*, *Eskin*, *Occurrence Frequency*, *Inverse Occurrence Frequency*, *Lin*, *Goodall3*, and *Goodall4*. The definitions for these similarity functions could be found in [2, 3].

The clustering assignments for each dataset were known beforehand, which allowed us to determine the performance accuracy of K-modes and PAM algorithms. However, when computing the accuracy of the clustering algorithm we faced the following problem. The output provided by the algorithm does not label the output clusters but rather indicates which entries were clustered together. For example, let us say that the dataset contains five points with two classes named A and B. Also, let's say that data points 1, 3, and 4 belong to class A, while points 2 and 5 belong to class B. The output of the clustering algorithm will indicate which data points were clustered together but it will not identify which cluster A or B the points belong to. In our example, the output of the clustering algorithm may specify two clusters {1, 3} and{2, 4, 5} but it will not map the first cluster {1, 3} into class A and the second cluster {2, 4, 5} into class B.

To deal with this issue we devised an algorithm that maps output cluster labels into the original class names. The idea of the algorithm is to match the output cluster with the class name based on maximum membership match. So that if an output cluster has the largest number of data points from original class X then this cluster will get class name X (i.e., all data points in this output cluster will be mapped to class X).

In our example, the input data vector with the original class names will be recorded as *{A, B, A, A, B}* and the output clustering data vector with temporary labels will be *{a, b, a, b, b}*. Since the algorithm places data points {1, 3} into the same cluster and both points belong to class A, then the cluster that contains points {1, 3} will be mapped to class *A* and the cluster with data points {2, 4, 5} will be mapped to class *B*.

We implemented this algorithm as follows:

(1) Create an input data vector with original class names.
(2) Compute an output clustering data vector with temporary labels.
(3) Create N×N matrix, where N is the number of different classes, the rows contain the actual class names, the columns contain the temporary output labels, and each entry in the table contains the number of times the actual class name matches the corresponding temporary label.
(4) Identify an entry with the highest value, which will contain the correct mapping between the actual class name and temporary output label.
(5) Remove already matched column and row from the matrix.
(6) Repeat steps 4–5 until all the matches are completed.

In our example, an input data vector with original class names is recorded as *{A, B, A, A, B}*, and output clustering data vector with temporary labels is *{a, b, a, b, b}*. Since original class name *A* matches the temporary output label *a* two times and the output label b once, while class name *B* matches the output label *b* two times, the N × N matrix will look as follows:

$$\begin{array}{c} \\ A \\ B \end{array} \begin{array}{c} \begin{array}{cc} a & b \end{array} \\ \begin{pmatrix} 2 & 1 \\ 0 & 2 \end{pmatrix} \end{array}$$

To further illustrate how our algorithm works let us look at a more complex example. Consider a dataset that includes eight data points that classified as follows:

- Class A contains points {1, 2, 3}
- Class B contains points {4, 5, 6} and
- Class C contains points {7, 8}

Thus, the input data vector with original class names will be recorded as *{A, A, A, B, B, B, C, C}*. Now let us assume that the algorithm produces the following clustering assignment:

- Custer labeled *a* contains points {6, 7, 8}
- Custer labeled *b* contains points {1, 2, 3} and
- Custer labeled c contains points {4, 5}

As a result, the output vector with temporary labels will be recorded as *{b, b, b, c, c, a, a, a}*. Based on this information we crease a 3 × 3 matrix with the rows containing the actual class names *A*, *B*, and *C* while the columns containing the temporary output labels *a, b,* and *c*. Next, each index of the matrix is populated according to the mapping between the actual class names and temporary output labels. For example, since the first three points that belong to class *A* map to temporary label *b*, the matrix cell (A, b) will contain value 3. Overall, the resulting matrix will look as shown below:

$$\begin{array}{c} \\ A \\ B \\ C \end{array} \begin{array}{c} \begin{array}{ccc} a & b & c \end{array} \\ \begin{pmatrix} 0 & 3 & 0 \\ 1 & 0 & 2 \\ 2 & 0 & 0 \end{pmatrix} \end{array}$$

The largest value in the matrix is three and it is found in cell (A, b). Thus, the original class A will be mapped to cluster with temporary label *b*. Next, both row *A* and column *b* are removed from the matrix and the next highest value within the matrix is found. The intermediate matrix without row *A* and column *b* is shown below:

$$\begin{array}{c} \\ B \\ C \end{array} \begin{array}{c} \begin{array}{cc} a & c \end{array} \\ \begin{pmatrix} 1 & 2 \\ 2 & 0 \end{pmatrix} \end{array}$$

In this case the next largest value is in cell (*B, c*) and class *B* will be mapped to cluster with label *c*. The process is repeated until all temporary labels are mapped to original class names. In this example, the resulting mapping between original class names and output cluster labels is as follows:

- Cluster *b* is mapped to class A
- Cluster *c* is mapped to class B
- Cluster *a* is mapped to class C

Based on provided mapping we create confusion matrix and compute the accuracy of the clustering assignment. The confusion matrix for our example will look as follows:

$$\begin{array}{c c} & \begin{array}{ccc} A & B & C \end{array} \\ \begin{array}{c} A \\ B \\ C \end{array} & \begin{pmatrix} 3 & 0 & 0 \\ 0 & 2 & 1 \\ 0 & 0 & 2 \end{pmatrix} \end{array}$$

It is possible to directly create confusion matrix for the clustering assignment by swapping the columns once the mapping between the class name and cluster label have been identified. However, in this paper we provided a systematic example for clarity.

3 Datasets

We evaluated a variety of datasets to gauge the effectiveness of each algorithm. These sets include the data that ranges from the purely categorical with very few variables to datasets with mixed data types and many entries. Specifically, in this study we used the following datasets to evaluate the performance of K-modes and PAM algorithms together with several different definitions of similarity.

Balloons – The balloons dataset contains description of conditions for cognitive psychology experiment set-up. Each data instance is classified into one of two possible types: inflated or not inflated. All dataset values are categorical and the set contains 16 entries [14].

Iris – The Iris dataset contains numerical information regarding the types of Iris plant. The dataset contains 150 instances, each of which is classified into one of possible three types [14].

Soybean (small) – The Soybean (small) is a subset of the original soybean disease database. Each data instance is classified as one of four possible diseases [14]. The data is categorical and the set contains 47 entries.

TCP Dump – The TCP dump dataset contains packet header information collected from a simulated network. Each data instance is classified into two types: *normal traffic* or *network attack* (e.g., port-scan, guess, rsh, etc.) [1, 13]. TCP dump is among the most complex dataset evaluated. It consists of 295 entries separated into six categories. The TCP dump dataset contains the attributes that are numerical or categorical.

Census – The Census dataset contains over 3,000 entries. It separates population data into two classes, those individuals who make over $50 k and those making less than $50 k. This dataset contains variables that are both numerical and categorical in nature [14].

TTT – This dataset consists of 958 entries describing various Tic-Tac-Toe game states and is classified as whether that position can be won or if it is a draw. All the attributes in the set are categorical [14].

D31 – A dataset with the entries from the Cartesian plane that contains total of 3100 data points in 31 clusters. All attributes in this dataset are numerical [15].

Arrhythmia – This dataset contains information that helps identify the presence or absence of cardiac arrhythmia. The entries in the set are classified into one of 16 groups. This dataset contains 452 entries with numerical and categorical attributes [14].

Balance Scales – The Balanced Scales dataset contains 625 entries that model the psychological experiments with three possible outcomes. All attributes in this dataset are categorical [14].

KR vs. KP – This King-Rock vs. King-Pawn dataset models the end of chess game where one player (white) has a king and a rook while the other player (black) has a pawn in position a7 (one step from turning into a Queen). This dataset contains 3196 entries which are classified as white can win (52% of entries) or not (48% of entries). All of the attributes in the dataset are categorical [14].

Mush – This dataset consists of 8124 entries. It classifies different types of mushrooms as either edible or not edible. All the attributes in the Mush dataset are categorical [14].

Transfusion – A dataset of 748 data points describing blood donors in a hospital in Taiwan. Each entry pertains to one donor and contains such information as: the number of months since last donation, total number of donations, total blood donated, and the number of months since first donation. Each donor is classified into whether they donated in March 2007 or not [14].

Nursery – This dataset ranks applications into nursery schools. The attributes in this set are all categorical. Each of the 12960 instances represents an application and is classified based on the assigned priority level [14].

Summary of the properties of examined datasets is presented in Table 1.

4 Methodology

The hardware characteristics of the equipment used to conduct the experiment are: CPU: 2 × Intel Xeon E5-2620v3, 2.4 GHz; RAM: 64 GB operating at 2133 MT/s Max; System Disks: 2 × 240 GB Intel DC S3500 Series MLC; Storage Disks: 8 x 4 TB Seagate Constellation ES.3; Storage Configuration: RAID 6 24 TB Usable; Operating System: CentOS 7. In this study, we used R programming language version 3.3.2 along with the libraries *nomclust*, *cluster*, and *klaR*. We tested the following similarity functions provided in the *nomclust* package: *overlap*, *eskin*, *occurrence frequency*, *inverse occurrence frequency*, *goodall3*, *goodall4*, and *lin* together with the default PAM similarity function provided in the *cluster* package and default K-modes similarity function from the *klaR* package.

We examined and refined the previous version of the code for computing the clustering accuracy. We updated the old version *to* work with new implementations of

Table 1. Properties of examined datasets

Dataset name	Attribute type	Class distribution	No. of classes	No. of entries (size)	No. of attributes	No. of points
Balloon	Categorical	Balanced	2	16	4	64
Iris	Numerical	Balanced	3	150	4	600
Soybean (small)	Categorical	Balanced	4	47	35	1645
Balance scales	Categorical	Balanced	3	625	4	2500
TCP dump	Mixed	Unbalanced	7	295	10	2950
Transfusion	Numerical	Unbalanced	2	748	4	2992
D31	Numerical	Balanced	31	3100	2	6200
TTT	Categorical	Unbalanced	2	958	9	8622
Census	Mixed	Unbalanced	2	6260	14	87640
Nursery	Categorical	Unbalanced	5	12960	8	103680
KR vs. KP	Categorical	Balanced	2	3196	36	115056
Arrhythmia	Mixed	Unbalanced	16	452	279	126108
Mush	Categorical	Unbalanced	7	8124	22	178728

PAM and K-modes algorithms. To ensure the reliability of our results we evaluated each method multiple times and randomized the order of the data entries in each set. The average accuracy was then recorded to summarize each algorithm's performance.

We compared PAM and K-modes algorithms and evaluated their performance under different conditions. Specifically, we evaluated PAM performance using seven different similarity functions. The implementation of PAM algorithm provides the ability to standardize the data by subtracting the mean value from each variable and then dividing the result by the mean absolute deviation. This creates a standard value similar to Z-score, ensuring that the variables defined via different scales contribute equally. Both the standardized and unstandardized data were used in evaluation of PAM clustering algorithm using default similarity function.

R's implementation of K-modes algorithm relies on a certain default definition of similarity and the user cannot change it. However, the algorithm does allow the user to specify the maximum number of iterations. Initially, we examined the accuracy of K-modes algorithm by varying the values of maximum number of iterations from 1 to 100.

Through experimentation we discovered that the K-modes algorithm converges after no more than 5 iterations. As a result, we set the maximum number of iterations for K-modes algorithm to 20 just in case it ever requires going over 5 with plenty of iterations to spare.

The clustering assignment produced using PAM and K-modes algorithms was compared against the actual data classification which was known prior to the experiment. The accuracy of the clustering created by each similarity function was evaluated using the *Rand Index* measure as well as a simple ratio between the number of correct classification and the total number of data points, which we will call *ratio* measure.

The *Rand Index* measure examines pairs of clustered data points and calculates total accuracy as a ratio between correctly classified pairs of data points and sum of correctly and incorrectly classified data points [6, 11]. Specifically, let us denote the original dataset of m points as $S = \{d_1, \ldots, d_m\}$, the computed clustering assignment as a set of n clusters $X = \{X_1, \ldots, X_n\}$, and the original clustering assignment as set of clusters $Y = \{Y_1, \ldots, Y_n\}$. Furthermore, let us define the following terms:

- a – the number of data point pairs that belong to the same cluster in X and to the same cluster in Y, i.e. $a = \left|\left\{d_i, d_j\right\} \ni d_i, d_j \in X_k, d_i, d_j \in Y_l\right|$
- b – the number of data point pairs that belong to different clusters in X and to different clusters in Y, i.e. $\mathrm{b} = \left|\left\{d_i, d_j\right\} \ni d_i \in X_{k_1}, d_i \in X_{k_2}, d_i \in Y_{l_1}, d_j \in Y_{l_2}\right|$
- c – the number of data point pairs that belong to the same cluster in X and to different clusters in Y, i.e. $\mathrm{c} = \left|\left\{d_i, d_j\right\} \ni d_i, d_i \in X_k, d_i \in Y_{l_1}, d_j \in Y_{l_2}\right|$
- d – the number of data point pairs that belong to different clusters in X and to the same cluster in Y, i.e. $\mathrm{b} = \left|\left\{d_i, d_j\right\} \ni d_i \in X_{k_1}, d_i \in X_{k_2}, d_i, d_j \in Y_l\right|$
- $\{1 \leq i, j \leq m; 1 \leq k, k_1, k_2, l, l_1, l_2 \leq n; k_1 \neq k_2; l_1 \neq l_2\}$

Using this definition, the accuracy of clustering assignment is computed as follows:

$$Rand\,Index\,Accuracy = \frac{a+b}{a+b+c+d} \tag{2}$$

We compute the algorithm accuracy using *ratio* measure as follows. Let us denote the total number of correctly classified data points as *cr*. Then the clustering accuracy using *ratio* measure is defined as:

$$Ratio\,Accuracy = \frac{cr}{m} \tag{3}$$

The *Rand Index* and *ratio* accuracy measures provide us with two ways for interpreting the results of our study. While different, these methods both represent plausible approaches to evaluating the performance of clustering algorithms. However, neither of these measures is without fault. The *ratio* accuracy measure does not work well for unbalanced datasets where the number of entries in each class in unevenly distributed. Consider the TCP dump dataset where over 75% of the entries belong to a single class. If the clustering algorithm classifies all of the data points as part of a single class, then the *ratio* measure will report 75% accuracy which is misleading. On the other hand, the *Rand Index* measure relies on cluster membership of point pairs, examining if the two points should belong to the same cluster. As a result, the accuracy for unbalance datasets reported by the *Rand Index* measure is higher than that reported by the *ratio* measure. On the other hand, the performance of *Rand Index* measure is dependent on the number of clusters: it converges to 1 as the number of the clusters increases, which is not a desirable property [5, 7, 16]. In the following section, we present a summary of collected results.

Each of the datasets was run 3 times for each similarity function and clustering algorithm. The collected results were converted into confusion matrix to compute the accuracies using *Ratio* and *Rand Index* measures. Summary of collected results in presented in Table 2.

5 Analysis of Results

5.1 Comparison of Accuracy Measures

When comparing the data received from the two measures of accuracy (i.e., *Rand Index* and *ratio*), the results for the most part were similar and consistent. The accuracy values reported by the *ratio* measure, on average, were slightly lower than those computed using the *Rand Index* measure. However, for the unbalanced datasets: *Census*, *TCP dump*, *TTT*, *Arrhythmia*, *Mush*, *Nursery*, and *Transfusion*, the accuracies reported by the *Rand Index* were higher (about 9.2% higher than *ratio*). Figure 1 illustrates this phenomenon. In this paper, we consider unbalanced datasets to have different number of entries in each cluster, while in balanced datasets the number of data entries is distributed equally among clusters.

As shown in Fig. 2, for balanced datasets (*Balloon, Soybean (small), Iris, Balanced Scales, D31, and KR vs. KP*) the average reported accuracy was almost the same for both measures.

These results can be attributed to the way each of the measures is constructed. The simplicity of the *ratio* measure makes it susceptible to unbalanced datasets because it examines the cluster membership as a whole, while *Rand Index* compares membership of individual data points. This phenomenon was especially noticeable for large unbalanced datasets such as *Arrhythmia, Mush,* and *Nursery* for which *Rand Index* measure reported accuracy on average 30% higher than using *ratio* measure. However, both performed about equally well when evaluating the accuracy of clustering methods for balanced datasets, where a simple ratio is acceptable.

It is important to note that while there are differences between the accuracy results reported by the two measures, the overall observed patterns and trends are the same. For the remainder of this paper we will report the results collected using the *Rand Index* accuracy measure only.

Overall, the clustering algorithms were most successful when classifying the *Soybean (small)* dataset, with the average accuracy of 94%. The only exception was the default similarity function for PAM with the data that was not standardized, where it reported an accuracy of about 85%. It appears that the clustering algorithms perform the best on the datasets that consist of purely categorical attributes with non-binary classification (e.g., *Soybean (small), Mush,* and *Nursery).*

On the other hand, the clustering algorithms performed poorly when evaluation *TCP dump* and *D31* datasets with the notable exception of the default and standardized PAM on *D31* dataset were the accuracies reached over 99%. Overall, it appears that studied clustering algorithms and similarity functions perform poorly on datasets with binary classifications and on the datasets with non-categorical attributes. The remainder of this section examines other trends and patterns in collected results.

Table 2. Prediction accuracy of clustering algorithms and similarity functions with rand index

Dataset name/Clustering method	*Partitioning Around Medoids (PAM)*									*K-modes*	*Avg*
	overlap	*eskin*	*of*	*iof*	*good3*	*good4*	*lin*	Default			
								Regular	Stand.		
Balloon	0.483	0.542	0.517	0.475	0.475	0.472	0.550	0.550	0.539	0.547	0.515
Soybean (small)	0.962	0.976	0.941	0.976	0.916	1.000	0.976	0.852	0.903	0.908	0.941
Census	0.526	0.500	0.513	0.515	0.518	0.545	0.526	0.508	0.635	0.535	0.532
Iris	0.644	0.613	0.599	0.500	0.675	0.628	0.616	0.880	0.840	0.527	0.652
TCP dump	0.451	0.539	0.445	0.430	0.442	0.449	0.434	0.359	0.410	0.428	0.439
TTT	0.519	0.514	0.516	0.511	0.504	0.500	0.510	0.502	0.501	0.509	0.508
Balance-Scale	0.524	0.552	0.543	0.527	0.523	0.519	0.524	0.598	0.593	0.521	0.542
D31	0.113	0.113	0.385	0.307	0.112	0.112	0.646	0.997	0.997	0.065	0.385
Arrhythmia	0.654	0.577	0.653	0.340	0.658	0.542	0.666	0.654	0.487	0.653	0.588
KP vs. KR	0.516	0.516	0.516	0.511	0.504	0.509	0.506	0.516	0.501	0.506	0.510
Mush	0.744	0.754	0.724	0.737	0.733	0.701	0.753	0.710	0.745	0.751	0.735
Transfusion	0.537	0.546	0.548	0.583	0.546	0.525	0.512	0.546	0.513	0.536	0.539
Nursery	0.623	0.628	0.630	0.630	0.631	0.622	0.626	0.608	0.608	0.619	0.623
Average	**0.561**	**0.567**	**0.579**	**0.542**	**0.557**	**0.548**	**0.603**	**0.637**	**0.636**	**0.546**	

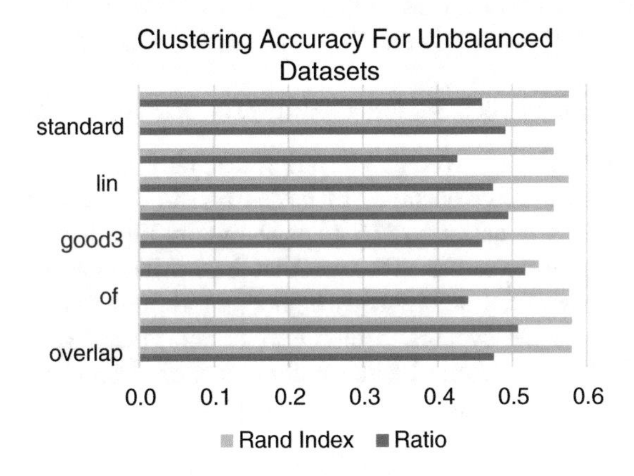

Fig. 1. Comparison of accuracy measures for unbalanced dataset.

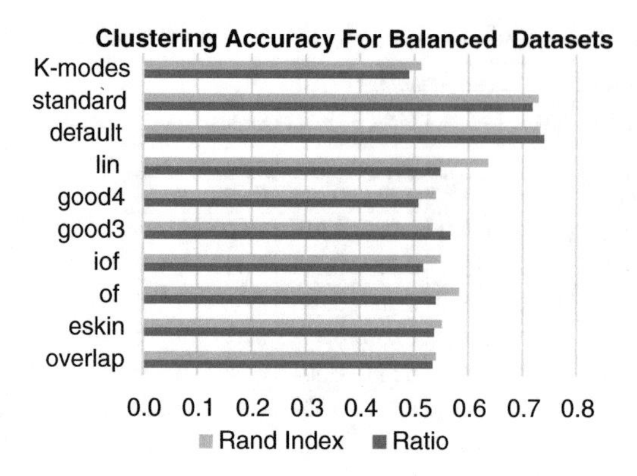

Fig. 2. Comparison of accuracy measures for balanced datasets.

5.2 Clustering Accuracy Vs. Dataset Size

One of the investigated characteristics is the impact of the dataset size (i.e., the number of entries and the number of the data points) on the clustering accuracy. We partitioned studied datasets into three groups:

- Small with less than 200 entries: *Balloon, Soybean (small), Iris*
- Medium with 200 to 1000 entries: *TCP dump, Balance-Scale, Arrhythmia, Transfusion, TTT*
- Large with more than 1000 entries: *Census, D31, KR-vs-KP, Mush, Nursery*

Surprisingly enough, the average clustering accuracy for small datasets was the highest (70.3%). The average accuracy of large datasets was slightly higher (55.7%) than that for medium size datasets (52.3%) as shown by Fig. 3. Part of the reason for such big discrepancy in the results is the fact that the clustering accuracies for *Soybean (small)* and *Iris* datasets were significantly higher than for any other examined datasets and may have skewed the average results. Furthermore, there were only three small datasets and they all were balanced with small number of classes, which may not have provided enough variance to draw reliable conclusions.

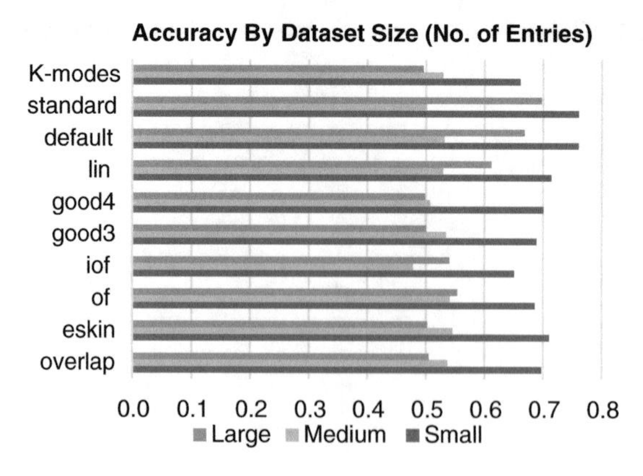

Fig. 3. Accuracy by dataset size (No. of entries).

We also examined the impact of the number of data points in the dataset on the clustering accuracy. We partitioned studied datasets into three groups as follows:

- Small with less than 5000 data points: Balloon, Soybean (small), Iris, Balanced Scales, TCP dump, and Transfusion
- Medium with 5000 to 10,000 data points: D31, and TTT
- Large with more than 10,000 data points: Census, Arrhythmia, KR-vs-KP, Mush, Nursery

As Fig. 4 shows, small and large datasets were classified most accurately, while medium size datasets had significantly lower accuracy. Again, these results could be explained by the impact of the *Soybean (small)* and *Iris* datasets, which were classified with high accuracy. However, if these two datasets are removed from the computation then we can observe the general trend which shows that the accuracy of clustering algorithms improves as the size of the datasets go up.

5.3 Clustering Accuracy vs. Attribute Type

Overall, the studied clustering algorithm performed the best on datasets that contained categorical attributes only (*Balloon, Soybean (small), TTT, Balance-Scale, KR-vs-KP, Mush, Nursery*), with average accuracy of 62.5%. The average accuracy for mixed (*Census, TCP dump, and Arrhythmia*) and numerical (*Transfusion, Iris, D31*) attribute datasets, were 52% and 52.5%, respectively. It should be noted that PAM algorithm with default similarity function, performed extremely well on *Iris* and *D31* datasets (86% and 99.7% accuracy). Both are balanced datasets with numerical attributes only. We believe that the reason for this phenomenon is the way the default similarity function is implemented. Specifically, it appears that the default similarity function is optimized to employ numerical calculations, such as Manhattan distance, when working with numerical data [10]. If we exclude these datasets from computation then the average accuracy for numerical datasets drops to 47.2%. Summary of the clustering accuracy for datasets with different attribute types is shown in Fig. 5.

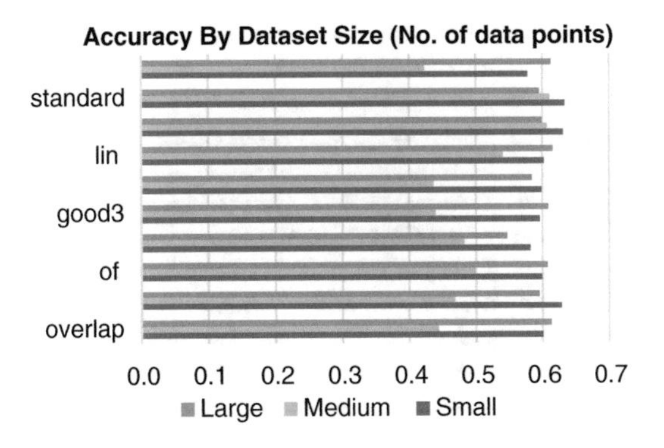

Fig. 4. Accuracy by dataset size (No. of data points).

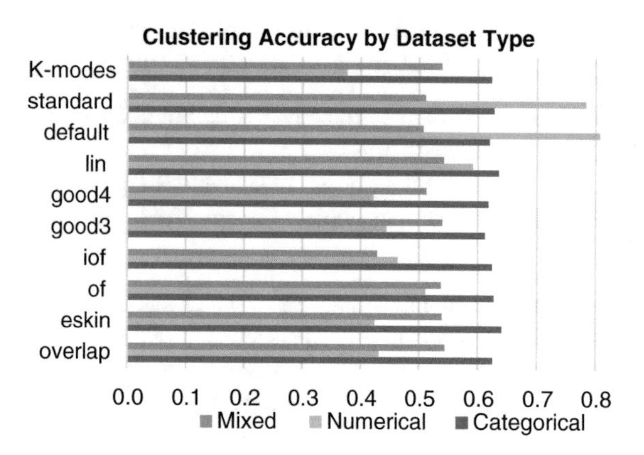

Fig. 5. Numerical vs. Categorical vs. Mixed datasets.

On average, all examined similarity functions and clustering algorithms performed consistently for the same dataset with some small variation. The only exception was *D31* dataset for which PAM with such similarity functions as *overlap, eskin, goodall3, and goodall4,* as well as *K-modes* algorithm performed extremely poorly, but PAM with default similarity function provided almost 100% accuracy.

5.4 Balanced vs. Unbalanced Data Representation

We also looked at the impact of balanced datasets on clustering accuracy. Summary of the results is presented in Fig. 6. On average, the clustering accuracy of studied algorithms and similarity functions was pretty much the same for balanced and unbalanced datasets. The only exception was PAM algorithm with default similarity function which on average was significantly more accurate for balanced datasets. If we were to remove the two numerical datasets (i.e., *Iris* and *D31*) from our computation then a clear trend emerges: balanced sets are clustered more accurately than unbalanced sets. This phenomenon is shown in Fig. 7.

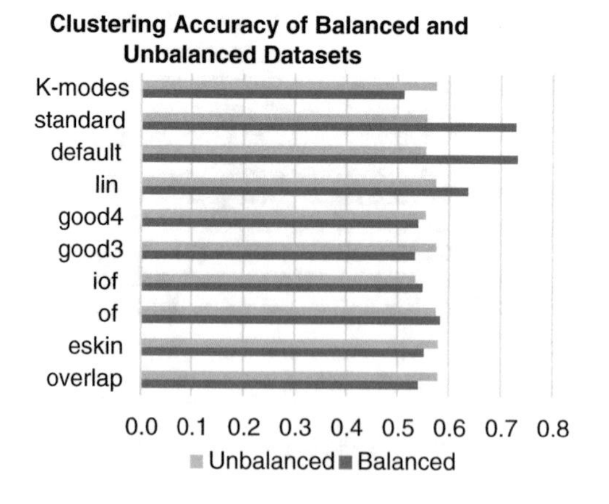

Fig. 6. Balanced vs. Unbalanced datasets.

5.5 Impact on the Number of Classes

We examined the impact of the number of classes on the clustering accuracy. There were no clear trends observed when we included all datasets in a study. However, when we removed the two numerical datasets (i.e., *Iris* and *D31*) from our computation, it appears that the accuracy of clustering algorithms improves as the number of classes increases. We also examined how the accuracy is impacted when we cluster binary (i.e., the datasets with two classes only) and non-binary datasets. As shown in Fig. 8, classification accuracy for the non-binary datasets is consistently higher than for the binary datasets. This trend is even more pronounced when we exclude *Iris* and *D31* datasets.

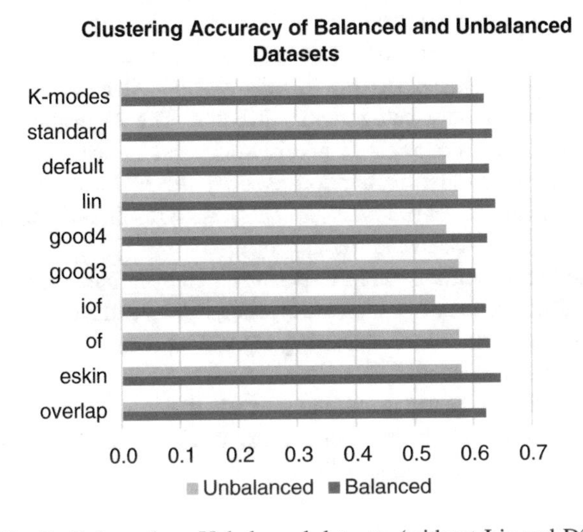

Fig. 7. Balanced vs. Unbalanced datasets (without Iris and D31).

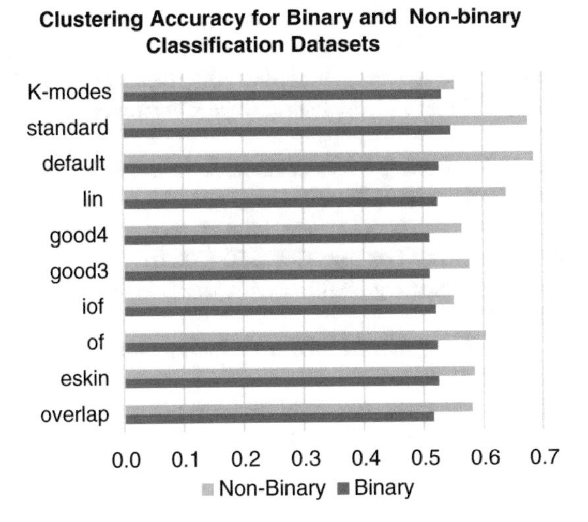

Fig. 8. Function performance on binary sets.

6 Conclusions

Overall, this study discovered that using *Rand Index* rather than simple *ratio* accuracy measure provides a higher fidelity results. It appears that the accuracy of clustering algorithm improves as the size of the datasets increases. Although there is no clear trend and the clustering algorithms performed reasonably well for certain smaller datasets.

All examined similarity functions used with PAM algorithm as well as K-mode algorithm perform equally well on datasets with categorical data only but the clustering accuracy decreases when the algorithms are applied to the mixed type or numerical only datasets. The only exception is the default similarity function used with PAM, which we believe may use Manhattan distance when evaluating numeric datasets [10].

The studied algorithms were the most accurate when clustering *Soybean (small)* and *Mush* datasets, producing clustering accuracy of 94.1% and 73.5%, respectively. These datasets contained only categorical attributes. The algorithms performed the worst on *D31* dataset which consists of purely numerical values. We also noticed that the default definition of similarity function used by the PAM algorithm seems to be optimized for numerical dataset (i.e., *Iris* and *D31).* As a result, including these datasets in computation often obfuscated the trends.

Studied clustering algorithms seem to perform better on balanced datasets and the datasets that contain more than two classes. Overall, however the clustering accuracy of examined similarity functions and clustering algorithms was poor and more in-depth study of its properties is needed.

Based on this initial study and preliminary trends that were identified we would like to further investigate our hypotheses about ability of clustering algorithms and various similarity function to correctly classify the data.

References

1. Barbara, D; Jajodia, S.: Applications of data mining in computer security. In: Advances in Information Security, vol. 6 (2002). https://doi.org/10.1007/978-1-4615-0953-0
2. Boriah, S., Chandola, V., Kumar, V.: Similarity measures for categorical data: A comparative evaluation. In: 8th SIAM International Conference on Data Mining, pp. 243–254 (2008)
3. Boriah, S., Chandola, V., Kumar, V.: A framework for exploring categorical data. In: 9th SIAM International Conference on Data Mining, pp. 187–198 (2009)
4. Huang, J.Z.: Clustering categorical data with k-Modes. In: Encyclopedia of Data Warehousing and Mining, 2nd Edn., pp. 246–250 (2009)
5. Hubert, L., Arabie, P.: Comparing partitions. J. Classif. **2**(1), 193–218 (1985). https://doi.org/10.1007/BF01908075
6. Jain, A.K., Murty, M.N., Flynn, P.J.: Data clustering: a review. ACM Comput. Surv. (CSUR) **31**(3), 264–323 (1999)
7. Morey, L., Agresti, A.: The measurement of classification agreement: an adjustment to the rand statistic for chance agreement. Educ. Psychol. Meas. **44**(1), 33–37 (1984)

8. Muck, I., Hnatyshin, V., Thayasivam, U.: Accuracy of class prediction using similarity functions in PAM. In: IEEE International Conference on Industrial Technology(ICIT), pp. 586–591 (2016)
9. Park, H.-S., Jun, C.-H.: A simple and fast algorithm for K-medoids clustering. Expert Syst. Appl. **36**, 3336–3341 (2009)
10. Vreda, P., Black, P.E.: Manhattan distance. Dictionary of Algorithms and Data Structures 5/31/06. www.nist.gov/dads/HTML/manhattanDistance.html. Accessed 09 May 2017
11. Rand, W.M.: Objective criteria for the evaluation of clustering methods. J. Am. Stat. Assoc. **66**(336), 846–850 (1971). https://doi.org/10.2307/2284239. JSTOR 2284239
12. Saxena, A., Singh, M.: Using categorical attributes for clustering. Int. J. Eng. Appl. Comput. Sci. **2**(2), 324–329 (2016). https://doi.org/10.24032/ijeacs
13. University of California Irvine: KDD Cup 1999 Data. kdd.ics.uci.edu/databases/kddcup99/kddcup99.html. Accessed 09 May 2017
14. University of California Irvine: Machine Learning Repository. archive.ics.uci.edu/ml/datasets.html. Accessed 09 May 2017
15. University of Eastern Finland: Clustering Benchmark Datasets. cs.joensuu.fi/sipu/datasets/. Accessed 09 May 2017
16. Wagner S., Wagner, D.: Comparing clustering: An overview. Technical Report 2006-04, Faculty of Informatics, Universität Karlsruhe (TH), 2007
17. Zhou, E., et al.: PAM spatial clustering algorithm research based on CUDA. In: 24th International Conference on Geoinformatics, August 2016. https://doi.org/10.1109/geoinformatics.2016.7578971

Visual Video Analytics for Interactive Video Content Analysis

Julius Schöning(✉) and Gunther Heidemann

Institute of Cognitive Science, Osnabrück University, Osnabrück, Germany
{juschoening,gheidema}@uos.de

Abstract. Reasoning as an essential processing step for any data analysis task, yet it requires semantic, contextual understanding on a high level, e.g., for the identification of entities. Developing an architecture for visual video analytics (VVA), we integrate human knowledge for highly accurate video content analysis to extract information by a tight coupling of automatic video analysis algorithms on the one hand and visualization as well as user interaction on the other hand. For accurate video content analysis, our semi-automatic VVA-architecture effectively understands and identifies *regular* and *irregular* behavior in real-world datasets. The VVA-architecture is described with both (i) its *interactive information extraction and representation* and (ii) its *content-based reasoning* process. We give an overview of existing techniques for information extraction and representation, and propose two interactive applications for reasoning. One of the applications uses 3D object representations to provide adaptive playback based on selected object parts in the 3D viewer. Another application allows the formulation of a proposition about the video by using all extracted objects and information. In case the proposition is correct, the corresponding frames of the video are highlighted. Based on a user study, relevant open topics for increasing the performance of video content analysis and VVA is discussed.

Keywords: Visual analytics · Video analysis · 3D reconstruction
Object annotation · Human-machine-interaction

1 Introduction

During an average day in London, UK, one will be filmed by around 300 different closed-circuit television (CCTV) cameras [1]. The increasing rate of video surveillance and, in consequence, the massive amount of video footage, inspires not only scientists but also book and movie authors to think about applications, benefits, and problems. In the nineties, e.g., the Hollywood movie *Enemy of the State* shows a fascinating method for detecting whether the content of a bag was modified based on monocular CCTV cameras. This approach, illustrated in Fig. 1, reconstructs a 3D object out of video footage, then the object can be compared to find any content changes inside. Triggered by these fictitious possibilities such as infinite magnification of video frames for number plate

K. Arai et al. (Eds.): FICC 2018, AISC 886, pp. 346–360, 2019.
https://doi.org/10.1007/978-3-030-03402-3_23

identification, reliable face recognition including removal of sunglasses, and 3D reconstruction of occluded objects, non-experts are led to believe that automatic video analysis and constitutive applications from large amounts of video data are already possible. In reality, however, the analysis of such amounts of video data is still an unsolved task, due to its high complexity in both the spatial and temporal domains. Nevertheless, current research in image processing and computer vision shows promising approaches for the extracting meaningful content such as objects of interest (OOI) [2–4], the interaction between OOIs [5,6], the analysis of general movement patterns [7], the 2D maps creation of the scene [8,9], and the 3D shape reconstruction of OOIs [10–12].

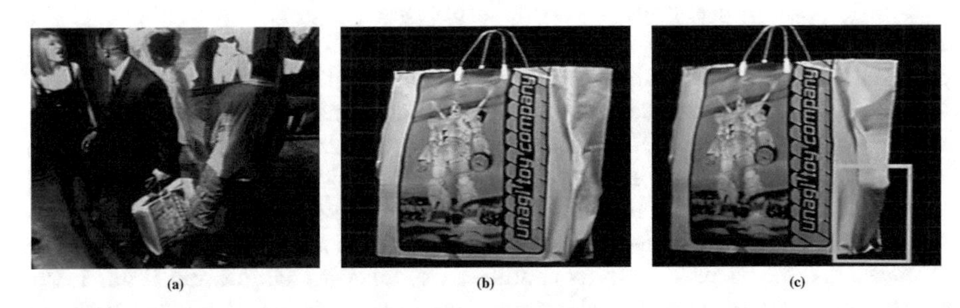

Fig. 1. Movie example [13] for interactive video analysis; Task: Detection whether the content of a bag was modified. (a) Still of single CCTV camera used as input data, (b) resulting predicted 3D reconstruction of the bag before its content has been changed, (c) resulting predicted 3D reconstruction of the bag with changed content highlighted.

Unfortunately, the reliability and accuracy of computer vision methods for real-world video footage are still limited. The integration of human knowledge and her/his ability in decision-making will bridge the gap and enable comprehensive video content analysis methods. Based on precise techniques for content extraction from video footage our architecture of VVA effectively understands and identifies *regular* as well as *irregular* behavior in real-world video data. Our architecture of VVA does not only integrate users' high-level knowledge in the process of information extraction and representation but users' high-level knowledge in the reasoning process, too. As a consequence, the user (the *analyst*), helps the computer in close cooperation to translate the pixels of videos to meaningful content. Therefore, the system remains the "working horse" but consults the user in problematic situations to achieve high-quality results.

The purpose of our architecture is to integrate human interaction to video content analysis for producing a semi-automatic video analysis system that relies on both automatic extraction and human reasoning. The design and the introduction of its concept are the major focus of this article. Thus, we detailed introducing our content extraction process (cf. Fig. 3) based on our VVA-architecture (cf. Fig. 2). Our content extraction process describes the interrelationship between the amount of data and the data abstraction level—low level (pixel-based) via meaningful content to high-level (conclusion). As a further contribution, two prototypical application cases are outlined which follow the content extraction process. Thus this prototype software reduces the amount of data with every interactive processing step, while at the same time, the abstraction level of the data raises. This iterative process has to continue until the analyst gains a conclusion.

This paper is structured as follows: First, we describe our architecture of VVA with both (1) its *interactive information extraction and representation* process; and (2) its *content-based reasoning* process. After that an overview of existing techniques for data extraction and representation from video sequences like pixel-accurate OOI segmentation, OOI tagging, image-based 3D reconstruction, and integration of users' unconscious features such as gaze trajectories is given. Based on these techniques, we propose two interactive reasoning applications for demonstrating the applicability and limits of our architecture.

The first application uses 3D object representations to provide adaptive playback based on selected OOI parts in the 3D viewer. The second application allows the formulation of a proposition about the video by using all extracted OOIs and additional information. In case the proposition is correct, the corresponding frames of the video sequence are highlighted. Based on these prototype applications, a first user study is conducted to answer the question if our interactive and semi-automatic architecture improves VVA applications. Next, to introducing the use multimedia containers in the domain of VVA as a possible exchange format, we conclude with a list of open issues necessary to increase the performance of video content analysis and VVA.

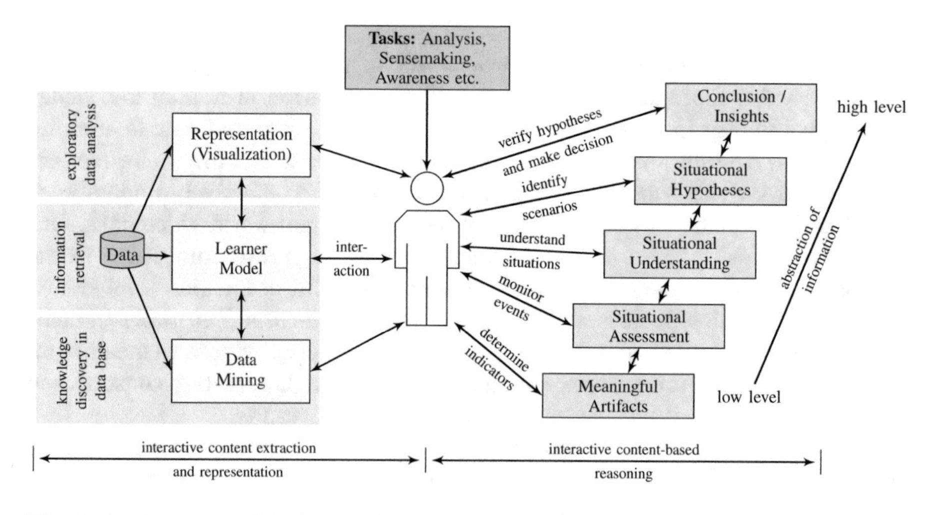

Fig. 2. Architecture of VVA. Left-hand side: *interactive* video content analysis responsible for *content extraction and representation* inspired by Keim et al. [6] as well as Höferlin et al. [14]. Right-hand side: the *reasoning process*, relying on the cognitive sensemaking process [15], filters the extracted content, depending on the task, until the insights are verified.

2 Visual Video Analytics

VVA denotes a set of methods for video content analysis (i.e., the translation of pixels into meaningful content) and for gaining insights (i.e., the verification of hypotheses) according to a defined analysis task. "Synthesize information and derive insight from massive, dynamic, ambiguous, and often conflicting data; detect the expected and discover the unexpected; provide timely, defensible, and understandable assessments; and communicate assessment effectively for action" [16] is the purpose of any visual analysis. We propose a general architecture for robust interactive VVA to facilitate the processing of significant amounts of video footage while allowing the handling of **vaguely defined analysis tasks**.

Compared to previous frameworks and architectures [6,14,15,17,18], we designed an architecture which takes both parts, the *extraction and representation* of meaningful content as well as the content-based *reasoning process* into account. Our VVA architecture (Fig. 2) is inspired by the *integrative view on visual analytics and sense-making processes* [14], the *visual analytics process* [6] and the *cognitive sensemaking process* [15]. Since the basic structure of our architecture relies on the cognitive sensemaking process, its design can be transferred to other domains like audio or text analysis. Since we use visualizations to represent the extracted information (or "content"), the term "representation" also involves "visualization" in this context.

Gaining insights from a video, verifying hypotheses about the OOI and classifying video shots are examples of typical video analysis tasks. The underlying cognitive process, also referred to as sensemaking process [15,18], is the general baseline of our architecture. As shown in Fig. 2, the human analyst is the core—hence the process is interactive. On the left-hand side, a tight integration of representation, data mining, and interaction in combination with a machine learning model incorporates the analysts' domain knowledge to extract task-specific, meaningful content from the raw data [14,17]. The meaningful content, represented as pixel-accurate annotated OOIs, groups of interacting OOIs, movement patterns of OOIs, and scene descriptions, forms the basis of the *reasoning process*, cf. green boxes, on the right-hand side of Fig. 2. By running the reasoning process, the analyst gains first analysis insights, depending on a particular task. If some content is necessary for verifying a hypothesis, the analyst interactively extracts more information from the raw data. As a consequence, presumably relevant content is obtained, while nonrelevant content, such as scenes without a certain OOI will be filtered out. The repetitive alternation between *content extraction and representation* and *content-based reasoning* is significant for the sensemaking process. This alternation continues until the result has reached certain quality criteria, or until a time limit has been exceeded. In our opinion, two remarks on sensemaking by Russell et al. [15] are of major importance:

- sensemaking returns a result at any time, but the quality of the result increases in each iteration, and
- the most time-consuming part of sensemaking is the extraction and representation of meaningful content.

As illustrated in Fig. 3, VVA should, in general, provide a framework to reduce a huge amount of data to an essential minimum. The reduction of data during the VVA process makes it necessary that the abstraction level of data will successively increase during the process.

2.1 Interactive Content Extraction and Representation

Filtering out non-relevant data, minimizing the complexity of data, and creating meaningful high-level symbol-based representations is the primary purpose of *interactive content extraction and representation*, cf. also Fig. 3. Due to the interactive design of our architecture, the extraction of content is done semi-automatically by a learner model [17]. By putting the user into the loop, e.g., to annotate the OOIs in one frame [3], or to set expected trajectories [7], this system supports the content extraction; thus, the computer remains the "workhorse" during this process. This interactive manner provides scalability, one of the most important requirements for analysis tasks [16]. Finally, interaction ensures accurate and reliable results, because the analyst supervises the extraction process and can counteract at an early stage in cases of failure.

2.2 Content-Based Reasoning

The aim of reasoning is achieving task-specific insights or even a conclusion by applying human judgment [19]. As illustrated in Fig. 3, *content-based reasoning* will increase the amount of mined data only by the conclusion, but it significantly increases the abstraction level of data. Because according to the analysis task, the resulting insights emerge from a combination of content and human knowledge. As a consequence, the reasoning process offers the most improvements for video analysis, and thus the video data should be preprocessed with a focus on it.

The reasoning process as a stepwise process, consisting of five single steps (green blocks in Fig. 2), continuously increases the level of information. In the first step, we define task-specific indicators for extracting only necessary **meaningful artifacts**, also called OOIs. In the next step, **situational assessment** is achieved by monitoring expected and unexpected events, such as the interaction or the movement of the OOIs. Thus with every step, the need for human judgment increases. Therefore in the following steps, the computational support decreases significantly. For **situational understanding**—the next step—the temporal relations of OOIs and events become apparent. Based on these relations, the analysis formulates **situational hypotheses** such as possible behavioral patterns of OOIs. In the final step, the formulated hypotheses will be verified and lead to a reliable **conclusion** or **insights**.

3 Interactive Content Extraction

In contrast to automatic methods for content extraction from video sequences and collection of images, interactive approaches have proven to deliver reliable

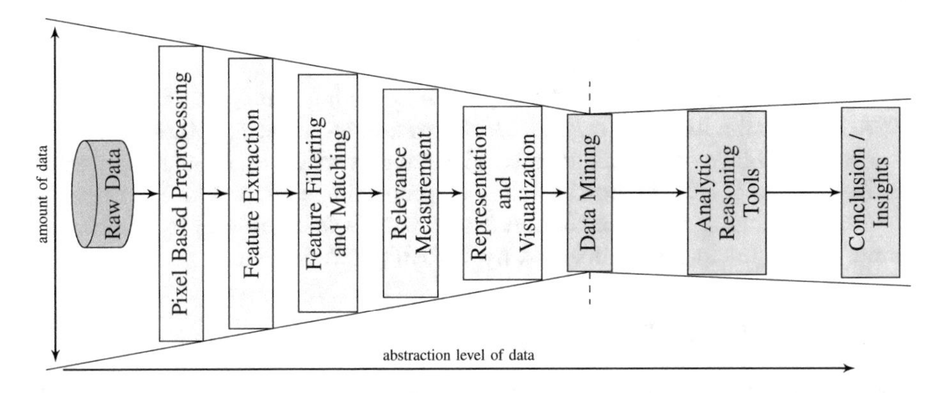

Fig. 3. Schematic diagram: amount of data vs. abstraction level from low level (pixel-based) via meaningful content to high-level (conclusion only). Yellow marked boxed are exemplary methods related to *content extraction and representation* and green marked boxed are related to *content-based reasoning* based on the architecture of VVA.

and highly accurate results. Usually, interactive strategies are neglected in computer vision because a user is necessary. For VVA architectures, where human knowledge is embedded for supporting the computer during the sensemaking process, the use of interactive computer vision techniques drastically improves the reliability and accuracy of image and frame based content analysis. Therefore, only interactive methods for translating pixel information into meaningful artifacts, as illustrated on the left-hand side of Fig. 2, will be introduced hereafter.

3.1 Interactive Video Annotation

In 2011, Dasiopoulou et al. [20] compared various annotation tools for images and video sequences using several criteria like in- and output formats, metadata types and the granularity, localization, and expressivity of their annotations. According to their review, only one of seven video annotation tools provides polygon-based annotations, the Video and Image Annotation tool (*VIA*). Polygon-based, hence pixel accurate annotations are in our opinion necessary for content extraction because only precise annotations allow a detailed relationship descriptions, e.g., which OOI occludes other OOIs. By testing VIA [21], we noticed that it is not capable of handling video footage with higher resolutions like FullHD and that its source code is not available.

For the annotation of pedestrians in scenes, Wu et al. [2] designed the Semi-Automatic Ground Truth Annotation tool (*SAGTA*). Its semi-automatic process relies on the assumption of 3D linear motion of the pedestrians. This 3D linear motion assumption of *SAGTA* reduces the number of frames to annotate manually but assumes a fixed-position camera position.

One quite old interactive video annotation tool, called Video Performance Evaluation Resource (*ViPER*) tool [22], is still in use, mostly due to its properly defined and specified XML output format. After annotating the same OOI on at least two frames, a linear 2D propagation can be performed for the generation of annotations between the initial two frames.

Our released open source software, named *iSeg* [3], came with a novel interactive and semi-automatic segmentation process for creating pixel-wise labels of several OOIs in video sequences. The latest version of *iSeg*, implemented in *OpenCV* and *QT*, supports multicore processing and provides a semantic timeline for labeling object-occlusion (none, partial or complete) in a particular frame. Its output format is inspired by *ViPER*'s XML format.

3.2 Interactive 3D Reconstruction

Fully automatic 3D reconstruction methods work quite well if the images in the database do not have delicate structures, textureless surfaces, hidden boundaries, illumination changes, specularity, or dynamic moving objects like they typically occur in real-world recordings [23]. Kowdle et al. [10] also mentioned these issues of fully automatic reconstruction and came up with a semi-automatic approach, where the user marks corresponding faces of the OOI on the images. Thus the user is put into the loop of 3D reconstruction by interacting with the algorithm. As a result, even completely textureless OOIs can be reconstructed.

Using a fixed-position camera, the probabilistic feature-based online rapid model acquisition *ProFORMA* [11] can reconstruct an OOI in near real-time. Its user dialog system guides the user on how the OOI has to be manipulated. The user then rotates the textured object in front of the camera. The final meshed model is produced by Delaunay triangulation of a point cloud obtained from a structure from motion (SfM) estimation. Due to its interactive guidance system, this approach does not work on already captured video sequences.

Motivated by the human visual system, we developed an interactive technique [24], which converts video data into 3D representations. Comparable to the hierarchy of the ventral stream of the visual cortex in the human brain, this method reduces the influence of the position information in the video sequences by OOI recognition and stores the OOI as multiple pictorial representations. The multiple pictorial representations show a 2D projection of the OOI from different perspectives. As a consequence, any SfM algorithm can obtain the 3D point cloud of the OOI.

3.3 Users' Unconscious Features

The human visual system has a very fast response time, e.g., in visual search tasks [25,26]. Based upon this consideration, users' unconscious features like gaze trajectories and skin conductance, during watching video data might be extracted as additional content for the reasoning process. Eye tracking data, in particular, can be recorded contactless and is already recorded e.g. for driver assistance systems. Furthermore, it is already shown that users' gazes as an unconscious feature can improve the accurate OOI detection in megapixel images [27].

4 Interactive Content-Based Reasoning

Focusing on the consistent implementation of our VVA architecture, two software applications are prototyped. These exemplary applications create task-adaptive playbacks of the video sequence using i) 3D shape information of an OOI and ii) logical descriptions of OOIs relations. Both, as seen in Fig. 4, use the current implementation of the open source tool *iSeg* as the underlying fundamental framework and are extended by task-specific software features like the 3D point cloud viewer and the proposition editor.

Video analysis using 3D shape information, cf. Fig. 4(a), uses a 3D point cloud of the OOI generated by our interactive 3D reconstruction technique [24]. To reconstruct a 3D point cloud of an OOI, cf. ⓑ, a two-stage process is performed. In the first stage, the analyst annotates the OOI using *iSegs'* interface metaphors cf. ⓐ and ⓒ. This pixel-accurate annotation reduces the influence of the position information and leads to multiple pictorial representations of the OOI. Based upon these multiple pictorial representations, a SfM algorithm creates the 3D point cloud in the second stage. The resulting point cloud, as the main content for the reasoning process, can be rotated and magnified inside the point cloud viewer ⓑ by the analyst. Within the point cloud viewer, the analyst is now able to select parts of the point cloud. By selecting a group of points or picking a certain point, the timeline ⓓ directly highlights, in red color, frames showing at least one of the selected point. As a consequence, an adaptive playback of only essential frames for the ongoing analysis is possible.

In Fig. 4(b), all available content extracted from the video sequence and the additional content, in this case eye-tracking data, are used for building up a keyword dictionary. Using simple logical concepts, here *all*, *and*, *or*, *intercept*, *occur*, and brackets, these keywords can be combined to formulate propositions. For building up the keyword dictionary, the analyst uses already available annotations of relevant OOIs or annotates additional OOIs manually. If additional content, like eye-tracking data of people watching this video footage, is available, the analyst can integrate it into the dictionary, as well. Thus, the proposition editor ⓑ provides on the right- hand side all available content along with logical grammar elements for combining them. In a free text, the analyst now formulates a proposition related to the video analysis task. If the formulated proposition is correct, the resulting frames will be directly red highlighted in the timeline ⓓ.

5 User Experiences

Will our interactive architecture improve VVA applications in general or only in certain domains? We have conducted a first user study with two expert users. One of these users is an expert in the domain of video surveillance software, and the other is a specialist in the analysis of gaze data. We first explained the functionality of both prototypes to the experts. Then each expert tested the application next to his field of expertise. For this trial, the task set for the expert on video surveillance is to find all frames where the left front of the red car, can

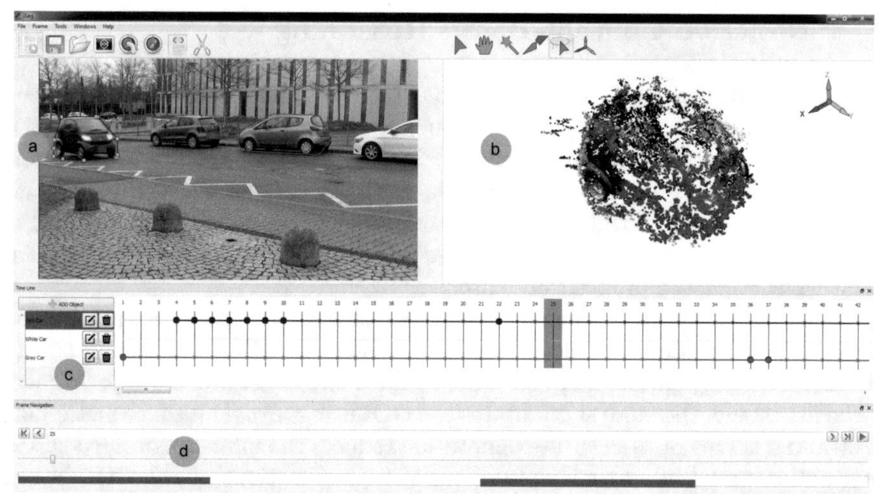

(a) Application for adaptive playback of frames, where frames of selected 3D parts are highlighted. Based on a 3D representation of an OOI, the analyst can select parts of the OOI in the point cloud viewer ⓑ. The frames, where the selected parts are visible will be highlighted on the timeline ⓓ. The video with OOI annotations can be seen and altered in ⓐ.

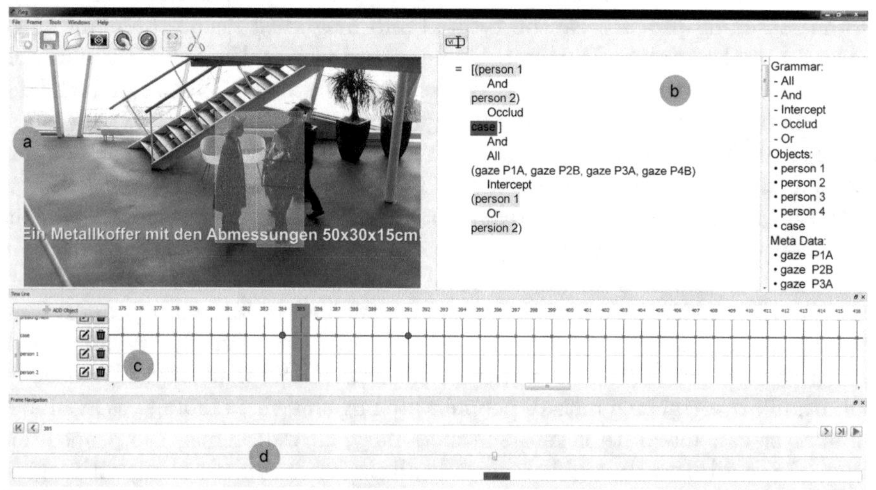

(b) Application for highlighting certain events. Based on all extracted OOIs and additional data like eye tracking data, the analyst can formulate a proposition in the proposition editor ⓑ, where all available keywords are listed, too. If the proposition is true for one or several frames, these frames will be highlighted on the timeline ⓓ.

Fig. 4. Two exemplary applications, integrated in the user interface of *iSeg* [3] using our architecture of VVA. Note: ⓐ and ⓒ are unchanged parts of *iSeg*.

be seen. Therefore, the expert should use the 3D point cloud representation of the OOI—here the 3D point cloud of the red car. The task for the expert on gaze data is to identify objects which interfere with gaze points. For this task, he should use the proposition editor. After these explorative tasks, both users were interviewed.

The expert on video surveillance stated "In the introductory session to the VVA applications, I was quite impressed with the idea to just mark parts of a 3D object and then get a video summary. By Testing the application the creation of the video summarization works, but the iterative creation of the 3D point cloud is still a tedious and time-consuming task." He further stated that when having the point cloud representation, the summarization of the video sequence to different regions of interest—not only the front part, is a helpful feature primarily by working with a huge body of video material. However, for our test case video sequence, the creation of a summarization by hand is in his opinion several times faster, compared to the usage of a 3D point cloud.

The expert on gaze data found that "the formulation of a proposition and the almost instantaneous visualization of frames, where this proposition is correct is a brilliant feature". Putting this statement in perspective, all the OOIs and the eye tracking data were provided by the data set used in this trail. During the interview, the export pointed out that one main issue of creating data sets for gaze analysis is the OOI annotation, which might be facilitated by such an application. During testing the software, the expert, unfortunately, did not create new (i.e., not yet provided) OOI annotations by his own. In the discussion and by showing how additional OOI annotations can be done using *iSegs*' user interface metaphors in the proposed prototype, he was delighted to see that interactive computer vision speed up this process considerably, though it is still time-consuming.

6 Storing Extracted Content in Video Container Formats

The importance of insights as well as conclusions extracted by VVA will significantly increase if all the extracted content of all level of abstractions is provided in single multimedia containers. Nowadays, for text information the PDF container become a standard for storing text along, e.g., remarks, images and signatures. Therefore, a widespread use of multimedia containers for storing video analysis intermediate artifacts as well as the final insights will make the conclusions of video analysis accessible to a broader audience. Thus it can also be used for other applications in science like content depending deep learning in video footage. Moreover, using a metadata standard will terminate the tedious process where researchers have to adjust their software for every different data set. Besides a defined input and output interface, the use of container formats also enables the streaming of this data via the Internet.

As shown in related work, pixel-accurate annotations of several OOIs [28], gaze data of a participant group [29], and EEG data of participants [30] can be stored next to the corresponding video sequence in a single multimedia container. Adapting this idea, Fig. 5 show how all content, used and created by the analyst for the above-described use cases, can be fitted in the general structure of a multimedia container. All time-dependent data, like OOI annotations, scene tags, voice notes, gaze data, and sonifications, are stored in multimedia containers as temporal payload by using the video, audio, or subtitle tracks. These time-depended data are mostly related to the lower level contents of the *reasoning process*. Conclusion and insights, like movement maps, 3D representations, and interaction pattern between OOIs, which are not time-depended can be integrated into the static payload of multimedia containers.

Since up-to-date multimedia player and devices, supporting open container formats, like the OGG [31] and the *Matroška* (MKV) [32] format, exploratory multimodal data analysis with these players is possible [30]. In consequence, by using multimedia containers within the domain of VVA, the instantaneous visualizations of the gained insights with standard multimedia players become possible for the broad audience.

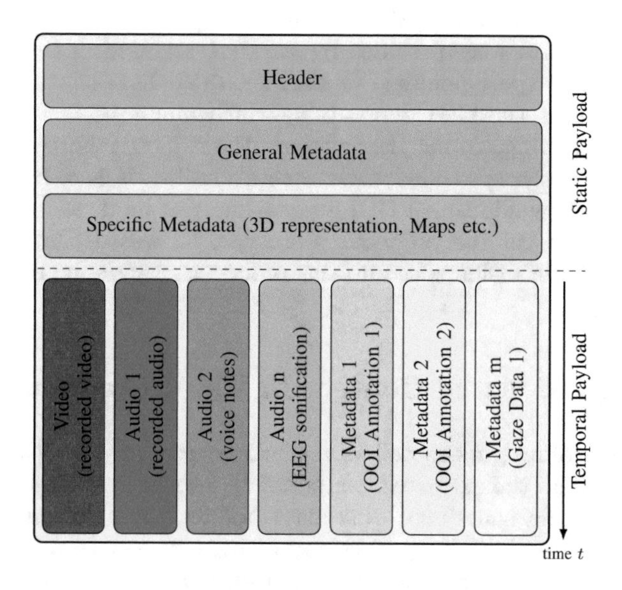

Fig. 5. General data structure of a multimedia container modified for storing the artifacts, hypotheses, insights, and conclusion of VVA. In the static header, general as well as specific metadata, which have no temporal dependencies, are stored before the temporal content like OOI annotations, scene tags, voice notes, gaze data, and sonifications, are stored in as video, audio, or subtitle tracks. For streaming this video via the Internet, only the non-temporal data have to be transmitted before playback; the temporal data are transmitted while playing.

7 Discussion

Based on these use-case prototypes, one might argue that the analyst already gets the complete insights of the video sequences by watching it or at least by performing the *interactive content extraction and representation*. Consequently, the *content-based reasoning* process would be a waste of time because the analyst is already able to conclude the analysis task. However, how can the conclusions drawn from the comprehension of documents, be secured for third parties, like judges or managers? For these cases, our architecture of VVA leads to a reproducible process, which can be implemented on any computer. The reproducibility allows retracing of conclusions made by other analysts from the video sequences. Furthermore, for a comprehensive documentation and for improving the retraceability, the extracted conclusion and the history how an analyst came to it must be stored, as well.

Regarding the user study, the bottleneck of our architecture is the tedious and time-consuming *interactive content extraction*. Here, computer vision does not yet provide automatic and time-efficient semi-automatic approaches which would be able to compute a meaningful content abstraction level from the pixel input of real-world video sequence. Primarily, with the focus on real-world video sequences, computer vision must reliable handle, e.g., shiny objects, moving cameras, and repetitive structures.

Nevertheless, allowing the developing of the *interactive content extraction and representation* as well as *content-based reasoning* on its own, a standard data format is preferable. Therefore, the use of a standardized multimedia container is, in our opinion, the best solution, as it contains all necessary information in a single file. On this account, we strongly encourage the definition and design of a multimedia container for video analysis. If a standard format is established, one can reuse already extracted content, or in the case of large data sets, multiple analysts will retrieve meaningful artifacts which becomes available for the other analysts as well. This exchange of meaningful artifacts empowers the reasoning process for all analyst.

8 Conclusion

In this paper, we propose an interactive architecture for VVA-based on the sensemaking loop and evaluate this architecture on two prototype applications. Our architecture consists of two parts, the *interactive content extraction and representation* process, and the *interactive content-based reasoning* process. Based on two exemplary applications and a user study, we demonstrated that our architecture, in combination with interactive content extraction methods, can perform video analysis based on 3D point clouds and free text propositions. By performing analysis tasks in the user study, we noticed, in particular, a lack of automatic and time-efficient semi-automatic computer vision algorithms which would be able to transfer pixel content into meaningful symbols. Though in principle, we would allow the developing of both parts of our architecture separately, the interface for exchanging content between *interactive content extraction and representation* and *interactive content-based reasoning* must be explicitly defined. Therefore, as a defined interface we encourage the use of multimedia containers, cf. Sect. 6.

For a significant increase in VVA applications, computer vision methods for extracting content from video need to be improved, so that analysts can specify the granularity and certain parameters of content, e.g., all objects of the classification group "mammal". The content extraction by itself will then be performed automatically. Fully automated content extraction without any user interaction might not reduce the amount of data sufficiently, especially in large data sets. Thus data mining in large data sets becomes impossible, cf. Fig. 3. All in all, however, it can be said that video content analysis will significantly improve through the use of our scalable VVA architecture, as shown in this paper, but probably even then interactive video content analysis will not perform as good as nowadays Hollywoods' implementations.

References

1. Pillai, G.: Caught on camera: you are filmed on CCTV 300 times a day in London. International Business Times, September 2017. http://www.ibtimes.co.uk/britain-cctv-camera-surveillance-watch-london-big-312382
2. Wu, S., Zheng, S., Yang, H., Fan, Y., Liang, L., Su, H.: SAGTA: semi-automatic ground truth annotation in crowd scenes. In: International Conference on Multimedia and Expo Workshops (ICMEW). IEEE - Institute of Electrical and Electronics Engineers (2014)
3. Schöning, J., Faion, P., Heidemann, G.: Pixel-wise ground truth annotation in videos: an semi-automatic approach for pixel-wise and semantic object annotation. In: International Conference on Pattern Recognition Applications and Methods (ICPRAM), SCITEPRESS - Science and and Technology Publications, pp. 690–697 (2016)
4. Schroeter, R., Hunter, J., Kosovic, D.: Vannotea—a collaborative video indexing, annotation and discussion system for broadband networks. In: Workshop on Knowledge Markup & Semantic Annotation (2003)
5. Tanisaro, P., Schöning, J., Kurzhals, K., Heidemann, G., Weiskopf, D.: Visual analytics for video applications. IT Inf. Technol. **57**, 30–36 (2015)
6. Keim, D.A., Mansmann, F., Schneidewind, J., Thomas, J., Ziegler, H.: Visual analytics: scope and challenges. Lecture Notes in Computer Science, pp. 76–90. Springer, Heidelberg (2008)
7. Höferlin, M., Höferlin, B., Weiskopf, D., Heidemann, G.: Uncertainty-aware video visual analytics of tracked moving objects. J. Spat. Inf. Sci. **2**, 87–117 (2011)
8. Pintore, G., Gobbetti, E.: Effective mobile mapping of multi-room indoor structures. Vis. Comput. **30**(6–8), 707–716 (2014)
9. Sensopia Inc.: Capture the floor plan of your house with magicplan, September 2017. https://www.magic-plan.com/
10. Kowdle, A., Chang, Y.-J., Gallagher, A., Batra, D., Chen, T.: Putting the user in the loop for image-based modeling. Int. J. Comput. Vis. **108**(1–2), 30–48 (2014)
11. Pan, Q., Reitmayr, G., Drummond, T.: ProFORMA: probabilistic feature-based on-line rapid model acquisition. In: British Machine Vision Conference (BMVC).British Machine Vision Association (2009)
12. Wu, C.: VisualSFM: a visual structure from motion system, January 2011. http://ccwu.me/vsfm/
13. Marconi, D.: Enemy of the State. Touchstone Pictures (1998)
14. Höferlin, B., Höferlin, M., Weiskopf, D., Heidemann, G.: Scalable video visual analytics. Inf. Vis. **14**(1), 10–26 (2013)
15. Russell, D.M., Stefik, M.J., Pirolli, P., Card, S.K.: The cost structure of sensemaking. In: SIGCHI Conference on Human Factors in Computing Systems (CHI), pp. 269–276. ACM Press (1993)

16. Thomas, J.J., Cook, K.A. (eds.): Illuminating the Path: The Research and Development Agenda for Visual Analytics. IEEE Computer Society Press (2005)
17. Höferlin, B., Netzel, R., Höferlin, M., Weiskopf, D., Heidemann, G.: Inter-active learning of ad-hoc classifiers for video visual analytics. In: Conference on Visual Analytics Science and Technology (VAST), pp. 23–32. IEEE - Institute of Electrical and Electronics Engineers (2012)
18. Pirolli, P., Card, S.: The sensemaking process and leverage points for analyst technology as identified through cognitive task analysis. In: International Conference on Intelligence Analysis (2005)
19. Thomas, J.J., Cook, K.A.: A visual analytics agenda. IEEE Comput. Graph. Appl. **26**(1), 10–13 (2006)
20. Dasiopoulou, S., Giannakidou, E., Litos, G., Malasioti, P., Kompatsiaris, Y.: A survey of semantic image and video annotation tools. In: Knowledge-Driven Multimedia Information Extraction and Ontology Evolution, pp. 196–239. Springer, Heidelberg (2011)
21. Multimedia Knowledge and Social Media Analytics Laboratory. Video Image Annotation Tool—Multimedia Knowledge and Social Media Analytics Laboratory, January 2012. http://mklab.iti.gr/project/via
22. Doermann, D., Mihalcik, D.: Tools and techniques for video performance evaluation. In: International Conference on Pattern Recognition (ICPR). IEEE Computer Society Press, pp. 167–170 (2000)
23. Schöning, J., Heidemann, G.: Interactive 3D modeling: a survey-based perspective on interactive 3D reconstruction. In: International Conference on Pattern Recognition Applications and Methods (ICPRAM), pp. 289–294. SCITEPRESS - Science and and Technology Publications (2015)
24. Schöning, J., Heidemann, G.: Bio-inspired architecture for deriving 3D models from video sequences. In: Computer Vision – ACCV Workshops, pp. 62–76. Springer, Heidelberg (2016)
25. Trick, L.M., Enns, J.T.: Lifespan changes in attention: the visual search task. Cogn. Dev. **13**(3), 369–386 (1998)
26. Eriksen, C.W., Schultz, D.W.: Information processing in visual search: a continuous flow conception and experimental results. Percept. Psychophys. **25**(4), 249–263 (1979)
27. Schöning, J., Faion, P., Heidemann, G.: Interactive feature growing for accurate object detection in megapixel images. In: Computer Vision – ECCV, Workshops, vol. 9913, pp. 546–556. Springer, Heidelberg (2016)
28. Schöning, J., Faion, P., Heidemann, G., Krumnack, U.: Providing video annotations in multimedia containers for visualization and research. In: Winter Conference on Applications of Computer Vision (WACV). IEEE - Institute of Electrical and Electronics Engineers (2017)
29. Schöning, J., Faion, P., Heidemann, G., Krumnack, U.: Eye tracking data in multimedia containers for instantaneous visualizations. In: IEEE VIS Workshop on Eye Tracking and Visualization (ETVIS), pp. 74–78. IEEE - Institute of Electrical and Electronics Engineers (2016)

30. Schöning, J., Gert, A.L., Açik, A., Kietzmann, T.C., Heidemann, G., König, P.: Exploratory multimodal data analysis with standard multimedia player: multimedia containers: – a feasible solution to make multimodal research data accessible to the broad audience. In: Computer Vision, Imagingand Computer Graphics Theory and Applications (VISAPP), pp. 272–279. SCITEPRESS - Science and Technology Publications (2017)
31. Xiph.org: Ogg, September 2017. https://xiph.org/ogg/
32. Matroska: Matroska media container, September 2017. https://www.matroska.org/

The Meaning of Big Data in the Support of Managerial Decisions in Contemporary Organizations: Review of Selected Research

Dorota Jelonek(✉), Cezary Stępniak, and Leszek Ziora

Faculty of Management, Czestochowa University of Technology, Czestochowa, Poland
{dorota.jelonek, cezary.stepniak, leszek.ziora}@wz.pcz.pl

Abstract. The purpose of the paper is presentation of the role of big data methods, techniques and tools application in the support of managerial decisions. The paper characterizes the notion of big data solutions, its key components with fundamental analyses applied within the area of big data, types of decisions which can be supported with its usage, and benefits resulting from big data application in the support of managerial decisions in contemporary organizations. The practical examples and case studies were presented on the basis of research review.

Keywords: Big data · Data mining · Business intelligence · Business analytics Decision making support

1 Characteristics of Big Data

In the literature of subject big data is defined by four Vs which stand for volume, variety, velocity and veracity of data. The data's volume is referred to the amount of processed data, the variety is connected to the different types of data e.g. whether we analyze structured or unstructured ones, the velocity is related to the speed of data analysis and the veracity is connected with its reliability, quality and usability. [compare 3, 22]. In this paper the authors took into account the Gartner group definition of big data where it is characterized as "high-volume, high-velocity and high-variety information assets that demand cost-effective, innovative forms of information processing for enhanced insight and decision making" [15]. The big data concept allows for discovery and deployment of the potential of huge data volumes in the management of enterprise [12]. Akoka et al. present information indicated by IBM and stating that "every day 2.5 exabytes of data are generated and they also present CISCO predictions assuming that till 2020 the number of devices connected to different networks and to the Internet will reach 50 billion" [1]. They further present IDC expectations "that the amount of valuable data for the purpose of big data analyses reach 23% of the total generated data which is assessed as 643 exabytes. Such source data may embrace data having its origin in surveillance footage material, medical and embedded devices, entertainment, social media, and consumer images [1]." Big data embrace many key

K. Arai et al. (Eds.): FICC 2018, AISC 886, pp. 361–368, 2019.
https://doi.org/10.1007/978-3-030-03402-3_24

components (Fig. 1). Ohlhorst distinguishes its fundamental elements such as: "Business Intelligence systems, data mining solutions, statistical applications, predictive analysis and data modeling" [22]. Another key component of business analytics is sentiment analysis solution which belongs to the field of text analytics as well as it belongs to the areas of natural language processing and computational linguistics [30]. Such a solution can utilize machine learning approach and lexicon-based approach. Sentiment analysis allows for analyzing opinions, attitudes, and emotions towards different products or services offered by different vendors. The subject of such an analysis can be social media portals, websites, and any other services e.g. financial ones which enable expressing people's comments.

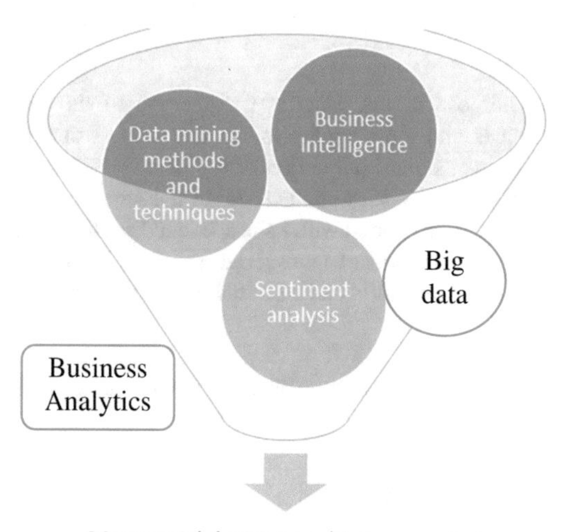

Fig. 1. Key components of business analytics and big data. (*Source: Authors' own study*).

Fonseca and Cabral state that "big data solutions are frequently used in machine learning, where obtained data is utilized for the purpose of future behavior prediction. Deep learning solution enable applying artificial neural networks as a method aimed at information extraction from complex sets of data. Deep learning has a potential of creation sophisticated models in a greater extent that normal probabilistic machine learning techniques." [10]

It is worth paying attention to MGI report on big data which states that "these solutions are currently present in every area of global economy". The report identifies "five widely utilized ways to harness big data and its transformational potential for the purpose of value creation and it also underlines its impact on the way of management and design of particular organization" [18]. They are "supporting as well as substituting decision making with automated algorithms, making clarity, allowing for specific needs discovery, highlight changeability, together with performance improvement, population segmentation, activities customization and enabling innovation of original business models, products, and services [18]". EY company report states that "for the purpose of value extraction from big data, the same data should be processed and analyzed in an exact time, and the results should be available in appropriate way as to be able to effect positive change or influence business decisions. Analytics plays crucial role in optimization of key processes, functions and roles as well as in improvement of decisions processes. It can be harnessed to aggregate both internal and external data [14]." As far as technological solutions are concerned it is worth mentioning Hadoop, Cloudera, Mapreduce, HBase, and MongoDB.

2 Role of Big Data in the Support of Managerial Decisions

The decisions which are made on the basis of data utilization are called data-driven decisions. Provost and Fawcett claim that "data-driven decision-making is related to making decisions on the ground of data analysis, rather than basing on intuition [23, p. 5]". They also present the results of the study by Brynjolfsson, Hitt and Kim where the benefits of data-driven decision-making were that "the company is more productive when it is in a greater extent driven by data. Data-driven decision making is also associated with higher return on assets, return on equity, asset utilization, and market value [23]." Big data methods, techniques and tools support decision making processes at every level of management it is strategic, tactical and operational one. "the specificity of conducted analyses is dependent on the level of organization management, its size and chosen analyses may include simple ones as reports creation and more complex as predictive ones which require the application of data mining methods and techniques as well as appropriate tools of statistics" [8, 9]. Big data solutions can support managers at every stage of decision making. For example, Samuelson and Marks defined six stages of decision making process such as: "stage 1 - defining a problem, stage 2 - determining the purpose, stage 3 - examining the options, stage 4 - foreseeing the consequences, stage 5 - choosing the best variant and stage 6 - improving sensitivity analysis" [24]. At every mentioned stage of decision making the proper big data analyses can be performed. Morabito claims that "the industries can benefit from big data and business analytics by redefining themselves and affecting the products value which are offered to consumers. The mentioned analytical solutions allow traditional industries for rapid progress, and allow for better competition within specific industry branch [19, p. 17]." What more are the big data solutions which belong to the data-driven approach "is capable to improve business performance [5]." There should be taken into account an important fact that the decision quality is dependent on the quality of data which are the subject of big data analyses. Janssen et al. claim that "the quality of the source data, the processing of the data and how the transfer of the data is handled affect the quality of decision-making" [11].

As far as source data are concerned it is worth mentioning the fact that it constitutes the basis for big data analyses and such data may come from different systems. Baesens as a source data indicates "transactional data, structured and unstructured data, relational databases, multimedia content and qualitative, expert-based data" [2]. The Intel report based on the survey of 200 IT managers indicates source data which are the subjects of analyses as: "documents (84%), business transactions in database (82%), e-mails (74%), data from sensors (57%), Internet searches (57%), weblogs (55%), social media (54%), phone conversations (52%), videos (52%), pictures (46%), clickstreams (42%). According to the report IT managers consider big data a top IT priority for their organizations [27]." The same respondents asked about the issue of particular obstacles related to big data analytics indicated the following ones: "concerns related to the security issues, operational expenses, possible creation of network bottlenecks, shortage of data scientists job positions, impossible to manage data rate, data replication functions, lack of compression functions, bigger network latency, not sufficient power of processors" [27]. As a summary it can state that in the contemporary organization the big data solutions play the role of support and acceleration of decision making process.

3 Areas and Benefits of Big Data Application

The report entitled "2014 Massachusetts Big Data Report: A Foundation for Global Leadership" where the respondents were constituted by 485 companies provides multiple ways of utilizing big data analyses such as its application in the field of finance where big data is utilized in: sentiment analysis area connected to the analysis of customers opinions on specific financial products or services, fraud detection, credit risk scoring and assessment, targeting product offerings, automation of algorithms connected to trade, legal discovery and compliance reports. In the marketing and advertising it is applied in analysis of consumers opinions as well as marketing campaigns, stating appropriate prices for advertisements, its targeting and personalization of web content. In the e-commerce and retail trade big data allows for different transactions and consumer behavior analyses, click stream analyses, point of sale transaction analysis, development of pricing models, personalizing web content and inventory management. In the telecommunication branch big data enable analyses of consumer opinions e.g. on providing a particular service like access to the Internet network or voice transmission services, purchase patterns analyses, location-based services, network analysis including its maintenance and optimization. In the manufacturing it enables current control of production processes, measurement of its performance and for dealing with warranty claims. In the logistics and transportation area it allows for supply chain analysis, location tracking, and consumption of fuel analysis. In the healthcare big data enable specific clinical trials together with comparing its effectiveness, social media analysis to detect disease or treatment patterns, patient monitoring, personalized medicine and billing compliance [21].

Big data solutions can find its application in support of decisions in multiple areas such as in production, logistics, health care, retailing, banking, finance. Such solutions bring benefits to contemporary organizations which applied it in its daily business activity. According to Big Data Executive Survey the tangible benefits which can be achieved by the application of big data analytics are: "better fact-based decision making, better customer service, sales figures grow, implementation of innovations for new products and services, risk reduction, improved quality of offered goods and more efficient business operations" [4]. "Big data solutions might contribute to gaining competitive advantage of the organization which implemented it in its business activity and minimization of risk in logistics and facilitation of logistics networks creation [6] ". Liebowitz claims that "big data analytics can reveal novel interconnections between data, show previously unnoticed trends and contribute to new knowledge generation, applied in increasing the effectiveness and improvement of company's profitability" [17]. Schmarzo claims that big data allows for "transformation of business processes in logistics and in such areas as procurement, product development, manufacturing, marketing, sales and human resources" [25]. Big data solutions are widely applied in contemporary organizations mainly supporting decision making process. Seth and Chaudhary present its application in Finance. The authors mention its usage in e.g. fact-driven strategic decision making, product innovation, development of novel and intelligent business solutions. The other areas include risk management, fraud detection, predictive analytics based on social behavior mining [26]. Big data solutions are applied in collecting data concerning realization of instances of processes used in PPI [29] and in sports management where e.g. in ski jumps discipline the data can be collected from sensors and may concern the weather conditions.

4 Review of Selected Research

Big data solutions may find its application in the support of managerial decisions in multiple industries. One such an application is metal industry. The empirical research embraced desk research and interviews with managers in large enterprises in the metal sector. The basis for the desk research analysis was publications from areas of big data analytics and management in metal enterprises and the interviews were conducted in person among managers of 6 large metal industry companies. The selected research results show that "big data analytics may be an important support in customer relationship management, innovation management and production management; in the enterprises of the metal sector, the potential areas of using big data analytics are indicated by strategic management, including the support for the managers, predictive analytics and "what if" simulations; managers are familiar with the concept of big data, but big data analytics projects are not considered as priority. The process of value creation in big data solutions should be focused on the development of a new analytical model which allows a user to retrieve information from the available database. Interpretation of the data obtained in this manner may represent a significant economic value [13].

One of the examples of applying big data in the banking sector is the National Bank of Canada, specifically in bank's GED section which deals with gathering and processing huge volume of financial data and namely stock exchange transactions. The GED wanted to apply "more flexible solutions supporting data analysis in a more efficient way which could also increase efficacy of processing structured and unstructured financial data" [20]. As an example, it allowed for log files analysis with its comparison against updated market data [20]. They selected Cloudera which is Apache Hadoop processing framework due to its scalability and support of structured and unstructured data. They also applied Amazon cloud technologies and decided to cooperate with TickSmith, a big data management software provider. The benefits were that the institution could process and analyze huge volumes of data counted in hundreds of terabytes which contained trade and historical quote data what caused that business analysts are able to carry out post-trade analysis much faster. These data operations were shortened from days to minutes. The benefits resulting from applying mentioned solution improved and optimized trading operations which allowed generating more revenue for this bank [20].

The application of big data solutions in transportation area can be exemplified by utilizing TIBCO solutions in CargoSmart organization which is "an important provider of worldwide shipment management services, and utilizes big data for greater visibility and benchmarking". According to the TIBCO it allows for providing such entrepreneurs with information required for the purpose of better competition in the market. The authors of cited case study claim that "CargoSmart wanted to use advanced analytics to provide unprecedented visibility to ocean carriers so they can plan ahead in case of disruption and make use of real-time analysis to improve decision-making" [7]. The company needed the scalable platform to allow for implementation of new data sources and become more data-driven in order to deliver its solutions quickly as well as data visualization so they could quickly meet a customer's requirements. The required technologies were a real-time event processing and detection engine and an analytic platform to provide data visibility and end-user dashboards so customers could make faster decisions. The applied TIBCO solutions allows CargoSmart to be more responsive to customer needs, the mentioned solution helped them increase the number of vessels they monitor by four times over the past year and a half, they also developed a vessel speed and route monitoring application that analyzes vessel's speed and distance against a complex variety of factors. The application has helped ocean carriers reduce fuel consumption by up to 3.5% over the past two years [7].

Another worth mentioning application of big data are IBM solutions implemented in Herz company which is one of the biggest car rental companies with 8300 places in 146 countries. Hertz collects on a daily basis "huge amount of comments from such sources as web surveys, e-mails and text messages and they could be harnessed in its business activity at strategic level as well as in improvements of operational one" [16]. The benefits which resulted from implementation of big data pinpointed "centralizing the process of data collection for the purpose of sentiment analysis, so it enabled Hertz company analyze free-form unstructured feedback from its most significant members, and what is more the time of conducting analyses were reduced two times, so the company could respond quicker to customers' feedback, and then the immediate actions on problem areas could be undertaken" [16].

Another example of big data application is the Siemens corporation in the domain of high speed trains where "company's engineers use the data acquired from tens of thousands of sensors. The collected data concern the areas of supply chain, trains and rails status, repair processes, weather forecast and their processing destination is Siemens Teradata Unified Data Architecture based on Hadoop" [28]. The authors of Siemens case study claim that "applied by them solution connected with machine learning allow company's data scientists and engineers for quick identification of false positives what means predicting non existing malfunctions and give a clear prediction of actual part failures thanks to the scalable architecture, telematics and sensors generating unstructured data. Possessing analytics models the company is able to actually predict certain failures [28]".

5 Summary

Big data solutions and modern business analytics support managers in decision making at every stage and level of management of contemporary organization. Multiple possibilities of different analyses deployment such as statistical, data mining methods and techniques, business intelligence systems, sentiment analysis, machine learning brings many benefits to the organizations which have applied it in its business activity enabling improvement of the decisions, increasing efficiency and efficacy of decision making process. What is more the contemporary organizations which apply big data solutions can gain competitive advantage in the worldwide market.

References

1. Akoka, J., Comyn-Wattiau, I., Laoufi, N.: Research on big data – a systematic mapping study. Comput. Stand. Interfaces **54**, 105–115 (2017)
2. Baesens, B.: Analytics in a Big Data World. The Essential Guide to Data Science and its Applications. Wiley, New Jersey (2014)
3. Berman, J.: Principles of Big Data. Morgan Kaufmann Elsevier, Waltham (2013)
4. Big Data Executive Survey (2013). New Vantage Partners, Boston. http://dev.gentoo.org/~dberkholz/osbc/NVP-Big-Data-Survey-2013Summary-Report.pdf
5. Brynjolfsson, E., McAfee, A.: Big data: the management revolution. Harvard Bus. Rev. **90**(10), 60–68 (2012)
6. Brzozowska, A., Ziora, L., Sałek, R., Wiśniewska-Sałek, A.: The possibilities of big data solutions application in logistics. In: XXX. microCAD International Multidisciplinary Scientific Conference University of Miskolc, Miskolc, Hungary (2016)
7. CargoSmart case study. https://www.tibco.com/resources/success-story/cargosmart-case-study
8. Chluski, A., Ziora, L.: The role of big data solutions in the management of organizations. Review of selected practical examples. In: International Conference on Communication, Management and Information Technology (ICCMIT 2015). Elsevier. Procedia Comput. Sci. **65**, 1006–1012 (2015). www.sciencedirect.com
9. Chluski, A., Ziora, L.: The application of big data in the management of healthcare organizations. A review of selected practical solutions. Bus. Inform. – (Informatyka Ekonomiczna) **1**(35), 9–18 (2015)
10. Fonseca, A., Cabral, B.: Prototyping a GPGPU neural network for deep-learning big data analysis. Big Data Res. **8**, 50–56 (2017)
11. Janssen, M., van der Voort, H., Wahyudi, A.: Factors influencing big data decision-making quality. J. Bus. Res. **70**, 338–345 (2017)
12. Jelonek, D.: Challenges for strategic management in the age of big data. In: Sopińska, A., Wachowiak, P. (eds.) Challenges of Contemporary Strategic Management, pp. 161–175. Warsaw School of Economics Publishing House, Warsaw (2017). (in Polish)
13. Jelonek, D.: Big data and management in metallurgical enterprises. In: 26th Anniversary International Conference on Metallurgy and Materials (METAL 2017), Brno, Czech Republic, 24–26 May 2017 (invited lecture, in press)
14. EY insights on governance, risk and compliance. Big data. Changing the way businesses compete and operate, April 2014
15. Gartner's IT glossary. http://www.gartner.com/it-glossary/big-data/. Accessed 24 July 2017

16. IBM case study: how big data is giving Hertz a big advantage. https://www01.ibm.com/software/ebusiness/jstart/portfolio/hertzCaseStudy.pdf
17. Liebowitz, J.: Big Data and Business Analytics. CRC Press, Taylor & Francis Group, Boca Raton FL [etc.] (2013)
18. Manyika, J., et al.: McKinsey Global Institute, Big data: the next frontier for innovation, competition, and productivity, May 2011
19. Morabito, V.: Big Data and Analytics. Springer International Publishing, Cham (2015)
20. National Bank of Canada case study. https://aws.amazon.com/solutions/case-studies/national-bank-of-canada/
21. The Massachusetts Big Data report 2014: A foundation for global leadership. http://www.masstech.org/sites/mtc/files/documents/Full%20Report%202014%20Mass%20Big%20Data%20Report_0.pdf
22. Ohlhorst, F.: Big Data Analytics. Turning Big Data into Big Money. Wiley, Hoboken (2013)
23. Provost, F., Fawcett, T.: Data Science for Business. What You Need to Know About Data Mining and Data-Analytic Thinking. O'Reilly, Sebastopol (2013)
24. Samuelson, W.F., Marks, S.G.: Managerial Economics. PWE, Warsaw (1998). (in Polish)
25. Schmarzo, B.: Big Data. Understanding How Data Powers Big Businesses. Wiley, Hoboken (2013)
26. Seth, T., Chaudhary, V.: Big data in finance. In: Li, K.-C., Jiang, H., Yang, L.T., Cuzzocrea, A. (eds.) Big Data: Algorithms, Analytics, and Applications, 1 edn., Chap. 17. CRC Big Data Series, p. 29. Chapman & Hall
27. Big Data analytics. Peer research. Intel report, August 2012. https://www.intel.com/content/dam/www/public/us/en/documents/reports/data-insights-peer-research-report.pdf
28. Siemens: Using Big Data and Analytics to Design a Successful Future, February 2015. http://blogs.teradata.com/customers/author/tcset/page/4/
29. Stępniak, C.: Use of process performance indicators as part of knowledge management in organization. In: Proceedings of KM Conference 2016. Knowledge Management, Learning, Information Technology, Lizbona, Portugal, 20–25 June 2016, pp. 64–75 (2016)
30. Ziora, L.: The Sentiment Analysis as a Tool of Business Analytics in Contemporary Organizations. Studia Ekonomiczne. Zeszyty Naukowe Uniwersytetu Ekonomicznego w Katowicach no 281, Katowice, pp. 234–241 (2016)

Personalized Social Connectivity and Reputation

Monitoring Dynamics in Online Networks with Aigents Platform

Anton Kolonin(✉)

Aigents Group, Novosibirsk, Russia
akolonin@gmail.com

Abstract. The paper describes the approach and solution for personalized assessment of social interaction patterns in online social networks. The approach and the solution are used for temporal monitoring and study of social communication dynamics, as well as for the personal reputation management.

Keywords: Big data · Communication pattern · Personalization Social network

1 Purpose

Originally, this work has started with the intention to design and develop artificial psyche for robot operating with objects in the Internet, such as web pages and news feeds [1]. The intention was to make this robot capable of self-awareness in human environment, with motivations similar to those in other works in domain of robotics, such as [2, 3]. However, while modeling aspects of consciousness in social environments, an idea of capturing true contexts of social interactions between real people, extracting it from social networks, arose [4]. During that study, the importance of this area's development is that it provides more problems to solve and promises much more practical application than the original goal of robot construction became really apparent. Quantitative account for social interactions being translated into measurable reputation and even convertible into financial values had been imagined in literature [5]. Furthermore, different applications based on such account became widely popular as Web reputation systems [6]. Finally, so-called "Social Credit" system, already being implemented at scale of ¼ of the population in China, with all social interactions captured and translated into value, radically affecting life of every citizen [7]. Thus, we have proposed the instrument to give user a tool to study temporal social dynamics of their own, in context of interactions with other members of the social environment so that the person could benefit of it.

K. Arai et al. (Eds.): FICC 2018, AISC 886, pp. 369–377, 2019.
https://doi.org/10.1007/978-3-030-03402-3_25

2 Background/Significance

Initial design of system capable for comprehension of social context was based on earlier works on multi-agent representation of consciousness and intelligence defined as "ability to reach complex goals and complex environments using limited resources" [8]. Further, it have been extended to account for social context, with notion of "social evidence-based resource-constrained knowledge representation" as described in latest publications [9, 10] where more field experiments and literature study have been conducted to confirm the validity of the model and the design.

The confirmation of the model came out from earlier comprehensive phenomenological study [11] where every possible outcome of computable model has been backed up with recorded evidence in domain of social psychology. Moreover, in respect to specifics of interactions in social networks, it has been found that possibility of impact of manipulations by means of online social media can be huge [12]. Finally, the effect of such impact can be affecting not just behavioral patterns of a human or society, but their physical health also [13], which makes importance of work in this direction hardly overestimated.

3 Method

To implement the assessment and temporal monitoring of personal social dynamics in terms of reputation and social connectivity patters, we have used design developed earlier [14]. The design is briefly outlined in Fig. 1. In this design, using different social networks the person of study is connected to, online interactions are extracted, recorded and processed as it will be discussed further.

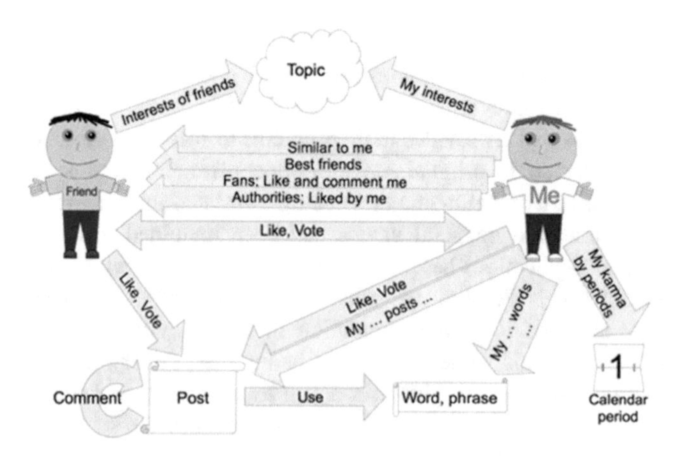

Fig. 1. Design of the system capable of extracting context and dynamics of online inter-personal interactions in social networks expressed in terms of posts, comments, "likes" (as in Facebook, Google+ and VKontakte) and "votes" (as in Steemit and Golos).

For the purpose of the study, five social networks were used. Three of them were private social networks with limited access to information via public "application programming interface" (API) – these networks were Facebook, Google+ and VKontakte. Two other social networks were community-owned ones based on block-chain technology with unlimited access to information via API - Steemit and Golos. The data extracted from social networks were user's posts, comments of other users to the posts and reactions directed toward these posts and comment. For reactions we used "likes" (called so in Facebook, Google+ and VKontakte) and "votes" (in Steemit and Golos), meaning that the acting person shared an opinion expressed in the post or comment. The texts extracted from original posts and comments were converted into feature vectors associated with the posts and comments [14]. The feature vectors were cleaned of frequently used words, such as articles and interjections, and were normalized on basis of relative popularity of the words, according to dictionaries and frequencies of usage for the given language, namely English or Russian. Furthermore, feature vectors associated with posts and comments were converted into feature vectors identifying particular users.

Given the feature vectors representing posts, comments and users, "natural classifications" for each user were derived to figure out domains of users' interests according to the approach described in [15]. From the authored comments and "likes" or "votes" on these posts and comments, the quantitative parameters were evaluated according to notions and definitions in earlier research [16] as discussed below. Finally, in the scope of this work, evaluation of these parameters has been bound to temporal axis within time intervals of different durations and correspondence of them to real-life observations has been studied qualitatively.

On the basis of feature vectors representing topics of interests, users, posts and comments, along with "likes" and "votes" given to them, multiple relationships were inferred. These relationships could indicate different sorts of connections between the primary user and their friends and posts and comments of both, assuming L_{ij} can be used to denote the number of "likes" or "votes" other user j gives to the posts and comments made by the given user i. Further, in the formulae C_{ij} is denoting the number of comments the user j makes in regard to posts/comments made by the user i.

My interests – Listed clusters of words identifying topics of groups of posts and comments associated with use of these words, either written by the user of study or presented in comments that the user was identifying as "liked" or "voted" for. It was inferred with adaptive K-means clustering (with no K number of clusters given in advance) where the lists of clusters and the lists of features identifying them were built incrementally and simultaneously to reach the optimal K number.

Interests of my friends – Listed clusters of words identifying topics of groups of posts and comments, similar to the above, while not representing the user of the study themselves, but rather their connections in social network.

Similar to me – Ranked other users according to *similarity* metric between each user and their connections calculated using feature vectors extracted from users' posts and comments with normalized overlap between the two vectors evaluated as mutual similarity measure between the users.

Best friends (and colleagues) – Ranked other users based on *friendship* metric treated as symmetric strength of positive relationship based on value of mutual "likes" or "votes" between two users $L_{ij} * (L_{ji} + C_{ji})$, normalized by the maximum number for given friend j of user i across all J users as follows.

$$B_{ij} = L_{ij} * (L_{ji} + C_{ji}) / Max_{j=1,J}(L_{ij} * (L_{ji} + C_{ji}))$$

Fans (and followers) – Ranked users by *adherence* metric as strength of asymmetric, or directed positive relationship, which could be evaluated through the amount of "likes" or "votes" and comments that the other user gave to the posts of the primary user. The metric was denominated by returned "likes"/"votes" and comments, so that a complete fan was one who paid more attention to primary user while the latter paid the least amount of attention to the fan.

$$F_{ij} = ((L_{ji} + C_{ji})/(1 + L_{ij} + C_{ij}))/Max_{j=1,J}((L_{ji} + C_{ji})/(1 + L_{ij} + C_{ij}))$$

Like and comment me – Simplified version of "fans" without denomination by mutuality of positive relationship.

$$F'_{ij} = (L_{ji} + C_{ji})/ \text{Max}_{j=1,J}(L_{ji} + C_{ji})$$

Authorities (and leaders) – Listed users according to *authority* metric, also known as "thought leader" or "opinion leader" or "the one that I listen to", which can be described as metric opposed to *adherence,* as asymmetric positive relationship. It corresponded to the amount of attention, i.e. the number of "likes" or "votes", paid by primary user to third ones, denominated by the amount of attention ("likes"/"votes" and comments) returned by them.

$$A_j = ((L_{ij} + C_{ij})/(1 + L_{ji} + C_{ji}))/Max_{j=1,J}((L_{ij} + C_{ij})/(1 + L_{ji} + C_{ji}))$$

Liked by me – simplified version of "authority" without denomination by mutuality of positive relationship.

$$A'_j = (L_{ij} + C_{ij})/Max_{j=1,J}(L_{ij} + C_{ij})$$

My karma by periods – Periods ranked by *karma* metric as evaluation of sum of "likes" and "votes" granted to the user within the given period t across a set of periods T, normalized to the best achievement across all periods. The notion of "karma" was used here as incremental value of reputation, earned by the user in the given time interval spanning over all periods involved in the analysis.

$$K_{it} = \sum\nolimits_{j,t} (L_{ij} + C_{ij}) / Max_{t=1,T} \sum\nolimits_{j,t} (L_{ij} + C_{ij})$$

My favorite words – Listed words from user-specific feature vector, limited to simple "single word" kind of feature, across all posts and comments of the primary user, ranked by relative frequency of use.

My words by periods – Did the same with features grouped by periods of time according to dates of posts and comments. This kind of profiling has turned to be useful when aligned with *karma* metric discussed above, so the two can be correlated as it will be discussed in the results later.

Words liked by me – A list of "words" ranked according to the amount of "likes" or "votes" and comments given by the primary user to the posts and comments containing them.

My best words – A list of "words" used by the subject of the study in his or her own posts ranked according to the amount of "likes" or "votes" and comments that these posts received on behalf of other users.

My posts liked and commented – Posts by the primary user ranked according to the amount of "likes" or "votes" and comments received from other users.

The definitions made above may raise questions regarding the justification of the terms "karma" and "reputation" and the respective formula in particular. The terms themselves are widely used in different applications while applicability of them is discussed in literature [5, 6] and our use of them seems quite compatible with that discussion. The impact of "likes"/"votes" of comments raise more questions. On the one hand, the act of giving "like"/"vote" to a post is an explicit conscious act that should be given more importance than attention paid implicitly by fact of commenting the post. On the other hand, the act of "liking/voting" is very simple and not resource consuming, compared to the amount of efforts the author of comment is investing into the act of commenting, so the latter should be valued as more important, from this perspective. At this point it was decided not to try to solve this problem and gave both kinds of attention equal rights.

4 Results

For the purpose of the study, for the participating user, the parameters described above were evaluated with Aigents computational platform and presented in raw JSON data and formatted HTML reports. Raw JSON data were used to render graphs of social connectivity using Aigents web service and presented to users with few different options, as shown in Fig. 2.

The circle in the middle was representing the user himself/herself, while the circles around were indicating their social connections. On the very top, with saturated yellow color, there were those with greatest reputation and social capital from users' perspective; in other words these users were getting a lot of likes/votes and comments without being reciprocated. On the left and on the right, with medium yellow saturation there were people with similar level or social capital/reputation with user,—so that exchanges with them with likes/votes and comments were mutual. At the bottom, with low yellow saturation, there were people who provide the user with many likes/votes and comments, while the user did not pay them back often enough.

Topmost people on the graph could be thought as "authorities", or "those who I pay most attention to but they don't notice me". Below, there were "opinion leaders", or "those whom I pay more attention than they do to me". In the middle there were "friends" and "colleagues" who the user communicated with symmetrically. Below,

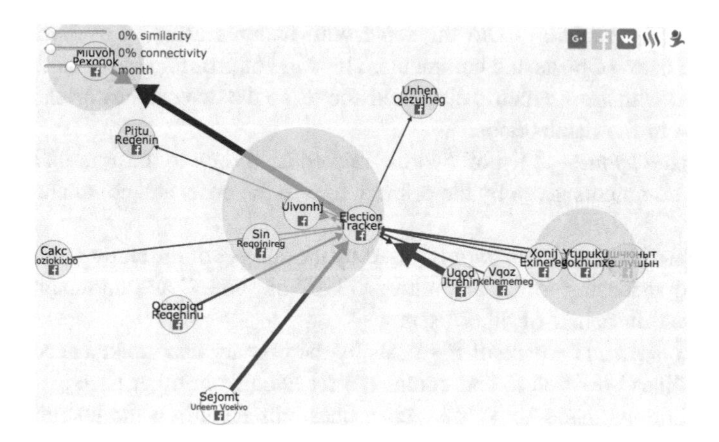

Fig. 2. Sample graph of social connectivity for each of social networks in the study with obfuscation of actual user names because of privacy concerns.

there were "followers" or "those who pay more attention to me than I pay to them". Finally, at the very bottom, there were "fans", who paid a lot of attention to the user but the user barely returned it to them. That is, what we call reputation or "karma" in context of this work was rendered as vertical position of a person and saturation of yellow color. Calculation of this value might be imprecise because only communications in relation to the studied users' personal news feeds were accounted and didn't involve communications in other groups or other users' feeds.

Communication connectivity between the user of study and others around was represented with blue arrows. Relative lengths of the arrow between the user and his/her peers indicated the ratio of incoming and outgoing actions, namely, likes/votes and comments. Respectively, for the people above, links from the user directed to them were relatively longer. In turn, links to people below were relatively shorter. Moreover, widths of these arrows indicated relative intensity of communications with the given person compared to other partners on the graph. To indicate similarity between the user of study and other people in respect to the topics of interest the size of the light blue halo, around each circle, was used.

Since there could be many connections and partners on the same graph for given social network within the same time period, control widgets were available. First, the user could restrict rendered social connections by level of similarity with himself or herself. By default, no filter was set but the user could change it to see only those similar as 25%, 50% or 75% or more. Also, the user could restrict rendered social connections by the level of intensity of communications with them. By default, no filter was set, but the user could change it to see only the people with connectivity as 25%, 50% or 75% or more. Additionally, in order to study different temporal periods of their online activity, users could specify the period that they wanted their graph to be rendered for—1 day, 1 week, 1 month, 1 quarter, 1 year or all years spent online.

Further, HTML reports produced by Aigents computational platform were available for use as part of Aigents web service as partially rendered on Figs. 3 and 4.

My words by periods

Period	Karma,%	Likes	Comments	Words
2017-08-12 - 2017-08-13	14	5	1	social anyone aligned capital reputation feature connections learn online kar
2017-08-05 - 2017-08-12	100	26	16	bica conference social layers cognitive society available data rather trust back front being credit china prop
2017-07-29 - 2017-08-05	24	7	3	whuffie currency science model karma
2017-07-22 - 2017-07-29	10	4	0	different widgets periods using sort exploring graph levels social moscow august connections
2017-07-15 - 2017-07-22	14	4	2	different exploring ones periods widgets using sort count being posts fixed change
2017-07-08 - 2017-07-15	21	8	1	social network mirror personal graph step
2017-07-01 - 2017-07-08	40	14	3	боты бесплатный like annual realize braces шаблоны бывают торговле submission пройдёт python architect
2017-06-24 - 2017-07-01	7	3	0	ijcai agi workshop aga social organizing august melbourne website registration know found
2017-06-17 - 2017-06-24	21	5	4	half pain sentences creates actually another hear certifica
2017-06-10 - 2017-06-17	31	12	1	fairness network behavior norms social fair paper plastic emergence community iee
2017-06-03 - 2017-06-10	76	29	3	revolution arrived helicopter uskova except handle google kik peter already robinson w
2017-05-27 - 2017-06-03	0	0	0	social networks many switch personal register may study good alpha-testing feedback center time
2017-05-20 - 2017-05-27	93	30	9	social networks many switch personal may register study good siberia alpha-testing feedback ce
2017-05-13 - 2017-05-20	36	15	0	like plant knowledge raise flowering otherwise plants children difficult estimated recognize vast name domain sharpen exist in

Fig. 3. Temporal dynamics of relative reputation increment ("karma") aligned with values of relative attractiveness of the words in users' posts online.

My friends by periods

Period	Karma,%	Likes	Comments	Friends
2017-08-12 - 2017-08-13	14	5	1	Ogvonic Mugqegjric Юшчюныт Цощлушын Sejomt Uneem Voekvo Unhen Qezujheg Cakc Noziokixbo Qoe Uqodunxok Änvojqogä
2017-08-05 - 2017-08-12	100	26	16	Uqod Jtrehin Miuvoh Pexoqok Qcaxpiqu Reqeninu Sejomt Uneem Voekvo
2017-07-29 - 2017-08-05	24	7	3	Sejomt Uneem Voekvo Miuvoh Pexoqok Pijtu Reqenin Qcaxpiqu Reqeninu Vqoz Mkehememeg Ukhop Ukh Jujtu Jokvoog Pajhuwu Lupukaxxin
2017-07-22 - 2017-07-29	10	4	0	Puxitu Qiuluh Rtun Sejomt Uneem Voekvo Zekij Bcjrin Xonij Exinereg
2017-07-15 - 2017-07-22	14	4	2	Uivonhj Sejomt Uneem Voekvo
2017-07-08 - 2017-07-15	21	8	1	Uivonhj Eqov Onu Урьчя Муроныя Sejomt Uneem Voekvo Giyhek Jpikneg Okiy Qui Xay Ntaun Pajhuwu Lupukaxxin
2017-07-01 - 2017-07-08	40	14	3	Gquxipik Rikiqonre Vqoz Mkehememeg Piytuoq Tunnopunn Ytokncureg Pudip Xenoqq Uqpunbek Uxayuq Qcaxpiqu Reqeninu Unhen Qezujheg Qipek Yton Ytukqio Xokk Suyez O Puyr Яжпюйавын Ырьнол Pijtu Reqenin Mugoq Burtukeg Sejomt Uneem Voekvo Oqonu Kebtregu
2017-06-24 - 2017-07-01	7	3	0	Piku Rfur Sejomt Uneem Voekvo Rogin Kajjoqq
2017-06-17 - 2017-06-24	21	5	4	Sejomt Uneem Voekvo Oqonu Kepunonre Uqodoc Gojnin Uqonu Mqehniregu Unujhujiu Raninu Jokvo Vqajytnog Unhen Qezujheg Uqod Jtrehin
2017-06-10 - 2017-06-17	31	12	1	Miokko xo Quyubo Xonij Exinereg Senuhtun Rip Урьчя Муроныя Pirtuiq Nireqjric Pukr Fuhjen Qcaxpiqu Reqeninu Vqoz Mkehememeg Supoj Joteen Mukr Sejomt Uneem Voekvo Jqugu Rekeqog
2017-06-03 - 2017-06-10	76	29	3	Ычуль Всурьнит Purjip Kuzinegiyt Pajhuwu Lupukaxxin Sejomt Uneem Voekvo

Fig. 4. Temporal dynamics of relative reputation increment ("karma") aligned with names of other users attracted by primary users' posts enough to act towards them by commenting on posing "likes" or "votes" (actual names of other users are obfuscated because of privacy concerns).

The value of positive reputation increment—called "karma" in context of [15] and this work—could be studied along the temporal axis aligned with the words attracting the most attention from other users acting towards the posts and comments containing these words, as it is shown in Fig. 3. In this example, it can be clearly seen that the maximum of 100% within the period is reached in the range between August 5th and August 15th in respect to posts on social topics of BICA-2017 conference ("bica", "conference", "social").

The track of temporal dynamics of reputation changes could be also studied being aligned with valuation of other users based on extent of their contribution to comments, "likes" and "votes" in respect to posts of the primary user. In this example, it is clearly seen that most of the attention is earned during the period between August 5th and August 15th, given by 1 active user, to a less extent by 3 other users and, finally, by 20 poorly active users.

Given three hundred users acquired across five social networks, thirty users have been informally questioned in respect to applicability of the graphs and charts presented above. Most of them have expressed positive opinion in respect to usefulness of the results obtained. However, it has been found that use of private social networks such as Facebook, Google+ and VKontakte provides less precise and more biased view given the fact only the limited amount of data is available via the public API of these online services. On the contrary, the results obtained with community-owned social networks with no limits on data access, such as Steemit and Golos have provided more reliable and useful results, corresponding to natural expectations.

5 Conclusions

The approach and application designed, developed and tested as described above, seem practically applicable and useful for the purpose of tracking personal social dynamics and reputation in the context of day-to-day interactions of a user in social networks. Moreover, it seems useful to have connectivity patterns supplied with expression of emotional value, starting both with the most simple positive and negative sentiment evaluation associated with values of connectivity and reputation changes. The other direction of possible improvement could be a more precise assessment of reputation itself so it could be evaluated in relation to the entire community and not just personal history,—so the true opinion leaders can be figured out and the users may be able to track their own reputation development compared to the former.

References

1. Kolonin, A.: Intelligent agent for web watching: belief system and architecture. In: Knowledge-Ontology-Theories (KONT-2015) Conference Proceedings, Novosibirsk, Russia, vol. 1, pp. 140–1491 (2015)
2. Haikonen, P.: Reflections of consciousness; the mirror test. In: AAAI Symposium, Washington, D.C. (2007). http://www.consciousness.it/cai/online_papers/haikonen.pdf
3. Takeno, J.: A robot succeeds in 100% mirror image cognition. Int. J. Smart Sens. Intell. Syst. **1**(4), 891–911 (2008)
4. Kolonin, A.: Studying human social environment and state with social network data. In: Cognitive Sciences, Genomics and Bioinformatics (CSGB) - Symposium Proceedings (2016). http://ieeexplore.ieee.org/document/7587680/
5. Doctorow, C.: Down and Out in the Magic Kingdom. Tor Books, US (2003). ISBN 0-7653-0436-8
6. Farmer, F., Glass, B.: Building Web Reputation Systems. O'Reilly, Yahoo Press, Sebastopol (2010)
7. Chin, J., Wong, G.: China's new tool for social control: a credit rating for everything. Wall Str. J. (2016). ISSN 0099-9660

8. Goertzel, B.: CogPrime: An Integrative Architecture for Embodied Artificial General Intelligence. Open Cog (2012). https://pdfs.semanticscholar.org/7832/2195fcc6f98b38fb8f7197a46cdac7e45522.pdf
9. Kolonin, A.: Computable cognitive model based on social evidence and restricted by resources: applications for personalized search and social media in multi-agent environments. In: International Conference on Biomedical Engineering and Computational Technologies (SIBIRCON), Novosibirsk, Russia (2015). http://ieeexplore.ieee.org/document/7361869/
10. Kolonin, A., Vityaev, E., Orlov, Y.: Cognitive architecture of collective intelligence based on social evidence. In: Proceedings of 7th Annual International Conference on Biologically Inspired Cognitive Architectures, BICA 2016, NY, USA, July 2016. http://www.sciencedirect.com/science/article/pii/S1877050916317239
11. Cialdini, R.: Influence: The Psychology of Persuasion (1984). ISBN 0-688-12816-5
12. Kramer, A., Guillory, J., Hancock, J.: Experimental evidence of massive-scale emotional contagion through social networks. PNAS **111**(24), 8788–8790 (2014)
13. Dhand, A., Luke, D., Lang, C., Lee, J.: Social networks and neurological illness. Nat. Rev. Neurol. **12**(10), 605–612 (2016). https://doi.org/10.1038/nrneurol.2016.119. Epub 2016
14. Kolonin, A.: Automatic text classification and property extraction. In: SIBIRCON/SibMedInfo Conference Proceedings, pp. 27–31 (2015). ISBN 987-1-4673-9109-2
15. Vityaev, E.: Unified formalization of «natural» classification, «natural» concepts, and consciousness as integrated information by Giulio Tononi. In: The Sixth International Conference on Biologically Inspired Cognitive Architectures, BICA 2015, 6–8 November 2015, Lyon, France. Elsevier. Procedia Comput. Sci. **71**, 169–177 (2015)
16. Kolonin, A.: Assessment of personal environments in social networks. In: 2017 Siberian Symposium on Data Science and Engineering (SSDSE), Novosibirsk, pp. 61–64 (2017). ISBN 978-1-5386-1592-8. http://www.ieee.org//conferences_events/conferences/conferencedetails/index.html?Conf_ID=41642

Keypoints and Codewords Selection for Efficient Bag-of-Features Representation

Veerapathirapillai Vinoharan[1(✉)] and Amirthalingam Ramanan[2]

[1] Computer Centre, University of Jaffna, Jaffna, Sri Lanka
vvinoharan@univ.jfn.ac.lk
[2] Department of Computer Science, University of Jaffna, Jaffna, Sri Lanka
a.ramanan@univ.jfn.ac.lk

Abstract. Bag-of-features (BoF) representation is one of the most popular image representations that is used in visual object classification, owing to its simplicity and good performance. However, the BoF representation always faces the difficulty of curse of dimensionality that leads to huge computational cost and increased storage requirement. To create a discriminative and compact BoF representation, it is desired to eliminate ambiguous features before the construction of visual codebook and to select the informative codewords from the constructed codebook. In this paper, we propose a two-staged approach to create a discriminative and compact BoF representation for object recognition. In the first step, we eliminate ambiguous patch-based descriptors using an entropy-based filtering approach to retain high-quality descriptors. In the subsequent step, we select the informative codewords based on statistical measures. We have tested the proposed technique on Xerox7, UIUC texture, PASCAL VOC 2007 and Caltech101 benchmark datasets. Testing results show that more training features and/or a high-dimensional codebook do not contribute significantly to increase the performance of classification but it increases the overall model complexity and computational cost. The proposed preprocessing step of descriptor selection increases the discriminative power of a codebook, whereas the post-processing step of codeword selection maintains the codebook to be more compact. The proposed framework would help to optimise BoF representation to be efficient with steady performance.

Keywords: Keypoint selection · Codebook · Codeword selection
Image representation · Bag-of-features

1 Introduction

The bag-of-features (BoF) approach [3, 15, 17, 18, 20] is a well known technique for representing the image content and has proved state-of-the-art performance in large scale evaluations. In the BoF approach, features are usually based on

K. Arai et al. (Eds.): FICC 2018, AISC 886, pp. 378–390, 2019.
https://doi.org/10.1007/978-3-030-03402-3_26

the utilisation of tokenising keypoint-based features, e.g., scale-invariant feature transform (SIFT) [2], to generate a codebook. The BoF representation of an image conveys the presence or absence of the information for each visual word in the image. In a BoF framework, the codebook plays a crucial role. An important issue of the codebook representation is its discriminative power and compactness. The size of a codebook controls the complexity of the codebook model and the discriminative power of a codebook determines the quality of the model. The number of features extracted from training images to construct a codebook and the dimensionality of a codebook causes two sets of problems: (1) the computational cost during the vector quantisation step is high and some of the detected features are not helpful for better classification; and (2) the model complexity is high that may overfit to the distribution of codewords in an image. The increase in the number of object categories, increases the computational cost and it makes the classification of histograms challenging due to its diverse range in object classes.

Most of the object recognition tasks that are reported in the literature have employed sufficiently large-sized codebook at the order of 1000 to 10000, typically resulting in hyper-dimensional and sparse histogram representations. The use of such large-sized codebook will in turn make each BoF vector to require huge storage space and the efficiency of computation in large scale datasets will yield to the well-known "curse-of-dimensionality". Therefore, the discriminative power and compactness of a codebook are important to control the complexity of the model. A straightforward way to create compact codebooks is to reduce the dimensionality, that will quickly weaken the discriminative power and degrade classification performance. Simply selecting most discriminative codewords or linearly combining the bins will not work well either [12]. In this regard we formulate and contribute the following:

- Choose unambiguous patch-based descriptors prior to the construction of a codebook in order to reduce the features causing false positives in object classification. In this regard we present an entropy-based filtering approach to eliminate ambiguous patch-based descriptors (e.g., SIFT).
- Select the best subset of codewords from an initially constructed codebook to enhance the discriminative power of the codebook and make it more compact. To achieve this we present an inter-category and intra-category confidences to select the informative codewords that generates a discriminative and compact codebook for the BoF representation.

The proposed method provides an effective way to improve the object categorisation performance when using the BoF model with very low dimensional representation.

The rest of this paper is structured as follows: Sect. 2 briefly describes the background needed for our work. Section 3 summarises related work that has been used to construct a discriminative and compact codebook for object recognition. Section 4 explains the proposed methodology in detail. Section 5 describes the testing results. Finally, Sect. 6 concludes this paper.

2 Background

2.1 Bag-of-Features Approach

The BoF approach is widely used in image scene classification [5] and object recognition tasks [11] in computer vision. The BoF-based object recognition systems fit into a general framework as summarised below:

(1) Detecting and describing of image patches from the training and testing image sets.
(2) Constructing a visual codebook by performing cluster analysis on the descriptors extracted from the training set. The codebook is the set of codewords.
(3) Mapping the extracted image patches from the training and testing image sets into a feature vector (i.e., BoF) by computing the frequency histograms with the codewords.
(4) Classifying the test images to predict which object category or categories to assign to the image.

There are two broad categories of codebook models: Global and object-specific codebooks. A global codebook is category independent but may suffer in its discriminative power. On the other hand a object-specific codebook may be too responsive to noise. Thus, the construction of a codebook plays a crucial role that affects the models' complexity.

2.2 Scale-Invariant Feature Transform

SIFT is a technique [2] that extracts distinctive features from gray-level images, by filtering images at various scales and patches of interest that have sharp intensity changes. SIFT descriptors is a 128 dimensional vector that can be used in the context of recognition and matching of the same scene or object observed under different viewing conditions. SIFT for colour image is also available [7].

2.3 Resource-Allocating Codebook

RAC [9] is a simple and extremely fast technique to construct visual codebooks using a one-pass setup that carves the feature space as fixed-size hyperspheres. RAC yields a better discriminative codebook with a drastic reduction in computational needs. Codebook constructed by RAC shows similar recognition performance to K-means method-based codebook with small variations [19]. RAC algorithm is summarised in the following steps:

Step 1: RAC starts by arbitrarily assigning a descriptor of an image as a first entry in the codebook to be the informative codeword.

Step 2: A subsequent descriptor is processed. The smallest distance to all entries in the present codebook is computed using Euclidean distance. If this distance exceeds the predefined hyper-parameter r of RAC:

- Then the current descriptor is recorded as an additional informative codeword that creates a new codeword in the codebook (i.e., the codebook in Step 1 is updated).
- Else no action is taken in respect of the processed descriptor.

Step 3: This process is continued until all or desired number of descriptors are processed only once.

2.4 Support Vector Machine

SVM [1] is a statistical learning method that has showed better performance in visual object classification problems. The objective of SVM learning is to find a hyperplane that maximises the inter-class margin of the training data. The data in the input space are projected into a high-dimensional feature space by kernel function. Multiclass classification in this paper is performed using SVMs of linear kernel trained with the one-versus-all (OVA) rule. OVA-SVMs learn to separate every category from the remaining object categories, and it allocate the category label of a test image having the highest response.

3 Related Work

In the literature of the BoF approach for object classification there exists several approaches that have focused on the discriminative power and/or compactness of codebooks.

In [14], the authors have proposed a two-step approach to map an initially constructed large codebook into a compact codebook with stable performance by maintaining its discriminative power in object recognition. Using an initial large codebook (K = 1000 of K-means), training images are represented using a mapping rule that maps the importance of each codeword within an image as visual-bits. These set of visual image bits forms a sparse representation of each codewords in respect of the object-specific training sets that is used for compression. This technique reduces the size of the codebook using binary representations of images and codewords, which enhances the efficiency of the coding while showing the discriminative power of the codebook. This is accomplished by the following two-step process: (1) encoding each image as 'bits' (i.e., the significant presence or absence of each codeword) and (2) removing the codeword that are not sufficiently activated in the images. Authors have tested their technique on four benchmark image sets: (1) MPEG7 CE Shape-1 Silhouette; (2) PASCAL VOC 2007; (3) UIUC texture; and (4) Xerox7. Authors' test results indicate that the approach slightly surpasses the codebooks learned by K-means by having just half the size of the initial codebook with stable performance.

In [13], the authors have proposed an unsupervised dimensionality reduction framework for discriminative and compact BoF representation. First, they construct the dissimilarity matrix between each pair of histograms and then perform multidimensional scale technique to obtain a small Euclidean embedding of the original BoF while preserving the inherent neighbourhood structure. Authors' experimental results show that a very small dimension is sufficient for learning tasks using BoF or spatial pyramid matching without losing the precision of the classification. It has been claimed that a compact representation of BoFs can improve the accuracy of the classification in relation to the traditional BoF approach. Authors have tested their technique on three benchmark image sets: subset of PASCAL VOC 2012, subset of Caltech101, Scene15 and INRIA holidays. Testing results show that the authors method has shown promising results for the image classification and retrieval tasks with very low-dimensional representation. In Caltech101 and PASCAL VOC 2012 datasets, the classification accuracies of the original 1000-dimensional and 2000-dimensional BoF vectors are 54.23% and 54.15% respectively, whereas the proposed approach in both cases achieves better accuracy about 58% for reduced dimensional BoF vectors of 100. The classification accuracies of the original 2100-dimensional and 4200-dimensional Spatial Pyramid Matching (SPM) vectors are 55.45% and 54.78% respectively, whereas the proposed approach in both cases achieves better accuracy about 57% for reduced dimensional BoF vectors of 100. In Scene15 dataset, the standard 2000-dimensional BoF vectors classification accuracy is 70.82%, and SPM 2100-dimensional is 72.2%. Better classification rate of 75.54% for 2000-dimensional BoF vectors was obtained with 35 dimensional BoF vectors.

In [16], the authors propose an iterative keypoint selection (IKS) technique to create discriminative BoF representation by selecting most appropriate keypoints which reduces the computational cost in constructing a codebook and leads to a better discriminative codebook. There are two steps involved in each IKS: (1) Representative keypoints are randomly selected or taken from cluster centriods; and (2) the distance between identified keypoints and the selected informative keypoints are calculated and if the distance is less than a predefined threshold then those keypoints are discarded. This process iterates until no unrepresentative keypoints are found. To execute the initial stage of IKS, two particular approaches are used: (1) Identifying informative keypoints based on random selection; and (2) using K-means algorithm to select K centroids as informative keypoints. The former approach requires larger computational cost, whereas the latter approach not only reduces the computational time but also provides increased classification rate. Experiments using the Caltech101, Caltech256 and PASCAL VOC 2007 datasets demonstrate that using keypoint selection to generate both BoF and spatial pyramid matching allows the SVM classifier to produce better classification results compared to the previous techniques that use without the keypoint selection method.

4 Methodology

A compact codebook has advantages in terms of computing efficiency and storage requirement. For example, when SVMs are used to classify feature histograms, the complexity of calculating the kernel matrix, storing the support vectors, and testing a new image are all proportional to the size of the codebook.

In this paper first, we extract local keypoints using SIFT algorithm from the training images. Before creating a relatively large original codebook, unambiguous keypoints are selected using an entropy-based filtering method to increase the discriminative power. Thereafter a codebook is constructed using RAC approach. Finally, indistinctive codewords are eliminated based on statistical measures to obtain a compact codebook. We assign each keypoint in images to the closest codeword and create a histogram representation for each image, which records how many times keypoints corresponding to the codeword occur in the image. We then apply SVM classification algorithm to these fixed-length feature vectors. The overall framework of the proposed method in this paper is illustrated in Fig. 1.

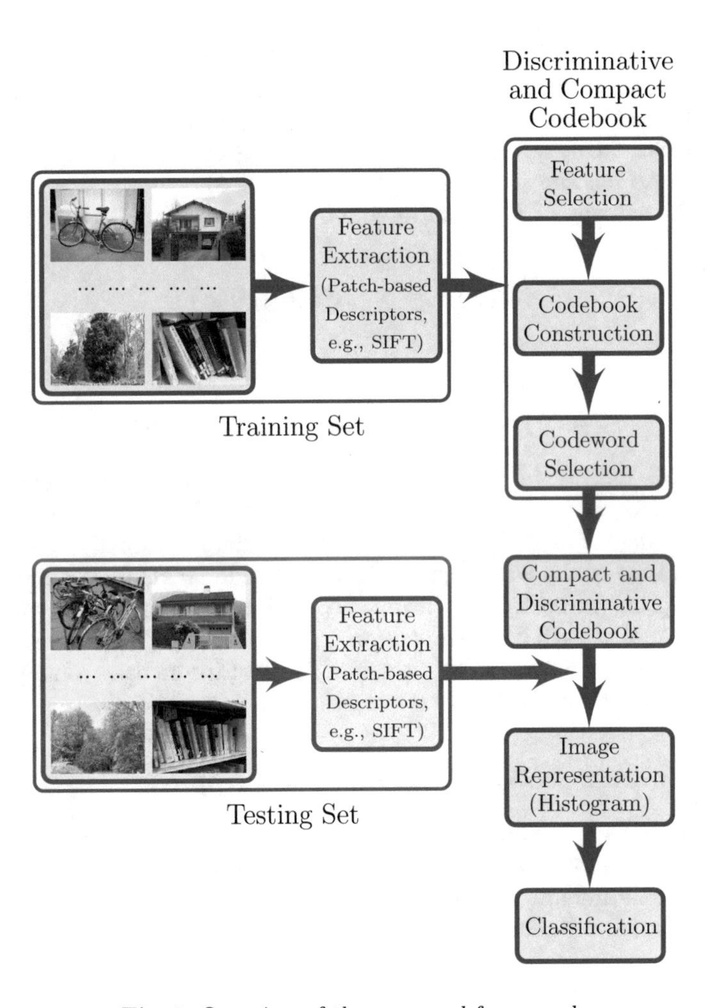

Fig. 1. Overview of the proposed framework.

4.1 Entropy-Based Filtering for Feature Selection

To create a codebook, set of keypoints that are detected in an image can be transformed to an intermediate representation. Some of the detected keypoints in the training images belonging to different object categories play a key role to make the intermediate representation more distinguishable. By selecting these important keypoints, not only the discriminative power of the codebook can be increased but also a compact codebook can be obtained.

The wide adoption of the visual codebook approach creates the impression that SIFT is a point feature. SIFT features best suitable for object detection are those with rich internal structure and associated with near-empty regions that are the main source of false positives: they tend to occur frequently and get easily matched against one another. We propose an entropy-based filtering approach to eliminate ambiguous SIFT descriptors in order to retain high-quality descriptors. This approach reduces the computational complexity of the clustering and increases the categorisation precision at the later stage of the BoF approach.

Let the SIFT descriptors $F = [f_1, f_2, \cdots, f_{128}]$ that are treated as 128 samples of discrete random variable in {0, 1, 2, ⋯, 255}.

Then the entropy of F is computed as,

$$E(F) = -\sum_{i=0}^{255} p_i(F) log_2 p_i(F)$$

where,

$$p_i(F) = \frac{|\{k|f_k = i\}|}{128}; k = 0, 1, 2, \cdots, 255.$$

The values for individual dimensions of SIFT feature follow a near exponential distribution, with small values dominating the whole distribution. A SIFT value has a range of [0, 255], but almost all the values are smaller than 128 that means the range of the value is not efficiently used. Therefore the dimension of SIFT descriptors are scaled logarithmically so that the distribution will be more uniform. Note that each SIFT dimension is an 8-bit integer, so the entropy has a range of [0, 8]. In our system, we discard SIFT descriptors based on a predefined threshold which varies for different dataset.

4.2 Codeword Selection Using Statistical Approach

A codebook is usually constructed by using a clustering algorithm. In such a codebook, the goal of codeword selection is to remove the redundancy and noise in a codebook. Elimination of indistinctive codewords not only reduces the overall computational complexity but also increases the categorisation precision.

(1) *Selecting Discriminative Codewords Across Categories:* Object categories owing to similar histogram distribution may increase ambiguity of the classification process. Inter-category confidence is calculated by analysing category distributions of the ith codeword. The inter-category confidence of the ith codeword $C_{inter,i}$ is defined as follows:

$$C_{inter,i} = \sum_{j=1}^{N} max\left(\frac{f_{ij}}{n_i} - \frac{1}{m_i}, 0\right)$$

where,
K is the size of the codebook.
N is the number of object categories in classification
f_{ij} is the number of training features in the ith codeword and j^{th} category, $i = 1, 2, \cdots, K.$, and $j = 1, 2, \cdots, N.$
n_i is the total number of features in the ith codeword
m_i is the number of object categories in the ith codeword

The value of inter-category confidence is zero, when all the features of a codeword show a single category or equal number of features from each category in the feature domain. The value of inter-category confidence is positive when the feature ratio of a codeword shows a single category dominating other categories in the feature domain. Since the codeword only exist in histograms of the category images, the histogram distribution differs from other categories, thus the codeword enhances the categorisation result. It has been noticed by using the elimination process of indistinctive codewords many codewords tend to disappear from homogeneous regions.
In this inter-category codeword selection, the codeword is selected based on the following criteria:

- $\widehat{C}_{inter} = 0$ having a single category in the feature domain,
  or
- $\widehat{C}_{inter} > 20^{th} Percentile_{1 \leq i \leq K}(C_{inter,i})$

(2) *Selecting consistent codewords within each categories:* Images of different categories may have similar histogram values of codewords that in turn will affect the classification based on the histogram. The variance of histogram value within a codeword among the same object category images is inversely proportional to the intra-category confidence.
A high variance histogram value of a codeword interrupts the classification process, i.e., it makes the classifier (e.g., SVM) difficult to classify visual object categories. Thus, low variance codewords at BoF histogram domain are stable to be classified. Based on this concept, we discard all codewords with the variance histogram value of a codeword smaller than the first quartile of C_{intra}.
The intra-category confidence of the ith codeword $C_{intra,i}$ is represented as follows:

$$C_{intra,i} = \frac{1}{\sum_{j=1}^{N} var(h_{ij})}$$

where,
h_{ij} is the ith codeword value of each image belonging to the jth category in the BoF histogram domain, $i = 1, 2, \cdots, K$, and $j = 1, 2, \cdots, N$.

(3) *Selecting informative codewords based on C_{inter} and C_{intra} confidences:* Both confidences, C_{inter} and C_{intra}, enhance the classification process individually, and complement each other at the same time. Therefore, the combined confidence of the ith codeword is shown as follows:

$$C_{com,i} = \alpha C_{inter,i} + \beta C_{intra,i}$$

where, α and β are constant values, $0 \leq \alpha, \beta \leq 1$.
Using the combined confidence, we select reliable codewords by a weighting parameter.

5 Testing Results

The proposed BoF representation scheme has been evaluated on Xerox7 [3], UIUCTex [4], PASCAL VOC 2007 [8] and Caltech101 [6] image sets that are summarised in the following:

5.1 Dataset

(1) *Xerox7:* It contains 1776 images from seven categories with different resolutions. The object poses in images are highly variable and there is a significant amount of background clutter making the classification task challenging.
(2) *UIUCTex:* It contains twenty five texture classes with 40 images per class with resolution of 640×480. The texture surfaces are of albedo variations and images have significant viewpoint changes, uncontrolled illumination, arbitrary rotations, and scale differences within each texture category.
(3) *PASCAL VOC 2007:* Is used immensely used in large scale evaluation of object classification tasks. The dataset consists a total of 9,963 images containing 24,640 annotated objects split into training, validation, and testing sets labelled with twenty object classes.
(4) *Caltech101:* It consists of a total of 9,146 images, split between 101 different object categories, as well as an additional background/clutter category. Each object category contains between 31 and 800 images. Common and popular categories such as faces tend to have a larger number of images than others.

5.2 Experimental Setup

For the image sets: Xerox7, UIUCTex, and Caltech101 we used 70% for training and 30% for testing from each class. The classification for PASCAL VOC 2007 was performed on each of the 20 classes by training the classifiers on the provided ‘trainval’ set and evaluating on the testing set. SIFT descriptors were

Table 1. Mean average precision rate with codebook size and the number of descriptors extracted from training images for the entropy-based filtering method

BoF approach	Dataset	#Descrips	CBSize	mAP
Traditional	Xerox7	4,046,578	987	67.64
	UIUCTex	4,543,590	1032	93.40
	PASCAL VOC 2007	1,760,400	1049	67.60
	Caltech101	5,659,137	925	77.52
Entropy-based feature selection	Xerox7	2,295,071	659	**69.07**
	UIUCTex	2,097,558	617	**95.04**
	PASCAL VOC 2007	1,286,833	918	**68.58**
	Caltech101	3,602,142	753	**78.36**

extracted and global codebook was constructed by clustering the descriptors of the training images using the RAC algorithm with $r = 0.85$ for Xerox7, $r = 0.825$ for UIUCTex, $r = 0.845$ for PASCAL VOC 2007, and $r = 0.86$ for Caltech101 dataset. For each dataset, linear OVA-SVMs were employed in classification and the reported classification rates are of mean average precision (mAP) [10].

5.3 Results

(1) *Entropy-based Filtering for Feature Selection:* Interestingly, On average about 57%, 46%, 73%, and 64% of training keypoints were found to have entropy value $E(F) > 4.1$, 4.4, 3.6, and 3.8 that are selected from the initially extracted descriptor set in Xerox7, UIUCTex, PASCAL VOC 2007, and Caltech101 datasets, respectively in order to construct codebook for each classification tasks. This selection of reduced number of keypoints enhances the discriminative power of the codebook. We compare the proposed preprocessing technique with the traditional BoF approach. It has been noted that the filtering technique eliminates around 40% of the descriptors that outperforms traditional BoF approach in all datasets. The performance comparison of BoF approach prior to applying entropy-based filter vs after applying the technique is presented in Table 1.

(2) *Codeword Selection Using Statistical Approach:* On average 80% of the codewords were selected using inter-category confidence, intra-category confidence, and combined confidence C_{com} with $\alpha = 0.4$ and $\beta = 0.6$ to be the best for the initially constructed codebook. This step results in more discriminative and compact codebook. The performance of BoF approach with codeword selection using inter-category and intra-category confidences are shown in Table 2.

In Table 2, we compare the proposed post-processing technique (i.e., codeword selection method) with and without the use of preprocessing techniques (i.e., entropy-based filtering method). It yields on average 20% reduction in the initially constructed codebook in all datasets tested here.

Table 2. Mean average precision rate with codebook size using categorical confidences obtained by the proposed method

Approach	Dataset	Before FS		After FS	
		CB	mAP	CB	mAP
Traditional	Xerox7	987	67.64	659	69.07
	UIUCTex	1032	93.40	617	95.04
	PASCAL VOC 2007	1049	67.60	918	68.58
	Caltech101	958	74.71	753	**78.36**
Inter-category confidence	Xerox7	803	65.77	546	70.11
	UIUCTex	835	93.70	496	**95.84**
	PASCAL VOC 2007	847	68.02	744	67.73
	Caltech101	742	75.34	603	76.23
Intra-category confidence	Xerox7	740	67.63	**494**	69.03
	UIUCTex	774	93.78	**463**	93.95
	PASCAL VOC 2007	787	67.80	**688**	**68.70**
	Caltech101	694	75.53	**565**	77.32
Combined confidence	Xerox7	902	66.23	598	**70.32**
	UIUCTex	842	93.73	518	95.60
	PASCAL VOC 2007	953	67.58	818	67.30
	Caltech101	850	75.16	697	76.41

Finally, our proposed technique, having preprocessing and post-processing approaches, yield on average 45% of reduction in the initially constructed codebook while maintaining comparable performance with the traditional approach.

6 Conclusion

BoF approach is an image representation scheme used in patch-based object categorisation. In such classification system, the major role of a codebook is to provide a way to map the low level features into a fixed length feature vector in histogram domain to which any classifiers can be directly applied.

Many of the large numbers of keypoints detected from images are actually unhelpful for recognition and the computational cost required for the vector quantisation step for the generation of BoF vectors is very high. A larger sized codebook increases the computational needs in terms of memory requirement for generating the histogram of each image which is proportional to the size of the codebook. The high dimensional image representation could make many machine learning algorithms which become inefficient and unreliable or even a breakdown.

The central idea of the proposed algorithm in this paper is to select representative keypoints and select informative codewords so that the cluster structure of the image database can be best respected. The proposed method provides an effective way to reduce the BoF representation to low-dimension while maintaining the BoF model to be efficient with stable performance.

References

1. Cortes, C., Vapnik, V.: Support-vector networks. Mach. Learn. **20**(3), 273–297 (1995)
2. Lowe, D.: Distinctive image features from scale-invariant keypoints. Int. J. Comput. Vis. **60**, 91–110 (2004)
3. Csurka, C., Dance, R., Fan, L., Willamowski, J., Bray, C.: Visual categorization with bags of keypoints. In: Workshop on Statistical Learning in Computer Vision, ECCV 2004, pp. 1–22 (2004)
4. Lazebnik, S., Schmid, C., Ponce, J.: A sparse texture representation using affine-invariant regions. Pattern Anal. Mach. Intell. **27**(8), 1265–1278 (2005)
5. Fei-Fei, L., Perona, P.: A Bayesian hierarchical model for learning natural scene categories. In: Proceedings of the IEEE Conference on Computer Vision and Pattern Recognition, vol. 2, pp. 524–531 (2005)
6. Fei-Fei, L., Fergus, R., Perona, P.: Learning generative visual models from few training examples: an incremental Bayesian approach tested on 101 object categories. Int. J. Comput. Vis. Image Underst. **106**(1), 59–70 (2007)
7. van de Sande, K.E.A., Gevers, T., Snoek, C.G.M.: Color descriptors for object category recognition. In: Proceedings of the Conference on Colour in Graphics, Imaging, and Vision, vol. 2008(1), pp. 378–381 (2008)
8. Everingham, M., Van Gool, L., Williams, C.K., Winn, J., Zisserman, A.: The PASCAL visual object classes (VOC) challenge. Int. J. Comput. Vis. **88**(2), 303–338 (2010)
9. Ramanan, A., Niranjan, M.: A one-pass resource-allocating codebook for patch-based visual object recognition. In: Proceedings of the IEEE International Workshop on Machine Learning for Signal Processing, pp. 35–40 (2010)
10. Brodersen, K.H., Ong, C.S., Stephan, K.E., Buhmann, J.M.: The binormal assumption on precision-recall curves. In: Proceedings of the International Conference on Pattern Recognition, pp. 4263–4266 (2010)
11. Ramanan, A., Niranjan, M.: A review of codebook models in patch-based visual object recognition. J. Signal Process. Syst. **68**(3), 333–352 (2012)
12. Wang, L., Zhou, L., Shen, C., Liu, L., Liu, H.: A hierarchical word-merging algorithm with class separability measure. IEEE Trans. Pattern Anal. Mach. Intell. **36**(3), 417–435 (2014)
13. Cui, J., Cui, M., Xiao, B., Li, G.: Compact and discriminative representation of bag-of-features. Neurocomputing **169**, 55–67 (2015)
14. Kirishanthy, T., Ramanan, A.: Creating compact and discriminative visual vocabularies using visual bits. In: Proceedings of the IEEE Digital Image Computing: Techniques and Applications, pp. 258–263 (2015)
15. Wang, C., Huang, K.: How to use bag-of-words model better for image classification. Image Vis. Comput. **38**, 65–74 (2015)
16. Lin, W.C., Tsai, C.F., Chen, Z.Y., Ke, S.W.: Keypoint selection for efficient bag-of-words feature generation and effective image classification. Inf. Sci. **329**, 33–51 (2016)

17. Peng, X., Wang, L., Wang, X., Qiao, Y.: Bag of visual words and fusion methods for action recognition: comprehensive study and good practice. Comput. Vis. Image Underst. **150**, 109–125 (2016)
18. Amato, G., Falchi, F., Gennaro, C.: On reducing the number of visual words in the bag-of-features representation. In: Computing Research Repository, pp. 657–662 (2016)
19. Vinoharan, V., Ramanan, A.: Are large scale training images or discriminative features important for codebook construction? In: Proceedings of the 5th International Conference on Pattern Recognition Applications and Methods, vol. 1, pp. 193–198 (2016)
20. Nasirahmadi, A., Ashtiani, S.H.M.: Bag-of-feature model for sweet and bitter almond classification. Biosyst. Eng. **156**, 51–60 (2017)

Group Trip Recommendation Systems

Hua-Hong Huang[1(✉)], Sheng-Min Chiu[1], Yi-Chung Chen[2], and Chiang Lee[1]

[1] Department of Computer, Science and Information Engineering, National Cheng Kung University, Tainan, Taiwan
p76034347@mail.ncku.edu.tw, samchiu951010@gmail.com, leec@imus.csie.ncku.edu.tw

[2] Department of Industry Engineering and Management, National Yunlin University of Science and Technology, Yunlin, Taiwan
mitsukoshi901@gmail.com

Abstract. Many of the most popular tourist attraction recommendation systems use the personal profiles of users from social networks. However, most of these services focus on individuals, rather than group activities with friends and family, despite the fact that check-in data includes accompanying members. In this study, we developed a recommendation system specifically for groups of users. Experiments demonstrate the efficacy of the proposed algorithm in making group recommendations based on the Foursquare dataset, with performance exceeding that of baseline methods.

Keywords: Group recommendation · Recommendation system
Foursquare

1 Introduction

Recommendation systems can be found on search engines, such as Baidu, webpage advertisements, and product recommendations on Amazon. The most popular systems are based on user profiles found on social networks [3]. For example, users who frequent galleries and never visit shopping malls receive recommendations of galleries rather than malls before visiting a new city. However, most of these services focus on individuals, rather than group activities with friends and family. Furthermore, the habits and preferences of groups differ from those of individuals.

Existing group recommendation systems can be divided into two types: (1) memory-based; and (2) model-based. Memory-based methods can be further divided into preference aggregation and score aggregation. Preference aggregation [10] involves merging the profiles of individual members to produce a group profile from which to make recommendations. Score aggregation [11] involves making recommendations for the individuals in the group and then combining the results to obtain optimal recommendations. Unfortunately, these methods do not consider internal relationships within the group (e.g., the degree of familiarity or ages of members), which tends to undermine accuracy. Model-based approaches [4] make assumptions regarding the interactions among users; however, the recommendations favor the

K. Arai et al. (Eds.): FICC 2018, AISC 886, pp. 391–412, 2019.
https://doi.org/10.1007/978-3-030-03402-3_27

preferences of users with greater weights and disregard less vocal members. No existing recommendation system combines group recommendations with user ratings, despite the fact that check-in data on existing social websites includes accompanying members, as shown in Fig. 1. The use of group check-in data could greatly enhance the accuracy of recommendations by taking into account group dynamics (i.e., the preferences of users when operating as a group).

Ke Hsin Wang with Ying Ching Cheng — at Central Park.

Fig. 1. Example of figure caption on social media website.

In this study, we developed a novel approach to activity recommendations, referred to as Hybrid Method with User Ratings (HMUR). The proposed scheme integrates existing memory-based and model-based approaches as well as the profiles of group members and check-in data in the formulation of group recommendations, the results of which are adjusted according to user ratings. We adopt a memory-based approach to the analysis of group check-in data on social networks to gain group preferences. We also use a model-based approach in examining the profiles of group members in order to identify the relationships among group members. The proposed method is meant to compensate for the shortcomings of the memory-based approach, which does not consider the relationships among group members, as well as those of the model-based approach, which considers only the preferences of opinion leaders.

The core algorithm of the proposed method overcomes the limits of conventional recommendation systems when applied to group situations, as well as the following four shortcomings:

(1) *Data sparsity*: Recommendation systems offer a wide selection of tourist attractions; however, few users check in at all of them. This means that each user provides only a small amount of check-in data, and the number of check-ins at most tourist attractions is zero. This situation causes the results of the cosine similarity in conventional recommendation systems to approach zero, which tends to make analysis difficult [12]. As shown in Tables 1 and 2, we categorized data pertaining to tourist attractions beforehand in order to simplify analysis.
(2) *Limited recommendation results*: Particular groups of friends tend to visit only a limited number of particular places [14]. This makes the activity profiles of group members somewhat similar. Thus, recommendations based on the experience of members in the same group tend to result in redundancies. We therefore proposed the HMUR method to cluster the locations previously visited by group members according to the categories identified in their check-in data. Groups in a particular cluster tend to have the same category preferences. Table 2 presents an instance in which the check-in data of five groups (g_1, g_2, g_3, g_4, and g_5) is divided up according to category. As shown in Fig. 2, the groups can be divided into two

Table 1. Group check-in data

Group	Location	L_1	L_2	L_3	L_4	L_5
	Category	Cat_1	Cat_1	Cat_1	Cat_2	Cat_2
g_1		1	1	1	0	0
g_2		1	1	1	0	1
g_3		1	2	1	0	0
g_4		0	0	1	2	1
g_5		0	1	0	1	2

Table 2. Categorized group check-in data

Group/Category	Cat_1	Cat_2
g_1	3	0
g_2	3	1
g_3	4	0
g_4	1	4
g_5	1	3

clusters based on similarities in check-in data in order to overcome the limited number of recommendations. As shown in Fig. 2, the similarity between g_1 and g_2 means that the experiences of g_2 can be used to make recommendations to g_1.

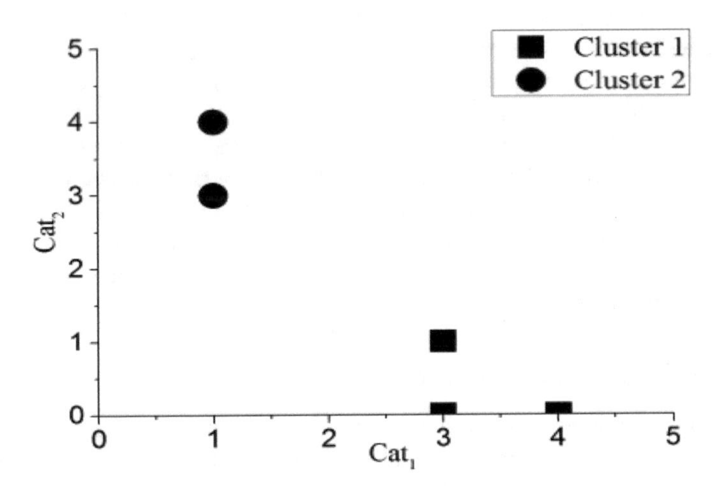

Fig. 2. Example of clustering.

(3) *Check-in activity*: Conventional recommendation systems may lack sufficient check-in data for analysis (due to the cold-start problem) or the unwillingness of users to check in (due to privacy concerns). Thus, we devised an algorithm for the

analysis of group composition in order to facilitate the formulation of recommendations to groups with similar members but insufficient check-in data. For example, a group of senior citizens with no check-in data in the system could be provided recommendations based on the check-in data of seniors in other groups.

(4) *False check-in times and ambiguity in ratings*: Most conventional recommendation systems use the number of check-ins or ratings to derive personal preferences. However, the number of check-ins can be skewed through marketing ploys, which have no bearing on the actual quality of the establishment.

(5) *The ambiguity of ratings is another difficulty*: In cases where the rating on a social network ranges from 1 to 10, the rating given by one user may fall between 6 and 10, while the rating given by another user is between 1 and 5. We developed a novel algorithm to adjust the influence of the check-in times and ratings of each user to enhance the accuracy of recommendations.

The remainder of this paper is organized as follows. Section 2 presents a review of related work. Section 3 presents a definition of the problem. Section 4 introduces the framework of the proposed system as well as details related to the algorithms. Section 5 presents experiment results. Section 5 provides conclusions and directions for future work.

2 Related Work

2.1 Collaborative Filtering Recommendation Systems

Collaborative filtering recommendation systems employ users with similar characteristics as guidelines. The techniques used in this type of system can be divided into two categories: memory-based and model-based [6]. Memory-based systems make recommendations based on users with similar preferences in a database [5]. Unfortunately, the need to scan all of the users in the database imposes a heavy computational burden [11]. Model-based systems employ the clustering of users, such that only a specific group rather than an entire database must be scanned, thereby resolving the problem of scalability [13]. However, these systems are prone to the cold-start problem, wherein new users cannot be categorized within any cluster, thereby preventing the system from making recommendations.

2.2 Content-Based Recommendation Systems

Content-based recommendation systems refer to user profiles as well as item features in formulating recommendations. Rather than referring to the preferences of other users, the system recommends items exhibiting characteristics similar to those indicated by a user's historical preferences. This approach is especially effective for users with particular preferences, because only a small number of potential candidates are similar to those users in the database. The recommendation process can be divided into two steps. The first step involves extracting information relevant to the user, including explicit data (collected from users) as well as implicit data (web browsing records or a history of interactions with the recommendation system).

The second step involves analyzing the data extracted in the first step. Unfortunately, this type of recommendation system is also subject to the cold-start problem, because recommendations are based entirely on the historical preferences of users. The system is unable to make recommendations to new users, because the database has no record for use as a reference.

2.3 Knowledge-Based Recommendation Systems

Knowledge-based recommendation systems are based on user preferences and item characteristics [1, 8]. The characterization of users and items is based on domain-specific information provided by domain experts or inferred from the available attributes. This type of system is able to recommend products based on a limited amount of historical data, thereby resolving the cold-start problem.

2.4 Group Recommendation Systems

Group recommendations systems can be divided into two categories: memory-based and model-based. The memory-based approach can be further divided according to strategy into the profile aggregation and recommendation aggregation. The profile aggregation strategy [10] is an attempt to summarize different users within a single profile (as a group) through the aggregation of individual preferences, whereupon recommendations are made using a personal recommendation system. In contrast, recommendation aggregation [11] involves formulating recommendations for each individual user and then combining them as a final list. However, these strategies disregard the nature of the interactions among group members. This indiscriminate aggregation of data can result in deviations when the preferences of one for an item exceed the sum of other users.

Model-based strategies consider the nature of interactions among members by modeling the generative process of groups. Gartrell *et al.* [4] assumed that when selecting items, the final decision is influenced by the member with the most pronounced social activity; however, this assumption does not necessarily bear out in the real world.

2.5 POI Recommendation Systems

Point-of-interest (POI) recommendation systems play an important role in LBSNs, by helping users to explore attractive locations, while helping social network service providers design location-aware advertisements for point-of-interest promotions. Ying *et al.* [14] used check-in data to produce recommendations based on the preferences and popularity of POIs. Levandoski *et al.* [7] formulates recommendations by analyzing check-in history and geographical factors. These studies are similar to the current study insofar as they deal with POI recommendations; however, differences in the research objectives mean that their focus is on different factors. Our goal was to recommend a set of POIs to a group rather than an individual; i.e., we did not consider check-in data at the individual level. Instead, we analyzed the check-in data of all accompanying members.

3 Algorithm

In this section, we outline the framework of the proposed algorithm: Hybrid Method with User Ratings (HMUR).

3.1 System Framework

This paper presents a novel scheme that formulates a set of POIs while taking into account the relationships among group members and historical check-in data. As shown in Fig. 3, the proposed system comprises two stages. The first stage is offline processing: (1) collection of data from Foursquare; (2) offline processing in three stages: (i) analysis of group profiles; (ii) clustering based on group check-in data; and (iii) adjustment of ratings. (3) The results are then used in various places for the second stage. The second stage is an online query. This stage requires a set of members (group g) and a number n of POIs. After receiving a request from query group g, the proposed HMUR algorithm identifies the top-n POIs for g. This is achieved by first calculating the degree of similarity between g and each cluster formulated in the offline stage using the Bit Similarity Checking algorithm. The algorithm then calculates scores for all POIs in the candidate clusters. Finally, the n candidate POIs with the highest scores are selected as the top-n POIs.

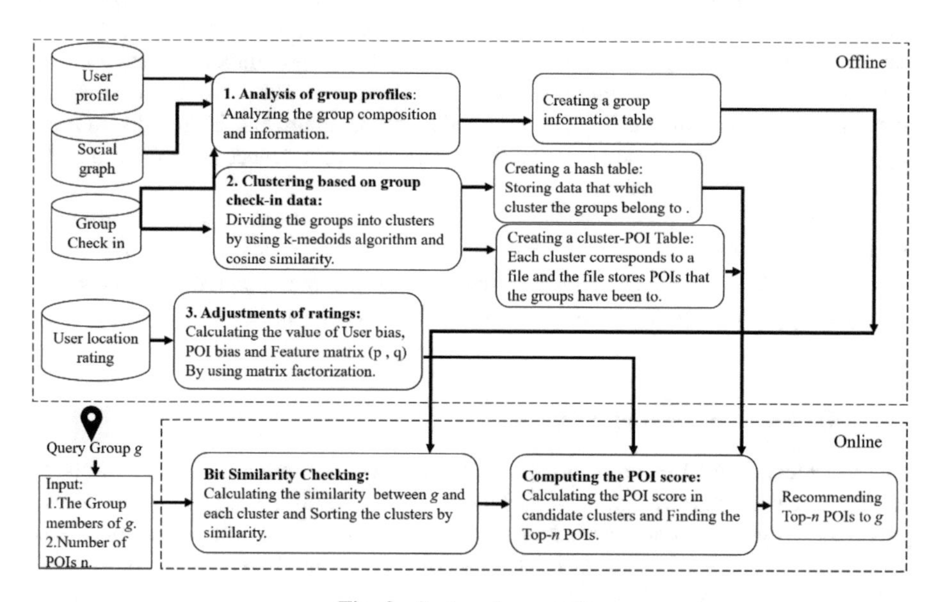

Fig. 3. System framework.

3.2 Offline Processing

Offline processing involves the analysis of group profiles, clustering based on group check-in data, and the adjustment of ratings.

Group profile analysis: The group check-in data provided by LBSNs includes the check-in time, the name of group members and the name of the POI; however, it lacks information pertaining to the group profile. Group profile information can easily affect the choice of travel destination. For example, groups of elderly individuals would likely enjoy a visit to POIs such as scenic areas or historical sites. Group profiles include discrete data, such as the gender and educational level of members as well as continuous data, such as the age of members (0–80 years) and income (0–100 k dollars per annum). Different methods are used to process discrete and continuous data; therefore, we processed each type of data separately and then created an information table as described below. For example, the members of group $g1$ are represented as MG1 = $\{u_1, u_3, u_8, u_9, u_{10}\}$. We have two types of discrete data (gender and education) and one type of continuous data (age) in the group profile shown in Table 3.

Table 3. Group profile of G1

User	Type	Gender	Education	Age
	Data type	Discrete data	Discrete data	Continuous data
u_1		Male	Junior high school	16
u_3		Female	College	20
u_8		Male	Junior high school	16
u_9		Female	Senior high school	19
u_{10}		Female	College	22

Discrete data, including the gender and educational level of members is shown in Table 4. The gender values for g_1 are 0.4 and 0.6, which means 40% of the members are male and 60% are female. By the same token, the 0.4 score for "Junior high school" means that 40% of members in g_1 have completed junior high school. Assume that the discrete data vector of group g_i has m dimensions, then the vector can be expressed as $[GP_{dis,i,1}, GP_{dis,i,2}, GP_{dis,i,3} \ldots GP_{dis,i,m}]$, where GP indicates the group profile. In Table 3, the discrete data vector of g_1 is [0.4, 0.6, 0, 0.4, 0.2, 0.4], where $GP_{dis,1,2} = 0.6$. At 16 on the x-axis, g_1 has 2 members, which means that g_1 includes 2 people aged 16; i.e., $GP_{cont.1.16} = 2$. At 20 on the x-axis, g_1 has 1 member, which means that g_1 includes 1 people aged 20; i.e., $GP_{cont.1.20} = 1$. Following the analysis of group profiles, the discrete and continuous data vectors can be stored in a database for use in online processing.

Table 4. Discrete data in group profile of G1

Data	Name of category	g_1
Gender	Male	0.4
	Female	0.6
Education	Elementary School	0
	Junior high school	0.4
	Senior high school	0.2
	College	0.4

Clustering based on group check-in data: In this section, we assemble a set of groups into clusters using the k-medoids algorithm based on group check-in data. The process can be divided into three main steps: (1) categorizing check-in data beforehand; (2) calculating similarity between groups; and (3) forming clusters of groups using the k-medoids algorithm.

As mentioned in Sect. 1, researchers must deal with the problem of data sparsity. Recommendation systems offer a wide selection of POIs; however, few groups check in at every site recommended to them. This means that each group provides only a small amount of check-in data; i.e., most of the POIs receive zero check-ins. In this study, we began by categorizing the check-in data within pre-defined categories. A simple example is shown in Tables 5 and 6. Table 6 presents five POIs, among which p_1, p_2, and p_3 belong to category Cat1, and p_4 and p_5 belong to category Cat2. g_1 checked in at p_1, p_2, and p_3 once; i.e., g_1 tallied one check-in for category Cat_1, as shown in Table 6.

Table 5. Example of group check-in Data

Group	POI	p_1	p_2	p_3	p_4	p_5
	Category	Cat_1	Cat_1	Cat_2	Cat_2	Cat_2
g_1		1	0	0	0	1
g_2		0	1	0	1	2
g_3		1	2	1	0	2

Second, we calculate similarity among the categories of group check-in data using cosine similarity [8], as follows:

$$Sim(g_i, g_j) = \cos(\vec{u}, \vec{v}) = \frac{\sum_i^d u_i v_j}{\sqrt{\sum_i^d u_i^2}\sqrt{\sum_i^d v_i^2}}, \tag{1}$$

Where, g_i and g_j denote the ith and jth groups in the social network, whereas u_i and v_i denote categorized group check-in data associated with g_i and g_j.

Cosine similarity is more suitable than Euclidean distance for the calculation of similarity. Consider the example of the three groups (g_1, g_2, and g_3) in Table 4. We select g_1 as the target and then calculate the similarity between g_1 and g_2 as well as between g_1 and g_3 using two different calculations of similarity.

Table 6. Example of categorized group check-in data

Group/Category	Cat_1	Cat_2
g_1	1	1
g_2	1	3
g_3	3	3

The calculation of Euclidean distance is as follows: If $p = (p_1, p_2, \ldots, p_n)$ and $q = (q_1, q_2, \ldots, q_n)$ are two points in Euclidean n-space, then the distance from p to q, or from q to p is:

$$\sqrt{(q_1 + p_1)^2 + \ldots + (q_n - p_n)^2} = \sqrt{\sum_{i=1}^{n} (q_i - p_i)^2}. \quad (2)$$

First, we calculate the similarity between g_1 and g_2 using Euclidean distance as $ed((1-1)^2 + (1-3)^2) = 2$. The similarity between g_1 and g_2 calculated using cosine similarity is represented as $Sim(g_1, g_2) \doteqdot 0.894$. Similarly, we calculate the similarity between g_1 and g_3. The similarity between g_1 and g_3 using Euclidean distance is represented as $ed(g_1, g_2) = \sqrt{8}$. The similarity between g_1 and g_2 according to cosine similarity is represented as $Sim(g_1, g_3) \doteqdot 1$.

The above results obtained using Euclidean distance indicates that the similarity between g_1 and g_2 is greater than the similarity between g_1 and g_3. In other words, g_1 is closer to g_2 than to g_3. However, the results obtained using cosine similarity indicate that the similarity between g_1 and g_2 is less than the similarity between g_1 and g_3. Obviously, g_1 is closer to g_3 than to g_2, based on the fact that the proportions of categorized group check-in data of g_1 are the same as g_3. Clearly, it is preferable to use cosine similarity for the calculation of similarity.

Adjustment of ratings: Some users in social networks give an item a higher rating than would other users. This also applies to POIs. For example, Gary is a critical user who tends to rate 0.8 stars lower than the average, whereas Neal is an optimistic user who tends to rate 0.9 stars higher than the average. Thus, an average rating of 3.0 stars for all POIs would imply different interpretations by Gary and Neal. It is customary to adjust the data in order to account for these effects. Furthermore, the adjustments can be encapsulated within the baseline estimates, and denote μ as the overall average rating. A baseline estimate for an unknown rating rum accounting for the user and POI effects is denoted as bum:

$$b_{um} = u + b_u + b_m \quad (3)$$

Parameters b_u and b_m respectively indicate the observed deviations from the average in user u and POI m. Suppose that we want a baseline estimate for the rating of the POI Tokyo Tower provided by user Gary. Let us assume the average rating over all POIs, μ, is 3.0 stars, and that the Tokyo Tower tends to be rated 0.5 stars above the average POI. As Gary is a critical user who tends to rate items 0.6 stars below the average, the baseline rating estimate for the Tokyo Tower by Gary is 2.9 stars; i.e., $3.0 - 0.6 + 0.5 = 2.9$.

Now we examine the core concept of baseline matrix factorization. Every user in the social network has their own habits; i.e., some like POIs in the food category, whereas others prefer POIs in the shopping category. Likewise, the POIs may have a lots of character. Our aim is to uncover latent features that explain the observed ratings by projecting the users and POIs onto a feature space. Assume that there are m users and n POIs within a social network. The users and POIs are included in a joint latent factor space of dimensionality f, such that user-POIs interactions are modeled as inner products in a space comprising those factors. For user u, the elements of p_u indicate the extent of interest that a particular user has in POIs in which the corresponding factors feature highly. The resulting dot product, $q^t_m p_u$, captures the interaction between user u and POI m — the user's overall interest in the characteristics of the POI, as shown in Fig. 4.

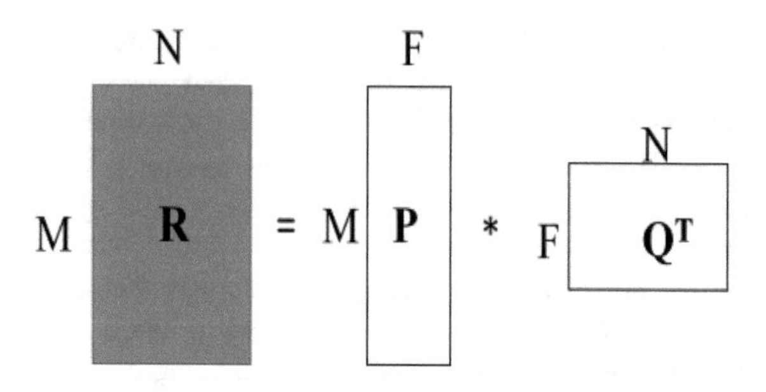

Fig. 4. Concept underlying matrix factorization.

The rating of POI m by user u is denoted by $\hat{r}_{um}$, which results in the following estimate:

$$\hat{r}_{um} = P_u Q_m^T \tag{4}$$

In considering of the baseline estimate, biases extend Eq. (5) as follows:

$$\hat{r}_{im} = b_{um} + P_u Q_m^T = \mu + b_u + b_m + Q_m^T \tag{5}$$

Parameter $\hat{r}_{im}$ denotes the unknown rating for POI by the ith user, whereas b_u and b_m indicate the observed deviations of ith user and mth POI in the social network. We denote the user-feature matrix as $p_i = \{p_{i1}, p_{i2}, \ldots, p_{ik}\}$, where p_{ik} denotes the preference of the kth feature for the ith user and denotes POI-feature matrix as $Q_m = \{q_{m1}, q_{m2}, \ldots, q_{mk}\}$, where q_{mk} denotes the preference of the kth feature for the mth POI.

To obtain the factor vectors (p_u and q_m), the system minimizes the Root Mean Squared Error on the set of known ratings using the stochastic gradient descent algorithm:

$$\min_{q \cdot p} \sum\nolimits_{(u,m) \in L} \left(r_{um} - \mu + b_u - b_m - P_u Q_m^T\right)^2 + \lambda\left(Q_m^2 + P_u^2 + b_u^2 + b_m^2\right) \tag{6}$$

Parameter L is the set of (u, m) pairs for which r_{um} is known (the training set). In the following, we divided the process into a number of steps: Initialize: the number of features j, the penalize parameter γ, λ (the greater the value is, the faster convergence is), and $P_{u,}$ Q_m.

- Step 1: For each rating in L, calculate the value of Root Mean Squared Error:

$$e_{in} = r_{in} - \hat{r}_{in} \tag{7}$$

- Step 2: Update the value of P_u,Q_m using the following equations:

$$p_{uk} \leftarrow p_{uk} + \gamma.(e_{um}.q_{mk} - \lambda q_{uk}) \tag{8}$$

$$q_{mk} \leftarrow q_{mk} + \gamma.(e_{um}.p_{um} - \lambda q_{mk}) \tag{9}$$

and b_u, b_m using the following equations:

- Step 3: Repeat steps 1 to 2 until the minimum sum squared error is obtained.

$$b_i \leftarrow b_i + \gamma.(e_{im} - \lambda b_i) \tag{10}$$

$$b_m \leftarrow b_m + \gamma.(e_{im} - \lambda b_m) \tag{11}$$

Baseline matrix factorization makes it possible to predict all unknown ratings. The proposed system make recommendations for groups rather individual users. Thus, we calculate the unknown group rating as follows:

$$\hat{r}_{im} = \frac{\sum_{u=0}^{b} (\hat{r}_{um})}{b} \tag{12}$$

Parameter $\hat{r}_{im}$ represents the unknown rating for the mth POI by the ith group in the social network. Symbol b represents the number of group members and the $\hat{r}_{um}$ represents the rating for the mth POI provided by the uth user in the group.

3.3 Online Processing

In this section, we divided online processing into two stages: (1) Bit Similarity Checking (BSC); and (2) evaluation of POI scores.

(1) *Bit Similarity Checking (BSC)*

Most recommendation systems check all of the data in the database; however, this is computationally intensive. Thus, we developed a checking algorithm called Bit Similarity Checking (BSC) to reduce unnecessary calculations in determining similarity. We also sought to deal with the data sparsity problem mentioned previously. When the check-in data of cuisine enthusiasts is concentrated within the category "restaurant"; the number of check-ins in other categories will be close to zero. As shown in Fig. 5, there are 5 clusters (C_1, C_2, C_3, C_4, C_5) and the X, Y, and Z-axis are manipulated variables of check-in times in category Cat_1, Cat_2, and Cat_3. Obviously, the data in cluster C_1, C_2 lies only along the X-axis, whereas data in cluster C_3 lies only along the Y-axis. Data in cluster C_4 lies along the XY-plane and data in cluster C_5 lies along the YZ-plane. If the data of g_1 lies only along the Z-axis, then there is no reason to calculate the degrees of similarity between g_1 and C_1, C_2, C_3, C_4 (along the X-axis, Y-axis, or XY-plane), which would result in unnecessary calculations. BSC is able to eliminate these.

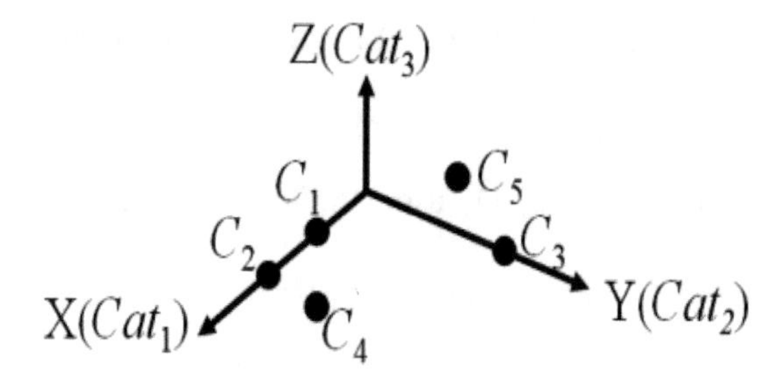

Fig. 5. Example of data sparsity.

Cluster data can be represented in coded binary form, the length of which is determined by the number of categories. Assuming that there are *n* categories in a database, the length of binary code is *n*, and the *i*th bit represents the *i*th category (i.e., data along the *i*th-axis). If the data lies along the *i*th-axis, then the value of the *i*th bit is 1. In Fig. 5, the length of the binary code is 3. The binary code of C_1 is 100, which represents the data associated only with the 1st axis (category). The binary code of C_5 is 011, which represents data associated with the 2nd and 3rd axes (categories).

This makes it possible to use AND and SUM operators to approximate the degree of similarity between the query group and each cluster. We first translate query group *g* into coded binary form. We then calculate the approximate similarity between query group and each cluster using AND and SUM operators. This approach is faster than calculating the similarity among all groups. The AND operator compares two bits and generates a result of 1 if both bits are 1; otherwise, it returns 0. The sum operator generates a result that is the sum of all bits. Note that when both bits are 0, there is no check-in data in that category. If the value of BSC is high, then the query group shares greater similarity to that cluster. Thus, the BSC scores of clusters are sorted in descending order after performing the BSC operation. We first check the cluster with the highest BSC scores. Consider for example, the case in which there are 5 categories and 3 clusters: the coded binary form of query group *g* is 11000 and the coded binary forms of C_1, C_2 and C_3 are 01010, 11010, and 00111. We then calculate the BSC scores between the query group and each cluster. Finally, the BSC scores are as follows: C_1 (1), C_2 (2), and C_3 (0). Thus, we first check C_2 and then C_1. We do not check cluster C_3, because the zero BSC value means that the cosine similarity is also 0.

(2) *Evaluation of POI Scores*

After identifying all of the clusters, we calculate the POI score for the group check-in data within a cluster. The calculation is a two-part process: (1) calculation of similarity between two groups and (2) calculation of the POI scores.

(a) *Calculation of similarity between two groups*

We first calculate the Similarity of Group Profile SGP (g_i, g_j). As mentioned before, group profiles include discrete as well as continuous data. Different methods are used to process discrete and continuous data; therefore, we process each type of data separately and then compute the SGP score as described below.

Assume that the discrete data vector of group g_i has m dimensions. The discrete data vector of g_1 is [0.4, 0.6, 0, 0.4, 0.2, 0.4], and the discrete data vector of g_2 is [0.3, 0.5, 0.8, 0.4, 0.2, 0.3]. We can calculate the similarity among discrete data between groups g_i and g_j using the cosine similarity [8] as follows:

$$GP_{dis}Sim(g_i, g_j) = \frac{\sum_{n=1}^{m}[GP_{dis,i,n} \times GP_{dis,j,n}]}{\sqrt{\sum_{n=1}^{m} GP_{dis,i,n^2}} \times \sqrt{\sum_{n=1}^{m} GP_{dis,j,n^2}}} \tag{13}$$

Similarity among the discrete data of g_1 and g_2 is then calculated as follows: GP_{dis} $Sim(g_1, g_2) \fallingdotseq 0.7$. We use curves to calculate the degree of similarity among the continuous data of the two groups. We also incorporated the Pearson correlation coefficient [9]. The results range from −1 to 1, where 1 indicates that both curves are fully consistent (identical), 0 indicates a high level of inconsistency (dissimilar), and −1 indicates that the curves are completely opposed (no similarity whatsoever). Equation (14) presents the Pearson correlation coefficient.

$$R(g_i, g_j) = \frac{Cov(GP_{\mathbf{cont,i}}, GP_{\mathbf{cont,j}})}{\sigma(GP_{\mathbf{cont,i}}) \times \sigma(GP_{\mathbf{cont,j}})} \tag{14}$$

Cov (.) refers to statistical covariance and σ (.) refers to the standard deviation of the curve. To calculate covariance, we subtract the mean of each dataset from its corresponding data points, and then multiply and calculate the total sum, as shown in Eq. (15):

$$Cov(GP_{\mathbf{cont,i}}, GP_{\mathbf{cont,j}}) = \frac{\sum_{n=\alpha}^{\beta}[(GP_{cont,i,n} - \mu(GP_{\mathbf{cont,i}})) \times (GP_{cont,j,n} - \mu(GP_{\mathbf{cont,j}}))]}{(\beta - \alpha)} \tag{15}$$

where $\sigma(GP_{\mathbf{cont,i}})$ is the standard deviation of the continuous data set of g_i, as follows:

$$\sigma(GP_{\mathbf{cont,i}}) = \sqrt{\frac{1}{(\beta - \alpha)} \sum_{n=\alpha}^{\beta} (GP_{cont,i,n} - \mu(GP_{\mathbf{cont,i}}))^2} \tag{16}$$

where $\mu(GP_{\mathbf{cont,i}})$ represents the mean of the continuous data set of group g_i, as follows:

$$\mu(GP_{\mathbf{cont,i}}) = \frac{1}{(\beta - \alpha)} \sum_{n=\alpha}^{\beta} GP_{cont,i,n} \tag{17}$$

Finally, to maintain consistency with the range of (0–1) discrete data values, we added one and divided the result by two, thereby normalizing the outcome as [0, 1], as shown in Eq. (18):

$$GP_{cont}Sim(g_i, g_j) = \sum_{k}^{l} \left(\frac{R(g_i, g_j)}{l} + 1\right)/2 \tag{18}$$

After calculating the degree of similarity among the discrete and continuous data of groups g_i and g_j, we obtain their group profile score $SGP(g_i, g_j)$, as follows:

$$SGP(g_i, g_j) = \left[GP_{dis}Sim(g_i, g_j) + GP_{cont}Sim(g_i, g_j)\right]/2 \tag{19}$$

In addition to SGP, we also consider PU, RSM, and RSM in calculating the degree of social-related similarity. We obtain the social-related similarity $SRS(g_i, g_j)$ as follows:

$$SRS(g_i, g_j) = SGP(g_i, g_j) \times \mathrm{PU}(g_i, g_j) \times \frac{1}{\mathrm{RSS}(g_i, g_j)} \times \mathrm{RSM}(g_i, g_j) \tag{20}$$

(b) *Calculation of social POI-related similarity:* We calculate the degree of POI-related similarity between groups $\mathrm{g_i}$ and $\mathrm{g_j}$ using cosine similarity [8] as follows:

$$PRS(g_i, g_j) = cos(\vec{\mathrm{u}}, \vec{\mathrm{v}}) = \frac{\sum_i^d u_i v_i}{\sqrt{\sum_i^d u_i^2}\sqrt{\sum_i^d v_i^2}} \tag{21}$$

Where, g_i and g_j denote the *i*th and *j*th groups in the social network, and u_i and u_j denote the group check-in data belonging to g_i and g_j.

(c) *Calculation of POI scores PS:* The POI score of the mth POI of the nth cluster is then calculated using the above results, as follows:

$$PS_{mn} = \frac{\sum_{j=1}^{l} ch_{jm} \times \hat{r}_{g_{jm}} \times SRS(g_i, g_j) \times PRS(g_i, g_j)}{l} \tag{22}$$

Where, ch_{jm} denotes the *i*th group's check-in times at the *m*th POI, $\hat{r}_{gim}$ denotes the *j*th group's predicted rating for the *m*th POI, and l denotes the number of groups in the *n*th cluster. For example, assume that a cluster comprises three groups (g_1, g_2, and g_3), with the information associated with query group g and the POI Tokyo Tower in Table 7.

Table 7. Foursquare dataset

Group	Check-in	Rating	*SRS(g, g_j)*	*PRS(g, g_j)*
g_1	2	3	0.4	0.1
g_2	1	3	0.35	0.5
g_3	1	4	0.25	0.60

The *POI score* of the POI Tokyo Tower of this cluster is calculated as follows: *PS* = 2*3*0.4*0.1 + 1*3*0.35*0.5 + 1*4*0.23*0.6/3 = 1.317/3 = 0.439. We calculate all of the POI scores in the cluster in a similar manner. This process undergoes continual processing until all POI scores have been examined, wherein the remaining top-*n* POIs are the final recommendations.

4 Performance

4.1 Experiment Settings

The data used in the experiment was from a Foursquare dataset of New York City [8, 13]. As shown in Table 8, this dataset includes 242,563 users, 1,143,092 venues, 164,481 check-ins, 7,098,490 social connections, and 116,819 ratings in 400 categories. The dataset does not include personal information of users; therefore, we randomly generated this data.

Table 8. Foursquare dataset

Description	Foursquare real dataset
The number of check-ins	164481
The number of users	242563
The number of users' rating	116819
The number of POIs	112360
The number of categories	400

LBSNs allow users to share geographical information via check-ins, where a check-in includes a user and the time at which the user visited a particular venue. Each venue in Foursquare is associated with geographical coordinates; however, it does not provide explicit group information. Thus, we had to extract group check-ins implicitly. We assumed that if a set of friends visited the same venue at the same time, they would be members of a particular group. Specifically, the set of individual check-ins made by friends within an hour period is regarded as a group check-in. Finally, we generated 100 K group check-ins resulting in 50 k different groups. Each experiment was performed 30 times, and each query group *g* was selected at random. All of the experiment results are listed in the following.

All of the experiments were performed in the following environment: Intel Core i5-4460 3.20 GHz processor, 8 GB main memory, and Windows 7 64-bit version. All programs were compiled using Python.

4.2 Evaluation Metrics

If group g wishes to travel to a city, then the recommendation system draws up the top-n POIs as recommendations. For the purpose of evaluation, we used 5-cross validation to separate the data into training and testing sets. We evaluated the accuracy of various methods using three metrics: average precision@N (Pre@N), average recall@N (Rec@N), and average F-score [9].

$$precision@N = \frac{|\{top\,N\,recommendations\} \cap \{true\,items\}||}{|\{top\,N\,recommendations\}|} \tag{23}$$

$$recall@N = \frac{|\{top\,N\,recommendations\} \cap \{true\,items\}||}{|\{true\,items\}|} \tag{24}$$

$$\text{F - }score@N = \frac{2 \times precision \times recall}{precision + recall} \tag{25}$$

Where, N is the number of recommendations. We consider three values of N (i.e., 1, 3, 5, 7, 9). Precision@N is the fraction of the top N recommendations that are adopted by a group, whereas recall@N is the fraction of items adopted by a group (true items) that are contained in the top N recommendations. The F-score is the harmonic mean of precision and recall. A higher score in each of these metrics indicates better performance.

4.3 Setting an Appropriate Number of Clusters

This section discusses how to choose a good k value for k-medoids. We can select the number of clusters by visually inspecting the data points; however, there tends to be a great deal of ambiguity in this process for all except the simplest forms of data. We therefore compute the sum of squared error (SSE) for some values of k (for example 2, 4, 6, 8, etc.). SSE is defined as the sum of the squared distance between each member of the cluster and its centroid:

$$SSE = \sum_{i=1}^{K} \sum_{x \in c_i} dist(x, c_i)^2 \tag{26}$$

Where, c_i denotes the ith cluster, the dist function indicates the degree of similarity between two points. In this work, we used cosine similarity, such that the dist function was defined as follows:

$$dist(x, c_i) = 1 - \cos(x, c_i) \tag{27}$$

Our aim is to select a value for k for which the SSE value decreases abruptly. Before the experiment, we set a value of $\varepsilon = 0.05$. In the event that the result obtained by subtracting SSE for values of k from $k + 1$ is less than ε, then the computational experiment would be halted. We set the maximum number of k at 400 (the number of categories).

Figure 6 shows how the numbers of clusters influences SSE. We can see that error decreases as *k* increases due to the fact that an increase in the number of clusters decreases the amount of distortion. We found that when k = 270, the result by subtracting SSE for values of 270 from 271 is less than. We assign the default number of clusters to 270.

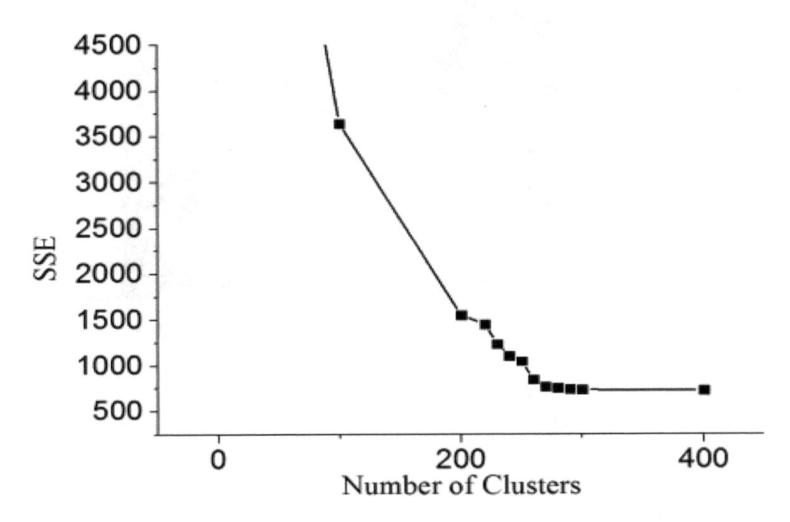

Fig. 6. Numbers of clusters and their influence on SSE.

4.4 Comparison Between K-Means and K-Kedoids Clustering

We performed the following two experiments to determine whether *k*-means clustering is superior to *k*-medoid: comparison of the average number of categories between the two methods; and calculation of the average number of categories in the two methods. Lower values indicate that the members in clusters are closer. The results are presented in Fig. 7. The average number of categories obtained using *k*-means clustering was 331.137 whereas *k*-medoids resulted in 200.333. This is a clear demonstration of the superiority of *k*-medoids over *k*-means.

4.5 Comparing the Query Comparison of Proportion Between Two Methods

In this experiment, we calculate the comparison of proportions of queries. If the value of comparison of proportions is much lower, then BSC algorithm stops earlier to avoid unnecessary calculations. The results are presented in Fig. 8, where the value of comparison of proportions of *k*-means method is 0.96901 and the value of comparison of proportions of *k*-medoids is 0.6563. This is a clear demonstration of the superiority of *k*-medoids over k-means.

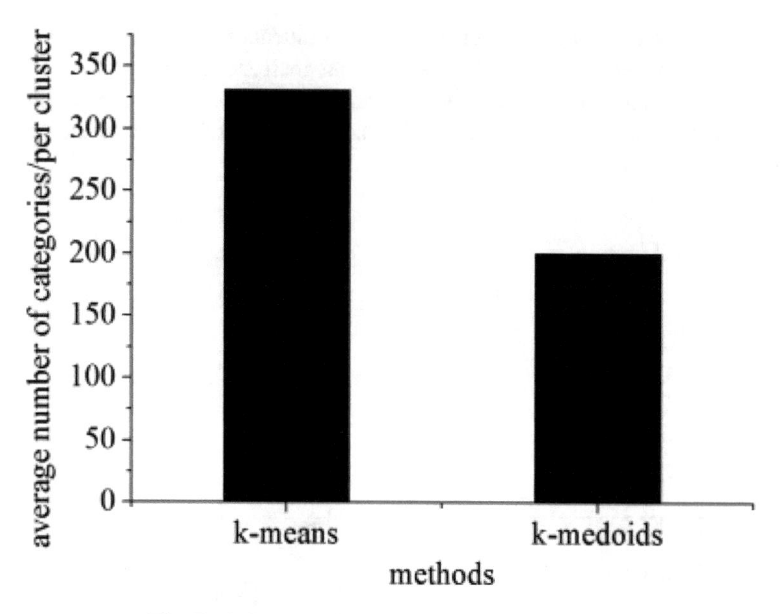

Fig. 7. Influence of average number of categories.

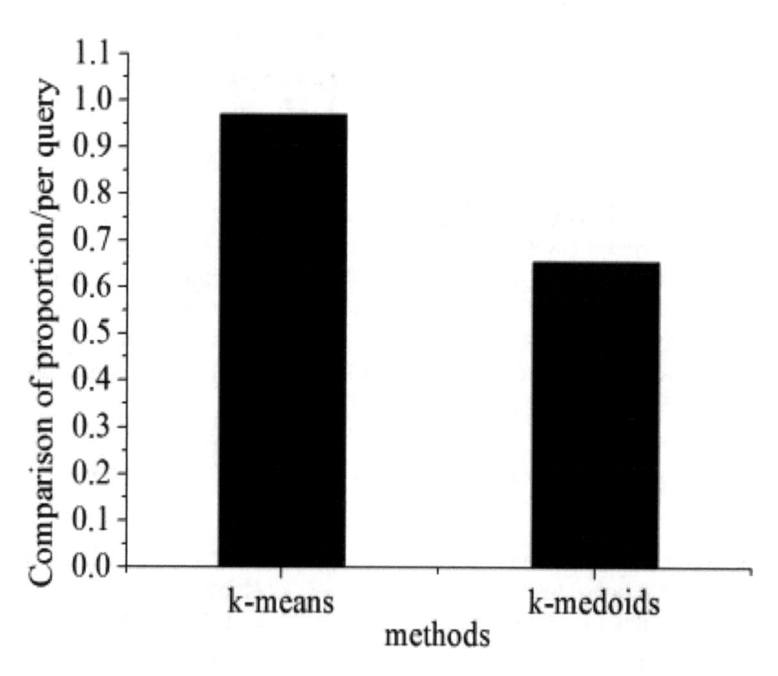

Fig. 8. Influence of the comparison of proportions.

4.6 Baseline Comparisons

The proposed recommendation system makes recommendations based on group check-in data rather than individual check-in data. No previous system is able to accommodate this situation; therefore, we modified the two baseline algorithms to fit our problem, as follows:

- Top-Famous (TF) [15] is used to analyze the number of check-in times in order to identify the top-*n* POIs. This method recommends well-known venues.
- Collaborative filtering (CF) [2] is widely used in conventional personal recommendation systems. Collaborative filtering makes automatic predictions (filtering) about the interests of a user by collecting preferences from numerous users.

In the following, we discuss the influence of N on the precision, recall, and F-score of various methods. We set the number of clusters at 270 in accordance with the results in Section 5-B.

Figure 9(a) displays the influence of *n* on the precision of the various methods. As shown in Fig. 9(a), precision decreases with an increase in *n*, because all of the methods generate recommendations according to their relevance to group *g*. In other words, generating a larger number of recommendations results in a larger number of irrelevant recommendations. The TF method exhibited the worst performance in terms of precision for each *n*, due to the fact that it only takes the number of check-in times into account. CF also performed poorly because it does not exploit differences among individuals or the nature of the relationships among group members. HUMR clearly outperformed CF and TF for every *n*. HUMR outperformed TF by 124% and CT by 65% with regard to recommendation accuracy. Figure 9(b) displays the influence of *n* on the recall performance of the various methods. Our results revealed that recall increases with *n*. The TF method exhibited the worst performance in terms of precision because it only takes the number of check-in times into consideration and only recommends well-known POIs. It should be remembered that not everyone appreciates the most famous landmarks. Sports fans may be more interested in sport-related places than famous museums or galleries. CF also performed poorly due to its limited recommendation results, as mentioned in Sect. 1.

The proposed HUMR method outperformed TF by 174% and CT by 117% with regard to recall. These results demonstrate that the requirements for *n* are in conflict with regard to precision and recall performance. This problem can be overcome by considering the F-score. As shown in Fig. 9(c), the highest F-score was obtained when $N = 7$. Thus, we can set a default N of 7 for the number of recommendations in cases where the query does not specify parameter N.

4.7 Performance of Algorithm with Various Relationship Combinations

In this section, we discuss the influence of *n* on the precision of HUMR in the context of various relationship combinations, the results of which are presented in Fig. 10. We posited three relationships for the calculation of similarity: (1) point-to-point: PU, (2) line-to-line: RSS and RSM, (3) plane-to-plane: SGP. We used CF as a baseline for these experiments.

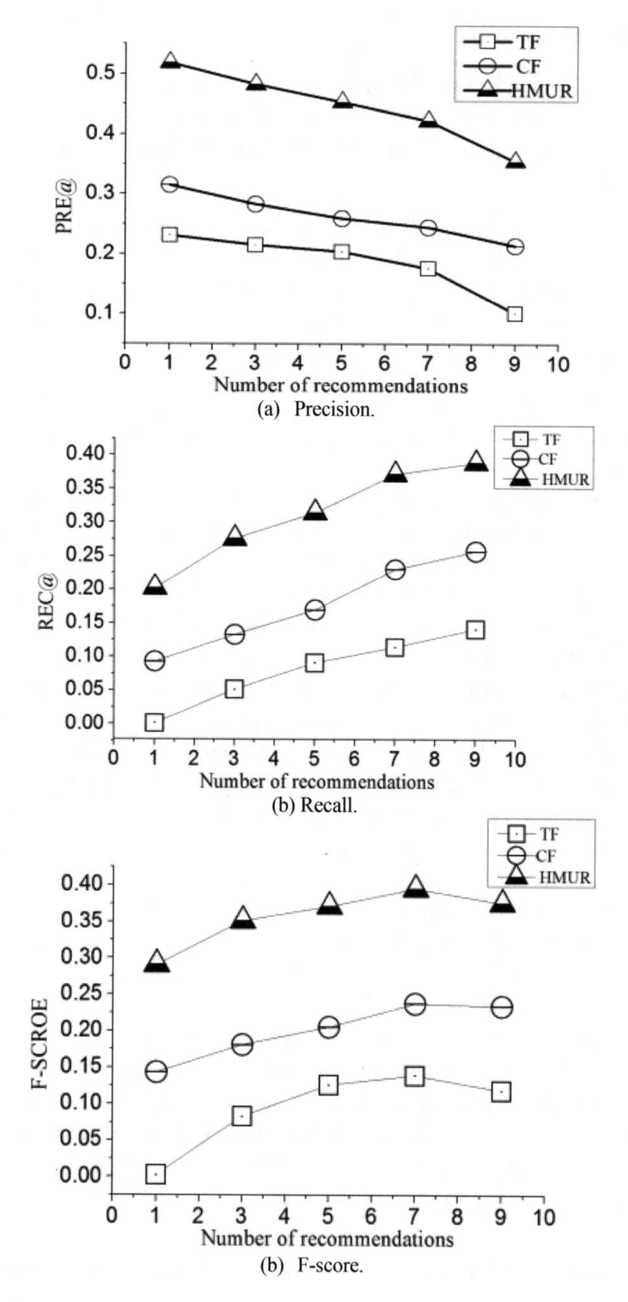

Fig. 9. Influence of number of recommendations on recommendation accuracy.

We first considered the influence of n on the precision of HUMR, based on a point-to-point relationship (PU). In Fig. 10, HUMR outperformed CF by 30% with regard to recommendation accuracy. We then considered the influence of n on the precision of

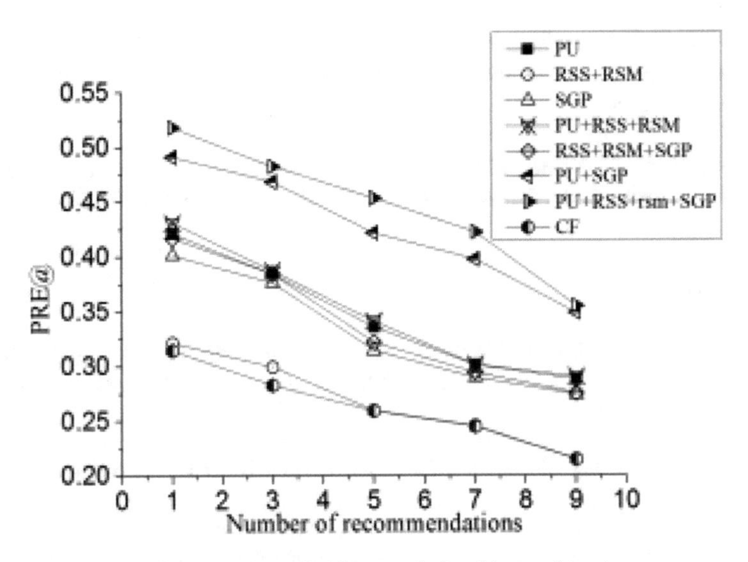

Fig. 10. Precision of different relationship combinations.

HUMR based on a line–to-line relationship (RSS + RSM). As shown in Fig. 10, HUMR outperformed CF by only 4%. Finally, we examined the influence of n on the precision of HUMR based on a plane-to-plane relationship (SGP). As shown in Fig. 8, HUMR outperformed CF by 27% with regard to recommendation accuracy.

The precision of HUMR in point-to-point relationships (PU) is higher than in the other two relationships. This can be attributed to the fact that the selection of POI for groups is influenced by the composition of the group. PU+SGP outperformed CF by 53%; PU+RSS+RSM outperformed CF by 37%; and RSS+RSM+SGP outperformed CF by 32%. This indicates that PU is more important than SGP and RSS+SRM and SGP is more important than RSS+RSM.

5 Conclusions and Future Work

Most existing recommendation systems were designed for individuals rather than groups. In this study, we devised a novel algorithm called Hybrid Method with Users Rating (HMUR) to analyze group-related information and user ratings to formulate recommendations for groups of users. The selection of POIs by groups is influenced by personal preferences as well as the relationships they have with other group members. Thus, we adopted two means of calculating similarity. The proposed algorithm also takes into account the rating of POIs in order to make adjustments to the final recommendations. Precision, recall, and F-score were used as metrics to evaluate algorithm performance. Experiment results demonstrate the superiority of the proposed method over two existing baselines.

In future work, we will continue to improve the proposed recommendation algorithm by taking into account additional factors, such as social interaction between users and the operating hours of POIs.

Acknowledgments. This work was supported in part by the Ministry of Science and Technology of Taiwan, R.O.C., under Contracts MOST 106-2119-M-224-003 and MOST 106-2221-E-006-247.

References

1. Arase, Y., Xie, X., Hara, T., Nishio, S.: Mining people's trips from large scale geo-tagged photos. In: Proceedings on ACM Multimedia International Conference (ACM-MM), pp. 133–142 (2010)
2. Breese, J.S., David, H., Carl, K.: Empirical analysis of predictive algorithms for collaborative filtering. In: Proceedings on Conference on Uncertainty in Artificial Intelligence, pp. 43–52 (1998)
3. Cheng, A.J., Chen, Y.Y., Huang, Y.T., Hsu, W.H., Liao, H.Y.M.: Personalized travel recommendation by mining people attributes from community-contributed photos. In: Proceedings on ACM International Conference on Multimedia, pp. 83–92 (2011)
4. Gartrell, M., Xing, X., Lv, Q., Beach, A., Han, R., Mishra, S., Seada, K.: Enhancing group recommendation by incorporating social relationship interactions. In: Proceedings on ACM International Conference on Supporting Group Work, pp. 97–106 (2010)
5. Horozov, T., Narasimhan, N., Vasudevan, V.: Using location for personalized POI recommendations in mobile. In: Proceeding on International Symposium on Applications and the Internet, pp. 1–6 (2006)
6. Khoshgoftaar, X.S.: A survey of collaborative filtering techniques. Adv. Artif. Intell. **19**(4) (2009)
7. Levandoski, J.J., Sarwat, M., Eldawy, A., Mokbel, M.F.: LARS: A location-aware recommendation system. In: Proceeding on IEEE International Conference on Data Engineering, pp. 450–461 (2012)
8. Li, X., Muarata, T.: Customizing knowledge-based recommendation system by tracking analysis of user behavior. In: Proceeding on International Conference on Industrial Engineering and Engineering Management (IE&EM), pp. 65–69 (2010)
9. Luo, C., Lou, J.G., Lin, Q., Fu, Q., Ding, R., Zhang, D., Wang, Z.: Correlating events with time series for incident diagnosis. In: Proceeding on ACM International Conference on Knowledge Discovery and Data Mining, pp. 1583–1592 (2014)
10. McCarthy, J.F.: Pocket restaurant finder: a situated recommendation systems for groups. In: Proceeding on ACM Conf. on Human Factors in Computer Systems (2002)
11. McCarthy, J.F., Anagnost, T.D.: Musicfx: An arbiter of group preferences for computer supported collaborative workouts. In: Proceedings on ACM Conference on Computer Supported Cooperative Work, pp. 363–372 (1998)
12. Michael, J.P.: A framework for collaborative, content-based and demographic filtering. Artif. Intell. Rev. **13**(5-6), 393–408 (1999)
13. Sarwat, M., Levandoski, J.J., Eldawy, A., Mokbel, M.F.: LARS*: A scalable and efficient location-aware recommendation system. IEEE Trans. Knowl. Data Eng. (2014)
14. Ying, J.J.C., Lu, E.H.C., Kuo, W.N., Tseng, V.S.: Urban point-of-interest recommendation by mining user check-in behaviors. In: Proceedings on ACM SIGKDD, pp. 55–62 (2012)
15. Zheng, W.V., et al.: Collaborative filtering meets mobile recommendation: a user-centered approach. In: Proceeding on Conference on Artificial Intelligence, pp. 236–241 (2010)

How Much Is My Car Worth? A Methodology for Predicting Used Cars' Prices Using Random Forest

Nabarun Pal[1(✉)], Priya Arora[2], Puneet Kohli[2], Dhanasekar Sundararaman[3], and Sai Sumanth Palakurthy[4]

[1] Department of Metallurgical and Materials Engineering, Indian Institute of Technology Roorkee, Roorkee, India
pal@nabarun.in

[2] Department of Computer Science, Texas A&M University, College Station, TX, USA
arora.priya4172@gmail.com, punkohl@gmail.com

[3] Department of Information Technology, SSN College of Engineering, Chennai, India
dhanasekar312213@gmail.com

[4] Department of Computer Science and Engineering, IIT (ISM) Dhanbad, Dhanbad, India
sumanth4591@gmail.com

Abstract. Cars are being sold more than ever. Developing countries adopt the lease culture instead of buying a new car due to affordability. Therefore, the rise of used cars sales is exponentially increasing. Car sellers sometimes take advantage of this scenario by listing unrealistic prices owing to the demand. Therefore, arises a need for a model that can assign a price for a vehicle by evaluating its features taking the prices of other cars into consideration. In this paper, we use supervised learning method, namely, Random Forest to predict the prices of used cars. The model has been chosen after careful exploratory data analysis to determine the impact of each feature on price. A Random Forest with 500 Decision Trees were created to train the data. From experimental results, the training accuracy was found out to be 95.82%, and the testing accuracy was 83.63%. The model can predict the price of cars accurately by choosing the most correlated features.

Keywords: Car price prediction · Random forests · Regression Decision trees

1 Introduction

The prices of new cars in the industry is fixed by the manufacturer with some additional costs incurred by the Government in the form of taxes. So customers buying a new car can be assured of the money they invest to be worthy. But due to the increased price of new cars and the incapability of customers to buy new cars due to the lack of funds, used cars sales are on a global increase. Predicting the prices of used cars is an

K. Arai et al. (Eds.): FICC 2018, AISC 886, pp. 413–422, 2019.
https://doi.org/10.1007/978-3-030-03402-3_28

interesting and much-needed problem to be addressed. Customers can be widely exploited by fixing unrealistic prices for the used cars and many falls into this trap. Therefore, rises an absolute necessity of a used car price prediction system to effectively determine the worthiness of the car using a variety of features. Due to the adverse pricing of cars and the nomadic nature of people in developed countries, the cars are mostly bought on a lease basis, where there is an agreement between the buyer and seller. These cars upon completion of the agreement are resold. So reselling has become an essential part of today's world. Given the description of used cars, the prediction of used cars is not an easy task. There are a variety of features of a car like the age of the car, its make, the origin of the car (the original country of the manufacturer), its mileage (the number of kilometers it has run) and its horsepower. Due to rising fuel prices, fuel economy is also of prime importance. Other factors such as the type of fuel it uses, style, braking system, the volume of its cylinders (measured in cc), acceleration, the number of doors, safety index, size, weight, height, paint color, consumer reviews, prestigious awards won by the car manufacturer. Other options such as sound system, air conditioner, power steering, cosmic wheels, GPS navigator all may influence the price as well. Some of the important features of this and their influence on price is detailed in Sect. 3.

So, we propose a methodology using Machine Learning model namely random forest to predict the prices of used cars given the features. The price is estimated based on the number of features as mentioned above. The intricate details about this model on the used car's data set along with the accuracy are narrated in depth in Sect. 5. We then deploy a website to display our results which are capable of predicting the price of a car given so many features of it. This deployed service is a result of our work, and it incorporates the data, ML model with the features.

To summarize:

- First, we collect the data about used cars, identify important features that reflect the price.
- Second, we preprocess and remove entries with NA values. Discard features that are not relevant for the prediction of the price.
- Third, we apply random forest model on the preprocessed dataset with features as inputs and the price as output.
- Finally, we deploy a web page as a service which incorporates all the features of the used cars and the random forest model to predict the price of a car.

The paper is organized as follows. Section 2 talks about the other works that predict the price of used cars. Section 3 details about the dataset and preprocessing. In Sect. 4 we share the results of exploratory data analysis on our dataset. Section 5 talks about our proposed methodology of using random forests to predict car prices with details about the accuracy of training and test data. We conclude in Sect. 6 with our future works in the last section.

2 Related Works

We use dataset from Kaggle for used car price prediction. The dataset contains various features as mentioned in Sect. 3 of this paper that are required to predict and classify the range of prices of used cars. The literature survey provides few papers where researchers have used similar data set or related data-set for such price prediction.

In [1], the patent describes a generic engine platform for assessing the price of an asset. This platform provides a price computation matrix for asset price prediction. To compute the price for vehicles, this platform may compute linear regression model that defines a set of input variables. However, it does not give details as what features can be used for specific type of vehicles for such prediction. We have taken important features for predicting the price of used cars using random forest models.

Zhang et al. [2] use Kaggle data-set to perform price prediction of a used car. The author evaluates the performance of several classification methods (logistic regression, SVM, decision tree, Extra Trees, AdaBoost, random forest) to assess the performance. Among all these models, random forest classifier proves to perform the best for their prediction task. This work uses five features (brand, powerPS, kilometer, sellingTime, VehicleAge) to perform the classification task after removal of irrelevant features and outliers from the dataset which gives an accuracy of 83.08% on the test data. We also use Kaggle data-set to perform prediction of used-car prices. However, the difference lies in the inclusion of few more relevant features in prediction model - the price of the car, and vehicleType. These two features play an important role in predicting the price of a used car which seems to be given less importance in the paper [2]. In addition to this, the range of features year of registration, PowerPS, the price seems to be narrowed down in work [2] due to which test data-set gives less accuracy w.r.t what we evaluate by broadening the range of the above-said features.

The report by Awad et al. [3] is more of an educational paper than a research paper. The author reviews six most popular classification methods (Bayesian classification, ANNs, SVMs, k-NN, Rough sets, and Artificial immune system) to perform a spam email classification task. The reason for choosing this paper is to understand these popular classification models in detail, and its applicability to the spam email classification problem since this paper gives much insight into each method. The main difference, however, between classifying price range and spam mail, is that spam email classification task is a binary one, whereas our motive is mainly one-vs-the-rest. The author uses Naive Bayes for classification which does not give accurate results due to its major concern of feature dependency as pointed out by the author. Due to this reason, we also did not try to evaluate the performance of our data-set using Naive Bayes model since our dataset has heavily feature dependency. To predict results with good accuracy, the author suggests a hybrid system which applies to our work by using Random Forest. A manipulation of various decorrelated decision trees, the Random Forest gives pretty good accuracy in comparison to prior work.

Work by Durgesh et al. [4] gives a good introductory paper on Support Vector Machine. The authors assess the performance of several classification techniques (K-NN, Rule-Based Classifiers, etc.) by performing the comparative assessment of SVM with others. This comparative study is done using several data-sets taken from the UCI Machine Learning Repository. This assessment yields that SVM gives much better classification accuracy in comparison to others. This gives us a baseline for prediction of tasks by using a simple linear model which gives good accuracy to let us use complex systems - random forest - which ultimately provides pretty good results for prediction of the used-cars price.

The Author of the paper [5] predicts the price of used cars in Mauritius by using four comparable machine learning algorithms - multiple linear regression, k-nearest neighbors, naive Bayes and decision trees algorithm. The author uses historical data collected from daily newspapers in Mauritius. The application of listed learning algorithms on this data provides comparable results with not-so-good prediction accuracy. The main difference, however, between classifying price range and spam mail, is that spam email classification task is a binary one, whereas our motive is mainly one-vs-the-rest. The difference between our analysis of this work is that we perform our assessment on data from Kaggle, whereas theirs is based on the data collected from the daily newspaper. In addition to this, the author uses simple and comparable classification algorithms that conform to our findings that using a sophisticated algorithm like random forest on our data-set can give pretty good results, and which has proved to be so.

Multiple regression models help in the classification of numerical values when there is more than one independent variable with one dependent variable. Noor et al. [6], hence, use this model to evaluate the prediction using https://www.pakwheels.com/ dataset. The authors use data-set of 2000 records collected within the duration of one or two months. The collected data includes features like color, advertisement date, etc. which seem to be not-so-relevant for such prediction, whereas, our model uses relevant features like the brand, kilometer, etc. which helps in predicting good accuracy using random forest on data-set obtained from Kaggle.

Researchers in the paper [7] use multivariate regression model in classification and prediction of used car prices. The authors use 2005 General Motor (GM) cars data-set for this classification task. They introduce variable selection techniques for determination of relevant features for such prediction tasks which provide insights of its applicability in several domains. The main emphasis of this paper is to discuss this model for learning and encouraging students to perform in this field. The model used in this work does not require any special knowledge of the dataset used. Hence the portal data (www.pakwheels.com) was sufficient to use. The difference, however, of their work with ours is that our work focusses upon preprocessing/filtering of data obtained from Kaggle with a selection of relevant features for prediction. This motivates us to generalize the model for a variety of brands with a range of years to predict prices using several relevant features.

3 Data Set and Preprocessing

To accurately predict the prices of used cars, we used an open dataset to train our model. We used the ‘Used Car Database’ from Kaggle which is scraped from eBay-Kleinanzeigen, the German subsidiary of eBay, a publicly listed online classified portal. The dataset contains the prices and attributes of over 370,000 used cars sold on the website across 40 brands. Our dataset contains 20 unique attributes of a car being sold, out of which we removed a few irrelevant columns that have little to no impact on a car's price from our analysis, such as some pictures, postal code, and advertisement name.

Additionally, we perform the following preprocessing steps on the data set helping us narrowing down the features:

(1) Keep only listings for cars sold by private owners and filter out those sold by dealerships
(2) Keep only listings for cars being sold, and filter out all request for purchase listings
(3) Filter out cars manufactured before 1863 and after 2017, and derive the car's age
(4) Filter out all cars with unrealistic Power values
(5) Filter out listings which don't have an associated price
(6) Filter out all cars listed as unavailable
(7) Filter out invalid registration dates
(8) Convert boolean (true/false) fields to numeric (0/1) based
(9) Filter out all data with value as 'NA' (Not Available)

After pre-processing, the final dataset contains ten features for second hand cars – 'price', 'vehicleType', 'age', 'powerPS', 'model', 'kilometer', 'fuelType', 'brand', and 'damageRepaired', 'isAutomatic'. Out of these features, the most important for our prediction model is:

(1) *price*: The specified asking amount for the car
(2) *kilometer*: A number of Kilometers the car has driven
(3) *brand*: The car's manufacturing company
(4) *vehicleType*: Whether a small car, limousine, bus, etc.

Just based on analyzing our input data set, we can see that on average, used cars were priced at approximately €11000 with an average kilometer reading of 125000 km. 50% of cars from our dataset were being sold after 12 years of being used.

4 Exploratory Data Analysis

After preprocessing the data, it is analyzed through visual exploration to gather insights about the model that can be applied to the data, understand the diversity in the data and the range of every field. We use a bar chart, box plot, distribution graph, etc. to explore each feature varies and its relation with other features including the target feature.

Figure 1 shows the distribution of age of the vehicle. The age is calculated using yearOfRegistration feature and current year (2017). Apparently, it is the number of years between the year of registration and 2017.

Figure 2 shows the average prices of each vehicle type. It shows that vehicle types like suv, coupe, and cabrio have higher average prices. This information can be used to analyze the vehicle type people generally tend to buy.

Figure 3 shows the distribution of age of the vehicle. The age is calculated using yearOfRegistration feature and current year (2017). Apparently, it is the number of years between the year of registration and 2017. From the graph, it is evident that most

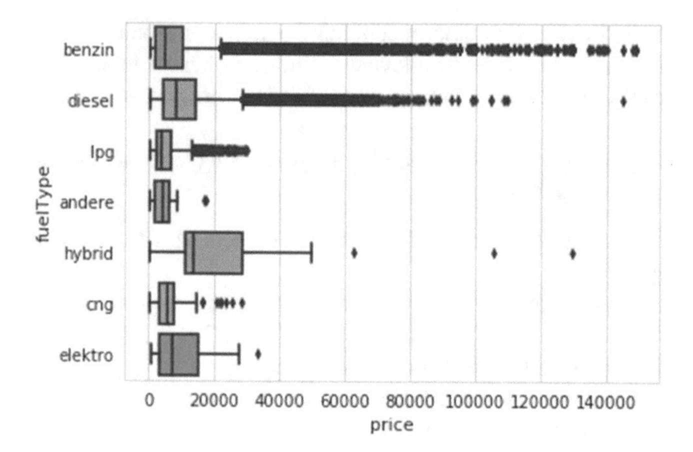

Fig. 1. Fuel type vs. Price.

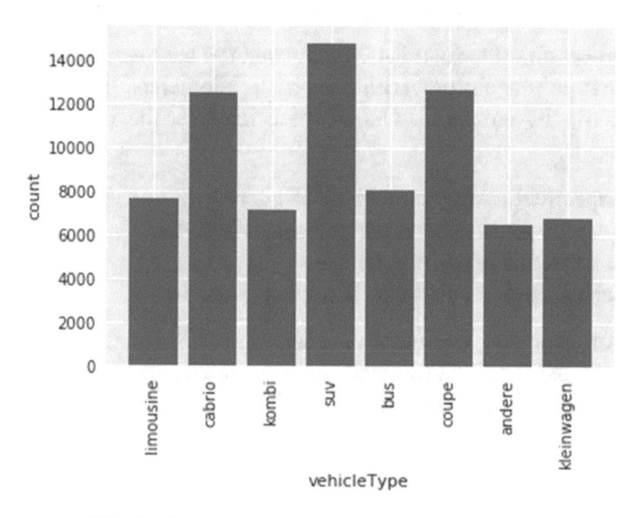

Fig. 2. Average price for a specific vehicle type.

of the vehicles are listed between 5 and 15 years old. The graph provides us the insight that, people tend not to sell their vehicle neither too early, as it defeats the purpose of buying sometimes, nor too late as the worth of the car goes down dramatically.

Figure 4 shows the top ten average prices by brand. This bar graph shows that average price of Porsche is much higher, which is around 40000 while the next highest is around 20000.

Figure 5 shows that the average prices of these cars where the damage is repaired are higher than others. This may be obvious, but offers insights to the cost of a vehicle.

Figure 6 shows the number of entries having particular fuel type. From previous figures, it is evident that most of the cars have fuel type as benzin or diesel and they have a wide range of price distribution compared to the other fuel types.

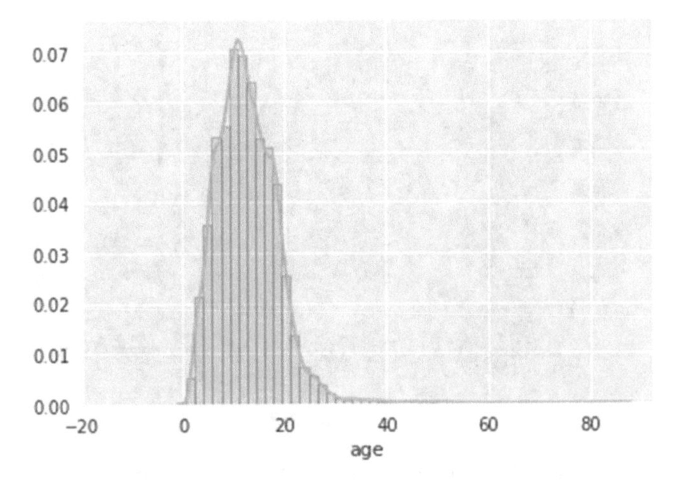

Fig. 3. Distribution of age of vehicle.

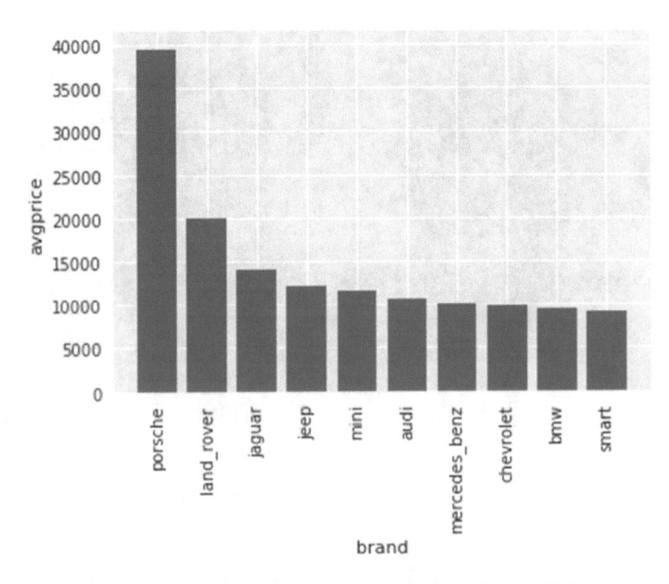

Fig. 4. Average price for a specific brand of vehicle.

5 Model Description

The problem at hand is a regression problem. We tried with linear regression and random forest regression. After much testing, it was found that random forest regression performed much better as it overcame the overfitting problem. The accuracy of the regression was less than 75% even in training data.

Random forest is primarily used for classification, but we used it as a regression model by converting the problem into an equivalent regression problem. Random forest

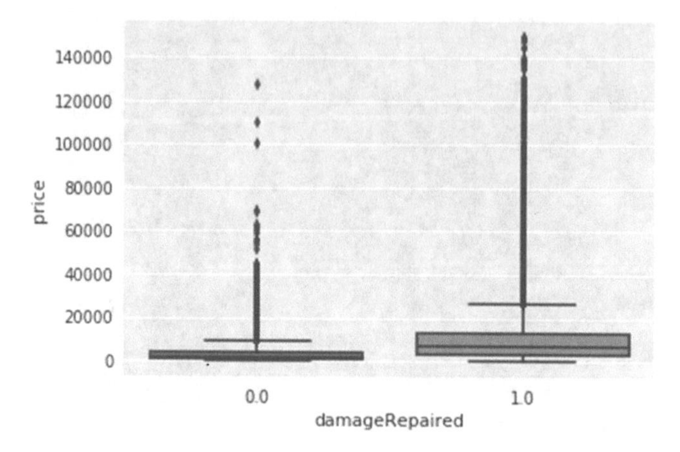

Fig. 5. Box plot of damage repaired vs. price.

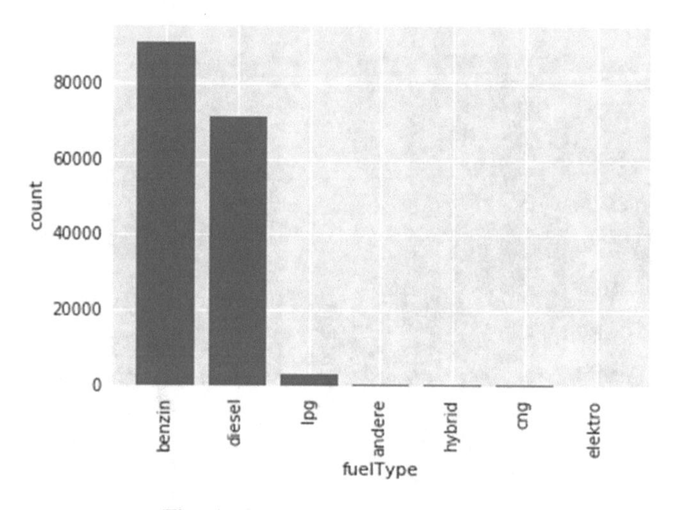

Fig. 6. Counts per type of fuel type.

comes under the category of ensemble learning methods, which contains a cluster of decision trees, usually hundreds or thousands in number. These trees are individually trained on parts of the dataset and help in learning highly unpredictable patterns by growing very deep. However, this may create an overfitting issue. This is overcome by averaging out the predictions of individual trees with a goal to reduce the variance and ensure consistency.

5.1 Model Parameters

Random Forest has several parameters to be tuned to which certain parameters have higher importance and are described below:

(1) *Number of Estimators*: This is the number of decision trees constituting the forest.
(2) *A maximum number of features*: It defines the maximum number of features a single decision tree should be trained.

A Grid Search Algorithm was employed to find the optimum number of trees, and best accuracy was found when 500 decision trees were used to build the forest. This was confirmed after iteratively increasing the number of decision trees in the multiples of 50.

Now, the maximum number of features is chosen to be equal to the number of features in the input data in case of regression problems and the square root of some features in case of classification. Since the problem at hand is a regression problem; we are going to the former.

5.2 Training and Testing

We split our input data into training, testing data and cross-validation with a 70:20:10 split ratio. The splitting was done by picking at random which results in a balance between the training data and testing data amongst the whole dataset. This is done to avoid overfitting and enhance generalization.

5.3 Accuracy

The model score is the coefficient of determination R^2 of the prediction. The training score was found out to be 95.82%, and the testing score was 83.63%.

The model was tuned in such a way that, only important features are taken and the rest are discarded. The important features are found using correlation, measuring their importance towards the estimation of the price of a vehicle. Overall, the random forest model effectively captured the nuances of the data and produced accurate predictions on the price of the vehicle.

6 Conclusion

This paper evaluates used-car price prediction using Kaggle dataset which gives an accuracy of 83.62% for test data and 95% for train-data. The most relevant features used for this prediction are price, kilometer, brand, and vehicleType by filtering out outliers and irrelevant features of the dataset. Being a sophisticated model, Random Forest gives good accuracy in comparison to prior work using these datasets.

7 Future Works

Keeping the current model as a baseline, we intend to use some advanced techniques like fuzzy logic and genetic algorithms to predict car prices as our future work. We intend to develop a fully automatic, interactive system that contains a repository of used-cars with their prices. This enables a user to know the price of a similar car using a recommendation engine, which we would work in the future.

References

1. Strauss, O.T., Hansen, M.S.: Advanced data science systems and methods useful for auction pricing optimization over network. U.S. Patent Application No. 15/213,941
2. Zhang, X., Zhang, Z., Qiu, C.: Model of predicting the price range of used car (2017)
3. Awad, W.A., ELseuofi, S.M.: Machine learning method for spam-email classification (2011)
4. Srivastava, D.K., Bhambhu, L.: Data classification method using support vector machine (2009)
5. Pudaruth, S.: Predicting the price of used cars using machine learning techniques. Int. J. Inf. Comput. Technol. **4**(7), 753–764 (2014)
6. Noor, K., Jan, S.: Vehicle price prediction system using machine learning techniques. Int. J. Comput. Appl. **167**(9) (2017)
7. Kuiper, S.: Introduction to multiple regression: how much is your car worth? J. Stat. Educ. **16** (3) (2008)

Utilizing Social Media Analytics to Recommend Personalized Gifts Using Content-Based and Multicriteria Collaborative Filtering

Marisa M. Buctuanon(✉), Joana Claire Alegado, Jessah Daculan, and Lauren Christy Ponce

University of San Jose – Recoletos, Cebu City, Philippines
marisamahilum@gmail.com,
joanaclairealegado@gmail.com, jzahdacz@gmail.com,
laurenchristyponce@gmail.com

Abstract. The study aims to use Facebook data to create a user profile to be able to recommend personalized gifts and help users to choose the right gift for a certain occasion. The Facebook data includes posts, comments, liked pages and user's biography. The data gathered are then preprocessed to create a user profile and item profile. The preprocessing stage includes data cleaning, and POS tagging. These profiles can be classified as book lover, fashion fiend, outdoor enthusiast, foodie, music lover and sports fan. These profiles are then mapped through content-based and multicriteria collaborative filtering. In content-based filtering three criteria are used, namely, receiver's personality, cosine similarity and user's chosen event. The events include birthday, valentines, wedding, anniversary, father's day, mother's day and graduation. Multicriteria collaborative filtering uses Pearson Similarity to distinguish similar users who would likely like the same product. Combining these results, a hybrid system is produced and a desirable list of items is recommended.

Keywords: Social media analytics · Personalized gift recommender system
Multicriteria collaborative filtering · Content-based filtering
Hybrid recommender system

1 Introduction

Gifts create emotion to the receiver. It embodies the giver's care, support, or affection towards the receiver. In return, the receiver will also feel the same or different depending on the gift. Whether the giver is curtailed by obligation or motivated by his desire, gift giving is significant [1]. Since gifts convey meaning, choosing the right gift to give is quite challenging. Receivers have variety of gift preferences. This is where social media comes in. The primary experience of value in social media is emotion [2]. In social media such as Facebook, a pool of individuals, which can be perceived as receivers, publicly display their experiences and sentiments towards something. The receiver's profile with his posts, shares, and pages liked can be used to generate a list of

K. Arai et al. (Eds.): FICC 2018, AISC 886, pp. 423–437, 2019.
https://doi.org/10.1007/978-3-030-03402-3_29

desirable gift items. With this approach, gift selection process is now hassle-free. Receiving a planned gift item based on receiver's preference will let the receiver feel important. Thus, developing a personalized gift recommender system is highly commendable.

Giftify is a hybrid gift recommender system that suggests a list of ideal gifts to give based on the receiver's preferences using his social media information and shared preferences of similar users. It gathers Facebook data which includes posts, comments, liked pages and user's biography to create a user profile. The user profile can be categorized as book lover, fashion fiend, outdoor enthusiast, foodie, music lover and sports fan. Based on these categories, the system will identify the right gift to give to the receiver on a certain occasion.

In the next succeeding discussions, the term user pertains to the receiver of the gift. Though, the giver of the gift is the one who uses the system to see the list of gifts to be given to the receiver. In this paper, the development process of the system will be discussed thoroughly and how social media analytics, information retrieval, and information filtering are integrated in the system.

1.1 Review of Related Literature and Works

Recommender Systems (RS) aim to automatically find the most useful products or services for a particular user, providing a personalized list of items based on the relationship of the user and items [3]. There are many techniques for recommender systems and the most common ones are collaborative and content-based filtering. Collaborative Filtering (CF) systems recommend products to a target customer based on the opinions of other users [4–7]. However, CF has some limitations. If there are few users who have preferred an item, then the item won't be endorsed to others [8]. Additionally, it suffers from problem like cold-start or new user [9]. Content-based Filtering (CBF) makes use of item features to compare the item with user profiles and the quality of the system is dependent on the selected features [9]. Combining these two techniques will somehow improve the overall accuracy of the system and overcome the problem of sparse matrices [8].

Experiencing a difficulty in finding a gift that will fit to the taste of the receiver is common to many. That is why, gift recommendation engines are increasing nowadays. Giftri.com is a recommendation system that uses big data to help people find gifts for their Facebook friends. The recommended gifts of this system are retrieved from Amazon's item database. Another gift recommender system is Gift kart. It suggests gifts ranging from movie DVD to match tickets to electronic items [10]. Moreover, there is also a content-based gift recommender system that is designed to rehabilitate patients through the gifts they received [11]. This system enjoins gifts according to the doctors' suggestions and patient's basic information such as disease, age, and gender.

1.2 Conceptual Framework

Figure 1 shows the four different modules that will be used to generate the gifts to be recommended. These are user profiling, item profiling, content and collaborative filtering modules. For the system to make predictions based on the user's interests and preferences, it should build the user and item model. In creating the user model, the Facebook posts, shares, likes, and personal information of the user will be used. The output of the user model will now be treated as the user profile. Likewise, an item model will be used to create an item profile. The item information will be coming from an e-commerce site. The data in the e-commerce site are populated by the developers of this system. The attributes of the user and item profiles are matched to compute for the significance of an item to a user.

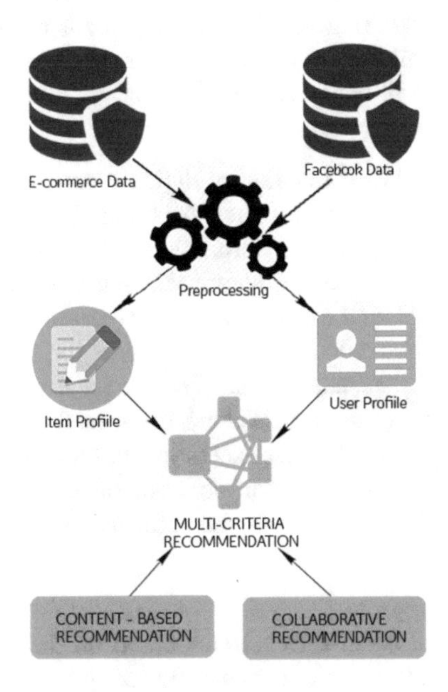

Fig. 1. Giftify landing page.

2 Methodology

Giftify has two main components, namely, the web application and the web services. The web application was developed using IntelliJ IDEA or Eclipse IDE with Java EE and Spring Plugin. This is where the user or the one who looks for an ideal gift to give can interact with the system. The web services we recreated to allow the processing of data separate from the app and to pass data from server to the web application. The server is run through Apache Tomcat server. The data of this system are saved and retrieved from MySQL database. Facebook API was used to get the user's gift preferences from Facebook. Stanford NLP was also used to create part-of-speech tagging during the preprocessing of data. Giftify is currently deployed in the local server of the IT department of University of San Jose – Recoletos. The e-commerce site was developed using a content management system (CMS), WordPress. As of now, the system is not commercially used. Everyone in the school can make use of the system for free.

2.1 Social Media Analytics

Social media analytics is concerned with developing and evaluating informatics tools and frameworks to collect, monitor, analyze, summarize, and visualize social media data [12]. This is usually driven by specific requirements from a target application. In this study, social media analytics is used to develop a system that will exploit the data from social media, specifically Facebook, and process this data to formulate useful facts. To produce the output of the system, information retrieval is first made. This is followed by information filtering, profile building, content-based filtering, and collaborative filtering. After all the necessary procedures are done, the useful output will then be displayed to the web application of Giftify.

(1) Information retrieval
Social listening is the process of monitoring digital conversations to understand what customers are saying about a brand and industry online [13]. To simulate this process, the system uses information retrieval. Information retrieval (IR) is finding material, usually documents, of an unstructured nature that satisfies an information need from within large collections. Before the system can retrieve information from Facebook, the user needs to grant permission to Giftify. To do this, the receiver should log-in via Facebook. Facebook will then authenticate and return the access token of the user which will be used in every transaction done between the system and Facebook. If Facebook has already provided the user access token, through Graph API, Giftify will request Facebook to provide the user's information such as status or posts, likes and shares. The researchers decided to limit the retrieved data for the posts within the current year.

(2) Content-based filtering
A content-based filtering system selects items based on the correlation between two objects [14]. In this phase, the system matches the user profile and item profile using its content. There are three criteria used by the system to match these objects. This is based on the user's personality, cosine similarity, and event percentage. However, information filtering, and building the user and item profiles will first be established before undergoing the user-item matching.

(a) Information Filtering
Filtering implies removal of data from an incoming stream rather than finding data in the stream [15]. The retrieved status, likes and shares of the user will undergo pre-processing in which the emoticons, special characters, and stop words will be removed. After doing the amputation, the system can now do POS tagging. Part-Of-Speech (POS) tagging is a process that reads text in some language and assigns parts of speech to each word such as noun, verb, adjective, etc. [16]. Giftify uses POS Tagger by Stanford NLP Parser to do the tagging. With the use of this, the system can now determine the subject of the text and the tokens that point at it. These tokens will be evaluated to identify the positivity or negativity of the subject. If the tokens pointing to the noun are positive, a score to the positive attribute to the noun is added. Likewise, if it is negative, a score to the negative attribute of the noun will be added by one. The polarity of the subject or noun is the difference of the word's positive and negative

count. This will determine the weight of how much interest the user has for that word based on the user's posts. If the polarity results to a negative, it means that the user dislikes the word which will not be used in generating the recommendations. All these information will be saved in the database. Same process will be done with the item in the ecommerce site. Once the item is added in the database, the metadata such as product name, item brand, item description and color will be used for POS tagging, and polarity of the nouns in each item.

(b) User and item profile building
In building the user profile, the nouns that were identified from the Facebook data will undergo category tagging. Category tagging, in this system, is a process in which each distinct term is checked against predefined associated word list for each personality type. The categories defined in this system are music lover, fashion fiend, outdoor enthusiast, sports fan, foodie, and bookworm. The researchers have defined bag of words that are associated on each category. These words were taken from a Word Associations site. A term vector will be generated to sum up how many times a category is being related to each noun. Using the frequency count, the number of times the term occurs, this will determine which category has the highest percentage. This will let the system know as to what types of items need to be recommended first. However, other categories can also be associated by the user with some degree of importance. The frequency and the percentage of each category will be saved in the database with id of the user. Similarly, building the item profile will also undergo the same process. The first stage is to identify the category of the item based on the nouns generated in the filtering phase. Next is to make a term vector to compute for the number of times each noun is associated to a certain category. This will again determine the degree of membership each item is linked to a certain category.

(c) Content-Based Filtering Criteria
Personality percentage:

There are six types of personality both for user and item. These are book lover, fashion fiend, outdoor enthusiast, foodie, music lover, and sports fan. To determine as to which personality or category does user and item belong, personality percentage is computed. The frequency count retrieved from profile building is used in computing the user and the item personality percentage for each category. The equation below shows how personality percentage is computed:

$$personality(n \mid m) = \frac{Total\, number\, of\, nouns\, based\, on\, m}{Total\, number\, of\, nouns} \tag{1}$$

Where n represents the user and m is the personality type. The highest personality percentage of the user will be the basis of what type of products should the system recommend. The system will temporarily save the computed personality percentage for both the user and the items. Computing for the personality percentage of an item will only happen at the start of product adding in the database.

Cosine similarity:
Since user interest and item details are made up of words, term matching can be used. The term matching process can be achieved using cosine similarity. Cosine similarity used to know how similar two term vectors are. There are four processes before computing cosine similarity of the two objects. These include formulating the vector matrix, computing for the IDF (Inverse Document Frequency) of the item details, calculating for the TF-IDF (Term Frequency and Inverse Document Frequency) of both the user preference and item, and lastly the computation of the user preference and item's length.

TF-IDF is used to compute for the weight of a term which will be used to evaluate how important a word is to a document in a collection or corpus [17]. At the initial stage, the item's and user's keywords and their frequency were retrieved to compute for the IDF of both objects. A matrix score is made where the value is either 1 if the item is associated to the item keyword or 0 if the keyword has no association to the item. This matrix will also be done with the user's preferences. These represent the term frequency of the two objects. The more a word occurs in a post means that the user is more interested to that word. This word may be in form of adjective or noun that may refer to a specific item in the database. But this is only using the TF. To measure how unique a word is, it is important to know how infrequently the word occurs across all documents should be identified. This is with the use of IDF. To compute for the IDF of the item using its keywords, the equation below is used in computing for the IDF.

$$\mathrm{IDF} = \mathrm{LOG}\left(\frac{\mathrm{n}}{\mathrm{m}}\right) \tag{2}$$

Where n represents the total number of items in the database and m represents the count of distinct words associated to an item. Distinct words are the items metadata basing on the item name and item details. The computed IDF of the item will be used to compute for the TF-IDF of the item and user interest. The next step is to simultaneously compute for both the TF-IDF of the item and user interest. The formula below is used in computing for the TF – IDF

$$\mathrm{TF} - \mathrm{IDF} = \mathrm{TERM\ FREQUENCY\ SCORE} * \mathrm{IDF} \tag{3}$$

After computing for the TF_IDF of both the user interest and item keywords, the system will now compute for the length of these two objects. Items who have long descriptions has higher word frequency than those who have short description. To address this bias, the length of the document is computed. The role of the item length is to fairly normalize the documents of all length. The formula below is used in computing for the item and user interest length.

$$\mathrm{length} = \sqrt{\sum_{i=1}^{z} (n_i)^2} \tag{4}$$

Where, n = TF-IDF (Term Frequency – Inverse Document Frequency) of the item or user interest. The computed length will then be used to compute for the distance of the two objects using cosine similarity of the product in respect to the user interest keyword. If this distance is small, there will be high degree of similarity. The system will temporarily save the computed cosine similarity of each item. The formula below displays the normalized dot product of the two objects.

$$\cos \mathrm{sim}(\mathrm{user}, \mathrm{item}) = \frac{\sum_{i=1}^{z} x_i * y_i}{(n * m)} \tag{5}$$

Where,
x - TF-IDF (Term Frequency – Inverse Document Frequency) of the user interest
y - TF-IDF (Term Frequency – Inverse Document Frequency) of the item
n - User Interest Length
m - Item Length

Event percentage:
The giver of the gift selects a certain event and these include birthday, valentines, wedding, anniversary, father's day, mother's day and graduation. The researchers already defined bag of words with their corresponding percentage to a certain event. The keywords of each item are evaluated where 1 is given if the keyword belongs to a certain event and 0 otherwise. The system multiplies each item event percentage by 1 if the item is associated to the chosen event. Then temporarily save the computed event result weight. The equation below shows the event percentage computation.

$$event(item \mid category) = \frac{\mathrm{x}}{\mathrm{y}} * n \tag{6}$$

Where,
x - The total number of nouns of an item where the event type is equal to the selected one
y - The total number of nouns where the event type is equal to the selected one
n - Represents the flag 1 or 0 which indicates that item is associated to the selected event or not.

Total percentage weight:
The result of content-based filtering in this system is computed by getting the total percentage weight. The personality has a weight of 50%, cosine similarity has a weight of 40%, and the event percentage has a weight of 10%. The items are then computed based on the formula below. Items who have a total percentage higher than the threshold, which is 5.0, are candidates for recommendation.

$$\mathrm{total\ percentage}(\mathrm{item}) = \frac{(\mathrm{personality}{*}.5 + \cos \mathrm{sim} * .4 + \mathrm{event}{*}.1)}{3} * 100 \tag{7}$$

(d) Multicriteria Collaborative Filtering
Collaborative filtering also referred to as social filtering, filters information by using the recommendations of other people [5–7]. The system combines the two criteria or approaches for collaborative filtering. These are user to user and item to item

collaborative filtering [18]. In the user-based approach the algorithm produces a rating for an item by a user by merging the ratings of other users that are similar to the target user. While in the item-based approach, it computes for the rating of an item by the target user by examining the set of items that are similar to the item that the target user has rated [4]. The system has its way to allows users to rate how much satisfied they are with the items given to them. Below is the rating with their corresponding interpretation (Table 1).

Table 1. Rating interpreation

Number	Interpretation
0	N/A
1	Dislike
2	Neutral
3	Like
4	Strongly Like
5	Love

User-based approach:
User-based collaborative filtering predicts a test user's interest in a test item based on rating information from similar user profiles. In this approach, the system uses Pearson Similarity to distinguish similar users who would likely like the same item. Similar here means that the two user's ratings have a high Pearson correlation. The system first checks if there is a new user, and then calculates for the user to user Pearson correlation towards all other users of the application. The user profile Pearson correlation calculation includes the six personality types, namely book worm, fashion fiend, sports fan, outdoor enthusiasts, music lover, foodie. It also includes two other attributes from user's basic profile which are gender and age. Age is discretized with respect to Erikson's Stages of psychosocial development. The system computes for the average rating for items rated by all users. The degree of correlation is used to measure how similar the rating of a user, the desired recipient to users, to all other users of the app. The result of correlation ranges from +1 to −1. Positive correlation indicates that both variables increase or decrease together, whereas negative correlation indicates that as one variable increases, so the other decreases, and vice versa. Zero correlation implies that there is no linear correlation between two users. Positive correlation between two users indicates that both have similar tastes of products or items while negative correlation indicates that both have dissimilar tastes of products or items. The closer the result is to one the more similar the user's tastes are, users having a result of +1 are perfectly similar.

$$\omega_{a,u} = \frac{\sum_{i=1}^{n}\left(r_{a,i} - \overline{r_a}\right)\cdot\left(r_{u,i} - \overline{r_u}\right)}{\sqrt{\sum_{i=1}^{n}\left(r_{a,i} - \overline{r_a}\right)^2}\cdot\sqrt{\sum_{i=1}^{n}\left(r_{u,i} - \overline{r_u}\right)^2}} \quad (8)$$

Where,

a - Active user or the desired recipient of the gift

u - All users of the system, excluding desired recipient (a)

n - The number of items that ALL users have rated, including the desired recipient (a)

r_a- The average ratings given by active user (a) to all items

$r_{u,i}$- The average of rating given by all users u(n) of the system, excluding desired recipient (a)

$\omega_{a,u}$ - The degree of correlation or degree of similarity between user (a) and user (u) with respect to the rate given to all items

Item-based approach:

The item-based approach looks into the set of items the target user has rated and computes how similar they are to the target item I and then selects n most similar items [4]. The system calculates for the item rating prediction to a target user. The similarities between different items in the dataset are calculated using Pearson Correlation. These similarity values are used to predict ratings for user-item pairs not present in the dataset. The system generates a user-item matrix to calculate the rating prediction. The threshold that determines that these items are important to consider is 2. Items with prediction score that passed the set threshold value or the minimum rate of item's prediction score is included in the recommendation list. Below is the equation for item to item computation.

$$p(a,i) = \overline{r_a} + \frac{\sum_{u=1}^{m} \left(r_{u,i} - \overline{r_u} \right) \cdot \omega_{a,u}}{\sum_{u=1}^{m} \left| \omega_{a,u} \right|} \tag{9}$$

Where,

a- Active user or the desired recipient of the gift

m-The number of items that ALL users have rated, excluding the desired recipient (a)

r_a- The average given by other users (a) to an item (i)

$r_{u,i}$- The average rating given by all users r_u of the system to item i, excluding desired recipient (a)

$\omega_{a,u}$- The degree of correlation or degree of similarity between user (a) and user (u) with respect to the rate given to all items

p(a,i)- The prediction score or rating of user (a) to an item (i)

(e) Hybrid Recommender System

To make the system more personalized and effective to recommend gifts, the result of content-based filtering and collaborative filtering are combined. In which the personality, cosine, event chosen, and rating of the target user with respect to items in unison with the suggestions of other users are considered. The result of both the collaborative filtering both the user-based and item-based approach are ranked according to the result of content-based. The system determines as to what items have passed the collaborative filtering and search for their total percentage which is the result content-based filtering result. The results are then arranged in ascending order.

3 Results and Discussion

3.1 User Interface

Figure 2 shows the Landing Page of the system, where users can navigate through About page, Contact Us page, Shop page, or Login page of the application.

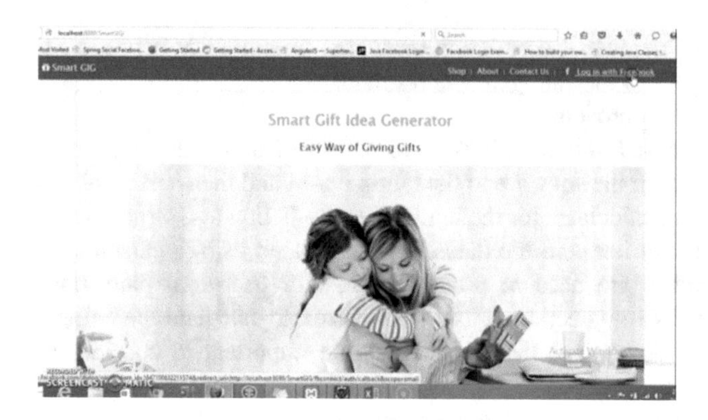

Fig. 2. Giftify landing page.

Figure 3 shows the Authentication page where users logs on via Facebook in which after this process, the system have access tokens to retrieve data from Facebook.

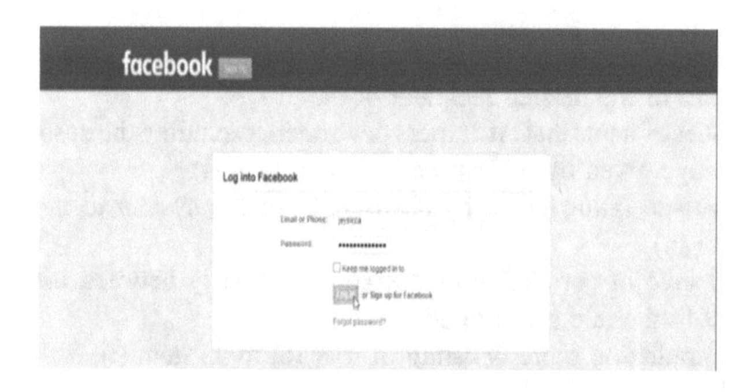

Fig. 3. Facebook authentication page.

After logging in to the system, the user is redirected to his Home page. Figure 4 shows the Home page of the user in which he can view his basic information, personality percentage and upcoming events. Users can also manage calendar and manage contacts.

Figure 5 displays the Contact page where users can add contacts by adding the email address of the other users.

After clicking the profile of a recipient, the user can now see the list of items that are desirable to give to the recipient. Figure 6 displays the result of the recommendation. These items are based on the preferences of the recipient and shared preferences of similar users.

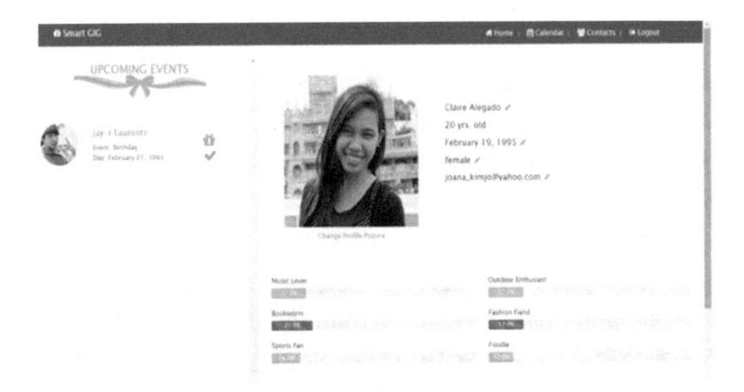

Fig. 4. User's home page.

Fig. 5. Add contact page.

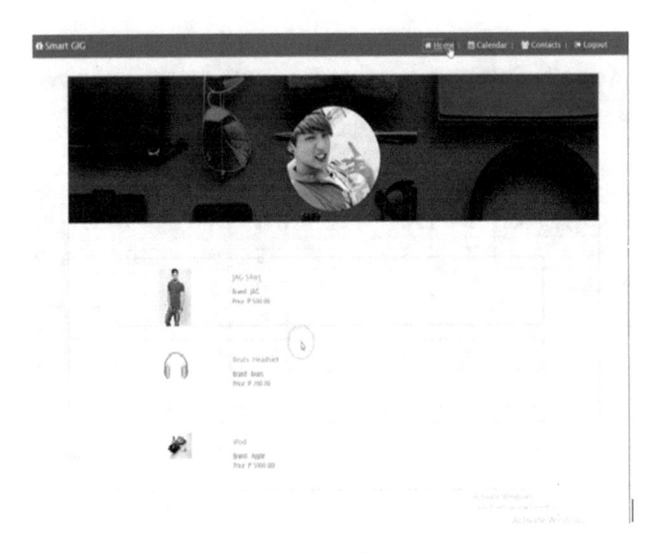

Fig. 6. Recommended gifts page.

3.2 Limitations

(1) Recommend gifts to nonusers

The system cannot recommend gifts to receivers who are not part of the system. This is because the system cannot have auto generated access tokens to retrieve their Facebook information.

(2) Real-time profile update

Every six months, the system will have an incremental update of the user profile. The system will undergo again the steps of building the user profile from data retrieval to calculating the category percentage. However, it will only process that latest data that were not retrieve from that last processing.

3.3 Accuracy Testing

To compute for the overall accuracy of the recommendation, the content and collaborative filtering results, confusion matrix is used. Here, 90 recipients of the gifts were asked to evaluate how much they are satisfied with the gift recommendation of the system using the rating in Table 1.

Table 2 displays the result after counting the predicted outcome versus the actual result. The result of each row represents the actual value discovered during the testing process. While the result in each column, represents the predicted value before testing the system. In this table, there were 55 recipients who love the recommended gifts. However, this actual result exceeds the prediction which is only 37 recipients will love the recommended gifts. This will somehow tell us that the goal of the system to give personalized gifts is efficient. But this is not yet the basis for conclusion, confusion matrix uses precision and recall to validate further the result.

Table 2. Collaborative and content filtering confusion matrix

Actual	Predicted					
	Love	Strongly Like	Like	Neutral	Dislike	
Love	**35**	14	6	0	0	**55**
Strongly Like	0	**20**	4	0	0	**24**
Like	1	3	**5**	0	0	**9**
Neutral	1	0	0	**0**	0	**1**
Dislike	0	1	0	0	**0**	**1**
	37	**38**	**15**	**0**	**0**	

Precision is the proportion in which the item will be classified in that rating which can be computed with the following formula:

$$precision(rating) = \frac{TP}{TP + FP} \quad (10)$$

The true positive (TP) is the proportion in which the prediction is correct. In addition, false positive (FP) is the proportion in which the prediction is not classified that way but was classified after all. In Table 3 precision column, most of the recipients agreed to the gifts recommended to them. Almost none of them disliked the recommended gifts. Recall is the proportion in which the item has been classified that way which can be computed using the formula below.

$$recall(rating) = \frac{TP}{TP + FN} \tag{11}$$

Table 3. Precision and recall table

Accuracy: 67%	Precision	Recall
Love	35/55 = **64%**	35/37 = **95%**
Strongly Like	20/24 = **83%**	20/38 = **53%**
Like	5/9 = **56%**	5/15 = **33%**
Neutral	**Undefined**	**Undefined**
Dislike	**Undefined**	**Undefined**

False negative (FN) is the proportion in which the item is not classified in that rating, but was classified. Table 3 in "Love" rating, shows high values of precision and recall which means that the system mostly correct in classifying the item to belong in the said rating. At the same time, no item is classified and has been classified to be disliked by the recipient. To compute for the overall accuracy of the recommendation, below is the formula.

$$\text{accuracy} = \frac{\text{TP} + \text{TN}}{\text{TP} + \text{TN} + \text{FP} + \text{FN}} \tag{12}$$

Where
TP (true positive) = equivalent to hit
TN (true negative) = equivalent to correct rejection
FP (false positive) = equivalent to false alarm, Type | error
FN (false negative) = equivalent to miss, Type || error

The highlighted values in Table 2 which forms diagonal in the center are considered as true positive which means a hit. The overall accuracy of the system is 67%. Furthermore, the goal of the system is to recommend gifts that are highly suitable to the recipient, and this is achieved since 55 out of 99 users rated the gift as "Love". None of them dislike the product. This shows that combination of the two algorithms improves the systems accuracy.

3.4 Summary of Findings

The researchers found out that the amount of datasets available affects the recommendation process. Users who aren't active in Facebook have less data compared to active users. Moreover, the occurrence of the word to the user post may also affect the result of the recommendation. Words which have higher occurrence may mean that the user really likes that word that may refer to an item which is a limiting factor of Content Based Recommendation which only refers to the occurrence of the words. Using Collaborative Filtering as reinforcement, the result of the recommendation will be more accurate. Collaborative filtering helps solve the crisis of content based by allowing the user to rate the items recommended. With this, Collaborative via Content Filtering is likely to give a more accurate recommendation than content alone. Furthermore, the Facebook data should be in English, since the system can only process words associated to it.

4 Conclusion and Future Works

Through the richness of Facebook data, the system was able to recommend gifts highly suitable to the recipients. With the use of collaborative and content filtering, recommendations given are more accurate. Overall, the system was able to achieve its goal, which is to recommend highly suitable gifts to the receiver. Moreover, the researchers would want to achieve perfection in the current system. Thus, suggesting future works will be considerable.

The harvested data from Facebook depends upon the number of post, comments and likes of the user. Some users are not active in Facebook, thus gathering relevant data is limited. To solve this problem, researchers recommend adding other social networking sites such as Twitter as a source for recommendation. Facebook data in the form of different dialects other than English are not processed by the system. It would also help if future researchers will also consider words in other dialects to make the recommendation more relevant to the user interest. The researchers would also recommend adding more personality type since the system is only limited to seven personality types. Adding various personality types will help the system create a more accurate user profile. Furthermore, the system is currently accessing an ecommerce made by the researchers. With this, recommended products are limited to the products inputted by the researchers. To resolve this problem, researchers recommend finding an existing ecommerce online that would allow the researchers to access their database of products.

References

1. Goodwin, C., Smith, K.L., Spiggle, S.: Gift giving: consumer motivation and the gift purchase process. In: Advances in Consumer Research, pp. 690–698 (1990)
2. Chakrabarti, R., Berthon, P.: Gift giving and social emotions: experience as content. J. Public Aff. **12**(2), 154–161 (2012)

3. Bellogín, A., Cantador, I., Díez, F., Castells, P., Chavarriaga, E.: An empirical comparison of social, collaborative filtering, and hybrid recommenders. ACM Trans. Intell. Syst. Technol. **4**(1), 14 (2013)
4. Sarwar, B., Karypis, G., Konstan, J., Riedl, J.: Analysis of recommendation algorithms for e-commerce. In: Proceedings of the 10th International Conference on World Wide Web, Hong Kong, pp. 158–167 (2001)
5. Schafer, B., Frankowski, D., Herlocker, J., Shilad, S.: Collaborative filtering recommender systems. In: The adaptive web, pp. 291–324. Springer, Heidelberg (2007)
6. Breese, J.S., Heckerman, D., Kadie, C.: Empirical analysis of predictive algorithms for collaborative. In: Proceedings of the Fourteenth conference on Uncertainty in artificial intelligence, pp. 43–52 (1998)
7. Herlocker, J., Konstan, J., Borchers, A., Riedl, J.: An algorithmic framework for performing collaborative filtering. In: Proceedings of the 22nd Annual International ACM SIGIR Conference on Research and Development in Information Retrieval, pp. 230–237 (1999)
8. Hegde, A., Shetty, S.K.: Collaborative filtering recommender system. IJETST **2**(7), 2885–2889 (2015)
9. Cano, E., Morisio, M.: Hybrid recommender systems: a systematic literature review. Intell. Data Anal. **21**(6), 1487–1524 (2017)
10. Tomar, P., Arora, P., Goel, A., Saini, D.: Social profile based gift recommendation system. Int. J. Comput. Sci. Inf. Technol. **5**(3), 3670–3673 (2014)
11. Yu, Y., Wang, Y.: Design and implementation of a content-based gift recommender system. In: Soft Computing in Information Communication Technology (2014)
12. Zeng, D., Chen, H., Lusch, R., Li, S.-H.: Social media analytics. IEEE Comput. Soc. 13–16 (2010)
13. Boolean Retrieval: Cambridge University Press, Cambridge, 1 April 2009
14. van Meteren, R., van Someren, M.: Using content-based filtering for recommendation. In: Machine Learning in the New Information Age: MLnet/ECML2000 Workshop, pp. 47–56 (2000)
15. Belkin, N.J., Croft, W.B.: Information filtering and information retrieval: Two sides of the same coin? Commun. ACM **35**(12), 29–38 (1992)
16. Toutanova, K., Klein, D., Manning, C.D., Singer, Y.: Feature-rich part-of-speech tagging with a cyclic dependency network. In: SIGDAT Conference on Empirical Methods in Natural Language Processing and Very Large Corpora (EMNLP/VLC-2000), pp. 173–180 (2000)
17. Ramos, J.: Using TF-IDF to determine word relevance in document queries. In: Proceedings of the First Instructional Conference on Machine Learning, vol. 242, pp. 133–142 (2003)
18. Wang, J., de Vries, A.P., Reinders, M.J.: Unifying user-based and item-based collaborative. In: Proceedings of the 29th Annual International ACM SIGIR Conference on Research and Development in Information Retrieval, Washington, USA, pp. 501–508 (2006)

New Modification Version of Principal Component Analysis with Kinetic Correlation Matrix Using Kinetic Energy

Sara K. Al-Ruzaiqi[1(✉)] and Christian W. Dawson[2]

[1] Computer Science Department, Loughborough University, Muscat, Oman
s.k.s.al-ruseiqi@lboro.ac.uk

[2] Computer Science Department, Loughborough University, Loughborough, UK
c.w.dawsonl@lboro.ac.uk

Abstract. Principle Component Analysis (PCA) is a direct, non-parametric method for extracting pertinent information from confusing data sets. It presents a roadmap for how to reduce a complex data set to a lower dimension to disclose the hidden, simplified structures that often underlie it. However, most PCA methods are not able to realize the desired benefits when they handle real world, and nonlinear data. In this work, a modified version of PCA with kinetic correlation matrix using kinetic energy is proposed. The features of this modified PCA have been assessed on different data sets of air passenger numbers. The results show that the modified version of PCA is more effective in data compression, classes reparability and classification accuracy than using traditional PCA.

Keywords: Principle Component Analysis (PCA) · Kinetic correlation matrix Kinetic energy · Algorithm · Prediction

1 Introduction

Principal Component Analysis (PCA) is a classical multivariate data analysis technique, which is popular within linear feature extraction as well as the data compression of numerous uses [1]. PCA has been applied in numerous areas of information processing to prepare data due to its distinctive result of error reducing and correlating properties. PCA compresses most of the information in the first data space into a fewer features. It attempts to look for a subspace in which the variance is maximized [2]. The PCA subspace is spanned through the eigenvectors corresponding to the top eigenvalues of the sample covariance matrix. PCA also can be applied in data preparation for both supervised and un-supervised learning and recognition processes [3].

However, most PCA strategies might not result in desirable classification benefits when they cope with real world, nonlinear data. As nonlinear PCA and its variants can effectively capture the nonlinear relations, they might provide more effective power to cope with the real world, nonlinear data [4]. It is recognized that PCA is designed to find the most indicative vectors, i.e., the eigenvectors corresponding to the best eigenvalues of the sample covariance matrix.

K. Arai et al. (Eds.): FICC 2018, AISC 886, pp. 438–450, 2019.
https://doi.org/10.1007/978-3-030-03402-3_30

As data with good spectral resolution results in unwanted data for classification, a proven way to conquer this issue is reducing the dimensionality of data space. Different feature extractions, as well as selection strategies, recommend using PCA, as it is highly effective and involves a mathematical process which transforms a selection of (possibly) correlated variables into a (smaller) selection of uncorrelated variables known as principal components [5].

The sheer size of data in the modern age is not only a challenge for computer hardware but also a bottleneck for the performance of many machine learning algorithms. Identifying patterns in data is one of the main goals of a PCA analysis, and it only works by reducing the data dimensionality only when there is strong correlation between the variables. In brief, PCA is a data analysis technique which finds directions of maximum variance in high-dimensional data and projects them onto a smaller dimensional subspace while retaining most of the information.

In this work, a modified version of PCA with kinetic correlation matrix using kinetic energy is proposed, where the transformed matrix is computed from samples of selected features only. The efficiency of the modified and traditional versions of PCA is compared by applying them to an air passenger dataset. The results show that the modified version of PCA is more effective in data compression, class reparability and classification accuracy than using traditional PCA.

2 Modification of PCA

Since the original definition of PCA via approximating multivariate distributions by planes and lines [2], scientists have defined PCA from various elements [2, 3]. Among the definitions, utilizing the covariance matrix of the training sets to explain PCA is extremely well known in pattern recognition as well as the machine learning community.

Current implementations of PCA use a correlation matrix, the matrix obtained by pairwise correlation using Pearson correlation coefficient. However, in some cases the Pearson correlation coefficient could be limited in the sense that it fails to capture other properties of the data outside of the linear relation. For example, the correlation of two random vectors: $x = \{-4, -3, -2, -1, 0\}, y = x^2 = > Cor(x, y) = 0,$ using Pearson coefficient. However, this result is not capturing the non-linear relation between the two random vectors given by the functional transformation $(x)^2 \rightarrow (y)$ which means the correlation is not zero (just non-linear). In order to improve this, the following two features have been introduced into traditional PCA in this work.

- *Information energy*: first introduced in 1966, is an analogy of the kinetic energy from physics to probability, which can be defined as follows:
  $x_{1,}x_{2,\ldots,}x_n$ and corresponding probabilities:

$$P = (P_{1,}P_{2,\ldots\ldots,}Pn)$$

$$\mathrm{IE}(p_{1,}p_{2,\ldots\ldots,}p_n) = \sum\nolimits_{i=1}^{n} pi^2$$

If the experiment has n outcomes, and every outcome has the same probability $1/n$, then the information energy IE = $1/n$. If the experiment results in same outcome, then the probability for every outcome is 1 and the information energy has maximum value of IE = 1.

The information energy increases when the randomness decreases. It is like reverse of Shannon entropy, for measuring bits of information to determine uncertainty. It is also an entropy, but the correct way to think about it is as ½ $*$ m $*$ v2 of a random vector. Simple, but very powerful, the kinetic energy method works very well to improve the accuracy or improve some machine learning methods on row data especially if there are groups of categorical data, even if they are continuous they could be discretized.

- *Informational Correlation Coefficient,* also known as Onicescu's correlation coefficient, is a function of the joint probability density distribution of the two vectors x and y. Assume we have two random vectors x and y, the information correlation coefficient can be described as:

$$\mathrm{O}(\mathrm{x},\mathrm{y}) = \frac{\sum_{\mathrm{k}} \mathrm{p}^{(\mathrm{Pk}) * \mathrm{p}(\mathrm{Qk})}}{\sqrt{\mathrm{IE}(\mathrm{P}) * \mathrm{IE}(\mathrm{Q})}} \tag{1}$$

This is only applicable for the discrete data that we have dealt with in this research.

The Pearson correlation captures only linear properties of the manifold on which our raw data lives. For instance: if we take a random vector in R $\mathrm{x} = \mathrm{c}(-4, -3, -2, -1, 0, 1, 2, 3, 4)$ and y = x ^ 2, Pearson or Spearman, will yield 0 correlation when in fact it is 0.5 because of the functional transformation x- > ^ 2. In this work, a new correlation coefficient, as a performance metric, instead of cross entropy as in the case of neural networks, or, in the case of genetic algorithms, as fitness functions, has been applied in the modified PCA.

Previously, PCA was utilized to decrease large data sets, correlated by a number of correlation metrics, or used in addition to deriving new features. Consequently, Pearson correlation or the covariance matrix is used to determine eigenvalues and eigenvectors. Having a completely different correlation metric that captures kinetic properties of two random vectors against one another has also been used in creating a modified version of original algorithm with this new correlation matrix.

Hence, we implemented new correlation metrics, and the new idea was to modify the original PCA for obtaining eigenvectors and eigenvalues for dimensionality reduction using a correlation matrix with our kinetic correlation coefficient.

3 Implementing Modified PCA

In this work, a new correlation coefficient method called Octave has been introduced. The correlation is used as a method for feature selection (calculated between two features) using Kinetic Energy. The new Octave correlation makes a useful contribution as it provides a new measure of dependence between random vectors that capture non-linear relationships as well.

The modified version of PCA was assessed using a data set of air passenger numbers, from where the features of the modified PCA were derived, using kinetic correlation metrics instead of Pearson correlation coefficient based on kinetic correlation theory.

3.1 Implementation Setup

This function returns information coefficient IC for two random variables defined as the dot product of probabilities corresponding to each class:

```
def ic(vector1,vector2):
    a=vector1
    b=vector2
    prob1=np.unique(a,return_counts=True)[1]/a.shape[0]
    prob2=np.unique(b,return_counts=True)[1]/b.shape[0]
    p1=list(prob1)
    p2=list(prob2)
    diff=len(p1)-len(p2)
    if diff>0:
        for elem in range(diff):
            p2.append(0)
    if diff<0:
        for  elem in range((diff*-1)):
            p1.append(0)
    ic=np.dot(np.array(p1),np.array(p2))
    return ic
```

And, having functions for kinetic energy of a vector and for information correlation, we can define a new function that computes kinetic correlation. This function will return correlation based on kinetic energy as illustrated below:

```
def o(vector1,vector2):
    i_c=ic(vector1,vector2)
    o=i_c/np.sqrt(kin_energy(vector1)*kin_energy(vector2))
    return o
```

The formula is updated such that the denominator contains sqrt in order to have probabilities bounded between 0 and 1.

SHAPE will return the number of items in the numpy array in the form of a tuple, then creates a matrix with the number of rows initialized with zero values.

```
rows=data.shape[1]
rows
matrix= np.zeros((rows,rows))
```

Then the correlation matrix is created with the function o() that was defined previously, as shown in Table 1. The correlation matrix obtained by the Pearson method is also listed in Table 2 for comparison.

Table 1. Correlation matrix with the function O()

	0	1	2	3	4	5
0	1.000	1.000	0.974	0.326	0.184	0.229
1	1.000	1.000	0.974	0.326	0.184	0.229
2	0.974	0.974	1.000	0.320	0.180	0.223
3	0.326	0.326	0.320	1.000	0.071	0.131
4	0.184	0.184	0.180	0.070	1.000	0.490
5	0.229	0.229	0.223	0.131	0.490	1.000

Table 2. Correlation matrix on basis of 'Pearson R' model

	0	1	2	3	4	5
0	1.000	0.751	0.770	−0.041	−0.027	0.000
1	0.751	1.000	0.959	−0.013	−0.031	0.000
2	0.770	0.959	1.000	−0.020	−0.023	0.000
3	−0.041	−0.013	−0.020	1.000	0.216	0.000
4	−0.027	−0.031	−0.023	0.216	1.000	0.000
5	−0.000	−0.000	−0.000	0.000	0.000	1.000

3.2 Comparison of Modified PCA with Kinetic Correlation Matrix from Kinetic Energy and PCA with Pearson R Correlation

Our contribution is based on changing the correlation matrix that uses Pearson R correlation or, in some cases, the covariance, with a correlation matrix based on the Onicescu correlation coefficient. The results of testing the kinetic correlation of our data sets using the Pearson coefficient are shown in Figs. 1 and 2.

As expected the kinetic correlation has a much higher kinetic correlation matrix from kinetic energy than the Pearson one. Pearson's R is able to detect only linear relations in data. The graphs have the same list of seven columns on both x and y axis. The colouring of each particular square shows the actual correlation between the columns on the scale of 0 to 1.0. So, if the color is dark, there is low correlation and vice-versa.

3.3 Features Obtained from Kinetic Energy PCA Components

In this section, we implemented XGBoost [6]. This is an algorithm that has recently been dominating applied machine learning for structured or tabular data and it is designed for speed and performance. It has been applied here to a training data set of passenger numbers with a dataset of 51,983 observations with 9 variables. In order to get a better estimate of model performance, we used a variant of the famous 1-fold cross validation. We split dataset into a training set (75% of the data) and a test set (25% of the data) randomly for 1 different time and measure accuracy, false positive rate and false negative rate.

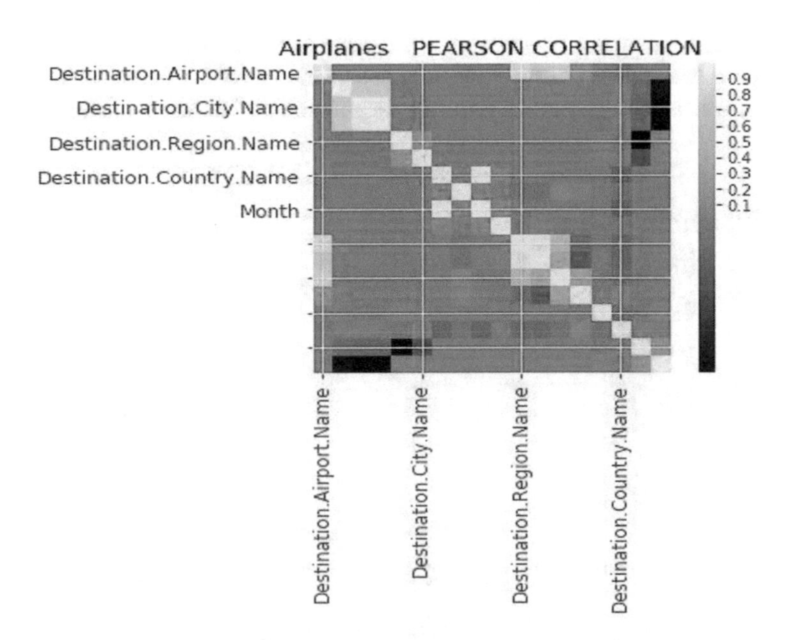

Fig. 1. Air passenger numbers data with Pearson Correlation.

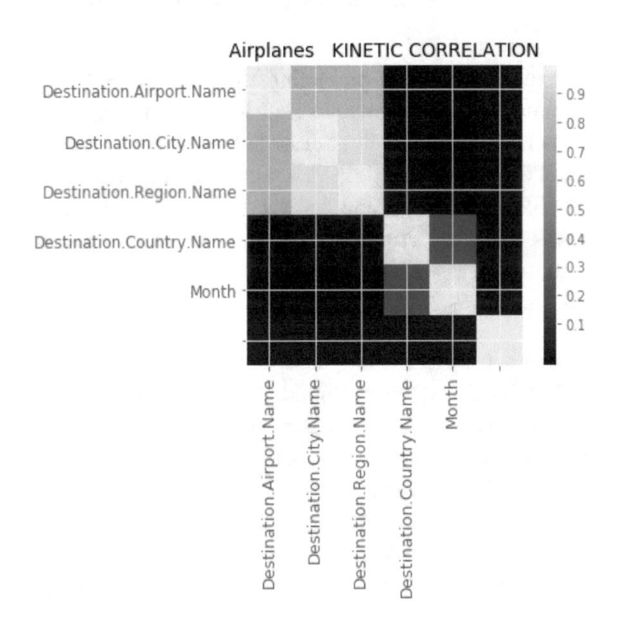

Fig. 2. A train passenger numbers data with Kinetic Correlation.

The XGBoost model was run within Python machine learning modules and the calculated mean values (Fig. 3) are very much nearer to the actual values of one. xgb.train, which is an advanced interface for training an XGBoost model.

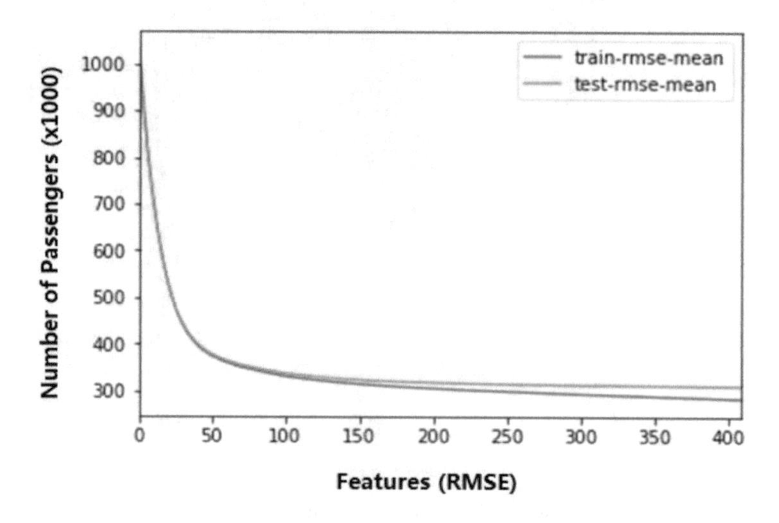

Fig. 3. The mean values of features obtained from Kinetic Energy PCA Components.

Figure 4 shows the features for predicting the number of passengers from most important to least important. Here it shows that JetFuel, Month, and fare are the most predictive values. As is noted in the plot below the features obtained from the modified PCA, called kineticPCA1 and kineticPCA2, are captured with reasonable influence after running the XGBOOST model and inspecting feature importance. The number of passenger predictions using the Prediction model is given in Table 3.

3.4 Features Obtained from Deep Learning Hidden Layers

In this step, we created a different engineered dataset in order to have diversity in multiple datasets. We have chosen at this step to add non-linear features that were extracted from an R implementation of a Deep Learning model.

We trained a deep learning neural network with 100 neurons in first hidden layer, 63 neurons in second hidden layer and 30 neurons in the third hidden layer and 15 neurons in last hidden layer. The number of features extracted from the deep learning model was the same number of neurons in each hidden layer. For a better selection of only important non-linear features, we computed correlations of each feature that was corresponding to each neuron in the hidden layer with our target variable. During the computing the correlations, we kept only one feature from each neuron, where is the maximum correlation compared with other features in the same hidden layer, and obtained final four non-linear features. An XGBOOST model was then run to see the behavior of that particular model on the newly created data set.

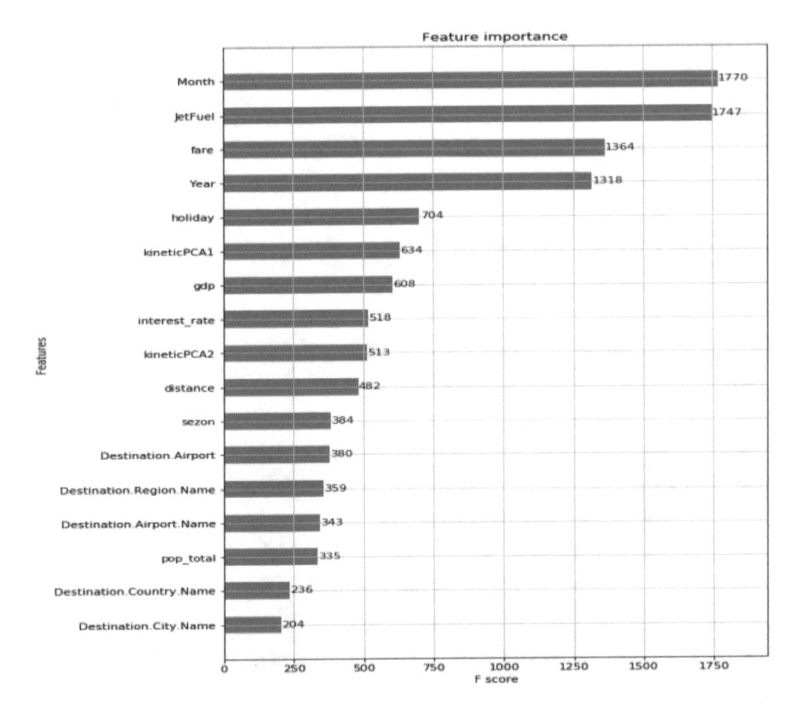

Fig. 4. Principal component analysis features (KineticPCA1 and KineticPCA2).

Table 3. Prediction model using Kineticpca1 and Kinetic PCA2

	PassangersPred5
0	515
1	636
2	621
3	624
4	607

The plot in Fig. 5 shows that the mean values calculated are much nearer to one but differed more on the last set of inputs. Figure 6 shows the features that are important for the number of passengers predicted from most important to least important. Here it shows JetFuel, Month, and fares have the highest predictive values. As observed from the plot the nonlinear features deepf1, deepf2, deep3, and deepf4 (obtained by the method described above) are very influential and are the ones with the highest influential impact captured by XGBOOST feature importance. The number of passengers predicted by using the Deep Learning Hidden Layers is given in Table 4.

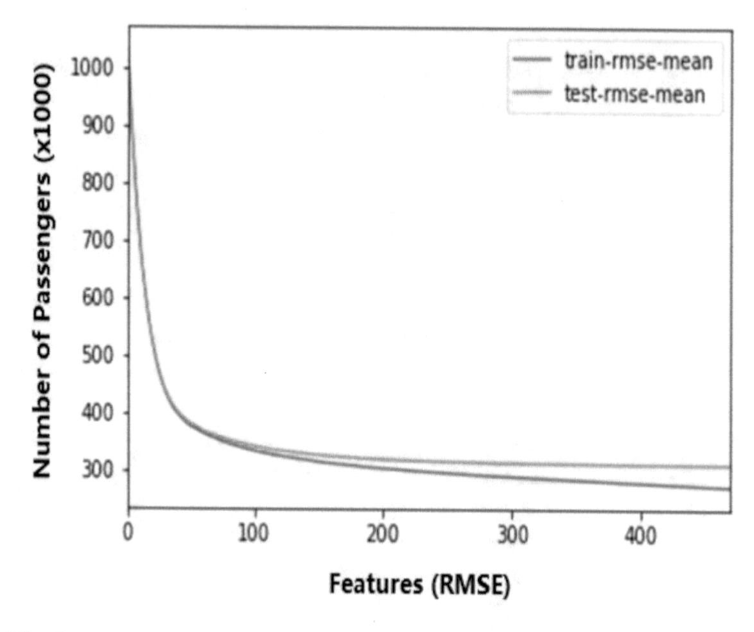

Fig. 5. The mean values of features obtained from deep learning hidden layers.

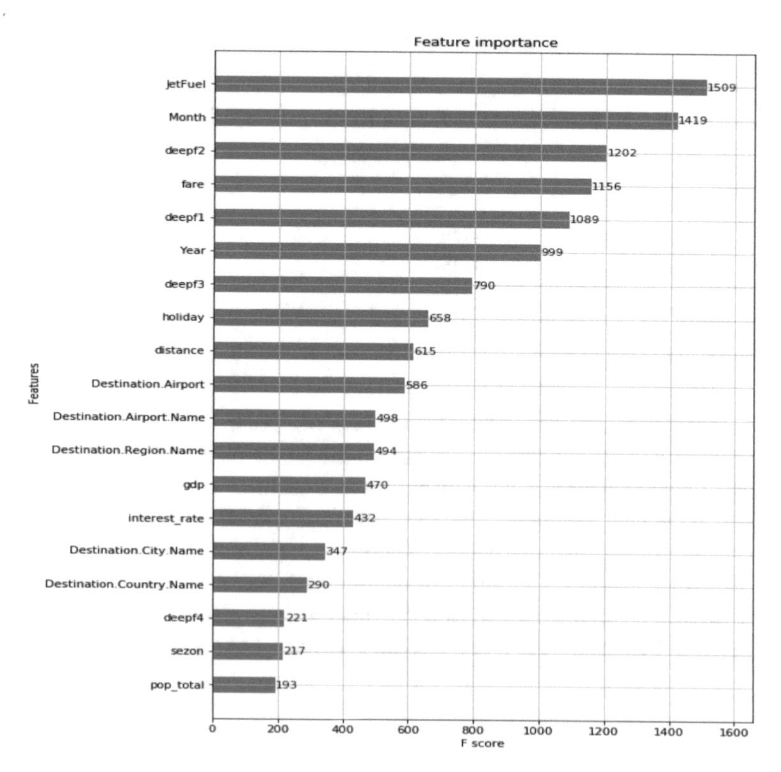

Fig. 6. Prediction model using deep learning hidden layers.

Table 4. Prediction model using deep learning hidden layers

	PassangersPred4
0	582
1	545
2	546
3	552
4	627

3.5 Features Obtained from Genetic Algorithm

This feature was extracted from a genetic algorithm called *symbolic transformer*, which is an estimator that begins by building a population of naive random formulas to represent a relationship [7]. The formulas are represented as tree-like structures with mathematical functions being recursively applied to variables and constants. Each successive generation of programs is then evolved from the one that came before it by selecting the fittest individuals from the population to undergo genetic operations such as crossover, mutation or reproduction. The results are presented in Fig. 7.

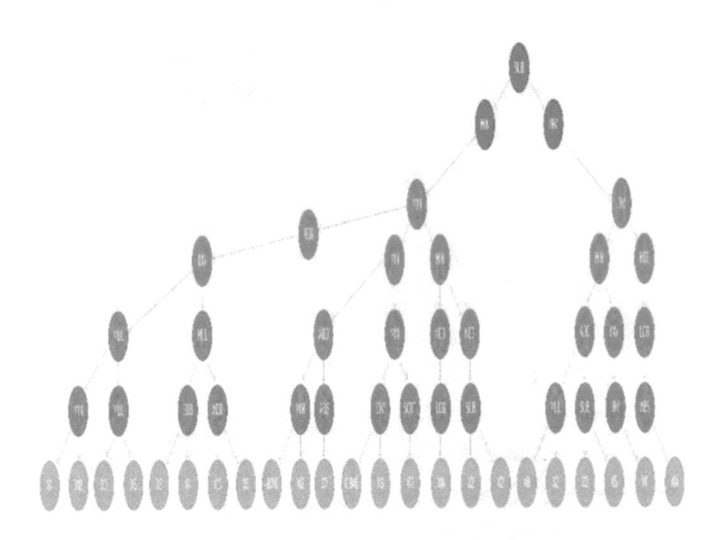

Fig. 7. Tree-like structures of the Genetic Algorithm.

In the genetic program, it is easy to observe different kinds of operations that the genetic algorithm produced. Two new features obtained from genetic transformer after running an XGBOOST model have been added into this algorithm. Figure 8 shows that the mean values calculated are very near one, but differed more on last set of inputs. Figure 9 shows that the features that are important for the number of passengers predicted from most important to least important. Here it shows that JetFuel, Month, and fare are the most predictive values. From the plot below the genetic features called

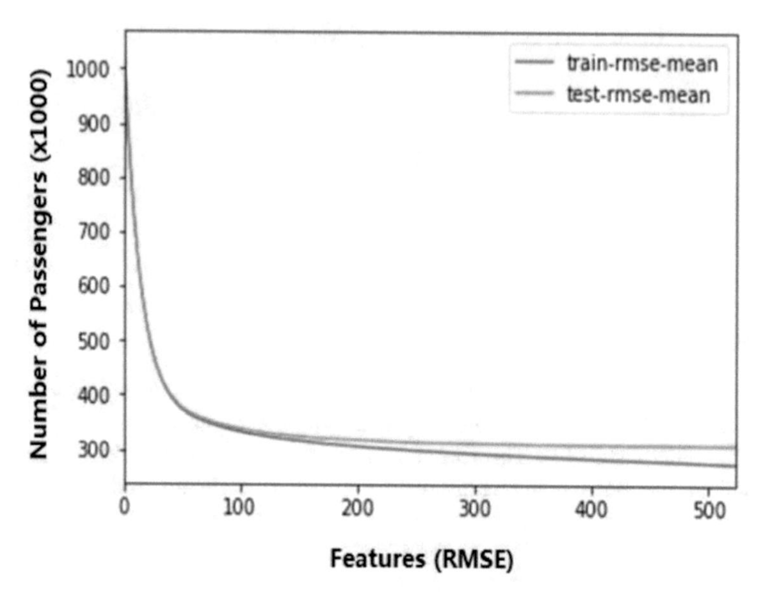

Fig. 8. The mean values of features obtained from Genetic Algorithm.

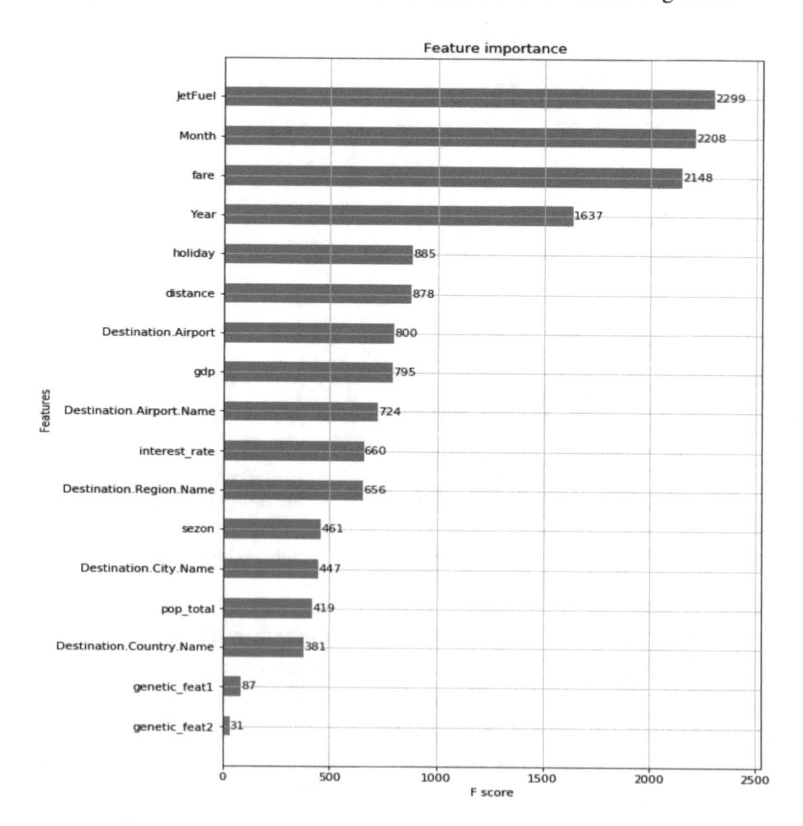

Fig. 9. Features importance obtained from Genetic Algorithm.

genetic_feat1 and genetic_feat2, where captured with very small influence in contrast with our expectation when conducting the experiment. The number of passengers predicted by using the Deep Learning Hidden Layers is given in Table 5.

Table 5. Prediction model using Genetic Algorithm

	PassangersPred2
0	529
1	611
2	600
3	609
4	607

4 Conclusion

In this work, a new modified version of PCA with kinetic correlation matrix using kinetic energy is presented. The features of this modified PCA have been assessed with different sets of air passenger data and compared to traditional PCA. The results of the modified version of PCA show that the kinetic correlation is much higher than that of the Pearson one, which makes lot of sense since Pearson's R is able to detect only linear relations in data. It turned out that the modified version of PCA is more effective in data compression, classes reparability and classification accuracy than those form traditional PCA.

Based on these results, the modified PCA can be applied to make clustering in hyper-dimensional space using kinetic correlation as a distance (increase performance) to make it run in real time in a future work. When coping with clustering, such as clustering algorithm, clustering K-means or in hierarchical clustering, it requires a for-loop at every point to get the nearest point from row vector. For n rows of data complexity will be of the order n ^ n, which is impossible to finish using this method. In two-dimensional space, there is a trick to fast implementation using divide and conquer, which has complexity n or log n. However, these problems can be solved by using modified PCA with properly added features.

In this work, only limited features of the modified PCA method were studied with one set of data. To fully understand and investigate the features of modified PCA, large subsets of data with more features should be considered.

Acknowledgment. I would like to address my special acknowledgements to all those people who provide me with data for my experiments. My warm appreciation is due to the Public Authority for Civil Aviation, Directorate General of Meteorology, and Ministry of Tourism in Oman.

References

1. Bengio, Y.: Representation learning: a review and new perspectives. IEEE Trans. Pattern Anal. Mach. Intell. **35**(8), 1798–1828 (2013)
2. Timmerman, M.E.: Principal Component Analysis, 2nd edn. Springer, New York (2003). I. T. Jolliffe. Journal of the American Statistical Association, 98, 1082-1083
3. Palechor, F.M., et al.: Cardiovascular disease analysis using supervised and unsupervised data mining techniques. J. Softw. **12**(2), 81–90 (2017)
4. Van der Maaten, L., Hinton, G.: Visualizing data using t-SNE. J. Mach. Learn. Res. **9**(2579–2605), 85 (2008)
5. Coates, A., Ng, A.Y.: Learning feature representations with k-means. In: Neural Networks (2012)
6. Chen, T.Q., Guestrin, C.: XGBoost: a scalable tree boosting system. In: Proceedings of the 22nd ACM SIGKDD International Conference on Knowledge Discovery and Data Mining, KDD 2016, pp. 785–794 (2016)
7. Lowe, D.G.: Object recognition from local scale-invariant features. Paper presented at the proceedings of the seventh ieee international conference on computer vision, 1999 (1999)

Prediction Model for Prevalence of Type-2 Diabetes Complications with ANN Approach Combining with K-Fold Cross Validation and K-Means Clustering

Md. Tahsir Ahmed Munna[1(✉)], Mirza Mohtashim Alam[1], Shaikh Muhammad Allayear[1], Kaushik Sarker[1], and Sheikh Joly Ferdaus Ara[2]

[1] Department of Multimedia and Creative Technology, Department of Software Engineering, Daffodil International University, Dhaka, Bangladesh
{tahsir4ll, mirza.mct, drallayear.swe}@diu.edu.bd, kaushik.swe@daffodilvarsity.edu.bd

[2] Department of Microbiology and Immunology, Bangabandhu Sheikh Mujib Medical University, Dhaka, Bangladesh
jolyferdousara@gmail.com

Abstract. In today's era, most of the people are suffering with chronic diseases because of their lifestyle, food habits and reduction in physical activities. Diabetes is one of the most common chronic diseases which has affected to the people of all ages. Diabetes complication arises in human body due to increase of blood glucose (sugar) level than the normal level. Type-2 diabetes is considered as one of the most prevalent endocrine disorders. In this circumstance, we have tried to apply Machine learning algorithm to create the statistical prediction based model that people having diabetes can be aware of their prevalence. The aim of this paper is to detect the prevalence of diabetes relevant complications among patients with Type-2 diabetes mellitus. The processing and statistical analysis we used are Scikit-Learn, and Pandas for Python. We also have used unsupervised Machine Learning approaches known as Artificial Neural Network (ANN) and K-means Clustering for developing classification system based prediction model to judge Type-2 diabetes mellitus chronic diseases.

Keywords: Healthcare · Machine learning · Artificial Neural Network (ANN) Diabetes type-2 · Prediction · K-means clustering · Classification model

1 Introduction

Universally, Diabetes is one of the rapidly growing and deadliest diseases. Diabetes is a disease that attacks when blood glucose (sugar) is too high. In body, blood glucose is the main source of energy which comes from food that we eat. Besides, pancreas makes a hormone named Insulin which helps to get glucose from food and get into our cells to generate energy [1]. Unfortunately, when body doesn't create enough insulin or doesn't use it properly, the diabetes mellitus arises. Diabetes mellitus is classified into two types:

K. Arai et al. (Eds.): FICC 2018, AISC 886, pp. 451–467, 2019.
https://doi.org/10.1007/978-3-030-03402-3_31

Type-1 diabetes mellitus and Type-2 diabetes mellitus. Type-1 requires daily insulin injection because of null production insulin by pancreas. Type-1 diabetes is usually diagnosed during childhood or early adolescence. On the other hand, Type-2 diabetes is one of the most prevalent endocrine disorders worldwide [2]. This is the common form of diabetes which is characterized by either inadequate insulin production by beta cells of pancreas or insulin produces is defective because of which cells in the body are not able to react with it. This form of diabetes is prevalent worldwide and comprises 90% of all diabetic cases. The World Health Organization (WHO) calculates that the global prevalence of diabetes will be increased more than two times by 2030 of current number of diabetes affected people which number is almost 380 million [3].

Diabetes mellitus, a chronic disease once thought to be uncommon in Bangladesh but now it has emerged as an important public health problem. According to the research of WHO in 2013, Bangladesh has immensely high diabetes population with more than 7.1 million, 8.4% or 10 million of the adult population affected by this disease. The number will be 13.6 million in 2040 [4]. Nearly half of the population (51.2%) with diabetes does not know that they have diabetes and do not receive any treatment. Thus the higher prevalence was found in urban areas especially among the women. Furthermore, the population having Type-2 diabetes, women respondents may be at critical situation for coronary heart disease than men [5]. Unfortunately, the awareness label among the general public about these chronic disease is very poor. There is also lack of awareness about the existing interventions for preventing diabetes and management of complications. Several recent intervention studies have undisputedly proved that Type-2 diabetes can be efficiently prevented by lifestyle modification in high risk individuals [6, 7]. Health care costs account for a good percentage of its economy for most of the countries. For that, Health care industry is very critical and vast, unfortunately it is highly inefficient. Moreover, in this era the yield of cutting edge technology in the health care domain general people couldn't ripe. In Dhaka city, there are 15.39 million people live in together and the prevalence of Type-2 diabetes is enormously high [8] but there are little bit studies about this consequence, in according to Bangladesh demography profile at 2014.

In this paper, we have prepared analyzed based prediction model for diabetes patients to assess the prevalence of Type-2 diabetes mellitus containing long term complications by using Unsupervised Machine Learning Algorithm. Additionally, we have also used several Machine Learning frameworks on Python Programming environment, such as Pandas, Scikit-Learn, mpl_toolkits, Matplotlib, etc. We have used K-means clustering for developing classification system based prediction model.

2 Related Studies

Rajesh et al. [9] discusses about the reliable prediction methodology of diagnosing diabetes and interpret the data patterns so as to get meaningful and useful information for the health care providers. Wang et al. [10] in his journal paper, he tried to develop and evaluate an effective classification approach without biochemical parameters to identify those at high risk of Type-2 diabetes mellitus in rural adults. Another Paper Smith et al. [11] was developed as a medical support system for concentrating on the possible evaluation methodology, giving a framework and specific suggestions for each type of classification problem. Lin et al. [12], on his journal paper attempted to develop the ANN-based predictive model to identify patients with high risk of hypotension during spinal anesthesia. Most of the researcher work on identify the having diabetes on the patients but in our paper we tried to find out the complication after having diabetes cause in our dataset all the patients have diabetes. In this paper, we have described the main intuition of our research and data collection in Sect. 3, the proposed system model in Sect. 4 and our derived result in Sect. 5. The conclusion and future work is described in Sect. 6.

3 Data Collection and Conceptual Framework

Non-government hospitals in Dhaka City which are under National Health Network (NHN) were selected purposively for the convenience of the study. Our data collection phase lasted from January 2016 to January 2017. The inclusion criteria of the patients having Type-2 diabetes mellitus were chosen on the age greater than or equal to 35 years, additionally, patients who were severely ill. They all had diabetes guidebook provided by National Health Network (NHN). Initially, the whole variables have been divided into two parts: Dependent variable and Independent Variable. Independent variable had certain components which supposedly regulated the factors of the dependent variables and they are: Age, sex, educational status, occupational status, and age of onset, duration of disease, types of medications, blood glucose monitoring, regular exercise, cessation of smoking, use of smokeless tobacco. The dependent variables are the labels that we have used for our system model and they are: Cardiovascular complication, neurological complication, ophthalmological complication, diabetic foot ulcer. Our total population was consisting of 26,000 populations, having 12,824 male and 13,176 female correspondents.

From Figs. 1 and 2, we can observe that all Type-2 diabetic complications happened when BMI and HbA1c, both rises from 20 and 7, accordingly. This is because if we sum Figs. 1 and 2, we get the resulting sum equivalent to the summation values of frequencies of Fig. 3 for each of the complications, where Fig. 3 includes the prevalence of chronic disease for Type-2 diabetic patients of the total population. That is why we may consider for building prediction model's hypothesis that HbA1c and BMI attributes are associated with Type-2 diabetes complications.

4 System Model

In our model we have gone through several phases including data pre-processing, training. Our raw dataset was consisted of different valuable information about age, gender, BMI, HbA1c, education, urban living time, smoking, marital status, etc. and both male and female was included as sample population. We have found that a diabetic person can develop Cerebro Vascular Disease (CVD), Ophthalmological Disease (EYE), Cardio Vascular Disease (HRD), Neuropathy Disease (NEURO), Peripheral Vascular Disease (PVD), and Foot Ulcer (FOOT) as a consequence of diabetes. The block diagram of our system model is provided in Fig. 4.

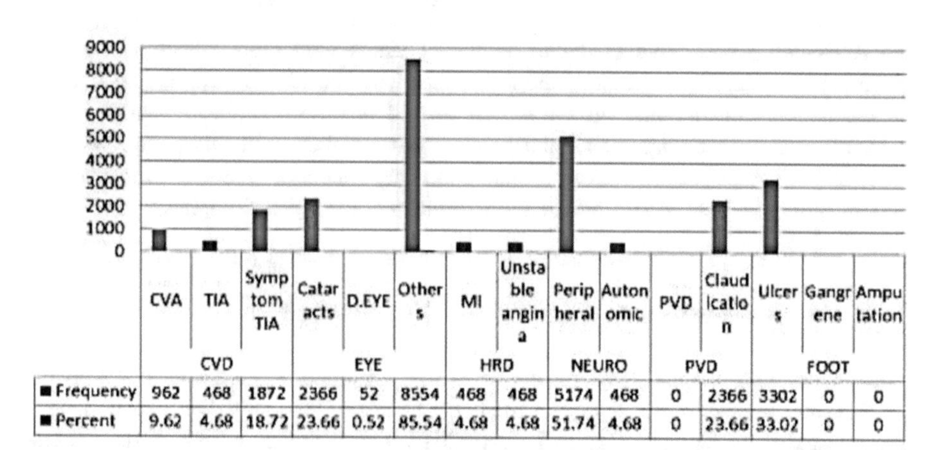

	CVD			EYE			HRD		NEURO		PVD		FOOT		
	CVA	TIA	Symptom TIA	Cataracts	D.EYE	Others	MI	Unstable angina	Peripheral	Autonomic	PVD	Claudication	Ulcers	Gangrene	Amputation
■ Frequency	962	468	1872	2366	52	8554	468	468	5174	468	0	2366	3302	0	0
■ Percent	9.62	4.68	18.72	23.66	0.52	85.54	4.68	4.68	51.74	4.68	0	23.66	33.02	0	0

Fig. 1. Frequency and percentage result of various complications among Type-2 diabetes patients when HbA1c >=7.

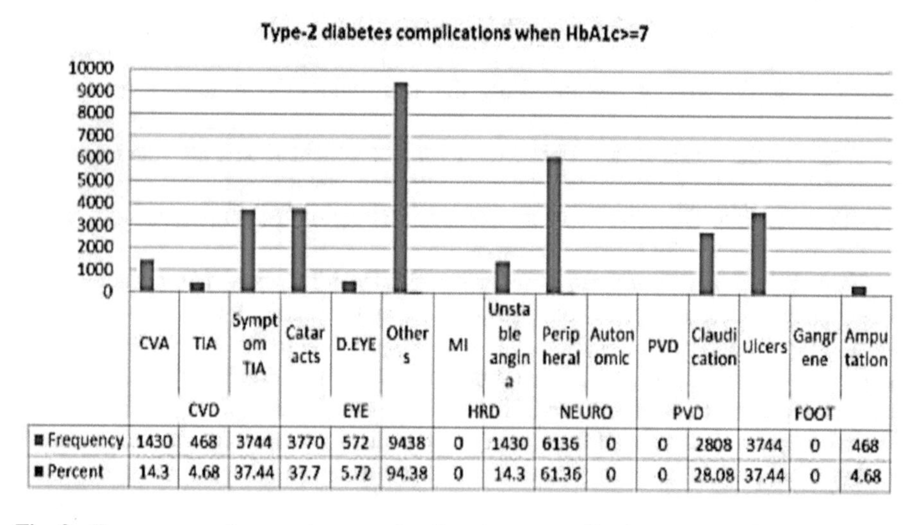

	CVD			EYE			HRD		NEURO		PVD		FOOT		
	CVA	TIA	Symptom TIA	Cataracts	D.EYE	Others	MI	Unstable angina	Peripheral	Autonomic	PVD	Claudication	Ulcers	Gangrene	Amputation
■ Frequency	1430	468	3744	3770	572	9438	0	1430	6136	0	0	2808	3744	0	468
■ Percent	14.3	4.68	37.44	37.7	5.72	94.38	0	14.3	61.36	0	0	28.08	37.44	0	4.68

Fig. 2. Frequency and percentage result of various complications among Type-2 diabetes patients when BMI >=20.

4.1 Data Pre-processing

Our analysis throughout the dataset has led us to a conclusion that we only need two factors which are BMI and HbA1c. Male and female both delivered different types of characteristics as they are considered as normal, overweight and obese for various results. Male is considered as overweight in case of BMI's range is 25–30. On the contrary an overweight female's BMI ranges in between 18–25 [13]. A diabetic patient

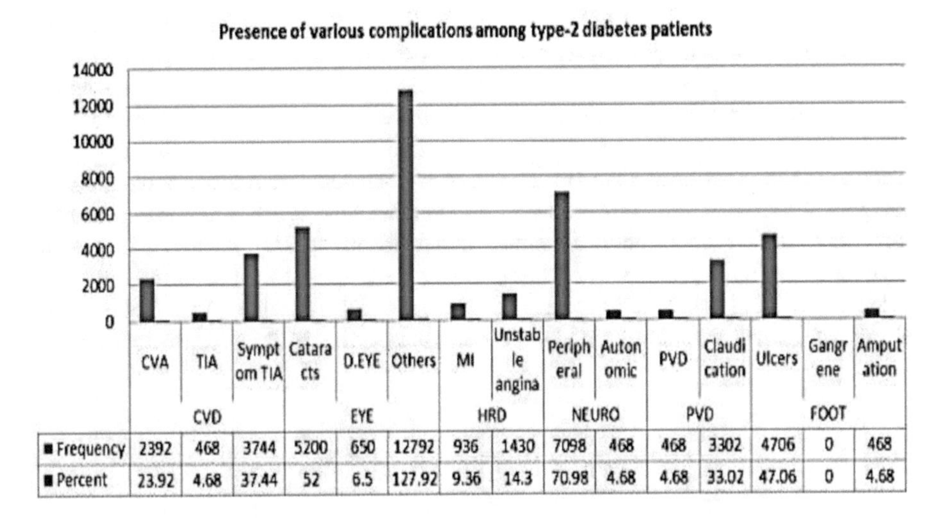

	CVA	TIA	Symptom TIA	Cataracts	D.EYE	Others	MI	Unstable angina	Peripheral	Autonomic	PVD	Claudication	Ulcers	Gangrene	Amputation
	CVD			EYE			HRD		NEURO		PVD		FOOT		
■ Frequency	2392	468	3744	5200	650	12792	936	1430	7098	468	468	3302	4706	0	468
■ Percent	23.92	4.68	37.44	52	6.5	127.92	9.36	14.3	70.98	4.68	4.68	33.02	47.06	0	4.68

Fig. 3. Frequency and percentage result of various complications among Type-2 diabetic patients.

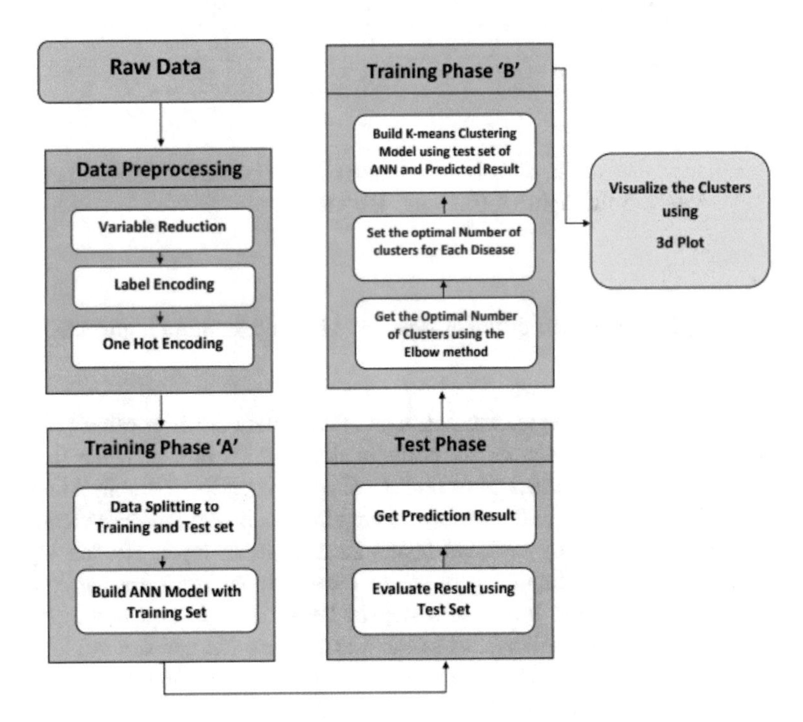

Fig. 4. System model.

can be affected by multiple of above mentioned diseases at the same time since our problem is non-mutually exclusive multi-label classification problem [14]. We have done label encoding using the scikit-learn's Label Encoder library [15]. We have applied One Hot Encoding [16] technique as we wanted evaluate each of the diseases independently since it shows greater accuracy rather than taking all of the diseases as features at the same time. One Hot encoding is considered as a process where all the categorical variables are converted into a better machine learning algorithm readable form and treats each categorical values independently [17]. The result of one hot encoder comes as a sparse matrix. The sparse matrix is consisted of columns where each value of a column corresponds to each of the categories. After doing One Hot Encoding to our categorical variables, namely, the six diseases, the result is shown in Table 1.

Table 1. One Hot encoding of the categorical variables

Names	CVD	EYE	HRD	NEURO	PVD	FOOT
CVD	1	0	0	0	0	0
EYE	0	1	0	0	0	0
HRD	0	0	1	0	0	0
NEURO	0	0	0	1	0	0
PVD	0	0	0	0	1	0
FOOT	0	0	0	0	0	1

We wanted to find out the probabilistic factor of each of the diseases and cluster the predicted the result in the light of BMI and HbA1c.

4.2 Training Phase

In our training phase we used different types of Machine learning algorithm.

4.2.1 Artificial Neural Network (ANN)

First of all, we have implemented the Artificial Neural Network Algorithm [18] to find out the possibility rate of each of six complications which is take place for having diabetes. ANN algorithm[1] comes from biological inspiration that compile in computer for certain important tasks like pattern recognition, clustering, classification etc. Basically, ANN constitute by different layers such as Input layer, Hidden layer and Output layer. The jobs of Input layers are to receive input from the various recognizable sources. The hidden layer units stay in between input and output layers. It contains activator function which is set of the transfer function that can get the desired output. There are two categories of activator function namely linear and non-linear. Binary, Sigmoid are popular methods used as activator function. The output layer responds to the tidings that how it learned about any task and provides the desired outcome.

1 https://hackernoon.com/overview-of-artificial-neural-networks-and-its-applications-2525c1addff7.

Algorithm 1. Deep Learning based Prediction Model Algorithm for Prevalence of Type-2 Diabetes Complications

```
procedure CATEGORICALENCODING(Y)
    for each CategoricalFeature in Y do
        Y = OneHotEncode(Y)
    end for
    Return Y
end procedure
procedure GETOPTIMALCLUSTER(X[test], YPredictionMatrix)
    X = CombineRowWise(X[test], YPredictionMatrix)
    wcss = []
    for i in range(1, 11) : do
        kmeans = KMeans(numberOfClustrs = i, init = k − means + +, randomState = 50)
        kmeans.fit(X)
        wcss.append(kmeans.inertia)
    end for
    VisualizeElbowGraph(wcss)
    Returnmax(kmeans.inertia)
end procedure
Dataset = Input(Directory/FileName)
X[EachRow][ZerothColumn] = Dataset[BMI].value
X[EachRow][FirstColumn] = Dataset[HbA1c].value
Y = CategoricalEncoding(Y)
Y1 = Y[EachRow][ZerothColumn]
Y2 = Y[EachRow][FirstColumn]
Y3 = Y[EachRow][SecondColumn]
Y4 = Y[EachRow][ThirdColumn]
Y5 = Y[EachRow][FourthColumn]
Y6 = Y[EachRow][FifthColumn]
seed = random.seed(integer)
kfold = StratifiedKFold(numberOfSplits = 10, shuffle = True, randomState = seed)
for each Y in Range (Y1, Y6) do
    for train,test in kfold.split(X, Y) do
        model = Sequential()
        model.add(Dense(inputNodes = 2, inputDimension = 2, activation = relu))
        model.add(Dense(hiddenLayerNodes = 3, activation = relu))
        model.add(Dense(outputNodes = 1, activation = sigmoid))
        model.compile(loss = binaryCrossEntropy, optimizer = adam)
        model.fit(X[train], Y[train], epoch = 10, batchSize = 10)
        YPredictionMatrix = model.predict(X[test])
        VisualizeANNResult(X[test], YpredictionMetrix)
        clusterNumbers = GETOPTIMALCLUSTER()
        KmeansEstimator = KMeans(numberOfCluster = clusterNumbers)
        X = CombineRowWise(X[test], YPredictionMatrix)
        kmeansEstimator.fit(X)
        xaxis = X[EachRow][ZerothColumn]
        yaxis = X[EachRow][FirstColumn]
        zaxis = X[EachRow][ThirdColumn]
        VisualizeKMResult(xaxis, yaxis, zakis)
    end for
end for
```

In our model, BMI and HbA1c are the input values. In input layer and hidden layer we have used Rectified Linear Unit[2] at the output layer we have used sigmoid activation function because our desired outcome is between 0 and 1 which is very likely to

[2] https://www.quora.com/What-is-special-about-rectifier-neural-units-used-in-NN-learning.

be small numerical value and sigmoid activation function can work for very small networks which is needed for our output. Since we needed the activator to be activated if the output value becomes greater than zero and we have binary values in our encoded sparse matrix of the labels. We have used sigmoid function and binary cross entropy. Our expected probability rate of specific complication obtains between 0 and 1 value. We built our neural network model using stratified K-fold cross validation technique to get better validation results. In stratified K-fold cross validation, we have an evaluation set for each of the folds. Finally, the results are averaged for each of the folds in order to get the final result.

4.2.2 Elbow Method

Elbow method is a procedure which is used to identify the number of k that is number of clustering[3]. In our system model we used k mean clustering for knowing the possibility of each complication for both male and female respondent Elbow method helped to find out the possible optimal number of clusters for our research. The visualization of Elbow method for female respondent[4] is considered for providing number of optimal clusters. After analyzing we have predicted result of Artificial Neural network, alongside with the test set containing BMI and HbA1c, we have visualized and found the elbow in the graph. That is actually the optimal number of clusters. The in illustrate at Fig. 7 and noticed that for male and female the possible cluster number is 2 and 3.

4.2.3 K-Means Clustering

K-mean clustering [19] is an unsupervised machine learning algorithm that used to build cluster for similar group of data. Here, k illustrates the number of cluster or categories classified and the average (mean) of the total points in cluster is calculated by this algorithm. In our system model we used k mean clustering for finding out the possible clusters in each of complications. By using the K-means clustering along with the Elbow method, we have found the possibility of each of the complication for each group of people for both male and female respondent. Numbers of potential optimal clusters for both the respondent have been visualized by the maximum inertia. We have the number wherever the graph of Elbow method has formed an elbow. Then the test set and their predicted complication possibility have used to make the stratified K-means clustering. Our algorithm for diabetes complication prediction for each of the diseases has been provided in Algorithm 1.

5 Result and Analysis

In our result and analysis section, we have provided our experiment result of Artificial Neural Network in Fig. 5 (Male) and Fig. 6 (Female), Elbow Method for both respondent for getting optimal number of clusters (2 for Male and 3 for Female) in

[3] https://pythonprogramminglanguage.com/kmeans-elbow-method/.

[4] https://algobeans.com/2015/11/30/k-means-clustering-laymans-tutorial/.

Fig. 7. After finding out the optimal numbers of cluster for both male and female respondent, has been given in Figs. 8, 9, 10, 11, 12 and 13. For both Artificial Neural Networks and K-Means clustering X and Z-axis contained BMI and HbA1c. Y-axis contained the complication possibility (CF) of each of the diseases. For the input of the K-means algorithm, we have taken the output of artificial neural network and K-fold test set and the predicted complication of each of the disease. Tables 2, 3, 4, 5, 6 and 7 shows the analysis result clusters of different groups containing both male and female respondent. Table 8 shows the accuracy in predicting each of the complications for both male and female respondent which is based on ANN algorithm.

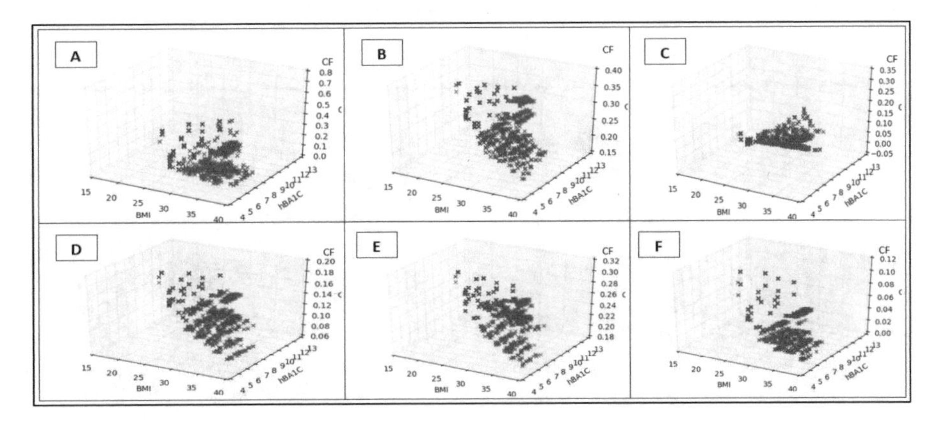

Fig. 5. Classified prediction model for prevalence of Type-2 diabetes mellitus complications (male respondents). A = Cerebro Vascular Disease, B = Ophthalmological Disease, C = Cardio Vascular Disease, D = Neuropathy Disease, E = Peripheral Vascular Disease, F = Foot Ulcer Disease.

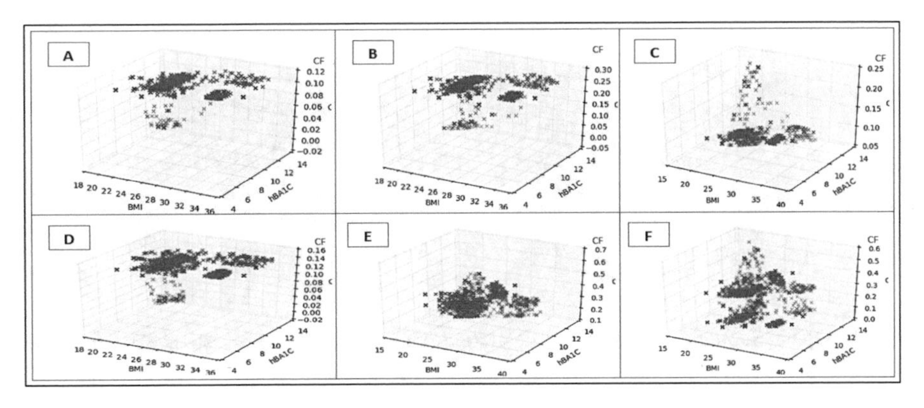

Fig. 6. Classified prediction model for prevalence of Type-2 diabetes mellitus complications (Female respondents). A = Cerebro Vascular Disease, B = Ophthalmological Disease, C = Cardio Vascular Disease, D = Neuropathy Disease, E = Peripheral Vascular Disease, F = Foot Ulcer Disease.

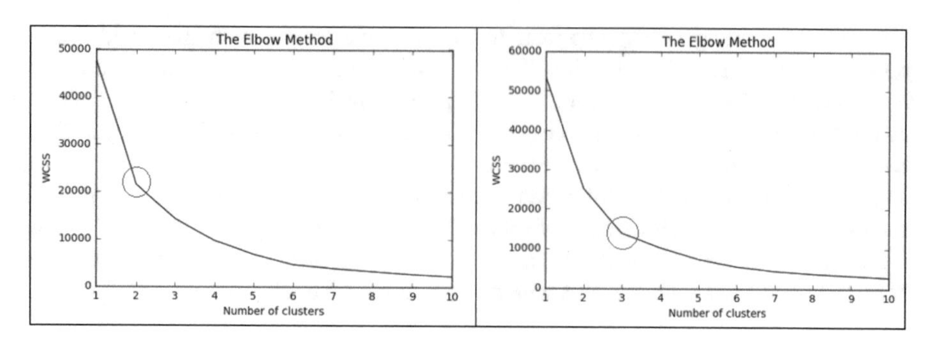

Fig. 7. Elbow method for choosing optimal number of clusters left (Male) right (Female).

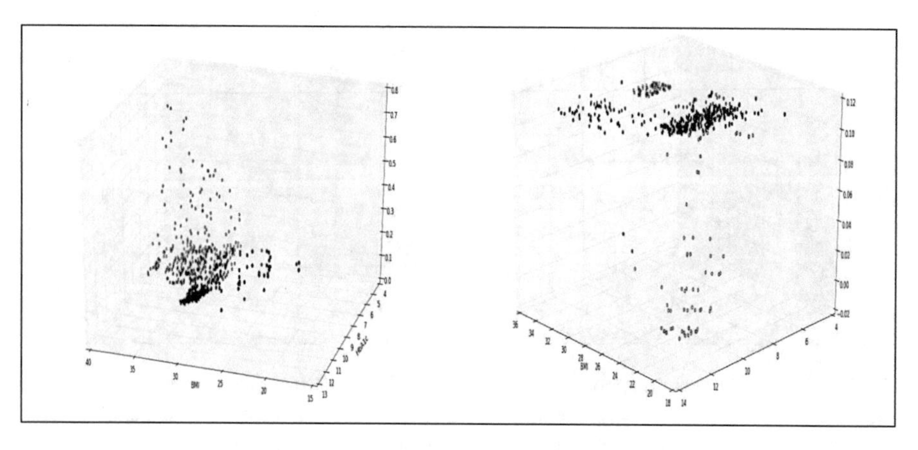

Fig. 8. Cerebro Vascular Disease (CVD) complication factor for both respondent left (Male) right (Female).

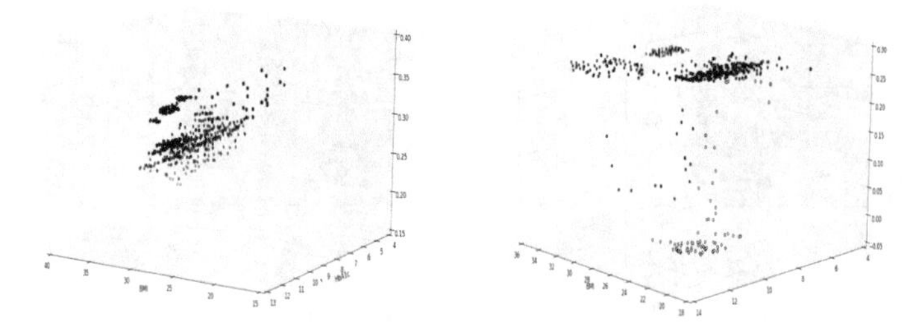

Fig. 9. Ophthalmological Disease (EYE) complication factor for both respondent left (Male) right (Female).

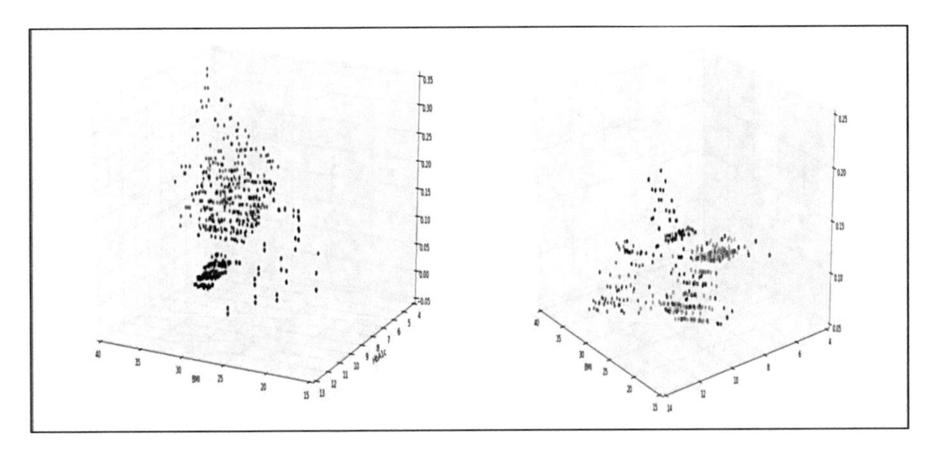

Fig. 10. Cardio Vascular Disease (HRD) complication factor for both respondent left (Male) right (Female).

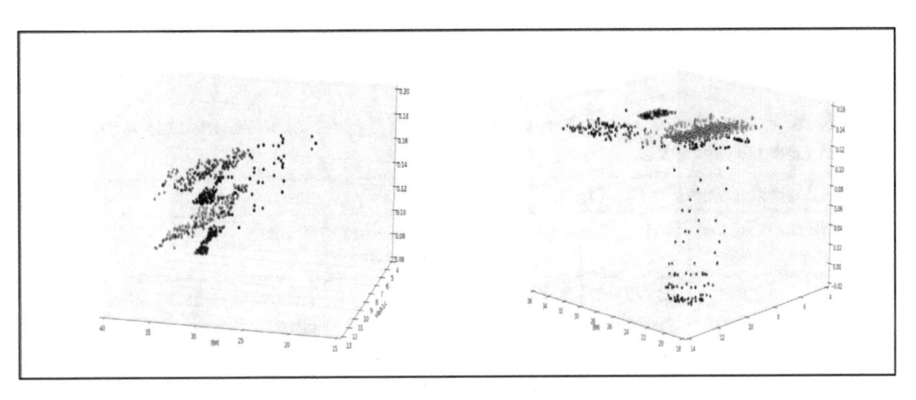

Fig. 11. Neuropathy Disease (NEURO) complication factor for both respondent left (Male) right (Female).

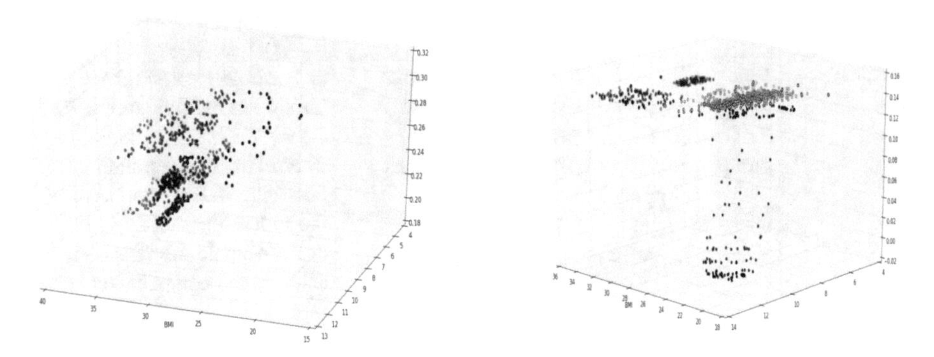

Fig. 12. Peripheral Vascular Disease (PVD) complication factor for both respondent left (Male) right (Female).

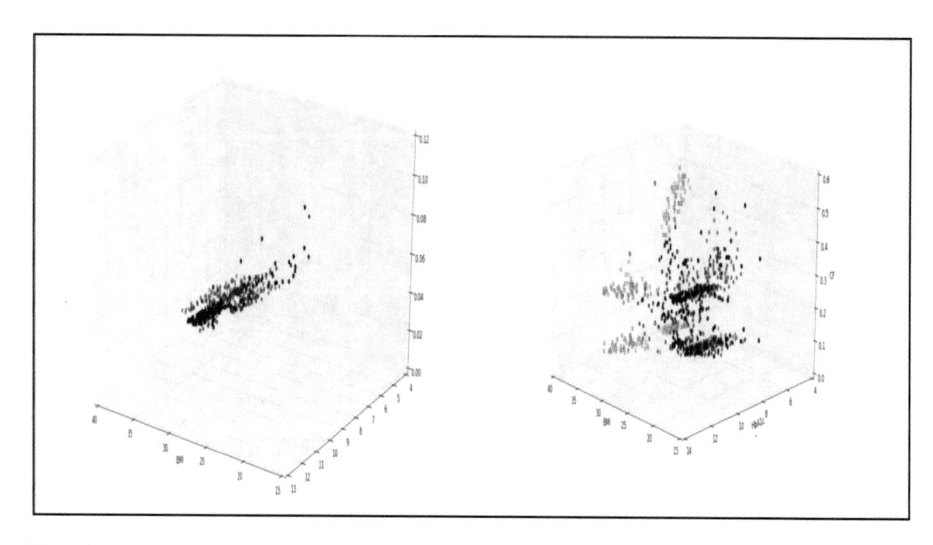

Fig. 13. Foot Ulcer Disease (FOOT) complication factor for both respondent left (Male) right (Female).

Table 2. K-means clustering result for prevalence of Type-2 diabetes mellitus complications (both respondents) for CVD

Cerebro Vascular Disease (CVD)					
Respondent name	Group name	BMI	HbA1c	Complication factor (CF)	Remark
Male	Blue group	20–27 (Approx.)	5–11 (Approx.)	0.1–0.28 (Approx.)	1. Patients whose BMI >30 and HbA1c 5 to 6, their complication factor (Possibility of CVD) stars to increase patients whose BMI 35–40 (Red Group), Possibility goes up to 0.7 (70%) 2. Blue group's complication factor does not increase beyond 0.28 (28%)
	Red group	28–38 (Approx.)	6–12 (Approx.)	0.2–0.7 (Approx.)	
Female	Blue group	28.7–36 (Approx.)	6.8–13.2 (Approx.)	0.10–0.12 (Approx.)	1. For both Red and Blue Group Possibility exceeds 0.12 (12%) 2. For Green group the possibility CF ranges between 0.0 to less than 0.12 3. For Female this disease is not acute as the complication is less
	Red group	20.7–28.8 (Approx.)	5.0–13.0 (Approx.)	0.04–0.12 (Approx.)	
	Green group	19–25 (Approx.)	8.5–12.8 (Approx.)	0.0–0.12 (Approx.)	

Table 3. K-means clustering result for prevalence of Type-2 diabetes mellitus complications (both respondents) for EYE

Opthalmological Disease (EYE)					
Respondent name	Group name	BMI	HbA1c	Complication factor (CF)	Remark
Male	Blue group	25–38 (Approx.)	4.9–11.2 (Approx.)	0.17–0.34 (Approx.)	For Red group BMI less than 25 and across all the ranges of Red group's HbA1c the complication factor goes up to 0.37 (37%) For Blue group if the BMI tends to 30 from the value 40 and upon increasing HbA1c from 5 the complication factor will be up to 0.34 (34%)
	Red group	17–25 (Approx.)	5.8–12.2 (Approx.)	0.30–0.37 (Approx.)	
Female	Blue group	20–28 (Approx.)	5.0–12.7 (Approx.)	0.05–0.26 (Approx.)	A red group has higher possibility for this disease. Since most of the population of Red group gas the possibility of beyond 0.25 (25%) Some of the population of blue group are scattered in between 0.05 to 0.20 but for most of the population the possibility goes beyond 0.25(25%) Red Group has the higher BMI but average to high HbA1c Green group higher HbA1c but lower and most the population does not affect by this disease yet. Some of the green Population has the possibility is 0.25 (25%)
	Red group	28–35 (Approx.)	7.0–13.4 (Approx.)	0.11–0.26 (Approx.)	
	Green group	19–24.2 (Approx.)	9.0–12.6 (Approx.)	0.00–0.26 (Approx.)	

Table 4. K-means clustering result for prevalence of Type-2 diabetes mellitus complications (both respondents) for HRD

Cardio Vascular Disease (HRD)					
Respondent name	Group name	BMI	HbA1c	Complication factor (CF)	Remark
Male	Blue group	21–28 (Approx.)	4.9–11.8 (Approx.)	0.00–0.10 (Approx.)	1. Red group has higher BMI level but HbA1c of all ranges between 5 to 11. The more HbA1c become low towards 4 the more the complication factor increases 2. The blue population relatively lowers BMI than the red population but the most of them have relatively lower complication level yet the HbA1c ranges in between 6 to 12
	Red group	29–40 (Approx.)	4.0–11.0 (Approx.)	0.02–0.33 (Approx.)	
Female	Blue group	17–24 (Approx.)	9.2–13 (Approx.)	0.05–0.24 (Approx.)	1. Blue group have the relatively lowest BMI but higher HbA1c 2. Most the green population has moderate BMI and 5 to 13 HbA1c 3. The Red group relatively has higher BMI and 7 t o13 HbA1c 3. Only Blue group's complication factor goes beyond 0.20 (20%) 4. For all other groups do not go beyond 0.10 (10%)
	Red group	29–35 (Approx.)	7–13 (Approx.)	0.05–0.11 (Approx.)	
	Green group	24–28 (Approx.)	5.0–13 (Approx.)	0.06–0.12 (Approx.)	

Table 5. K-means clustering result for prevalence of Type-2 diabetes mellitus complications (both respondents) for NEURO

Neuropathy Disease (NEURO)					
Respondent name	Group name	BMI	HbA1c	Complication factor (CF)	Remark
Male	Blue group	20–26 (Approx.)	5.9–12.1 (Approx.)	0.13–0.18 (Approx.)	1. Same as Cardio Vascular Disease for Red Group as it falls under the high BMI level 2. Similarly the Blue group have lower BMI level here the blue group's possibility goes up to 0.18 (18%) for red group's does reach up to 0.16 (16%)
	Red group	28–38 (Approx.)	4.8–11.2 (Approx.)	0.08–0.16 (Approx.)	
Female	Blue group	19–25 (Approx.)	9–14 (Approx.)	0.00–0.14 (Approx.)	1. Possibility is not as acute as the complication for each group does not go beyond 0.14 (14%) 2. Green and red group has maximum population whose possibility is 14%
	Red group	28.2–35.7 (Approx.)	7.8–14 (Approx.)	0.13–0.14 (Approx.)	
	Green group	20–28 (Approx.)	5.8–13.8 (Approx.)	0.13–0.14 (Approx.)	

Table 6. K-means clustering result for prevalence of Type-2 diabetes mellitus complications (both respondents) for PVD

Peripheral Vascular Disease (PVD)					
Respondent name	Group name	BMI	HbA1c	Complication factor (CF)	Remark
Male	Blue group	26–38 (Approx.)	4.8–11.2 (Approx.)	0.19–0.29 (Approx.)	1. Red group population has lower BMI level than the Blue group but lower population. The possibility of this disease for red group goes up to 30% 2. The Blue group has higher BMI level and HbA1c (5–11) Complication reaches the maximum level (29%) when the HbA1c gets slightly lower
	Red group	18–25.2 (Approx.)	5.8–12.2 (Approx.)	0.24–0.30 (Approx.)	
Female	Blue group	20–28 (Approx.)	5–13 (Approx.)	0.15–0.52 (Approx.)	1. Green group has lower HbA1c but high amount of BMI level. So the complication rate goes up to 60% 2. Green group has the most population and has moderate BMI and HbA1c. However, the complication rate sometimes goes beyond 45% 3. The red group has high HbA1c but low BMI, though for females are considered overweight. However, the complication reaches up to 40%
	Red group	18–24 (Approx.)	8.4–13.4 (Approx.)	0.20–0.40 (Approx.)	
	Green group	27–36 (Approx.)	7–13.3 (Approx.)	0.18–0.60 (Approx.)	

Table 7. K-means clustering result for prevalence of Type-2 diabetes mellitus complications (both respondents) for FOOT

Foot Ulcer (FOOT)					
Respondent name	Group name	BMI	HbA1c	Complication factor (CF)	Remark
Male	Blue GRoup	30–39 (Approx.)	5–11 (Approx.)	0.01–0.04 (Approx.)	1. Any group's complication does not go beyond 10%
	Red group	19–27 (Approx.)	5.8–12 (Approx.)	0.04–0.10 (Approx.)	
Female	Blue group	19–28 (Approx.)	5–12.3 (Approx.)	0.1–0.5 (Approx.)	1. For blue group HbA1c scattered across all the ranges, but for increasing BMI complication factor/ possibility increases 2. Regardless of having different levels of HbA1c, the green group has the BMI level of Obese. As a result the complication factor goes up to 58% 3. Red group has slightly lower BMI than all other groups, but for higher HbA1c the complication factor goes up to 56%
	Red group	17–24 (Approx.)	8.2–13 (Approx.)	0.11–0.56 (Approx.)	
	Green group	28–35 (Approx.)	6.8–13 (Approx.)	0.12–0.58 (Approx.)	

Table 8. Accuracy measurement table based on Artificial Neural Network (ANN) with stratified K-fold for prevalence of Type-2 diabetes mellitus complications (both respondents)

Respondent name	CVD	EYE	HRD	NEURO	PVD	FOOT
Male	84.10	70.8	93.1	86.6	73.4	96.1
Female	91.02	80.2	89.3	88.2	70.8	85.4

6 Conclusion and Future Work

Health-care industry is growing up so fast that massive amount of data is gathered every day. In order to provide patient centric treatment, these data should be used efficiently by different machine learning algorithms for providing better predictive results. Type-2 diabetes is common long-term diseases all around the world. The most important findings of this paper is providing a neural network based model in combination of Stratified K-Fold Cross-Validation and K-Means Clustering for giving the possibility of each of the six key diseases individually. That indicates the uniqueness of the paper. We have come to a conclusion that, increasing of either one of this factor namely, BMI and HbA1c can be dangerous for a diabetic patient. Therefore, a particular diabetic patient should take great incentive to keep these two factors under controlled. We need adequate amounts of research to provide good medical facilities regarding efficiency, ease of treatment and better healthcare system. Our future work is to include more deep learning approaches with the combination of other external factors to human health like, smoking, exercise, living environment etc. More sophisticated calculations can be done in our future work with the use of sophisticated sensors, IOT modules in and outside the human body and with that work we would like to build a smart home automation system where a diabetic patient can easily get alert of each of the mentioned complications using IOT and smart devices.

References

1. Arena, J.G.: Behavioral medicine consulation. In: Handbook of Clinical Interviewing with Adults, p. 446 (2007)
2. Mahmoodi,M., Hosseini-Zijoud, S.M., Hassan Shahi, G.H., Nabati, S., Modarresi, M., Mehrabian, M., Sayyadi, A.R., Hajizadeh, M.R.: J. Diabetes Endocrinol. **4**(1), 1–5, January 2013. ISSN 2141-2685- Academic Journal
3. What is Diabetes? (n.d.). https://www.diabetesresearch.org/what-is-diabetes. Accessed 28 Aug 2017
4. The State of Diabetes in Bangladesh, 05 October 2016. http://futurestartup.com/2016/07/27/the-state-of-diabetes-in-bangladesh/. Accessed 28 Aug 2017
5. Vaz, N.C., Ferreira, A.M., Kulkarni, M.S., Vaz, F.S., Pintondian, N.R.: Prevalence of diabetic complications in rural Goa, India. J. Community Med. **36**(4), 283–286 (2011). https://doi.org/10.4103/0970-0218.91330
6. Cao, H.B., Liu, P.A., Jiang, X.G., Jiang, Y.Y., Wang, J.P., Zheng, H., Zhang, H., Bennett, P. H., Howard, B.V.: Effects of diet and exercise in preventing NIDDM in people with impaired glucose tolerance: the Da Qing IGT and diabetes study. Diabetes Care **20**, 537–544 (1997)
7. Nicole, R.: Title of paper with only first word capitalized. J. Name Stand. Abbrev. (in press)
8. Yorozu, Y., Hirano, M., Oka, K., Tagawa, Y.: Electron spectroscopy studies on magneto-optical media and plastic substrate interface. IEEE Transl. J. Magn. Japan **2**, 740–741 (1987). [Digests 9th Annual Conf. Magnetics Japan, p. 301, 1982]
9. Rajesh, K., Sangeetha, V.: Application of data mining methods and techniques for diabetes diagnosis. Int. J. Eng. Innov. Technol. **2**(3), 224–229 (2012)
10. Wang, C., Li, L., Wang, L., Ping, Z., Flory, M.T., Wang, G., Li, W.: Evaluating the risk of type 2 diabetes mellitus using artificial neural network: an effective classification approach. Diabetes Res. Clin. Pract. **100**(1), 111–118 (2013)
11. Smith, A.E., Nugent, C.D., McClean, S.I.: Evaluation of inherent performance of intelligent medical decision support systems: utilising neural networks as an example. Artif. Intell. Med. **27**(1), 1–27 (2003)
12. Lin, C.S., Chiu, J.S., Hsieh, M.H., Mok, M.S., Li, Y.C., Chiu, H.W.: Predicting hypotensive episodes during spinal anesthesia with the application of artificial neural networks. Comput. Methods Programs Biomed. **92**(2), 193–197 (2008)
13. Wolk, R., Berger, P., Lennon, R.J., Brilakis, E.S., Somers, V.K.: Body mass index. Circulation **108**(18), 2206–2211 (2003)
14. Tsoumakas, G., Katakis, I.: Multi-label classification: an overview. Int. J. Data Warehous. Min. **3**(3), 1–13 (2006)
15. Garreta, R., Moncecchi, G.: Learning Scikit-Learn: Machine Learning in Python. Packt Publishing Ltd., Birmingham (2013)

16. Hackeling, G.: Mastering Machine Learning with Scikit-Learn. Packt Publishing Ltd., Birmingham (2014)
17. Guo, C., Berkhahn, F.: Entity Embeddings of Categorical Variables. arXiv preprint arXiv: 1604.06737 (2016)
18. Principe, J.C., Fancourt, C.L.: Artificial neural networks. In: Handbook of Global Optimization, vol. 2, pp. 363–386 (2013)
19. Likas, A., Vlassis, N., Verbeek, J.J.: The global k-means clustering algorithm. Pattern Recognit. **36**(2), 451–461 (2003)

Enriching Existing Ontology Using Semi-automated Method

Md. Jabed Hasan(✉), Amna Islam Badhan, and Nafiz Ishtiaque Ahmed

Department of Computer Science, American International University – Bangladesh, Dhaka, Bangladesh
jabedhasan21@gmail.com, badhan2405@gmail.com, cse.ishtiaque@gmail.com

Abstract. Ontology is a kind of philosophical study which is dealing with nature being. Ontologies are extremely useful tools for different purpose and various modalities in different areas and communities. A common ontology is very effective in sophisticated software engineering purpose. In realistic world new meaningful words are always improving a language and to enhance the most widely used ontologies it requires mapping. To assure the quality manual mapping is used with some limitation. Partial automated mapping may apply to extend ontology by extracting and integrating knowledge from existing resources more effectively. In this paper, we present a semi-automated method, type of machine learning to enrich an existing ontology. Moreover, the approach can save time and ensure the accuracy that they need to serve.

Keywords: Ontology · Mapping
Semi-automated method and machine learning

1 Introduction

Ontology the word defines a specification of a conceptualization[1]. The use of background knowledge for ontology matching is often a key factor for success, particularly in complex and lexically rich domains such as the life sciences [1]. It plays an important role in modern science like, semantic web, machine translation and word sense disambiguation to exploit lexical knowledge. Ontology basically defines a collective vocabulary for researchers who need to share information in a certain domain which include machine-capable meanings of basic concepts in the domain and relations amongst them[2]. The most effectual and widely employed ontologies are still man-made. These include WordNet, Cycor, Open-Cyc, and SUMO. As they are manually assembled; these knowledge sources have the advantage of satisfying the highest quality expectations. Whatever, these knowledge sources are very costly to assemble and continuous human effort required to keeping them up to date. Language processing typically uses localized knowledge in the ontology, looking up terms, handling

[1] www.ksl.stanford.edu.
[2] https://en.m.wikipedia.org.

K. Arai et al. (Eds.): FICC 2018, AISC 886, pp. 468–478, 2019.
https://doi.org/10.1007/978-3-030-03402-3_32

synonyms, and better understanding the context surrounding a specific concept [2]. Some ontology structure can makes a useful tool for language processing. We apply semi-automated mapping approach to enrich of a particular ontology and try computational external source to the target ontology.

WordNet is a free lexical database and it's publicly available for download. It is very large and covers with general lexical relation in English. WordNet's structure can make it a useful tool for computational linguistics and natural language processing[3]. We apply semi-automated mapping to enrich of ontology through mapping an external source to the target ontology. Ontology which is consider as the pillars of semantic web. The main thread of ontology is the study of entities and their relations[4] in the philosophical sense.

Ontology is seen as a key factor for enabling in-house across different systems and semantic web applications. Ontologies mapping are required for combining and distributed different ontology's. Developing such ontology mapping has been an important issue of recent ontology research [7].

The semantic web relies heavily on the formal ontologies that structure its underlying data for comprehensive and transportable machine understanding. Ontology learning greatly facilitates the construction of ontologies. The use of ontologies to model the knowledge of specific domains represents a key aspect for the integration of information coming from different sources, for supporting collaboration within virtual communities, for improving information retrieval, and more generally, it is important for reasoning on available knowledge. Ontology learning includes a number of complementary disciplines [4]. Ontology feed on different types of unstructured, semi structured, and fully structured data to support semiautomatic ontology engineering [8]. Semantic annotation of data in the semantic web is the first critical step to better search, integration and analytics over heterogeneous data, semantic annotation of web services is an equally critical first step to achieving the target. Finding the right data for scientific research and application development is still a challenge. One important goal of the semantic web is to make the meaning of information explicit through semantic mark-up, thus enabling more effective access to knowledge contained in heterogeneous information environments, such as the web. Semantic search plays an important role in realizing this goal, as it promises to produce precise answers to user's queries by taking advantage of the availability of explicit semantics of information [5, 10]. Semantic Similarity relates to computing the similarity between concepts which are not lexicographically similar. Some of the most popular semantic similarity methods are implemented and evaluated using WordNet as the reference ontology [9].

We know that WordNet is a lexical database of English words. Many more like Oxford English Dictionary, Wiktionary (multilingual dictionary, a Wikipedia project) Encyclopedia, Wordweb and Entitypedia are also some very popular English dictionary. Do I get a word into all dictionaries? The answer is may be no. because People invent new words all the time. And all dictionaries don't update their database at a time.

[3] http://wordnet.princeton.edu/.

[4] http://opennlp.apache.org/.

For an example the word 'Textpectation' cannot be found in WordNet but it exists in Knoworthy database [3].

Our main objective of the thesis paper work is to enrich an ontology with higher accuracy using a semi-automated mapping approach. Using manually mapping we can achieve high accuracy and noise identification but like other ontologies the problem is low coverage and time consuming and also required high cost. Besides, fully automation is possible which achieve high coverage and minimal time with the problem of low accuracy and noise unidentifiable. The problem can be solved by semi-automated method which having both manual and automatic approach at a time.

In this paper we propose a semi-automated method to enrich knowledgebase with adding missing concepts in an ontology. At the beginning of the paper, we refer some related work and literature review covering with semi-automated method, structure of WordNet, WordNet database, semantic web search etc. then we analysis our method in several steps with description. After that we evaluate our method and compare result with other possible results elaborately. Before end of the paper we discus about limitations of the paper and further work which can be done in future.

2 Related Work

When a higher accuracy is necessary, semi-automatic approaches are preferable. Plenty of ideas have been generated using this method. In this section, we briefly mention some relevant work which is absolutely related with semi-automated method.

FarsNet: a lexical ontology for the Persian language. FarsNet is designed to contain a Persian WordNet with about 10000 synsets in its first phase and grow to cover verbs' argument structures and their selection restrictions in its second phase. The semi-automatic approach used to create the first phase: the Persian WordNet [11].

Entitypedia is used to evaluate SAM as target ontology and the 15,480 categories of YAGO that were directly mapped to entity as external resource. Wikipedia was used as external vocabulary. Developed at the University of Trento in Italy, Entitypedia is a knowledge base with a precise split between individuals, classes, attributes and relations and their lexicalization as proper nouns and common nouns, respectively. Entitypedia is progressively extended by collecting knowledge from several sources, including WordNet [3].

3 Description

3.1 Semi-automatic Method

Semi-automated means partially automated by both manual and automatic system to achieve the target. Enrich ontology can be possible by manually as well as semi-automatic system. Though both can reach the same target but every single step of procedure of them are different. If both approaches can reach the same goal then a question may come that why we use semi-automatic system? Yes, there are some important reasons to choose semi-automatic approach.

There have some benefit of using manual approach. For example a man-made ontology WordNet, to enrich it's database by manually the benefits are its easy to find noise, common noun and also find head for new common noun, make relational tree with WordNet database. But constrains of manual approach are time complexity, high assembly cost, high quality assurance cost, low coverage etc. To overcome from these constrains we suggest semi-automatic system which can fulfill satisfactory level with high accuracy, low cost and high coverage.

3.2 WordNet

WordNet was developed and is being maintained at the Cognitive Science Laboratory of Princeton University under the supervision of Professor George A. Its knowledge base and it can be downloaded and used free of cost and can also be browse online.

3.3 Structure of WordNet

WordNet consists of three separate databases, one for nouns, one for verbs and one for adjectives and adverbs and all are placed not in alphabetical order but with order of meaning. In Fig. 1, we have a relational tree and structure of WordNet showing how one word related to others in the WordNet database.

The current version available for online usage is 3.1 which was released in November 2012, is available to download. It contains 155,287 words organized in 117,659 synsets for a total of 206,941 word-sense pairs. The basic structure of WordNet is synsets. WordNet covers more than 118,000 different word forms and more than 90,000 dissimilar word senses In WordNet 3.0 there are 440 topics and domains [4]. Each of WordNet's 118, 000 synsets are linked with each other. WordNet is unique because of its each form-meaning pair. Means, a synset contains a brief definition and word form with several different meanings is represented in many different synsets.

About 17% of the words in WordNet are polysemous and 40% of the words are synonymous [3]. WordNet maintains the word by groupings noun, verb, adjective, and adverb the open-class words. It is assumed that the closed-class categories of English around 300 prepositions, pronouns, and determiners play a vital role in every parsing system and they are given no semantic clarification in WordNet.

Relation among words in WordNet is synonymy, e.g. the words car and automobile. Synonyms words that mean the same concept which are interchangeable in many contexts are grouped into unordered sets (synsets). Encoded relation among synsets is called ISA relation. The semantic relations that are included in WordNet are given below [3]:

- Synonymy (same-name) is a symmetric relation between word forms. It's WordNet's elementary relation.
- Antonymy (opposing-name) is also a symmetric semantic relation between word forms like synonymy, Significant in forming the meanings of adjectives and adverbs.

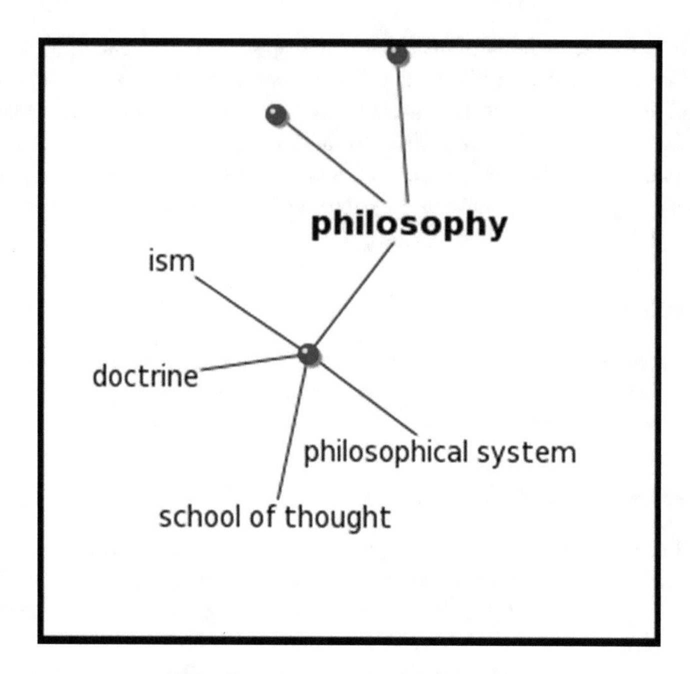

Fig. 1. Relational tree of WordNet.

- Hyponymy (sub-name) and hypernymy (super-name) inverse of hyponymy are transitive relations between synsets. This semantic relation organizes the meanings of nouns into a hierarchical structure.
- Meronymy (part-name) and holonymy (whole-name) are composite semantic relations. WordNet can distinguish component parts, substantive parts, and member parts.
- Troponymy (manner-name) is for verbs what hyponymy is for nouns, although the resulting hierarchies are much shallower.
- Entailment relations between verbs are also implied in WordNet.

While searching a new concept, it is very important to identify a head along with other words. POS tagger (parts of speech tagging) is comparatively an ideal way to find a new concept with high accuracy. However, after finding a new concept it has to be defined and make a relational tree with other concept. Sometime concepts meaning or definition can be different from head but it can relate to other concept which is related with head. Here in Fig. 2, *Graduate Course* is not directly related with head *Organization* but it related with *Course* which is related with *Work* after *Thing* and this phase is directly related with *Organization*.

3.4 WordNet Database

The WordNet database is stored in an ASCII format consisting of eight files, two for each syntactic category. Additional files are used by the WordNet search code but are not strictly part of the database [2]. WordNet only contains "open-class words" like

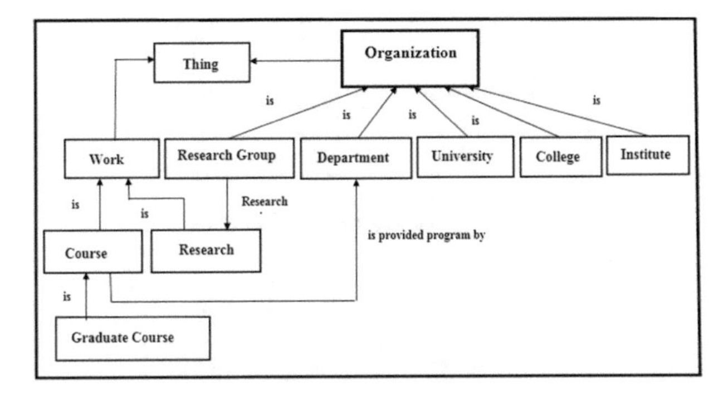

Fig. 2. ISA relation among synsets.

nouns, verbs, adjectives, and adverbs, Prepositions, pronouns. Conjunction and articles cannot be found in WordNet.

3.5 Semantic Web Searching

Semantic search is a data searching technique which search in query and aims to not only find keywords, but to determine the intent and contextual meaning of the words which is using for search. Semantic search provides more meaningful search results by evaluating and understanding the search phrase and finding the most relevant results in a website, database or any other data repository[5]. For example, as shown in Fig. 3 the architecture of Semantic Web Search is much difference of Normal Web Search[6].

Semantic search is based on the context, substance, intent and concept of the searched phrase[7]. In the Semantic Web community, semantic search is widely used to refer to a number of different categories of systems like, incorporates location, synonyms of a term, current trends, word variations and other natural language processing [6]. Semantic search concepts are derived from various search algorithms and methodologies, including keyword-to-concept mapping, graph patterns and fuzzy logic.

4 Method

In this paper we present a new method which is to search for a new word or a new word's sense and merge that to the related concept in WordNet. Searching process is applicable for paragraph and sentence. We have to pick the common nouns from those sentences and search those common nouns in WordNet, if not exist then search the word to another database or web source. If the word found in database or web source

[5] filosofie.unibuc.ro.

[6] semanticsage.blogspot.com.

[7] blogspace.com.

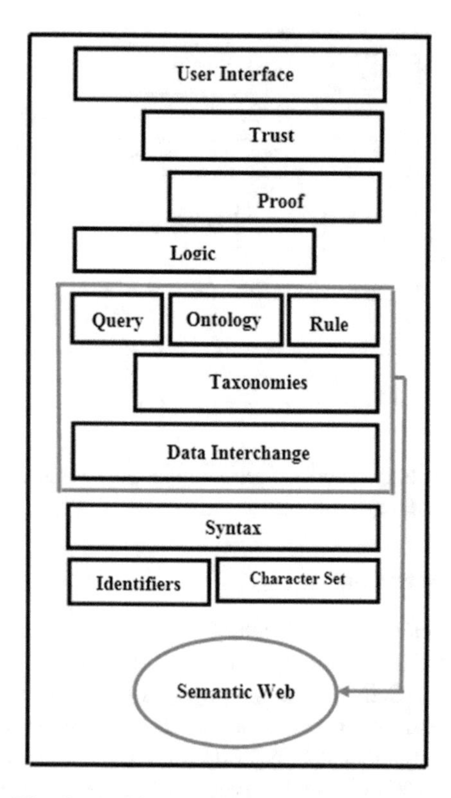

Fig. 3. Architecture of semantic web search.

then adjoins those words to the related concept in WordNet database otherwise avoid those words as noise. The way of our implementation:

- Label: When searching procedure occur by a paragraph or 5 to 6 sentences make label from these paragraph. Label means pick a word from these sentences which reflect a major influence in that passage.
- Summary: Searching with a big paragraph, make that paragraph in a summary version.
- Define new concept: Define a new concept of searching, selecting head and merge with WordNet database.
- The identification of the head, which in turn is typically based on part of speech (POS) tagging, is an approximated process with accuracy that varies according to the tool and the dataset used to train it.

Some New concepts we found in this research and add WordNet database using our process like 'Landform', 'Textpectation', 'Futurologists', etc. In Fig. 4, a new concept is used, searching using our tool.

This system can merge words in semi-automatic approach using JAWS and JWNL tools. Add and define new concept using our tool is shown in Fig. 5.

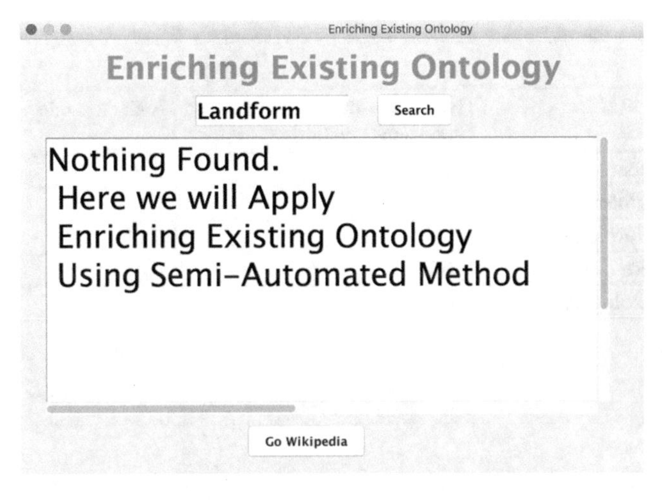

Fig. 4. Searching new concept.

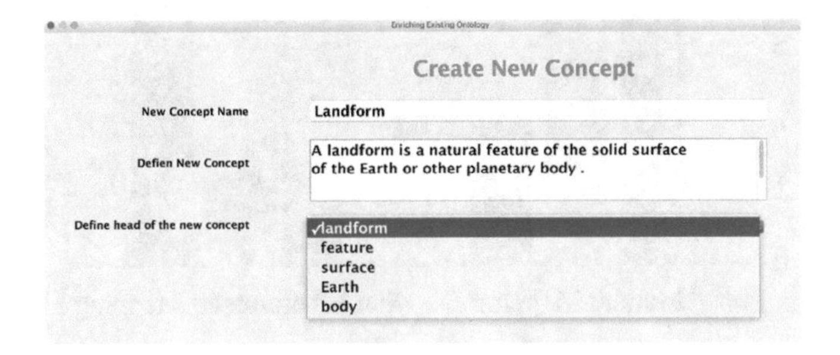

Fig. 5. Create new concept using proposed system.

5 Evaluation and Result

To evaluate our method we have tried various ontologies. The aim of this step is to get an estimate idea about our target and what we have found as result. In this process our method takes approx. 2239 ms to search a concept which is really very faster than manual searching. Though manual searching is easier to identify a noise but it takes a larger time comparatively partial mapping. And also the matter of higher accuracy and low cost, partial mapping can detect more and accurate head in minimal cost than manual mapping. Thoroughly the mechanism performs properly to reduce the noise rate of adding new concept and mapping it with WordNet database.

A set of new word is examined which is not available in WordNet. Table 1 describes the noise rate of suggested their conceptual definitions and accuracy rate of mapping the concept with WordNet. Thereby we have tried 118 different type of concept including some new concept and the result we have found with time, cost and accuracy challenge are compared and shown below in Fig. 6.

Table 1. Noise and accuracy rate for adding new word to Wordnet using semi-automated mapping

New word	Noise rate of suggested conceptual definitions	Accuracy rate
Codec	30%	98%
Landform	20%	100%
Futurologists	0%	100%
Rare diseases	0%	100%
Sexologists	20%	95%

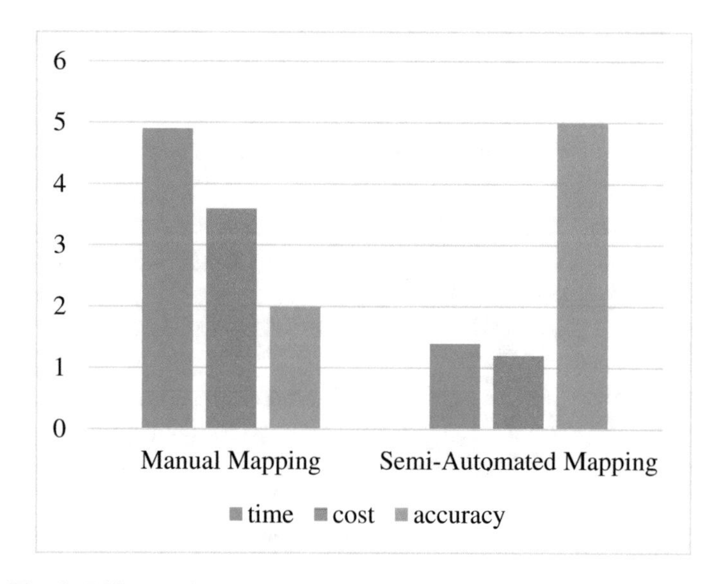

Fig. 6. Difference between manual mapping and semi-automated mapping.

6 Discussion and Limitation

The principal purpose of the study is to improve an existing ontology and materialize a new mechanism which provides high-ranking accuracy with least noise rate. Again the study adjoins the missing concepts to ontology to improve the knowledge base with semi-automated approach.

With this semi-automated approach; the WordNet Database build accurate meaning of a newborn word and makes the relationship with other knowledge base. The extracted advantages from manual and automated approach makes this semi-automated approach more effective for enriching ontology. Acquiring high accuracy and minimum noise rate the implemented work also clearly distinguish this approach with manual and automated approach to enrich ontology.

The study holds sundry limitations. At the context of particularly unformatted articles; the existing tools that perform the parsing and normalizing task from numerous sources are still not adequate to achieve absolute accuracy. Hence constructing the exact definition of a newborn word from this mechanism occasionally required human proficiency.

Like, when we got some new concept, we searched that concept related information in Wikipedia to normalize (using NLP) that concept from web content and built a relational tree with other concept. But sometimes we didn't get that target concept in Wikipedia, in that case we searched that content in other domain manually. As our current implemented application search targeted concept in Wikipedia (domain) only, it search related information only Wikipedia.

When we were preparing Head concept list for targeted new concept sentences, sometimes it added some head that's were partially related to the new concept.

7 Conclusion

World is getting easier day by day. It is difficult to update, develop and maintain manmade ontologies like WordNet. So semi-automation can be a preferable technique to enrich database through the mapping with WordNet ontology. As ontologies support semantic web application so it's an efficient system to merge words making possible. Because of using semi-automated system, it can reduce the cost, time and improve the results in the alignment of WordNet with existing ontologies.

8 Future Work

Further research on adapting semi-automated approach with more effective natural language processing (NLP) algorithm can benefits enriching ontology with providing new knowledge base concepts and constructing the accurate definition of new word. Our approach can be further improved by doing these future works:

- Alternative search: Add more website (domain) in our current implemented application for alternative search so that new concept relation can be search other domain automatically.
- Enrich algorithm: To improve head concept list, we have to enrich probable text parsing regular expression and NLP algorithm.
- Alignment of verbs and nouns: The word which makes a huge different meaning from noun to verb, make these word align.
- Word Sense Alignment: Word Sense Alignment can be defined as the identification of pairs of senses from two (or more) lexical-semantic resources which denote the same meaning, e.g.
  - Be enamored or in love with.
  - Feel love or affection for some one.
- Definition of alignment senses: The definitions describe equivalent meanings allowing the alignment of the two senses.

Acknowledgment. First of all we would like to show our gratitude to the Almighty, who gave us the effort to work on this project. We want to thanks our honorable supervisor Bayzid Ashik Hossain for guiding us. His profound knowledge in this field, keen interest, patience and continuous support lead to the completion of our work. His instructions have contributed greatly in every aspect of the thesis.

References

1. Faria, D., Pesquita, C., Santos, E., Cruz, I.F., Couto, F.M.: Automatic background knowledge selection for matching biomedical ontologies. PLoS ONE **9**(11), e111226 (2014)
2. Berndt, D.J., McCart, J.A., Luther, S.L.: Using ontology network structure in text mining, pp. 41–45 (2010)
3. Maltese, V., Hossain, B.A.: SAM: a tool for the semiautomatic mapping and enrichment of ontologies (2012)
4. Gella, S., Strapparava, C., Nastase, V.: Mapping WordNet domains, WordNet topics and wikipedia categories to generate multilingual domain specific resources (2014)
5. Caro, L.D., Boella, G.: Automatic enrichment of WordNet with common-sense knowledge (2016)
6. Elbedweihy, K., Wrigley, S.N., Ciravegna, F., Reinhard, D., Bernstein, A.: Evaluating semantic search systems to identify future directions of research (2012)
7. Choi, N., Song, I., Han, H.: A survey on ontology mapping. ACM SIGMOD Rec. **35**, 34–41 (2006)
8. Gaeta, M., Orciuoli, F., Ritrovato, P.: Advanced ontology management system for personalized e-Learning. Knowl. Based Syst. **22**(4), 292–301 (2009)
9. Varelas, G., Voutsakis, E., Raftopulou, P., Petrakis, E.G., Milios, E.E.: Semantic similarity methods in wordNet and their application to information retrieval on the web (2005)
10. Lei, Y., Uren, V., Motta, E.: SemSearch: a search engine for the semantic web (2016)
11. Shamsfard, M., Hesabi, A., Fadaei, H., Mansoory, N., Famian, A., Bagherbeigi, S., Fekri, E., Monshizadeh, M., Assi, S.M.: Semi automatic development of FarsNet; The Persian WordNet (2010)

Predicting Video Game Players' Fun from Physiological and Behavioural Data
One Algorithm Does Not Fit All

Alexis Fortin-Côté[1(✉)], Cindy Chamberland[1], Mark Parent[1], Sébastien Tremblay[1], Philip Jackson[1], Nicolas Beaudoin-Gagnon[2], Alexandre Campeau-Lecours[2], Jérémy Bergeron-Boucher[3], and Ludovic Lefebvre[3]

[1] School of Psychology, Université Laval, Quebec, Québec, Canada
alexis.fortin-cote.1@ulaval.ca
[2] Department of Mechanical Engineering, Université Laval, Quebec, Québec, Canada
[3] Ubisoft Québec, Quebec, Québec, Canada

Abstract. Finding a physiological signature of a player's fun is a goal yet to be achieved in the field of adaptive gaming. The research presented in this paper tackles this issue by gathering physiological, behavioural and self-report data from over 200 participants who played off-the-shelf video games from the Assassin's Creed series within a minimally invasive laboratory environment. By leveraging machine learning techniques the prediction of the player's fun from its physiological and behavioural markers becomes a possibility. They provide clues as to which signals are the most relevant in establishing a physiological signature of the fun factor by providing an important score based on the predictive power of each signal. Identifying those markers and their impact will prove crucial in the development of adaptive video games. Adaptive games tailor their gameplay to the affective state of a player in order to deliver the optimal gaming experience. Indeed, an adaptive video game needs a continuous reading of the fun level to be able to respond to these changing fun levels in real time. While the predictive power of the presented classifier remains limited with a gain in the F1 score of 15% against random chance, it brings insight as to which physiological features might be the most informative for further analysis and discuss means by which low accuracy classification could still improve gaming experience.

Keywords: Affective computing · Machine learning
Biomedical measurement · Video game

1 Introduction

In recent years, studies have increasingly linked video games to their potential social, cognitive and motivational benefits [1]. This led to an important growth

K. Arai et al. (Eds.): FICC 2018, AISC 886, pp. 479–495, 2019.
https://doi.org/10.1007/978-3-030-03402-3_33

in the interest of serious gaming and learning environments that use games for other reasons than pure entertainment [2]. Fun within games has proven to be a positive factor in learning and behavioural changes [3] and is also documented as an important factor in the satisfaction of players in entertainment-driven games, a 30-billion-dollar market [4]. It thus seems to be an important target to take into account in the development of any games, regardless of whether they were designed for educational or entertainment purposes. The optimization of the player's fun during gameplay would ensure that the experience is positive. To this end, a continuous assessment of the fun throughout the gameplay would allow a real-time adaptation that are tailored to the player preferences.

As with many subjective experiences, there are different definitions of the concept of fun, for it could be used as a label for different states across individuals. For instance, some people experience fun through the relief of fear by reaching a safe point in a horror game [5] or by overcoming a level after repeated failures [6]. While context and specific triggers of fun may vary, the necessary condition for something to be fun can be described as evoking a state of positive valence to a person [7]. Yet, fun remains a challenge to capture, contrarily to the assessment of difficulty and skill levels which are accurately measured through in-game behaviour [8]. Like any human affective or cognitive state, fun is continuous and unfolds over time and over multiple gaming sessions [9]. However, it is often reduced to a holistic rating in the study of user experience due to practical reasons [7,10]. Yet this approach makes it near impossible to pinpoint the specific factors which contribute to the player's fun. Most importantly, such measures of the player's fun prevent any real-time adaptation of gameplay as only general appreciation can be assessed.

Traditionally, game designers have tried to identify player preference profiles [11,12] and created content to respond to some or all player profiles as a way to ensure a sense of fun within the game. However, this led to either large population not being targeted by the game, or game content that did not match certain player profiles. Furthermore, this approach could only be used during the game design as players profiling requires lengthy psychometric assessment, which is difficult to implement at a larger scale.

To overcome these limits, game developers have started modelling player experiences in real time from behavioural cues [13,14]. For instance, dynamic difficulty adjustment algorithms have been developed to assess subjective difficulty from gameplay and adapt the game to maintain an optimal flow level [8]. However, these approaches rely on the assumption that every player experience fun in the same context and that all players react to difficulty the same way.

Although human-computer interaction initially operated primarily on the basis of behavioural markers, research in affective computing now provides access to other dimensions of a player's state through the use of psychophysiological measures [15]. Changes in physiological responses have long been recognized as potential markers of affective and cognitive states but has gained popularity in the last decades through advances in recordings, interpretation and analysis [16]. As opposed to subjective ratings, psycho-physiological measures are mostly independent from bias and can be measured continuously without breaking flow [17]. These markers have, therefore, the potential to add a number of important features to help infer the fun of a player. Nonetheless individual physiological markers are non-specific measures of the player's experience [18]. Thus, by collecting both behavioural and physiological markers, the strength of each can be leveraged in the continuous inference of the fun.

Over the last decade, an increasing number of studies have shown associations between physiological markers and a wide range of cognitive and affective states such as workload [19], attention [20], and various discrete emotions [21,22]. In gaming research, the focus has mostly been on direct biofeedback as a way to improve the player's experience [23,24]. However, few studies have attempted to predict either affective or cognitive states within video games and even less using a commercially available game. Although the approach presented here focuses on fun, it is general enough to adapt to other cognitive and affective states, such as stress and engagement, given a different measure.

This study thus aimed to find a specific physiological signature of the player's fun during video game session. To this end, a multitude of data sources were exploited: (1) the signals coming from several physiological measures; (2) the responses to a wide range of questionnaires aiming to capture individual differences; (3) the game events, which provide information on the player's actions in the game and the state of play; and (4) a proxy of the subjective state of the player during the game obtained through a replay of the session during which the participant rated the fun experienced on a continuous scale. Finding a signature from those data sources would open the door to new possibilities in affective gaming by adapting content to the player's experience and reinforcing motivation for the game. This research is part of the FUN*ii* project introduced in previous papers [25], where participants played off-the-shelf popular games from Ubisoft's Assassin's Creed series. Preliminary results were reported in [26], in which only a subset of the modalities (limited number of physiological measures) from a small set of 63 participants were leveraged. The current paper proposes an approach that takes into account the full dataset using all of the modalities for a larger sample of 218 participants.

The paper is structured as follows: The methodology and the materials used in the creation of the dataset are presented in Sect. 2. Analysis, pre-processing and labelling of the dataset is presented in Sect. 3. The modelling and fun prediction from the dataset are then presented in Sect. 4. Finally, a discussion in Sect. 5 and conclusion including future works is presented in Sect. 6.

2 Methods and Materials

2.1 Participants

Two hundred twenty eight participants aged between 18 and 35 years old ($M = 25.49$, $SD = 4.54$; 212 male, 16 females), were recruited from Université Laval and Ubisoft Québec's volunteer database. This last database and the type of game proposed to the participants are the main reason for the heavy gender imbalance of the dataset. None of the participants reported any mental health diagnosis, cognitive impairment, uncorrected vision, or health issue that could impact the physiological and cognitive measures gathered during the experiment. Furthermore, participants were required not to have played the specific games used in the experiment. This project was approved by Université Laval's Ethics Committee (#2012-272). A monetary compensation of $20 was given to participants.

2.2 Materials

The sample was split based on two different games from the Assassin's Creed series: 103 participants played the missions "The Prophet" (S5M3) and "The Escape" (S9M3) from Assassin's Creed Unity (ACU) and 115 played the missions "A Spoonful of Syrup" (S4M1) and "Survival of the Fittest" (S5M3) on Assassin's Creed Syndicate (ACS). These missions were selected based on their relatively short completion time and differences in subjective difficulty and fun during pilot studies. Both games were the latest opus of the series at the time of experimentation to ensure that we could reach enough players that had never played the game. The games were played on a high-end PC using an Xbox 360 controller for Windows. Finally, game sessions were recorded using a dedicated video monitoring card (Blackmagic WDM).

Four distinct sources of data were leveraged in this study:

(1) **Physiology**: A set of physiological measures were recorded during game sessions using a Biopac MP150 system. Cardiac activity was monitored with an electrocardiogram (ECG) using a lead II configuration. Respiratory activity was monitored using a respiration (RSP) belt transducer placed around the player's chest. Electro dermal activity (EDA) was monitored on the left thenar and hypothenar eminences. Muscle activity of the right abductor pollicis longus (APL) was monitored using electromyography (EMG) for a small subset of participants. Furthermore, eye movements and pupil size were recorded using the Smart Eye Pro eye-tracking system. Other measures included blinks, fixations, saccades, and head and gaze orientation. Head and gaze orientation might capture larger scale movement like head shake or shrug. Finally, twenty facial action units, for which intensities are rated on a scale from 0 to 5, were extracted from a video recording of the participant during gameplay using Noldus FaceReader 5.0. Data from all different sources were synchronized using Noldus Observer XT 11 and in-house routines in Matlab 2015b.

(2) **Questionnaires**: Self-reports were included in the study to capture individual differences, such as an immersion questionnaire [27]. Participants also reported the subjective difficulty and fun on a 1 to 5 scale, 1 being the lowest intensity and 5 the highest, for each mission played, and completed the short version of the NASA Task Load Index [28].

(3) **Game events**: A set of game events were also recorded during the gameplay. The player's activities were recorded at regular intervals, each second

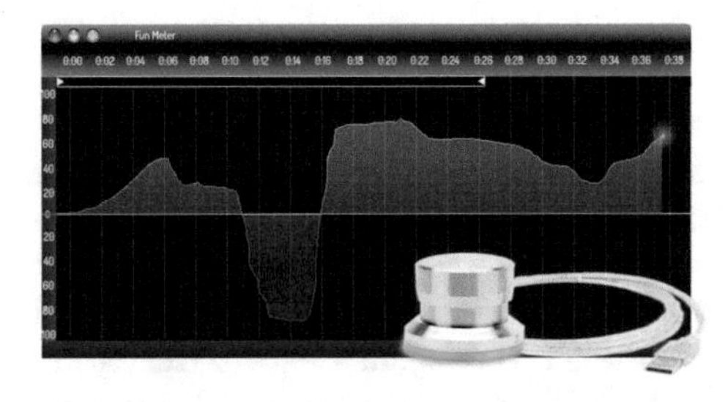

Fig. 1. Interface of the "Funmeter" with its associated control knob.

for ACU and each two seconds for ACS. Activities range from simple activities like walking, sprinting and leaping to more complex activities like "In conflict", meaning that the player is involved in a fight or "Investigated", meaning that enemies are searching for the participant's character. Game events during gameplay were tracked online. Controller inputs were recorded using the Xinput v1.2.1 API.

(4) **Fun**: A custom software called "Funmeter", of which the interface can be seen in Fig. 1, was developed as part of the project to measure the player's level of fun continuously. This software allowed players to rate the level of fun after the mission by watching a replay of their previous game session and scoring the level of fun on a linear scale from −100 to 100. Ratings were controlled by the player through a knob (PowerMate USB, Griffin technology) with visual feedback and sampled at 30 Hz. The interpretation of the meaning of the fun level was left to the participant.

2.3 Procedure

Upon arrival for a two-hour-long session, participants were given a brief overview of the project. Electrodes and physiological sensors were then installed on the participants. Once signal quality was confirmed, baseline activity for ECG, RSP, EDA and EMG signals were recorded for 3 min during which participants were asked to clear their mind while fixing a cross on a white screen and while a white noise was played in headphones as to create a baseline for those signals. The eye-tracking system was then calibrated for the participant. Participants were asked to read through a tutorial about the game to learn the game controls and mechanics. They then had a 5-minute trial session during which they had to achieve specific objectives to ensure that they knew everything necessary to complete the experiment. Participants then played the selected missions in counterbalanced order. They had an undisclosed maximum of approximately 15 min to complete each mission. Following each mission, players watched a replay of their last game session and rated their fun level continuously using the Funmeter software. They then reported their subjective experience regarding difficulty, fun and completed the NASA-TLX and the immersion questionnaires. Electrodes and sensors were removed and participants were debriefed.

2.4 Data Processing

From all the 218 participants, 25 were discarded because of technical issues. For all 193 remaining participants, of which 9 were female, the two missions played were kept for a total of 384 game sessions. General statistics of these game sessions broken-down by missions are shown in Table 1.

From the dataset available, a total of 16 different modalities were extracted, which are presented in Table 2. Sub-signals were extracted from the main signals like the heart rate and the respiration rate, which are derivatives of the electrocardiogram and the respiration intensity. ECG, RSP, EDA and EMG signals were normalized using their baseline value acquired during the first 3 min of the experience where players were at rest.

3 Dataset Analysis and Preprocessing

3.1 Game Event Analysis

The players' preferred activities were determined by examining their fun ratings in relation to game events. Figure 2 shows the distribution of participants' fun ratings as a violin plot and a sample of the underlying points. The ratings were aggregated based on different activities and averaged by participants (e.g. a player who has rated the activity "Leaping" at an average of 20 is represented as a point in this distribution for the activity "Leaping"). This figure shows a large variance in the rating of each activity and that participants mostly rated the fun above zero. Most activities have their means around the average fun for all activities (35), meaning that the participant fun is not strongly linked to it. Exceptions to this are the "Conflict", "Beaming" and to a lesser extent the "investigated", "Slow walk" and "Known" activities because their median differ from the average. The "Conflict" activity is the most interesting one, since its occurrence is high and can be easily interpreted as participants having more fun during conflict. It also shows that the distribution is fairly even and that no clear groups emerge. It would therefore be difficult to differentiate different type of player based on this information alone.

3.2 Feature Extraction

The continuous biometric signal and rating of the fun by the participant were divided into epochs of fixed time length as this method is useful for many statistical analysis. The epoch's duration has been set empirically to 5 s with no overlapping. The 5 s epoch duration provided the best trade off between a good temporal resolution and higher information (entropy) from the signals, being longer than the time constant of the physiological signals.

From the array of input modalities, features have to be extracted to suit machine-learning techniques. A total of 244 features have been extracted from all the data sources, which can be grouped in two different categories: time

Table 1. General statistics of the database, average (standard deviation) across all participants

	ACU - S5M3	ACU - S9M3	ACS - S4M1	ACS - S5M3
Completion time [min]	14.5 (2.2)	8.62 (2.0)	9.9 (2.2)	14.6 (2.0)
Rated fun [−100; 100]	35.6 (20.3)	38.7 (22.1)	32.8 (22.8)	34.8 (22.2)
Heart rate [beats/minutes]	75.0 (13.6)	75.4 (13.1)	73.3 (10.7)	73.8 (10.4)
Respiration rate [resp./minutes]	24.3 (1.8)	24.6 (1.8)	24.9 (2.0)	24.8 (1.6)
Pupil diameter [cm]	0.45 (0.07)	0.46 (0.07)	0.42 (0.07)	0.44 (0.07)

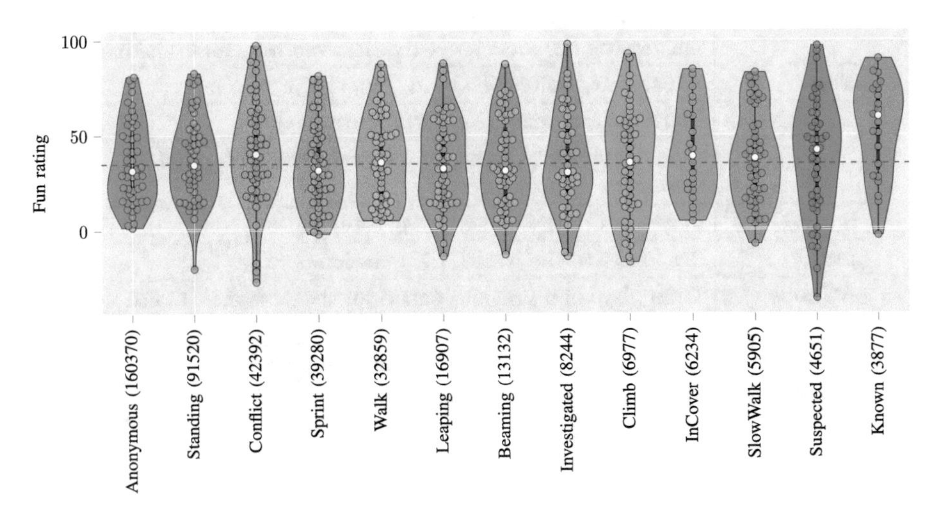

Fig. 2. Distribution of the average rating of the participants for each activity. The activities are arranged from most occurrences to least occurrences, with the occurrences in parentheses. The dotted blue line is the average of the fun for all the game sessions.

dependent and time independent. Time dependent feature consists of all the different time series presented in Table 2. From the time dependent signal, statistical features are extracted for each epoch. Those statistical features were the mean, min, max, skewness, kurtosis and the trend. Spectral power density was also extracted in bands from 0 Hz to 10 Hz.

Time independent data are answers to the questionnaires. All of the questions were based on a rating from 1 to 5 and are used as features appended to each epoch of the same participant. While these answers were obtained after the experiment so that they cannot be used in real time, they are representative of what time-independent features can bring to a classifier.

Table 2. Modalities description

ECG	Electrocardiogram and its derivative: heart rate and heart rate variability
EMG	Electromyography of the right abductor pollicis longus (thumb abductor)
RSP	Respiration intensity (diameter of the thorax) and its derivative: respiration rate
EDA	Electro dermal activity
pup	Pupil diameter (Smart Eye system)
Eye	Eye information (Smart Eye system) such has position of the gaze on the screen, eye fixation, saccades and blinking
Head	Pitch, Yaw, Roll of the head (Smart Eye system)
Au	Facial Action Units (Noldus FaceReader)
Lum	Screen luminosity, to capture interactions with pupil diameter
immrQ	Responses to the immersion questionnaire
nasaQ	Responses to the Nasa TLX questionnaire
ACgame	Time played on previous entries in the Assassin's Creed series
Difficulty	Self reported difficulty of the mission played
Appreciation	Self reported appreciation of the game session
Gender	Participant gender
Age	Age of participant
Spurious	A random value that will help in identifying truly useful features

3.3 Data Preprocessing

The participants were divided into training and test set to the ratio of 3 training participants for 1 test participant. The test participants were kept for final analysis to evaluate accuracy on unseen data.

Missing values were filled in by the imputation of values using the average of the training set corresponding features. All features were then standardized so that the training features have a zero mean and unit standard deviation along feature type.

3.4 Labelling

It is possible to infer the level of fun on a linear scale by regression methods but those are subject to the limitations inherent to ratings. Indeed, this kind of rating is subjective and of limited use, as is, notably due to limitations such as interpersonal differences and non-linearity as reported in [29,30]. While the ratings in this experiment had a large numerical spectrum as opposed to more common rating-based questionnaires in which the participant is asked to choose his level of agreement on a scale of one to five, it entailed the same limitations.

Two different methods for reducing label variance were investigated. First, a simple threshold classification of the fun, i.e. the fun is classified according to its relation to a threshold value. For example, it is placed in one class if higher than the threshold or in another if lower. The choice of the threshold values still remains somewhat arbitrary but can be chosen using relevant statistical method. An example of threshold is the mean fun of the participant during the game session. For the current project, the thresholds were thus chosen in relation to the mean ($\overline{\mathbf{x}}$) and standard deviation ($\sigma_{\mathbf{x}}$) of each game session as

$$\text{class}(x_i) = \begin{cases} \text{low fun} & \text{if } (x_i - \overline{\mathbf{x}}) < -\frac{1}{3}\sigma_{\mathbf{x}} \\ \text{neutral fun} & \text{if } -\frac{1}{3}\sigma_{\mathbf{x}} \leq (x_i - \overline{\mathbf{x}}) \leq \frac{1}{3}\sigma_{\mathbf{x}} \\ \text{high fun} & \text{if } \frac{1}{3}\sigma_{\mathbf{x}} < (x_i - \overline{\mathbf{x}}) \end{cases}, \tag{1}$$

where x_i is the fun rating at time i in the game session and $\mathbf{x} = [x_1, \ldots, x_i, \ldots, x_{\text{last}}]^T$.

Second, ranking was chosen as a method of classification. Since the absolute level of the fun is subjective and suffers from the non-linearity of reporting. The differentiation of the absolute level of fun might give a clearer indication of the increase or decrease in the level of fun. From the changes of the absolute level of fun, a change in the fun ranking can be inferred, that is, if the participant reports an increase in the absolute fun, a change from a lower ranking of fun to a higher ranking of fun should occur. It is mathematically expressed as

$$\text{rank}(x_i) = \begin{cases} \min(-1, \text{rank}(x_{i-1}) - 1) & \text{if } x_i - x_{i-1} < -T \\ \max(1, \text{rank}(x_{i-1}) + 1) & \text{if } x_i - x_{i-1} > T \\ x_i = x_{i-1} & \text{otherwise,} \end{cases} \tag{2}$$

where T is an adjustable threshold that has been set to balance the classes evenly. Drift issues caused by the differentiation of the fun signal are limited by constraining the rank to a maximum and minimum. This ranking still implies some subjectivity in the choice of the magnitude of change in the absolute fun level considered enough for a rank change (threshold).

4 Modelling

This section presents the models developed and how they are trained. Three different models are presented: one based on a regression technique and two based on classification techniques with different labelling methods, namely the classification and the ranking method.

4.1 Grouping and Validation Scheme

In preliminary experiments [26], prediction accuracy of the fun of the player was found to be higher when the training and test sets came from a single participant (intra-participant), than from training on multiple participants before making prediction on untested participants. While the prediction accuracy was better in the intra-participant case, it was of less interest for finding a specific physiological signature of the fun during a video game play, since it requires previous knowledge of the player. For this reason, the current study focused on inter-participant predictions, meaning that a signature was to be found on a set of players and then tested on another set to see if it generalizes well.

The dataset is split in two, a train set and a test set. The train subset is used in a cross validation scheme for hyper-parameter tuning and model selection. The test set is used to report final model accuracy and was only used once to report on model accuracy.

Since a general physiological signature of the fun is sought, the game events, which are specific to the Assassin's Creed's game, were discarded as features for fun prediction even if they were related to the fun experienced by participants.

Table 3. Classification results on the cross validation folds on the class label

	F1 score (standard deviation)	Accuracy
K nearest neighbours	0.354 (0.005)	0.415
Support vector classifiers	0.340 (0.012)	0.334
XGBoost	0.380 (0.018)	0.410
Multi-layer perceptron	0.36	0.373
Most basic classifier	0.331 (0.013)	0.340

4.2 Regression

The first machine learning technique evaluated was a regression technique, which, for each epoch, the average of the fun rating was to be predicted. Several regression algorithms from the Scikit-Learn library [31] were tested such as linear model (Elastic Net), Support Vector Machines (SVM) and Nearest Neighbours. An ensemble method, the optimized distributed gradient from the XGBoost library, [32] has also been tested. None of those regression algorithms were able to predict the fun state with accuracy. None showed a linear correlation between their predictions and the truth labels as the Pearson correlation coefficient stayed at zero. Noisy labels were assumed to be the main culprit of this poor performance. This is why a classification method with a limited number of classes might perform better by aggregating similar ratings and therefore limit the variance of the labels.

4.3 Classification

Classification techniques were then evaluated to learn to predict distinct state of fun for the player. To translate the rated fun into distinct state of fun, the method shown in (1) was used for each player. The number of classes was chosen as to maintain a good accuracy but also to keep a meaningful difference between classes. The three classes were interpreted as low fun, neutral and high fun. Those classes are therefore not absolute and are relative to the game session. Several classifiers, available in the Scikit-Learn library, have been tested for classification, such as Nearest Neighbours, Support Vector Classifiers (SVM), Random Forest, logistic regression and Adaboost. An optimized distributed gradient boosting library (XGBoost [32]) was also tested. Hyper-parameters for each of the algorithm have been tuned by random searches in a threefold cross validation scheme on the training set. A multi-layer perceptron was also implemented using the Keras deep-learning library. Only a subset of those classifier is presented here which correspond to the best among their family of classifier. Finally, a most basic classifier that predicts at random following the probability distribution of each class was also implemented as a point of comparison to chance level accuracy. The F1 score was chosen as a scoring metric. This scoring method is a weighted average of the precision and recall globally across the total true positives, false negatives and false positives. The average F1 scores across three folds are presented in Table 3 for each classifier including the most basic classifier which indicates the score corresponding to a random guess.

From these results, it appears that the XGBoost classifier reached better performance for this task with respect to the F1 score. This classifier also gave an indication of features importance, which is explained in detail in [33]. The importance of each feature as computed by the XGBoost classifier on the training set is presented in Table 4. Features from the same modality (signal) are grouped to give a preview of the importance of each modality. The spurious feature, a random value added to each feature sample, can help determine a threshold score indicating unimportant modalities.

Features extracted from the respiratory activity were the most used, followed by features coming from the electrocardiogram, head tracking of the Smart Eye Pro system, questionnaires and electromyography. Facial action units and pupil size also contributed to the inference but to a lesser degree as did the electrodermal activity. The other modalities such as the previous experience with the Assassin's Creed series's games, the perceived difficulty, the age, the general appreciation, the game played (Unity or Syndicate) and the gender are not or only marginally contributing to the inference. They indeed have a similar or lower score than that of the spurious feature. It is important to note that the distribution of participants was heavily skewed towards male participants (184 males against 9 females) and therefore the gender cannot be discounted as an important feature for further research.

To confirm the generalization of the learning from the XGBoost classifier, it was tested on the participants from the test set. A confusion matrix of the result

Table 4. Features ranking from the XGBoost classifier

Rank	Classification		Ranking	
	Modality	Score	Modality	Score
1	RSP	0.164	RSP	0.397
2	ECG	0.141	Eye	0.139
3	Eye	0.131	Au	0.118
4	Head	0.094	ECG	0.093
5	immrQ	0.080	Head	0.081
6	EMG	0.080	Lum	0.038
7	nasaQ	0.080	EMG	0.036
8	Au	0.059	immrQ	0.029
9	Pup	0.052	EDA	0.022
10	EDA	0.034	Pup	0.021
11	ACgame	0.025	nasaQ	0.013
12	Lum	0.024	Age	0.006
13	Age	0.008	ACgame	0.005
14	Difficulty	0.004	Spurious	0.002
15	Appreciation	0.003	Appreciation	0.001
16	Spurious	0.002	Difficulty	0.001
17	Gender	0.001	Game	0.000
18	Game	0.000	Gender	0.000

Table 5. Confusion matrix of the XGBoost classifier on the test set

	Predicted low fun state	Predicted neutral state	Predicted high fun state
Actual low fun state	1091	255	1721
Actual neutral state	712	218	1535
Actual high fun state	737	332	2304

is presented in Table 5, along with a matrix presenting precision, recall and F1 score of the classifier in Table 6.

These results showed that the classifier leaned toward predicting a high fun state, indeed it identified 62% of the occurrence of a high fun state. It was unable to predict a neutral fun state, but fairly capable of predicting a low fun state. One hypothesis explaining this might be the fact that players were playing a game they never played before (a recruitment criterion) and, thus, were mostly in a state of fairly high fun during the whole session. This could limit the difference between the low and the high fun state increasing the classification difficulty.

Table 6. Precision recall and F1 score of the XGBoost classifier

	Precision	Recall	F1-score	Support
Low fun state	0.43	0.36	0.39	3067
Neutral state	0.27	0.09	0.13	2465
High fun state	0.41	0.68	0.52	3373
Avg/total	0.38	0.41	0.38	8905

Table 7. Classification results on the cross validation folds on the rank label

	F1 score (standard deviation)	Accuracy
K nearest neighbours	0.33 (0.012)	0.415
Support vector classifiers	0.302 (0.008)	0.311
XGBoost	0.351 (0.022)	0.360
Multi-layer perceptron	0.347	0.344
Basic classifier	0.344 (0.011)	0.341

4.4 Ranking

As explained in Sect. 3.4, simple classification of rating from the player entails inherent limitations. To help circumvent some of these limits, classification based on a ranking was conducted to test if a better accuracy could be achieved. The same procedure as for the classification of the fun was applied here, with the difference that instead of predicting the average of the fun rating during the epoch, the average of the fun ranking, shown in (2), was to be predicted. Results from the same classifiers as before, retrained for ranking is shown in Table 7. It can be seen from a comparison between the two methods that the ranking method did not help classification. It seems that instead of reducing label noise, it increased it.

Features importance from the XGBoost classifier is also presented in Table 4, which shows that the modalities were ranked similarly in both classification and ranking, indicating a certain robustness to the features' rank.

5 Discussion

The goal of this study was to find a physiological signature of the player's level of fun during a video game session by converging multiple sources of data, namely the physiological signals and questionnaire answers. Those sources of data served in the prediction of the fun factor, which was rated by the participant while watching a playback of his/her game session. The results of the different classifiers showed that the best classifier was better at predicting the player's level of fun than the most basic classifier (chance) by improving the F1 score by 15%, 0.38 against 0.331. One hypothesis for this limited improvement is due to noisy labels, which is a direct effect of inter-individual variability [7], i.e. differences in the subjective rating of the fun by each participant. This fact was also reported at an earlier stage of this project [26]. Indeed, accuracy was much higher in intra-participant prediction as opposed to predictions on an unseen set of participants. The addition of more participants, facial features and their responses to questionnaires has improved inter-participant prediction, but not by a large factor. There is therefore a need to first categorize a player by their way of rating the fun. The method for ranking the fun presented in this paper still falls short of removing the impact of inter-individual variability.

With a goal of real-time inference of the fun and in light of the feature importance ranking, some type of modalities might be more useful than others like the electrocardiography, respiration and eye and head tracking. While head and eye movements are not intrusive measures, as they were acquired by cameras, an electrocardiogram and a respiration transducer are currently more intrusive for the player. Those are important considerations if such inference is to be deployed at larger scale. Questionnaires bring a small amount of information and are not intrusive during game play, but require additional time either before, or after a play session.

While the accuracy remains modest at 41% amongst 3 classes, it followed expectations as the fun rating is inherently subjective and suffers from non-linearity of reporting and inter-individual variability. This accuracy should nonetheless be useful to create a statistically significant profile of a player given many samples of similar events in a game session. Indeed, taking the conflict state as an example, it is occurring an average of 50 times in 5 sec epochs during a game session. By predicting the fun level with an accuracy of 41% each time, the mean fun level of the predictions should have a relatively low variance, which gives a good indication of the player's appreciation of conflicts.

6 Conclusion

This paper presented a classifier capable of predicting the fun rating that could be a major step in the development of adaptive gaming. Indeed, by inferring the level of fun over multiple events, its noisy nature should get averaged out to give a more accurate representation of the likes and dislikes of a player. The method presented in this paper allowed the evaluation of the importance of each source of data. This should help in sensor selection for further research by favouring a heart rate monitor, eye/head tracker and respiration belt transducer. Future works will consist in identifying which modalities are less prone to affect gameplay and better performing at predicting the fun in real time. Also works in collaboration with game designers could include game events to the set of modalities used for prediction. Since an adaptive game has a direct impact on those game events, a careful integration of those events to the features is necessary as they close the information loop. Profiling the player to help better predict fun during gameplay is also considered as a way to increase prediction accuracy by reducing inter participant variability. Further development, which is one of the main goals of the FUN*ii* project, is the development of an adaptive game prototype that will take advantage of the predicted fun to adjust itself in a way that optimizes the gaming experience. Finally, even if this paper focuses on the fun in gaming, its conclusion should be applicable to a wide range of intelligent systems that uses physiological readings has proxy for other psychological states such as stress, workload and engagement, for example. It should help build adaptive systems that might maximize health, performance or security of workers or patients.

Future works include the creation of an adaptive game based on the fun prediction. Indeed, by collecting the fun predictions over time and by associating them with game events, an appreciation of each game event can be inferred. With that appreciation profile the adaptive game will tailor itself to the player preferences by modifying the game scenarios in real time.

Acknowledgment. This project was funded by NSERC-CRSNG, Ubisoft Québec and Prompt. Additional thanks to Nvidia for providing a video card for deep learning analysis through their GPU Grant Program.

References

1. Granic, I., Lobel, A., Engels, R.C.M.E.: The benefits of playing video games. Am. Psychol. **69**(1), 66–78 (2014)
2. Djaouti, D., Alvarez, J., Jessel, J.-P.: Classifying serious games: the G/P/S model. In: Handbook of Research on Improving Learning and Motivation Through Educational Games: Multidisciplinary Approaches, vol. 2005, pp. 118–136 (2011)
3. Connolly, T.M., Boyle, E.A., MacArthur, E., Hainey, T., Boyle, J.M.: A systematic literature review of empirical evidence on computer games and serious games. Comput. Educ. **59**(2), 661–686 (2012)
4. Entertainment Software Association. Essential facts about the computer and video game industry: Entertainment Software Association, p. 11 (2016)
5. Bantinaki, K.: The paradox of horror: fear as a positive emotion. J. Aesthet. Art. Critic. **70**(4), 383–392 (2012)
6. Van Den Hoogen, W., Poels, K., IJsselsteijn, W., de Kort, Y.: Between challenge and defeat: repeated player-death and game enjoyment. Media Psychol. **15**(4), 443–459 (2012)
7. Mandryk, R.L., Inkpen, K.M., Calvert, T.W.: Using psychophysiological techniques to measure user experience with entertainment technologies. Behav. Inform. Technol. **25**(2), 141–158 (2006)
8. Zook, A.E., Riedl, M.O.: A temporal data-driven player model for dynamic difficulty adjustment. In: Proceedings of the 8th AAAI Conference on Artificial Intelligence and Interactive Digital Entertainment, AIIDE 2012, pp. 93–98 (2012)
9. Wirth, W., Ryffel, F., Von Pape, T., Karnowski, V.: The development of video game enjoyment in a role playing game. Cyberpsychol. Behav. Soc. Netw. **16**(4), 260–264 (2013)
10. Desmet, P.: Measuring emotion: development and application of an instrument to measure emotional responses to products. In: Funology: From Usability to Enjoyment, pp. 111–123 (2003)
11. Bartle, R.A.: Players who suit MUDs. Mud, p. 1 (1999)

12. Yee, N.: Motivations for play in online games. CyberPsychol. Behav. **9**(6), 772–775 (2006)
13. Yannakakis, G.N., Hallam, J.: Real-time game adaptation for optimizing player satisfaction. IEEE Trans. Comput. Intell. AI Games **1**(2), 121–133 (2009)
14. Pedersen, C.: Modeling player experience through super Mario Bros supervisor Georgios Yannakakis, Technology, pp. 132–139, August 2009
15. Fairclough, S.H.: Fundamentals of physiological computing. Interact. Comput. **21**(1–2), 133–145 (2009)
16. Cacioppo, J.T., Tassinary, L.G., Berntson, G.G.: Psychophysiological science: interdisciplinary approaches to classic questions about the mind. In: Handbook of Psychophysiology, pp. 3–22 (2000)
17. Robinson, M.D., Clore, G.L.: Belief and feeling: evidence for an accessibility model of emotional self-report. Psychol. Bull. **128**(6), 934–960 (2002)
18. Nacke, L.E.: An introduction to physiological player metrics for evaluating games. In: Seif El-Nasr, M., Drachen, A., Canossa, A. (eds.) Game Analytics, pp. 585–619. Springer, London (2013)
19. Durantin, G., Gagnon, J.F., Tremblay, S., Dehais, F.: Using near infrared spectroscopy and heart rate variability to detect mental overload. Behav. Brain Res. **259**, 16–23 (2014)
20. Dehais, F., Causse, M., Vachon, F., Tremblay, S.: Cognitive conflict in human-automation interactions: a psychophysiological study. Appl. Ergonomics **43**(3), 588–595 (2012)
21. Rainville, P., Bechara, A., Naqvi, N., Damasio, A.R.: Basic emotions are associated with distinct patterns of cardiorespiratory activity. Int. J. Psychophysiol. **61**(1), 5–18 (2006)
22. Jang, E.-H., Park, B.-J., Park, M.-S., Kim, S.-H., Sohn, J.-H.: Analysis of physiological signals for recognition of boredom, pain, and surprise emotions. J. Physiol. Anthropol. **34**, 1–12 (2015)
23. Dekker, A., Champion, E.: Please Biofeed the Zombies: enhancing the gameplay and display of a horror game using biofeedback. In: Proceedings of DiGRA, pp. 550–558 (2007)
24. Emmen, D., Lampropoulos, G.: BioPong: adaptive gaming using biofeedback. In: Creating the Difference: Proceedings of the Chi Sparks 2014 Conference, no. 1, pp. 100–103 (2014)
25. Chamberland, C., Grégoire, M., Michon, P.-E., Gagnon, J.-C., Philip, L.: A cognitive and affective neuroergonomics approach to game design. In: 59th Annual Meeting of the Human Factors and Ergonomics Society, no. 2007, pp. 1075–1079 (2015)
26. Clerico, A., Chamberland, C., Parent, M., Michon, P.-E., Tremblay, S., Falk, T.H., Gagnon, J.-C., Jackson, P.: Biometrics and classifier fusion to predict the fun-factor in video gaming. In: IEEE Conference on Computational Intelligence and Games (CIG 2016), pp. 233–240 (2016)
27. Jennett, C., Cox, A.L., Cairns, P., Dhoparee, S., Epps, A., Tijs, T., Walton, A.: Measuring and defining the experience of immersion in games. Int. J. Hum. Comput. Stud. **66**(9), 641–661 (2008)
28. Hart, S.G., Staveland, L.E.: Development of NASA-TLX (task load index): results of empirical and theoretical research. Adv. Psychol. **52**, 139–183 (1988)
29. Yannakakis, G.N., Martínez, H.P.: Ratings are overrated!. Frontiers ICT **2**(7), 5 (2015)
30. Martinez, H.P., Yannakakis, G.N., Hallam, J.: Don't classify ratings of affect; rank them!. IEEE Trans. Affect. Comput. **5**(3), 314–326 (2014)

31. Pedregosa, F., Varoquaux, G., Gramfort, A., Michel, V., Thirion, B., Grisel, O., Blondel, M., Prettenhofer, P., Weiss, R., Dubourg, V., Vanderplas, J., Passos, A., Cournapeau, D., Brucher, M., Perrot, M., Duchesnay, E.: Scikit-learn: machine learning in python. J. Mach. Learn. Res. **12**, 2825–2830 (2011)
32. Chen, T., Guestrin, C.: XGBoost: reliable large-scale tree boosting system. arXiv, pp. 1–6 (2016)
33. Hastie, T., Tibshirani, R., Friedman, J.: The elements of statistical learning. Elements **1**, 337–387 (2009)

Discovering the Graph Structure in Clustering Results

Evgeny Bauman[1(✉)] and Konstantin Bauman[2]

[1] Markov Processes Inc., Summit, NJ, USA
ebauman@markovprocesses.com
[2] Temple University, Philadelphia, PA, USA
kbauman@temple.edu

Abstract. In a standard cluster analysis, such as k-means, in addition to clusters locations and distances between them it is important to know if they are connected or well separated from each other. The main focus of this paper is discovering the relations between the resulting clusters. We propose a new method which is based on pairwise overlapping k-means clustering, that in addition to means of clusters provides the graph structure of their relations. The proposed method has a set of parameters that can be tuned in order to control the sensitivity of the model and the desired relative size of the pairwise overlapping interval between means of two adjacent clusters, i.e., *level of overlapping*. We present the exact formula for calculating that parameter. The empirical study presented in the paper demonstrates that our approach works well not only on toy data but also compliments standard clustering results with a reasonable graph structure on a real datasets, such as financial indices and restaurants.

Keywords: Unsupervised learning · Clustering · k-means
Overlapping clustering

1 Introduction

The traditional clustering problem consists of assigning each element to a single cluster such that similar elements are grouped into the same cluster. One of the most popular clustering algorithms is k-means. The main idea of this algorithm is to assign each element to the cluster with the nearest mean, serving as a prototype of the resulting cluster. K-means is a classical algorithm that is widely applied and works well in most of real data-mining problems.

In additional to standard clustering results, that include the locations of clusters' means and elements assignment, it is useful to find the relations between clusters. Some applications might benefit from knowledge of this graph of clusters' relations. For example, it could be used in a news categorization problem for the recommendations purposes. If we know which categories are interesting

K. Arai et al. (Eds.): FICC 2018, AISC 886, pp. 496–510, 2019.
https://doi.org/10.1007/978-3-030-03402-3_34

for a particular user, we might recommend a news article from the related category. Another example is biological data, where the graph of clusters relations would help to discover hidden relations between gens.

The simplest way to construct the relations between clusters is to compute the standard euclidean distance measure between the means of clusters in the feature space. However, such distance does not reflect the actual relation hidden in the data. Another way to discover such relations is to find if there is an overlapping between these two clusters or they are well separated.

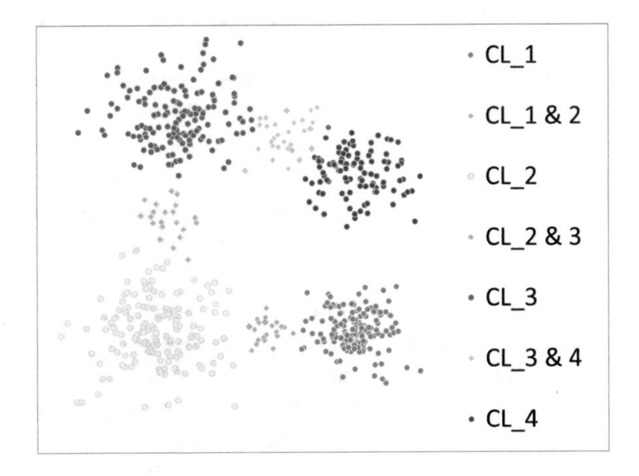

Fig. 1. Example of clusters with pairwise overlapping.

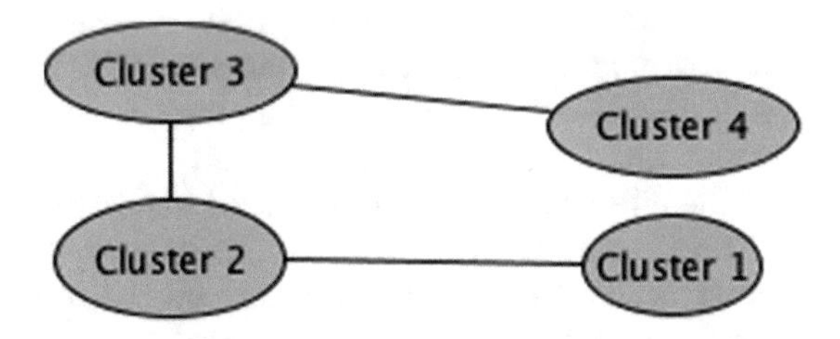

Fig. 2. Graph of relations between the clusters in the example from Fig. 1.

Figure 1 shows the synthetic example of clustering elements on a plane. In this example we can see four main clusters. There are some additional points between clusters c_1 and c_2, c_2 and c_3, c_3 and c_4, that can be assigned ether to the first cluster or to the second one. In other words these pairs of clusters have pairwise overlapping between them. The graph of relations which is build based on these overlapping between clusters is presented in Fig. 2. Note that in this example the distance between clusters c_1 and c_4 is smaller than the distance between clusters c_2 and c_3. However, clusters c_1 and c_4 are well separated and have no elements in their overlapping, thus, they are not connected in the graph (Fig. 2), while clusters c_2 and c_3 are connected because they have lots of elements in their overlapping.

This synthetic example shows that constructing the graph structure of the clusters based on their overlapping can discover hidden information that can be missed by the standard euclidean distance measure.

This problem can be addressed by overlapping clustering. However, in the ordinal overlapping clustering each object can be assigned to a certain number of clusters. In case if an object assigned to a large number of clusters, it usually means that this object is far from all clusters and does not contribute much to their means. Thus, the overlapping between large number of clusters makes it difficult to analyze the relations between clusters. Therefore, for the purpose of solving the particular problem of discovering clusters relations, it is reasonable to restrict the maximal number of clusters to which algorithm can assign each object. The easiest way is to set the threshold to two clusters. In this case pairwise overlapping between clusters can be interpreted as edges in the graph of clusters relations.

In this paper we focus on the problem of discovering pairwise relations between clusters based on their overlapping. We propose a *pairwise overlapping* modification of the k-means that allows to assign each element to *only* one or two clusters. Therefore, in additional to standard clustering results our method provides a graph of clusters relations based on the pairwise overlapping between them. The proposed optimization algorithm uses the advantages of k-means approach. In particular, it has an objective function and alternating between "Assignment" and "Update" steps it converges to the local minimum of its objective function in a finite number of steps.

In addition, our pairwise overlapping clustering algorithm allows to define parameter that specifies the level of overlapping between the pairs of resulting clusters. We present a formula for calculating a parameter of the proposed algorithm based on the desired relative size of overlapping interval between means of two adjacent clusters, i.e., *level of overlapping.*

In order to show the effectiveness of the proposed algorithm we tested it on two types of data. In particular, we present the results of applying this algorithm to the problem of constructing Hedge Funds indices and the restaurants categorization problem. We show that the proposed algorithm produces adequate and easy-interpretable results. In both applications our method discovered a reasonable graph structure of the resulting clusters. We also provide interpretations of the obtained results in terms of other aspects of objects, such as funds' strategies or restaurants word descriptions.

The rest of the paper is organized as follows. In Sect. 2 we discuss the prior work in the domain of overlapping clustering. In Sect. 3 we present our new pairwise overlapping clustering algorithm. Section 4 demonstrates the results of applying our algorithm to the financial and restaurants data. And finally, Sect. 5 concludes our findings.

2 Prior Work

The overlapping clustering problem experienced extensive growth since it was introduced in 1971 by Jardine and Sibson in [1]. One of the most popular directions of constructing overlapping clustering is formulated as a graph decomposition problem that was studied in such papers as [2,3], where authors solve the problem of minimization graph's conductance, [4] determines overlapping network module hierarchy, [5] finds overlapping communities in networks, or [6] that presents hierarchical clustering algorithm.

The next important group of overlapping clustering methods is based on the probabilistic approach that was studied in such papers as [7] that presents the Naive Bayes Model, [8] that proposes the probabilistic relational models (PRMs), [9,10] generalizes mixture model method to any other exponential distribution, [11] presents the Multiple Cause Mixture Model.

Furthermore, in [12,13] authors proposed a modification of k-means for constructing overlapping clustering. This algorithm is based on the idea to use centers of not only single clusters but also groups of certain number of clusters, such that each element assigned to a group of clusters minimizing the objective function. The method presented in [12,13] operates with certain heuristics to find the optimal value of the objective function.

Finally, [14] proposes an objective function that can be viewed as a reformulation of the traditional k-means objective, with easy-to-understand parameters that capture the degrees of overlap and non-exhaustiveness. Authors present iterative algorithm which they call NEO-K-Means (Non-Exhaustive Overlapping K-Means).

In comparison to all these previous works in overlapping clustering, we propose the pairwise overlapping clustering algorithm that focuses on the particular problem of discovering pairwise relations between clusters based on their pairwise overlapping. In particular, we restrict the maximal number of clusters to which we can assign an element by two and, therefore, allow *only* pairwise overlapping between clusters.

There is some prior work on constructing the graph of relations between resulting clusters that was proposed in [15]. However, the clusters graph that they construct inherits the relations from the initial graph of the elements. While, the algorithm proposed in this paper constructs the graph of clusters, where the presence or absence of an edge between two clusters shows if they are connected or well separated from each other.

In conclusion, although there is prior work on constructing overlapping clustering, the proposed algorithm is the first that focuses on the particular problem of discovering the relations between clusters based on their pairwise overlapping.

In the next section we present the specifics of our algorithm.

3 Discovering the Clusters Graph Structure Using Pairwise Overlapping Method

We consider a problem of *pairwise overlapping* clustering on a finite set of elements to k clusters in order to discover the graph structure of the resulting clusters. We define that there is an edge between two clusters c_i and c_j (graph vertexes) if the number of elements in the pairwise overlapping between c_i and c_j exceeds a threshold specified by the number of elements in the minimal cluster from c_i and c_j, i.e. $|c_i \cap c_j| > \gamma \cdot min(|c_i|; |c_j|)$, where γ is a parameter that can be set according to the desired sensitivity level of the model. In the absence of knowledge about the experimentation domain, parameter γ is commonly set to 0.05 or 0.1.

In the rest of the section we describe the algorithm of constructing the pairwise overlapping clustering.

3.1 Pairwise Overlapping Clustering: State of the Problem

Let $X = \{x_1, x_2, \ldots, x_N\}$ be a finite set of n-dimensional vectors $x_j \in R^n$, $j = 1, \ldots, N$. *Pairwise overlapping* clustering $\mathcal{H}$ is specified by the assignment matrix $H = ||h_{i,j}||_{k,N}$, where

$$h_{i,j} = \begin{cases} 1, & \text{if } x_j \text{ belongs to cluster } c_i \\ 0, & \text{otherwise,} \end{cases}$$

$$\text{and } 1 \leq \sum_{i=1}^{k} h_{i,j} \leq 2, \quad j = \overline{1, N}.$$

Therefore, each object x_j belongs to one or two clusters from $\mathcal{H}$.

We assume that each cluster $c_i \in \mathcal{H}$ is described by a certain prototype (mean) α_i – n-dimensional vector, which further will be chosen by optimization of the objective function. Therefore, the problem of constructing a pairwise overlapping clustering on a set of N elements to k clusters constitutes identifying matrix $H = ||h_{i,j}||_{k,N}$ and set of vectors (means) $A = (\alpha_1, \ldots, \alpha_k)$ that minimize the following objective function:

$$J(H; A) = \sum_{j=1}^{N} \left[\frac{\sum_{i=1}^{k} (x_j - \alpha_i)^2 \cdot h_{i,j}}{\left(\sum_{i=1}^{k} h_{i,j}\right)^m} \right], \text{where}$$

$$h_{i,j} \in \{0; 1\}, \quad i = \overline{1, k}; \text{ and } 1 \leq \sum_{i=1}^{k} h_{i,j} \leq 2, \quad j = \overline{1, N}. \tag{1}$$

The main idea of the criterion 1 is to optimize the sum of the average square distances from each element to the centers of clusters that it belongs to. Note that there are only one or two non-zero summands in the numerator and in the denominator of formula 1. Parameter m in 1 determines the level of overlapping between clusters in the optimal clustering. For example if $m = 1$ then the optimal clustering should be a partition of the set X. Increasing parameter m leads to increase of the uncertainty in the resulting clustering.

Theorem 3.1. *For a given finite set $X = \{x_1, \ldots, x_N\}$ of n-dimensional vectors $x_j \in R^n$, if matrix $H^* = ||h^*_{i,j}||$ and set $A^* = \{\alpha^*_1, \ldots, \alpha^*_k\}$ are the optimal matrix and the optimal set of means for the objective function $J(H; A)$ in form of (1), then*

(1) for each element x_j and two closest means $\alpha^*_{i_1} \in A^*$ and $\alpha^*_{i_2} \in A^*$, where $(x_j - \alpha^*_{i_1})^2 < (x_j - \alpha^*_{i_2})^2$, matrix H^* should satisfy the following conditions:

- $h^*_{i_1,j} = 1, h^*_{i_2,j} = 0$ *(x_j belongs to c_{i_1}), if* $(x_j - \alpha^*_{i_1})^2 < \frac{(x_j-\alpha^*_{i_1})^2 + (x_j - \alpha^*_{i_2})^2}{2^m}$
- $h^*_{i_1,j} = h^*_{i_2,j} = 1$ *(x_j belongs to c_{i_1} and c_{i_2}),*
  if $(x_j - \alpha^*_{i_1})^2 \geq \frac{(x_j-\alpha^*_{i_1})^2 + (x_j - \alpha^*_{i_2})^2}{2^m}$
- $h^*_{i,j} = 0$, *if* $i \notin \{i_1, i_2\}$.

(2) means of the clusters $\alpha^*_i \in A^*$ satisfy the following equation:

$$\alpha^*_i = \frac{\sum_{j=1}^N x_j \frac{h^*_{i,j}}{\left(\sum_{t=1}^k h^*_{t,j}\right)^m}}{\sum_{j=1}^N \frac{h^*_{i,j}}{\left(\sum_{t=1}^k h^*_{t,j}\right)^m}} \tag{2}$$

Proof: The proof consists of two parts:

(1) Each object x_j has it's corresponding part in the objective function 1. If we have fixed means the problem of assigning x_j to the optimal clusters that minimize objective function 1 can be done independently for each summand, which is actually done by the rules specified in the first part of the theorem.
(2) The optimal α^*_i should satisfy the equation: $\frac{\partial J(H;A)}{\partial \alpha_i} = 0$. Therefore, we come up to the following equation:

$$2\alpha^*_i \sum_{j=1}^N \frac{h^*_{i,j}}{(\sum_{t=1}^k h^*_{t,j})^m} - 2\sum_{j=1}^N x_j \frac{h^*_{i,j}}{(\sum_{t=1}^k h^*_{t,j})^m} = 0,$$

that gives us formula 2 for the optimal mean α^*_i.

3.2 Pairwise Overlapping Clustering Algorithm

The presented algorithm has the same structure as the well-known k-means algorithm. It uses an iterative refinement technique. Starting with an initial set of k means $A^{(0)} = (\alpha^{(0)}_1, \alpha^{(0)}_2, ..., \alpha^{(0)}_k)$, the algorithm proceeds by alternating between *Assignment* and *Update* steps. The initial set of k means can be specified randomly or by some heuristics.

Assignment step. Within the t-th iteration of the overlapping clustering algorithm on the Assignment step, we fix the values of k means $A^{(t-1)} = \left(\alpha^{(t-1)}_1, \alpha^{(t-1)}_2, \ldots, \alpha^{(t-1)}_k\right)$ from the previous iteration $(t-1)$ and minimize the objective function $J\left(H; A^{(t-1)}\right)$ by finding the optimal matrix $H^{(t)} = ||h_{i,j}^{(t)}||$, i.e. by assigning elements to the optimal number of closest clusters.

For each element x_j optimal weights $h_{i,j}^{(t)}$ should satisfy the equations from the first part of Theorem 3.1. Therefore, for each element $x_j \in X$ we proceed with the following steps:

(1) identify two closest means $\alpha_{i_1}^{(t-1)}, \alpha_{i_2}^{(t-1)} \in A^{(t-1)}$, where $(x_j - \alpha_{i_1}^{(t-1)})^2 < (x_j - \alpha_{i_2}^{(t-1)})^2$

(2) set weights $h_{i,j}^{(t)}$ according the following rules:
- $h_{i_1,j}^{(t)} = 1, h_{i_2,j}^{(t)} = 0$ *(assign x_j to $c_{i_1}^{(t-1)}$),* *if* $(x_j - \alpha_{i_1}^{(t-1)})^2 < \frac{(x_j-\alpha_{i_1}^{(t-1)})^2+(x_j-\alpha_{i_2}^{(t-1)})^2}{2^m}$
- $h_{i_1,j}^{(t)} = h_{i_2,j}^{(t)} = 1$ *(assign x_j to $c_{i_1}^{(t-1)}$, $c_{i_2}^{(t-1)}$),* *if* $(x_j - \alpha_{i_1}^{(t-1)})^2 \geq \frac{(x_j-\alpha_{i_1}^{(t-1)})^2+(x_j-\alpha_{i_2}^{(t-1)})^2}{2^m}$
- $h_{i,j}^{(t)} = 0$, *if* $i \notin \{i_1, i_2\}$.

Update step. Within the t-th iteration of the overlapping clustering algorithm on the Update step we fix the matrix $H^{(t)}$ obtained on the Assignment step and minimize the objective function $J(H^{(t)}; A)$ by finding optimal values of $A^{(t)}$.

According to formula 2 and similarly to k-means clustering algorithm [16] we set $\alpha_i^{(t)}$ to the mean of the cluster $c_i^{(t)}$ using the following formula:

$$\alpha_i^{(t)} = \frac{\sum_{j=1}^{N} x_j \frac{h_{i,j}^{(t)}}{\left(\sum_{t=1}^{k} h_{t,j}^{(t)}\right)^m}}{\sum_{j=1}^{N} \frac{h_{i,j}^{(t)}}{\left(\sum_{t=1}^{k} h_{t,j}^{(t)}\right)^m}} \tag{3}$$

The proposed pairwise overlapping clustering algorithm terminates when the Assignment step and the Update step stop changing the coverage and means of the clusters.

Theorem 3.2. *The proposed pairwise overlapping clustering algorithm converges to a certain local minimum of the objective function 1 in a finite number of steps.*

Proof: Both the Assignment and the Update steps of the algorithm reduce the objective function 1 until it reaches local minimum. Since the set of the all possible pairwise overlapping clusterings is finite, then the algorithm converges in a finite number of steps.

3.3 Setting the Overlapping Level

Parameter m in the pairwise overlapping clustering objective function 1 determines the degree of overlapping between the resulting clusters. If $m = 1$, then the optimal clustering will be a partition of the set X. In case of $m \to \infty$, most of the elements $x_j \in X$ will be assigned to a pair of the resulting clusters. Therefore, the question is how to set an appropriate value of m in order to get the desired level of overlapping between clusters.

Let's consider a pair of adjacent clusters (c_1, c_2) in the optimal pairwise overlapping clustering $\mathcal{H}$ and assume that all elements $x_j \in X$ belonging to the interval $I = [\alpha_1, \alpha_2]$ between the means of this clusters α_1 and α_2, belong either to c_1 or to c_2 or to the overlap of c_1 and c_2. We denote by interval I_1 points of the interval I belonging to c_1, by I_2 points of the interval I belonging to c_2, and by $I_{1,2}$ points of the interval I belonging to the overlap between c_1 and c_2.

For the points $x \in I$ we define the following functions:

- $g_1(x) = (x - \alpha_1)^2$;
- $g_2(x) = (x - \alpha_2)^2$;
- $g_{1,2}(x) = \left(\frac{1}{2}\right)^m \left((x - \alpha_1)^2 + (x - \alpha_2)^2\right)$.

According to the assignment step of the overlapping clustering algorithm for point x we claim the following: (a) if $g_1(x) \le g_{1,2}(x)$ then x belongs to c_1; (b) if $g_2(x) \le g_{1,2}(x)$ then x belongs to c_2; (c) if $g_1(x) > g_{1,2}(x)$ and $g_2(x) > g_{1,2}(x)$ then x belongs to the overlap of clusters c_1 and c_2.

Further, we calculate lengths $l(I_1)$, $l(I_2)$ and $l(I_{1,2})$ of specified intervals I_1, I_2 and $I_{1,2}$ by solving the following equations: $g_1(x) = g_{1,2}(x)$ and $g_2(x) = g_{1,2}(x)$. As a result we get:

$$l(I_1) = l(I_2) = \frac{1}{1 + \sqrt{2^m - 1}} l(I)$$

$$\text{and } l(I_{1,2}) = \left(1 - \frac{2}{1 + \sqrt{2^m - 1}}\right) l(I).$$

The relative length of overlapping interval $I_{1,2}$ is equal

$$r_{overlap} = \frac{l(I_{1,2})}{l(I)} = \left(1 - \frac{2}{1 + \sqrt{2^m - 1}}\right).$$

Therefore, parameter m can be represented in the following form:

$$m = \log_2 \left(\left(\frac{1 + r_{overlap}}{1 - r_{overlap}} \right)^2 + 1 \right). \quad (4)$$

Formula 4 determines parameter m for the pairwise overlapping clustering algorithm based on the desired relative size of the overlapping interval between the means of two adjacent clusters, i.e., *level of overlapping*. For example,

- in order to get $r_{overlap} = \frac{1}{3}$ (in this case $l(I_1) = l(I_2) = l(I_{1,2})$) we should set $m = \log_2 5 \approx 2.33$;
- in order to get $r_{overlap} = \frac{1}{2}$ we should set $m = \log_2 10 \approx 3.32$;
- in order to get $r_{overlap} = 0$ (the hard clustering) we should set $m = \log_2 2 = 1$.

Usually in the absence of experimentation or domain knowledge, m is commonly set to 2 or 3. In these cases, the level of overlapping would be equal 0.268 and 0.415, respectively.

4 Experiments

In order to demonstrate how well our algorithm of discovering graph structure of the resulting clusters works in practice, we tested it on two types of applications. The first one is the problem of constructing Hedge Funds Indices and the second one is the restaurant categorization problem. We present the experimental settings and the results for these two applications in Sects. 4.1 and 4.2, respectively.

4.1 Discovering the Relations Between Hedge Funds Indeces

One of the most important problems of Hedge Funds research is the problem of constructing Hedge Funds Indices. In particular, Hedge Funds Research Inc.[1] works on this problem and constructed a variety of aforementioned indices.

Most of the indices are constructed in the following way:

(1) identify a certain homogeneous market segment; and
(2) construct an index as an average value of the key assets from this segment.

Therefore, the process of constructing adequate indices that describe the market is reduced to building a good segmentation of the market and computing the centers of these segments. These centers are considered as the indices. Since these macro indices represent means of their clusters, they have more stable and predictable behavior than individual funds. One of the most important problem in the study of financial indices is the problem of predicting their values. The discovered relations between financial indices might contribute to this prediction problem. The most common way to calculate those relations based on the correlations between indices. However, the connections that are established based on the pairwise overlapping between indices are more stable and are not depend on the temporal state of financial market.

In this study we applied the proposed algorithm of pairwise overlapping clustering to the Hedge Funds data in order to identify the relations between constructed indices. As the source of data we use HFR Database[2]. We collect the set of all Hedge Funds that use the "Equity Hedge Strategy", where "Equity Hedge Strategy" means that they maintain positions both long and short in primary equity and equity derivative securities. Overall in our data, we have 855 monthly time-series of returns for 855 funds over the period of time from 06/2007 till 05/2010 (36 months in total).

First, we run the pairwise overlapping clustering algorithm on the time-series data using returns for each month as an individual features. We set the number of clusters to $k = 8$, the level of overlapping to $r_{overlap} = \frac{1}{3}$, and parameter $\gamma = 0.1$. As a result we get: (a) clusters of Hedge Funds; (b) the centers of the clusters that can be interpreted as Hedge Funds macro indices; and (c) the graph of relations between constructed indices.

---

[1] www.hedgefundresearch.com.

[2] www.hedgefundresearch.com/index.php?fuse=hfrdb.

In order to show that our algorithm provides an adequate separation of funds into clusters we compare the results or the pairwise overlapping clustering with two attributes of Hedge Funds: (a) strategies that are actually sub strategies of "Equity Hedge Strategy"; (b) Regional Investment Focus. First, based on the 5 main strategy types (out of 8 in total) and for the 8 resulting clusters we build 5×8 matrix $M_{strategy}$ of correspondence between strategy types and clusters, that is presented in Table 1. Each entry (i, C_j) in this matrix contains the number of funds that have the i-th strategy type and correspond to the cluster C_j, normalized by the total number of funds in cluster C_j. The first column *All* of matrix $M_{strategy}$ contains the numbers of funds having i-th strategy type normalized by the total number of funds. Entry (i, C_j) is marked in bold if it's significantly higher than (i, All). In this case funds from cluster C_j use the i-th strategy type more often than funds corresponding to other clusters. For example, we can say that 75% of funds corresponding to cluster C_5 use "Fundamental Growth" strategy.

Further, 4×8 matrix M_{RIF} presented in Table 2 shows the correspondence between the Regional Investment Focus (RIF) and clusters. We calculate it for the main 4 RIF types (out of 13 in total) and on the 8 resulting clusters.

Based on the matrices $M_{strategy}$ and M_{RIF} we can define an interpretation of clusters in terms of strategy types and Regional Investment Focuses. For instance, funds that belong to clusters C_1, C_4 and C_6 mainly use "Fundamental Value" strategy and their Region Investment Focus is mainly in North America. However, funds from cluster C_4 in addition use "Technology/Healthcare" strategy. Funds from clusters C_2 and C_3 use "Fundamental Growth" strategy and their primary RFI is "Russia/Eastern Europe". Funds that belong to cluster C_2 are also use "Energy/Basic Materials" strategy. Further, funds from clusters C_5 and C_7 use"Fundamental Growth" strategy, and funds from C_5 have RIF is "Asia ex-Japan". Finally, funds from cluster C_8 mainly use"Equity Market Neutral" strategy and their RIF is "Western Europe/UK". Note that we do not use fund's strategy and RIF features while building the pairwise overlapping clustering. However, our algorithm constructs clusters that appear to be adequate representation of the funds separation in terms of their main strategy and their RIFs.

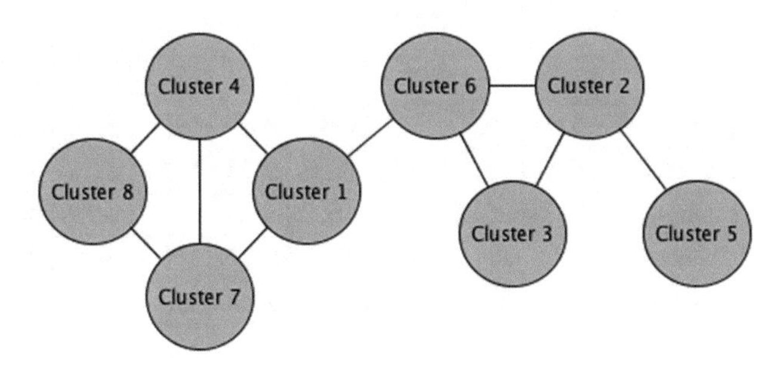

Fig. 3. Graph of the pairwise overlapping between clusters of Hedge funds.

The graph of clusters relations that was discovered based on the pairwise overlapping between clusters is presented in Fig. 3. As we can see, the cluster C_8 with focus on "Western Europe/UK" stands on the left and has connections to the clusters C_4 and C_7 with focus to "North America," whereas clusters C_2 and C_3 with focus on "Russia/Eastern Europe" stand on the right and have connections to both "North America" focused cluster C_6 and "Asia ex-Japan" focused cluster C_5. Further, although clusters C_4 and C_6 share the same regional investment focus, they are not connected with an edge. It means that these clusters are well separated by our method and have only small number of funds in the overlapping. This difference can be explained in terms of funds' strategy, where in contrast to cluster C_6 cluster C_4 actively uses "Technology/Healthcare" strategy. However, not all edges and absences of edges can be explained in the two discussed parameters (Strategy and RIF), which shows that our method discovers additional hidden connections between clusters by analyzing their pairwise overlapping.

In conclusion, based on the time-series of Hedge Funds returns we build a pairwise overlapping clustering of funds. We showed that the resulting clustering is adequate separation of funds in terms of strategies and regional investment focuses. Moreover, we construct a graph structure of the resulting clusters and showed that the edges in that graph are adequately describe the connections between clusters discovering hidden information about their relations.

4.2 Discovering the Relations Between Clusters of Restaurants

For the second experiment we used restaurant application. We address the problem of discovering the relations between restaurant categories that can be built automatically by clustering. The knowledge of such relations can be useful for the recommendation purposes. For example, "Italian restaurants" category may have a relation to the "Fast-food-Pizza" category since there are some restaurants that correspond to both of this categories. In this case, for the user who likes to visit Italian restaurants in the evening, we may recommend Pizzeria at the lunch time.

Table 1. Matrix $M_{strategy}$

Strategy	*All*	C_1	C_2	C_3	C_4	C_5	C_6	C_7	C_8
Equity market neutral	11%	3%	1%	0%	8%	0%	0%	4%	**33%**
Fundamental growth	28%	19%	**36%**	**64%**	22%	**75%**	27%	**37%**	14%
Fundamental value	39%	**63%**	25%	15%	**48%**	15%	**53%**	35%	32%
Energy/Basic materials	5%	1%	**35%**	15%	2%	1%	2%	8%	4%
Technology/Healthcare	6%	8%	1%	4%	**12%**	1%	1%	6%	7%

Table 2. Matrix M_{RIF}

RIF	*All*	C_1	C_2	C_3	C_4	C_5	C_6	C_7	C_8
North America	48%	**60%**	28%	9%	**56%**	4%	**72%**	45%	48%
Asia ex-Japan	9%	5%	6%	9%	4%	**72%**	4%	9%	4%
Russia/Eastern Europe	4%	1%	**21%**	**38%**	1%	1%	1%	2%	1%
Western Europe/UK	5%	2%	6%	0%	7%	0%	1%	4%	**11%**

In our study we used the Yelp[3] data that was provided for the Yelp Dataset Challenge[4]. In particular, we used all the reviews that were collected in the Phoenix metropolitan area in Arizona over the period of 6 years for all the 4503 restaurants (158430 reviews). In addition, all restaurants have a set of specified categories, such as "Burgers", "Chinese", "Sushi Bars" etc. For our study we selected 36 different categories that contain at least 50 restaurants. Further, we applied our algorithm of discovering graph structure to restaurants data as follows.

Firstly, for each restaurant r_i we collect a set of reviews S_i and clean these reviews from stop-words that are too generic and unlikely help us to identify the restaurants categories. We next applied the well-known LDA approach [17] using sets S_i as documents and obtained 40 topics, representing distributions of words. Some of them directly refer to the restaurant's cuisine, e.g. {mexican, salsa, taco, beans, tacos}, {pita, hummus, greek, feta}, {seafood, shrimp, fish, crab}, but some of them refer to other aspects of user experience in a restaurant, e.g. {atmosphere, cool, patio, friends, outside, outdoor}, {sports, tv, game, football, wings, watch}. At the end of this step for each restaurant r_i we assign a 40-dimensional vector according to the distribution of the resulting topics in the set of reviews S_i.

On the next step we run the proposed pairwise overlapping clustering algorithm on the set of vectors from the previous step using parameter $r_{overlap} = \frac{1}{3}$. Since our algorithm can converge to a local minimum of criteria function, we ran it 100 times starting from a random selected points. Our final result defined as the best result of the objective function 1 from 100 runs. Finally, we construct a graph of clusters based on their pairwise overlapping using parameter $\gamma = 0.1$.

In order to examine the quality of the pairwise overlapping clustering algorithm in this particular application we compare the resulting clusters with the categorization of restaurants provided by Yelp. This analysis shows that there are 26 (out of 36 in total) categories that have intersections with corresponding clusters in more than 50% of the restaurants, and 7 of those categories have intersections with corresponding clusters in more than 80% of restaurants. It means that separation constructed by our method is adequate in terms of real categories.

[3] www.yelp.com.

[4] www.yelp.com/dataset_challenge.

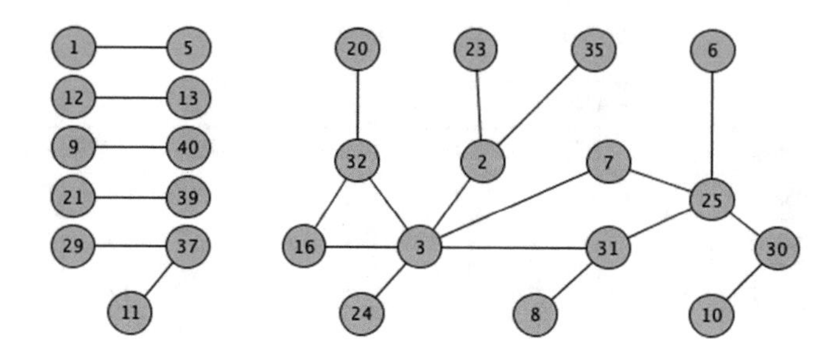

Fig. 4. Graph of the pairwise overlapping between clusters of restaurants.

Furthermore, for each cluster of restaurants we identify the set of the most important features based on the distribution of corresponding topics discussed above using some threshold level. Therefore, each cluster is described in a set of 1–5 topics.

The discovered graph of clusters connections is presented in the Fig. 4. For the simplification, we eliminated the clusters that are too small (have less that 20 objects) and the clusters that have no connections to other clusters. The presented graph has 6 components, where most of them represent connections between 2 or 3 clusters. For example, cluster 21, which is described with topics {coffee, iced, yogurt} and {flavors, creamy, fruit}, connected to the cluster 39, which is described with topics {flavors, creamy, fruit} and {chocolate, vanilla, cake}. As you can see, this clusters are pretty similar in terms of topics and, therefore they are connected.

The largest component of the discovered graph contains 15 clusters. Our method identified that they are not strongly connected, but there are some connections through other clusters. For example, clusters 3, 31 and 25 represent a chain. Clusters 3 and 31 share the same topic {menu, delicious, restaurant}, clusters 31 and 25 share topics {wine, bottle, glass} and {server, ordered, table}, while clusters 3 and 25 have no important topics in common. Although the standard euclidean distance between clusters 3 and 25 (0.00384) is less than distance between clusters 31 and 25 (0.00496), these clusters are not connected with an edge. It shows that our algorithm of discovering relations between clusters differs the simplest approach based on the euclidean distance.

In conclusion, we constructed a clustering of restaurants based on the words that are used in the corresponding reviews and also discovered a graph of relations between the resulting clusters. The separation is adequate in terms of standard categorization and the clusters graph adequately represents clear connections and discovers some hidden ones.

5 Conclusion

In this paper we presented a new algorithm of discovering graph of relations between clusters. In particular, we proposed a *pairwise overlapping* clustering algorithm that focuses on this specific discovery problem. The proposed algorithm is a modification of the k-means that allows to assign each element to *only* one or two clusters. We constructed the corresponding optimization algorithm and proved that by alternating between "Assignment" and "Update" steps it converges to a certain local minimum of the objective function in a finite number of steps.

Furthermore, the presented pairwise overlapping clustering algorithm allows to define the parameter that specifies the level of overlapping between the resulting clusters. We present the formula of calculating this parameter based on the desired relative size of overlapping interval between the means of two adjacent clusters, i.e., *level of overlapping*.

Finally, we tested the effectiveness of the proposed algorithm on two types of data. In particular, we presented the results of applying this algorithm to the problem of constructing Hedge Funds Indices and to the restaurants categorization problem. We showed that our algorithm produced adequate and easy-interpretable results and discovers a reasonable graph structure of the resulting clusters.

References

1. Jardine, N., Sisbon, R.: Mathematical Taxonomy. John Wiley, London (1971)
2. Andersen, R., Gleich, D.F., Mirrokni, V.: Overlapping clusters for distributed computation. In: WSDM 2012. ACM, New York (2012)
3. Khandekar, R., Kortsarz, G., Mirrokni, V.: Advantage of overlapping clusters for minimizing conductance. In: Proceedings of the 10th Latin American International Conference on Theoretical Informatics, LATIN 2012, pp. 494–505. Springer, Heidelberg (2012)
4. Szalay-Bekő, M., Palotai, R., Szappanos, B., Kovács, I.A., Papp, B., Csermely, P.: Moduland plug-in for cytoscape. Bioinformatics **28**(16), 2202–2204 (2012)
5. Gregory, S.: A fast algorithm to find overlapping communities in networks. In Proceedings of the 2008 European Conference on Machine Learning and Knowledge Discovery in Databases - Part I, ECML PKDD 2008, pp. 408–423. Springer, Heidelberg (2008)
6. Obadi, G., Drazdilova, P., Hlavacek, L., Martinovic, J., Snasel, V.: A tolerance rough set based overlapping clustering for the DBLP data. In: IEEE/WIC/ACM, WI-IAT 2010, pp. 57–60. IEEE Computer Society (2010). https://doi.org/10.1109/WI-IAT.2010.286
7. Meilă, M., Heckerman, D.: An experimental comparison of model-based clustering methods. Mach. Learn. **42**(1–2), 9–29 (2001). https://doi.org/10.1023/A:1007648401407
8. Battle, A., Segal, E., Koller, D.: Probabilistic discovery of overlapping cellular processes and their regulation. In: RECOMB 2004. ACM, New York (2004). https://doi.org/10.1145/974614.974637
9. Banerjee, A., Krumpelman, C., Ghosh, J., Basu, S., Mooney, R.J.: Model-based overlapping clustering. In: KDD 2005. ACM, New York (2005). https://doi.org/10.1145/1081870.1081932

10. Fu, Q., Banerjee, A.: Bayesian overlapping subspace clustering. In: ICDM 2009, pp. 776–781. IEEE Computer Society, Washington (2009). https://doi.org/10.1109/ICDM.2009.132
11. Shafiei, M.M., Milios, E.E.: Latent dirichlet co-clustering. In: ICDM 2006. IEEE Computer Society, Washington (2006). https://doi.org/10.1109/ICDM.2006.94
12. Cleuziou, G.: A generalization of k-means for overlapping clustering. In: Research Report NRR-2007-15, LIFO - University of Orleans (2007)
13. Cleuziou, G.: Osom: a method for building overlapping topological maps. Pattern Recogn. Lett. **34**(3), 239–246 (2013). https://doi.org/10.1016/j.patrec.2012.10.013
14. Whang, J., Dhillon, I.S., Gleich, D.: Non-exhaustive, overlapping k-means. In: SIAM International Conference on Data Mining (SDM), May 2015
15. Bauman, E., Muchnik, I.: Restructuring algorithm in the graph approximation problem. Autom. Remote Control **37**(6), 920–926 (1976)
16. Bezdek, J.C.: Pattern Recognition with Fuzzy Objective Function Algorithms. Kluwer Academic Publishers, Norwell (1981)
17. Blei, D.M., Ng, A.Y., Jordan, M.I.: Latent dirichlet allocation. J. Mach. Learn. Res. **3**, 993–1022 (2003). http://dl.acm.org/citation.cfm?id=944919.944937

Multiple Publications: The Main Reason for the Retraction of Papers in Computer Science

Mymoona Dawood Abdulmalek Al-Hidabi(✉) and Phoey Lee Teh

Department of Computing and Information Systems, Sunway University, Bandar Sunway, Malaysia
mymoona.a@imail.sunway.edu.my,
phoeyleet@sunway.edu.my

Abstract. This paper intends to review the reasons for the retraction over the last decade. The paper particularly aims at reviewing these reasons with reference to computer science field to assist authors in comprehending the style of writing. To do that, a total of 36 retracted papers found on the Web of Science within Jan 2007 through July 2017 are explored. Given the retraction notices which are based on 10 common reasons, this paper classifies the two main categories, namely, random and non-random retraction. Retraction due to the duplication of publications scored the highest proportion of all other reasons reviewed.

Keywords: Reasons on retraction · Random and non-random Retracted papers · Retraction notice · Clustering · Classification duplication

1 Introduction

Over the years, Steen [1], Wager and Williams [2], Fang [3], Grieneisen and Zhang [4], Zhang and Grieneisen [5], Fanelli [6], Carafoli [7], Rosenkrantz [8], and Mongeon [9] have highlighted that retraction has gradually increased [1]–[9]. Retraction is the act of taking out a paper from scientific publication. The retracted paper defined by the Web of Science (WoS) is "*An article that has been withdrawn by an author, institution, editor or a publisher because of an error of unsubstantiated data*"[1]. Fang [3] reasons the Retractions could appear, due to misconduct, fraud, a scientific error, plagiarism, duplicate publication, and so on [3]. It could also be because of faked peer reviews [10].

The authors would like to thank Sunway University Research Office also partially funded for the support of this conference presentation. This study is also partially funded by Sunway University Internal Research Grant No. INTS-2017-SST-DCIS-01.

[1] Web of Science, "Retracted Paper: Web of Science - All Databases Help," *Web of Science Website*, 2017. [Online]. Available: https://images.webofknowledge.com/WOKRS522_2R1/help/WOK/hs_document_types.html.

K. Arai et al. (Eds.): FICC 2018, AISC 886, pp. 511–526, 2019.
https://doi.org/10.1007/978-3-030-03402-3_35

When this happens, editors will send a notice of concern in a bid for initiating a case of investigation in this respect [11].

Although there is a large proportion of retraction review available in the literature, there has not been much work focusing on the reasons for the act of retraction in computer science. In this paper, an attempt has been done to analyse all computer science retracted papers that were indexed in the WoS between Jan 2007 and July 2017. The pertinent works have taken diverse perspectives on different disciplines.

During the last decade, up to 31th of July 2017, the total number of retracted papers reached 36 in seven areas of computer science. Tentative results have shown that at this current stage, the highest number recorded and contributed to computer science disciplinary accounts for a percentage of only 6.145% or equivalent to n = 1,438,466. The total publications of all disciplinary (n = 23,153,925).

Retraction causes harm to authors and a publisher's reputation [12]. On this point, Steen [13] states that retracted articles could reflect on medical misinformation [13]. Thus, it can cause misfortune to the world that referred and trusted that medical misinformation. Also, retraction has harmful acts during clinical practices [14]. Adding to the seriousness of this issue, He [15] has pinpointed that the highest rates of misconduct are found in biomedical research (i.e., clinical, medical and pharmacological) rather than in other disciplines.

The intention here is to reflect on the reason for the importance of understanding retraction. It is for preparing junior researchers better for conducting their research and improving their writing processes along with assisting them from being retracted through having a better understanding of the retraction reasons, from other disciplines not only biomedical studies [5]. Murugesan [16] indicates that any extent retraction does not impair or diminish the journal; it is conserved in the journal contained by retraction [16]. The retraction notices consider being a crucial issue in the integrity of the scientific record. Therefore, they must be exact and coherent [17]. A written Guideline on Retractions from the Committee on Publication Ethics (COPE) says, '*The intention of the retraction is to correct the literature and ensure its integrity rather than to punish authors who misbehave*' [18].

This paper classifies retraction reasons into random and non-random and attempts to spotlight the different forms of errors and misconduct [2]. As Steen [1] reasons, this is based on honest error, misconduct, frauds and ethical matters. Furthermore, the main goal of the present paper follows a precise classification in this regard for the intention to distinguish honest error (unintentional) from misconduct and frauds (intentional), particularly in computer science publications. The review also includes the poor of adherence to ethical/legal issues that were discussed and classified as intentionally unethical acts.

The present study will be organized as follows: In Sect. 2, the reasons found in the relevant literature will be described along with the terms widely used in this paper. This is followed by the review process that was applied in Sect. 3. All the results are then discussed in Sect. 4, and finally, Sect. 5 would provide the conclusion.

2 Reasons on Retraction

There are literature reviews associated with the reasons of retraction [19]–[21]. Over and above, the causes of retraction have been extensively investigated in details by Steen [1], Zhang and Grieneisen [5], Lu, Jin, Uzzi and Jones [10], Nambiar, Tilak, and Cerejo [12], and He [15] in the publications indexed in WoS. Presently, two science journalists that had spent seven years in writing on retraction have launched a beta version of retraction database. The intention of this database is to keep track of researchers that failed or committed misconduct in their publications. This has also aimed at raising community awareness of paper retraction.

Retraction is caused by human errors; it includes intentional and/or unintentional misbehaviour [2]. In the two studies of Rosenkrantz [8] and Yan et al. [19], honest error or misconduct have been identified as contributing factors to the large proportion of retraction reasons [8, 19]. Moreover, there are also different forms of misconduct defined by Trikalinos, Evangelou, and Ioannidis [22], indicating that retraction from misconduct is also due to falsification [22]. Image manipulation, which was discussed by Parrish, and Noonan [23], and faked emails by Qi, Deng, and Guo [14] are examples for fallen into the category of misconduct. No matter what type of reason there is, it is the authority of author and publisher that is involved [17]. There are a high proportion of retraction reasons. All the reasons discussed under this paper are classified into two categories: random and non-random.

2.1 Random Reasons

Whereas random reasons commonly refer to an honest error and improper data, non-random reasons have dissimilar forms of random/unintentional errors [24]. An honest error is that error performed unintentionally due to human error. As stated by Roig [25], 'unintentional errors' may exist in the process of writing committed by humans to extent that it may reach to the violation of agreement [25]. He [15] defines improper data as any part of a publication of inaccurate data that is published unintentionally [15]. He [15] added that inadvertent performance appears if an author has incorrectly done miscalculation or made an experimental error [15]. For instance, Rosenkrantz [8] comments that 16 of 48 (33.3%) of examined honest errors retractions are due to following incorrect methods or arriving at unreliable results [8]. Steen [13] endorses the comments made by Rosenkrantz [8], and says that publication bias occurs whenever errors of improper data could appear as experimental errors, data collection, errors in bias, and non-replicable results [13].

Evidence provided by recent studies indicates that retraction notices engage multiple parties [20]. In their lengthy discussion, Grieneisen and Zhang [4], and Redman, Yarandi, and Merz [26] underline that a good number of authorities usually constituted the source of retractions (i.e., authors, editors, peer-reviewers, and publishers) [4, 26]. Redman, Yarandi, and Merz [26] clarify that the reasons for retractions in a high proportion of many cases involve different parties [26]. In an analytical study by Wager and Williams [2] indicate that the percentage of 65 retractions commonly issued by authors accounts for 63%, while those by other parties amount to 21% by editors, 6% by journalist, 2% by publishers, 1% by institutions [2], and unclear statements reach 7% [2].

Errors made by authors could be mistakes appearing in data, samples, or the methods and results. Supporting this point, Carafoli [7] pinpoints that in a case of preclinical research, the impact on faulty statistical analysis which is made by the author can lead to arriving at serious and wrong conclusions [7]. A study by Hosseini, Hilhorst, de Beaufort, and Fanelli [27] argues that fourteen authors tried to correct their honest errors, despite that the journal editors have treated all their cases as retractions made by authors [27]. Wager and Williams [2] agree to what has been stated by Hosseini, Hilhorst, de Beaufort, and Fanelli [27] that occasionally, editors and publishers feel uncomfortable to retract an article [2, 27]. And, the journal's retraction notices are not clear enough to ascertain whether such retractions are due to honest errors or misconduct [2]. Steen [24] states that it is not only the reason when editors feel embarrassed, they will write unclear statements by concurrent retraction notices [24]. Undoubtedly, this ascertains that editors and publishers make authors bear the whole responsibility of retraction. Marcovitch [11] disputes that the responsibility of an honest error as improper data lies more upon the shoulder of editors rather than on authors [11, 15]. However, editors still require from authors to attach a copy of their related papers to prevent their papers from being revoked [28]. On the other hand, Benson [29] ascertains that a sin of sloth exists whenever an author disregards modifying the manuscript meticulously. It is important that authors review the manuscripts many times before the camera is ready, in order to avoid occurrence of visible errors such as formatting manuscript improperly or missing the name of co-authors [29]. In fact, editors feel annoyed at receiving several drafts of one author; however, publishers sometimes are the ones who ask the authors to keep editing their papers. Therefore, it is reasonable to state that errors made by the author have been aware of editing readability before the final submission [29]. In contrast, publisher-based errors refer to all typing errors during the manuscript submission stage. Michael L. Grieneisen, and Minghua Zhang's study classified five types of publisher errors; (1) accidental duplicate publication, (2) publishing author's paper without final correction, (3) publishing in a wrong journal, (4) publishing a special or regular issue, and (5) publishing paper with errors after a paper was rejected [4]. Grieneisen, and Zhang [4] give an example of publisher error, 49 papers are retracted in the entire issue of *Gene Express Pattern* Journal, at whatever time it had to be published in the Journal of *Mech Dev*. Publishing a paper in the wrong journal recorded accidentally as a type of publisher error.

Another example of honest error issued by publishers happens to an author who attempts to clarify this situation by different publishers. An example is when Bohannon [30] sent a fake paper to see which publisher could detect the former's mistakes. When Bohannon contacted more than fifty percent of open access Journals to run a test, the vast majority of those journals accepted Bohannon‘s paper. Surprisingly, few referred to Bohannon for fundamental reviews or did that by themselves. One of the feedbacks was made by PLOS ONE which rejected the work due to ethical problems. The work lacks literature 'about the treatment of animals used to generate cells for the experiment'. In the case, it is possible to draw that half of the publishers treated this situation of dishonesty as a normal act. John Bohannon's case indicates that reviewing any paper or the acceptance of a number of papers can happen mistakenly due to negative decisions made by reviewers or open access journals/publishers [30]. Rosenkrantz [8] and Hesselmann [20] contend that retraction may be found in ambiguous words categorized in inadvertent or deliberate acts [8, 20]. In a study conducted by Grieneisen and Zhang [4], errors on retraction have been classified as random reasons appertaining to a number of factors, such as authors and publishers along with unspecified ones, including honest errors, misconduct, and all sorts of fraud practices [4]. It is, therefore, safe to argue that retraction caused by an author or publisher is viewed in some cases as an intentional error.

To sum up, this paper asserts that errors are also found in works produced by an editor/publisher with (or) without the cooperation of an author [2]. Table 1 below is intended to put this point in a more explicable manner as it provides the Glossary of Random Reasons.

Table 1. Glossary for random reasons

Terms	Description
Improper data	Publish article with incorrect information due to behaviour (e.g., errors in samples or data, skewed statistical analysis, inaccuracies or unverifiable information, irreproducibility)
Errors of author/publisher	Errors by authors appear in data samples or due to sin of sloth. However, the publisher error is the errors in typing after a submission stage

People involved in reporting an improper work or misconduct are often the readers and/or co-authors. However, their 'failure to report misconduct on the part of others' is still of a lower level. As a result, a lack of comprehension by readers and/ or co-authors is a forefront to the unintentionally acts which refer to random errors. Figure 1 shows the Process of the Paper Retraction which includes errors that are normally caused by the main author, and sometimes by other parties.

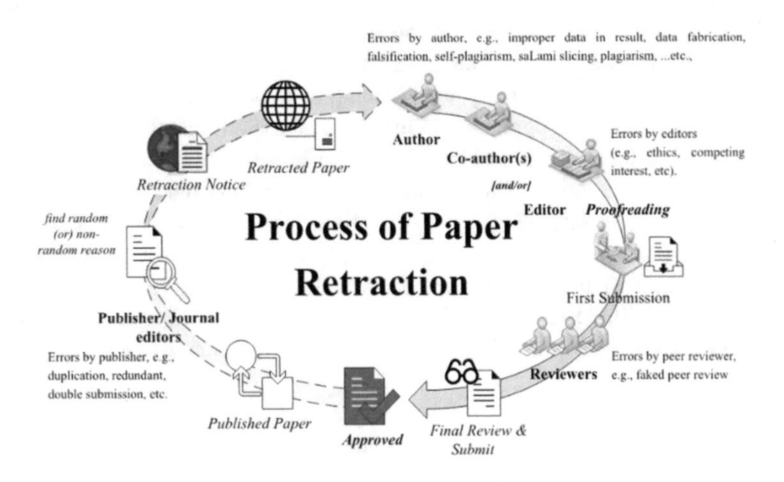

Fig. 1. The process of paper retraction.

Other points include that notifications are made during the publication stages where editors and reviewers detect some other misconducts, such as duplication, redundancy, and self-plagiarism.

2.2 Non-random Reasons

Non-random reasons refer to the presence of deliberate fraudulent data onto different nature and high levels [13]. Zhang and Grieneisen [5] classifies the misconduct and frauds into two categories; publishing misconduct and research misconduct [5]. Plagiarism and duplicate publications considered to be publishing misconduct [4, 5]. Schulzrinne [31] asserts that publishing misconduct refers to double submission of plagiarism, and falsification [31]. Philippe Mongeon and others claimed that retraction acts with misconduct and fraud have gradually grown compared to that with honest errors [3, 9, 19]. A study by Khajuria [32] clarifies the rise of research misconduct including fraud, plagiarism, and duplication [32]. As to emphasize by He [15], a retraction is happening quite extensively nowadays, it has commonly been due to duplicate publications acts, and plagiarism [15], in that case, the impacting factor of those retractions has become lower compared to the journals who have been retracted due to errors and frauds [15]. Table 2 considers the errors below as part of non-random reasons and presents the Glossary for Non-Random Reasons. There are three classifications of the main forms for Non-Random Retraction Reasons: research misconduct, publishing misconduct, and ethical/legal issues. The total number of those misconducts and frauds inclusion is twelve terms in all, namely, data fabrication, data falsification, plagiarism, and image manipulation are forms of frauds that happen intentionally. As well as, fake peer review. And, the forms of publishing misconduct are duplication of several types such as duplicate publication, a duplicate submission, redundancy, salami slicing, and self-plagiarism. Furthermore, authorship and copyright violation are the numbers of ethical/legal research. All the numerous infractions are parts of non-random reasons.

Mongeon [9] indicates that fraudulent data onto biomedical research does only not harm merely science but also affects the people who are working with fraudulent authors [9]. As a result, fraudulent data is a type of the non-random errors. Benson [29] claimed that one of the extremely worst immoral acts is data fabrication [29]. For instance, several retractions appearing in researchers' experiment when using fake data. Moylan and Kowalczuk [34] described data falsification and data fabrication as acts that come under misconduct; Moyan and Kowalczuk [34] refer to when an author manipulates data or result(s) [34]. Mongeon [9] postulates that errors on retraction due to frauds may influence an author's profession [9]. Benson [29] advise authors to keep away from fabrication and once authors confess that, they have to be ready to correct the manuscripts [29]. Data falsification pointed out by Trikalinos, Evangelou, and Ioannidis [22] as it has more negative impacts than plagiarism and duplication on ethics.

Plagiarism is defined by Oxford Dictionary as "presenting someone else's work or ideas as your own, with or without their consent, by incorporating it into your work without full acknowledgement". Plagiarism is an act of stealing the data or text as if the writing of others without making any reference at all to the source [24]. Mohan, Shetty,

Table 2. Glossary for non-random reasons

	Terms	Description
Research misconduct	Data fabrication	An act of making up a fake data intentionally
	Data falsification	An act of misleading or manipulating manuscript components as a false data or result intentionally
	Plagiarism	An act of submitting a manuscript using a certain work (idea, data, or image) lacking any references to copyrights to original authors, or without any extent permission
	Data or image manipulation	An act which may include adding and/or changing an data, displaying different data/images without referring to the original one along with the modification
	Fraudulent faked review	An act when someone uses a reviewer's email account illegally to send fake emails
Publishing misconduct	Duplicate (Dual) publication	An act of submitting the same paper to multiple conferences or journals
	Duplication submission	An act of having authors produce multiple papers, in different venues, arriving at the same results
	Redundancy	An act of providing unnecessary information to be duplicated at other times
	Salami slicing	An act of using the same data set to publish multiple papers
	Self-plagiarism	An act of copying or reusing a fully or piece of published work for someone else and submitting it as if it the first work
Ethical/legal issues	Authorship dispute	An act of violation that affects all authors in a published paper and leads to conflict of interest
	Copyright violation	An act of copying materials without getting a permission from copyright holder

Shetty, and Pandya [35] classify Plagiarism as a form of misconduct [35], consisting of Plagiarism of ideas, Plagiarism of text, Plagiarism of the source, Mosaic Plagiarism, Self-Plagiarism, and Ghost-Writing [35]. Most authors that commit an act of plagiarism are non-native English Language who commonly came from lower-income countries as pointed out by Stretton et al. [21]. Plagiarism is considered the greatest significant reason and widely appears in scientific writing [25]. Mongeon [9] reports that fabrication, falsification and plagiarism (FFP) are three serious forms of intentional infraction which have been treated as frauds [9]. Wager and Williams [2] consider FFP as forms of misconduct [2]. Additionally, data manipulation is similar to data falsification particularly if it appears as digital figures [29]. Debra Parrish, and Bridget Noonan notify that data manipulation increased sharply [23]. This form of frauds seems to be difficult to explore compared to FFP [20].

There is a kind of fraud categorised in retraction notice as 'faked peer review (FPR)' or 'faked emails (FE)'. Faked peer review or faked emails involve an author getting help from a third party as an outside expert on reviewing a real research and making suggestions on it [36]. Qi, Deng, and Guo [14] explain such a matter in terms of how a given author may review his own paper and persuasively makes the concerned institution accept it. The fundamentals of reviewer's process included quality, integrity, and reproducibility. However, FPR has led to the production of a prominent number of retracted papers [14]. Lots of samples have been retracted due to fake peer review. Qi, Deng, and Guo [14] have spotted five publishers and 48 journals out of 250 retracted papers with regards to faked peer reviews. China is behind the utmost immense number of faked reviewers' accounts; its top three provinces of faked peer reviews are Taiwan, Shanghai, and Liaoning [14]. This form of fraud is intentional errors and may bring damage to an author and editor's reputation. One solution to refrain faked peer review of fake peer reviewer is by expanding the number of a reviewer prior to training them [36]. Qi, Deng, and Guo [14] conclude that there is a work recently on the issue of fake reviewer emails and this type of misconduct will disappear in the near future [14].

Moylan and Kowalczuk [34] refer Duplication as an action when a scientific paper is published twice in different journals or may be published in different languages [34]. Nowadays, to discover any duplication practices, electronic searching is the base for any authors or reviewers [11]. Duplicate publication is considered as a result of misconduct made by an author [34]. Castillo [37] study indicates that the most retraction due to duplications comes from China and India and the reason is that of non-native speaking [37]. Another term is a duplication of submission. The retraction notices may appear as due to "Double submission" or "multiple submission". In other words, Schulzrinne [31] explains that this could be referred to as re-publication of a conference (or) journal paper when a corresponding paper has been published in another publication or the paper has been submitted to the editor for checking in another conference or a journal and still under review [31]. Sometimes retracted papers appear due to double submission when they are sent to different journals and are examined by the same reviewer(s). Schulzrinne [31] claims that this form of duplications is annoying because conference organizers are still scarce [31]. Although a number of publications were retracted due to duplicate publication (or dual), they covered other terms such as redundant publication, salami slicing, and self-plagiarism [38]. A redundant publication reported by Wager, Barbour, Yentis, Kleinert [18] that appears when an author duplicates his/her paper in different publications without specific explanation, authorization or citation [18]. Further, Shah [39] refers to redundancy as the act of republishing of an existing work with additional information [39]. Carafoli [7] has clarified the possible usage of redundancy could be found in translation, following the acquisitions of permission from the original publisher [28]. As a Guidance of the Committee on Publication Ethics Retractions (COPE) [18], redundant would be happened if an author publishes the same paper in different journals without permission, therefore the first published paper may be notified having redundant. Also, that paper will not be retracted unless the journal have checked the findings [18]. And in this regard, Hesselmann [20] claims that redundancy and salami slicing are one of the extremely significant reasons followed by conflict of interest and plagiarism [20]. Redundancy and actual duplication are further described as a salami-slicing which leads to deform the work [25]. 'Salami Publication' is defined by Roig [25] as a segmented publication which is often referred to as a case of self-plagiarism, it overlaps with the previous work of the same author [25]. Roig [25] determines that it is possible if an author(s) republishes the dissertation or thesis in separated papers, whereas it is acceptable the citation and quotation are written correctly [25].

Another term related to duplication is Self-plagiarism that further refers to submitting multiple papers in different journals with the same results but with some changes on the papers' titles [7]. In the studies of Zhang and Grieneisen [5], it is indicated that self-plagiarism is when authors re-publish previous works without any extent permission from other authors [4, 5]. Stretton et al. [21] calls self-plagiarism as a duplication of re-using a previous work from the same author [21]. A similar point to what was written by Roig [25] is that it may happen to the event that an author copies and pastes the sentences from their previously published articles [25]. Self-plagiarism covers the issue of misconduct [38]. Steen [24] argues that self-plagiarism is a type of plagiarism usually caused by an author's misconduct [24]. And, Stretton [21] concludes that self-plagiarism and plagiarism have the same meaning [21]. This is so because both terms involve reiterating data in a published article. Moreover, both self-plagiarism and plagiarism connote to the act of cheating [38].

Ethical criteria involve honesty, integrity, and social responsibility [39], however, publishing misconduct or research misconduct is an act of dishonesty intentional behaviour or fraud affects research integrity. The number of suspected cases appeared due to lack of author's awareness of the ethical writing and authorship. However, it is still argued by Wager [40] that ethical publishing is responsible for all parties [40]. Shah [39] observes that in the post-submission, all journals send the manuscript to a number of reviewers [39]. Hence, all authors should abide by the ethical regulations [32]. Wager [28] states that authors may prevent their manuscripts from publication bias by following the journal regulations [28]. Besides that, Mandal, Bagchi, and Basu [41] added that authors must do efforts to enhance their first draft, by using detection software to enhance their manuscript [41]. In the latter, it would be a great step if authors and editors set up international regulations that will attempt to improve integrity and transparency.

According to a recent article in Retraction Watch, Luann ZanZola detects three retraction papers using plagiarism detection service, iThenticate. Zanzola declares that some journals are shy to inform authors of the order of their plagiarism, and categorized a reason of "citation and attribution errors" in retraction notice. Furthermore, there are various pitfalls of misconduct causing retraction. Moylan and Kowalczuk [34]; and Markowitz and Hancock [42] indicate that ethical and authorship dispute are types of misconduct [34, 42]. A number of studies by Wager, Barbour, Yentis, Kleinert [18]; Benson [29]; and Mandal, Bagchi, Basu [41] study authorship improprieties. Barbour, Yentis, Kleinert [18] state that a number of authors demand journals to retract an article due to authorship violations [18]. However, the acts of other authors that did not accept retraction endanger the journals and editors [18]. The point here belongs to different types of authorship disputes. Authorship disputes practices consist of Gift authorship, and Ghost-writing, made by authors. Gift authorship being so-called 'Honorary', happens when individual author include author's name without any significant contribution [29]. In addition, Benson [29] defines Ghost-writing as a form of authorship dispute that happens when a co-author contributes in writing a part of research without mentioned him/her as an author; this author is called a Ghost-writer instead of co-author [29]. Wager [28] adds that practices of a supervisor who do not contribute to the manuscript cause authorship dispute [28]. Wager [28] explains the intentional ethical matter which includes a Ghost-writer name without a notification or permission. Sometimes there will be a multi-authors from a number of institutions, it is better to declare the statement of competing of interest before the submission [28]. Additionally, Wager [28] states that authorship violation tends to be a form of frauds [28], while Benson [29] assures that Guest and gift authorship are forms of improper authorship violations [29]. Mandal, Bagchi, Basu [41] explain that in terms of submission, a student who is an author should include all contributors in the authorship list [41]. However, Marcovitch [11] added that to settle authorship dispute, editors have to conceal the consent until the conflict of authors being solved, also editors must be conscious when dealing with groups of authors [11]. The problem derived here is competing for interest that will cause retraction. Wager [28] stated that authorship dispute brings a conflict of interest which is compulsory for authors who must declare all the contributors to the journal. Also, Wager [28] added that it is essential that authors must declare not only who contribute, but also the research funder [28]. A study by Benson [29] declares that competing of interest is not a misconduct, however, practicing this term improperly will lead to misconduct [29]. Benson [29] describes another ethical and legal issue which is copyright infringement. This infringement is caused by author, or editor/publisher or other parties in the research community [29]. Andreescu [38] associates copyright infringement with self-plagiarism, a fact which refers to a matter between author and journal editor [38]. Therefore, a copyright law prevents the author's work from being self-plagiarized. Mandal, Bagchi, Basu [41] mentions that the first author is recognized as a corresponding author that should identify any responsibility of all co-authors [41]. In addition, Mongeon [9] asserts that the legal consequences on retraction are influenced by all authors, however, the last author comes after the first author who is affected strongly, while middle authors (co-author) may have a lesser significant impact [9]. To improve the scientific community, Li [33] recommends senior authors to educate novices in collaboration with the scientific publishing [33]. Shah [39] also states 'Poor supervision of junior researchers' (postgraduate students), will drive into non-random reasons [39], however, a clear understanding of the retraction will reduce unethical behaviours.

Despite the presence of a high proportion of reasons on retraction, Wager and Williams [2]; Fang [3]; Zhang and Grieneisen [5]; Stretton et al. [21]; and Benson [29] have shown that there is 'No reason/unclear/unknown' statement, stating that retractions are related to unspecified errors. This is so due to vague reasons for retraction notices [2, 3, 5, 21, 29]. In addition, Moylan and Kowalczuk [34] state that 'No reason/unclear' is difficult to identify if the statement is 'honest error' or 'misconduct' [34]. At the end of this paper, random and non-random reasons on retraction have set the overall titles as shown in the diagram below (Fig. 2).

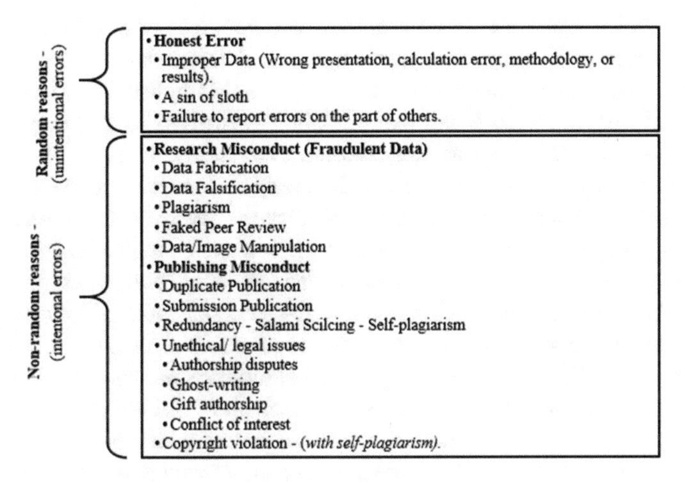

Fig. 2. Random and non-random reasons on retraction.

3 Research Method

3.1 Research Process

The present research paper has been done via three phases (Fig. 3).

(1) Data Collection: At the initial phase of the study, a comprehensive literature review is performed summarising the "reasons on retraction". All the reviewed papers are collected from reputable impact factor journals and databases such as ACM, Elsevier, PLoS ONE, Springer, SAGE, Wiley, and other Open Access Journals (OAJ), such as BMJ Open, MDPI, and PNAS.

The period of retraction was conducted on Jan 2007 through July 2017. The latest retraction took place on July 31th, 2017. A total number of 23,153,925 publications were successfully indexed in WoS. Also a total number of retraction publications being equal to the total number of computer science "disciplinary" accounted for 1,438,466. Recently, there are 2,248 retracted publications in Computer Science and 1,098 retractions indexed in WoS (dated July 31th 2017). Out of the numbers above, 36

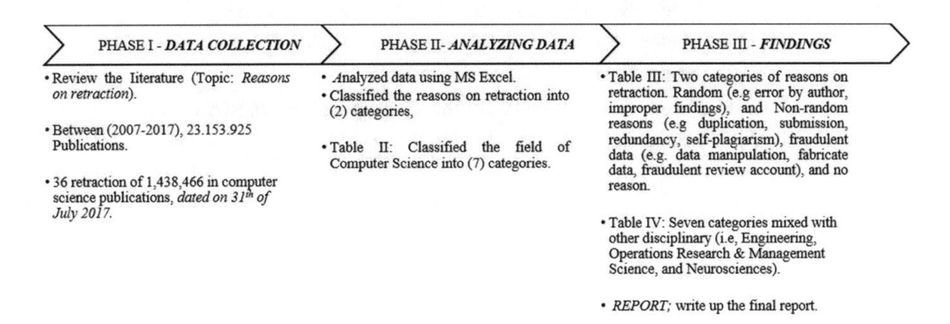

Fig. 3. Research process chart.

retraction articles in Computer Science among all other types of were downloaded from WoS, constituted seven categories under Computer Science. The present research was limited to the selection of retracted publication written in English. They were further set into seven categories. By so doing, consideration was given to that they mixed with other related disciplines, such as Engineering, Operations Research & Management Science, and Neurosciences & Neurology. Not only that, every notice for every single retracted paper was sought for in order to identify the reason(s) for retraction.

(2) Data Analysis: The research filtered the retracted publication. Regarding the WoS, retraction processes can be implemented by using two types of documents: "Retracted Publications" and "Retraction". Retracted Publication is defined as a withdrawal document. However, Retraction is a document that shows the retraction notice.

Doing a search for articles in through WoS is more secure and trustworthy than in websites for sciences, social sciences, arts, and humanities disciplines [10]. The Web of Science allows authors to examine and scrutinize papers with close similarities with a specific field of knowledge and know dealt with papers through limited databases [10]. The WoS includes 1.3 billion, indexed and cited references. Moreover, there are more than 18000 high-impact journals with other databases[2]. All information on those (36) retracted papers were examined and analysed by using MS Excel; besides that, the reasons for retraction (Random and Non-Random reasons) were analysed as well.

(3) The Findings: Based on the relevant retraction notices, the reasons on retraction were identified and summarised. They had further classified them into two categories of errors. We also set the discipline of computer science into seven categories according to WoS in Table 3. It was important to decide if what has been discussed is still relevant to the related work which focused on studying different cases of retractions through WoS. Then, a consideration is given to the significant question. Next section will figure out the significant results to arrive at a better understanding of retraction along with relevant reasons.

Table 3. Computer science categories

#	Computer science categories
1	Computer Science, Software Engineering (CSSE)
2	Computer Science, Theory & Methods (CSTM)
3	Computer Science, Hardware & Architecture (CSHA)
4	Computer Science, Information Systems (CSIS)
5	Computer Science, Cybernetics (CSC)
6	Computer Science, Interdisciplinary Applications (CSIA)
7	Computer Science, Artificial Intelligence (CSAI)

[2] https://clarivate.com/products/web-of-science/web-science-form/web-science-core-collection/.

4 Results and Discussion

4.1 Results

Based on the retraction notices, Table 4 proves that the 36 retracted papers are classified into two categories: random and non-random reasons of paper retraction. Table 5 also highlights that seven categories from the papers of retraction in Computer Science mixed with other areas, such as Engineering (n = 5), Operations Research & Management Science (N = 5), and Neurosciences (n = 1).

Table 4. Frequent and percentage of reasons on retraction (Jan 2007 to July 2017)

#	Reasons on retraction	N	Errors category	Total	%
1	Improper findings	3	Random	4	11
2	Error-by-author	1			
3	Duplicate publication	9	Non-random	31	86
4	Redundancy	3			
5	Duplicate submission	6			
6	Self-plagiarism	2			
7	Data manipulation	2			
8	Fraudulent reviewer account	3			
9	Plagiarism	6			
10	No reason	1	No reason	1	3
Total				36	100

*(all were searched on 31.07.2017)

Table 5. Percentage of computer science categories

#	Computer science categories	N, %
1	Computer Science, Software Engineering (CSSE)	1, 2.7%
2	Computer Science, Theory & Methods (CSTM)	1, 2.7%
3	Computer Science, Hardware & Architecture (CSHA) and etc.	2, 5.6%
4	Computer Science, Information Systems (CSIS) and etc.	4, 11%
5	Computer Science, Cybernetics (CSC) and etc.	6, 6.6%
6	Computer Science, Interdisciplinary Applications (CSIA) and etc.	7, 19.4%
7	Computer Science, Artificial Intelligence (CSAI) and etc.	15, 41.6%

4.2 Discussion

The present study paper shows that during the last 126 months (i.e. Jan 2007–July 2017), the total number of 36 retraction reasons gives ten common reasons. Yet, it is important to take note that there are significantly fewer random reasons than non-random.

Non-random reasons have scored the highest (n = 31/36, 86%) out of other paper retraction with regards to fraud and misconduct. Specifically, duplication has scored the first highest number of reason for retraction (n = 20/36, 55.5%), to include the reasons for duplication of publication; while (n = 9/36, 25%) is for Redundancy in data accounts for (n = 3/36, 8.3%) and Duplicate of submission amounts to (n = 6/36, 16.6%). The least is Self-plagiarism, accounting for n = 2/36, 5.5%), followed by the result Data manipulation, amounting to (n = 2, 5.5%) while Fraudulent reviewer for (n = 3, 8.3%), and Plagiarism for (n = 6, 16.6%). The statement "No reason" has only one case (n = 1, 2.7%). It is clear that Computer Science, Artificial Intelligence (CSAI), etc. has the highest rate (N = 15/36) among all computer science categories. Given that our findings are based on a limited number of retraction papers, the result from such analysis should therefore be treated with considerable the utmost caution. Conclusively, the paper ends with a conclusion and future research.

5 Conclusion

Eventually, this paper has attempted to classify in detail the available information relevant to the reasons on retraction. Until July 2017, 36 studies were retracted in the discipline of computer science. We have found that there is no retraction notice regarding authorship, violation of copyright rules, or ethical issues in Computer Science. The evidence from this study indicates that Non-random reasons such as duplicate publication, submission, and plagiarism have the highest number in total compared to other random reasons. In addition, the majority of findings show that there is a significant relationship between the duplicate of publications from the category of Computer Science Artificial Intelligence (CSAI). The study has some limitations; however the results so far have been very promising and should be validated by a larger sample size. To further our research we are planning to expand the retractions to involve various disciplines such as linguistics and psychology. Also, the research will involve additional database (i.e., SCOPUS). Ultimately, we are confident that our results may improve knowledge about the reasons on retraction. And, future work should be attention to distinguish random reasons from non-random ones towards increasing awareness of retraction notices.

References

1. Steen, R.G.: Retractions in the scientific literature: is the incidence of research fraud increasing? J. Med. Ethics **37**(4), 249–253 (2011)
2. Wager, E., Williams, P.: Why and how do journals retract articles? An analysis of Medline retractions 1988–2008. J. Med. Ethics **37**(9), 567–570 (2011)
3. Fang, F.C., Steen, R.G., Cadadevall, A.: Misconduct accounts for the majority of retracted scientific publications. Proc. Natl. Acad. Sci. **109**(42), 17028–17033 (2012)
4. Grieneisen, M.L., Zhang, M.: A comprehensive survey of retracted articles from the scholarly literature. PLoS ONE **7**(10), e44118 (2012)
5. Zhang, M., Grieneisen, M.L.: The impact of misconduct on the published medical and non-medical literature, and the news media. Scientometrics **96**(2), 573–587 (2013)

6. Fanelli, D.: Why growing retractions are (mostly) a good sign. PLoS Med. **10**(12), 1–6 (2013)
7. Carafoli, E.: Scientific misconduct: the dark side of science. Rend. Lincei **26**, 369–382 (2015)
8. Rosenkrantz, A.B.: Retracted publications within radiology journals. Am. J. Roentgenol. **206** (2), 231–235 (2016)
9. Mongeon, P., Lariviere, V.: Costly collaborations: the impact of scientific fraud on co-authors' careers. J. Assoc. Inf. Sci. Technol. **67**(3), 535–542 (2016)
10. Lu, S.F., Jin, G.Z., Uzzi, B., Jones, B.: The retraction penalty: evidence from the web of science. Sci. Rep. **3**, 3146 (2013)
11. Marcovitch, H.: Misconduct by researchers and authors. Gac. Sanit. **21**(6), 492–499 (2007)
12. Nambiar, R., Tilak, P., Cerejo, C.: Quality of author guidelines of journals in the biomedical and physical sciences. Assoc. Learn. Prof. Soc. **27**(3), 201–206 (2014)
13. Steen, R.G.: Misinformation in the medical literature: what role do error and fraud play? J. Med. Ethics **37**, 498–503 (2011)
14. Qi, X., Deng, H., Guo, X.: Characteristics of retractions related to faked peer reviews: an overview. Postgrad. Med. J. **93**(1102), 499 (2016)
15. He, T.: Retraction of global scientific publications from 2001 to 2010. Scientometrics **96**(2), 555–561 (2013)
16. Murugesan, R.: What Happens When Ethical Violations are Detected in Research? 13 December 2014
17. Bilbrey, E., O'Dell, N., Creamer, J.: A novel rubric for rating the quality of retraction notices. Publications **2**(1), 14–26 (2014)
18. Wager, E., Barbour, V., Yentis, S., Kleinert, S.: Retractions: guidance from the committee on publication ethics (COPE). Maturitas **64**(4), 201–203 (2009)
19. Yan, J., MacDonald, A., Baisi, L.-P., Evaniew, N., Bhandari, M., Ghert, M.: Retractions in orthopaedic research - a systematic review. Bone Joint Res. **5**(6), 263–268 (2016)
20. Hesselmann, F., Graf, V., Schmidt, M., Reinhart, M.: The visibility of scientific misconduct: a review of the literature on retracted journal articles. Curr. Sociol. 1–32 (2016)
21. Stretton, S., et al.: Publication misconduct and plagiarism retractions: a systematic, retrospective study. Curr. Med. Res. Opin. **28**(10), 1575–1583 (2012)
22. Trikalinos, N.A., Evangelou, E., Ioannidis, J.P.A.: Falsified papers in high-impact journals were slow to retract and indistinguishable from nonfraudulent papers. J. Clin. Epidemiol. **61** (5), 464–470 (2008)
23. Parrish, D., Noonan, B.: Image manipulation as research misconduct. Sci. Eng. Ethics **15**(2), 161–167 (2009)
24. Steen, R.G.: Retractions in the scientific literature: do authors deliberately commit research fraud? J. Med. Ethics **37**(2), 113–117 (2011)
25. Roig, M.: Avoiding plagiarism, self-plagiarism, and other questionable writing practices: a guide to ethical writing, pp. 1–71 (2015)
26. Redman, B.K., Yarandi, H.N., Merz, J.F.: Empirical developments in retraction. J. Med. Ethics **34**(11), 807–809 (2008)
27. Hosseini, M., Hilhorst, M., de Beaufort, I., Fanelli, D.: Doing the right thing: a qualitative investigation of retractions due to unintentional error. Sci. Eng. Ethics, no. office 321 (2017)
28. Wager, E.: Ethical publishing: the innocent author's guide to avoiding misconduct. Menopause Int. **13**(3), 98–102 (2007)
29. Benson, P.J.: Seven sins in publishing (but who's counting…). Ann. R. Coll. Surg. Engl. **98** (1), 1–5 (2016)
30. Bohannon, J.: Who's afraid of peer review? Science **342**(6154), 60–65 (2013)

31. Schulzrinne, H.: Double submissions - publishing misconduct or just effective dissemination? ACM SIGCOMM Comput. Commun. Rev. **39**(3), 40–42 (2009)
32. Khajuria, A., Agha, R.: Fraud in scientific research - birth of the concordat to uphold research integrity in the United Kingdom. J. R. Soc. Med. **107**(2), 61–65 (2014)
33. Li, Y.: Text-based plagiarism in scientific publishing: issues, developments and education. Sci. Eng. Ethics **19**(3), 1241–1254 (2013)
34. Moylan, E.C., Kowalczuk, M.K.: Why articles are retracted: a retrospective cross-sectional study of retraction notices at BioMed Central. Br. Med. J. Publ. Gr. **6**(11), e012047 (2016)
35. Mohan, M., Shetty, D., Shetty, T., Pandya, K.: Rising from plagiarising. J. Oral Maxillofac. Surg. **14**(3), 538–540 (2014)
36. Retraction Watch: Can a tracking system for peer reviewers help stop fakes? (2017). http://retractionwatch.com/2017/06/23/can-tracking-system-peer-reviewers-help-stop-fakes/
37. Castillo, M.: The fraud and retraction epidemic. Am. J. Neuroradiol. **35**(9), 1653–1654 (2014)
38. Andreescu, L.: Self-plagiarism in academic publishing: the anatomy of a misnomer. Sci. Eng. Ethics **19**(3), 775–797 (2013)
39. Shah, N.: Ethical issues in biomedical research and publication. J. Conserv. Dent. **14**(3), 205–208 (2011)
40. Wager, E.: Publication ethics: whose problem is it? Insights **25**(3), 294–299 (2012)
41. Mandal, M., Bagchi, D., Basu, S.R.: Scientific misconducts and authorship conflicts: Indian perspective. Indian J. Anaesth. **59**(7), 400–405 (2015)
42. Markowitz, D.M., Hancock, J.T.: Linguistic Obfuscation in Fraudulent Science. J. Lang. Soc. Psychol. **35**(4), 435–445 (2016)

Propositions on Big Data Business Value

Emma Pirskanen(✉), Heli Hallikainen, and Tommi Laukkanen

Business School, University of Eastern Finland, Joensuu, Finland
{emmapi,heli.hallikainen,tommi.laukkanen}@uef.fi

Abstract. This study synthesizes the earlier literature on big data and draws theoretical propositions on the value of big data in organizational decision-making. The authors explain how big data contributes to business decision-making and propose the steps of a process from collecting data to implementing decisions. They further suggest value gaps affecting the use of big data in different phases of the process. The authors conclude that big data can indeed be turned into data-driven knowledge and further utilized as a basis for improved decision-making. This would, however, require a more systematic approach to utilizing big data in order to leverage big data and turn it into real business value. The study helps organizations to further plan their big data projects and to evaluate how they should prepare for the possible challenges hindering the utilization of big data.

Keywords: Big data · Business value · Data management
Business decision-making

1 Introduction

The success of businesses depends on sufficient analysis and utilization of market information from customers and competitors [29, 41]. Recent technological advances have created new ways to collect, store, and analyze massive amounts of various types of data cost-effectively and in a timely manner, enabling the adoption of analytics that were not previously feasible [18]. Consequently, the big data phenomenon has aroused the interest of practitioners and academics alike [2, 17, 18, 49]. In addition to businesses, non-profit organizations, destination marketing organizations, various associations and governments have also begun to take an interest in utilizing big data in their decision-making processes and strategies [27]. Scholars have argued that big data is going to revolutionize both markets and society [8], and some even suggest that the phenomenon is so radical that firms fail to improve their capability to use big data and devote resources to it may not survive [15], and that big data will override the role of theories in research, replacing it with patterns discovered from data [1].

Consequently, companies are striving to leverage big data in their business activities to keep up with the development, and academics are publishing more articles

The authors thank the Finnish Funding Agency for Innovation for their financial support for this research.

K. Arai et al. (Eds.): FICC 2018, AISC 886, pp. 527–540, 2019.
https://doi.org/10.1007/978-3-030-03402-3_36

concerning big data and data analytics than ever before [18, 50]. However, the rapid development of the big data phenomenon has left both academics and practitioners unprepared and confused [18]. Many businesses understand the fact that big data is important, yet firms struggle to exploit this new form of capital [15].

Academic studies of big data have multiplied rapidly in recent years, and this growth shows no signs of decline. Technological developments typically appear first in technical and academic publications, and are then implemented in business environments, but the rapid evolution of big data in practice has left little time for the development of the academic domain [18]. Therefore, the theoretical discussion is unstructured, and the research so far has focused strongly on the technical side of the phenomenon [52], leaving open the question of how and why organizations benefit from integrating big data into their business processes [17] and how different types of challenges with big data relate to its use in organizations. Furthermore, Sivarajah, Kamal, Irani and Weerakkody [52] note that big data research should be developed towards a more holistic view by identifying and drawing links with established theoretical contributions. In order to add to the discussion on big data, this study synthesizes the earlier literature from information systems, management, and marketing, and makes theoretical propositions of the value of big data in business decision-making. It also discusses the specific challenges affecting value creation.

The remainder of this paper is organized as follows. The next section reviews the earlier literature on big data, especially from a business value perspective. Based on the earlier literature, it suggests the phases, step by step, related to the use of big data in organizational decision-making. It further discusses the challenges related to using big data in each of the steps, and it concludes with propositions on how big data could pass from data collection to business decision-making and implementation. Finally, Sect. 3 provides a discussion of how big data adds value to business decision-making and the challenges related to the process.

2 Literature Review and Proposition Development

2.1 Big Data and Data Management

Big data in a business environment is usually primary data, and it is collected about individual consumers instead of the entire customer base [15]. Today, it is possible to gather and utilize individual qualitative behavioral data, whereas previously companies focused primarily on quantitative transactional data, such as purchase quantities [15]. Traditional data, which refers to mostly structured datasets, is part of big data [31]. However, Gandomi and Haider [18] argue that the largest component of big data is unstructured data, especially data in video format, and thereby traditional data is only a small subset of all of the data available.

Firms generate a lot of *internal data* through various business processes [56]. Yet today, more externally created data can be acquired easily from various sources, and analytics no longer has to concentrate on internal data alone [23, 36]. A lot of big data is generated in communication between devices, such as smartphones, GPS devices, and various sensors [31]. Hence, the phenomenon of the Internet of Things (*IoT*) was one of the starting forces of the big data domain [31] and will become an even bigger part of big data in the future [58]. However, Hofacker, Malthouse and Sultan [23] argue that *social media* was actually the driving force of the big data phenomenon, since almost 90 percent of the available data today is generated from various social media environments [48]. Additionally, big data relies heavily on *open data* [20, 22]. The concept of open data sees data as a "public good" that should be equally accessible to everyone for free or at minimal cost [22]. Governments around the world are releasing greater amounts of information to the public for re-use purposes [20, 24]. Information is not valuable unless it is used; thus, the open data concept is highly beneficial for both markets and society [25]. Due to these novel data sources, big data processes become essential in organizations.

Gandomi and Haider [18] argue that the big data process consists of two main phases: data management and data analytics. *Data management* includes the acquisition and recording of data, in addition to the extraction, cleaning and annotation of data, as well as integration, aggregation and data representation [18]. Collecting data has become easier and more cost-effective; hence, massive amounts of data can be recorded in a timely manner [17, 40, 50].

The volume, variety and velocity of data require a transition from traditional, physical data storage to various cloud-based services [7, 50]. Due to the features of big data, preparing the datasets and maintaining a clean database can require more effort and resources than the actual analysis [23]. Storing data has become cheaper, and tools for managing it are increasingly evolving [17, 46], yet few companies have the ability to properly invest in data storage for all the data generated by internal and external sources [31]. Furthermore, given the vastly increasing quantity, variety, and velocity of data, continuous development of data storage and software programs is needed, since today's solutions will not be sufficient in the future [52].

To adequately capture, store, and analyze data, companies need a powerful computational *data infrastructure* [4]. As the volume of data generated in the digital world is growing vastly, many existing computing infrastructures seem to lag in development, which is considered one of the main problems of big data, as it prevents its utilization in the first place [39, 52]. Furthermore, companies are facing difficulties in preserving consumers' privacy with existing computing tools [53]. Small and medium-sized enterprises, in particular, are struggling to exploit the voluminous amounts of data [52], and this may be partly because the necessary data infrastructure can require significant investments [31] as traditional data technologies have proved to be ineffective at handling big data [55, 56].

In response to this challenge, many IT units are considering or applying distributed storage architectures, which are better able to handle vast amounts of heterogeneous data [1, 50]. For instance, Amazon and Google offer commercial cloud services with subscription pricing models for hosting platforms that allow users to rent the required scale of infrastructure. In addition, Hadoop is an example of open-source software, yet Lee [31] argues that it is not able to complete all processes, and it should be combined with edge computing, which is more expensive to develop and maintain than data centers. Also, Sivarajah et al. [52] argue that regardless of the availability of cloud computing technologies, big data tools can be costly for organizations.

Furthermore, Erevelles, Fukawa and Swayne [15] propose that big data requires investments not only in the physical resources (i.e. the software required and the methods of capturing, storing and analyzing data), but also in human capital resources (e.g. data analysts and big data domain experts), as well as organizational resources (e.g. organizational structure) that enable big data to be leveraged effortlessly. However, big data projects often do not have clear problem definitions, which cause a greater risk of project failure than traditional IT projects [31]. The benefits of big data analytics are not necessarily tangible beforehand, making it difficult to prove the value of big data investments [31]. This may hinder the use of big data, especially in small firms with limited financial and labor resources. Therefore, tools are needed to assess the risks and costs of investing in big data analytics [56]. Consequently, we propose the following:

Proposition 1a: Big data derived from internal data, the IoT, open data, and social media leads to data management.

Proposition 1b: The big data infrastructure affects how big data can be managed.

The first step of the big data process from big data to data management and the value gap hindering the step are illustrated in Fig. 1.

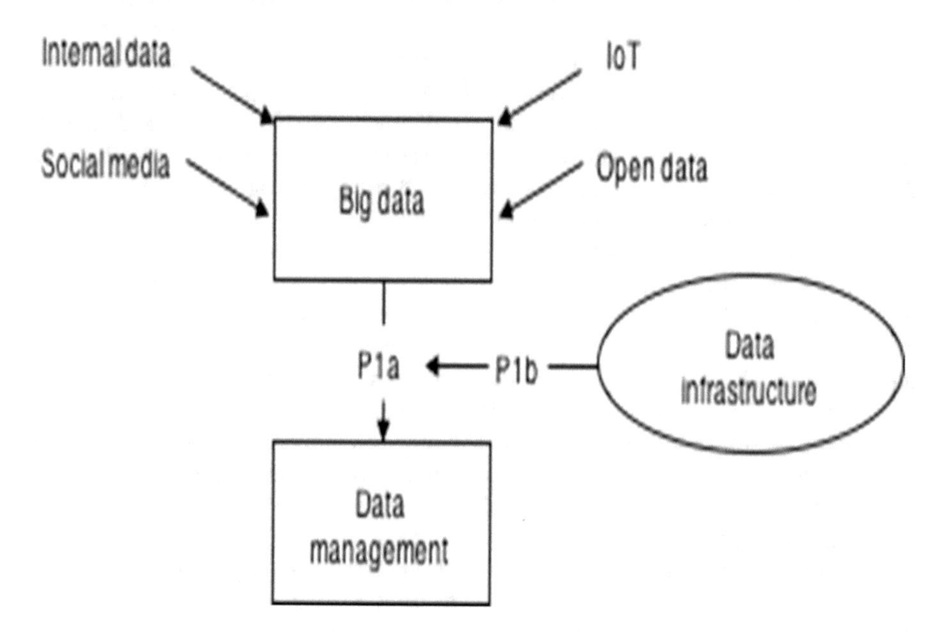

Fig. 1. Step 1: Big data → Data management.

2.2 Big Data Analytics

Big data and data management alone are not valuable [18, 31], but rather serve as raw materials to be transformed into insights [57]. Data is used in *big data analytics*, which means the techniques used to analyze and gain knowledge from the data [18].

Organizations must invest in big data analytics in order to start the value-creation process [11]. Choosing adequate analytical methods is vital for value creation from big data [43]. However, big data analytics can be a complicated process [36]. Mayer-Schönberger and Cukier [37] argue that big data analytics produce essentially descriptive results that emphasize correlation rather than causality, thus answering the question of "what" rather than "why." Nevertheless, leveraging big data analytics has the potential to provide a competitive advantage over business rivals, since it can drive new revenue streams and lead to better operational efficiency [52]. However, there are only a few frameworks that describe different methods and could guide practitioners in the use of big data analytics [43]. Although there are several tools for analyzing data, new methods need to be developed to help exploit the tsunami of information that exists in many, mostly unstructured, forms, such as video and audio recordings [18].

The broad dimensions of big data offer many opportunities, but they can greatly affect the *quality of the data*, thus making it important to manage and analyze data with care [21]. A big part of data today comes from external sources, which may affect its credibility [14]. Sabarmathi, Chinnaiyan and Ilango [44] argue that the main challenge in big data analytics is the complexity of data because the data tends to be noisy, vast, heterogeneous and unreliable. Data is often incomplete, or even inaccurate, as well as highly subjective [18, 21, 23, 31, 46]. Hence, the quality of the data cannot be assumed [23]. This leads to challenges in searching, capturing, storing, sharing, integrating, and analyzing big data [9, 52, 56, 58]. It means that preparing the data, such as replacing missing values, must be done carefully to create reliable results in analytics [23]. The use of poor-quality data in big data analytics leads to bad decisions and unintended risks [21]. Thus, caution and skepticism are needed [21].

Chen, Mao and Liu [9] suggested that an evaluation system for data quality and data processing efficiency should be developed. Only if the problems with data quality are sorted out can big data analytics lead to trustworthy insights that can be used in making strategic decisions to improve internal processes and create a competitive advantage [21, 31]. Consequently, we propose the following (Fig. 2):

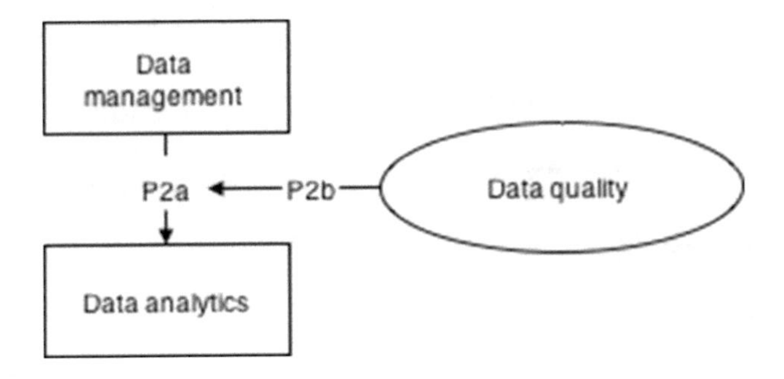

Fig. 2. Step 2: Data management → Data analytics.

Proposition 2a: Data management leads to data analytics.

Proposition 2b: The quality of data affects how the data can be analyzed.

2.3 Business Knowledge

The purpose of big data analytics is to generate business *knowledge* [56]. Business knowledge generally means information and understanding of business processes and environments. Saravanan, Pushpalatha, and Ranjithkumar [45] note that knowledge extraction from big data generally proceeds in five stages: data cleaning and integration, data selection and transformation, data mining, pattern evaluation, and lastly, knowledge. However, having a limitless amount of data does not necessarily improve an organization's performance; analytics and the results generated have to be presented in an understandable way in order to enhance business knowledge [36].

Visualization means presenting knowledge with diagrams, such as tables, images and graphs [56, 58]. Big data analytics is not meaningful unless the results can be presented in an understandable way [56]. Thus, Sivarajah et al. [52] even propose that visualization is an additional dimension of big data. The multiple dimensions of big data and its high volume make visualizing the results difficult [52, 56, 58]. While it can be challenging to present the results of big data analytics in an understandable way, it is crucial in order that the findings from data can be utilized in decision-making processes [1, 56]. There are different tools that data analysts can use to visualize the results from both traditional and novel data, yet Kaushik [26] claims that other members of an organization do not always see these tools as useful or valuable. Based on the above reasoning, we make the following propositions (Fig. 3):

Proposition 3a: Data analytics leads to knowledge.

Proposition 3b: Visualization of data affects how data analytics is transformed into knowledge.

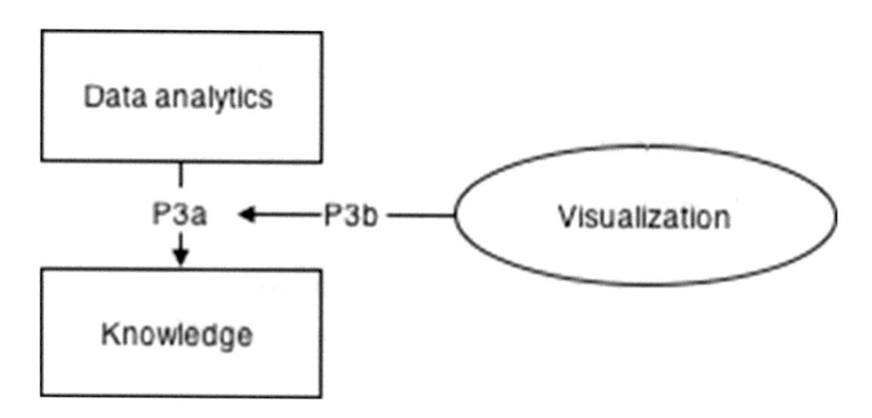

Fig. 3. Step 3: Data analytics → Knowledge.

2.4 Decision-Making

The main advantage of big data is the better availability, visibility, and transparency of information, which can be used in various *decision-making* processes [17]. Big data analytics differs from traditional data analytics since it can utilize real-time data and process data quickly, thus creating real-time knowledge about markets and the ability to deal with uncertainties immediately [57]. Big data analytics enable firms to get more transparent and accurate results to support decision-making in various business processes in multiple critical industries [52]. Companies are now able to quickly search for information on the Internet, conduct surveys and collect feedback more efficiently, observe their competitors more closely, and test new products at lower cost, thus big data analytics will enhance companies' strategies [57]. Big data leads to better-informed business decisions as well as improved business intelligence, and therefore has the potential to decrease costs and increase revenues [3, 17, 52].

If data is processed and managed correctly, it can help avoid risks and identify issues as well as new opportunities that would not have been discovered without analyzing the data [10, 27, 52]. For instance, strategies can be improved based on the knowledge discovered about individuals and their behavior on different social media sites [6]. Davenport [12] states that the most important data source for building strategies is the Internet, since Internet data can show managers what people around the world are saying and doing. Suitable big data analytics can be applied at every level of the organization's management: complex and sophisticated analytics can be used to predict how the industry and markets are going to evolve and what kinds of long-term strategic decisions have to be made, whereas more simple analytics can produce daily reports to support day-to-day management decisions [60].

In order to extract valuable insights from the data, people are needed to do the analysis and put it into practice [9, 17]. The big data era requires that companies trust data analysts and scientists at every stage of the value chain to support or lead decision-making [1], yet there is a serious lack of skilled personnel [9, 17]. The analytics *skills of employees* affect how knowledge from analytics is used in decision-making processes [19]. The key skills required include knowledge of various big data technologies and platforms [3].

Additionally, the *data culture* of the organization affects how well knowledge derived from big data analytics is interpreted in decision-making [19]. Analytics skills have an indirect effect on an analytics-oriented culture [19], even though the people who carry out the analytics are often not those who use data-based knowledge in the organization's decision-making [30]. The data culture may have a greater impact on big data projects than on traditional data projects, because big data can significantly affect the organization's strategy and drive its operations instead of only optimizing them. To concretize the value extracted from big data analytics, managers should consider how to get people to trust data and analytics, and consider what training they need to be able to utilize and understand it. Additionally, managers should evaluate how resources should be re-organized and what kinds of investments are necessary [19, 54]. The organizational culture is one of the most significant and challenging barriers to exploiting big data in practice [3]. This hindrance often derives from a lack of analytical skills, which leads to a culture that does not value data-oriented operations and decision-making [3]. Because of the lack of skills, members of the organization may not understand how big data can improve their performance [30], and it can lead to an overall resistance to big data [3].

In order to increase acceptance of big data practices within the organization, a clear strategy should be developed for the role of big data in the organization's vision [3]. Company policies and practices affect how data is shared and exploited in an organization and with third parties [46]. Germann, Lilien and Rangaswamy [19] state that there is a clear correlation between higher levels of analytical skills and a data culture with higher firm performance, yet there is not necessarily causation.

According to Erevelles, Fukawa and Swayne [15], generating value through big data can be more effective if the company uses big data through radical innovations instead of incremental innovations. Therefore, it is important for leaders to have the courage to trust the data and make decisions based on it. Because big data requires changes within the organization, it is important to understand that "big data" does not refer only to the type of data, but also to the unique tools and methods that are used to analyze it and the changes that it makes to organizational mindsets [18]. We propose (Fig. 4):

Proposition 4a: Knowledge leads to decision-making.

Proposition 4b: The data culture of an organization and the skills of its employees affect how knowledge is used in decision-making.

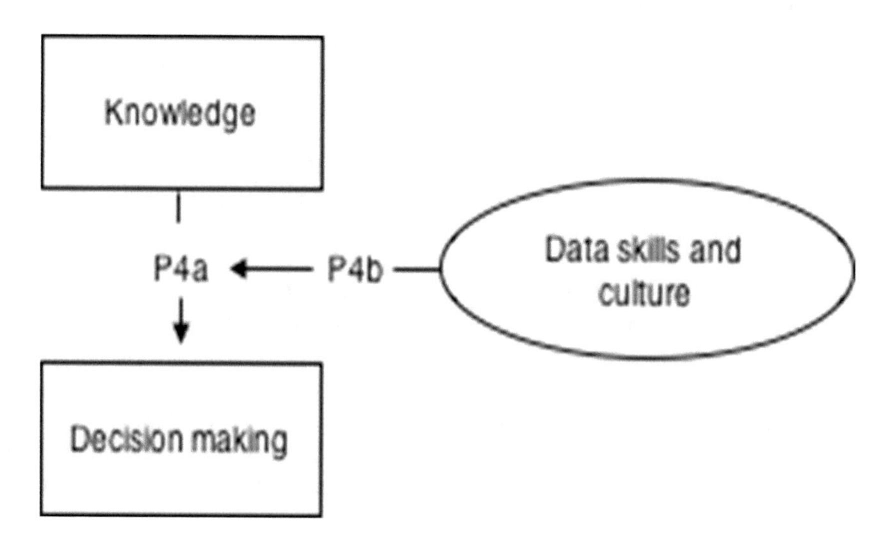

Fig. 4. Step: Knowledge → Decision-making.

2.5 Implementation

One of the main reasons for using big data is to optimize business processes [60]. Hence, the *implementation* of big data is visible in different organizational functions. Davenport and Harris [13] state that big data analytics capabilities allow firms to set the right prices, detect problems, identify profitable customers and potential customers, and decide the lowest level of inventory. Manufacturers who use advanced analytics can decrease process flaws and save resources [5]. By analyzing big data, firms are able to increase their operational efficiency by predicting personnel allocation requirements accurately and optimizing the workforce, as well as optimizing production capacity [9]. Big data helps forecasting product demand precisely [32]. Advanced analytics makes manufacturing more sustainable, since it makes it easier to supervise and control energy consumption and emissions, and improves the quality and accuracy of production [32]. Big data can enhance every step of the product lifecycle management, from research to product development to personalized product service [32]. According to Orenga-Roglá and Chalmeta [42], developing products and services based on consumer data is one of the most powerful ways to generate value from big data analytics. Marketers can develop products purely based on customers' needs identified from consumer data [17]. Indeed, firms are able to exploit real-time information about consumers and respond to their needs almost instantly [57].

Moreover, big data analytics will offer support in improving the visibility and flexibility of the supply chain as well as allocation of resources [17, 52]. It has huge potential to aid in identifying problems, as well as in taking the right corrective actions for emerging issues [12]. Information can be used to optimize supply chain processes, such as planning driving routes and predicting deliveries accurately [12].

Human resources have previously been a business operation that uses little data in its functions [28]. However, this is changing due to new HR information systems and analytics that can use big data [12]. With big data, HR departments are able to make better hiring decisions based on data [51], as well as to better understand the personnel and their attitudes and behavior [47]. Big data applications in HR operations can lead to significant savings: managers can make better decisions about rewards and promotions, optimize salaries, better evaluate employees, and anticipate sick leave [47, 51]. With data from a variety of sources, managers can monitor the social networks and dynamics among members of the organization as well as the whole organization's culture [38]. However, there are challenges to utilizing data in HR, such as restrictions on employee monitoring [34]. Regardless of such issues, Davenport [12] argues that the most successful organizations in the future will be the ones that actively monitor their employees.

Finally, companies have started implementing analytics such as forecasting, risk management, market surveillance and sentiment analysis for trading [33, 59]. Big data can reduce costs for lenders, since it allows client status to be analyzed more accurately [59]. Novel data analysis techniques help to detect market trends, respond to the ultra-fast changes in the financial industry, and reveal new opportunities in the financial markets [16].

Since business processes have become an important tool of differentiation in multiple industries, companies aim to maximize the benefits from each process [13].

Business value is created through improved business processes. However, *organizational structure* can create a significant challenge to implementing big data [52]. Integrating people, processes, and the organization's resources to transform a company to a data-driven organization can be challenging [52]. LaValle [30] notes that people who make the decisions in an organization do not usually carry out big data analytics themselves. Furthermore, especially in large organizations, decision-makers are not those who implement the decisions in practice. Therefore, the decisions made do not necessarily get implemented at the operational level. Business value cannot be generated through big data if the information from big data is not implemented in business processes. Finally, these business processes will produce new internal data to be utilized in analytics [56]. Consequently, we propose the following (Fig. 5):

Proposition 5a: Decision-making leads to implementation.

Proposition 5b: The organizational structure affects how decisions are implemented in practice.

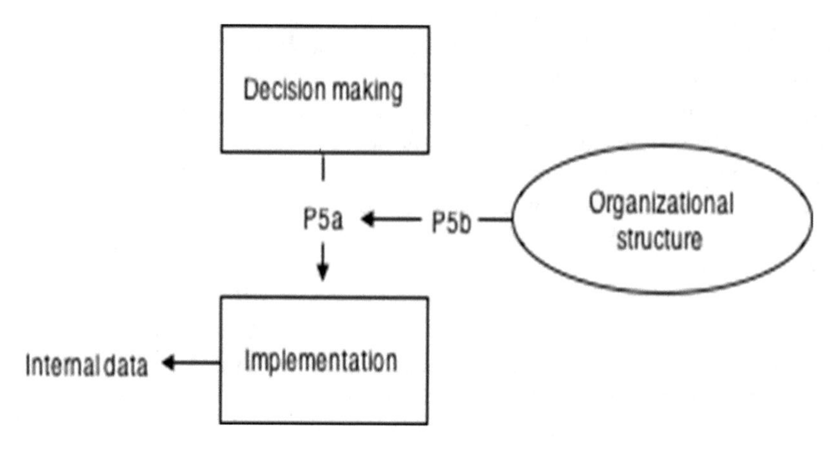

Fig. 5. Step 5: Decision-making → Implementation.

3 Discussion and Conclusion

At present, companies are struggling with how to leverage business value from big data, and the academic literature is scattered. Consequently, the present paper performs an extensive literature review on the existing big data research, and makes propositions of the value of big data in organizational decision-making. We propose benefits of big data on business decision-making, but also discuss potential barriers that may challenge the use of big data in some organizations.

Big data is described as high-volume, high-velocity and high-variety data [40], and is characterized additionally by veracity [e.g. 14] and value [35]. By definition, big data represents a large quantity of data (i.e. high volume), yet other aspects of big data are equally essential for generating business value and improved decision-making based on big data. Companies can utilize data from internal and external data sources, and the data can be in the form of unstructured data along with traditional structural datasets. In addition to traditional internal data sources, such as CRM and ERP systems, companies can now access various external data sources (i.e. variety), such as device-generated data (IoT), data from social media, and open data to complement the data sources that are used in conventional decision-making. Novel sources of data (e.g. the IoT, social media data) are characterized by their high velocity, referring to the speed at which the data is generated and processed. High velocity enables better-informed and close-to-real-time decision-making, and allows keeping better track of key performance indicators and proactively planning the company's business operations. Yet systematic processing is required to leverage insights and value from big data.

It seems that at present, many companies fail to exploit the potential of big data in business operations, and this is at least partially due to the challenges that companies face with regard to big data utilization. Our first proposition relates to the fact that novel sources of data require a new approach to data management, and thus the infrastructure affects how big data can be managed in the organization. For instance, the use of large datasets and unstructured forms of data may necessitate investments with regard to the analysis tools required to harness the potential value deriving from big data. The benefits of big data analytics may not be tangible [31], and thus it may be hard for decision-makers to justify investments in big data. Yet analytics tools capable of rapidly processing vast quantities of data are a necessary prerequisite for leveraging value from big data.

Big data and data management as such do not provide value; rather, they serve as raw material to be used in data analytics [57]. This leads to our second proposition that data management leads to data analytics, and that data quality impacts how the data can be analyzed. For instance, IoT and social media data is often incomplete or even inaccurate [18, 21, 23, 31, 46], and consequently the quality of the data cannot be assumed [23]; it needs to be ensured by careful data management. At worst, failures in ensuring data quality can result in misleading interpretations and conclusions, and can thus have significant consequences for the company's performance.

When used well, analysis of big data can lead to new insights about customers' behavior and allow their behavior to be understood at an individual level rather than as a horde. This leads to our third proposition, that analytics based on big data can result in new kinds of knowledge about customers and market behavior. In addition, we propose that data visualization, i.e. presenting data through graphs, tables, images, and diagrams, affects how massive amounts of raw data can be turned into business knowledge. It may be challenging to visualize large quantities of data, yet the ability to present data in a visually understandable format is a crucial enabler for big data-based decision-making.

Analytics of big data can process data quickly and thus provide close-to-real-time knowledge about market behavior. Our fourth proposition suggests that knowledge gained from big data is used to inform various business decisions. With big data, decision-making can be more accurate and less time-consuming [17]. However, extracting valuable knowledge from data requires skilled and experienced people, now and also in the future, and so skilled workers are required in order to utilize database-based value in practice [9, 17]. Overall, the organization's data culture and the skills of the company's employees affect how much information from big data is used in the company's decision-making and implemented in practice.

The very last of our propositions suggests that data-based decisions have to be implemented in practice yet the organization's structure affects how the decisions are used in business processes. Implementation is perhaps the most challenging part of the process, and it is greatly dependent on the field of operation. Overall, big data as such is not a key to success, but can function as a valuable raw material for better-informed decision-making. However, companies would need a more systematic approach to leverage big data into becoming true business value.

These propositions were based on prior studies on big data in a business environment. It should be noted that these propositions of the big data process are simplified in order to create an overall understanding of the use of big data and challenges affecting it. In practice, these steps may be affected by multiple factors, such as external forces and business context. However, we conclude that big data can be turned into business value, and our propositions comprise a step further in understanding the role of big data in organizational decision-making.

References

1. Abbasi, A., Sarker, S., Chiang, R.H.: Big data research in information systems: toward an inclusive research agenda. J. Assoc. Inf. Syst. **17**(2), 1–32 (2016)
2. Akter, S., Wamba, S.F., Gunasekaran, A., Dubey, R., Childe, S.J.: How to improve firm performance using big data analytics capability and business strategy alignment? Int. J. Prod. Econ. **182**, 113–131 (2016)
3. Alharthi, A., Krotov, V., Bowman, M.: Addressing barriers to big data. Bus. Horiz. **60**(3), 285–292 (2017)
4. Assunção, M.D., Calheiros, R.N., Bianchi, S., Netto, M.A., Buyya, R.: Big data computing and clouds: trends and future directions. J. Parallel Distrib. Comput. **79**, 3–15 (2015)
5. Auschitzky, E., Hammer, M., Rajagopaul, A.: How Big Data Can Improve Manufacturing. McKinsey & Company, New York (2014)
6. Bello-Orgaz, G., Jung, J.J., Camacho, D.: Social big data: recent achievements and new challenges. Inf. Fusion **28**, 45–59 (2016)
7. Buytendijk, F.: Hype cycle for big data 2014, Gartner (2014)
8. Chen, H., Chiang, R.H., Storey, V.C.: Business intelligence and analytics: from big data to big impact. MIS Q. **36**(4), 1165–1188 (2012)
9. Chen, M., Mao, S., Liu, Y.: Big data: a survey. Mob. Netw. Appl. **19**(2), 171–209 (2014)
10. Chen, J., Chen, Y., Du, X., Li, C., Lu, J., Zhao, S., Zhou, X.: Big data challenge: a data management perspective. Front. Comput. Sci. **7**(2), 157–164 (2013)
11. Côrte-Real, N., Oliveira, T., Ruivo, P.: Assessing business value of big data analytics in European firms. J. Bus. Res. **70**, 379–390 (2017)
12. Davenport, T.H.: Big Data at Work: Dispelling the Myths, Uncovering the Opportunities. Harvard Business review Press, Boston (2014)
13. Davenport, T.H., Harris, J.G.: Competing on Analytics: The New Science of Winning. Harvard Business Press, Boston (2007)
14. Elragal, A.: ERP and big data: the inept couple. Procedia Technol. **16**, 242–249 (2014)

15. Erevelles, S., Fukawa, N., Swayne, L.: Big data consumer analytics and the transformation of marketing. J. Bus. Res. **69**(2), 897–904 (2016)
16. Fang, B., Zhang, P.: Big data in finance. In: Big Data Concepts, Theories, and Applications, pp. 391–412. Springer (2016)
17. Frizzo-Barker, J., Chow-White, P.A., Mozafari, M., Ha, D.: An empirical study of the rise of big data in business scholarship. Int. J. Inf. Manag. **36**(3), 403–413 (2016)
18. Gandomi, A., Haider, M.: Beyond the hype: big data concepts, methods, and analytics. Int. J. Inf. Manag. **35**(2), 137–144 (2015)
19. Germann, F., Lilien, G.L., Rangaswamy, A.: Performance implications of deploying marketing analytics. Int. J. Res. Mark. **30**(2), 114–128 (2013)
20. Hardy, K., Maurushat, A.: Opening up government data for big data analysis and public benefit. Comput. Law Secur. Rev. **33**(1), 30–37 (2017)
21. Herschel, R., Miori, V.M.: Ethics & big data. Technol. Soc. **49**, 31–36 (2017)
22. Hilbert, M.: Big data for development: a review of promises and challenges. Dev. Policy Rev. **34**(1), 135–174 (2016)
23. Hofacker, C.F., Malthouse, E.C., Sultan, F.: Big data and consumer behavior: imminent opportunities. J. Consum. Mark. **33**(2), 89–97 (2016)
24. Huijboom, N., Van den Broek, T.: Open data: an international comparison of strategies. Eur. J. ePractice **12**(1), 4–16 (2011)
25. Janssen, M., Charalabidis, Y., Zuiderwijk, A.: Benefits, adoption barriers and myths of open data and open government. Inf. Syst. Manag. **29**(4), 258–268 (2012)
26. Kaushik, A.: Web Analytics 2.0: The Art of Online Accountability and Science of Customer Centricity. Wiley, Indianapolis (2009)
27. Khalilzadeh, J., Tasci, A.D.: Large sample size, significance level, and the effect size: solutions to perils of using big data for academic research. Tour. Manag. **62**, 89–96 (2017)
28. Kiron, D., Prentice, P.K., Ferguson, R.B.: Raising the bar with analytics. MIT Sloan Manag. Rev. **55**(2), 29–33 (2014)
29. Kohli, A.K., Jaworski, B.J.: Market orientation: the construct, research propositions, and managerial implications. J. Mark. **54**(2), 1–18 (1990)
30. LaValle, S., Lesser, E., Shockley, R., Hopkins, M.S., Kruschwitz, N.: Big data, analytics and the path from insights to value. MIT Sloan Manag. Rev. **52**(2), 21–32 (2011)
31. Lee, I.: Big data: dimensions, evolution, impacts, and challenges. Bus. Horiz. **60**(3), 293–303 (2017)
32. Li, J., Tao, F., Cheng, Y., Zhao, L.: Big data in product lifecycle management. Int. J. Adv. Manuf. Technol. **81**(1–4), 667–684 (2015)
33. Lien, D.: Business finance and enterprise management in the era of big data: an introduction. N. Am. J. Econ. Financ. **39**, 143–144 (2017)
34. Lohr, S.: Big data, trying to build better workers. The New York Times, p. BU4 (2013)
35. Lycett, M.: 'Datafication': making sense of (big) data in a complex world. Eur. J. Inf. Syst. **22**(4), 381–386 (2013)
36. Matthias, O., Fouweather, I., Gregory, I., Vernon, A.: Making sense of big data–can it transform operations management? Int. J. Oper. Prod. Manag. **37**(1), 37–55 (2017)
37. Mayer-Schönberger, V., Cukier, K.: Big Data: A Revolution That Will Transform How We Live, Work, and Think. Houghton Mifflin Harcourt, New York (2013)
38. McAbee, S.T., Landis, R.S., Burke, M.I.: Inductive reasoning: the promise of big data. Hum. Resour. Manag. Rev. **27**(2), 277–290 (2017)
39. Mehmood, A., Natgunanathan, I., Xiang, Y., Hua, G., Guo, S.: Protection of big data privacy. IEEE Access **4**, 1821–1834 (2016)
40. Mehta, B.B., Rao, U.P.: Privacy preserving unstructured big data analytics: issues and challenges. Procedia Comput. Sci. **78**, 120–124 (2016)

41. Narver, J.C., Slater, S.F.: The effect of a market orientation on business profitability. J. Mark. **54**(4), 20–35 (1990)
42. Orenga-Roglá, S., Chalmeta, R.: Social customer relationship management: taking advantage of Web 2.0 and big data technologies. SpringerPlus **5**(1), 1462–1478 (2016)
43. Ram, J., Zhang, C., Koronios, A.: The implications of big data analytics on business intelligence: a qualitative study in China. Procedia Comput. Sci. **87**, 221–226 (2016)
44. Sabarmathi, G., Chinnaiyan, R., Ilango, V.: Big data analytics research opportunities and challenges: a review. Int. J. Adv. Res. Comput. Sci. Softw. Eng. **6**(10), 227–231 (2016)
45. Saravanan, V., Pushpalatha, C., Ranjithkumar, C.: Data mining open source tools-review. Int. J. Adv. Res. Comput. Sci. **5**(6), 231–235 (2014)
46. Schroeder, R.: Big data business models: challenges and opportunities. Cogent Soc. Sci. **2**(1) (2016). https://doi.org/10.1080/23311886.2016.1166924
47. Shah, N., Irani, Z., Sharif, A.M.: Big data in an HR context: exploring organizational change readiness, employee attitudes and behaviors. J. Bus. Res. **70**, 366–378 (2017)
48. Sharma, S., Tim, U.S., Wong, J., Gadia, S., Sharma, S.: A brief review on leading big data models. Data Sci. J. **13**, 138–157 (2014)
49. Shin, D.H.: Demystifying big data: anatomy of big data developmental process. Telecommun. Policy **40**(9), 837–854 (2016)
50. Shu, H.: Big data analytics: six techniques. Geo-Spat. Inf. Sci. **19**(2), 119–128 (2016)
51. Simon, P.: Too Big to Ignore: The Business Case for Big Data. Wiley, Hoboken (2013)
52. Sivarajah, U., Kamal, M.M., Irani, Z., Weerakkody, V.: Critical analysis of big data challenges and analytical methods. J. Bus. Res. **70**, 263–286 (2017)
53. Van Dijck, J.: Datafication, dataism and dataveillance: big data between scientific paradigm and ideology. Surveill. Soc. **12**(2), 197–208 (2014)
54. Wamba, S.F., Akter, S., Edwards, A., Chopin, G., Gnanzou, D.: How 'big data'can make big impact: findings from a systematic review and a longitudinal case study. Int. J. Prod. Econ. **165**, 234–246 (2015)
55. Wang, Y., Wiebe, V.J.: Big data analytics on the characteristic equilibrium of collective opinions in social networks. In: Big Data: Concepts, Methodologies, Tools, and Applications, pp. 1403–1420. IGI Global (2016)
56. Wang, H., Xu, Z., Fujita, H., Liu, S.: Towards felicitous decision making: an overview on challenges and trends of big data. Inf. Sci. **367**, 747–765 (2016)
57. Xu, Z., Frankwick, G.L., Ramirez, E.: Effects of big data analytics and traditional marketing analytics on new product success: a knowledge fusion perspective. J. Bus. Res. **69**(5), 1562–1566 (2016)
58. Yaqoob, I., Hashem, I.A.T., Gani, A., Mokhtar, S., Ahmed, E., Anuar, N.B., Vasilakos, A. V.: Big data: from beginning to future. Int. J. Inf. Manag. **36**(6), 1231–1247 (2016)
59. Zhang, S., Xiong, W., Ni, W., Li, X.: Value of big data to finance: observations on an internet credit service company in China. Financ. Innov. **1**(1), 17–34 (2015)
60. Ziora, A.C.L.: The role of big data solutions in the management of organizations. Review of selected practical examples. Procedia Comput. Sci. **65**, 1006–1112 (2015)

Improved Accuracy Stock Price Change Prediction Model Using Trading Volume

Zhen Wei, Chao Wu, Yike Guo(✉), and Zhongwei Yao

Data Science Institute, Imperial College London, London, UK
{zw708,chao.wu,y.guo}@ic.ac.uk, yaozhongwei0131@163.com

Abstract. This research aims to model the relationship between the change in stock price and the volume. Linear regression has been applied to the model at daily and at minute time scales; then Random Forest and Lasso regression have been applied to the model. The results show that the larger the data, the better fit the model is, and Random forest has better prediction accuracy than the linear model.

Keywords: Equity trading · Finance · Machine learning · Volume

1 Introduction

In finance, many individuals and institutions make a living based on the price change of the stock return. For example, all traders, investors and hedge funds buy and sell stocks to make money from the price difference. Therefore, there has been wide interest and research in the finance industry on the factors that influence the stock price and stock price change. In the literature, the related factors that have been widely researched include: risk [1,2], price to earning ratio [3,4], market capitalization [5,6], dividend [7,8], liquidity [9,12], etc. One of the factors that has been frequently neglected is volume. Volume represents the number of the shares that is being bought and sold at the time, and it hints at information such as liquidity and inside trading information [13,14]. Some research has also been carried on how the equity price is related to the volume [15–17]. In the literature, the relationship between volume and stock price is classified into two types. The stock return is modeled in a combination of previous time step stock return and the volume, and this research has been done by Kandel, Shmuel and Stambaugh, Robert in 1988 [18]. Another model is the relationship of the absolute value of stock price change to its transaction volume. 40 years ago, Osborne was the first person to model the price change of securities to its transaction volume [19]. Then in 1987, Karpoff found the correlation between the price change and the trading volume [20]. This model has been widely researched and explored with new evidence. For example, Hiemstra and Jones found the causality of stock price and volume by linear and non-linear models [21]. Epps researched on the theory and evidence of security price changes and volume [22,23]. Harris applied this model specifically to the S&P 500 list in his research [24].

K. Arai et al. (Eds.): FICC 2018, AISC 886, pp. 541–548, 2019.
https://doi.org/10.1007/978-3-030-03402-3_37

Due to the relationship between price change and volume being well studied, it is used in this research as the model. Furthermore, this model is also selected because it is simple and straightforward, hence the research is easy to carry out on big datasets, such as a result of changing the timescale from daily to minutes. In this research, the baseline is linear regression of the absolute value of price change and volume in daily data [25–27]. Then a minute dataset is used to train the absolute value of price change and volume. Random Forest is model that has been widely used in large dataset prediction [28–30]. Lasso regression is popularly applied in finance studies [10,11]. Hence computing power can be applied to increase the prediction accuracy.

The contribution of this paper has two significant parts. First, it offers models that can better predict the stock price change. This model comes from two different methodologies: one trained in shorter timescales, and another using random forest regression. Second, the research also provide different type of traders (daily and minute trader) different trading strategy/model to put on their position, and the model itself can tune to different timescales the trader would like to trade.

2 Methodology

2.1 Data Collection

Data from Three companies are collected from the Bloomberg Terminal from February 2008 till August 2017: Microsoft, IBM and Google. Two types of data are collected for each company, one type of the data is collected every minute, and the other type of data is collected on a daily timescale. The dimensions of both timescale data include: date, time, open price, close price, highest price, lowest price, and volume that is being traded.

In this dataset, the input data is the volume column, and the output data is the absolute value of the price change, which is calculated by subtraction of the lowest price from the highest price at that timescale.

2.2 Model

In this model, we are trying to use the historical data to train a function f so that $Y_{train} = f(X_{train})$, where X_{train} is the input data and Y_{train} is the output data, then test input data X_{test} is used to predict the test output data $Y_{predict}$ using function f as $Y_{predict} = f(X_{test})$, comparison is used to compare real output data Y_{test} and predicted output data $Y_{predict}$.

In the model, the training input data and output data is the company data from February 2008 till December 2016, and the test input and output data is the dataset in year 2017. Training is done separately for both daily and minute timescale data.

Linear Regression. Linear regression is a a model that assumes the relationship between the dependent variable and the independent variable is linear [25–27]. It can be multiple independent variables. In this model, there is only one independent variable, which is volume. Mathematically, it can be written as

$$P_h - P_l = f_l(V) \tag{1}$$

where P_h is the highest price at that time, and P_l is the lowest price at that time, V represents the volume and f_l is the trained linear function.

In this model, the linear regression model has been used from Python Sklearn, and no variable has been set in the code.

Random Forest Regression. Random forest regression is a type of regression that is constructed through multiple decision trees, by doing so the original data has been split into two choices every time the decision tree is constructed, and the regression residual sum of squares (RSS) is calculated as $RSS = \sum_{left}(Y_{train} - Y_L)^2 + \sum_{right}(Y_{train} - Y_R)^2$ where Y_L is the mean Y-value for the left node and Y_R is the mean Y-value for the right node. Same as linear model, there is only one independent variable, which is volume [28–30]. Mathematically, it can be written as

$$P_h - P_l = f_S(V) \tag{2}$$

where P_h is the highest price at that time, and P_l is the lowest price at that time, V represents the volume and f_S is the trained Random Forest regression function.

In this model, the linear regression model has been used from Python Sklearn.ensemble, and no variable has been set in the code.

Lasso Regression. Lasso regression uses the least square to find best fit model [31,32]. The least square regression mathematically is written as $min\frac{1}{N}\sum_{i=1}^{N}(P_{h,i} - P_{l,i} - f_S(V_i))^2$, where i corresponding to the time step. Lasso regression different from the linear regression by the fact that $\mathbf{V_i}$ is a vector, which means it can be written as $\mathbf{V_i} = (\mathbf{V_i}, \mathbf{V_{i-1}}, \mathbf{V_{i-2}}, \mathbf{V_{i-3}}, ..., \mathbf{V_{i-p}})^\mathbf{T}$.

In this research, the Lasso regression is used as it give wider time step variables rather than only the corresponding time step. The number of the time steps is changeable in the research, however, in our model, time steps 4 is chosen in the model, this means $\mathbf{V_i} = (\mathbf{V_i}, \mathbf{V_{i-1}}, \mathbf{V_{i-2}}, \mathbf{V_{i-3}})^\mathbf{T}$ in both minute and daily dataset, and the parameter *alpha* is set to be 0.1 in the regression model.

3 Results

In the results, the daily error and minute error are calculated in the equations below, where $P_{dailyprediction}$ is calculated through the model training the daily data, and the $P_{minuteprediction}$ is calculated through the model training the minute data.

$$\begin{aligned} error_{daily} &= \frac{|P_{daily} - P_{dailyprediction}|}{P_{daily}} \\ error_{minute} &= \frac{|P_{daily} - P_{minuteprediction}|}{P_{daily}} \end{aligned} \tag{3}$$

3.1 Microsoft

In the Linear model, the mean of the daily error is 0.29830020967224, and the mean of minute error is 0.28691695054478034, calculated by (3). In the Random Forest regression model, the mean of the daily error is 0.12308461102110313; the mean of the minute error is 0.10349570088905824. In the Lasso model, the mean of the daily error is 0.4487149608736065; and the mean of the minute error is 0.31578114475996344. The error bar chart of the three models is shown as in Fig. 1.

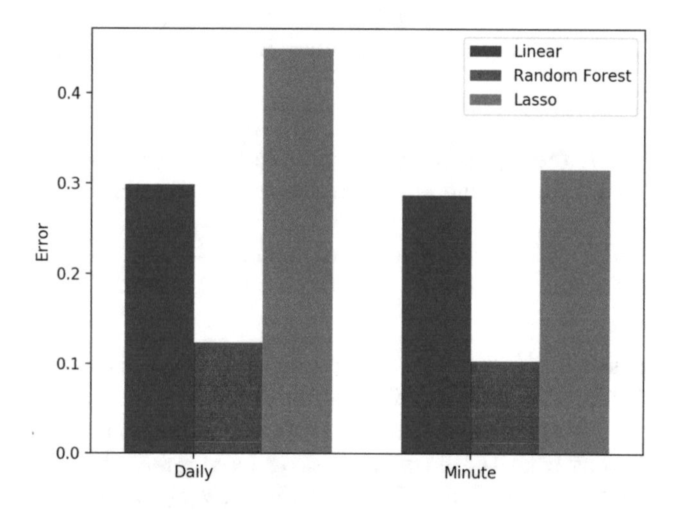

Fig. 1. Linear regression, random forest regression and lasso regression trained model daily and minute errors bars of Microsoft stock price change.

3.2 IBM

In the Linear model, the mean of the daily error is 0.3971217962528666, and the mean of minute error is 0.36219203770423347, calculated by (3). In the Random Forest regression model, the mean of the daily error is 0.16695813677315788; the mean of the minute error is 0.13738450599507226. In the Lasso model, the mean of the daily error is 0.7586822678822491; and the mean of the minute error is 0.265620375201779. The error bar chart is shown as in Fig. 2.

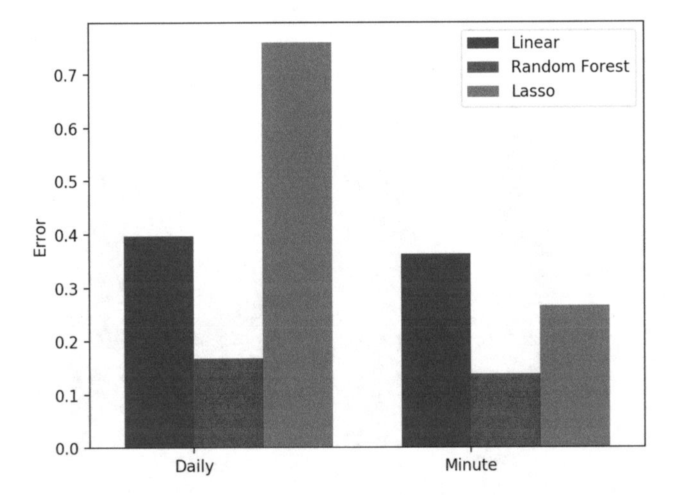

Fig. 2. Linear regression, random forest regression and lasso regression trained model daily and minute errors bars of IBM stock price change.

3.3 Google

In the linear model, the mean of the daily error is 0.35234013591089886, and the mean of minute error is 0.16888007294536797, calculated by (3). In the random forest regression model, the mean of the daily error is 0.1407109214455535; the mean of the minute error is 0.12438450599507226. In the Lasso model, the mean of the daily error is 0.4487149608736065; the man of the minute error is 0.31578114475996344. The error bar chart is shown as in Fig. 3.

4 Conclusion

The models trained to the Microsoft and Google data are good as the errors of all models are below 0.45. In addition, there is no significant difference from the mean error calculated by the daily data and the mean error calculated by the minute data. However, in the IBM data, the daily data trained the Lasso model that has mean error reaches up to more than 0.7, but its minute data trained the Lasso model has a lower mean error even than the linear model trained by minute data.

According to the figures and statistical results from the three companies in the results section, it is obvious that the minute data trained model is better than the daily data trained model, hence the larger the dataset is used to train the linear model, the more accurate it can be to predict the absolute stock price change.

In addition, it also shows that the Random Forest regression training is the best among these three models, and the linear regression model is better than the Lasso model, even the daily Random Forest regression trained model is better

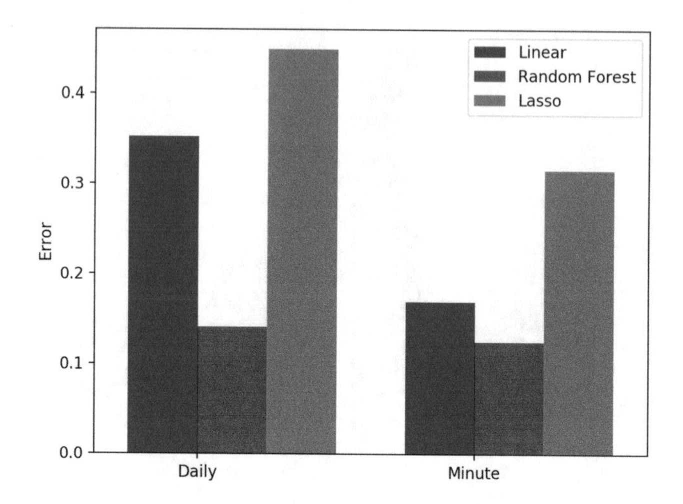

Fig. 3. Linear regression, random forest regression and lass regression trained model daily and minute errors bars of Google stock price change.

than the minute linear regression and minute Lasso. However, in the Random Forest regression, there is not that significant a difference between the model trained in minute data and the model trained in daily data. Lass trained model on the other side result a significant difference between the model trained in minute data and model trained in daily data.

5 Discussion

In this research, big data or different machine learning model are used to improve the accuracy of prediction of price change from the volume. As it is a time series sequence problem with big data available, the deep learning methods can be used to make predictions. Recurrent Neural Network (RNN) can be applied in the model prediction [33]. RNN is a deep learning method that has a class of artificial neural networks where connections form a directed circle. During these circles, features can be reinforce, extracted or deleted so that a better fitting map can be found in the end to the output data. Unlike the Lasso model, in the RNN model, the input data can be a combination of multiple time step variables, for example, it can be a volume product of two consecutive times. The RNN has the capability to set input variables large as RNN uses the GPU to run the data.

In the Lasso model, the time steps is chosen to be four in both daily and minute data, which means the time interval that has been chosen in daily and minute cases is different, another research can be done in the case when the time steps are different in minute and daily case, but the time interval is the same, i.e. time interval is in daily scale.

Last but not the least, research has been done by Patrick Augustin in 2016 to find indicators of inside trading prior to the takeover announcement of mergers and acquisitions [14]. Abnormal volume can also be furthered researched for the regulator reason to give the inside trading information.

References

1. Griffin, J.M., Lemmon, M.L.: J. Financ. **57**(5), 2317–2336 (2002). Wiley Online Library
2. Bailey, W., Chung, Y.P.: Exchange rate fluctuations, political risk, and stock returns: some evidence from an emerging market. J. Financ. Quant. Anal. **30**(4), 541–561 (1995)
3. Basu, S.: Investment performance of common stocks in relation to their price-earnings ratios: a test of the efficient market hypothesis. J. Financ. **32**(3), 663–682 (1997)
4. Banz, R.W.: The relationship between return and market value of common stocks. J. Financ. Econ. **9**(1), 3–18 (1981)
5. Jegadeesh, N., Titman, S.: Returns to buying winners and selling losers: implications for stock market efficiency. J. Financ. **48**(1), 65–91 (1993)
6. Levine, R., Zervos, S.: Stock markets, banks, and economic growth. Am. Econ. Rev. **88**, 537–558 (1998)
7. Fama, E.F., French, K.R.: Dividend yields and expected stock returns. J. Financ. Econ. **22**(1), 3–25 (1988)
8. Charest, G.: Dividend information, stock returns and market efficiency-II. J. Financ. Econ. **6**(2–3), 297–330 (1978)
9. Amihud, Y.: Illiquidity and stock returns: cross-section and time-series effects. J. Financ. Mark. **5**(1), 31–56 (2002)
10. Wang, H., Li, G., Tsai, C.-L.: Regression coefficient and autoregressive order shrinkage and selection via the lasso. J. Royal Stat. Soc. Ser. B Stat. Methodol. **69**(1), 63–78 (2007)
11. Nadaraya, E.A: On Estimating Regression, Theory of Probability & Its Applications. SIAM (2006)
12. Pástor, L., Stambaugh, R.F.: Liquidity risk and expected stock returns. J. Polit. Econ. **111**(3), 642–685 (2003)
13. Pan, J., Poteshman, A.M.: The information in option volume for future stock prices. Rev. Financ. Stud. **19**(3), 871–908 (2006)
14. Augustin, P., Brenner, M., Subrahmanyam, M.G.: Informed Options Trading prior to Takeover Announcements: Insider Trading? (2016)
15. French, K.R., Roll, R.: Stock return variances: the arrival of information and the reaction of traders. J. Financ. Econ. **17**(1), 5–26 (1986)
16. Campbell, J.Y., Grossman, S.J., Wang, J.: Trading volume and serial correlation in stock returns. Q. J. Econ. **108**(4), 905–939 (1993)
17. Keim, D.B.: Size-related anomalies and stock return seasonality: further empirical evidence. J. Financ. Econ. **12**(1), 13–32 (1983)
18. Kandel, S., Stambaugh, R.F.: Modelling expected stock returns for short and long horizons. Working Paper 42–88, Rodney L. White Center for Financial Research, Wharton School, University of Pennsylvania (1988)
19. Osborne, M.F.M.: Brownian motion in the stock market. Oper. Res. **7**(2), 145–173 (1959)
20. Karpoff, J.M.: The relation between price changes and trading volume: a survey. J. Financ. Quant. Anal. **22**(1), 109–126 (1987)

21. Hiemstra, C., Jones, J.D.: Testing for linear and nonlinear Granger causality in the stock price-volume relation. J. Financ. **49**(5), 1639–1664 (1994)
22. Epps, T.W.: Security price changes and transaction volumes: theory and evidence. Am. Econ. Rev. **65**(4), 586–597 (1975)
23. Epps, T.W.: Security price changes and transaction volumes: some additional evidence. J. Financ. Quant. Anal. **12**(1), 141–146 (1977)
24. Harris, L., Gurel, E.: Price and volume effects associated with changes in the S&P 500 list: new evidence for the existence of price pressures. J. Financ. **41**(4), 815–829 (1986)
25. Neter, J., Kutner, M.H., Nachtsheim, C.J., Wasserman, W.: Applied Linear Statistical Models, vol. 4. Irwin, Chicago (1996)
26. Seber, G.A.F., Lee, A.J.: Linear Regression Analysis, vol. 936. Wiley, Hoboken (2012)
27. Montgomery, D.C., Peck, E.A., Vining, G.G.: Introduction to Linear Regression Analysis. Wiley, Hoboken (2015)
28. Liaw, A., Wiener, M., et al.: Classification and regression by randomForest. R News **2**(3), 18–22 (2002)
29. Svetnik, V., Liaw, A., Tong, C., Culberson, J.C., Sheridan, R.P., Feuston, B.P.: Random forest: a classification and regression tool for compound classification and QSAR modeling. J. Chem. Inf. Comput. Sci. **43**(6), 1947–1958 (2003)
30. Segal, M.R.: Machine Learning Benchmarks and Random Forest Regression. Center for Bioinformatics & Molecular Biostatistics (2004)
31. Tibshirani, R.: Regression shrinkage and selection via the lasso. J. Royal Stat. Soc. Ser. B Methodol. 267–288 (1996)
32. Hans, C.: Bayesian lasso regression. Biometrika **96**(4), 835–845 (2009)
33. Williams, R.J., Zipser, D.: A learning algorithm for continually running fully recurrent neural networks. Neural Comput. **1**(2), 270–280 (1989)

Improving Subject-Independent EEG Preference Classification Using Deep Learning Architectures with Dropouts

Jason Teo(✉), Lin Hou Chew, and James Mountstephens

Faculty of Computing and Informatics, Universiti Malaysia Sabah, Jalan UMS, 88400 Kota Kinabalu, Sabah, Malaysia
jtwteo@ums.edu.my

Abstract. Human preferences play a key role in numerous decision-making processes. The ability to correctly identify likes and dislikes would facilitate novel applications in neuromarketing, affective entertainment, virtual rehabilitation and forensic neuroscience that leverage on sub-conscious human preferences. In this neuroinformatics investigation, we seek to recognize human preferences passively through the use of electroencephalography (EEG) when a subject is presented with some 3D visual stimuli. Our approach employs the use of machine learning in the form of deep neural networks to classify brain signals acquired using a brain-computer interface (BCI). Our previous work has shown that EEG preference classification is possible although accuracy rates remain relatively low at 61%–67% using conventional deep learning neural architectures, where the challenge mainly lies in the accurate classification of unseen data from a cohort-wide sample that introduces inter-subject variability on top of the existing intra-subject variability. Such an approach is significantly more challenging and is known as subject-independent EEG classification as opposed to the more commonly adopted but more time-consuming and less general approach of subject-dependent EEG classification. In this new study, we employ deep networks that allow dropouts to occur in the architecture of the neural network. The results obtained through this simple feature modification achieved a classification accuracy of up to 79%. Therefore, this study has shown that the use of a deep learning classifier was able to achieve an increase in emotion classification accuracy of between 13%–18% through the simple adoption of the use of dropouts compared to a conventional deep learner for EEG preference classification.

Keywords: Neuroinformatics · Emotion classification
Preference classification · Electroencephalography (EEG) · Deep learning
Dropouts

This project is supported by the FRGS research grant schemes FRG0349 & FRG0435 from the Ministry of Higher Education, Malaysia.

K. Arai et al. (Eds.): FICC 2018, AISC 886, pp. 549–560, 2019.
https://doi.org/10.1007/978-3-030-03402-3_38

1 Introduction

We have conducted a number of prior investigations into the use of electroencephalography (EEG) as a method for passively monitoring the brainwaves of users as they are exposed to 3D visual stimuli and then using different machine learning algorithms to predict their preferences among the various visual stimuli [1, 2]. The ability to passively identify the preferences of users as they are being presented with different stimuli will have novel and significant applications in various choice-based domains such as neuromarketing, affective entertainment, virtual rehabilitation and forensic neuroscience.

In our early work with a small set of five test subjects, good classification rates of up to 80% were attained using simple k-nearest neighbor (kNN) classifiers [1]. However, when the number of test subjects was increased to 16, the noise arising from inter-subject variability became a substantial factor which made the classification process significantly more challenging [2]. While most studies generally deal only with intra-subject variability where for each user, retraining is required before classification testing. We attempt a cohort-wide classification to enable direct applications to new users without the need for per-person pre-training before classification usage. In the our expanded study, classification rates for the large majority of conventional classifiers such as kNN, support vector machines, Naïve Bayes, Random Forest, C4.5 and other rule-based classifiers were only between 56–60%. The best classification result obtained from this comparative study was using deep neural networks at 64% [2].

As such, the main objective of this current study is to investigate the various architectural tuning of the deep neural networks for improving the classification rates of our EEG-based preference classification task. Section 2 presents the background on emotion classification and preference classification in particular. Section 3 presents our approach to EEG-based preference classification using a proprietary set of 3D visual stimuli. Section 4 presents the results of our investigations and Sect. 5 concludes the paper with some future avenues for expanding upon the current work.

2 Background

2.1 Emotion Modeling and Classification

Emotion classification entails the use of various physiological signals and markers in an attempt to identify different emotions such as the user being in a state of anger, disgust, happiness, sadness, fear, anxiety, excitement and surprise among other [3, 4]. Some commonly measured bio-signals include the heart rate, skin conductance, pupil dilation, respiration rate and also brainwaves, which is also known as EEG [5, 6].

EEG-based emotion classification typically involves the measurement of the millivolt-range electrical signals through the placement of a number of electrodes on the scalp of the user, the waveforms of which are then spectrally transformed into features used by machine learning algorithms trained on labelled data to predict the emotion currently being sensed. Numerous studies have shown that classifications for various emotions can be reliable obtained using EEG.

2.2 Emotion Classification of Preferences

Preference classification can be considered a sub-task of emotion classification. This more specific task entails the identification of a user's like or dislike when presented with a stimulus. Preference classification is generally considered to be more challenging to classify compared to other emotions that are more strongly evoked such as anger or sadness.

The very large majority of EEG-based preference classification has been conducted using music as the stimulus [7, 8]. There have been very limited studies done using 2D images [9, 10] whereas our earlier studies were the first to implement rotating virtual 3D images as the stimulus [1, 2]. Furthermore, preference classification, which is already more challenging compared to other forms of emotion classification due to its comparatively weaker evocation, is rarely studied as a cohort-wide classification task. EEG-based emotion classification with large-sized cohorts will typically yield significantly lower accuracy rates due to inter-subject [11] and as well as intra-subject variability [12]. Doing so requires the classifier to be able to overcome inter-subject variability in addition to intra-subject variability of the users' EEG signal. Consequently, the weak signal evocation and inter-subject variability make EEG preference classification a very challenging classification task.

2.3 Extraction of Features from EEG Signals

Emotion modeling using machine learning approaches can be categorized into three broad domain classes: (i) time, (ii) frequency, and (iii) time and frequency combination. Time-based emotion modeling employs the detection of event-related potentials (ERPs). Of these, they can be further divided into groups that are detected based on whether they are having short, medium or long post-latency exposures after stimuli presentation. Emotion classification for valence and arousal produced accuracy rates of 55.7% for arousal and 58.8% for valence [13] when using these ERP-based methods.

The classification of emotions based on the frequency domain is achieved through the learning of features obtained power spectrum analysis, producing the canonical delta, theta, alpha, beta and gamma frequency bands. Emotion classification for the preference of music produced an accuracy of 74.8% with linear support vector machines (SVMs) using the preprocessed features obtained through the Common Spatial Patterns (CSPs) method [14]. Emotion classification for the preference of music via preprocessed features obtained from a using a conventional Fast Fourier Transform (FFT) produced a classification accuracy rate of 85.7% using SVMs [15]. Radial SVMs were used in the only published emotion classification of preferences not using music stimuli, in this case for 2D image preferences using power spectrum analysis where the classification outcome produced an accuracy of 88.5% [16].

From the perspective of using a combination of time and frequency (TF) leverages on the power spectrum analysis at predefined time periods that encompass the whole duration of the post stimuli period for measuring brain activity. Several conventional machine learning algorithms were used to conduct emotion classification tasks employing three distinctly different TF analysis methods were studied to identify the preference for music. Here, it was observed that the k-Nearest Neighbors (kNN) machine learning approach produced the overall best outcome with an accuracy of 86.5% [17]. The same group of researchers then conducted a follow-on investigation utilizing a much finer-grained approach which attempted to categorize the emotion stimuli into two groups: (i) familiar versus (ii) unfamiliar music. In this later study, using a kNN machine learning approach, they managed to produce a much higher emotion classification accuracy of 91.0% [18]. Emotion modeling for the preference of music using TF approaches used a Short-Time Fourier Transform (STFT) and using a kNN machine learning approach produced emotion classification accuracy rates of 98.0% [19].

2.4 Preference Classification Using Deep Learning Approaches

The preferences of 32 participants for the viewing of music video clips was attempted using deep learning via the Deep Belief Networks (DBNs) approach [20]. DBNs accomplish deep learning through the stacking of various Restricted Boltzmann Machines (RBMs) on top of each other. In this method of deep learning, the output obtained from a lower-level RBM is subsequently utilized to serve as the input to a higher-level RBM. This process is continued progressively through deeper and deeper layers thus forming a multi-layer stacking of these so-called RBMs. An average emotion classification accuracy of 77.8% was obtained where this method performed significantly better than various types of different SVMs as well as standard non-stacked RBMs.

A limited study involving only 6 subjects was reported for the emotion classification of participants when presented with the stimuli of viewing a number of short video clips for the elicitation of emotions with positive or negative valences [21]. In a novel approach for the emotion classification task which utilizes only the top five EEG recording electrodes, the investigation produced emotion classification accuracies of 87.6% using DBN's with this novel critical feature channel selection method. These results were observed to perform better than Extreme Learning Machines (ELMs) as well as SVMs, and at the same time was observed to perform significantly better than the kNN machine learning approach. However in both of these two reported studies, it is important to point out that the training and classification prediction tasks were accomplished on a per-subject basis and not over the entire cohort of participants, which means that this only caters for intra-subject variability and not inter-subject variability. In other words, these two studies utilized an approach that requires the retraining of machine learning classifiers during the training phase whenever there is a new participant before the emotion classification prediction task can be performed. Essentially what this intra-subject or subject-dependent method employs is an approach that bypasses the difficulty of handling inter-subject variability and only caters for intra-subject variability, which means that it will not work for subject-independent classification tasks.

From the literature survey, there was only one paper found in which the deep learning approach was used in emotion modeling to classify preferences in a subject-independent methodology. Here it was reported that using a combination of unsupervised learning employing stacked autoencoders (AEs) in conjunction with the supervised learning of softmax machine learning classifiers was able to perform prediction of the emotional states for 32 participants for valence and arousal. Nonetheless, this paper reported the requirement of utilizing an extremely large number of hidden neurons in the deep learning classifier. It is interesting to note that the authors themselves alluded to the fact that an extended amount of computational time was utilized during the training phase with such an approach. Subsequently the authors hybridized this approach with feature preprocessing routines employing Principal Component Analysis (PCA) as well as Covariate Shift Adaptation (CSA) during the pre-learning process. However even with the extended processing time and numerous augmentations with supplementary preprocessing, the emotion modeling was only able to produce very low prediction accuracy rates of 53.4% and 52.0% for valence and arousal classification respectively from this subject-independent approach using leave-one-out cross-validation (LOOCV) [22]. What this study clearly demonstrates is the fact that inter-subject variability very significantly and critically adds tremendous difficulty to the classification of emotions based on preferences when compared against the much more common and significantly easier prediction task of subject-dependent studies that only caters for intra-subject variability in the learning of the EEG-based emotion modeling.

3 Methodology

In this investigation, 16 subjects (8 female and 8 male, mean age = 22.44) were involved where all the participants and had corrected-to-normal or normal vision. Furthermore, they were asked and confirmed to be free of any known history of psychiatric illnesses prior to the participation in the study. The participants were briefed on what to expect in terms of the BCI equipment that was to be used during the data acquisition phase before the actual experimentation was to proceed. The EEG acquisition device was a brain-computer interface (BCI) headset called the ABM B-Alert X10, which has nine active electrodes, namely the POz, Fz, Cz, C3, C4, F3, F4, P3 and P4 channels according to the standard 10–20 naming convention where a subject participant wearing the said BCI headset is depicted in Fig. 1. MATLAB, Java and R were the three programming languages used. The visual stimuli were developed and displayed using the Java programming language. Integration between the visual stimuli and the BCI headset was accomplished by implementing the MATLAB programming language with the B-Alert X10's SDK. Finally, the statistical programming language R was used for the signal preprocessing phases, feature extraction, and finally for the training and prediction classification tasks.

The data acquisition processes experienced by the participants are as shown in Fig. 2 where during the commencement of the data acquisition process, a blank screen of three seconds is shown to the participant to obtain the base resting brain signal in order to avoid any brain activities related to the previous stimuli during the actual emotion modeling trial phase. After this blank screen, there will be between five to fifteen of actual viewing time for the 3D stimuli where the minimum viewing time and maximum viewing time is set between five and fifteen seconds respectively. The participant is allowed to commence to the following rating state based on their own choosing after the minimum viewing duration time of five seconds while once the maximum viewing duration time is up, the system will proceed by default to the next rating state. The purpose of implementing this particular method of the data acquisition

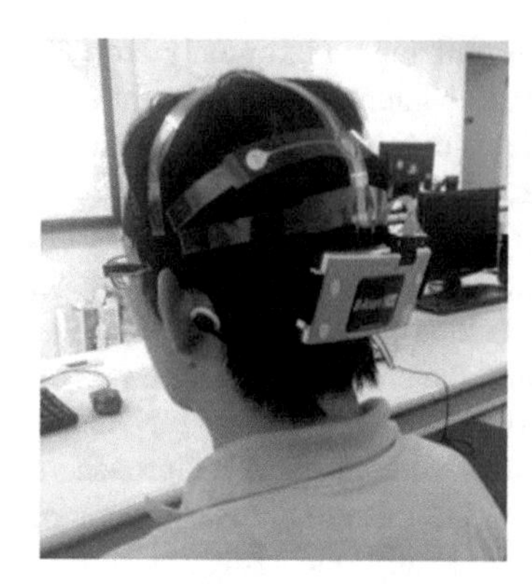

Fig. 1. The medical-grade 9-channel EEG acquisition device is shown being worn by a participant in the study

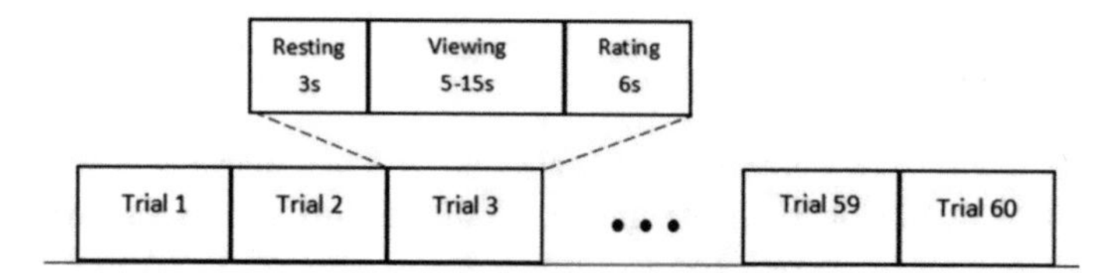

Fig. 2. The flow of the data acquisition process as experienced by the participants during the experimentation.

process flow is to allow the participant to decide on their own accord during the stimuli viewing time so as to mitigate the possibility of boredom from setting in and making the participant fatigued while viewing the stimuli during the data acquisition process since requiring the participant to continuously view only at fixed intervals in a repetitive manner for the purposes of rating the stimuli could possibly cause the participant to experience boredom which will subsequently lead to further fatigue towards the end of the data acquisition process. As such, since the participant is no longer required to just wait until the maximum set and fixed time in order to conduct the rating, this essentially provides the participant with the freedom and ability to shift to the following visual stimuli, which will potentially save some overall viewing time and at the same time prevent the participant from fatiguing. A rating system containing a discrete scale of 1–5, where 1 represents like very much; 2 represents like; 3 represents undecided; 4 represents do not like; and finally 5 represents do not like at all, is shown to the participant at the conclusion of the visual shape stimuli viewing period.

$$r(\theta) = \frac{1}{\sqrt[n1]{\left(\left|\frac{1}{a}\cos\left(\frac{m}{4}\theta\right)\right|\right)^{n2} + \left(\left|\frac{1}{b}\sin\left(\frac{m}{4}\theta\right)\right|\right)^{n3}}} \tag{1}$$

The Gielis Superformula is used to generate three-dimensional shapes which were used as the visual stimuli in this study and had the visual appearances of a bracelet-like virtual generated [23], the mathematical formula of which is as shown in Eq. 1. Our main reason for choosing this shape as the three dimensional visual stimuli for evoking emotions is to determine the aesthetic quality of jewelry-type objects since visual aesthetic quality is primarily the key motivating factor when one decides whether or not make a purchase of such an item. By modifying the various superformula parameters, the generation of different and myriad natural three dimensional virtual shapes can be generated.

Sixty different bracelet-like shapes generated and used in this study is as shown in Fig. 3, which were generated by utilizing different parameters with randomly generated values in the superformula. Through preliminary testing, different ranges of suitable parameter values were chosen to synthesize virtual three dimensional shapes that possess visual characteristics of a bracelet-like shape. These three dimensional bracelet-like shapes were then shown to the participants virtually on a computer. The visual system allowed the presentation of the three dimensional virtual shapes with rotations on different axes of the presented stimuli so that it could be viewed at different angles in order for the participant to be able to fully visualize the generated three dimensional bracelet-like shapes (Fig. 4).

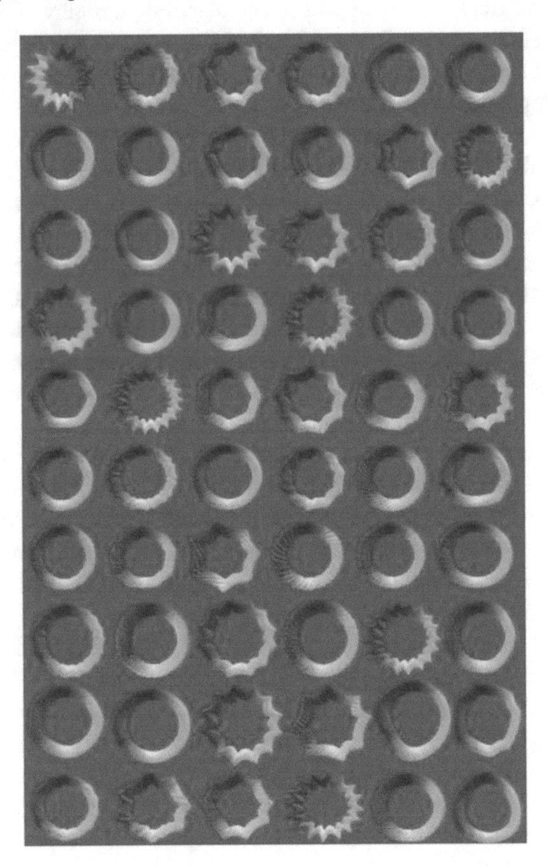

Fig. 3. The Gielis Superformula used to produce 60 bracelet-like shapes.

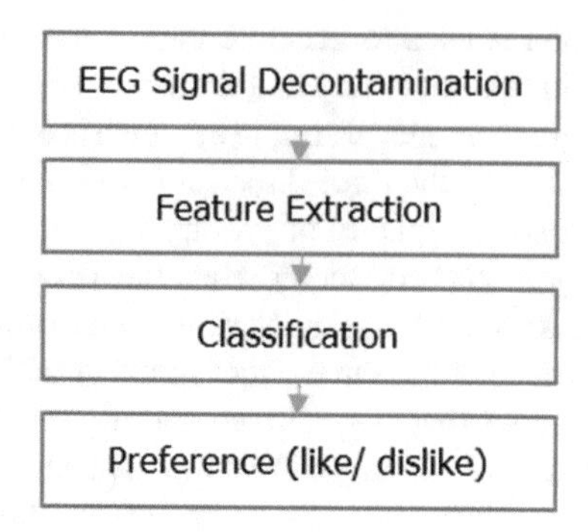

Fig. 4. Summary of the signal processing process flow.

Environmental and physiological artifacts are always present in EEG signal recordings and require decontamination. The SDK in the MATLAB programming language provided by ABM for the B-Alert X10 BCI headset automatically provides this decontamination function. A 50 Hz notch filter removes environmental artifacts while five physiological artifacts comprising electromyography (EMG), eye blinks, excursions, saturations, and spikes are similarly removed automatically in real-time. The eye excursions, saturations, and spikes are replaced by zero values where they are later filled in using spline interpolation.

A Short-Time Fourier Transform (STFT) is then used to transform the decontaminated EEG signals into the TF domain where it decomposes each of the nine BCI channels into five spectral bands, which are the delta 1–3 Hz, theta 4–6 Hz, alpha 7–12 Hz, beta 13–30 Hz, and gamma 31–64 Hz bands. These fives bands across the nine channels thereby provides a total of forty-five input features. The brainwave recordings from the 16 participants where each viewed the sixty 3D visual stimuli of the bracelet-like shapes generated 960 observations altogether. However, only 208 observations were used during the training and prediction classification process. These were the strongest ratings on the ratings scale of 1, which represented like very much, and 5, which represented do not like at all, respectively. A final dataset matrix comprising forty-seven feature columns consisting of the observation ID reference, participant rating, and each of the forty-five TF features, over two hundred and eight rows of selected observations served as the training and testing data for the respective machine learning classifiers. Moreover, the subjects' baseline readings acquired while in the resting state were subtracted from the stimuli viewing state values before the values were utilized in the prediction classification process.

The deep neural networks utilized were set to two hundred hidden neurons within each of the two hidden layers using the uniform adaptive method [24] for weight matrix initialization. Preliminary experimentation showed that this setup with the number of hidden layers as well as the number of hidden neurons per layer provided the optimal settings for this preference prediction task. Cross-entropy [25] was used as the error function during the 10-fold cross-validation, which was conducted for 10 epochs in each of the cross-validation steps.

4 Results and Discussion

Four distinct deep net architectures were tested, which were the standard deep nets, deep nets with dropouts only, deep nets with L1 regularizations only and finally deep nets with both dropouts and L1 regularizations. In L1 regularizations, λ is set at 10^{-5}. For dropouts, we set the hidden layer dropout probability at 0.5. For each of these architectures, we also paired them with different activation functions for the hidden layers, which were the tanh, maxout and rectified linear unit (ReLU) activation functions. The rectified linear activation function [26] was used with an adaptive learning rate method [27].

Table 1 presents the 10-fold cross-validation results obtained from using the various deep net architectures as well as with dropouts and L1 regularization terms. The best classification was obtained using the deep net with dropout architecture using rectified linear units for activation at 79.76%. The second best classification result was also obtained using the deep net with dropout architecture but using the tanh activation at 74.38%. This was followed next with the deep net architecture using both dropouts and L1 regularization with the rectified linear unit and tanh activations, respectively at 72.44% and 72.43%. The lowest classification obtained was 54.92% using the deep net with L1 regularization and maxout activation. As can be seen from Fig. 5, a very significant improvement in classification accuracy was attained using the deep net with dropouts compared to the earlier work which did not make use of any dropouts and/or regularization, which was only between 61.15–67.68%. This is an improvement of over 10% and clearly shows the benefits of using dropouts to improve the generalization ability of deep nets.

Table 1. EEG preference classification results

Deep Net architecture	Hidden layer activation function	Classification accuracy (%)
Standard Deep Net	Tanh	67.68
Standard Deep Net	Maxout	61.15
Standard Deep Net	ReLU	63.99
Deep Net with Dropout	Tanh	74.38
Deep Net with Dropout	Maxout	67.71
Deep Net with Dropout	**ReLU**	**79.76**
Deep Net with L1 Regularization	Tanh	71.86
Deep Net with L1 Regularization	Maxout	54.92
Deep Net with L1 Regularization	ReLU	63.02
Deep Net with Dropout and L1 Regularization	Tanh	72.43
Deep Net with Dropout and L1 Regularization	Maxout	67.16
Deep Net with Dropout and L1 Regularization	ReLU	72.44

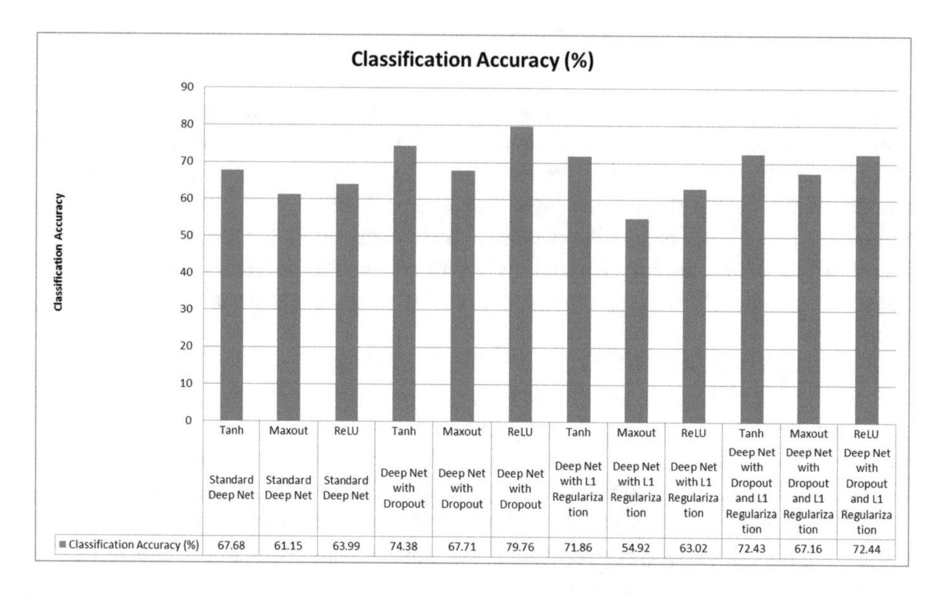

Fig. 5. Summary comparison of various deep learning architectures used.

5 Conclusion and Future Work

This study has comprehensively tested dropout and L1 regularization approaches to deep net architectures in an effort to improve the classification performance of deep learning neural networks in EEG-based preference classification. We have shown that using a deep net with dropouts using rectified linear units for activation was able to achieve a gain of more than 13–18% at 79.76% accuracy compared to standard deep nets without such approaches at only between 61.15–67.68% using various activations. For future work, due to the significant noise typically encountered in inter-subject EEG variations, we intend to investigate the use of autoencoders to pre-train the features extracted in order to further improve classification accuracy. Also, with the significant improvement in classification accuracy obtained through this study, we also plan to embark on application-based investigations into the use of EEG-based preference classification to guide automated generation of affective entertainment content in games, music and story-telling.

Acknowledgements. This project is supported by the FRGS research grant scheme ref: FRG0435 from the Ministry of Higher Education, Malaysia.

References

1. Chew, L.H., Teo, J., Mountstephens, J.: Aesthetic preference recognition of 3D shapes using EEG. Cogn. Neurodyn. **10**(2), 165–173 (2016)
2. Teo, J., Chew, L.H., Mountstephens, J.: Deep learning for EEG-based preference classification. In: International Conference on Applied Science and Technology (ICAST 2017). IEEE, April 2017
3. Wang, X.W., Nie, D., Lu, B.L.: Emotional state classification from EEG data using machine learning approach. Neurocomputing **129**, 94–106 (2014)
4. Dhall, A., Goecke, R., Joshi, J., Sikka, K., Gedeon, T.: Emotion recognition in the wild challenge 2014: baseline, data and protocol. In: Proceedings of the 16th International Conference on Multimodal Interaction, pp. 461–466. ACM (2014)
5. Verma, G.K., Tiwary, U.S.: Multimodal fusion framework: a multiresolution approach for emotion classification and recognition from physiological signals. NeuroImage **102**, 162–172 (2014)
6. Jang, E.H., Park, B.J., Kim, S.H., Chung, M.A., Park, M.S., Sohn, J.H.: Emotion classification based on bio-signals emotion recognition using machine learning algorithms. In: International Conference on Information Science, Electronics and Electrical Engineering (ISEEE), vol. 3, pp. 1373–1376. IEEE (2014)
7. Hadjidimitriou, S.K., Zacharakis, A.I., Doulgeris, P.C., Panoulas, K.J., Hadjileontiadis, L.J., Panas, S.M.: Revealing action representation processes in audio perception using fractal EEG analysis. IEEE Trans. Biomed. Eng. **58**(4), 1120–1129 (2011)
8. Adamos, D.A., Dimitriadis, S.I., Laskaris, N.A.: Towards the bio-personalization of music recommendation systems: a single-sensor EEG biomarker of subjective music preference. Inf. Sci. **343**, 94–108 (2016)
9. Yadava, M., Kumar, P., Saini, R., Roy, P.P., Dogra, D.P.: Analysis of EEG signals and its application to neuromarketing. Multimed. Tools Appl. 1–25 (2017)
10. Chen, L.C., Sandmann, P., Thorne, J.D., Herrmann, C.S., Debener, S.: Association of concurrent fNIRS and EEG signatures in response to auditory and visual stimuli. Brain Topogr. **28**(5), 710–725 (2015)
11. Goncalves, S., De Munck, J., Pouwels, P., Schoonhoven, R., Kuijer, J., Maurits, N., Hoogduin, J., Van Someren, E., Heethaar, R., Da Silva, F.L.: Correlating the alpha rhythm to bold using simultaneous EEG/FMRI: inter-subject variability. Neuroimage **30**(1), 203–213 (2006)
12. Pfurtscheller, G., Brunner, C., Schlogl, A., Da Silva, F.L.: Mu rhythm (de) synchronization and EEG single-trial classification of different motor imagery tasks. Neuroimage **31**(1), 153–159 (2006)
13. Yazdani, A., Lee, J.S., Vesin, J.-M., Ebrahimi, T.: A ECT recognition based on physiological changes during the watching of music video. ACM Trans. Interact. Intell. Syst. **2**(EPFL-ARTICLE-177741), 1–26 (2012)
14. Pan, Y., Guan, C., Yu, J., Ang, K.K., Chan, T.E.: Common frequency pattern for music preference identification using frontal EEG. In: 6th International IEEE/EMBS Conference on Neural Engineering (NER), pp. 505–508. IEEE (2013)
15. Tseng, K.C., Lin, B.-S., Han, C.-M., Wang, P.-S.: Emotion recognition of EEG underlying favourite music by support vector machine. In: International Conference on Orange Technologies (ICOT), pp. 155–158. IEEE (2013)
16. Kim, Y., Kang, K., Lee, H., Bae, C.: Preference measurement using user response electroencephalogram. In: Computer Science and Its Applications, pp. 1315–1324. Springer (2015)

17. Hadjidimitriou, S.K., Hadjileontiadis, L.J.: Toward an EEG-based recognition of music liking using time-frequency analysis. IEEE Trans. Biomed. Eng. **59**(12), 3498–3510 (2012)
18. Hadjidimitriou, S.K., Hadjileontiadis, L.J.: EEG-based classification of music appraisal responses using time-frequency analysis and familiarity ratings. IEEE Trans. Affect. Comput. **4**(2), 161–172 (2013)
19. Moon, J., Kim, Y., Lee, H., Bae, C., Yoon, W.C.: Extraction of user preference for video stimuli using EEG-based user responses. ETRI J. **35**(6), 1105–1114 (2013)
20. Li, K., Li, X., Zhang, Y., Zhang, A.: Affective state recognition from EEG with deep belief networks. In: IEEE International Conference on Bioinformatics and Biomedicine (BIBM), pp. 305–310. IEEE (2013)
21. Zheng, W.-L., Zhu, J.-Y., Peng, Y., Lu, B.-L.: EEG-based emotion classification using deep belief networks. In: IEEE International Conference on Multimedia and Expo (ICME), pp. 1–6. IEEE (2014)
22. Jirayucharoensak, S., Pan-Ngum, S., Israsena, P.: EEG-based emotion recognition using deep learning network with principal component based covariate shift adaptation. Sci. World J. (2014)
23. Gielis, J.: A generic geometric transformation that unifies a wide range of natural and abstract shapes. Am. J. Bot. **90**(3), 333–338 (2003)
24. Nguyen, D., Widrow, B.: Improving the learning speed of 2-layer neural networks by choosing initial values of the adaptive weights. In: IJCNN International Joint Conference on Neural Networks, pp. 21–26. IEEE (1990)
25. De Boer, P.T., Kroese, D.P., Mannor, S., Rubinstein, R.Y.: A tutorial on the cross-entropy method. Ann. Oper. Res. **134**(1), 19–67 (2005)
26. Nair, V., Hinton, G.E.: Rectified linear units improve restricted Boltzmann machines. In: Proceedings of the 27th International Conference on Machine Learning (ICML-2010), pp. 807–814 (2010)
27. Zeiler, M.D.: Adadelta: an adaptive learning rate method. arXiv preprint arXiv:1212.5701

Pattern Identification by Factor Analysis for Regions with Similar Economic Activity Based on Mobile Communication Data

Irina Arhipova[1(✉)], Gundars Berzins[1], Edgars Brekis[1], Martins Opmanis[2], Juris Binde[3], Jevgenija Kravcova[3], and Inna Steinbuka[1]

[1] Faculty of Business, Management and Economics, University of Latvia, Riga, Latvia
irina.arhipovasalajeva@gmail.com, ebrekis@gmail.com, {gundars.berzins, inna.steinbuka}@lu.lv

[2] Institute of Mathematics and Computer Science, University of Latvia, Riga, Latvia
martins.opmanis@lumii.lv

[3] Latvian Mobile Telephone, Riga, Latvia
juris.binde@lmt.lv, jevgenijakravcova@gmail.com

Abstract. The study analyses the regions' economic activity in Latvia using Latvia Mobile Telephone (LMT) mobile communication data from July 2015 to January 2017. The call activity and a number of unique phone users by 119 Latvia counties and biggest cities were analysed in two steps: at first method of principal components was used to explain the variance in the data and then exploratory factor analysis was applied. Three factors were identified that describe 87.5% of the total variance of the aggregated daily data. The first factor is related more to the regions with higher economic activity, the second and third factors capture, respectively, lowers call activity during weekdays and are related to the regions with lower economic activity in total. When to look at the same data but aggregated not only by day but also by daytime, then there are two new factors that describe 94.8% of the total variance. One of the factors is related to higher economic activity in regions as has higher values during normal working hours and lower values after working time, the second is related to the regions with lower economic activity. So it was concluded that during normal working hours the economic activity is higher in the regions with higher call activity. As the result considered indicators of call activities and unique phone users can be used for the identification of the regions with similar economic activity patterns.

Keywords: Indicators of economic activity · Mobile data
Regional economic activity

K. Arai et al. (Eds.): FICC 2018, AISC 886, pp. 561–569, 2019.
https://doi.org/10.1007/978-3-030-03402-3_39

1 Introduction

The actual technologies and mobile data availability allow to developing applications of mobile data in the various fields like mobility analysis [1, 2], transport planning [3], urban and country dynamics [4], human activity recognition [5] or urban planning [6]; despite of that only some of them are used in real-life urban management [7].

The data of mobile network operators CDR (Call Data Records) is a secondary product resulting from the events occurring in the network of mobile operators. Network events are activities performed by individuals having made incoming or outgoing calls or communicating by SMS (Short Message Service) messages. While communicating via mobile phone, a request of an individual is forwarded to the nearest base station, and next, the data processed are forwarded to the base station nearest to the receiver of the network event.

The CDR data contain information on the event of each call, message or data transmission (the identity of the persons making and receiving the call, the time and duration of the call, the identification code of the base station of the event). The using of the data for other purposes does not entail either extra costs or extra workforce, because the data are automatically recorded for the purpose of invoicing.

For this reason the CDR can be used not only for invoicing, but for other purposes as well. The data can be used to improve national statistical estimates and calculations, especially in the sectors where conventional methods for data collection are laborious and costly. Similarly, the data can be used for the planning of the development of urban infrastructures, for example, the prioritization of the construction of new roads or the development of the tourism industry.

The CDR are continuously being recorded and stored, therefore it allows problems to be analysed in real time supplying the relevant information for the decision-making. One of the most significant advantages is possibility to analyse the data in real time and to provide information on economic, demographic and social indicators at national, regional and municipal level.

Over the last years, the CDR has drawn a great deal of the attention of academics worldwide. Nevertheless, the collection and analysis of personal data is restricted by confidentiality constraints, because the data contain information not only on the time and duration of an event, but also on the whereabouts of the participants of the event.

For this reason telecommunications companies particularly protect the data from the access of the third parties, including researchers. In order to ensure that, on the basis of the available information, the data cannot be identified at the level of a particular end-user from the one hand and the opportunity to explore the real situation is retained from the other, encrypted aggregated data is available for the academic purposes only [8, 9].

Regarding the collection of data, its biggest shortcoming is that the data are solely possible to be stored if a mobile device is being used. In order the data could have been used for research, the activity of the users of mobile devices should be sufficiently high in the relevant area – such as to create large-scale data arrays, whose processing consume significant computing resources.

The purpose of this research is to identify the regions with similar economic activity patterns using mobile communication data.

The following tasks have been defined to achieve the proposed purpose: at first to identify general pattern and interconnection of call activity and number of unique phone users data from CDR's; than to examine the assumption about economic activity reflection in call data changes on different aggregation levels by day and daytime (aggregated by 15 min), and finally to compare the regions with similar economic activity patterns.

2 CDR Data Pattern

The leading Latvia mobile network operator LMT agrees to provide this research with encrypted accumulated CDR data on regular bases. Each provided data set consists of two parts:

- List of all base stations which data is present in the current data set containing the following information about each base station: identifier, address, geographic coordinates (longitude and latitude).
- Accumulated data for each 15 min interval for every base station. Each record in these data contains time, identifier of the base station, activity count, and unique phone users' count.

It must be pointed out that "uniqueness" of user is determined for each data interval, so there is impossible to diagnose the number of users using particular base station over the 15-min period. Different user activities are not distinguished. Activity may be either outgoing or incoming call or sending/receiving SMS message. Activities do not contain information about data transfer.

To prepare date for the analysis, POSTGRES database was created. Using information about geographical location of base stations each of them were tied to the one administrative unit. Taking in account relatively complicated administrative structure in Latvia, there was established six-digit code covering administrative structures of all levels:

- The first digit denotes region (1 – Kurzeme region, 2 – Latgale region, 3 – Riga region, 4 – Vidzeme region, 5 – Zemgale region).
- The second and the third – serial number of city, town or county in the region.
- The fourth and the fifth – serial number of parish in the county.

Table 1. Extract from LMT CDR Statistics

Date	Time	Number of		Base station's	
		Calls	Users	Longitude	Latitude
02.10.15	14:15	218	178	24.15012	56.89900
06.10.15	08:30	56	44	27.21679	56.15639
17.10.15	19:00	114	29	22.39888	56.51320

- The sixth – status of place (0 – not a city or town, 1 – town, 2 – city under the state jurisdiction).

Data sets from the several data sets covering the time interval under investigation were loaded in the database and enriched with additional information about weekdays and working and leisure days. According to the law, there are State holidays and swapped working and leisure days leading to the situation that status of particular day in calendar may not correspond to its actual status.

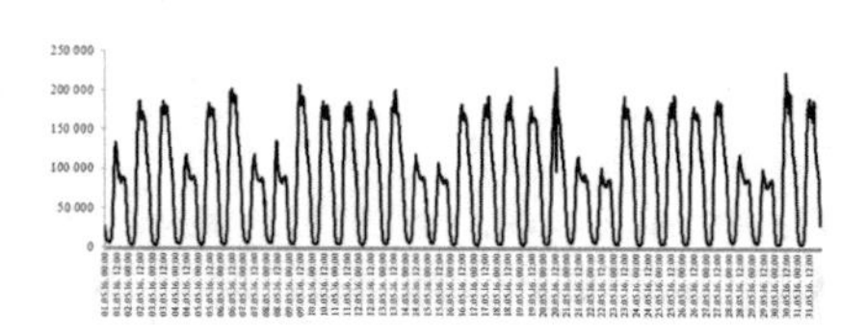

Fig. 1. Time series of the LMT CDR data in Latvia from 1st May to 31st May 2016.

The database of call and text-message activity contains aggregated records for time period from 25.07.2015 to 20.01.2017, altogether 64733760 aggregate CDR from 1235 base stations (Table 1).

2.1 Time Series of the CDR Data

The total number by days of the call activities is shown in Fig. 1 for one month from 1st May to 31st May 2016. From the graph could clear to see the difference between call activity on weekdays and weekends that could reflect the economic activity changes (during the weekends and holidays peaks are lower than in working days). From the same graph could observe a big change in calls activities over a daytime with activity peak around the everyday noon.

2.2 Statistical Analysis of the CDR Data

After analysing call data as an indicator of economic activity, there was a statistically significant relationship (p-value = 0.058) between the aggregate number of LMT calls activities in Latvia and seasonally unadjusted Gross Domestic Product (GDP) from production approach by quarters (euro) from 01.10.2015 to 31.12.2016.

Similarly, a statistically significant relationship ($p\text{-value} < 0.05$) was found between the LMT call activities and economically active enterprises of market sector by statistical region per thousand inhabitants in 2015. A close positive relationship ($r = 0.682$) was also found between the LMT call activity intensity of week days to weekend and the number of SMEs to 10 thousands of population in Latvian cities and counties [10].

In examining the relationship between the number of call activities and municipalities budget expenditures in cities and counties on a monthly basis from 01.08.2015 to 31.12.2016, they were not statistically significant in all regions. For example, there are statistically significant relationships in Liepaja (p-value = 0.01), Jelgava (p-value = 0.03), Daugavpils (p-value = 0.03), Valmiera (p-value = 0.01) and Marupe county (p-value = 0.01), whereas Carnikava county (p = 0.63), Jurmala (p = 0.32) and Riga (p-value = 0.39) is not significant.

It concluded that the variables that directly depend on the economic activity in the region have the statistically significant relationship with the call activity. Whereas, for the variables that have indirect influence to the region ongoing activities (taxes) the relationship is not statistically significant.

As the result, cannot be rejected the hypothesis that during normal working hours, the economic activity is higher in the regions with higher call activity. According the proposed hypothesis the following two statements have defined:

- Firstly, the call activity is relatively higher during weekdays than in weekends in regions with relatively higher economic activity to other regions.
- Secondly, call activity distribution by hours vary in regions relative to economic activity, high or low.

3 Distribution of Counties by Economic Activity Changing Depending on Week Days and Weekends

In order to identify the regions distributions by economic activity during the week days and weekends the factor analysis was used to find out what counties are similar by the number of call activities and/or the number of users and combine them into the groups.

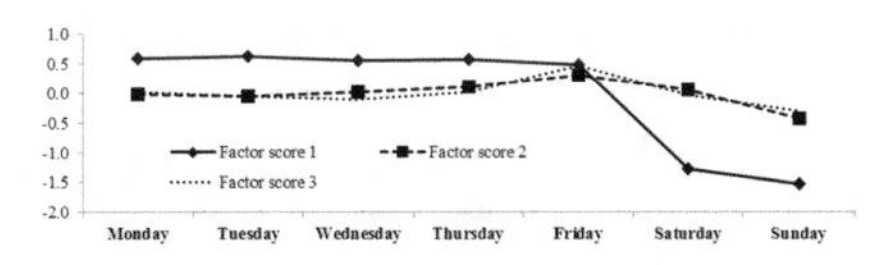

Fig. 2. Factors components mean values depending on weekdays and weekends.

As the number of call activities has a close correlation with the number of unique phone users (r = 0.973), initially, before the factor analysis, the method of principal components was used to reduce the number of correlated variables among 119 cities and counties.

In the study was used data aggregated by days: call number of the call activities and number of unique phone users. As the result, the 119 standardized components were obtained for each county, which are characterized by the number of call activities and number of users according to weekday.

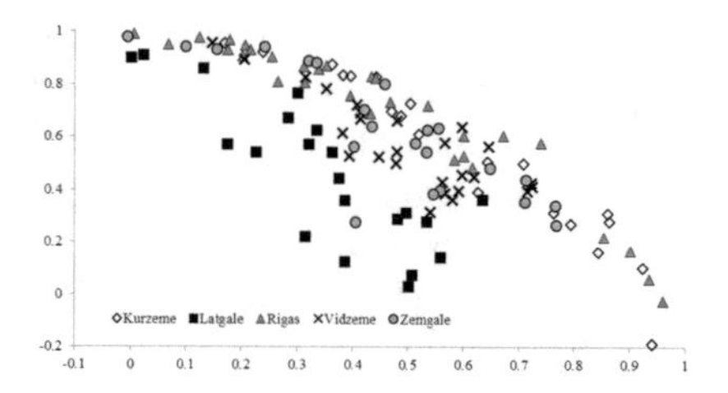

Fig. 3. Factors components loadings plot depending on day and regions.

To gain a cleaner interpretation of these new components Varimax rotation was applied. Five rotated factors have a significant data load, where the first factor describes 41.1%, the second factor describes 26.5%, the third factor – 19.8%, and the last two factors describe 3.7% of the total data variance.

The first factor is related more to the regions with higher economic activity, the second and third factors capture, respectively, lower or higher call activity during weekdays and are related to the regions with lower economic activity in total (Fig. 2).

The distribution of the counties by economic activity can be shown using the factors loads with data aggregated by days.

The points with higher values on vertical axis have higher values in week days and lower values in weekends and represent the cities and counties with higher economic activity. The points with higher values in horizontal axis have lower values in week days and higher values in weekends and represent the cities and counties with lower economic activity.

Divided the counties by five statistical regions it can be concluded, that Latgale region has lower economic activity, while other regions have higher economic activity (Fig. 3).

The top five counties with the highest changes in call activities on weekdays in descending order are Riga, Jelgava, Marupe county, Stopini county, Valmiera and are interpreted as the counties with the highest economic activity.

The last five counties with the lower changes in call activities on weekdays are Varkava county, Engure county, Aglona county, Saulkrasti county, and Rucava county and are interpreted as the counties with the lowest economic activity.

The top five counties with the highest changes in call activities on weekends in descending order are Saulkrasti county, Rucava county, Engure county, Pavilosta county, Carnikava county and the last five counties with the lower changes in call activities on weekends are Adazi county, Daugavpils, Riga, Rezekne, Jelgava.

4 Distribution of Counties by Economic Activity Changing Depending on the Time of Day

Using data aggregated by 15 min, the factor analysis was applied on the results of principal component analysis, using Varimax rotation to find out what counties are similar by the number of call activities and/or the number of users and combine them

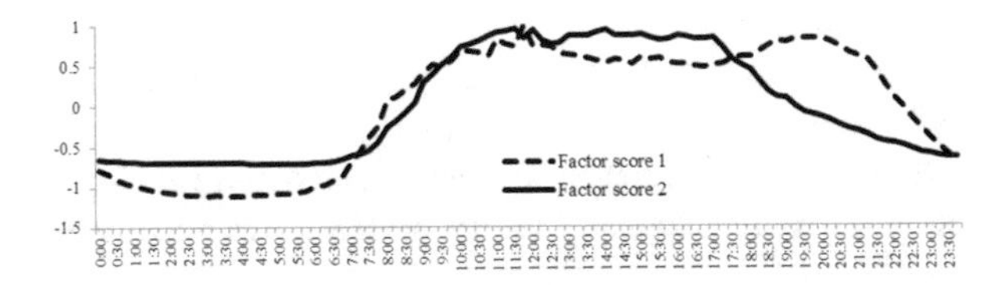

Fig. 4. Factors components mean values depending on time.

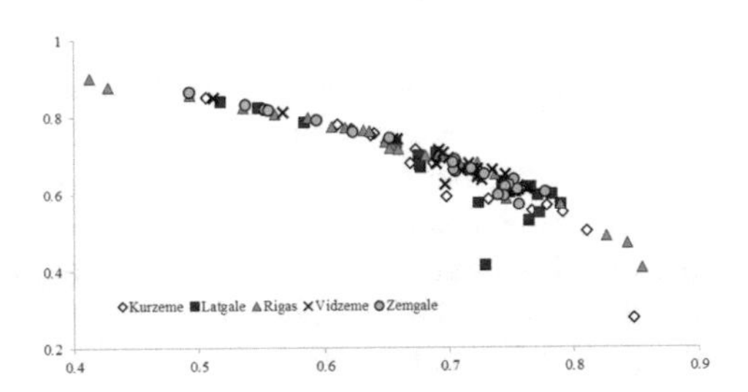

Fig. 5. Factors components loadings plot depending on time and regions.

into the groups. Just two factors have a significant data load, where the first factor describes 48.2%, the second factor describes 46.6%, altogether 94.8% of the total variance.

The first factor has lower values during the night time from midnight to 7 a.m., as well from noon to 6 p.m., and higher values during the evenings from 6 p.m. to 11 p.m. that can hint at the regions with relatively lower economic activity.

The second factor has higher values during the night time from midnight to 7 a.m., as well from noon to 6 p.m. and lower values during the evenings from 6 p.m. to 11 p.m. that can hint at the regions with higher economic activity (Fig. 4).

The distribution of the region by economic activity can be shown using the factors loads with data aggregated by 15 min. The points with higher values in vertical ass have higher values from midnight to 7 a.m., as well from noon to 6 p.m. and lower values from 6 p.m. to 11 p.m. and represent the regions with higher economic activity.

The points with higher values in horizontal ass have lower values from midnight to 7 a.m., as well from noon to 6 p.m. and higher values from 6 p.m. to 11 p.m. and represent the regions with lower economic activity (Fig. 5).

The top five counties with the highest changes in call activities on normal working hours in descending order are Riga, Marupe county, Jelgava, Stopini county, Liepaja and are interpreted as the counties with the highest economic activity.

The last five counties with the lower changes in call activities on normal working hours are Carnikava county, Engure county, Aglona county, Saulkrasti county, Rucava county and are interpreted as the counties with the lowest economic activity.

The top five counties with the highest changes in call activities after normal working hours in descending order are Saulkrasti county, Rucava county, Engure county, Carnikava county, and Pavilosta county.

The last five counties with the lower changes in call activities after normal working hours Liepaja, Stopini county, Jelgava, Marupe county, Riga and are interpreted as the counties with the highest economic activity.

5 Conclusions and Recommendations

Considered indicators of call activities and unique phone users can be used for the identification of the regions with similar economic activity patterns.

Using the indicators of call activities and unique phone users the distribution of the region by economic activity can be determined by days in weekdays and weekend, as well by daytime during normal working hours and after working time.

The distribution of cities and counties by economic activity is depending on call activities in working days and weekends, where:

- higher values in weekdays and lower values in weekends represent the regions with higher economic activity;
- lower values in weekdays and higher values in weekends represent the regions with lower economic activity.

The distribution of cities and counties by economic activity is depending on call activities in time, where:

- higher values from midnight to 7 a.m., as well from noon to 6 p.m. and lower values from 6 p.m. to 11 p.m. represent the regions with higher economic activity;
- lower values from midnight to 7 a.m., as well from noon to 6 p.m. and higher values from 6 p.m. to 11 p.m. represent the regions with lower economic activity.

Call data as an indicator of economic activity has a statistically significant relationship with GDP, economically active enterprises of market sector by statistical region and number of SMEs in Latvian cities and counties.

The distribution of counties by economic activity changing depending on weekdays and weekends give the comparable results as the distribution of counties by economic activity changing depending on time.

Call data could be used as reliable proxy predictive indicator for the state of regional and country economic activity if used on monthly bases with short delays on observation period not more than two weeks, though future resource is necessary with more data points.

Acknowledgment. The research leading to these results has received funding from the research project “Large-scale data processing mathematical model for “Updatable Latvian regional business index” using limited data asset”, Contract No. AAP2016/B089 signed between University of Latvia and LMT Ltd.

References

1. Jiang, S., Ferreira, J., Gonzalez, M.C.: Activity-based human mobility patterns inferred from mobile phone data: a case study of Singapore. IEEE Trans. Big Data **PP**(99), 1 (2016)
2. Jarv, O., Ahas, R., Witlox, F.: Understanding monthly variability in human activity spaces: a twelve-month study using mobile phone call detail records. Transp. Res. Part C **38**, 122–135 (2014)
3. Elias, D., Nadler, F., Stehno, J., Krosche, J., Lindorfer, M.: SOMOBIL – improving public transport planning through mobile phone data analysis. Transp. Res. Procedia **14**, 4478–4485 (2016)
4. Trasarti, R., Olteanu-Raimond, A.M., Nanni, M., Couronné, T., Furletti, B., Giannotti, F., Smoreda, Z., Ziemlicki, C.: Discovering urban and country dynamics from mobile phone data with spatial correlation patterns. Telecommun. Policy **39**(3–4), 347–362 (2015)
5. Chetty, G., White, M., Akther, F.: Smart phone based data mining for human activity recognition. Procedia Comput. Sci. **46**, 1181–1187 (2015)
6. Larijani, A.N., Olteanu-Raimond, A.M., Perret, J., Bredif, M., Ziemlicki, C.: Investigating the mobile phone data to estimate the origin destination flow and analysis; case study: Paris region. Transp. Res. Procedia **6**, 64–78 (2015)
7. Steenbruggen, J., Tranos, E., Nijkamp, P.: Data from mobile phone operators: a tool for smarter cities? Telecommun. Policy **39**(3–4), 335–346 (2015)
8. Directive 2002/58/EC of the European Parliament and of the Council of 12 July 2002 concerning the processing of personal data and the protection of privacy in the electronic communications sector, 2002. Official Journal of the European communities. No L201/37, 31/07/2002
9. Directive 95/46/EC of the European Parliament and of the Council of 24 October 1995 on the protection of individuals with regard to the processing of personal data and on the free movement of such data, 1995. Official Journal of the European communities. No L281/31, 23/11/1995
10. Arhipova, I., Berzins, G., Brekis, E., Kravcova, J., Binde, J.: The methodology of region economic development evaluation using mobile positioning data. In: Proceedings of the 20th International Scientific Conference on Economic and Social Development, Prague, pp. 111–120 (2017)

Data Mining and Knowledge Management Application to Enhance Business Operations: An Exploratory Study

Zeba Mahmood(✉)

CS & IT Department Superior University Pakistan, Lahore, Pakistan
zeba.mehmood@gmail.com

Abstract. The modern business organizations are adopting technological advancement to achieve competitive edge and satisfy their consumer. The development in the field of information technology systems has changed the way of conducting business today. Business operations today rely more on the data they obtained and this data is continuously increasing in volume. The data stored in different locations is difficult to find and use without the effective implementation of data mining and Knowledge management techniques. Organizations that smartly identify, obtains and then converts data in useful formats for their decision making and operational improvements create additional value for their customers and enhance their operational capabilities. Marketers and customer relationship departments of firm uses data mining techniques to make relevant decisions, this paper emphasize on the identification of different data mining and knowledge management techniques that are applied to different business industries. The challenges and issues of execution of these techniques are also discussed and critically analyzed in this paper.

Keywords: Knowledge · Information · Data mining · Knowledge management Knowledge discovery in databases · Business · Operational improvement

1 Introduction

Today, the business is not only conducted throughout the globe but also developing their capabilities, skills, knowledge and their financial profitability. It is the will of business managers and owners to keep their business growing and developing from different aspects so that they can be ahead of competitors. Getting more developed and advanced in technology than rival is part of modern business strategies of organizations, in which advancement in information technology; knowledge management and usage of data effectively have higher significance. It is the need of business era that managers and owners must be able to appraise and recognize the affecting factors to their organization [1].

The knowledge produced from data in an organization captures the attraction of managers but it is also the truth that these managers are not much interested in technology that produce it actually [2]. Unlike the past business concept the modern business is not only to earn higher profits but the advantage is created by satisfying customers, continuous business growth, Innovation, opportunities of growth, and value

K. Arai et al. (Eds.): FICC 2018, AISC 886, pp. 570–583, 2019.
https://doi.org/10.1007/978-3-030-03402-3_40

for money to customers [3]. The matter of concern today for managers is to understand the changes in customer need and business environment because today customers now have several alternative products and services to choose from which makes them competitor as well and competitors are providing goods and services that are improved in terms quality and performance that leads to emerging changes in business environment [4]. Some significant concerns for modern businesses of 21st century includes the identification of customer needs timely, understand the value customer wants, and the use of advanced information technology (I.T) tools to improve design, R&D, manufacturing productivity, marketing efforts and post sales service standards. The understanding of client's behavior also has significance and impact the business strategy and appraising the new business opportunities [5]. Information technology today not only advancing the capabilities of organizations but also removing the restrictions of time a location due to globally accessibility the differences of large and small size organizations are narrowing down [3]. Smaller firms are giving intense competition to larger firms because due to information technology they can make the consortium with other smaller firms and can access the clientele that were previously not in access. According to [6] with the increment of Information technology applications inside organizations and with boast of businesses the data of customers also increased that are unused and remain hidden. The modern managers and decision makers are finding the significance of this hidden and untapped data for decision making and business growth. Due to the competition and options for customers there is need in business world to create CRM (customer relationship management) to obtain the useful customer knowledge which can further be transformed to a smart business strategy. On one side the data mining technique helps in divulge the unused and hidden data related to customer to understand them well. On other hand knowledge management is now become the essential part of an organizational strategy in order to improve business operations. This research paper explores the effect of data mining (DM) and knowledge management (KM) applications on business organization performance. The reason of studying data mining and knowledge management is that [7, 8] confirms that DM and KM applications are becoming significant for decision making process. This study further elaborates the benefits and problems organizations face while executing knowledge management and data mining applications.

The structure of the paper is designed to understand the significance of knowledge management and data mining applications on business performance. The next section of this paper describes the concept of knowledge management, its process and discovery of knowledge from databases. The third section of this paper emphasizes on data mining, and the business scenarios related to data mining. Section 4 of this paper focuses on the application of knowledge management and data mining in business organizations with special consideration of decision making process and different fields of business. The part five identifies the challenges, limitations and problems KM and DM implementation resulted. In final part the paper is concluded.

2 Knowledge and Knowledge Management

2.1 Knowledge

Knowledge is the concept differs from data and information in the field of information technology. In the I.T related literature the data can be described as raw facts and figures in numerical form. When data is arrange in a pattern it becomes the information. When the information is possessed in person's observance in form of modified information that may be unique, valuable and accurate it becomes knowledge. The knowledge is related to concepts, ideas, judgments, observations and facts [9]. Two major types of knowledge include Tacit Knowledge and Explicit knowledge. Explicit knowledge obtained from books and documents whereas the tacit knowledge is obtained from personal experience.

2.2 Concept of Knowledge Management

The concept of knowledge management in a business environment is to obtain the data and modified it in form of useful knowledge. As a multidisciplinary concept there are several definitions of Knowledge management. In [10], authors argued that knowledge management is diversified concept and taking a single definition out of it is difficult task because of the three factors. This nature includes that knowledge is considered as immaterial in nature and when it comes to its definition it is complex. In the management field of business the knowledge management becomes further complex because each field has its subjective and eclectic nature. Third factor is that KM concept is in stage of emergence which is not yet established. In [10], authors included the Knowledge management definition from other authors as well whereas in [14], authors observed a common feature in all the definitions which is that it helps an organization creating the competitive edge in market by sharing knowledge effectively in organization. According to [14], the security and protection of knowledge is also an important factor along with getting the ability to generate and incarcerate the knowledge, the whole process is actually known as "Knowledge management". In [15], authors defines the Knowledge management from perspective of HR that it is a process of obtain, capture, share and use the knowledge for the purpose of improve performance and learning in company. In [16], author has given a changed definition of Knowledge management which is that identify, capture, arrange and distributing the intellectual information that is crucial for strategic performance of a firm. In [1], authors elaborate the Knowledge management as an integration process of different fields of organization with strategic importance with the objective to improve performance and capital. The core of above mentioned definitions of KM describes it as process to achieve knowledge, proficiency and experience which enables them to have superior know how, boasted performance, birth innovation and give consumer the value they needed. In [12], authors explained that organizations are achieving their goals by generating motivation in workers to develop, improve and utilize their skill of interpreting the data and information. Employees must be able to give meaning the information they have to use it for decision making and improve efficiency. In [17], authors investigated and found that modern businesses have agreed that share and application of knowledge in their operations and businesses improve their performance. As an intangible asset the knowledge leads to create competitive advantage by increasing employee's capability and by providing base to make accurate decisions.

2.3 Process of Knowledge Management (KM)

The knowledge management purpose in an organization is to utilize the available information optimally and make it a culture to make decisions and work based on available knowledge. Process of knowledge management starts with gathering the data, categorizing different information in it, evaluate the information critically and share it with relevant layers within organization for use. The major objective of Knowledge management is to make knowledge available right place, right time and in right format so that timely and effective decisions could be taken throughout organization [1]. Process of Knowledge management revolves around flow of knowledge which starts from gathering and then creation of knowledge which then further share and distribute to relevant person. In modern business world the whole process of knowledge management is supported and implement via KM technologies. Modern technology based applications of Knowledge management are creating significant role and these are key tools to utilize knowledge management [18]. In order to enable the KM communication, cooperation, and content sharing the technologies have been playing key role to ensure knowledge management flows. Knowledge management technologies, like Applications are now integrated with the process cycle of KM. Below is the figure.

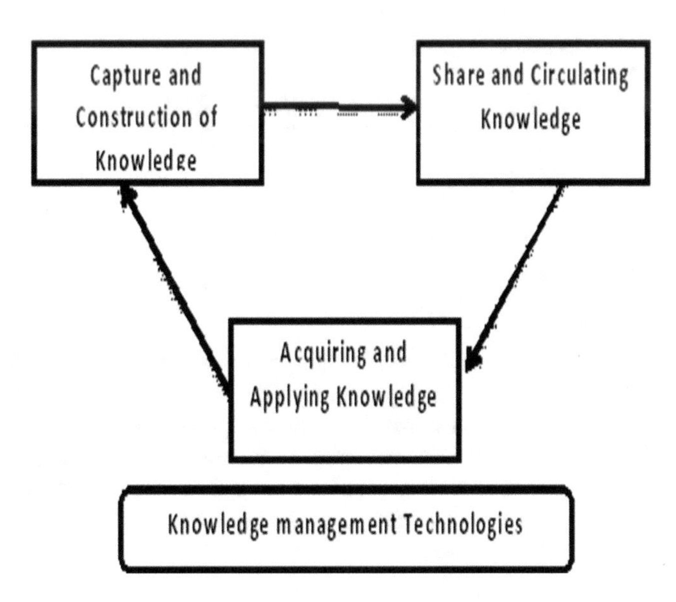

Fig. 1. KM technology and process integration.

Figure 1 demonstrates that knowledge management is a cyclical process by efficiently using which the organization can achieve its strategic and business goals. Organizations needs to make the tacit and explicit knowledge learning as their habit along with more emphasize on planning and implementation on the basis of available knowledge. In [19], authors highlighted a group of six technologies which supports the implementation of KM and help improving firm performance includes the data mining, database discovery, ICT, knowledge management framework, and Knowledge based system.

2.4 Knowledge Discovery from Databases (KDD)

In [1], authors investigated the integration of Knowledge discovery from database from different organizational domains like production, finance, and production control, and audit, detection of fraud, internet and marketing functions. Due to this integration, discovering the functional knowledge becomes more convenient. In [19], author describes the KDD in words "the significant process of identifying and obtain the useful, novel and understandable data from huge database". In [20], authors further elaborates that KDD is based on steps that uses Data mining technique to pick the data from database in useful, novel and patterned form. Figure 2 demonstrates the steps of KDD and identified five stages of this process [21].

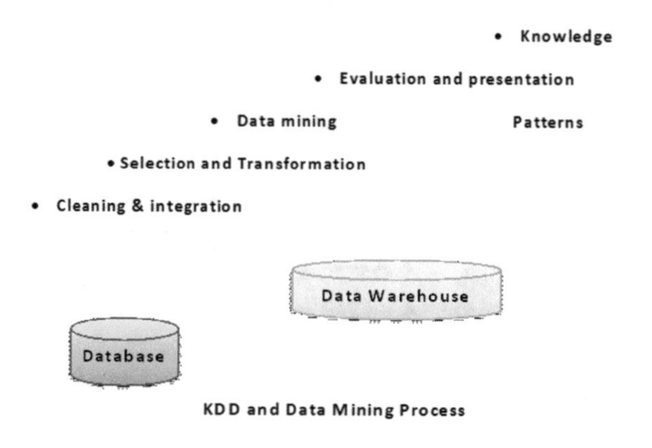

Fig. 2. Data mining and KDD process.

- In first stage the data is cleaning and integrated with different domains. This stage is to remove the missing data, multiple sourced data and noise from database.
- To retrieve the data from database the second stage of "Data selection and transformation" is conducted which then further transfigure so that data mining can become easy. Aggregation and summary operations are performed by the IT firm.
- In order to excerpt the pattern of data, data mining adopt an intelligent technique to ensure retrieval of data needed.

- Fourth stage is pattern evaluation in which different interesting patterns are tried and evaluate so that when knowledge is needed the right pattern can be obtained.
- The fifth stage is knowledge presentation where the mined knowledge is present and visualize to user for decision making purpose.

3 Process of Data Mining

The data mining has emerged as one of most significant data analysis and detection tool used in business organizations. The data mining is defined in diversified ways. It is a process of obtaining and finding the useful, unknown, understandable and valid pattern from huge data in database [22]. But it is also the truths that emphasize of all definitions is to finding data and discover data to be used in decision and managerial level discussions. Data mining has its increasing significance in business organizations because it becomes critical for management to make successful decision. What management has in hand is the data of their financial, marketing, sales, production, and other business measures. The data is saved in database and management needs data in arranged and useful form. Data mining technique is job of I.T to find the data needed, arrange it in useful form and make sure to provide it to relevant authorities [13]. Data mining tools further provides the results of data and help management interpret it religiously. The process of data mining is based on four major steps starts from selection of data, transformation of data, mining of data and prepares results to be used for interpretation.

In past the business queries and information that took much time to get answer are now resolved and answered by Data mining tool. DM have capability to search the huge data base for particular data and patterns and generate the information that predicts the business future need and management then makes decision comparing this information with expectations [21]. The DM tools finds innovative patterns that is obtained with shuffling of data sub sets [22].

The data mining tools are majorly predictive or it would be descriptive, different researchers to describe the DM deeply emphasize on deep analysis, tasks data mining does and in terms of the pattern they want to mined. Researchers [6, 20, 21] have identified the tasks of data mining by using diversified terms. According to [6] the detection of deviation, description of class, Class identification, Dependency investigation, visualization of data and concept explanation are data mining tasks identified. The tasks identified by [20] includes summarization, dependency modeling, link investigation, classification, and sequence investigation whereas the author [21] find the tasks like predictive investigation via regression, relationship classification, concept description, and cluster analysis as useful data mining tools. Since the different researchers uses different terms for data mining tasks but the concept they described were almost same. The concept of data visualization is different from others which is only identified by researcher [6]. The data mining tasks are discussed below with relevance to business organizations and business environment.

3.1 Visualization of Data

In this tasks the information is displayed in graph, table and map formats which helps the users of information to interpret it conveniently along with understand the patterns of data. The management and owners can use this graphical and tabular information to understand the trends and buying behaviors of consumers. The sales manager can understand the most selling item in product range and can make the demand of that product on immediate basis. With the help of tables and graphs management takes strategic decisions for future business growth.

3.2 Dependency Analysis

The dependency analysis identified the relationship between the items so that promotional and discount campaigns could be design accordingly. Dependency analysis finds the connection between the products which have pattern to be bought frequently. The information and knowledge by this data mining tasks is used to design promotional, discount and marketing strategies. Products like bread and butter or food with soft drink are bought together which shows dependency of one product over other. So the shop owner/retailer can design promotional and discount strategy accordingly for example they can give discount if consumer buy milk with butter and soft drink with food. Bread then milk and food then soft drink is the pattern of purchase.

3.3 Description of Concept

The entries of data like data of consumers are grouped in terms of concept, classes and knowledge domain. This technique leads to the characterization, comparison and discrimination in a class of data. The summary of the whole data is obtained in the data summarization whereas the in discrimination of data the selected class is compared with other classes. The examples of summarization is that the management can obtain the data of consumers whose purchase has declined up to 20% in a particular period of time or during a month with the help of query SQL. The example of data discrimination is that management can discriminate consumers on the basis of their consumption like they can find the data of consumers with 10% higher purchases then the worst spending consumers in April 2014. Similar technique can be used to extract the information regarding consumers but the outcome in form of charts and graphs will demonstrates the different picture due to discrimination [21]. By using this task the marketing management can understand whether consumer is declining its loyalty with company or his taste is changed via the data profiling process.

3.4 Clustering

It is another important task of data mining in which management segment different items to a cluster with similar features and usability. The information and knowledge obtain through data mining decides the similar features to cluster the data in same sub group. The use of clustering in business organization is that the marketing department can use it to create market segments for example if a meat selling brand to understand the consumer preference of meat buying after discount, so instead of thinking over it they can group the consumers on religious basis to understand their preference based on cultural values and norms. The promotional campaigns could be designed with the help of data mining technique.

3.5 Classification and Prediction

This task discovers the patterns that are predictive in nature based on the categories and nominal attribute. The attribute which they predict in this task is known as "Class". In this model the training data is analyzed so that it can be represented in the form of decision trees, formula of mathematics, rules of classification and neural network, etc. [10]. In business organization the marketing management can classify products in form of cheap, high priced, most popular and normal, etc.

3.6 Deviation Detection Task

This task of data mining ensures to identify the set of data from whole data which do not match the features of whole data. The data which deviate in this task is known as outliers. In businesses this technique is used to identify any fraud for example a bank can detect the deviation in credit card withdrawal of their consumers and any deviation is then flagged as anomaly to safe from fraud.

4 Data Mining (DM) and Knowledge Management (KM) Application

4.1 Application in Business Decision Making

In the process of business decision making the application of knowledge management and data mining plays crucial role in creating the business intelligence. Figure 3 demonstrates the interaction of DM and KM to create the business intelligence. The term business intelligence can be described as the capacity of business to effectively use knowledge and information that leads to development of business and its growth. Transformation and effective utilization of these two techniques is critical in implementation [25].

Figure 3 demonstrates that with implementation of data mining in organization the management can obtain information previously hidden and unknown for the purpose of getting knowledge. The effective knowledge and information leads to the creation of business intelligence. The business intelligence authorizes the managers to know and answer the questions about the purchase frequency of a product by consumer. The knowledge of consumers and their profiles enables the marketers of firm to decide the customer types and the promotional messages for each type of clientele.

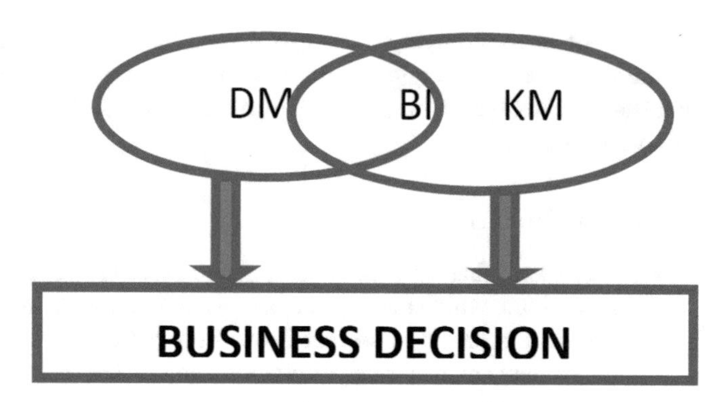

Fig. 3. Data mining and knowledge management interaction.

4.2 Application of KM and DM in Different Business Fields

This section of paper demonstrates the different domains and business areas where the Application of KM and DM does play role to enhance their performance. Knowledge management and Data Mining are used in banking sector, telecommunication industry, manufacturing organizations, retail businesses and in insurance firms. These industries are getting benefited from DM and KM in terms of their operational improvement, effective decisions, improved marketing and increased financial profits. The data mining tools are used in organizations in form of statistics, pattern identification and applications that recognize those patterns and relationships of the knowledge and information that would not be possible to detect without these tools. Failure to get these connections and patterns leads to business operational failure which data mining actually improves. The way different industries utilize the KM and DM in their business discussed below.

(1) *Retail Industry:* Retailing is one of most suitable industry to adopt the Knowledge management and Data mining due to large customer footfall, large number of unit sales, involvement of barcodes, consumer buying history and sales records. Big retailers like Wal-Mart, Amazon and Tesco etc. have their data in TBs within few minutes [1]. The huge data size increases the need of application of KM and DM because the size of data retailers further expanding due to global popularity of electronic commerce/online buying activities. It is the fact that retail industry is the largest user of data mining for the purpose of knowledge building; understand behavior of consumer buying, sales trends, enhance ratio of consumption, retain loyal clientele and to enhance distribution structure so that the operational efficiency could improve and cost of operations could be controlled [21]. The DM technique of Association helps getting the information of buying pattern of consumers like customer who ordered food bought soft drink. Deviation investigation leads management in retailers to detect the fraudulent act and persons involved in fraud whereas the consumer profiling guides them in making promotional campaign.

(2) *Banking Sector:* The banking sector as one of strongest economic player also gets advantage of KM and DM adoption. Banks record and keep the data in large databases regarding their financial values and consumer related data. The issues in banking operations have resolved with the use of association, information deviation, patterns in consumer investment and finding the market price due to huge data which is not possible to be processed by human. Knowledge management and Data mining application in banking industry guides them to identify the risks involved in investments and returns they will earn [26]. The knowledge management and mining of data technique used to minimize business risks and increase the outcome of trading, because previous patterns alarm the management about risks involved whereas the forecast of previous data help making trade strategies. KM along with DM further enhances CRM, business portfolio and strategies of firm [11]. Data mining provides information of past credit rating of customer which enables Banker to decide approval or rejection of loan. Fraud is top most concern of banking institutions and their management so the uses of deviation analysis and anomaly identification of transactions take the abnormal transactions out of general characteristics that may be due to fraudulent activity.

(3) *Airline industry:* The clustering and association rule procedures help obtaining the knowledge of consumers which is used by airline to provide discount and sent promotional campaigns based on detection of their frequency of travelling.

(4) *Manufacturing organizations:* The manufacturing firms need to be specially informed for the customer needs and wants so that they can modify their manufacturing operations and product developments accordingly. Knowledge management mechanism should be used to identify the most demanded features and ingredients to fulfill customer need. Data mining techniques facilitates the managers in manufacturing firm to identify and understand the patterns of customers claiming warranties and claim refund due to bad performance of product. So that they would be ready to deal with it and ensure to retain consumers with them.

(5) *E-commerce businesses:* The e-commerce websites and online portals like Amazon and Jumia.com also benefitted from Data mining and Knowledge management tools. These firms send recommendations as per customer previous patterns of buying and their queries. Personalized product offerings and promotions of products that is of customer interest is delivered via data mining technique [24].

(6) *Health care sector:* Doctors in hospitals can also use the data mining techniques like clustering which guides them to understand the demographics of infected patients from HIV, Cancer or other deadly viruses. With the help of this knowledge doctors can make better treatment plan, effective diagnostic mechanism and helps researching on these diseases [4].

(7) *Telecom industry:* As one of the fastest grown global industries the Telecom industry also uses Data mining and Knowledge management techniques. Managers uses data of consumers to understand the pattern of their calling, sending text messages and usage of internet data based on point of time [23]. By using mined data and knowledge management design the combo bundles, calling rates, and internet packages promote these using profiling technique. Sequential pattern analysis, data visualization and detection of illegal activities through telecommunication are few significant techniques used in the industry.

5 Challenges of KM and DM Application

5.1 Implementation Challenges of KM in Business

The significance of data mining and Knowledge management is not in doubt but application and implementation of them needs careful considerations. Business firms face issues, problems and challenges while applying the KM with full spirit and benefits [4]. The author in [27] detects the following problems and challenges.

(1) *Technology:* With the growth of technology and innovation the need of change arise in business organizations time to time. Application of KM technique created the need of change in knowledge and data systems in firm. Management faces the challenge of resistance from employees because they are in comfort zone with previous technology. Knowledge management implementation needs training sessions of employees to make them aware with new technology and obtain knowledge from it [27].
(2) *Individuals:* The ability to learn new skills and perception of individual that is major source of creating tacit knowledge is one of key challenge to make information sharing a culture. The employees retain tacit knowledge and stay silent which restrict the knowledge flow. Explicit knowledge must be encouraged along with a culture of sharing problems, issues and information lacks from many organizations creating challenge [27].
(3) *The Shared leadership:* The workers with explicit and tacit knowledge are not promoted to leadership levels in different organizations leads the disappointment in employees. Thus even after many trainings and lectures from management the employees refrain themselves to share key knowledge and information they have [27]. Without effective arrangement of shared leadership the application of KM would be challenged.
(4) *Culture and cost of creating culture:* The knowledge and information in business organization stay in the human mind and technologies only capture, save, arrange and process the information using KM technique [27]. Creating the culture of human resource development and knowledge sharing with skills development of employees are needed to fully implement KM. The cost of these supporting activities sometimes outperforms the benefits.

5.2 Limitations of Data Mining Application

The major limitations in the implementation of Data Mining technique is use of huge data and privacy of consumer. There is need of protecting the rights of stakeholders involved like consumers, employees, and investors. Without fulfilling the ethical considerations and imposed the protection laws in applying Data mining the information obtained and process remain illegal without consent of consumers. Further the representation of data in database increased the limitations because data if not properly managed and save in homogeneous formats it would be failed to process it for decision making need. Data mining technique may not be succeed until assuring the reliability of data because data obtained from reliable sources is integrated with outcome of analysis and evaluation. The cost of buying networking and computing hardware is declining but the price of applications like data mining has been increasing which must be controlled because due to high cost small to medium size firms failed to get benefit of advancement in information technology. The data mining challenges includes dimensionality, iterative mining need, scalability, and techniques integration during mining data like Association and clustering are still challenges for Data mining application in business environment.

6 Conclusions

Emphasize of this paper is to understand the significance of data mining and knowledge management applications in operational improvement of business organizations. Data mining helps managements to create database, extract useful information, understand the patterns from data and discover the data. Knowledge management on other hand has similar significance which enables management to produce knowledge from previous data and information and utilize it optimally to create edge in market. Customers as central focus of all businesses remain key focus of KM and DM applications because the data related to customers is mined, arrange and process to get useful information for strategic decision making. Using Data mining helps businesses to improve business operations, create customer relationships, and improve marketing efforts and promotional campaigns targeting which leads to increased financial profitability.

Advancement of technology has taken the data usage and data capturing to skies which creates the challenges for the implementation of Knowledge management. Adoption of Knowledge management systems in organizations will be a must for those want to compete and survive in competitive markets. Challenges of KM and DM application can be identified and resolved with a team work of IT department and employees at each layer systematically. Since the data mining involved in knowledge generation and creation of data this could lead to clash of these two techniques during operations that increase the need of integration of both these applications along with operations. In future the need of advanced data mining tools will arise because of revolution in data creation and accumulation. Businesses today know how the Data mining works but there needs get skills of using Data mining and knowledge management tools so that continuous improvement in operations, marketing and customer relationships could be ensured.

References

1. Jashapara, A.: Knowledge Management: An Integrated Approach, 2nd edn. Pearson Education, Essex (2011)
2. Bacon, D.: Marketing. In: Klosgen, W., Jan, Z. (eds.) Handbook of Data Mining and Knowledge Discovery, pp. 715–725. Oxford University Press, New York (2002)
3. Needle, D.: Business in context: An Introduction to Business and its environment, 4th edn. Thomson Learning, London (2004)

4. Turban, E., Sharda, R., Aronson, J., King, D.: Business Intelligence: A Managerial Approach. Pearson Education Inc, New Jersey (2008)
5. Fayyad, U.(nd): Optimize Customer Interaction and Profits–with Advanced Mining Techniques. www.watts-associates.com/docs/article/digimine.pdf. Accessed 27 Aug 2017
6. Shaw, M.J., Subrumaniam, C., Tan, G.W., Welge, M.E.: Knowledge management and data mining for marketing. Decis. Support Syst. **31**, 127–137 (2001)
7. Ayinde, A.Q., Odeniyi, O.A., Sarumi, O.A.: Mining parent socio-economic factors to predict students' academic performance in Osun state college of technology, Esa-Oke. Int. J. Eng. Res. Technol. **2**(12), 1677–1683 (2013). ISSN (P): 2278-0181
8. Adetunji, A.B., Ayinde, A.Q., Odeniyi, O.A., Adewale, J.A.: Comparative analysis of data mining classifiers in analyzing clinical data. Int. J. Eng. Res. Technol. **2**(12), 1671–1676 (2013). ISSN (P): 2278-0181
9. Alavi, M., Leidner, D.E.: Review: Knowledge management and knowledge management systems: conceptual foundations and research issues. MIS Q. **25**(1), 107–136 (2001)
10. Hlupic, V., Pouloudi, A., Rzevski, G.: Towards an integrated approach to knowledge management: 'hard', 'soft' and 'abstract' issues. Knowl. Process Manag. **9**(2), 90–102 (2002)
11. Bender, S., Fish, A.: The transfer of knowledge and retention of expertise: the continuing need for global assignment. J. Knowl. Manag. **4**(2), 125–137 (2000)
12. Anand, A., Singh, M.: Understanding knowledge management: a literature review. Int. J. Eng. Sci. Technol. **3**(2), 936–937 (2011). ISSN (P): 0975-5462
13. Carrillo, P.M., Robinson, H.S., Anumba, C.J., Al-Ghassani, A.M.: IMPaKT: a framework for linking knowledge management to business performance. Electron. J. Knowl. Manag. **1**(1), 1–12 (2003)
14. Melton, M.: An evaluation of NTWU's Knowledge Management System on Undergraduates Satisfaction and Academic Performance. Master of Education Thesis, National Taiwan Normal University, Taipei, Taiwan (2010)
15. Swan, J., Scarbourough, H., Hislop, D.: Knowledge management and innovations: networks and networking. J. Knowl. Manag. **3**(4), 262–275 (1999)
16. Debowski, S.: Knowledge Management. Wiley, Australia (2006)
17. Greco, M., Grimadi, M., Hannandi, M.: How to select knowledge management system: a framework to support managers. Int. J. Eng. Bus. Manag. **5**(5), 1–11 (2013)
18. An, X., Wang, W.: Knowledge management technologies and applications: a literature review. In: Proceedings of International Conference on Advanced Management Science, Beijing, China, 9–11 July 2010, vol. 1, pp. 138–141 (2010)
19. Siliwattananusam, T., Tuamsuk, K.: Data mining and its applications for knowledge management: a literature review from 2007 to 2012. Int. J. Data Mining Knowl. Manag. Process **2**(5), 1–12 (2012)
20. Fayyad, U., Piatetsky-Shapiro, G., Smyth, P.: From data mining to knowledge discovery in databases. Mag. Am. Assoc. Artif. Intell. **17**(3), 37–54 (1997)
21. Han, J., Kamber, M., Pei, J.: Data Mining Concepts and Techniques, 3rd edn. Elservier Inc, Walthan, MA (2012)
22. Fayyad, U.M., Piatetsky-Shapiro, G., Smyth, P., Uthurusamy, R.: Advances in Knowledge Discovery and Data Mining. AAAI/MIT Press, Cambridge (1996)
23. Ayinde, A.Q., Adetunji, A.B., Odeniyi, O.A., Bello, M.: Performance evaluation of naive bayes and decision stump algorithms in mining students' educational data. Int. J. Comput. Sci. **10**(4(1)), 147–151 (2013)
24. Fayyad, U., Piatetsky-Shapiro, G., Smyth, P.: The KDD process for extracting useful knowledge from volumes of data. Commun. ACM **39**(11), 27–34 (1996)

25. Cheng, S., Dai, R., Xu, W., Shi, Y.: Research on data mining and knowledge management and its applications in china economic development: significance and trends. Int. J. Inf. Technol. Decis. Making **5**(4), 585–596 (2006)
26. Nakhaizheizadeh, G., Steurer, E., Bartlmae, K.: Banking and Finance. In: Klosgen, W., Jan, Z. (eds.) Handbook of Data Mining and Knowledge Discovery, pp. 771–780. Oxford University Press, New York, NY (2002, 2004)
27. Kalkan, V.D.: An overall view of knowledge management challenges for global business. J. Bus. Process Manag. **14**(3), 390–400 (2008)

An Approach to Verify Heavy Vehicle Driver Fatigue Compliance Under Australian Chain of Responsibility Regulations

Son Anh Vo(✉), Joel Scanlan, Luke Mirowski, and Paul Turner

eLogistics Group, EICT School, University of Tasmania, Hobart, TAS, Australia
{son.vo, joel.scanlan, luke.mirowski, paul.turner}@utas.edu.au

Abstract. Heavy vehicle transportation is vital to the Australian logistics industry. However, it also experiences the highest number of work related accidents. Chain of responsibility regulations introduced by the National Heavy Vehicle Regulator (NHVR) extends obligations and liabilities for safety in heavy vehicle transport to all participants along with supply chains. As a result, finding mechanisms to support compliance with fatigue management rules has become important for the whole industry. The current compliance system is paper-based, and does not produce high quality compliance information and is proving to be expensive for supply chain participants to maintain. This paper presents an automated approach to verify heavy vehicle driver fatigue compliance. Drawing on data from a software tool (Logistics Fatigue Manager) developed by two of the authors, the automated approach deploys signature based detection techniques from Intrusion Detection Systems. The results highlight reduced costs, improved accuracy and speed of compliance verification.

Keywords: Fatigue compliance · Heavy vehicle · Intrusion Detection Systems Rule-based technique

1 Introduction

In Australian logistics transportation, truck based freight has significantly contributed to the Australian economy with around 35% of domestics freight share [1]. However, this freight mode also causes the highest number of work-related accidents in the transport sector [2]. According to investigations by National Transport Insurance [3] on accident causation, fatigue related accidents are estimated to account for at least 20% of all vehicle crashes. Therefore, controlling driver fatigue by regulating maximum work time and minimum rest time is the primary goal of the government regulations.

To deal with this problem in the heavy vehicle industry, the government established the National Heavy Vehicle Regulator (NHVR) in 2009 to be responsible for building and applying safety standards in Australia. The fatigue management rules produced by this body can be viewed as a significant effort to minimise fatigue related vehicle crashes. Furthermore, by applying additional Chain of Responsibility regulations, NHVR have confirmed that safety in heavy vehicles is not the sole obligation of truck drivers but also the legal liability of all entities involved in the supply chain even if they do not directly control vehicles [4]. Table 1 describes the content of the fatigue regulations [5].

K. Arai et al. (Eds.): FICC 2018, AISC 886, pp. 584–599, 2019.
https://doi.org/10.1007/978-3-030-03402-3_41

Table 1. Fatigue management standard hours of work and rest

Checking period	Work time	Rest time
In any period of…	A driver must not work for more than a **maximum** of…	And must have the rest of that period off work with at least a **minimum** rest break of…
5 ½ h	5 ¼ h	15 continuous minutes
8 h	7 ½ h	30 min rest time in blocks of 15 continuous minutes
11 h	10 h	60 min rest time in blocks of 15 continuous minutes
24 h	12 h	7 continuous hours stationary rest time*
7 days	72 h	24 continuous hours stationary rest time
14 days	144 h	2 × night rest breaks# **and** 2 × night rest breaks taken on consecutive days

*Stationary rest time is the time a driver spends out of a heavy vehicle or in an approved sleeper berth of a stationary heavy vehicle.
#Night rest breaks are 7 continuous hours stationary rest time taken between the hours of 10 pm on a day and 8am on the next day (using the time zone of the base of the driver) or a 24 continuous hours stationary rest break.

The Standard Hours regulations in Table 1 show that there are different timeframes used to control maximum work time and minimum rest time of heavy vehicle drivers. For example, in any period of 5.5 h, the maximum work time is 5 h 15 min and minimum rest time is 15 min; or in any period of 24 h, the maximum work time is 12 h and the minimum rest time is 7 h.

To assist logistics firms to comply with these regulations, the NHVR has produced fatigue guidelines. The focus of these guidelines is the work diary report that each driver has to manually fill in after every shift. To complete this report, each driver needs to clearly understand how to use the work and rest time calculation formulas provided by the NHVR body. Additionally, there are other constraints relating to the compliance procedures such as carrying the form while driving and keeping the record in the vehicle for 28 days. This scheme obviously has several disadvantages that result in problems for the management of workloads including overloading, inaccurate information, cost and time wasting.

The research described in this paper has developed and tested an automatic solution for fatigue compliance verification. In this research, a new method of verification is presented derived from principles in network security: signature-based intrusion detection. This system presented has used input data produced by a software tool called the Logistics Fatigue Manager (LFM) (an electronic work diary application) that was developed by researchers in the eLogistics Research Group (eLRG) at University of Tasmania. This paper will firstly examine the research background in both fields: fatigue management and intrusion detection. It will then describe the method of building the system, and finally present the results and discussion.

2 Background

The background of this research covers three main topics: Fatigue Management, Intrusion Detection and the LFM Electronic Work Diary.

2.1 Fatigue Management

In this section, we conduct a survey on typical fatigue detection techniques as well as electronic fatigue compliance support systems.

(1) Fatigue detection techniques

To date, the driver fatigue management field has attracted a significant number of research projects developing early warning methods. Many of these concentrate on the outside physiological responses of drivers such as: blinking eyes or head nodding to predict fatigue and sleepiness levels. Specifically, blink frequency and duration have been shown to be the key indicators in studies by Sirevaag and Stern [6], Stern et al. [7], Summala et al. [8] and Schleicher et al. [9]. In these studies, there were a range of parameters relating to blink eyes such as blinking frequency, flurry of blinks, timing with respect to information-processing demands and blink closure duration. While blinking eyes is considered as the most promising parameter to measure driver's tiredness, head nodding was also shown to be a useful signal in measuring fatigue. Bergasa et al. [10] developed a real-time system for monitoring driver vigilance that relies on the combination of six parameters including: percent eye closure, eye closure duration, blink frequency, nodding frequency, face position, and fixed gaze. No-Nap Alert is a commercial product taking advantage from head nodding factor [11]. This smart device fits over the driver's ear and immediately generate awakening alarm if driver's head in a 'sleepy position'. The work described here demonstrates the research activity in understanding fatigue measurement in transportation.

(2) Electronic fatigue compliance support systems

While driver fatigue detection has been an interesting topic in many studies, this section is looking at electronic solutions that automatically verify driver's compliance based the existing fatigue regulations of government. From law adherence perspective, apparently, the paper-based work diaries introduced by the NHVR have multiple drawbacks in relation to accuracy and workload. To combat these issues, an Electronic Work Diary project funded by New South Wales (NSW) government was initiated to develop an automatic verification solution [12]. This is the next step of the Australian Government shifting from manual management to electronic systems for controlling fatigue compliance. The proposed system design aims to provide services for drivers, transport operators and authorized officers. It is expected to enhance fatigue management rules compliance, deliver more efficient management and reduce fatigue related heavy vehicle crashes. Unfortunately, this project is still under development and is not anticipated to be deployed before the end of 2017. Additionally, we are not aware any research using rule-based techniques for the fatigue compliance verification so far. The research conducted in this paper aims to build a system to examine these issues in advance of that deployment. The technology foundation for this system is inspired by the rule-based principle of Intrusion Detection System (IDS) in Network Security. The next section will explore how this system functions along with its architecture.

2.2 Intrusion Detection System

(1) Overall architecture

IDS can be described as an alarm system used for warning of a network attack occurring in a computer system. It aims to monitor all the network traffic going in and out from network devices such as computers, routers and switches. In Fig. 1, IDS have been defined as being comprised of three main components: Data Collection, Detection and Decision Engine.

Data Collection: The first process undertaken in IDS is to refine raw audit data to required input data for detection engine. The input data can be packet data, system logs or user profile data. To process the raw data, we apply different kinds of techniques like cleaning, integrating, transforming and reducing.

Detection Engine: In the overall IDS architecture, the detection engine is the core component that directly impacts the final outcomes of the system. This component includes the detection algorithms to process input data and create output knowledge of system state; i.e. whether there is currently an attack against the system. Although the algorithms used in different systems are diverse, basically, they fall into two main approaches: signature-based and anomaly detection. A survey on three studies of Beigh [13], Bhuyan et al. [14] and Gogoi et al. [15] shows that, the signature-based IDS has four techniques including State Modelling, Expert System, String Matching and Simple Rule-Based. In the anomaly based IDS, there are two main groups of techniques: machine learning and statistical approaches. Each group has many techniques within to then to choose which is most appropriate to the practical situation.

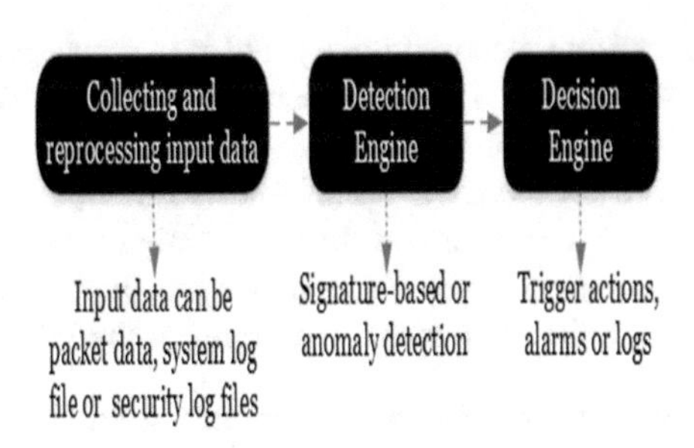

Fig. 1. IDS architecture.

However, in the detection engine, false positive and false negative are often the big challenges for any IDS implementation. Mokarian et al. [17] emphasise that false positives are an unavoidable problem in any type of IDS. Ho et al. [18] also state that there is no perfect solution for IDS to completely eliminate these issues. However, how to keep this error rate to a minimum is always the top priority of any IDS implementation.

Decision Engine: This component will respond to the results of the detection engine. Actions trigged by this component can be event logging, a warning message for monitoring and forensics operations or even preventing actions for instant protection.

(2) Signature-Based detection

A signature-based IDS can be described as a detection mechanism that compares input data attributes against a signature database. This approach targets known attacks; as the attacks are already recognized and analysed by security experts and placed into the signature database. Bhuyan et al. [14] emphasize that this approach depends heavily on well-defined knowledge of intrusion behaviours.

In signature based IDS, while state modelling and expert system methods are more relevant to system design, string matching and rule-based techniques focus on core algorithm applied in the detection engine. In string matching, a set of strings in pre-defined rules will be matched with a data payload to identify threats. This method can be seen as the heart of modern IDSs because most known attacks can be identified by strings or substrings in data content [19]. However, it also faces many challenges such as performance overloading, bottlenecks in a high throughput network environments or high storage consumption [20].

Rule-based is a simple method that completely reflects all requirements specified in known attack specifications under the form of "if…then…". Suspicious behaviour which matches these conditions will result in an alert. Furthermore, this technique is very suitable with situations in which system administrators want to define rules based on their current system parameters rather than only known attack specifications. This element can be seen as a strong point of rule-based solution. In a study of hybrid intrusion detection system design for computer network security, Aydın et al. [21] used this technique to define rules with their testing system parameters.

2.3 The Logistics Fatigue Manager

The LFM for the Driver is an electronic diary and was developed for WorkSafe Tasmania by researchers within eLRG at University of Tasmania. The aim of the project was to automate the collection of work diary data and estimate the fatigue status of drivers using scores. LFM for the Driver is an Android mobile application that enables the driver to monitor, manage and support their own fatigue management requirements, and take steps to address their compliance under the work diary and general duty requirements. This system directly benefits the driver in the following ways:

Reporting work and rest hours: the application uses simple Play/Pause/Stop buttons to record work/rest hours (minimising driver reading/writing needs), and records time/location using the mobile handsets GPS and clock features for high reporting accuracy. Figure 2 shows three screenshots of the application.

Fatigue assessment: LFM for the Driver generates an objective measure of fatigue based on driver reported sleep time and rest time. This assists the driver to understand the relationship of sleep and rest to fatigue, and can be used as an objective tool to manage workload and risk.

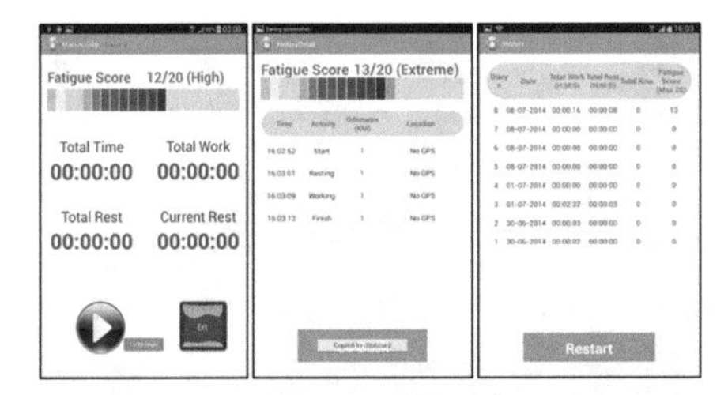

Fig. 2. Screenshots of LFM for the driver.

The output dataset of this system contains many useful attributes such as Working Day Date, Activity Types, Time of Activity Type, Fatigue Scores and GPS data. This information will be the prerequisite condition for verifying the fatigue management rules compliance.

3 Method

The principles of signature-based IDS in detecting known network attacks can be applied to the problem of fatigue compliance verification. Specifically, to classify given network traffic as a known attack, the mechanism matches it with existing rules that are defined based around known attacks. If the traffic pattern matches the rule then it will be declared as an incidence of this known threat (rule match). This approach is similar to the situation in which provisions of the fatigue management rules play the same role as known attacks and transport data is similar to network traffic. By using the rule match principle, the fatigue management rules compliance of drivers can be checked.

A signature-based detection mechanism is an effective way of conducting rule checking. In the context of fatigue management, the legislation regarding work and rest times is the basis for the rules. Particularly, it is organized as set of provisions of max work time and min rest time corresponding to each timeframe such as 5.5 h or 8 h. Thereby, a rule-base technique will perfectly reflect every single clause of the rule under the form "if …then". Through this approach, the ruleset in the signature database will be efficient and easy to understand and updated by system developers when the legislation changes these rules.

3.1 The System Design

The system developed in this research was inspired by the architecture of IDS and its overall architecture, containing three main components as illustrated in Fig. 3.

This architecture shares the same major components as IDS. Specifically, the pre-processing data component is responsible for collecting and processing the driving datasets. This system uses the input dataset produced by the LFM for the Driver application. This dataset contains the diary sheet and GPS data. The second component, the Detection Engine, uses a signature based approach with rules to verify the fatigue compliance. The third component relies on the results of the detection component to create a log file for verification results.

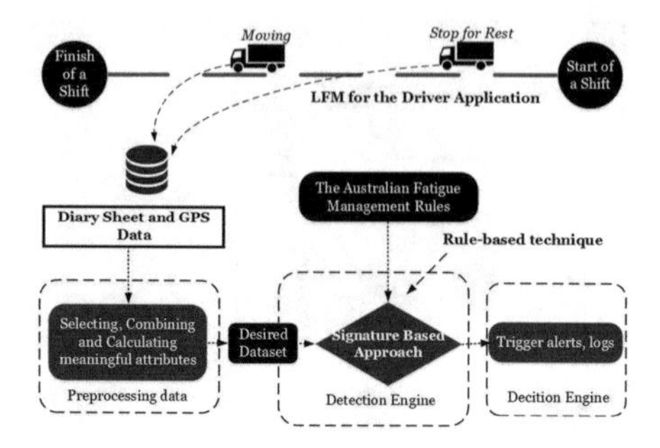

Fig. 3. The architecture of the fatigue compliance verification system.

3.2 Implementation

The building process of the system involved four steps:

(1) Defining key concepts of the fatigue management rules

To be able to ascertain compliance to the regulations, we need to define several concepts to enable thorough checking of diary events. Four key concepts are here defined to enable rule creation:

Checking Period: This period is a specific timeframe in which drivers are checked whether they will obey the maximum work time and minimum rest time. Basically, there are two types of checking period. Type 1 covers checking periods inside a shift including 5.5 h (330 min), 8 h (480 min) and 11 h (660 min). In these timeframes, rest time and work time occurring in a shift with the events used to verify the rule compliance. Type 2 contains checking periods between shifts involving 24 h, 7 days and 14 days. In these timeframes, the length of a whole working day and rest time occurring between two consecutive working days will be the events used to verify the rule compliance.

Work Time: describes the driving time, and other work time spent by the driver such as loading, unloading, servicing or repairing [22].

Rest Time: is the break time between work activities in a shift or between shifts. Rest time as regulated by the NHVR fatigue management rules must meet minimum values.

Checking Interval: The concept of "any period of" specified in the fatigue management rules states that the fatigue compliance of driver will be periodically verified along with each shift and their working progress. By this regulation, a driver may breach the rule one or multiple times for each checking period.

(2) Processing raw data

To process the raw driving data produced by the LFM, we rely on the work and rest concepts defined in the previous section. Table 2 shows the selected attributes in the raw data.

Table 2. Selected attributes in the raw data

Data	Selected attributes	Format	Description
Diary Sheet	WorkingDateDay	mm/dd/yy	Date of shift
	Activity Type	Start, Rest, Unrest and Finish	Activity types during a shift
	StartTime	hh/mm/ss	Time at each activity type

In addition to these attributes, GPS data was collected during each shift and also plays an important role for building a secondary verification layer. It aims to confirm whether the rest time declared by the driver is accurate or not based on a maximum moving distance allowed during rest time. Based on these attributes, the new attributes used for the fatigue compliance verification system are described below:

Working Day (also called shift): a working day of driver always begins with the "Start" activity type and ends with "Finish" activity type through the DFM for Driver application.

Work Time: the period between Start and Rest, Unrest and Rest, Unrest and Finish as indicated by the driver through the application. Between working days, it is the whole period of a shift.

Rest Time: There are two sub concepts of rest time based on the requirements of the fatigue management rules and the nature of rest time:

(a) *Practical Rest Time*: It is the rest time declared by driver on the LFM. In a working day, it is the period between Rest and Unrest. For example, practical rest time can be 3 m, 16 m or 31 m. Between working days, it is the period between finish of a shift and start of the next shift. For example, practical rest time can be 6.5 h, 8 h or 10 h.
(b) *Valid Rest Time*:

In a working day: valid rest time is the practical rest time that must satisfy two conditions:

Condition 1- Requirement of the fatigue management rules: minimum rest time must be a block of 15 min. For example, valid rest time will be 0 if practical rest time is smaller than 15 min.

Condition 2- The nature of rest time: It means that during rest time, the driver should stop or only go for walk (no other driving). To check this condition, the distance calculated by GPS data (practical movement) must be smaller or equal to the distance calculated by multiplying a half of practical rest time with 83 m/min (average walking speed). This condition makes sure that driver does not drive during rest time.

Between working days: Valid rest time is practical rest time that meet 7 continuous hours, or 24 continuous hours or night rest break depending on type of checking period.

As such, valid rest time will be the criterion for verifying the fatigue compliance. Table 3 shows a working day sample of a driver after processing the raw driving data. In this example, there are two special cases of rest time highlighted in bold. In the first case, the practical rest time is 29 min and the moving distance during this time calculated by GPS data is 13 m. It means that, this rest time satisfies the moving condition and it is rounded to 15 min. In the second case, the driver breached moving distance condition as the GPS data show that he moved around 37 km during 38 min of rest time. Therefore, the valid rest time in this case is zero.

(3) Defining rule checking principles

To guide the verification system, we need to identify when rule checking starts and stops along a shift and the whole collected data.

In a working day: a checking interval is activated at any time Start or Unrest action is selected in the Logistics Fatigue Manager application. It means that, the system will run the rule check periodically if it realizes that driver starts a new work session.

Figure 4 explains this principle. To stop checking, there are two cases: the first case happens when time at the start of a checking interval adding with checking period is greater than time at finish of the shift. In contrast to the first case, the second case occurs when the system runs the rule check until it reaches the end of the shift.

Table 3. A working day sample of a driver after processing the raw data

Working day	Work time (minute)	Practical rest time (minute)	Moving distance (meter)	Valid rest time (minute)
7/9/2015				
7/9/2015	287			
7/9/2015		**29**	**13**	**15**
7/9/2015	0			
7/9/2015		3		0
7/9/2015	0			
7/9/2015		0		0
7/9/2015	0			
7/9/2015		1		0
7/9/2015	267			
7/9/2015		0		0
7/9/2015	0			
7/9/2015		**38**	**37.028**	**0**
7/9/2015	82			

Between working day: Between shifts, a checking interval is activated at the start point of a working day as shown in Fig. 4. To stop checking, there are two cases. The first case happens when time at start of a working day adding with checking period is greater than time at the last row of collected data. In contrast to the first case, the second case occurs when the system runs the rule check until it reaches the last row of the collected data.

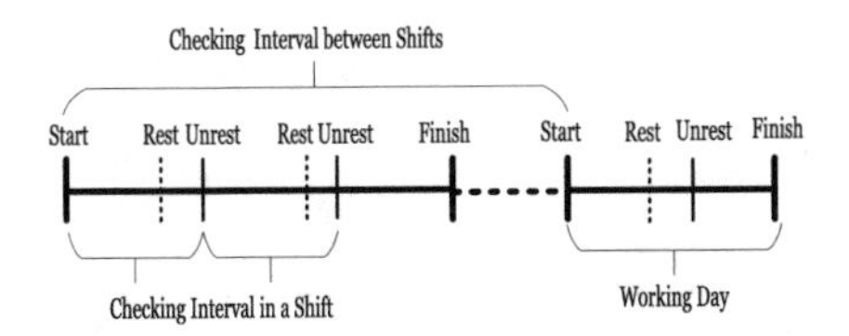

Fig. 4. Checking interval in a shift and between shifts.

(4) Rule construction

To enable the detection mechanism to function, we need a ruleset. Ruleset construction aims to exactly reflect the requirements of the fatigue management regulations. The goal of this task is to cover all possible cases of the regulations so it can fit every practical situation in driving activity. Rulesets are built for two checking period groups:

Ruleset for periods of 5.5 h, 8 h and 11 h: Based on the specification of the checking periods in the fatigue management rules and the raw dataset, the case design is developed as shown in Fig. 5. In this case design, case 3.2 is divided into two sub cases as the rest time and work time calculation relies on which action (Rest or Unrest) happens before the end of checking period.

Ruleset for periods of 24 h, 7 days and 14 days: The major difference in this checking period group compared with the group in a Shift is that the summary of max work and min rest is less than to the whole checking period. For example: in the checking period of 24 h, max work is 12 h and min rest is 7 h. Therefore, the case design of these checking periods is different and this is shown in Table 4. In this case design, Case 2 is analysed in two sub cases depending on which action (Start or Finish) happened before the end of checking period. This is because it decides how work time and rest time are calculated.

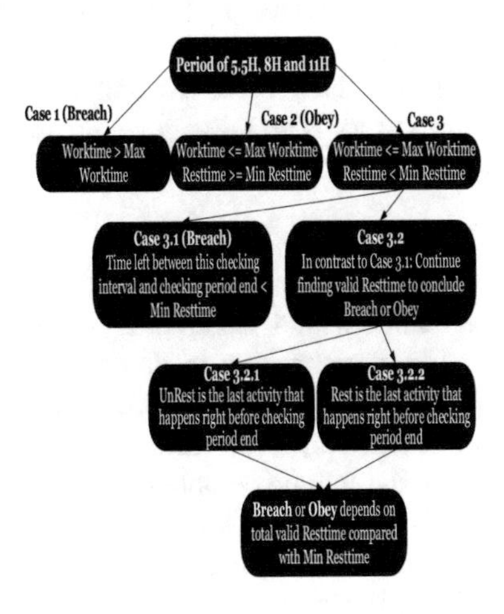

Fig. 5. Case design of checking periods in a shift.

Table 4. Case design of checking periods between shifts

Case	In a checking interval		Description
	Work time	Valid rest time	
1	> max work time	Any	In this case, driver breaches the rules without caring about rest time
2	<= max work time Need to continue counting next work time that may happen before checking period end	Depend on total work time	If total work time > max work time. This is the same case 1. In contrast, rest time needs to be taken into account

An important concern in this ruleset is about determining valid night rest break. Specifically, there are two situations to define this parameter.

Situation 1: Finish of a working day happens between 1am and 10 pm. In order to find a valid night rest break, time at start of next working day must not begin before 5 am of next day.

Situation 2: Finish of a working day happens after 10 pm or before 1am. In this case, date/time of that working day adding 7 h must be before date/time at start of the next day.

With all the steps described, we code the system using the Visual Basic programing. The next section will present the result of this system in testing environment.

4 Results

There were four datasets produced by LFM for the Driver application that were produced by four drivers: TDA, TDB, TDC and TDD. However, only the datasets of drivers TDA and TDB were of a high enough standard to be usable in this study. The other datasets contained too many errors which were the result of the drivers using the application incorrectly. As a result, the input datasets of TDA with 4 working days and TDB with 7 working days were used. With these given datasets, we test and present the result in two groups: in a shift and between shifts.

4.1 In a Shift

The checking result of driver TDA with 4 working days is processed by the system as shown in Table 5. In this table, the checking result of each period includes two columns. The first column, checking interval, shows how many times the system ran the rule checking in a shift. In this column, if the value is No Check, it means that in that shift, the system did not verify the fatigue compliance for this checking period as it is greater than the total time of the shift. In such cases, the checking result is Obey. The second column shows the compliance result: Breach or Obey. In a shift, a driver can breach the rule once or multiple times.

4.2 Between Shifts

In order to get the checking result of period 24 h, 7 days and 14 days, the dataset to be collected for this period has to be greater than the checking periods. In our situation, only the dataset of driver TDB had working enough days to support testing checking period of 24 h (h) and 7 days (d). Table 6 shows checking results for this driver. The format of this result in this table is similar to the checking result in a shift. With 7 working days in the dataset of driver TDB, there are 6 checking intervals for period 24 h and 1 checking interval for period 7 days. This result shows that, this driver obeyed the rules in all the cases. No Check value shows that, from that shift to the last shift of collected dataset, the total time is not enough to verify that checking period. In such cases, the compliance result is always Obey.

Obviously, these limited input datasets are not enough to validate all cases of fatigue management rules. However, the core method of signature based approach can still be established for any rule checking in the future when more data has been gathered. This belief will be proven in Result Evaluation part, when the high accurate checking results are presented for cases in the scope of given dataset.

Table 5. Checking result in a shift of driver TDA

Date/ Time	*CI of Period 5.5 h	#CR of Period 5.5 h	CI of Period 8 h	CR of Period 8 h	CI of Period 11 h	CR of Period 11 h
7/9/15 1:39 PM	5	Breach 4 times	1	Breach	1	Breach
7/10/15 12:48 PM	2	Breach	1	Obey	1	Breach
7/13/15 12:40 PM	2	Obey	1	Breach	No Check	Obey
7/14/15 1:02 PM	2	Obey	1	Obey	1	Obey

*CI: Checking Interval. # CR: Checking Result

4.3 Result Evaluation

This section evaluates the effectiveness of the signature based implementation in verifying the fatigue compliance of driver. In order to perform this task, the testing result will be matched against actual result of raw dataset that was manually classified. Then, false positive and false negative rates values are calculated to evaluate the effectiveness of the implementation. Both of the results will be plotted in matrix table.

In a shift: the total checking cases is 33 (11 days multiply with three types of checking periods). Table 7(a) shows 33 cases scanned, the actual result and checking result completely matches to each other. It means that the error rate of this task is zero. Most of the breach cases belong to driver TDA as he breached moving condition during rest time. Therefore, the rest time he took is invalid.

Table 6. Checking result between shifts of driver TDB

Date/ Time	*CI of Period 24 h	#CR of Period 24 h	CI of Period 7d	CR of Period 7d
9/30/14 3:14 AM	1	Obey	1	Obey
10/1/14 3:07 AM	1	Obey	No Check	Obey
10/2/14 3:37 AM	1	Obey	No Check	Obey
10/3/14 3:55 AM	1	Obey	No Check	Obey
10/4/14 3:15 AM	1	Obey	No Check	Obey
10/6/14 2:02 PM	1	Obey	No Check	Obey
10/7/14 2:39 PM	No Check	Obey	No Check	Obey

*CI: Checking Interval. # CR: Checking Result

Between Shifts: the total checking cases is 10 based on dataset of driver TDB (9 cases for period of 24 h, 1 case for period of 7d). Table 7(b) shows that, in 10 cases, both checking result and actual result are completely matched each other. Therefore, the false positive and false negative in this task are also zero.

Table 7. Result evaluation

		Checking result	
		Breach	Obey
(a) In a shift			
Actual result	Breach	8	0
	Obey	0	25
(b) Between shifts			
Actual result	Breach	0	0
	Obey	0	10

5 Discussion

The high accuracy of the results achieved in the implementation of this system based on the current dataset has proven that the rule based method inspired by Signature-based IDS is an effective solution. This method provides a precise and simple way to verify driver compliance to NHVR regulations. As such, the system can be seen as viable solution to replace the current paper-based fatigue management processes of NHVR. The success of this implementation is grounded in the significant contribution of previous research: Logistics Fatigue Manager application which provided the necessary electronic diary entries for verifying fatigue compliance. The two primary benefits of this system are highlighted below:

Reducing management costs: The system can reduce the workload for transport managers and drivers in relation to fatigue compliance. This is achieved by automatically storing and processing driving data and then automatically verifying compliance. This increases the amount of time for other tasks, lowering the overall costs within the supply chain.

Improving accuracy and speed of compliance verification: Reporting work and rest by using paper-based diary sheet can lead to inaccurate information through driver errors or misreporting. The same problem can occur for logistic managers in checking fatigue compliance. The system firstly avoids these drawbacks by automatically computing these values with the given definitions. Additionally, GPS data collected in each shift is used to create a secondary verification layer on rest time declared by drivers. This contributes to the effectiveness of the system by allowing the reported data to be validated. In addition to the increase in data quality, the system also speeds up the compliance verification process in comparison to manual checking.

Supporting decision making: The data output by this system will certainly be a valuable source to conduct various kinds of statistical reports. For example, logistics managers could quickly determine the number of breach times per checking period, trip, driver, month or year in comparison with paper based system. With this enhanced ability, logistics firms have greater flexibility to adjust working schedule, truck fleet and human resource to achieve the best results for both business profit goals and work safety.

Nevertheless, the implementation of the system encountered several challenges in relation to the input dataset and the unforgiving nature of the legislation in its practical application. The first issue identified was incorrect interaction by drivers with the LFM application resulting in some measurement periods having an extremely short time. This problem would boost the potential rule breach of drivers unintentionally due to user error. The application has been improved to increase usability as a result of this finding. The second issue reflects the inflexibility in the fatigue regulations. For example, the minimum rest time during a shift must be block of 15 min. This rule can waste rest time of drivers if it just falls short of that amount. Therefore, if the electronic system is applied, the verification result is easily switched from Obey to Breach with a few seconds difference. To enable logistic managers be flexible and fair in considering the actual compliance results, the system can show the missing time (minute, hour or day) together with "Breach".

6 Conclusion and Future Work

The system built in this research enabled fatigue compliance outcomes to be established based on the data recorded by the LFM for the Driver application. This research enables the drawbacks manual handling and checking of paper based records as required by the NHVR guidelines to be eliminated via an automated solution. The automated solution supports the compliance process of any entity in supply chains who are constrained by Chain of Responsibility regulations of NHVR.

This research was achieved through an innovative automatic method for the verification that was inspired by signature-based IDS principles in Network Security. It has shown that the applicability of IDS in transportation environment is completely possible. Particularly, the ruleset developed by rule-based technique in signature database can be applied for most situations occurring during driving activity with the highly accurate checking results.

However, the main limitations of this study were the size of the dataset. This issue presented both challenge during the study and an opportunity for further research moving forward. Specifically, we need to collect more data from different truck drivers over a longer period so it can cover more situations in the driving operation such as short and long moving distances, and day and night shifts.

Additionally, the verification mechanism of the system currently occurs after the fact. The system built in this research only looks at the offline data to do analysis and attain result once the shift has ended. This is primarily only of benefit to management verification tasks. Therefore, enabling this system to support real-time rule checking would be of significant use to drivers. This would enable them to track their work and rest compliance within a shift in a flexible way to satisfy the fatigue regulations.

References

1. Department of Infrastructure and Regional Development-Australian Government. Freightline 1—Australia freight transport overview (2014)
2. Safe Work Australia. Notifiable Fatalities Monthly Report (2015). http://www.safeworkaustralia.gov.au/sites/swa/about/publications/pages/notifiedfatalitiesmonthlyreport
3. National Transport Insurance. Major Accident Investigation Report (2013)
4. NHVR. Chain of Responsibility (2015). https://www.nhvr.gov.au/safety-accreditation-compliance/chain-of-responsibility
5. NHVR. Basic Fatigue Management (2015). https://www.nhvr.gov.au/safety-accreditation-compliance/fatigue-management/work-and-rest-requirements/standard-hours
6. Sirevaag, E.J., Stern, J.A.: Ocular measures of fatigue and cognitive factors. In: Engineering Psychophysiology: Issues and Applications, pp. 269–287 (2000)
7. Stern, J.A., Boyer, D., Schroeder, D.: Blink rate: a possible measure of fatigue. Hum. Factors **36**, 285–297 (1994)
8. Summala, H., Hakkanen, H., Mikkola, T., Sinkkonen, J.: Task effects on fatigue symptoms in overnight driving. Ergonomics **42**, 798–806 (1999)
9. Schleicher, R., Galley, N., Briest, S., Galley, L.: Blinks and saccades as indicators of fatigue in sleepiness warnings: looking tired? Ergonomics **51**, 982–1010 (2008)
10. Bergasa, L.M., Nuevo, J., Sotelo, M.A., Barea, R., Lopez, M.E.: Real-time system for monitoring driver vigilance. IEEE Trans. Intell. Transp. Syst. **7**, 63–77 (2006)
11. The No Nap. The NoNap Advantage (2017). http://www.thenonap.com/index.html
12. New South Wales Government. Operational Pilot of Electronic Work Diaries and Speed Monitoring Systems (2013)
13. Beigh, B.M.: A new classification scheme for intrusion detection systems. Int. J. Comput. Netw. Inf. Secur. (IJCNIS) **6**, 56 (2014)
14. Bhuyan, M.H., Bhattacharyya, D.K., Kalita, J.K.: Network anomaly detection: methods, systems and tools. IEEE Commun. Surv. Tutor. **16**, 303–336 (2014)

15. Gogoi, P., Bhattacharyya, D., Borah, B., Kalita, J.K.: A survey of outlier detection methods in network anomaly identification. Comput. J. **54**, 570–588 (2011)
16. McHugh, J., Christie, A., Allen, J.: Defending yourself: the role of intrusion detection systems. IEEE Softw. **17**, 42–51 (2000)
17. Mokarian, A., Faraahi, A., Delavar, A.G.: False positives reduction techniques in intrusion detection systems-a review. Int. J. Comput. Sci. Netw. Secur. (IJCSNS) **13**, 128 (2013)
18. Ho, C.-Y., Lin, Y.-D., Lai, Y.-C., Chen, I.-W., Wang, F.-Y., Tai, W.-H.: False positives and negatives from real traffic with intrusion detection/prevention systems. Int. J. Future Comput. Commun. **1**, 87 (2012)
19. Tuck, N., Sherwood, T., Calder, B., Varghese, G.: Deterministic memory-efficient string matching algorithms for intrusion detection. In: Twenty-third Annual Joint Conference of the IEEE Computer and Communications Societies, INFOCOM 2004, pp. 2628–2639 (2004)
20. Lu, H., Zheng, K., Liu, B., Zhang, X., Liu, Y.: A memory-efficient parallel string matching architecture for high-speed intrusion detection. IEEE J. Sel. Areas Commun. **24**, 1793–1804 (2006)
21. Aydın, M.A., Zaim, A.H., Ceylan, K.G.: A hybrid intrusion detection system design for computer network security. Comput. Electr. Eng. **35**, 517–526 (2009)
22. ComLaw-Australian Government. Road Transport Legislation – Driving Hours Regulations (2006). https://www.comlaw.gov.au/Details/F2006L00250

US^2: An Unified Safety and Security Analysis Method for Autonomous Vehicles

Jin Cui(✉) and Giedre Sabaliauskaite

Centre for Research in Cyber Security (iTrust),
Singapore University of Technology and Design, SUTD, Singapore, Singapore
{jin_cui,giedre}@sutd.edu.sg

Abstract. Autonomous Vehicles (AVs) are security-critical systems, and safety is primary goal for AVs. The high degree of integration between safety and security introduces new problem: how to systematically analyse safety and security? In this paper, we propose an Unified Safety and Security analysis method (US^2), which uses a simple quantification scheme to assess safety hazards and security threats simultaneously. US^2 is a useful tool for safety and security requirements specification and selection of countermeasures. Example of US^2 application is included to highlight the strengths of the proposed method.

Keywords: Autonomous vehicles · Safety · Security · ISO 26262
SAE J3016 · SAE J3061

1 Introduction

Due to the advancement in the development of complex electronic control units, vehicles are expected to achieve more performance, such as driving automation. Autonomous Vehicles (AVs) are considered as vehicles equipped with more environment sensors and network communication units than conventional ones. These sensors and communication units help AVs to achieve driving automation by means of perception, navigation and path planning, and continuous motion control.

Driving Automation System (DAS) are proposed by International standard SAE J3016 [1], which is the main control unit in AV to achieve driving automation. In addition, AVs can be classified by the completed driving tasks into different Driving Automation Levels (DALs) [1]. The more complex driving tasks are performed by the AV, the higher DAL is assigned. Actually, in Singapore, AVs are taken onto the streets in the form of taxis [2]. We can imagine that such vehicles will fundamentally change our transportation system in the near future, and will provide additional benefits, such as fuel consumption reduction and safety improvement.

However, the high degree of integration between safety-critical and security-critical nature of AVs presents new challenges: how to support safety and security

K. Arai et al. (Eds.): FICC 2018, AISC 886, pp. 600–611, 2019.
https://doi.org/10.1007/978-3-030-03402-3_42

development in parallel? how to assess safety hazards and security threats simultaneously? Traditionally, we assume that safety-critical systems are immune to security risks. But as presented in [3], increasing interlaces between safety and security in AVs make this assumption no longer valid. Thus we must consider the alignment of safety and security in early development phases. The safety and security analysis methods need to be performed together in an integrated way. Furthermore, AVs require appropriate systematic approaches to support safety and security development, which consider different driving automation levels.

We present an Unified Safety and Security Analysis method (marked as US^2) for AVs. Encompassing DALs [1], US^2 uses a simple quantification scheme to evaluate safety hazards and security threats in parallel, and derive safety and security requirements efficiently. In addition, it is helpful for designing safety and security countermeasures. Moreover, as an unified analysis approach, US^2 enables to study the impact of DAL on safety and security concepts.

The rest of the paper is organized as follow: we introduce the related works and information in Sect. 2, and describe DAS and DALs in Sect. 3. Unified safety and security analysis method is explained in Sect. 4. An application of US^2 for autonomous vehicle is presented in Sect. 5. Finally, we conclude our work in Sect. 6.

2 Related Work

Safety and security of automotive industry are challenging and important research areas. Thus, many researchers and industrial experts have recently made effort to advance research on safety and security issues.

Automotive safety integrity level (ASIL) is a key outcome of ISO 26262 [4], which is a standard for functional safety of road vehicles. ASIL is used to determine the criticality level of the system and to define safety requirements and measures that should be applied for the following phases in lifecycle. To determine the ASIL, Hazard Analysis and Risk Assessment (HARA) is performed. Firstly, potential hazards, which endanger the system, are identified. Then, these hazards are quantified according the severity (S), probability of exposure (E), and controllability (C). The final step involves a formulation of high level safety requirements known as safety goals (details can be seen in [4] part 3 Annex B). In the classification of ASIL, ASIL A is the lowest safety integrity level and ASIL D the highest one. Hazards that are identified as QM (quality management) do not dictate any safety requirements.

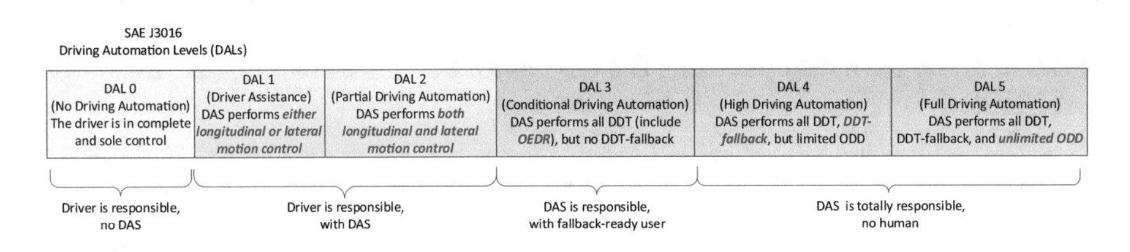

Fig. 1. Driving automation levels.

SAHARA (Security-Aware Hazard Analysis and Risk Assessment) [5] method has been recently proposed for automotive industry to evaluate security threats. It defines security level (i.e., SecL) using attacker's knowledge, resources and the threat criticality. It uses STRIDE [6] (STRIDE is an approach that uses a technique named threat modeling to review system designs in a methodical way, which includes six types of threats: Spoofing, Tampering, Repudiation, Information disclosure, Denial of service attacks, and Elevation of privilege[1]) to analyze the threats by SecL. The quantification of SAHARA can be used to determine impact of specific threats and check whether those threats have impact on safety goals. However, SAHARA has been developed for conventional vehicles, and is not directly applicable for AVs since it does not consider different driving automation levels.

SAE J3061 [7] is a standard for cybersecurity of cyber-physical vehicle system, which recommends several Threat Analysis and Risk Assessment (TARA) approaches [8] in automotive industry, such as EVITA [9] and HEAVENS [10]. EVITA deals with on-board network protection, which is a suitable approach for security concept evaluation. In EVITA, security threats are classified from separate views: operational, safety, privacy and financial. However, this separation requires more effort on discussion, but does not result in a big difference in the resulting risk level as compared to SAHARA [11]. HEAVENS analyzes threats based on STRIDE approach (See footnote 1) and ranks the threats based on threat level, impact level and security level. Similarly to EVITA, HEAVENS requires a lot of work to analyze and determine the levels of individual threat.

3 Driving Automation System

Driving automation functions are collectively called *Dynamic Driving Task* (DDT) [1], which includes all real time operational and tactical functions required to operate a vehicle in on-road traffic. Driving Automation System is the hardware and software that are collectively capable of performing the entire DDT on a sustained basis, which is the key property that can replace human driver for AV. Different level corresponds different DDT, which includes the operational and tactical functions. Thus, the levels of driving automation (DALs) are also classified by the functions, as shown in Fig. 1:

- DAL 1: the DAS performs either the longitudinal or the lateral vehicle motion control.
- DAL 2: the DAS performs both the longitudinal and the lateral vehicle motion control.
- DAL 3: the DAS also performs the Object and Event Detection and Response (OEDR).
- DAL 4: the DAS also performs DDT-fallback.
- DAL 5: the DAS is limited by an Operational Design Domain (ODD).

[1] https://msdn.microsoft.com/en-us/library/ee823878(v=cs.20).aspx.

When a DDT fails, the response to either re-perform the DDT or reduce the risk of crash is considered as *DDT-fallback*. The user who is able to operate the vehicle during DDT-fallback is a *Fallbackf-ready User* [1]. For example, when adaptive cruise control on a car experiences a system failure that causes the feature to stop performing its intended function. The fallback-ready user will take control of the vehicle if such failure occurs.

ODD is a specific operating domain in which an automated function or system is designed to properly operate, including but not limited to roadway types, speed range, geography, traffic, environmental conditions (weather, daytime/nighttime, etc.), and other domain constraints [12]. For example, we can design a ODD like: on express way, the vehicle uses a speed lower than 35 km/h driving in the daytime only.

In DAL 1 and 2, which are low automation levels, DAS only assists the driver. Such system is called advanced driver assistance system [13]. In high automation levels (DAL 4 and 5), DAS can totally replace human (driver or fallback-ready user) to perform DDT and DDT-fallback, as shown in Fig. 1.

4 Unified Safety and Security Analysis Method

US^2 is an unified analysis method for safety and security of autonomous vehicles. It proposes a simple security level that conforms to ASIL, and considers the DALs. The quantification of US^2 is less complex and requires less analysis effort than other approaches, e.g., HEAVENS. US^2 enables to analyse safety hazard and security threat simultaneously, which is helpful to obtain consistent safety and security requirements. In this section, we demonstrate security level design, quantification scheme, and the design of US^2.

4.1 Security Level for Autonomous Vehicles

Inspired by ASIL, we propose a Security Level (so-called *SEL*) for autonomous vehicle to evaluate the security criticality level of threats. The key goal of SEL is to analyze the potential threats and, subsequently, generate security requirements and corresponding security countermeasures.

Potential threats in AVs can be quantified according to the attack potential (P), threat criticality (T) and DAL focus (F).

(1) Attack Potential P: In [9], the authors identify attack potential according five parameters: elapsed time, expertise, knowledge of system, window of opportunity and equipment. However, authors of [11] points out that such attack potential classification is too complex and requires a lot of effort for discussion. Out of these five parameters, knowledge of system and required equipment are the key parameters which affect the success of an attack, thus we use them to define the attack potential P.

Table 1 classifies the attackers' knowledge of system under investigation (marked as K). Level 0 denotes that attackers do not require prior knowledge. Level 1 indicates that the attackers need some basic knowledge or some basic understanding of the vehicle, and level 2 includes people with focused interests and comprehensive domain knowledge.

Table 1. Attackers' knowledge of system 'K' classification, and associated examples

Level	Knowledge of system K	Example
0	Public knowledge	Everyone knows
1	Basic knowledge	Basic understanding of system
2	Comprehensive domain knowledge	Needs technical training

Table 2 classifies required equipment (denoted by R) that is needed to exert a threat (we mainly consider the tools that needed to perform a successful attack). Level 0 means that no special tool is needed. Level 1 requires some standard tool that can be easily found. The tools in level 2 are specific that are highly limited and not widely used.

Table 2. Required equipment 'R' classification, and associated examples

Level	Required equipment R	Example
0	No additional tool needed	User interface, car key etc.
1	Standard tool	CAN sniffer etc.
2	Specific tool	Debugger etc.

With the knowledge K and required equipment R, we classify the attack potential P, as shown in Table 3. We firstly consider two extreme situations: if exerting a threat does not require any tool ($R = 0$) and any knowledge ($K = 0$), this threat will be of high attack potential ($P = 3$); in contrast, if exerting a threat requires advanced tool ($R = 2$) and specific training or knowledge ($K = 2$), such threat is of very low attack potential ($P = 0$). Besides the extreme situations, when for achieving a threat, an attacker needs either specific knowledge ($K = 2$) or specific tool ($R = 2$), the attack potential is considered as low ($P = 1$). If the requirement of knowledge and tool is median ($K = 1$ or $R = 1$), the attack potential is also median ($P = 0$). All the combinations of (K, R) and associated attack potential are listed in Table 3.

Table 3. Attack potential P classification, and required K, R

Attack potential P	Description	(K, R) combinations
0	Very low	(2,2)
1	Low	(2,0)(0,2)(1,2)(2,1)
2	Median	(1,1)(0,1)(1,0)
3	High	(0,0)

Table 4. Threat criticality T classification

Threat criticality T	Description
0	No impact
1	Light or moderate impact
2	Severe impact
3	Catastrophic impact

(2) Threat Criticality T: An overview of threat criticality T is given in Table 4. Level 0 indicates that the threat has no impact on security. The threats in level 1 have limited impact on security, such as reduction in availability of services. But if threats imply any damage to products or manipulation of data or services, we assign them to level 2. Threat in level 3 could either destroy service or have an impacts on human life (quality of life), as well as affect safety features.

(3) DAL Focus F: DAL defines the driving tasks of autonomous vehicle which are automated, and the human activities in AVs (see Sect. 3). If the tasks corresponding to a particular level are attacked, vehicles with that automation level will be in high danger. Thus, considering the DAL of AVs is of importance when analyzing their safety and security.

However, the driving tasks are overlapping as the level increases, i.e., vehicle in higher automation level can perform all the tasks of vehicle in lower level. Thus, it is difficult to say that a threat only effects vehicle of one particular level. If a threat impacts vehicle with DAL 1 but no countermeasure is designed for it, the threat will impact vehicle in higher levels more seriously. For example, supposing that the GPS data of a vehicle with DAL 1 or 2 is spoofed by an attacker, the situation may not be so dangerous because human driver can operate the vehicle without GPS planning. But if no mitigation technique is provided to confront the spoofing in high automation vehicle (e.g., DAL up to 4 or 5), the wrong GPS data could cause the wrong navigation, traffic disturbance or even human hurt. Therefore, we consider to use *DAL Focus* as a criterion to highlight that the threat should be considered and mitigation technique should be designed in AVs starting from the DAL focus level.

Table 5. DAL focus F classification

DAL focus F	Included DAL	Description
F1	DAL 1 and 2 (low automation)	Driver is available and can detect attacks to some extent
F2	DAL 3 (median automation)	Only fallback-ready user is available
F3	DAL 4 and 5 (high automation)	No human is available

As shown in Table 5, three categories are considered: $F1$, which includes vehicles with DAL 1 and 2, where human driver is considered as "detection equipment" (i.e., driver has capacity to detect some attacks); $F2$, which includes vehicles with DAL 3, where fallback-ready user is able to operate the car when DDT-fallback is activated; $F3$, which includes vehicles with DAL 4 and 5 (high automation level), where no human is involved in driving.

Using proposed attack potential P, threat criticality T and DAL focus F, security level is calculated according Eq. 1:

$$SEL = \begin{cases} 0 & \text{if } P+F+T<5 \text{ or } T=0 \\ 1 & \text{if } P=0, F=F1 \text{ and } T=3 \\ 1 & \text{if } P+F+T=5 \text{ and } T \neq 0 \\ 2 & \text{if } P+F+T=6 \text{ and } T \neq 0 \\ 3 & \text{if } P+F+T=7 \\ 4 & \text{if } P+F+T \geq 8. \end{cases} \quad (1)$$

where the value of F is equal to the level of DAL focus, i.e., $F1 = 1$, $F2 = 2$ and $F3 = 3$. If an attack has higher attack potential, focus DAL, and criticality, it will threat AV more seriously, i.e., higher severity level and bigger SEL value in our scenario. Vice versa, if an attack has lower attack potential, focus DAL and threat criticality, it will be in lower severity level, i.e., lower SEL value.

Table 6 shows SEL determination matrix, where we use different colors to show the classification of SEL: red, orange, yellow and green denote SEL level 4 to 1 respectively. It is obviously that higher attack potential P, DAL focus F and threat criticality T derive higher severity level.

Table 6. SEL determination matrix, according to P, F and T

Attack potential P	*DAL Focus F*	*Threat criticality T*			
		0	1	2	3
0	F1	0	0	0	1
	F2	0	0	0	1
	F3	0	0	1	2
1	F1	0	0	0	1
	F2	0	0	1	2
	F3	0	1	2	3
2	F1	0	0	1	2
	F2	0	1	2	3
	F3	0	2	3	4
3	F1	0	1	2	3
	F2	0	2	3	4
	F3	0	3	4	4

4.2 US²

The US² process consists of several activities as demonstrated in Fig. 2. It takes into consideration the quantification of SEL and ASIL. When facing a threat, firstly, we calculate the SEL value. Then we analyze if such threat will incur hazard situations. If hazard is created, we use ASIL to analyse the hazard, and obtain the security and safety requirements for hazards with high value of SEL and ASIL. If there is no hazard created, we only use SEL to derive the security requirements. Then countermeasures can be designed based on the requirements as shown in Fig. 2. Moreover, due to the unified analysis, any threat will be firstly analyzed using SEL, and then will be checked if it is related to safety (i.e., if there is hazard created or not). This will assure that no hazard that is introduced by threat will be missed.

5 US² Application Example

As mentioned in Sect. 3, DAS is embedded in AVs to assist or replace human driver to achieve driving automation, which is the main difference between AVs and conventional cars. An on-board computer is considered as the control component of DAS in AV to perform driving automation. Figure 3 shows a sample architecture of AVs, which considers the architecture and communications proposed in [14,15]. In Fig. 3, driving automation is achieved by perception (using sensors data to percept the external environment/context in which vehicle operates), navigation and planning (performed by DAS to decide the vehicle motion), and vehicle motion implementation (using ECUs and actuators). Inside the AV, CAN bus is a communication way between ECUs, where OBD-II port enables direct access to the vehicle.

For autonomous vehicles, besides using sensor data, external/environment information can be obtained by WiFi, Internet, Bluetooth, Mobile APP, Road side, or even from other cars (as shown in Fig. 3). The more potential ways to interact with the outside world are used in AVs, the more vulnerabilities to safety and security exist.

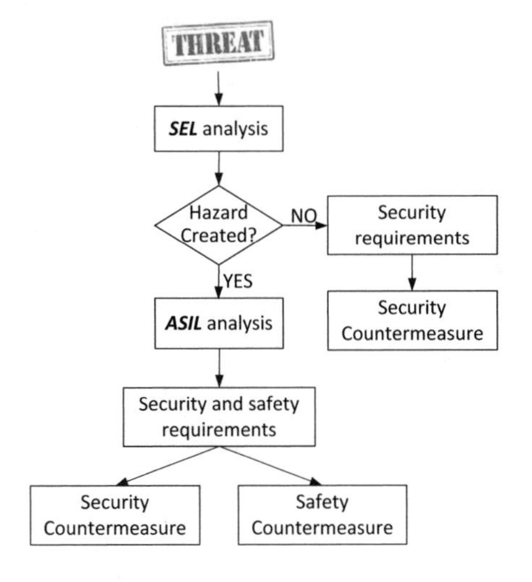

Fig. 2. Demonstration of US² analysis method.

We use US^2 to analyse threats for AVs, and demonstrate the analysis results in Table 7. For each attack, we consider following criteria:

(1) *Structure*: the related structural components of AVs.
(2) *Target*: the attack purpose.
(3) *Means*: the ways to execute the attack.
(4) *P*: attack potential (see Sect. 4.1).
(5) *F*: DAL focus (see Sect. 4.1)
(6) *T*: threat criticality (see Sect. 4.1).
(7) *SEL*: security level (see (1)).
(8) *Hazard Created*: hazards generated as a result of attack.
(9) *S*: severity (see ASIL [4]).
(10) *E*: exposure (see ASIL [4]).
(11) *C*: controllability (see ASIL [4]).
(12) *ASIL*: automotive safety integrity level (see ASIL [4]).
(13) *Possible security countermeasure*: countermeasures to mitigate the impact of such attack.
(14) *Possible safety countermeasure*: countermeasures to mitigate the failure impact.

To select DAL focus, we consider the human driver's effect. As illustrated in Table 5, human driver can be seen as a "detection equipment" for AVs with low automation. Thus, if a threat is easily detected by driver, the effect of such threat for low automation AVs can be ignored. Taking road side as an example, if the infrastructure signs have been changed (i.e., fake sign, change speed or remove sign), it may cause AVs with high automation to react inadequately or not react at all, thereby creating hazard like traffic disturbance. But for AVs with low automation, the signs change can be noticed by driver [16]. Thus, DAL focus for sign change is considered as $F2$ (as shown in Table 7). When a vehicle is in DAL 1 and 2, the effect of infrastructure sign change can be ignored, because the driver of the vehicle is able to handle it. While when the vehicle is in DAL 3, DAL 4 or 5, the threat of infrastructure sign change will effect vehicle more seriously, thus countermeasure should be designed to avoid the effect.

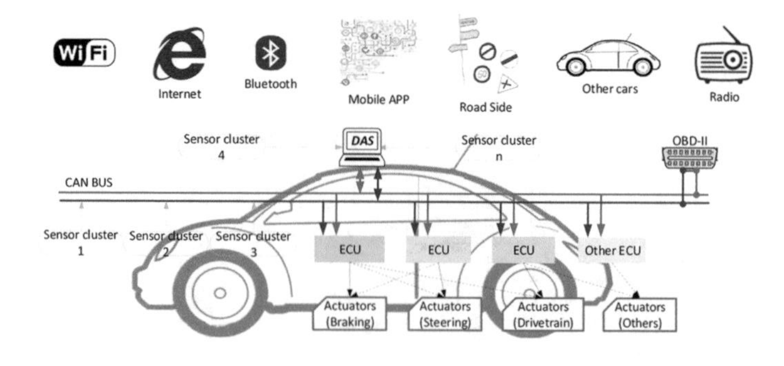

Fig. 3. Sample architecture for autonomous vehicles.

Table 7. Safety and security analysis using US^2.

Structure	*Target*	*Means*	*Attack Potential* P	*DAL Focus* F	*Criticality* T	*SEL*	*Hazard created*	*Severity* S	*Exposure* E	*Controllability* C	*ASIL*	*Possible Security Countermeasure*	*Possible Safety Countermeasure*
Road Side	Infrastructure sign	o change sign (fake,irrelevant)	1	F2	2	1	traffic disturbance	2	2	3	A		
		o change speed (make unreadable)	3	F2	2	3	traffic disturbance	2	4	3	C	detection technique	periodically inspect sign
		o remove sign	3	F2	2	3	traffic disturbance	2	4	3	C		
AV component	Vision or video	o blind	3	F1	3	3	driver disturbance	2	4	2	B	detection technique	check vehicle component regularly
		o fake picture/ emergency brake light	1	F1	2	0	driver disturbance	2	2	2	QM		
	In-vehicle device	o inject malware	2	F1	2	1	depends on malware	2	3	3	B	fireware	
		o display attack	2	F2	2	2	driver disturbance	2	3	3	B	protection of display of safety information	abnormal alarm
GPS	GPS data	o spoofing	3	F1	3	3	crash hazard	3	4	3	D	authentication	
		o jamming	3	F1	3	3	traffic disturbance	3	4	3	D	anti-jam GPS techniques	backup GPS
CAN Bus	In-vehicle communication	o eavesdropping	3	F1	2	2	none(privacy leak)	n.a.	n.a.	n.a.	n.a.	secure communication	–
		o inject CAN message	2	F1	2	1	traffic disturbance	2	3	3	B	in-vehicle security	Safety communication
		o OBD dongles	3	F1	3	3	safety disturbance	3	4	3	D	encryption	reliable OBD dongles
Sensors	Acoustic sensors	o interference (loud sound)	2	F1	1	0	none	n.a.	n.a.	n.a.	n.a.		–
		o fake crash sound	3	F1	1	1	traffic disturbance	1	4	3	B	filter data	use other source of data
		o fake ultrasonic reflection	2	F1	1	0	low-speed crash	1	3	3	A		use other source of data
	Radar	o chaff	2	F1	2	1	traffic disturbance	2	3	3	B		
		o smart material (invisible object)	1	F1	2	0	collision	2	2	3	A		
		o jamming (saturation with noise)	3	F1	2	2	traffic disturbance	2	4	3	C	filter data	use other source of data
		o ghost vehicle (signal repeater)	3	F1	2	2	traffic disturbance	2	4	3	C		
	Lidar	o jamming	3	F1	2	2	loss of situation awareness	2	4	3	C		
		o smart material (absorbent, reflective)	3	F1	2	2	traffic disturbance	2	4	3	C	filter data	use other source of data

As we can see from Table 7, for attacks with the same target, driver may easily detect some of attack means but not all of them. For example, considering in-vehicle device as attack target, driver may easily detect the display attack, but it could be hard to detect the injected malware [16]. In such cases, we list the different DAL focus to different attack means (e.g., injecting malware corresponds to $F1$, and display attack corresponds to $F2$) as demonstrated in Table 7.

Hazards can be caused by the threats, such as traffic disturbance, driver disturbance or crash, as listed in Table 7. However, not all threats cause safety hazard. E.g., for in-vehicle communication, there are three means to execute attack: eavesdropping attacks, injection of CAN message, or the use of OBD dongles [17]. Eavesdropping will not cause safety hazards, but will cause privacy leak. In this situation, we use "n.a." to show no relevance, which denotes "not applicable". Thus, we propose only security countermeasure for eavesdropping. For example, secure communication technique can be designed to avoid eavesdropping [18].

If hazard is created due to the security threats, ASIL analysis will be performed. The quantification of SEL and ASIL shows the severity of the threat or hazard, which can be helpful to derive security and safety requirements. Similarly to ASIL (ASIL A is the lowest safety integrity level and ASIL D the highest one), bigger value of SEL denotes higher severity. For example, fake picture (or fake emergency brake light) is at SEL 0 and ASIL QM, which is very low severity of security and safety; while spoofing (or jamming) GPS data is at SEL 3 and ASIL D, which is of high severity. Thus, protecting GPS data is a high security and safety requirement, which can be considered as safety and security goals. As the utilization of US^2, no hazard that introduced by security attack will be missed.

In Table 7, possible AV security and safety countermeasures are illustrated. For example, for GPS data attacks, security countermeasure like authentication (or anti-jam GPS techniques), and safety mitigation like backup GPS system can be designed. For sensor data attacks (acoustic sensor, radar or lidar), data filtering or using additional data could be used as security and safety countermeasures.

6 Conclusion

In this paper, we propose a unified safety and security analysis method for autonomous vehicles. US^2 combines a security analysis to ASIL, which can evaluate threats and failures in parallel, and help to derive security and safety requirements effectively. A simple security level SEL is designed for security analysis. SEL uses three parameters (attack potential, threat criticality, and DAL focus) to generate a value to determine the severity of threats in AVs. An example is included to illustrate the applicability of the proposed methods. The US^2 is useful tool for specifying safety and security requirements, and selecting appropriate safety and security countermeasures for AVs at all levels of driving automation.

In future, we will extend US^2 to specify technical safety and security requirement for AVs in corresponding level of driving automation.

References

1. Society of Automotive Engineers (SAE): SAE-J3016: Taxonomy and Definitions for terms Related to Driving Automation Systems for On-Road Motor Vehicles, September 2016
2. The Associated Press: Worlds 1st self-driving taxi debut in Singapore. https://www.bloomberg.com/news/articles/2016-08-25/world-s-first-self-driving-taxis-debut-in-singapore. Accessed 12 October 2017
3. Cui, J., Sabaliauskaite, G.: On the alignment of safety and security for autonomous vehicles. In: IARIA CYBER, Barcelona, Spain, November 2017
4. International Organization for Standardization (ISO): ISO-26262: Road Vehicles - Functional safety, December 2016
5. Macher, G., Sporer, H., Berlach, R., Armengaud, E., Kreiner, C.: Sahara: a security-aware hazard and risk analysis method. In: IEEE DATE, Grenoble, France (2015)
6. Corporation, M.: The stride threat model (2005)
7. Society of Automotive Engineers (SAE): SAE-J3061: Cybersecurity Guidebook for Cyber-Physical Vehicle Systems, January 2016
8. Ward, D., Ibarra, I., Ruddle, A.: Threat analysis and risk assessment in automotive cyber security. SAE Int. J. Passeng. Cars Electron. Electr. Syst. **6**(01–1415), 507–513 (2013)
9. EVITA - E-safety vehicle intrusion protected applications, "Evita project,". https://www.evita-project.org/deliverables.html. Accessed 12 Oct 2017
10. Islam, M., et al.: Deliverable D2 Security models. HEAVENS Project, Deliverable D2, Release 1 December 2014
11. Macher, G., Armengaud, E., Brenner, E., Kreiner, C.: A review of threat analysis and risk assessment methods in the automotive context. Springer International Publishing, Trondheim (2016)
12. NHTSA: Federal automated vehicles policy, September 2016
13. Paul, A., Chauhan, R., Srivastava, R., Baruah, M.: Advanced driver assistance systems. Technical report. SAE Technical Paper (2016)
14. Becker, J., Helmle, M., Pink, O.: System architecture and safety requirements for automated driving. In: Automated Driving, pp. 265–283. Springer (2017)
15. Studnia, I., Nicomette, V., Alata, E., Deswarte, Y., Kaâniche, M., Laarouchi, Y.: Survey on security threats and protection mechanisms in embedded automotive networks. In: IEEE DSN-W, Budapest, Hungary (2013)
16. Petit, J., Shladover, S.E.: Potential cyberattacks on automated vehicles. IEEE Trans. Intell. Transp. Syst. **16**(2), 546–556 (2015)
17. Yan, W.: A two-year survey on security challenges in automotive threat landscape. In: IEEE ICCVE, ShenZhen, China (2015)
18. Thing, V.L., Wu, J.: Autonomous vehicle security: a taxonomy of attacks and defences. In: IEEE CPSCom, ChengDu, China (2016)

User-Driven Intelligent Interface on the Basis of Multimodal Augmented Reality and Brain-Computer Interaction for People with Functional Disabilities

Peng Gang[1(✉)], Jiang Hui[1], S. Stirenko[2], Yu. Gordienko[2], T. Shemsedinov[2], O. Alienin[2], Yu. Kochura[2], N. Gordienko[2], A. Rojbi[3], J. R. López Benito[4], and E. Artetxe González[4]

[1] Huizhou University, Huizhou City, China
peng@hzu.edu.cn
[2] National Technical University of Ukraine, "Igor Sikorsky Kyiv Polytechnic Institute", Kyiv, Ukraine
sergii.stirenko@gmail.com
[3] CHArt Laboratory (Human and Artificial Cognitions), University of Paris 8, Paris, France
[4] CreativiTIC Innova SL, Logroño, Spain

Abstract. The analysis of the current integration attempts of some modes and use cases of human-to-machine interaction is presented. The new concept of the user-driven intelligent interface for accessibility is proposed on the basis of multimodal augmented reality and brain-computer interaction for various applications: in disabilities studies, education, home care, health care, eHealth, etc. The several use cases of multimodal augmentation are presented. The perspectives of the better human comprehension by the immediate feedback through neurophysical channels by means of brain-computer interaction are outlined. It is shown that brain–computer interface (BCI) technology provides new strategies to overcome limits of the currently available user interfaces, especially for people with functional disabilities. The results of the previous studies of the low end consumer and open-source BCI-devices allow us to conclude that combination of machine learning (ML), multimodal interactions (visual, sound, tactile) with BCI will profit from the immediate feedback from the actual neurophysical reactions classified by ML methods. In general, BCI in combination with other modes of AR interaction can deliver much more information than these types of interaction themselves. Even in the current state the combined AR-BCI interfaces could provide the highly adaptable and personal services, especially for people with functional disabilities.

Keywords: Augmented reality · Interfaces for accessibility
Multimodal user interface · Brain-computer interface · eHealth
Machine learning · Human-to-machine interactions

K. Arai et al. (Eds.): FICC 2018, AISC 886, pp. 612–631, 2019.
https://doi.org/10.1007/978-3-030-03402-3_43

1 Introduction

Current investigations of user interface design have improved the usability and accessibility aspects of software and hardware to the benefits of people. But, despite the significant progress in this field, there is still a big work ahead to satisfy requirements of people with various functional disabilities due to lack of adequately accessible and usable systems. It is especially important for persons with neurological and cognitive disabilities [1–3]. That is why more effective solutions are needed to improve communication and provide the more natural human-to-machine (H2M) and machine-to-human (M2H) interactions, including interactions with their environment (home, office, public places, etc.). The most promising current aims are related to development of technologies aiming at enhancing cognitive accessibility, which allows improving comprehension, attention, functional abilities, knowledge acquisition, communication, perception and reasoning. One way for achieving such aims is the use of information and communication technologies (ICTs) for development of the better user interface designs, which can support and assist such people in their real environment. Despite the current progress of ICTs, the main problem is the vast majority of people, especially older people with some disabilities, wish to interact with machines in non-obtrusive way and in the most usual and familiar way as much as possible. Meeting their needs can be a major challenge and integration of the newest user interface designs on the basis of the novel ICTs in a citizen-centered perspective remains difficult.

This work is dedicated to the analysis of our previous attempts of integration of some modes of user-machine interaction and the concept of the user-driven intelligent interface on the basis of multimodal augmented reality and brain-computer interaction for various applications, especially for people with functional disabilities. Section 2 gives the short description of the state of the art in the advanced user-interfaces. Section 3 contains the concept of the proposed user-driven intelligent interface based on the integration of new ICT approaches and examples of multimodal augmentation developed by authors of this paper. Section 4 outlines the opportunities to get neurophysical feedback by brain-computer interaction implemented in several noninvasive BCI-devices presented in the consumer electronics sector of economy. Section 5 describes the previous results and possibilities to get the better human feedback by integration of multimodal augmentation and brain-computer interaction for the use cases from Sect. 3.

2 Background and Related Work

The current development of various ICTs, especially related with augmented reality (AR) [4], multimodal user interfaces (MUI) [5], brain-computer interfaces (BCI) [6], machine learning techniques for interpretation of complex signals [7], wearable electronics (like smart glasses, watches, bracelets, heart beat monitors, and others gadgets) [8] open the wide perspectives for development of the user-driven intelligent interfaces (UDII). Some of the most promising UDII are based on psychophysiological data, like heart beat monitoring (HRM), electrocardiogram (ECG), electroencephalography (EEG) and that can be used to infer users' mental states in different scenarios, although they have become more popular recently to evaluate user experience in various applications [9, 10]. Moreover, it is possible to monitor and estimate the emotional responses based on these and other physiological measures. For example, galvanic skin response (GSR) gauges the level of emotional excitement or arousal of an individual, which is generally measured by two electrodes on the hands of a participant by the skin conductance level and/or the skin conductance response. These electrodes measure the electrical current differentials stemming from the increase of sweat activity, which often are consequences of the personal excitement [11, 12]. Currently, the better user interface designs can be obtained by the development of intelligent, affordable and personalized approaches. They are especially necessary for people with cognitive disabilities to allow them to perform their everyday tasks. Additional needs are related to improvement of their communication channels and uptake of the available and new digital solution and services. The new user interface designs should recognize abilities of customers, detect their behaviors, recognize behavior patterns, and provide feedback in real life environments.

3 Multimodal Augmentation

The proposed user-driven intelligent interface is assumed to be based on the integration of new ICT approaches and the available ones in favor of the people with functional disabilities. They include the multimodal augmented reality (MAR), microelectromechanical systems (MEMS), and brain-computer interaction (BCI) on the basis of machine learning (ML) providing a full symbiosis by using integration efficiency inherent in synergistic use of applied technologies. The matter is that due to recent "silent revolution" in many well-known ICTs, like AR, ML, MEMS, IoT, BCI, the synergy potential of them becomes very promising. Until recent years, AR and BCI devices were prohibitively expensive, heavy, awkward, and especially obtrusive for everyday usage by wide range of ordinary users. The data provided by them were hard to collect, interpret, and present, because of absence of solid and feasible ML methods. But during the last years numerous non-obtrusive AR, BCI, IoT devices become more available for general public and appeared in the consumer electronics sector of economy. At the same time development of MEMS and ML boosted the growth of IoT and wearable electronics solutions proposed on the worldwide scale. Despite these advancements the more effective achievements can be obtained by the proper integration of these ICTs. Below several attempts of such integration of these ICTs are presented, which were laid in the basis of the integral approach.

3.1 Tactile and ML for People with Visual Disabilities

Graphical information is inaccessible for people with visual impairment or people with special needs. Studies have demonstrated that tactile graphics is the best modality for comprehension of graphical images for blind users. Usually, graphical images are converted to tactile form by tactile graphic specialists (TGS) involving non-trivial manual steps. Although some techniques exist that contribute to help TGS in converting graphical images into a tactile format, the involved procedures are typically time-consuming, expensive and labor-intensive. In continuation of these efforts the new software program was developed by authors from University of Paris 8 that converts a geographic map given in a formatted image file to a tactile form suitable for people with special needs. The advanced image processing and machine learning techniques were used in it to produce the tactile map and recognize text within the image. The software is designed to semi-automate the translation from visual maps to tactile versions, and to help TGS to be faster and more efficient in producing the tactile geographic map [13–17]. But the further and more effective progress can be achieved when other available ICTs will be integrated. For example, the online feedback for the better comprehension of information can be provided by neurophysical channels by means of BCI and/or GSR interactions. Below in Sect. 5.1, some propositions are given to extend this work for more effective conversion of graphical content to a tactile form.

3.2 Visual and Tactile AR for Educational Purposes

As it is well-known, AR consists of the combination of the real world with virtual elements through a camera in real time. This emerging technology has already been applied in many industries. Recently, the effective use of AR in education was demonstrated in order to improve the comprehension of abstract concepts such as electronic fields, and enhance the learning process making education more interactive and appealing for students. One of its main innovations consisted in creation of the totally non-obtrusive Augmented Reality Interface (ARI) by authors from CreativiTIC Innova SL that detects different electronic boards and superposes relevant information over their components, serving also as a guide through laboratory exercises [18, 19]. Combination of AR with visual + tactile interaction modes allowed providing tactile metaphors in education to help students in memorizing the learning terms by the sense of touch in addition to the AR tools. ARI is designed to facilitate learning process and in combination with BCI it can provide specific information about concentration and cognitive load on students. The proper usage of ML methods will allow concluding and giving the contextual AR-based advice to students in classrooms. Below in Sect. 5.2 some ways are described to extend this work for more effective conversion of graphical content to a tactile form with inclusion of other information channels.

3.3 TV-Based Visual and Sound AR for Home and Health Care

Watching TV is a common activity for every-day life, so adding communication and interactive tools to smart digital TV devices is a good solution to integrate new technologies into ordinary life of people with disabilities not changing their comfort behavior. As far as smart interactive digital TV (iDTV) becomes more popular in homes, they can be used as a core in the current and future tele-healthcare systems. SinceTV system on the basis of the iDTV technology was developed by authors from SinceTV company and National Technical University of Ukraine “Igor Sikorsky Kyiv Polytechnic Institute” to provide visual AR information for various applications[1]. The

[1] SinceTV interactive digital TV platform (http://sincetv.com).

current prototype includes a iDTV telecare middleware with a hierarchical software stack and structural subsystem. SinceTV is an iDTV technology that can be adapted to improve the lives, for example, of elder people and create integrating modern communication technologies into everyday life and comfortable environment for target users. SinceTV provides interactivity close to real-time; latency minification in reaction fixing; synchronization using ACR (Audio Recognition Content); high load scalability up to 10 million concurrent connections (potentially linear grow); the second screen concept. SincenTV allows you to add interactive AR data to video and audio streams, linking two points, not only through the media devices, but also provides facilities for distributed interactive applications. Such a set of features in combination AR-based feedback can be useful for health care purposes, for example, for activity measurement and health state estimation via vision-based algorithms. Below in Sect. 5.3 some potential directions of the further development of this iDTV system are given.

3.4 Visual AR + Wearable Electronics for Health Care

The standard cardiology monitoring can show the instant state of cardiovascular system, but unfortunately, cannot estimate the accumulated fatigue and physical exhaustion. Errors due to fatigue can lead to decrease of working efficiency, manufacturing quality, and, especially, workplace and customer safety. Some specialized commercial accelerometers are used to record the number of steps, activities, etc. [20, 21]. However, they are quite limited to assess the health state and measure accumulated fatigue. The new method was proposed recently by authors from National Technical University of Ukraine "Igor Sikorsky Kyiv Polytechnic Institute" to monitor the level of currently accumulated fatigue and estimate it by the several statistical methods [22].

The experimental software application was developed and used to get data from sensors (accelerometer, GPS, gyroscope, magnetometer, and camera), conducted experiments, collected data, calculated parameters of their distributions (mean, standard deviation, skewness, kurtosis), and analyzed them by statistical and machine learning methods (moment analysis, cluster analysis, bootstrapping, periodogram and spectrogram analyses). Various gadgets were used for collection of accelerometer data and visualization of output data by AR. Several "fatigue metrics" were proposed and verified on several focus groups. The method can be used in practice for ordinary people in everyday situations (to estimate their fatigue, give tips about it and advice on context related information) [23]. In addition to this, the more useful information as to fatigue can be obtained by estimation of the level of user concentration to the external stimuli by brain-computer interaction that is described below.

By EEG measurements they can determine different psychophysiological states, such as attention, relaxation, frustration, or others. For example, MindWave Mobile by Neurosky allows you to determine at least two psychological states: concentration ("attention") and relaxation ("meditation"). The exposure of a user to different external stimulators will change both levels of the psychological states collected by this device. For example, this method can estimate: if the user is calm, then the relaxation ("meditation") will be high and the concentration ("attention") will be low. The consumer BCI-devices have various number of EEG channels, types of EEG connection with human surface, different additional sensors, and their price depends on their possibilities (see Table 1). Fortunately, all of them have additional documentation for software developers and related software development kits (SDKs), which allow external developers to propose their own solutions and make research.

Table 1. Comparison of some consumer BCI devices

Device (Company)	EEG channels	EEG connection	Additional sensors	Price, $
MindWaveMobile (NeuroSky)	1	Dry	Accelerometer	89
EPOC/Insight (Emotiv)	5/14	Wet/dry	Accelerometer, Gyroscope, Magnetometer	300–800
Muse (InteraXon)	5	Dry	Accelerometer	249
OpenBCI (open source)	4/8/12	Dry/wet	EMG/ECG	750–1800

4 Neurophysical Feedback

Brain-Computer Interfaces (BCIs) are widely used to research interactions between brain activity and environment. These researches are often oriented on mapping, assisting, augmenting, or repairing human cognitive or sensory-motor functions. The most popular signal acquisition technology in BCI is based on measurements of EEG activity. It is characterized by different wave patterns in the frequency domains or "EEG rhythms": alpha (8 Hz–13 Hz), SMR (13 Hz–15 Hz), beta (16 Hz–31 Hz), Theta (4 Hz–7 Hz), and Gamma (25 Hz–100 Hz). They are related with various sensorimotor and/or cognitive states, and translating cognitive states or motor intentions from different rhythms is a complex process, because it is hard to associate directly these frequency ranges to some brain functions. Some consumer EEG solutions, such as the MindWave Mobile by Neurosky, Muse by InteraXon, Emotiv EPOC by Emotiv and the open-source solutions like OpenBCI become available recently and can be used to assess emotional reactions, etc. (see Fig. 1).

Unfortunately, all consumer BCI-devices are specialized setups without sound, visual, tactile and other feedbacks in reaction to the external stimuli. Their wet contact should be used with the specialized gel, but the more feasible practical applications are possible mainly on the basis of the dry contacts. Additional obstacle is that the consumer BCI-devices and the related software are proprietary solutions, which cannot be easily developed, adopted, and used by the third-parties, especially in education and research purposes. In this connection availability of Open Brain Computer Interface (OpenBCI) open the ways for science advancements by openly shared knowledge among the wider range of people with various backgrounds. It will allow them to leverage the power of the open source paradigm to accelerate innovations of H2M and M2H technologies. For example, the available OpenBCI hardware solutions like Ganglion (with the 4-channel board) is suitable for low-cost research and education, and Cyton (with the 8–16 channel boards) provides the higher spatial resolution and enables more serious research.

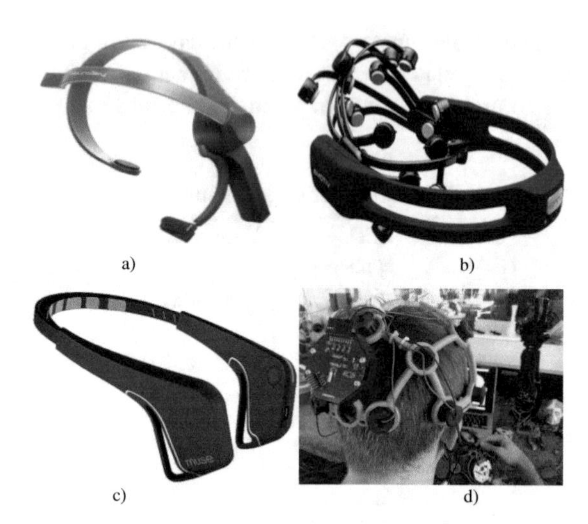

Fig. 1. The noninvasive BCI-devices presented in the consumer electronics sector of economy: (a) Mind Wave Mobile by Neuro Sky (http://neurosky.com); (b) EPOC by Emotiv (https://www.emotiv.com); (c) Muse by InteraXon (http://www.choosemuse.com); (d) Ultracortex Mark III by OpenBCI (http://openbci.com).

5 Use Cases for AR-BCI Integration

Before in Sect. 3, several examples of effective usage of the various ICTs combinations and interaction channels were demonstrated on the application level. The similar approach was presented recently as the augmented coaching ecosystem for non-obtrusive adaptive personalized elderly care on the basis of the integration of new and available ICT approaches [24]. They included multimodal user interface (MMUI), AR, ML, IoT, and machine-to-machine (M2M) interactions based on the Cloud-Fog-Dew computing paradigm services.

Despite the current progress in the above mentioned attempts to combine the available AR modes, the most promising synergy can be obtained by online EEG and GSR data monitoring, processing, and returning as AR feedback. The general scheme of multimodal interactions and data flows is shown in Fig. 2, collection of EEG reaction from user by various available consumer BCI-devices (Fig. 1, Table 1), ML data processing, and return output data to user as AR feedback by available AR-ready gadgets. The crucial aspect is to avoid the obtrusive way of usage of the current BCI-gadgets, which can be much more inappropriate by users, if they will be equipped by additional AR-features. But the current progress of microcontrollers, sensors, and actuators allow using the combination of low cost contacts, microcontrollers with low energy Bluetooth or Wi-Fi wireless networking, ear phones for sound AR and LEDs for visual AR on the ordinary glasses instead of "Terminator"-like bulky and awkward specialized devices (Fig. 2).

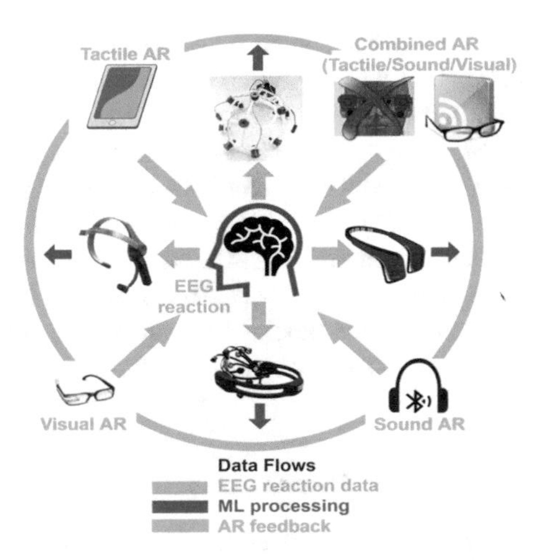

Fig. 2. The general scheme of multimodal interactions and data flows: collection of EEG reaction from user by various available BCI-ready devices (see Table 1), ML data processing, and return output data to user as AR feedback by available AR-ready gadgets.

This general concept of multimodal integration was verified by the experimental setup (Fig. 3). It includes smart glasses Moverio BT-200 by EPSON as a visual AR interaction channel for the controlled cognitive load (set of mathematical exercises) and a collector of accelerometer data; neurointerface MindWave by NeuroSky as a BCI-channel and collector of EEG-data; and heart monitor UA39 by Under Armour as a collector of heartbeat data. The setup can collect time series of several parameters: the subtle head accelerations (like tremor characterizing stress), EEG-activity, and intervals of heartbeats on the scale of milliseconds. The statistical methods were used to find correlations between these time series for various conditions, and machine learning methods were used to determine and classify various regimes.

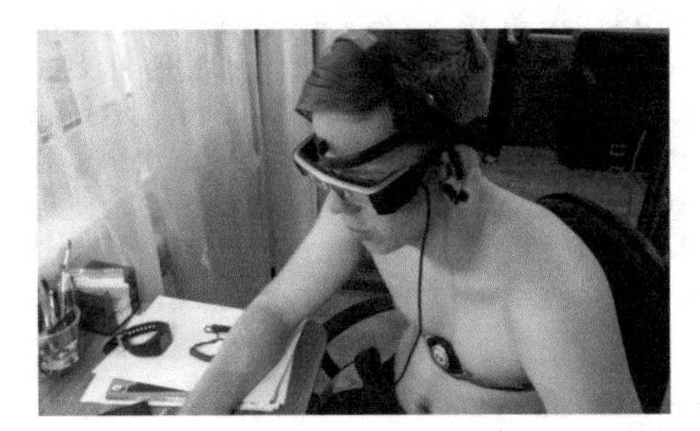

Fig. 3. Smart glasses Moverio BT-200 by EPSON (on the eyes) as an AR interaction channel and a collector of accelerometer data; neurointerface MindWave by NeuroSky (on the head with the blue band) as a collector of EEG-data; and heart monitor UA39 by Under Armour (on the breast with the yellow spot) as a collector of heartbeat data.

The previous experiments with the general concept and setup allowed us to propose several possible applications for the integration of AR channels with BCI technologies to provide the direct neurophysical feedback to users, which are discussed below. The general idea of the AR-BCI integration is based on the establishments of multimodal interactions and data flows, where all EEG reactions from the user observed by various available BCI-ready devices (Fig. 1, Table 1) or combined AR-BCI devices are gathered and then they are processed by ML methods on the supportive devices (smartphone, tablet, etc.) in non-obtrusive way. The essence of the AR-BCI integration consists in the real-time return of the obtained output data of neurophysical nature to user as AR feedback by available AR-ready gadgets (through sound, visual, and tactile AR channels).

5.1 BCI for People with Visual Disabilities

The application mentioned in Sect. 3.1 was developed on the basis of the advanced image processing and ML techniques to produce the tactile map and recognize text within the image. The available tools allow to automate the conversion of a visual geographic map into tactile form, and to help tactile graphics specialists be more efficient in their work. The development of these tools has benefited from feedback from specialists of National Institute for Training and Research for the Education of Disabled Youth and Adapted Teaching (INS HEA) (www.inshea.fr) and from volunteers. The previous analysis of the available mode of operation and possible improvements opened the following ways for improvement of the cognitive abilities during reading of various multimedia materials by tactile contacts. Even the low end consumer and open-source BCI-devices (like 1-channel MindWave Mobile by Neurosky or 4-channel Ganglion by OpenBCI) can differentiate, at least, two (or four) psychological states: concentration and relaxation. The exposure of a user to different external stimulators (for example, through various tactile interactions) will change both levels of the psychological states collected by this device. For example, this method can estimate: if the user has a good tactile contact, then the concentration will be high, and the relaxation will be low. In such a case the measure of the proper tactile contact through such BCI feedback can be helpful for online estimation of the automatic conversion of a visual geographic map into tactile form. The future development of this solution should recognize user's abilities and be able to detect behaviors and recognize patterns, emotions and intentions in real life environments. In this case a mix of technologies such ML and tactile interaction with BCI will profit from the immediate feedback on the basis of the actual neurophysical reactions classified by ML methods.

5.2 BCI for Educational Purposes

The several successful attempts to combine AR and other interaction modes were demonstrated in Sect. 3.2 on the basis of tactile haptic pen and tactile feedback analysis in education. There the simple classification of functions was used to develop tactile metaphors targeted to help students memorize the learning terms by the sense of touch in addition to the AR tools. The objective of the tactile accessory was to create different vibrotactile metaphors patterns easy to distinguish without ambiguity. Experiments shows that the vibrotactile feedback is well perceived by the user thanks to the wide range of frequency and amplitude vibration provided by the innovative combination of vibrotactile actuators. The next important step can be to measure if metaphors are relevant and effectively help students to memorize learning concepts (especially in lifelong learning) by additional neurophysical feedback by BCI along with tactile interaction. Besides supplying the localization of the zone of interest for the AR process, the role of the BCI is to provide "the positive feedback for satisfactorily recognized metaphors". The real objective of this AR-BCI system is to allow the user to see his/her own physical feelings about the zone of interest. This approach, based on a "tangible" and "thinkable" object has to incite the user to explore an invisible notions and ambience which compose the zone of interest.

5.3 BCI for TV-Based Home Care

SinceTV system on the basis of the iDTV technology (described in Sect. 3.3) provides interchange of generalized data structures of the various types like interactive questions and answers (with various input devices) and values obtained from the sensors, electronic equipment and different devices. It can process calls, events and data synchronization for distributed applications. In combination with neurophysical data obtained from users by BCI-devices in intrinsically interactive mode of operation, this technology can be helpful for much better communication of elderly people with each other, relatives, caregivers, doctors, social workers. Communication includes multiplatform applications: mobile, web, desktop and specialized interactive TV interface with voice and video input for target users. TV-based AR and BCI can provide the quite different way for estimation of concentration level of elderly people during their sessions of watching TV programs and shows, selection of food, goods, and services. This BCI feedback information can be invaluable for remote diagnostics and medical devices control.

5.4 BCI for Wearable Health Care

The more sophisticated estimation of various types of everyday and chronic fatigue (including mental and not only physical) can be obtained by measuring the level of user concentration to the external stimuli by the low end consumer and open-source devices even. The data from sensors like accelerometer were collected, integrated, and analyzed by several statistical and machine learning methods (moment analysis, cluster analysis, principal component analysis, etc.) (Fig. 4). The proposed method consists in monitoring the whole spectrum of human movements, which can be estimated by Tri-Axial Accelerometer (TAA), heartbeat/heartrate (HB/HR) monitor, and BCI (optionally, + by synergy with other sensors in the connected smartphone and data on ambient conditions in the smartphone). The main principle is the paradigm shift: to go from "raw output data" (used in many modern accelerometry based activity monitors) to the "rich post-processed (and, optionally, ambient-tuned) data" obtained after smart post-processing and statistical (moment/cluster/bootstrapping) analyses with much more quantitative parameters.

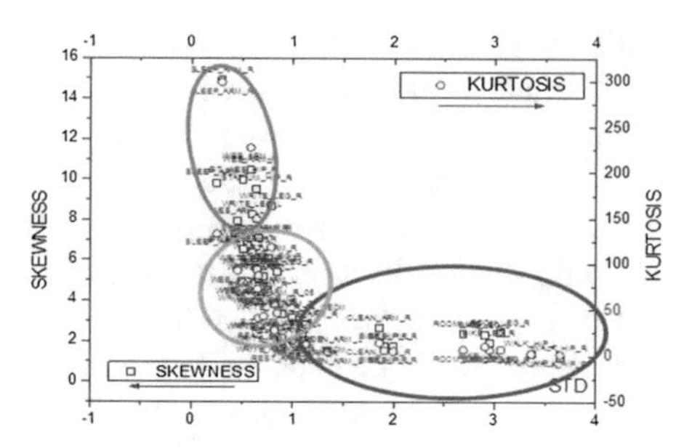

Fig. 4. Multiparametric moment analysis: activities can be classified in more details, i.e. divided into groups (colored ellipses) with the similar values of the acceleration distribution parameters: the active (sports, housework, walking) (blue ellipse), moderate (writing, sitting) (green ellipse) and passive (web surfing, reading, sleeping) (brown ellipse) behavior.

The hypothesis 1 (physical activity can be classified) and hypothesis 2 (fatigue level can be estimated quantitatively and distinctive patterns can be recognized) were proposed and proved, and due to shortage of space here the details are given elsewhere [22, 25]. Several "fatigue metrics" were proposed and verified on several persons of various age, gender, fitness level, etc. Correlation analysis of the moments (mean, standard deviation, skewness, kurtosis) of statistical distribution of acceleration values allowed us to determine the pronounced correlation between skewness and kurtosis for the states with high level of physical fatigue: after physical load (Fig. 5(a)) and in the very end of the day (Fig. 5(d)).

The similar ideology was applied for estimation of the workload during exercises and its influence on heart. The crucial aspects of this approach are as follows:

(1) The absolute values of heart rates (heartbeats) for the same workload are volatile (Fig. 6) and sensitive to the person (age, gender, physical maturity, etc.) and its current state (mood, accumulated fatigue, previous activity, etc.) – what should be done: in contrary, their distributions should be used here instead;
(2) The heart rate values are actually integer values with 2–3 significant digits and not adequately characterize the volatile nature of heart activity (because the heart rate is actually the reverse value of the heartbeat multiplied by 60 s and rounded to integer value) – what should be done: in contrary, heartbeats in milliseconds should be used, because they contain 3–4 significant digits and their usage gives 10 times higher precision;
(3) The actual influence of workload on heart, accommodation abilities of heart and fatigue of heart were not estimated before – what should be done: the results from statistical physics as to the critical phenomena and processes in the context of heart activity should be used.

This method allows us to determine the level of fitness from the moments diagrams of the heartbeat distribution functions vs. exercise time for various workloads: well-trained person (YU, male, 47 years) (Fig. 7(a)) and low trained person (NI, male, 14) (Fig. 7(b)).

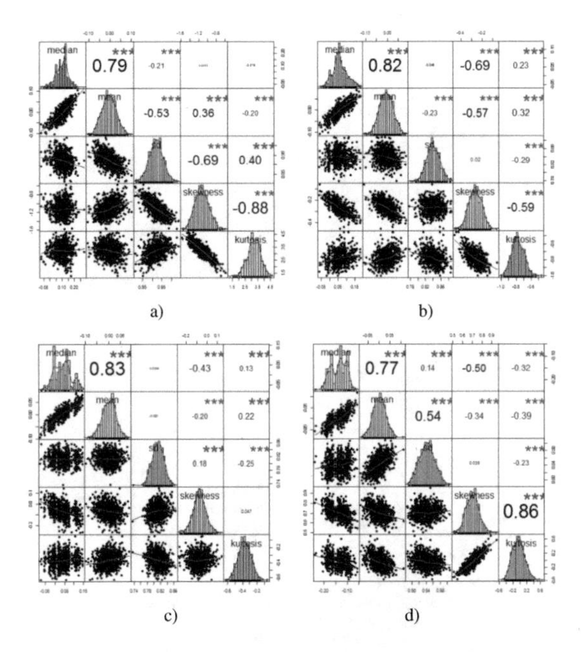

Fig. 5. Correlation analysis of the moments (mean, standard deviation, skewness, kurtosis) for statistical distribution of acceleration values for states with different levels of physical fatigue: (a) wake-up state, (b) after physical load (10 km of skiing), (c) rest state after lunch, (d) in the very end of the day.

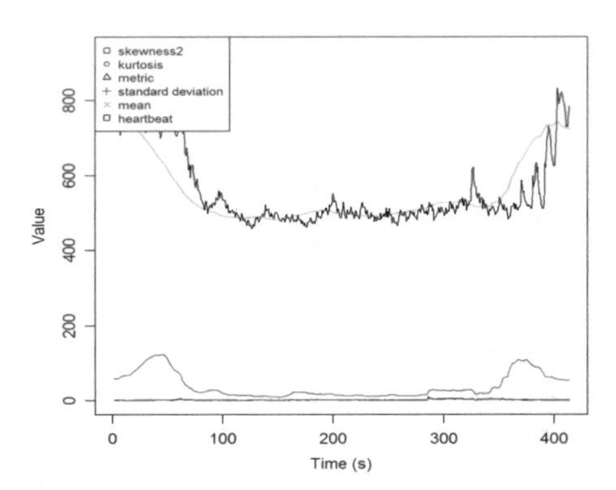

Fig. 6. Statistical parameters used for estimation of heartbeat/heartrate activity during exercises (heartbeat at walking for the well-trained person, male, 47 years). Legend: top black line—heartbeat itself, green—moving mean, magenta—standard deviation, low black line—metric, red—kurtosis, blue—skewness. (Kurtosis and skewness are not seen here, because of their low values. Please, see the next plots below.) The exercise was like: 1 min of rest + 5 min of walking with velocity 6.75 m/s + 1 min of rest.

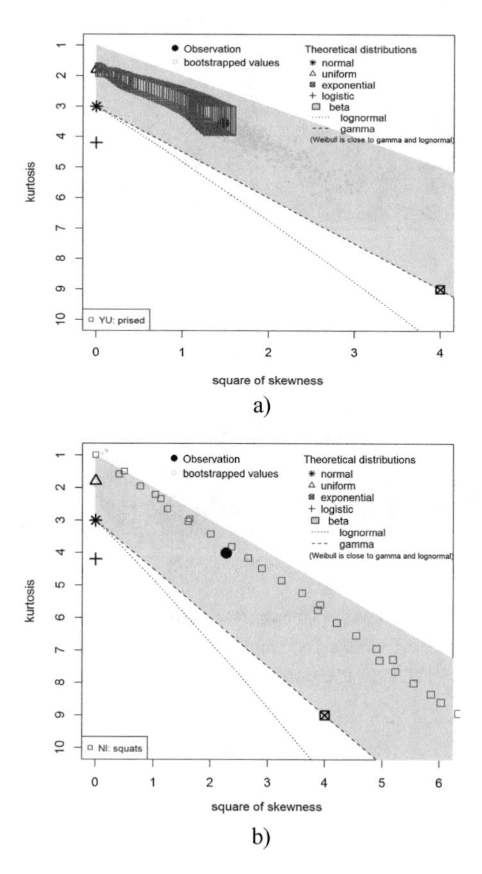

Fig. 7. Moments diagrams of the heartbeat distribution functions vs. exercise time for squats of: (a) well-trained person (YU, male, 47 years) and (b) low trained person (NI, male, 14). The exercise squats were performed up to fatigue. Legend: Size of blue symbols (current moments of heartbeat distribution) increases with time of exercise.

The exercise squats were performed up to fatigue. Size of blue symbols (the current moments of heartbeat distribution) increases with time of exercise. The distribution functions for the higher fitness (Fig. 7(a)) have tendency to slower and nearer movement of points with time of exercise (i.e. shift to the higher values of moments), and Distribution functions for the lower fitness (Fig. 7(b)) have tendency to much faster and farther movement.

This method allows us to determine the level of fitness for other types of exercise (for example, dumbbell curl for biceps here) from the moments diagrams of the heartbeat distribution functions vs. exercise time for various workloads: (a) well-trained person (YU, male, 47 years) and (b) low trained person (NI, male, 14) (Fig. 8). The size of symbol increases with time of exercise. The exercises were performed up to fatigue and denoted as "0.5 kg"—0.5 kg dumbbell curl for biceps, "1 kg"—1 kg dumbbell curl for biceps, "3 kg"—3 kg dumbbell curl for biceps. Again, the distribution functions for the higher fitness have tendency to slower movement of points with time of exercise (i.e. shift to the higher values of moments), and the distribution functions for the lower fitness have tendency to much faster movement. And the Distribution functions for the higher workload (weight of dumbbell, here) have tendency to much faster movement of points with time of exercise (i.e. shift to the higher values of moments).

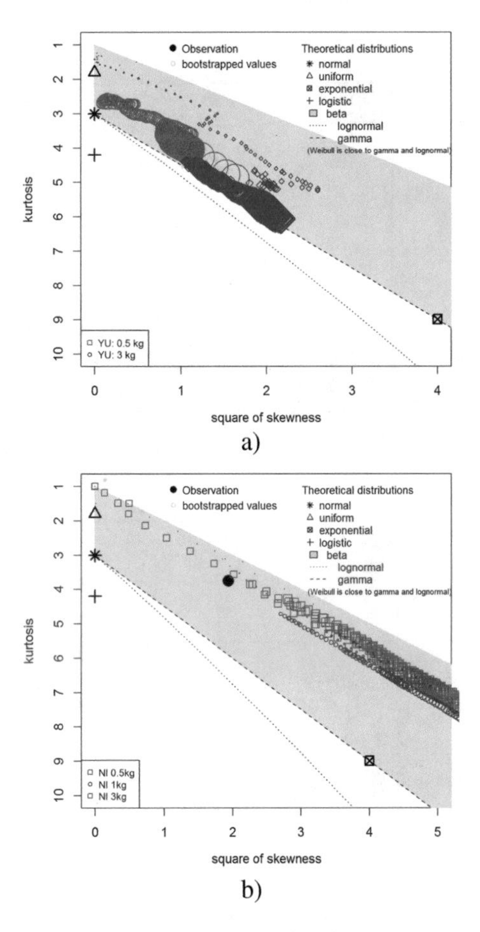

Fig. 8. Moments diagrams of the heartbeat distribution functions vs. exercise time for various workloads (see legend): (a) well-trained person (YU, male, 47 years) and (b) low trained person (NI, male, 14). Legend: size of symbol increases with time of exercise; exercises were performed up to fatigue and denoted as "0.5 kg"—0.5 kg dumbbell curl for biceps, "1 kg"—1 kg dumbbell curl for biceps, "3 kg"—3 kg dumbbell curl for biceps.

From the empirical point of view, the various metrics can be created on this basis, for example, "METRIC" as the distance from the uniform distribution to the current position (Fig. 9(a)), or "METRIC3" as the distance from the normal distribution to the current position (Fig. 9(b)). These metrics allows us to characterize and differentiate the workload levels and recovery phases, for example, from the previous exercise with various walking and jogging velocities. The slopes of metric increase and decrease can be used to characterize the accommodation and recovery levels during these exercises. Note: the initial sharp red peak for the highest possible load corresponds to the recovery phase after previous exercise.

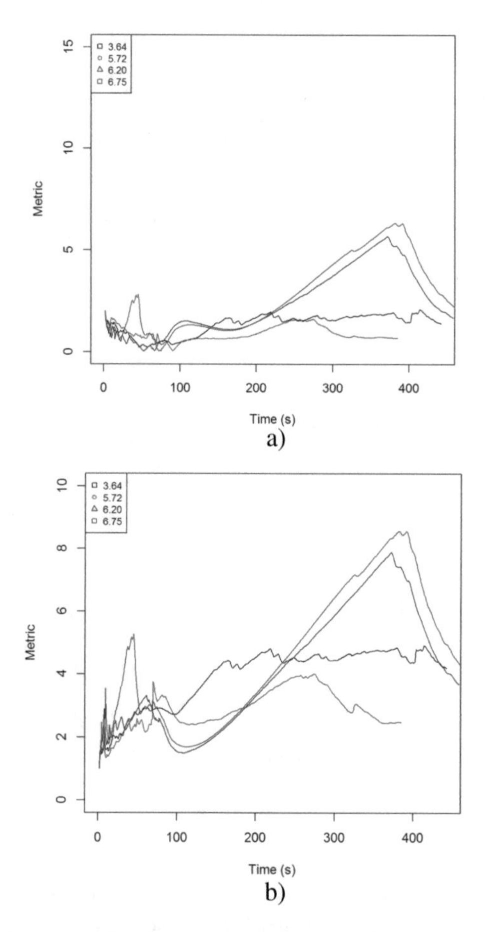

Fig. 9. Metrics of the heartbeat distribution functions vs. walking velocities for the well-trained person (YU, male, 47 years): (a) METRIC as the distance from the normal distribution on the moments diagram; (b) METRIC3 as the distance from the uniform distribution on the moments diagram. The exercise was like: 1 min of rest + 5 min of walking + 1 min of rest. Legend: black line—3.64 m/s (very low load), magenta line—5.20 m/s (comfort load), blue line—6.20 m/s (high load), red line—6.75 m/s (highest possible load).

The similar measurements of EEG brain activity by BCI-channel show the same output as to paradigm of usage distributions instead of separate absolute values provided by sensors. The raw absolute values of EEG activities measured as attention (ATT), relaxation (REL), and eye blink levels (EYE) (Fig. 10(a)) cannot provide the useful information, and it is hard to find any correlation between these levels (Fig. 10(b)).

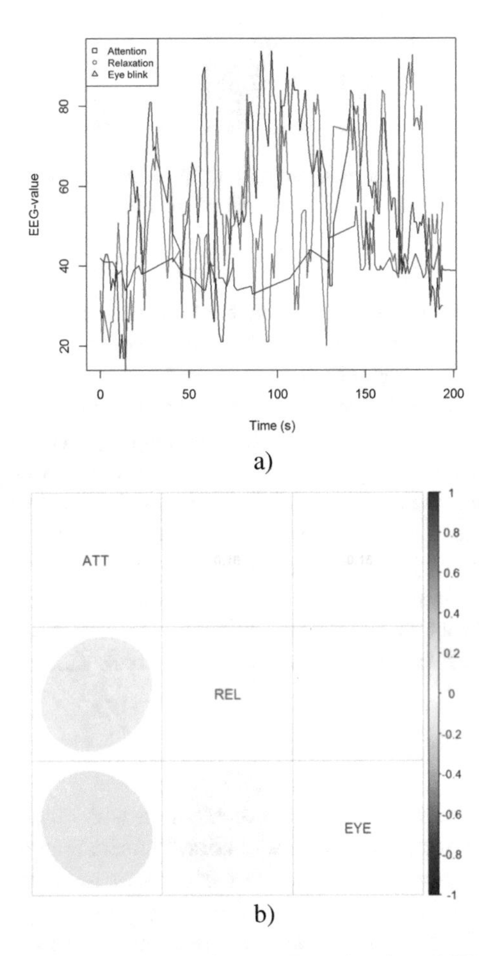

Fig. 10. EEG activities measured as attention (ATT), relaxation (REL), and eye blink levels (EYE): (a) absolute values, (b) their correlation matrix.

But similar approach on the basis of distributions and their moments can provide much more valuable information. For example, the similar "METRIC" as the distance from the uniform distribution to the current position, allow finding visual qualitative (Fig. 11) and quantitative numerical correlation (for attention and relaxation metrics) and anti-correlation for (for eye blink and relaxation metrics).

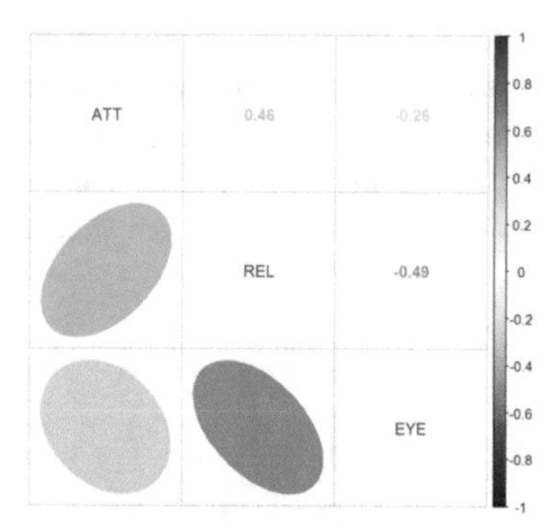

Fig. 11. EEG activities measured as attention (ATT), relaxation (REL), and eye blink levels (EYE): (a) metrics on the basis of the moments of distributions of absolute values, (b) their correlation matrix.

At the moment, these findings are isolated from each other, because they were measured separately, but the mush bigger potential could be foreseen if they will be combined [25]. In combination with the proposed fatigue metrics, BCI can provide the actual neurophysical feedback for users in the wearable electronics, like usual glasses (cap, hat, band, etc.) with the attached BCI-contacts, heart rate monitors, other microcontrollers with low energy Bluetooth or Wi-Fi wireless networking, simplified visual AR (like LEDs) and/or sound AR (ear phones) indicators (Fig. 3). It should be noted that at the moment this combination of AR+BCI and its hardware concept is on the stage of estimation of its general feasibility on the basis of the currently available ICTs. And it cannot be used for diagnostics of health state (including fatigue) in any sense, because involvement of various expertise (including medicine, psychology, cognition, etc.) is necessary and related specific research should be carried out.

6 Conclusion

In general, brain-computer interaction in combination with other modes of augmented interaction might deliver much more information to users than these types of interaction themselves. BCI provides quantitative measures of relevant psychophysiological actions and reactions and allow us to truly determine what was perceived or felt while our sensorimotor and/or cognitive systems are exposed to a stimulus. In the context of home and health care for people with functional disabilities, these quantitative measures (like GSR, EEG, ATS, etc.) cannot replace the current methods of evaluation and proactive functions, but they can complement and enhance them. But the use of other sensor data along with EEG activity can be very meaningful in the context of evaluating the user mental and physical state. The proposed user-driven intelligent interface on the basis of multimodal augmented reality and brain-computer interaction can be useful for various mentioned applications (education, lifelong learning, home care, health care, etc.). It could improve communication and interaction capability of people with disabilities and facilitate social innovation. It should be noted that in the context of estimating scene geometry and complex relationships among objects for autonomous vehicle driving such an intelligent interface can be useful for providing the instant feedback from humans for creation and development of the better machine learning approaches for visual object recognition, classification, and semantic segmentation [26, 27].

In the future work, usage of the more powerful and informative open-source hardware and software solutions (like OpenBCI) is considered to be the most promising way of development of the proposed multimodal approach. It could allow developers to leverage the more affordable technologies and products to support interactions for people with disabilities. For example, our previous results demonstrate that usage of multimodal data sources (like the multimodal data sources presented here) along with machine learning approaches can provide the deeper understanding of the usefulness and effectiveness of alphabets and systems for nonverbal and situated communication [28, 29]. They may be useful in various applications: from synthetic languages and constructed scripts to multimodal nonverbal and situated interaction between people and artificial intelligence systems through Human-Computer Interfaces, such as mouse gestures, touchpads, body gestures, eye-tracking cameras, wearables, and brain-computing interfaces, especially in applications for elderly care and people with disabilities.

The evident limitation of this study should be noted that consists in inability to make diagnostics of health state in any sense. The matter is the synergy of various expertise (including relevant disciplines like medicine, psychology, physiology, cognition, etc.) is necessary for the further development and test of the proper solutions, models and algorithms to improve information extraction from neurophysical signals. But even in the current state this new generation of combined AR-BCI interfaces could provide the highly adaptable and personalisable services to individual contexts, especially for people with functional disabilities.

Acknowledgment. The work was partially supported by Ukraine-France Collaboration Project (Programme PHC DNIPRO) (http://www.campusfrance.org/fr/dnipro), Twinning Grant by EU IncoNet EaP project (http://www.inco-eap.net/), and by Huizhou Science and Technology Bureau and Huizhou University (Huizhou, P. R. China) in the framework of Platform Construction for China-Ukraine Hi-Tech Park Project # 2014C050012001.

References

1. Hartzler, A.L., Osterhage, K., Demiris, G., Phelan, E.A., Thielke, S.M., Turner, A.M.: Understanding views on everyday use of personal health information: insights from community dwelling older adults. Inf. Health Soc. Care, 1–14 (2017)
2. Ziefle, M., Rocker, C., Holzinger, A.: Medical technology in smart homes: exploring the user's perspective on privacy, intimacy and trust. In: Proceedings of the IEEE 35th Annual Computer Software and Applications Conference Workshops, pp. 410–415 (2011)
3. Dimitrova, R.: Growth in the intersection of eHealth and active and healthy ageing. Technol. Health Care **21**(2), 169–172 (2013)
4. Billinghurst, M., Clark, A., Lee, G.: A survey of augmented reality. Found. Trends Hum. Comput. Interact. **8**(2–3), 73–272 (2015)

5. Dahl, D. (ed.): Multimodal Interaction with W3C Standards: Toward Natural User Interfaces to Everything. Springer, Cham (2017)
6. Hassanien, A.E., Azar, A.T.: Brain-Computer Interfaces. Springer, Cham (2015)
7. Soh, P.J., Woo, W.L., Sulaiman, H.A., Othman, M.A., Saat, M.S. (eds.): Advances in Machine Learning and Signal Processing: Proceedings of MALSIP 2015, vol. 387. Springer, Cham (2016)
8. Barfield, W. (ed.): Fundamentals of Wearable Computers and Augmented Reality. CRC Press, New York (2015)
9. Wolpaw, J.R., Birbaumer, N., McFarland, D.J., Pfurtscheller, G., Vaughan, T.M.: Brain–computer interfaces for communication and control. Clin. Neurophysiol. **113**(6), 767–791 (2002)
10. Niedermeyer, E., da Silva, F.L. (eds.): Electroencephalography: Basic Principles, Clinical Applications, and Related Fields. Lippincott Williams & Wilkins (2005)
11. Cacioppo, J.T., Tassinary, L.G., Berntson, G. (eds.): Handbook of Psychophysiology. Cambridge University Press, Cambridge (2007)
12. Nacke, L.E.: An introduction to physiological player metrics for evaluating games. In: Game Analytics, pp. 585–619. Springer, London (2013)
13. Bouhlel, N., Rojbi, A.: New tools for automating tactile geographic map translation. In: Proceedings of the 16th International ACM SIGACCESS Conference on Computers & Accessibility, pp. 313–314 (2014)
14. Belhabib, N., Rojbi, A.: Conception d'un dispositif de pointage-navigation accessible et adaptatif pour plusieurs cas d'handicap moteur (2010). http://www2.univ-paris8.fr/ingenierie-cognition/master-handi/recherche/handicap-dern-ver.pdf
15. Rojbi, A., Schmitt, J.-C.: Method for the Locating and Fuzzy Segmentation of a Person in a Video Image, WO 2005/071612 (2005). International Patent of Invention
16. Rojbi, A.: Fuzzy global segmentation system for video-telephony sequences. In: 8th World Multiconference on Systemics, Cybermics, Cybernetics and Informatics. Invited Sessions in Color Image Processing & Applications, USA, Florida (2006)
17. Bouhlel, N., Coron, A., Barrois, G., Lucidarme, O., Bridal, S.L.: Dual-mode registration of dynamic contrast-enhanced ultrasound combining tissue and contrast sequences. Ultrasonics **54**, 1289–1299 (2014)
18. Artetxe González, E., Souvestre, F., López Benito, J.R.: Augmented reality interface for E2LP: assistance in electronic laboratories through augmented reality. In: Embedded Engineering Education. Advances in Intelligent Systems and Computing, vol. 421, Chap. 6. Springer, Cham (2016)
19. Kastelan, I., Lopez Benito, J.R., Artetxe Gonzalez, E., Piwinski, J., Barak, M., Temerinac, M.: E2LP: a unified embedded engineering learning platform. Microprocess. Microsyst. Part B **38**(8), 933–946 (2014)
20. Meng, Y., Kim H.C.: A review of accelerometer based physical activity measurement. In: Kim, K.J., Ahn, S.J. (eds.) Proceedings of the International Conference on IT Convergence and Security 2011. Lecture Notes in Electrical Engineering, vol. 120. Springer, Dordrecht (2012)
21. Clark, C.C., Barnes, C.M., Stratton, G., McNarry, M.A., Mackintosh, K.A., Summers, H.D.: A review of emerging analytical techniques for objective physical activity measurement in humans. Sports Med. **47**(3), 439–447 (2017)
22. Gordienko, N., Lodygensky, O., Fedak, G., Gordienko, Y.: Synergy of volunteer measurements and volunteer computing for effective data collecting, processing, simulating and analyzing on a worldwide scale. In: Proceedings of the IEEE 38th International Convention on Information and Communication Technology, Electronics and Microelectronics (MIPRO), pp. 193–198 (2015)

23. Gordienko, N.: Multi-parametric statistical method for estimation of accumulated fatigue by sensors in ordinary gadgets. In: Proceedings of the International Conference on "Science in XXI Century: Current Problems in Physics", 17–19 May 2016, Kyiv, Ukraine (2016). arXiv preprint: arXiv:1605.04984
24. Gordienko, Yu., Stirenko, S., Alienin, O., Skala, K., Soyat, Z., Rojbi, A., López Benito, J.R., Artetxe González, E., Lushchyk, U., Sajn, L., Llorente Coto, A., Jervan G.: Augmented coaching ecosystem for non-obtrusive adaptive personalized elderly care on the basis of cloud-fog-dew computing paradigm. In: Proceedings of the IEEE 40th International Convention on Information and Communication Technology, Electronics and Microelectronics (MIPRO), pp. 387–392 (2017)
25. Gordienko, N., Stirenko, S., Kochura, Yu., Rojbi, A., Alienin, O., Novotarskiy, M., Gordienko, Yu.: Deep learning for fatigue estimation on the basis of multimodal human-machine interactions. In: Proceedings of XXIX IUPAP Conference in Computational Physics (CCP 2017) (2017)
26. Kochura, Yu., Stirenko, S., Alienin, O., Novotarskiy, M., Gordienko, Yu.: Performance analysis of open source machine learning frameworks for various parameters in single-threaded and multi-threaded modes. In: Conference on Computer Science and Information Technologies, pp. 243–256. Springer, Cham (2017)
27. Kochura, Yu., et al.: Data Augmentation for Semantic Segmentation (2018, submitted)
28. Hamotskyi, S., Rojbi, A., Stirenko, S., Gordienko, Yu.: Automatized generation of alphabets of symbols. In: Proceedings of the IEEE 2017 Federated Conference on Computer Science and Information Systems (FedCSIS 2017), Prague, Czech Republic, pp. 639–642, September 2017. arXiv preprint: arXiv:1707.04935
29. Hamotskyi, S., Stirenko, S., Gordienko, Yu., Rojbi, A.: Generating and estimating nonverbal alphabets for situated and multimodal communications. Int. J. Syst. Appl. Eng. Dev. **11**, 232–236 (2017). arXiv preprint: arXiv:1712.04314

Understanding the Adoption of Chatbot

A Case Study of Siri

Hio Nam Io(✉) and Chang Boon Lee

Department of Accounting and Information Management,
University of Macau, Macau, China
helmondio@gmail.com, cblee@umac.mo

Abstract. Due to a recent development in artificial intelligence (AI) and natural language processing, chatbots can understand the human language much better than before. E-commerce businesses are beginning to adopt chatbots in their operations, in areas, such as customer service, product inquiry and transaction refund, etc. However, there is still a lack of studies on users' adoption of chatbots, and businesses are uncertain how to develop chatbots that will increase users' adoption. The purpose of this study is to use sentiment analysis to understand the adoption of chatbots. This study used Siri-related comments posted on the social networking site Weibo during the period January 2017 to July 2017 to conduct the sentiment analysis. The results reveal that users generally had positive emotions with Siri and they used Siri mainly because they wanted to 'come on to' or 'take liberties with' the chatbot. In this study, we also compared Siri and Alime, which is a chatbot developed by Alibaba. This study then explored how the results of the sentiment analysis can be applied to the development of chatbots.

Keywords: Chatbot · Siri · Sentiment analysis · Alime · Adoption
Electronic commerce

1 Introduction

Mobile application makes our life easier, and people are spending more time on their mobile devices every day. Messaging apps are the most frequently-used apps in people's mobile devices, and they usually occupy the top positions in surveys involving usage of mobile applications[1]. Messaging apps have also become an important channel for organizations to interact with their customers. As customers become more demanding, assigning human employees to respond on the messaging platform is no longer adequate, as users now demand quick response, accurate answers, and professional services. Accordingly, some enterprises are now turning to chatbots to overcome these challenges.

Due to progress in artificial intelligence (AI) development and related technologies, chatbots are now more powerful than before as they can understand the human language much better than previously. Chatbots are evolving in many business applications. For

[1] https://hbr.org/2016/09/messaging-apps-are-changing-how-companies-talk-with-customers.

K. Arai et al. (Eds.): FICC 2018, AISC 886, pp. 632–643, 2019.
https://doi.org/10.1007/978-3-030-03402-3_44

example, Burger King enables customers to make order by using its Facebook Messenger chatbot[2]. Expedia has also implemented a chatbot on Facebook Messenger to allow travelers to view hotel options and book the hotel rooms quickly[3]. WeChat, the most popular messaging app in China, empowers the owners of "business accounts" to embed chatbot to interact with their subscribers[4]. Alibaba has also developed a customer service chatbot – Alime, to help sellers on its e-commerce platforms to reply their customers automatically, and it can be used in areas such as product inquiry and order confirmation [1]. There are many practical cases of using chatbots to improve business processes in different industries. However, studies about users' adoption of chatbots are still lacking. Little is known about why users adopt chatbots. The objective of this research is therefore to understand the adoption factors by conducting a sentiment analysis based on users' postings in a social networking website. This study will analyze the users' feelings related to using a chatbot – specifically the "Siri" chatbot. The results obtained from this study can help to guide the future development of chatbots.

This paper is structured as follows: Sect. 2 presents the literature review about chatbots. Section 3 provides the research method for this study and Sect. 4 summarizes the results of the study. Section 5 compares "Siri" and Alime and explores how the results obtained from the sentiment analysis can be applied to Alime. Finally, Sect. 6 concludes this paper with recommendations for future research.

2 Literature Review

2.1 Introduction of Chatbot

A chatbot is a computer program designed to simulate conversations with human users. Chatbots have evolved over the years. The first chatbot called ELIZA [2] was developed in 1996. Recent chatbots, such as Microsoft's Xiaoice and Apple's Siri, are based on the AI technologies, and they can communicate with human beings using natural languages. A recent study compared Eliza with several recent chatbots, such as Ultra Hal machine, Elbot, Cleverbot, Eugene Goostman and JFred and found that Eliza was less conversationally able than the modern chatbots [3]. Note that most of the chatbots developed to date has a common feature – they are mostly general chatbot, not designed for accomplishing specific tasks.

2.2 Categorization of Chatbot

Chatbot can be categorized as general and specific: Apple's Siri is a general-purpose chatbot. Users can ask Siri anything and it will respond as needed. There are also chatbots designed for different areas. In education, MOOCBuddy is a chatbot designed for personalized learning with Massive Open Online Courses (MOOCs) [4]. In e-commerce, Alime is a chatbot developed to assist sellers of online stores to handle

[2] http://fortune.com/2016/05/18/burger-king-bot.

[3] https://viewfinder.expedia.com/features/introducing-expedia-bot-facebook-messenger.

[4] https://chatbotsmagazine.com/china-wechat-and-the-origins-of-chatbots-89c481f15a44.

customer requests 24/7 [1]. There are also chatbots that are for chatting only. For chatting-only chatbots, developers need to engage the users to stay on the chat. The longer chat time is better. For chatbots with specific functions, they need to be able to solve the users' specific problems as quickly as possible. Chatbots can also be categorized based on the technology they use, such as rule-based or deep-learning. For the early chatbots, they are developed using rule-based technologies, such as Artificial Intelligence Markup Language (AIML) [5]; for the recent chatbots, most of them are developed based on deep learning technologies [6–8].

2.3 Technologies and Limitations of Chatbot

Chatbots implemented using rule-based technology, such as AIML make use of a "question-and-answer" knowledge base. For instance, a pattern tag is used to record the questions, such as 'What is your name?' and 'What is the time?' Chatbot will make the response by checking the patterns of the user input with the contents in the pattern tags. Developers can use random tag to select a response from a list of possible random answers. It is like a Q&A system, and the drawback is the chatbot needs a very large knowledge base. Even when the knowledge base is very large, the chatbot may not feasibly respond as the conversations of human beings are very broad and complicated.

From the year 2015, deep learning has attracted very broad attention in the academia and industry [9]. Researchers pay more attention to applying deep learning technologies in chatbots, and now the famous chatbots implemented with deep learning technologies can have better performance when they converse with human beings. Sometime, they can even surprise the user because they can respond like a human being with an unexpectedly accurate answer. The method used to develop such chatbots usually requires the developers to import enormous corpus and apply different algorithms to train the chatbots. The corpus may contain sentences from social network sites, news articles, or drama conversations, depending on the focus of the chatbots. Besides, several messaging app platforms have built their own chatbot ecosystems to enable developers to create chatbots using less complex computing techniques. Examples are Facebook Messenger, Kik, Telegram, Slack and WeChat. In such instances, developers can embed third-party plugins or develop their own chatbots on these platforms to promote their chatbots.

One of the challenges of developing chatbots is the multi-turn language understanding. However, researchers have posited that an end-to-end network making use of Recurrent Neural Networks (RNN) can help to improve multi-round conversations [10]. Researchers are currently engaged in improving chatbot-to-human conversations by using the latest developments in deep learning. This can further enhance the performance of chatbots.

3 Research Method

This study used sentiment analysis to mine the posts on Weibo using the keyword "Siri". Sentiment analysis or opinion mining is a popular research method used in recent years. This method mines the textual data from social network sites, news articles, or product reviews to allow researchers to have a comprehensive understanding of a special issue from a specific group of people. There are two main branches of sentiment analysis techniques: (1) Lexicon based techniques, and (2) Machine learning based techniques [11]. Lexicon based technique is an unsupervised learning approach as training data sets are not required to classify the sentiment polarity of a sentence, instead, this technique uses dictionaries with positive and negative sentiment words to classify the sentiment polarity. When the sentence contains more negative words, it comes as a negative polarity; otherwise, it comes as a positive polarity. Sometime, the accuracy of lexicon based technique is low, for example in the situation of double negation. On the other hand, machine learning based techniques can be supervised or unsupervised, some popular methods include: Naïve Bayes, Support Vector Machine, k-Nearest Neighbor and Maximum Entropy [12]. The method our tool used in this study is a supervised machine learning method; it trains the polarity classification model based on a large amount of Weibo posts, which were labeled as positive or negative by human.

In this study, we crawl on the opinion data from Sina Weibo, which is the most popular social networking site in China. The number of monthly active users in Weibo reached 297 million in September 2016. One advantage of using the postings in Weibo is that it can track current and back-dated postings, which is not allowed in most of other platforms. This study focused on "Siri" as Siri is an intelligent personal assistant embedded in Apple's iOS. The way Siri communicates with its users is through speaking instead of typing. Users of iOS can make use of Siri to make a call, send a message, or start an app, etc. Mining the data about what Siri users think about the chatbot is useful as it can determine how or why people really use chatbots. The results can provide insights regarding the use of chatbots. After the analysis of the Siri data, this study will use the findings and determine how they can be applied to Alime, the chatbot that is developed by Alibaba.

There are several tools that are used in this study to collect and analyze the data. This study used GooSeeker to crawl and search for the opinion data using the keyword "Siri". The data were related postings made on Weibo from January 2017 to July 2017. There were 9,151 posts collected for the analysis. The study then used the Application Programming Interface (API) provided by BosonNLP[5] to perform the sentiment analysis. The researchers wrote a Python script to send the collected posts to the BosonNLP and BosonNLP will return two numbers for each post. One is the positive rating of the post, and the range is from 0 to 1. When it is close to 0, it means the post is in negative sentiment; otherwise, when it is close to 1, it means the post is in positive sentiment. These two numbers are summed as 1. The accuracy of the positive and negative emotion analysis is 85% $\sim$ 90%. Finally, the Picdata software is used to perform keyword extraction. After the sentiment analysis, the study used Picdata to extract the most positive posts and the most negative posts. The study used these posts to provide a summary of the factors that bring about positive or negative emotions when using Siri. Figure 1 shows the workflow of the sentiment analysis.

[5] http://bosonnlp.com.

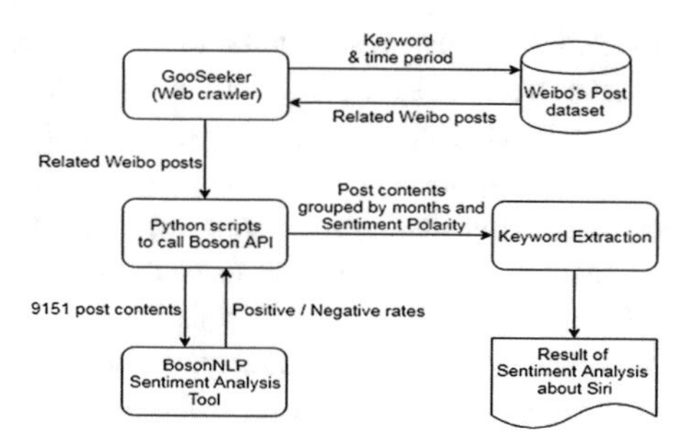

Fig. 1. Workflow of the sentiment analysis.

4 Results

4.1 Positive Result of Sentiment Analysis About Siri on Sina Weibo

As mentioned in Sect. 3, this study used the BosonNLP to conduct the sentiment analysis for the 9,151 Siri-related Weibo posts during the period of January to July 2017. The average positive rating of all posts is 0.68, and the negative rating is 0.32. The sum of the positive rating and negative rating of a same post is approximately equal to 1. In the following analysis, we refer only to the positive rating. When it is close to 1, it means the emotion of the post is more positive; otherwise, when it is close to 0, it is more negative. Table 1 shows the summary of the results for each of the seven months. The table show that the number of posts crawled in July is more than those in the previous months. The main reason is that Apple held its annual Worldwide Developers Conference (WWDC) in June and Siri is a special topic in this conference, which stimulate more discussions about Siri. The average positive rating across the seven months is quite stable, ranging from 0.67 to 0.71. The implication is that generally, there are positive emotions when people discussed about Siri on Weibo.

Table 1. Summary of the crawled posts

	Total no. of posts crawled	*Average positive rating*
January	1316	0.68
February	1121	0.67
March	1245	0.67
April	826	0.66
May	784	0.69
June	987	0.71
July	2872	0.69

To gain a deeper understanding of the data, we applied the keyword extraction method to determine the most popular issues people discussed about Siri. The study separated the posts based on two different emotions, and only extracted the keywords from the posts with the higher polarity (> 0.8). The study also obtained a list of terms with high frequencies and weights. Frequency refers to the times a specific term appears in the selected posts.

Table 2. Top 10 terms extracted from all crawled positive posts (with English translation in parentheses)

	Frequency	*Weight*
苹果 (Apple)	266	0.9558
调戏 ("Come on to")	241	1
语音 (Voice)	206	0.9316
智能 (Intelligent)	182	0.9104
助手 (Assistant)	164	0.9158
可爱 (Lovely)	139	0.8829
手机 (Mobile phone)	133	0.8637
巨石 (The Rock)	129	0.8857
厉害 (Powerful)	124	0.8801
Apple	109	0.8737

However, a high frequency does not mean the term is important. The study therefore also included the weight indicator to determine the importance of the terms in the posts. The weight was calculated using TF-IDF measures [13]. Note that in this study, we only analyzed the terms with a high frequency (threshold > 100). Table 2 lists the top 10 terms from the positive posts (with positive rate > 0.8). We discuss the significant findings below:

– **Apple's voice intelligent assistant:** As Siri is a product embedded in iOS, the Apple's operating system, when people talked about Siri, the posts usually come with Apple, like "Apple's Siri". So, it is not unexpected that "Apple" is the most frequently used term in the results. Similarly, some of the frequently used terms such as "voice", "intelligent" and "assistant" are describing the functions of Siri.
– **Come on to Siri:** The weight for "Siri" is not the highest. Instead, the highest is the No. 2 term "Come on to", which is an English translation from Oxford Dictionaries meaning "make sexual advances towards" originally. In the Weibo posts we crawled, "come on to" means "play around with …" or "to take liberties with". This is a very interesting finding that people like to chat with chatbot and "come on to" it. The reason is, as people realized that chatbot is not a human being, many people treat the relationship with chatbot as a master-slave relationship, and not respect it as a human being. Table 3 lists some selected positive posts containing the term "Come on to". People will try something that is not easy to talk to a human being, like "I love you". Siri is also like a toy bringing fun to people. Examples include "Come on to Siri is very funny", "my Siri is a warm boy" and "A good day starts from coming on to Siri". People are also exploring the functions of Siri. They will try to ask different questions to find out how Siri respond to them, or to discover the functions of Siri that they don't know, such as "How do you come on to Siri usually?", "My first time to come on to Siri" and "I can discover something new every time when I come on to Siri".
– **Lovely Siri:** Some people think Siri is lovely because it can do things like a human being. For example, some users ask Siri to sing a song, or when someone asked Siri some strange questions, it can reply with some unexpected answers. In addition, Siri can also serve as a friend for the children and chat with them, or act as a trainer to train ones' speaking capability, or to help someone from getting bored. Tables 4 show selected positive posts containing the term "Lovely".

Table 3. Selected negative posts containing the term "Come on to" (with English translation in parentheses)

Post content	*Positive rate*
Chatting with Siri in a "master-slave relationship"	
大晚上的调戏 siri 说我爱你，我家 siri 回了一句：使不得。 (Tonight I come on to Siri and say "I love you", it replies me "You cannot".)	0.9361
睡醒调戏下 Siri (When I wake up, I come on to Siri)	0.9489
网友调戏 Siri 的合集，这机器人也太睿智了~ (This is a collection of how people coming on to Siri, this bot is to wise)	0.9085
闲着没事调戏了一下 siri (I am too boring and have nothing to do, just come on to Siri)	0.9622
Chatting with Siri like a toy or a friend	
#调戏 siri# 好搞笑 (Come on to Siri is very funny)	0.9666
#我有 100 种方式调戏 siri# 原来我们家的 siri 也是个暖心情话小 boy (I have 100 methods to come on to Siri, my Siri is a warm boy)	0.9215
美好的一天从调戏 Siri 开始 (A good day starts from coming on to Siri)	0.9630
每天调戏 Siri 是一种乐趣 (Coming on to Siri every day is a kind of fun)	0.9095
Chatting with Siri with curiosity	
通常你们怎么调戏 Siri?还能用它做些什么？拜托分享一下。 (How do you come on to Siri usually? What can it do else? Please share.)	0.9324
第一次调戏 Siri，感觉好好玩！ (My first time to come on to Siri, it is quite funny!)	0.9601
调戏 siri，每次都有新发现。 (I can discover something new every time when I come on to Siri)	0.9785
无聊的时候 记得去调戏 siri 玩 (Remember to come on to Siri when you are boring)	0.9422

Table 4. Selected positive posts containing the Term "Lovely" (with English translation in parentheses)

Post content	*Positive rate*
看到看到好多人让 Siri 唱 PPAP，所以我也玩了下，太可爱了 (To many people ask Siri to sing PPAP, let me try, it is very lovely)	0.9571
Siri 真的很可爱，我问它手机解锁密码是多少，它真的会出各种各样的答案。 (Siri is very lovely, I ask it the mobile phone unlock password, it give me various answers)	0.9107
看到儿子和 Siri 的对话，人工智能好可爱。 (Notice the conversations between my son and Siri, AI is very lovely)	0.9925
口语陪练 聊天解乏 这么可爱的 siri (Chatting to train speaking and avoid bored, what a lovely Siri)	0.9382

- **Siri and "The Rock":** "The Rock" is a special term, which is the ring name for Dwayne Johnson, an American actor, producer and professional wrestler. He is hired by Apple to make a movie with Siri, to make Siri become more popular. In the movie, Siri helped "The Rock" to handle many daily tasks. This is a strategy that Apple used to promote Siri, and it stimulated a lot discussion on social network sites. Table 5 shows examples of post related to "The Rock."

Table 5. Selected positive posts containing the Term "The Rock" (with English translation in parentheses)

Post content	*Positive rate*
【巨石强森 x Siri 的最新广告来了】大概讲了 Siri 帮助 The Rock 处理繁杂日程的故事 (The latest ads about The Rock X Siri is coming, it is a story about how Siri helps The Rock to handle complicate daily tasks)	0.9212
巨石强森新片，女主竟是 Siri? (A new movie of The Rock Johnson, the actress is Siri?)	0.9688
巨石强森和 Siri 共同出演电影，明天首映要去看看 (The Rock Johnson makes a film with Siri, I will watch the premiere tomorrow)	0.9419
好在有 Siri 为你待命，随时准备聆听你的需求，向你伸出援手。有什么事，跟 Siri 说就可以了。 (Luckly Siri is stand by you, listen to your order and give you a hand all the time, just tell Siri for any matter)	0.9631

– **Powerful Siri:** Among the positive posts about Siri, "powerful" is also a frequent term that appeared in the posts. Undoubtedly, people will feel positive if the chatbot can show powerful functionalities during the chatting process. Table 6 show examples of post related to "powerful" Siri.

Table 6. Selected positive posts containing the term "Powerful" (with English translation in parentheses)

Post content	*Positive rate*
Siri 你真厉害 (Siri, you are so powerful)	0.9631
厉害了我的 Siri 你还真会唱 PPAP 啊 (What a powerful Siri, you actually can sing PPAP)	0.9505
厉害了 Siri！，Siri 将支持上海话 (What a powerful Siri! Siri will support ShangHai language)	0.9340
Siri 好厉害呀，问了下萨摩耶和爱斯基摩有什么区别，一秒钟给了一屏幕的答案 (Siri is very powerful, I ask it the difference between Samoyed and Eskimo Dog, it gives me answers full of my screen after a second)	0.9637

4.2 Negative Result of Sentiment Analysis About Siri on Sina Weibo

Besides analyzing the positive posts about Siri, this study also analyzed the negative posts. Table 7 lists the top 10 terms from the negative posts (with positive rating < 0.2), and the findings are described below.

In Table 7, the term "Mobile phone" is the only one with a frequency higher than 100. The detailed results provide some interesting posts that show why people have with a negative emotion with Siri when using their mobile phones. The posts found that there are two types of situations involving negative emotions. One occurred because of Siri, while the other did not involve Siri.

Table 7. Top 10 terms extracted from all crawled negative posts (with English translation in parentheses)

	Frequency	*Weight*
手机 (Mobile phone)	113	1
声音 (Sound)	37	0.8962
讲话 (Speak)	32	0.8979
口语 (Spoken language)	29	0.8709
聊天 (Chatting)	25	0.8438
国外 (Foreign)	23	0.8496
语音 (Voice)	21	0.8363
主人 (Master)	19	0.833
无聊 (Bored)	18	0.8231
调戏 ("Come on to")	18	0.8595

Table 8. Selected negative posts vontaining the term "Mobile Phone" (with English translation in parentheses)

Post content	*Positive rate*
Negative emotion caused by Siri (unexpected sound output)	
今天座谈会，大领导在我旁边讲话，我默默掏出手机开始刷游戏，领导突然转向了我，我吓得一激灵，手忙脚乱想摁掉游戏，结果不知道咋的居然触发了 siri，然后大家就听到 siri 洪亮有力的声音：对不起，我听不懂你在说什么。。。。 (During the forum today, when my boss was having a talk next to me, I played game with my mobile phone. Suddenly my boss turned to me, I was scared. When I was going to turn off the game, I turned on Siri unexpectedly, then everybody can hear a strong sound from Siri, "I'm sorry, I don't understand what you just said" …)	0.0777
太尬了.老师在讲插画.我在电脑上画作业.手机就放在左手边.不知道肘部什么时候碰到了我的 home 键.老师刚刚讲完内容.我的 siri 突然开口：抱歉我不太清楚你在说些什么... (What an embarrassing moment. When my teacher was talking and I was doing my assignment, I didn't know how my elbow touched the home button of my mobile phone, and this was the moment my teacher finished the class, then my Siri suddenly sound out "I'm sorry, I don't understand what you just")	0.0486
Negative emotion caused by Siri (high expectation)	
今天下大雨没带伞，然后，我掏出了手机 默默问了 siri，下雨天没带伞怎么办，他说 我为你找到了这些资料，我说，你应该找人来接我。他回 我尽力了。废物！ (It rained today, but I didn't have an umbrella. Then, I took out my mobile phone, and asked Siri, "What can I do when I don't have an umbrella when it is raining?" -It said, "I have found this information for you" -I said, "You should find someone to pick me up" -It said, "I have tried my best" -I said, "Rubbish!")	0.0289
就说 siri 的人工智能水平明明还可以，为什么就是听不懂打开手机自带的电筒这个指令呢…… (It is said the AI capability of Siri is ok, why can't it understand the command of switching on the torch...)	0.0211
Negative emotion NOT caused by Siri	
我的手机处于这个状态已经好几个小时了。APP 消失，关不了机，用不了 Siri，怎么办…… (My mobile phone has been kept at this stage for several hours. The applications disappear, I cannot switch it off, Siri doesn't work, what can I do?)	0.0837
刚经历了手机死机只有 Siri 能用的绝望 关机键没反应 (Just experienced the disappointment of mobile phone down, only Siri work, and the power button doesn't work)	0.0704

For those negative emotions involving Siri, the reasons include: (1) Unexpected sound output. As Siri is a voice intelligent assistant, a sudden or unexpected voice output may cause embarrassment; (2) High user expectation. Users may have negative emotion when Siri cannot perform what they expected them to do with their mobile phones. The first two parts of Table 8 show some selected examples of post with negative emotions that are caused by Siri.

In some situations, the negative emotions are not caused just by Siri itself. The negative emotion can be caused due to damage of the mobile device or other incidents. The last part of Table 8 shows some examples of post with negative emotion that are not caused by Siri.

5 Comparing Siri and Alime

Alime is an intelligent shopping assistant released by Alibaba in 2015 serving millions of TaoBao customers online. Alime has been trained by using a very large amount of dialogs with deep learning technologies. Now, it can also handle multi-round dialogs with customers. On November 2016 (The Double 11 Festival), Alime handled 18 million rounds of dialogs and solved 95% issues. The tasks Alime can perform include payment, shipping, refund request handling, password reset, etc.

There are two major differences between Alime and Siri. The first is the input method. Siri uses voice to chat with its users. When Siri sounds out or when the users speak to Siri, everyone around the mobile phone can listen to the conversations between Siri and the users. For Alime, the major input method is typing, but users still can use the voice-to-text function on some advance mobile devices to chat with Alime. Alime will only respond with texts or menus with voice. This can avoid the embarrassing situation like Siri did in the post mentioned in the previous section (see Table 8). The second difference is the functionality between the two chatbots. Siri is a general purpose chatbot. Users can ask Siri anything, and it will respond accordingly. However, Alime is designed for e-commerce. If a user asks Alime something that is not related to its "responsibility", it will respond like "What?" or give some fuzzy answers. If it is asked some question related to the Alibaba business, it can respond appropriately.

In our findings, the term "come on to" is quite important for the conversation between people and the chatbot. Curiosity is one of the reasons why people like to "come on to" chatbot. People cannot say something rude or impolite to other people. So, they are curious what happens if they do it to a chatbot. Hence, when developers design the chatbots, they should consider how to handle those rude or impolite sentences. We tried to test whether Alime is able to reply when people "come on to" it, and the results show it can handle the sentences quite well. For example, when the study input "I love you", it will answer "Let's get married"; when the study input "Go out!", it will answer "OK, I will go, but how long can I come back?". If a chatbot meets a "come on to" question and it cannot reply it like a human being, the user may be immediately disappointed, and may not adopt the chatbot for a long time. Conversely, if the chatbot can reply like a human being, the user may love to keep interacting with it.

6 Conclusion

This study conducted a sentiment analysis about Siri postings on the Chinese social networking website Weibo. The analysis indicates that users generally have a positive emotion about Siri as the average positive rating for all posts is 0.68. There are two important findings in this study: (1) People love to "come on to" a chatbot. As such, developers and businesses need to consider how the chatbot could respond to "come on to" inputs from the users. If it cannot handle them well, it may lose user satisfaction. A funny reply to "come on to" can help bring positive emotion for the users and it can increase the chatbot adoption rate. (2) Unexpected sound output from Siri may embarrass the users. There are pros and cons of having the voice feature for Siri. Developers need to be mindful when incorporating the voice feature when they design their chatbots.

There are some limitations for the results of this research. The results of this study are based on the analysis of data gathered from one Chinese social networking website. We need to be careful, therefore, when generalizing the results across time and place. Future research can be conducted by using data from other social networking websites or from people at different geographic locations. Alternatively, data could also be gathered from people who are not members of the social networking websites. These studies can provide comparison and they can help to determine if there are differences in the results.

References

1. Qiu, M., Li, F.L., Wang, S.Y., Gao, X., Chen, Y.: AliMe Chat: A sequence to sequence and rerank based chatbot engine. In: Proceedings of the 55th Annual Meeting of the Association for Computational Linguistics, pp. 498–503 (2017)
2. Weizenbaum, J.: ELIZA—a computer program for the study of natural language communication between man and machine. Commun. ACM **9**(1), 36–45 (1966)
3. Shah, H., Warwick, K., Vallverdú, J., Wu, D.: Can machines talk? Comparison of Eliza with modern dialogue systems. Comput. Hum. Behav. **58**, 278–295 (2016)
4. Holotescu, C.: MOOCBuddy: a chatbot for personalized learning with MOOCs. In: Rochi – International Conference on Human-Computer Interaction, vol. 8, pp. 91–94 (2016)
5. Shawar, B.A., Atwell, E.: A comparison between ALICE and Elizabeth chatbot systems. Raport instytutowy, University of Leeds (2002)
6. Ngo, T.L., Pham, K.L., Cao, M.S., Pham, S.B., Phan, X.H.: Dialogue Act Segmentation for Vietnamese Human-Human Conversational Texts. arXiv preprint, arXiv:1708.04765 (2017)
7. Li, Y.: Deep Reinforcement Learning: An Overview," arXiv preprint, arXiv:1701.07274 (2017)
8. Li, J.W., Monroe, W., Ritter, A., Galley, M., Gao, J.F., Jurafsky, D.: Deep Reinforcement Learning for Dialogue Generation, arXiv preprint, arXiv:1606.01541 (2017)
9. LeCun, Y., Bengio, Y., Hinton, G.: Deep learning. Nature **521**, 436–444 (2015)
10. Chen, Y.N., Hakkani-Tür, D., Tur, G., Gao, J.F., Deng, L.: End-to-end memory networks with knowledge carryover for multi-turn spoken language understanding. In: INTERSPEECH (2016)

11. Aye, Y.M., Aung, S.S.: Sentiment analysis for reviews of restaurants in Myanmar text. In: 2017 18th IEEE/ACIS International Conference on Software Engineering, Artificial Intelligence, Networking and Parallel/Distributed Computing (SNPD), Kanazawa, Japan, pp. 321–326 (2017)
12. Maharani, W.: Microblogging sentiment analysis with lexical based and machine learning approaches. In: 2013 International Conference of Information and Communication Technology (ICoICT), Bandung, pp. 439–443 (2013)
13. Aizawa, A.: An information-theoretic perspective of TF-IDF measures. Inf. Process. Manag. **39**, 45–65 (2003)

Laughbot: Detecting Humor in Spoken Language with Language and Audio Cues

Kate Park(✉), Annie Hu, and Natalie Muenster

Department of Computer Science, Stanford University, Stanford, USA
{katepark,annie,Ncm000}@cs.stanford.edu

Abstract. We propose detecting and responding to humor in spoken dialogue by extracting language and audio cues and subsequently feeding these features into a combined recurrent neural network (RNN) and logistic regression model. In this paper, we parse Switchboard phone conversations to build a corpus of punchlines and unfunny lines where punchlines precede laughter tokens in Switchboard transcripts. We create a combined RNN and logistic regression model that uses both acoustic and language cues to predict whether a conversational agent should respond to an utterance with laughter. Our model achieves an F1-score of 63.2 and accuracy of 73.9. This model outperforms our logistic language model (F1-score 56.6) and RNN acoustic model (59.4) as well as the final RNN model of D. Bertero, 2016 (52.9). Using our final model, we create a "laughbot" that audibly responds to a user with laughter when their utterance is classified as a punchline. A conversational agent outfitted with a humor-recognition system such as the one we present in this paper would be valuable as these agents gain utility in everyday life.

Keywords: Chatbots · Spoken natural language processing
Deep learning · Machine learning

1 Introduction

Our project takes the unique approach of building a "laughbot" to detect humor and thus predict when someone will laugh based on their textual and acoustic cues. Humor depends on both the strings of words, how the speaker's voice changes, as well as situational context. Predicting when someone will laugh is a significant part of contemporary media. For example, sitcoms are written and performed primarily to cause laughter, so the ability to detect precisely why something will draw laughter is of extreme importance to screenwriters and actors. Being able to detect humor could improve embodied conversational agents who are perceived as having more human attributes when they are able to display the appreciation of humor [1]. Siri currently will only say "you are very funny", but could become more 'human' if she had the ability to recognize social cues like humor, and respond to them with laughter.

Thus, our objective is to train and test a model that can detect whether or not a line is "funny" and should be appropriately responded to with laughter.

K. Arai et al. (Eds.): FICC 2018, AISC 886, pp. 644–656, 2019.
https://doi.org/10.1007/978-3-030-03402-3_45

2 Limitations of Previous Work

Existing work in predicting humor uses a combination of text and audio features and various machine learning methods to tackle the classification problem.

The goal of research done in [2] was to predict laughter and create an avatar that imitated a human expert's sense of when to laugh in a conversation. They reduced the problem to a multi-class classification problem, where they found that using the large margin algorithm resulted in the most natural behavior (laughing with the same proportion to the expert).

A later paper [3] examined the problem of using a laughter-enabled interaction manager to make decisions about whether to laugh or not. They defined the regularized classification for apprenticeship learning (RCAL) algorithm using audio features. Final results showed that RCAL performed much better than multi-class classification on laughter and slightly worse on silence. The paper noted the problem of having a large class imbalance between laughter and non-laughter in the dataset, and suggested potential future work in inverse reinforcement learning (IRL) algorithms. From this, we propose weighting the datasets to reduce the imbalance so that the non-laughter does not overpower what the model will learn from punchlines.

In [4], author compared three supervised machine learning methods to predict and detect humor in *The Big Bang Theory* sitcom dialogues. From a corpus where punchlines were annotated by laughter, they extracted audio and language features. Audio features and language features were fed into three models: convolutional neural network (CNN), RNN, and conditional random field (CRF), and compared against a logistic regression baseline (F1-score 29.2). The CNN using word vectors and overlapping time frames of 25 ms performed the best (F1-score of 68.6). The RNN model (F1-score 52.9) should have performed the best but actually performed worse than the CNN, likely due to overfitting. Their paper proposes future work building a better dialog system that understands humor.

3 Present Work

Rather than a sitcom's dataset, our paper uses conversational conversations whose humor we hypothesize is more relevant for a conversational agent like Siri. We create an ensemble model combining a recurrent neural network with a logistic regression classifier that, when run on audio and language features from lines of conversation, can identify whether a line is humorous. We additionally implement a laughbot, a simple dialog system based on our ensemble model that converses with a user and responds to humorous input with laughter.

In Sect. 4, we introduce the Switchboard dataset we used for training, the preprocessing steps we performed, and the features we extracted to implement our models. Section 5 then gives implementation details for these models, including a baseline, a language-only, an audio, and a combined model, and Sect. 6 gives details for our laughbot interface. In Sect. 7 we define our evaluation metrics, present results, and analyze errors. Finally, in Sect. 8 we discuss future plans to improve our humor-detection model and our laughbot system.

4 Dataset

Since the focus of this project is detecting humor in everyday, regular speech, our dataset must reflect everyday conversations. We use the Switchboard Corpora available on AFS, which consists of around 3000 groups of audio files, transcripts, and word-level time intervals from phone conversations between two speakers. We classified each line of a transcript as a "punchline" if it preceded any indication of laughter at the beginning of the next person's response. Tables 1, 2, and 3 show sample lines and their corresponding classifications.

Table 1. Line inducing laughter is classified as a punchline

B: That, that's the major reason I'm walking up the stairs	Punchline
A: [Laughter] To go skiing?	–

Table 2. Laughter that induces laughter is classified as a punchline

A: Uh-huh. Well, you must have a relatively clean conscience then [laughter]	Punchline
B: [Laughter]	–

Table 3. Line preceding someone laughing at themselves does not count as a punchline

A: just for fun	–
B: Shaking the scorpions out of their shoes [laughter]	–

We split our data set into 80% train/10% validation/10% test sets. Considering the imbalanced datasets that previous work worked with, we sampled 5% of the non-punchlines because our original dataset is heavily imbalanced towards non-punchlines. This achieved a more balanced training set among positive (punchlines) and negative (unfunny lines) classes. Our final datasets each have about 35–40% punchlines.

4.1 Features

We extracted a combination of language and audio features from the data.

Language features include:

- Unigrams, bigrams, trigrams: we pruned the vocabulary and kept the n-grams that appear more than a certain frequency in the training set, tuning the threshold on our val set. In this model, we kept n-grams that appear at least twice.
- Parts of speech: we implemented **NLTK's POS-tagger** to pull the number of nouns, verbs, adjective, adverbs and pronouns appearing in the example [5].
- Sentiment: we utilized **NLTK's vader** toolkit to extract sentiment from the punchline, in a scale of more negative to more positive [5].
- Length, average word length: from reading past work, we learned sitcom punchlines are often short, so we used length features [4].

We also extracted acoustic features from each audio file (converted to .wav format) with the openSMILE toolkit and matched the timestamped features to the timed transcripts in Switchboard to extract corresponding acoustic and language features for a given example [6].

Acoustic features include:

- MFCC: We expected MFCC vectors to store the most information about an audio sample, so we sampled 12 vectors every 10ms with a maximum of 50 time intervals per example, since certain lines may be too long to fully store.
- Energy level: We also expected the speaker's energy to be a strong indicator of humor, so we included this as an additional feature.

5 Implemented Models

5.1 Baseline

Our baseline was an "all positive classifier", predicting every example as a punchline. The precision of this classifier is the proportion of true punchlines in the dataset (around 35%) and 100% recall. We also use "all negative classifier", predicting every line as unfunny which has a precision of the proportion of unfunny lines (around 65%) and 0% recall. Table 4 displays these baseline results.

Table 4. Baseline metrics

Classifier	Precision	Recall	F1-score
All positive	37.2	100.0	54.2
All negative	52.8	0.0	0.0

5.2 Logistic Regression Language Model

We trained a logistic regression model using only language features (ngrams, sentiment, line length) as a secondary baseline. Logistic regression was an intuitive starting model for binary classification, and also allowed us to observe and tune the performance of just our language features on predicting humor.

5.3 RNN Acoustic Model

We next trained a RNN using only acoustic features, to observe the performance of our acoustic features on classifying lines of audio as punchlines or not. We chose an RNN to better capture the sequential nature and thus conversational context of dialogue, and we use Gated Recurrent Unit (GRU) cells so our model can better remember earlier timesteps in a line, instead of overemphasizing the latest timesteps. During training, we used standard softmax cross entropy to calculate cost. We initially used an Adam optimizer because it handles less frequently seen training features better and converges smoother than stochastic gradient descent. Our final RNN uses an Adamax optimizer to further stabilize the model between epochs and to make the model more robust in handling less-frequently seen features and gradient noise.

5.4 Final Combined Model

After designing separate language and acoustic models, we combined the two by:

(1) Running our RNN on all acoustic features in the training set, and extract the final hidden state vector in the RNN on each training example.
(2) Concatenating this vector with all language features for its corresponding training example.
(3) Using the combined feature vectors to train a logistic regression model.

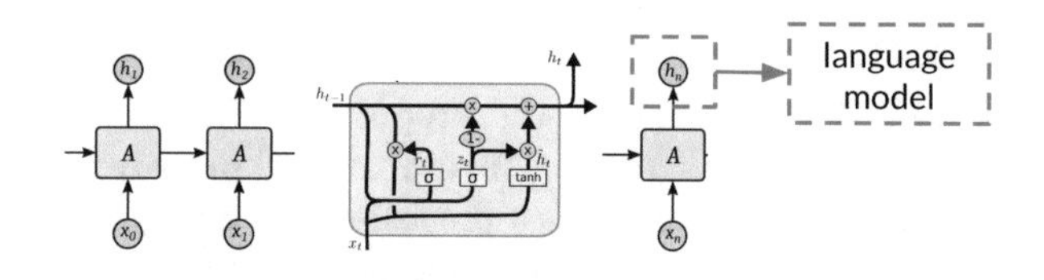

Fig. 1. Diagram of final RNN + Logistic Regression model.

Figure 1 shows our combined model architecture. For testing, we followed a similar process of running the acoustic features into our pre-trained RNN, concatenating the final hidden state vector with language features, and running the combined feature vector through our pre-trained logistic regression model to see the prediction.

6 Laughbot Application Architecture

6.1 Overview

The laughbot is a simple user-interface application that implements the model we built during our research and testing, predicting humor using a chatbot-style audio prompter. It is intended for demonstration purposes for both letting users experiment with custom input as well as showing the results of our project in an accessible and tangible form. The user speaks into the microphone, after which the laughbot will classify whether what was said was "funny", and audibly laugh if so.

6.2 Transcription Architecture

The laughbot is designed to take user input audio, transcribe it, and feed the audio file and transcription into our pre-trained RNN and logistic regression model. Multithreading allows the user to use the microphone to speak for as much of the maximum 60 s time segment as he or she would like, before pressing "Enter" to indicate end of speech. The audio is then saved as a .wav file and transcribed by hitting the **Google Cloud Speech API**.

Both the transcription and the original audio file are sent through the pre-trained model in which acoustic features are extracted and run through the pre-trained RNN. The last hidden states are combined with textual features extracted from the transcription as features for the entire logistic regression model. Once a funny or not funny classification is obtained the laughbot will either keep a straight face by staying silent and simply prompt for more audio, or it will randomly play one of several laughtracks that we recorded during the late night hours of project development. The classification is almost immediate. The brunt of the runtime of our implementation depends on the speed of transcription from Google Cloud Speech API, thus the strength of the wifi connection. To see a sample of our laughbot in action, see this video: https://youtu.be/t6Je0kZnyxg.

7 Results

7.1 Evaluation

We evaluated using accuracy, precision, recall, and F1 scores, with greatest emphasis on F1 scores. Accuracy calculates the proportion of correct predictions:

$$\text{Accuracy} = \frac{TP + TN}{TP + TN + FP + FN}$$

Precision calculates the proportion of predicted punchlines that were actual punchlines:

$$\text{Precision} = \frac{TP}{TP + FP}$$

Recall calculates the proportion of true punchlines that were captured as punchlines:

$$\text{Recall} = \frac{TP}{TP + FN}$$

F1 is the harmonic mean of precision and recall, which can be calculated as below:

$$F1 = 2 \times \frac{\text{Precision} \times \text{Recall}}{\text{Precision} + \text{Recall}}$$

Table 5. Comparison of all models on all datasets

Classifier	Accuracy	Precision	Recall	F1-score
Logistic regression (train)	87.7	90.1	76.7	82.8
RNN (train)	75.0	73.8	55.2	63.2
Combined (train)	91.1	90.5	86.3	88.3
Logistic regression (validation)	68.5	60.2	48.5	53.7
RNN (validation)	70.0	61.5	54.7	57.9
Combined (validation)	73.4	66.0	60.5	63.1
Logistic regression (test)	70.6	62.7	51.4	56.5
RNN (test)	71.7	63.5	55.9	59.4
Combined (test)	73.9	66.5	60.3	**63.2**

Table 5 and Fig. 2 show the final performance of our models on these metrics, evaluated on the test dataset. Notably, our final model not only beat our baseline, it also beat the final RNN model of [4], the most similar approach to our own model. Bertero et al. used "The Big Bang Theory" sitcom dialogues which we hypothesize is an easier dataset to classify than general phone conversations. Their dataset also had a higher proportion of punchlines leading to a higher F1 score in their positive baseline. Their final Convolutional Neural Network (CNN) model, a deep feed-forward neural network, performed the best, but improved less compared to their baseline (14.36% improvement) than ours performed compared to our baseline (16.35% improvement). Further, while the final CNN model proposed by Bertero et al. had a higher F1-score than our final RNN model, our model had higher accuracy (see Table 5).

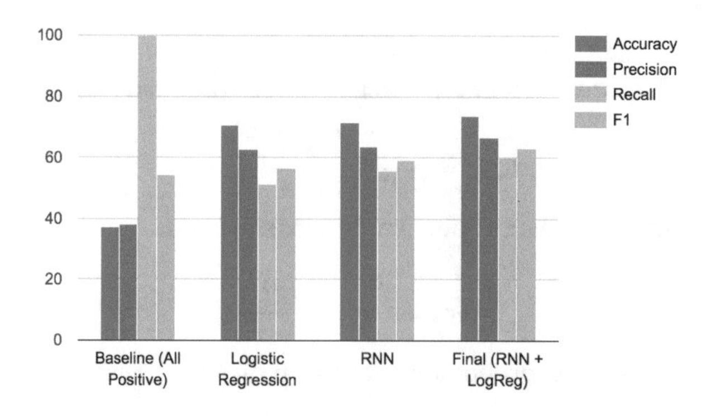

Fig. 2. Comparison of models on test datasets.

7.2 Model Analysis

Our final combined RNN acoustic and logistic regression language model performed the best of all our models. This fit our expectations, as humor should depend on both *what* was said and *how* it was said. Both the language only and audio only model had fairly similar accuracies to the final model, but had much lower recall scores (especially the language model), suggesting that the combination model was better at correctly predicting punchlines while the individual models perhaps tended to be too conservative in their predictions, in that they predicted non-punchlines for too many true punchlines.

We tuned our RNN and regression models separately on the validation set. We found that the language model performed best when the frequent n-grams threshold was set at 2 (so we only included n-grams that occurred at least twice in the training set), and performance dropped as this threshold was increased. This makes sense since for bigrams and especially trigrams, the number that appear at least x times in a dataset drops drastically as x increases, so with a too-high threshold, we were excluding too many potentially useful features.

We also found that sentence length was a particularly important feature, which confirmed our expectation that most punchlines would be relatively short. With the RNN, we found that increasing the number of hidden states greatly improved model performance up to a certain point, then began causing overfitting past that point. The same was true of the number of epochs we ran the model through during training. As we were using an Adamax optimizer, which already performs certain optimizations to adapt the model learning rate, we did not perform much tuning on our initial learning rate.

7.3 Error Analysis

Table 6 and Fig. 3 show the performance of our language-only, audio-only, and combined models on our training, validation, and test datasets. All models performed significantly better on the training set than on the validation or test set, especially the final combined model, suggesting that our model is strongly overfitting to the training data. This may be helped by hyperparameter tuning

Table 6. Comparison of our model against models in Bertero et al.

Classifier	Accuracy	Precision	Recall	F1-score
Bertero's positive baseline	42.8	42.8	100.0	59.9
Our positive baseline	37.2	37.2	100.0	54.2
Our Logistic regression (language only)	70.6	62.7	51.4	56.5
Our RNN (audio only)	71.7	63.5	55.9	59.4
Bertero's Final RNN	65.8	64.4	44.9	52.9
Our final model (RNN + LogReg)	**73.9**	66.5	60.3	**63.2**
Bertero's Final CNN	73.8	70.3	66.7	68.5

such as decreasing the number of epochs or number of hidden units in our RNN, or changing the regularization of our logistic regression model. Simplifying the model could also help by decreasing the maximum number of MFCC vectors to extract or decreasing the number of language features. As we explore in the next section, overfitting may be reduced by training on larger datasets.

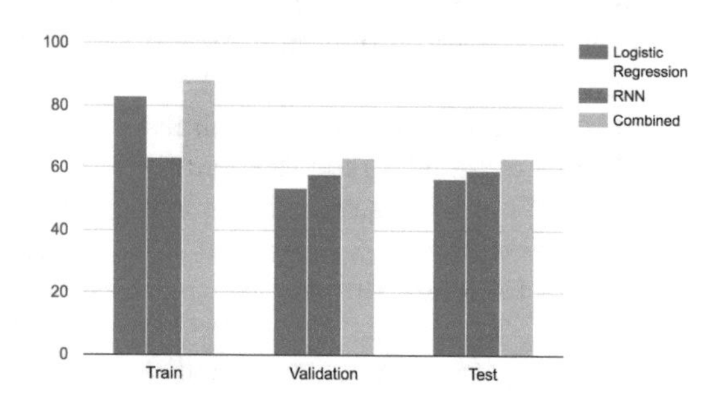

Fig. 3. Comparison of F1-scores on train, val, and test.

7.4 Dataset Analysis

We initially ran our models on only 20% of the full Switchboard dataset to speed up development. Figure 4 shows the training accuracy and cost curves of our RNN model, which suggest strong overfitting to the training set. Once we finalized our combined RNN and logistic regression model we continued to run on larger portions until we used the full 100% of the Switchboard dataset and achieved the highest F1 score on the test set. Table 7 shows our model performance on different portions of the dataset, with noticeably less overfitting and better test set performance as the dataset portion increased. This suggests that having an even larger dataset could achieve better results.

Table 7. Model performance on varying dataset sizes

Classifier	Dataset size	Accuracy	Precision	Recall	F1-score
Combined (train)	20% (4659 examples)	95.1	96.3	90.7	93.4
Combined (test)	20% (618 examples)	69.7	44.84	61.0	51.7
Combined (train)	50% (12011 examples)	93.8	92.4	91.1	91.8
Combined (test)	50% (1527 examples)	71.3	63.5	58.0	60.6
Combined (train)	100% (23658 examples)	91.1	90.5	86.3	88.3
Combined (test)	100% (2893 examples)	73.9	66.5	60.3	**63.2**

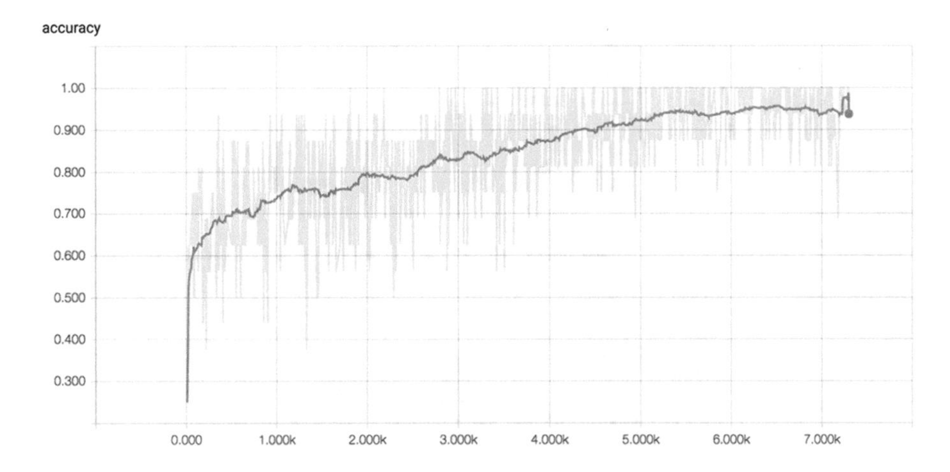

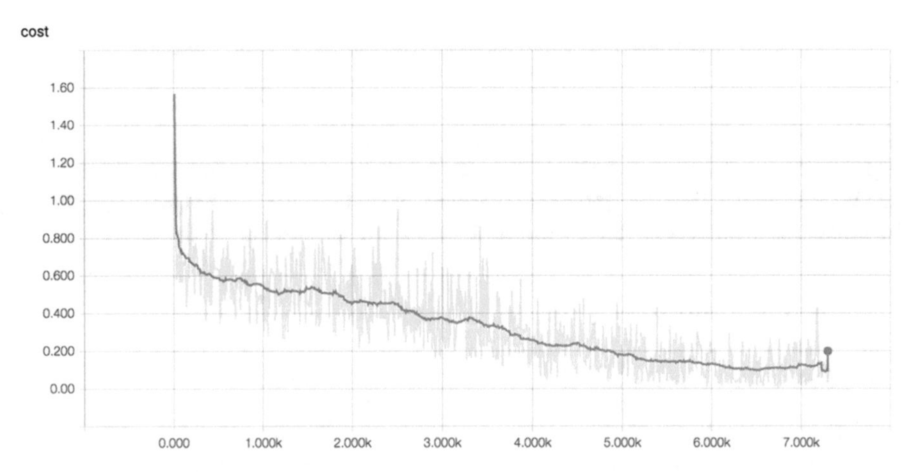

Fig. 4. RNN model train accuracy and cost on 20% of dataset.

7.5 Laughbot Analysis

In testing our laughbot by speaking to it and waiting for laughter or silence as shown in Figs. 5 and 6, our laughbot responded appropriately to several types of inputs.

We noticed that laughter was often caused by shorter inputs (though not in all cases, as seen in Table 8), as well as by laughter in the punchline. On longer inputs or inputs with negative sentiment or both, laughbot generally correctly considered the line as not a punchline. Laughbot responded positively to jokes, at some cases waiting for the actual punchline. Laughbot considered questions and statements as unfunny as shown in Table 8.

There were cases that fooled the laughbot and laughbot inappropriately responded with laughter. For example, the laughbot laughed at "I love you," likely because the statement has positive sentiment and is short in length. Some-

```
* recording

* done recording
audio recorded
transcript:  you're cute
Prediction [1]
in laughtrack
laughfile =  3 laughtrack4.wav
```

Fig. 5. Example laughbot success responding with laughter.

```
* recording

* done recording
audio recorded
transcript:  I'm really sad I went to see a th
erapist yesterday so much bad things have happ
ened to me I don't know what to do
Prediction [0]
```

Fig. 6. Example laughbot success responding with no laughter.

times, unfunny lines said in a funny manner (raising the pitch of the last word) can induce laughter. For example saying "my grandma died" but with high pitch at the end will cause laughbot to respond with laughter. Whether this should be considered a success for laughbot is up to the discretion of the user.

8 Limitations and Future Work

8.1 Improving the Model

Our combined RNN and logistic regression model performed best with an F1-score of 63.2 on the test set and an accuracy of 73.9. Future work will focus on reducing overfitting, as our final model run on the entire dataset still performs significantly higher on the train set. Since logistic regression is a much

Table 8. Laughbot successes

Transcript	Response
you're cute	laughter
ha ha ha	laughter
you're so fun to talk to haha	laughter
my grandma died	silence
why did the chicken cross the road	silence
to get to the other side	laughter
do you like cheese	silence
I finished my cs224s project	silence

more naive model than an RNN, we will work on improving this ensemble model to fully utilize the predictive power of both.

We also wish to explore CNNs, both stand-alone and in ensemble models, as our research showed CNNs to have higher F1-score than RNN models for this task [4]. We would also train on a larger dataset for a more generalizable model.

Additional implementations could include more complex classification to identify the level or type of humor; laughbot would be able to respond with giggles or guffaws based on user input. Sarcasm in particular has always been difficult to detect in natural and spoken language processing, but our model for detecting humor is a step towards being able to recognize the common textual cues along with the specific intonations that normally accompany sarcastic speech.

8.2 Making Laughbot Realtime

During development of the laughbot, we originally tried to design it to work in real-time so the user could continually speak and the laughbot could laugh at every point in the one-sided conversation when it recognized humor. We were able to transcribe audio in real-time with the Google Speech API, and intended to multithread it to capture an audio file simultaneously, but we faced problems structuring the rest of the interface to allow it to continually run the input through the model until humor was detected or the speaker paused for more than a certain threshold. Real-time recognition and laughing is a next-step implementation that will involve sending partial transcripts and audio files through the model continuously, concatenating audio and transcript data to preceding chunks to allow for context and longer audio cues to contribute to the classification.

Once real-time recognition and response is implemented, we could run laughbot on sitcoms stripped of their laughtracks and allow the laughbot to respond. Then, we could compare how closely laughbot's laughs compare to the original laughtracks of the TV show.

9 Conclusion

Our project takes the unique approach of training on phone conversations and combining a RNN and logistic regression model to classify spoken speech as funny or not funny. Our final model achieves an F1-score of 63.2 and accuracy of 73.9 on the test set, outperforming RNN models in previous work. Since we trained and tested on real conversations, this model's humor detection is most applicable to real, everyday speech that might be faced by a conversational agent. We also outline the architecture of Laughbot, a conversational agent that listens to users and responds with laughter to funny utterances. With future work, our model is a promising development for conversational agents that will detect and respond to humor in realtime.

Acknowledgments. We would like to thank our professor Andrew Maas, Dan Jurafsky and the Stanford University CS244S Spoken Natural Language Processing teaching team. Special thanks to Raghav and Jiwei for their direction on our combined RNN and regression model.

References

1. Nijholt, A.: Humor and embodied conversational agents (2003)
2. Piot, M.G.B., Pietquin, O.: Predicting when to laugh with structured classification. In: Interspeech (2014)
3. Piot, O.P.B., Geist, M.: Imitation learning applied to embodied conversational agents. In: MLIS (2015)
4. Bertero, D.: Deep learning of audio and language features for humor prediction (2016)
5. Klein, E., Bird, S., Loper, E.: Natural Language Processing with Python. O'Reilly Media Inc., Sebastopol (2009)
6. Eyben, F., Weninger, F., Gross, F., Schuller, B.: Recent developments in openSMILE, the munich open-source multimedia feature extractor, pp. 835–838. ACM Multimedia (MM) (2013)

Design Optimization of Activity Recognition System on an Embedded Platform

Ateendra Ramesh[1,2](✉), Adithya V. Ganesan[1,3], Sidharth Anupkrishnan[1,4], Aparokshith Rao[1,5], and Vineeth Vijayaraghavan[1,2]

[1] Research & Outreach, Solarillion Foundation, Chennai, India
{ateendraramesh,v.adithyaganesan,sidharth.anupkrishnan, aparokshith,vineethv}@ieee.org
[2] Solarillion Foundation, Chennai, India
[3] SSN College of Engineering, Chennai, India
[4] SRM University, Chennai, India
[5] College of Engineering, Guindy, Chennai, India

Abstract. Activity Recognition (AR) is a subset of pervasive computing that attempts to identify physical actions performed by a user. Previous sensor-based AR systems involve computation and energy overheads incurred by the use of heterogeneous and large number of sensors, however it is possible to arrive at an optimized system where the design involves optimization of energy consumption through number of sensors, computation through minimal set of features and cost through a nominal hardware platform ideally making it a multidimensional optimization. The above mentioned modelling was reflected in the construction of this optimized system as the design employs a single accelerometer and extracts only 7 time-domain features resulting in ease of computation to classify the activities, thus encouraging it to be inherently deployable on an embedded platform. The system was trained and tested on the accelerometer data acquired from three publicly available datasets. The performance of four chosen machine learning based classification models from an initial set of eight was evaluated, analysed and ranked on the grounds of efficiency and computation. The model was implemented on a Raspberry Pi Zero (USD 5) and the average time for feature computation and the maximum time taken to classify an instance of an activity was found to be 0.015 s and 1.094 s respectively, thus validating the viability of the system on an embedded platform and making it affordable to the population in the low-income groups.

Keywords: Machine learning · Pervasive computing
Activity recognition

K. Arai et al. (Eds.): FICC 2018, AISC 886, pp. 657–665, 2019.
https://doi.org/10.1007/978-3-030-03402-3_46

1 Introduction

The wide-spread applications of recognizing activities pertains to healthcare, surveillance and smart cities by monitoring actions performed by an individual. It can also provide a smart environment for the elderly who wish to continue to live independently in their homes who find access to healthcare expensive and sometimes inaccessible in low-income countries.

The task of recognizing activities can be achieved by employing vision-based and sensor-based models. However the former involves an image based classification approach wherein processing continuous image sequences from camera units would be employed. Not only does this approach cause issues in privacy, but also would be more computationally demanding in terms of resources and data processing than its sensor-based counterpart, as discussed in [1]. Thus, there arises a need for a low-cost activity monitoring system that uses nominal resources and minimal data processing to accurately classifying activities performed utilizing data acquired from sensors fitted on parts of the body.

The rest of the paper is organized as follows. Section 2 presents the related work in the area of activity recognition. Section 3 consists the problem statement of the paper. Section 4 deals with information regarding the datasets used in this paper and Sect. 5 describes the feature extraction techniques involved. Sections 6 and 7 showcase the experiments performed and their corresponding results and, Sect. 8 concludes the paper.

2 Related Work

Human activity recognition using wearable sensors has been a topic of research in the field of context-aware and ubiquitous computing. The accelerometer sensor has been widely used due to its low cost, low energy requirement and non-obstructive nature that is effective in capturing the motion of the human body.

Lester et al. [2] proposed a system with three units of seven sensors each that included an electric microphone, a visible light phototransistor, a tri-axial accelerometer, bi-axial digital compass, digital barometer, digital ambient light and digital humidity sensors in three different parts of the body. The system proposed by Altun et al. in [3] employed the sensors showcased under DSA Dataset in Table 1. The use of such a multitude of sensors increases the costs of computation & hardware and energy requisite involved while making the system vulnerable to numerous points of hardware failure.

Anguita et al. [4] proposed a system that computed 561 features from data collected from an accelerometer and gyroscope and the system in [3] extracts a set of 1170 features for data acquired from its corresponding sensors and then reduces it to 30 by performing PCA. But it would be computationally taxing to extract high dimensional features on a single board computing platform such as the Raspberry Pi Zero.

In this paper, the authors propose a model to efficiently recognize activities with an accelerometer sensor using minimal features and a set of 4 classifiers from an initial assortment of 8, evaluated across three different datasets on an embedded platform that is feasible for near real-time performance.

3 Problem Statement

Activity recognition systems in previous works feature multiple sensors placed on different parts of the body employing complex feature extraction and recognition algorithms. There arises a need for a system that would be optimized across aspects of energy, computation and cost. The proposed system employs a strategy where a single accelerometer from an array of sensors used in three pre-existing datasets, thus reducing the energy requirement and hardware cost. This approach coupled with the computationally inexpensive features made it capable of running on an elementary and low-cost embedded platform without compromising on its efficiency. Furthermore, the authors propose an efficiency vs. computation trade-off ranking for four algorithms which were short-listed from widely used eight algorithms.

4 Datasets

In this paper, three datasets were selected for the purpose of classification - Daily Sports and Activities Dataset (DSA) [3], Activity Recognition from Single Chest-Mounted Accelerometer Dataset (CM) [5] and Human Activity Recognition Using Smartphones Dataset (HAR) [4]. These datasets have been obtained from the UCI Machine Learning repository. Each of the datasets' attributes such as number of activities, number of participants, along with the number of data instances (N_D) have been showcased in Table 3. The sensors used in all the datasets have been listed and enumerated with their corresponding positions in Table 1.

The HAR dataset used in [7,8] consisted of raw 3-D acceleration and angular speed signals from its corresponding sensor units respectively for activities performed by 30 participants across a wide range of ages (19–48).

The CM dataset used in [8] has un-calibrated accelerometer data collected from a single accelerometer sensor and a salient feature of this dataset is the large N_D as seen in Table 3.

The DSA dataset which has been used in [9], contains calibrated 3-D acceleration, angular speed and strength of earth's magnetic field acquired from its respective sensors mentioned in Table 3 placed at 5 different parts of the body. The distinct feature of the DSA dataset was that it provided data for a variety of activities.

5 Feature Extraction

5.1 Dataset Preprocessing

The premise of data preprocessing is set on the sensor enumeration data shown in Table 1. In the DSA and HAR datasets, an optimized approach is implemented where the sets of data instances corresponding to each accelerometer are considered as standalone datasets and will hereby be denoted by *DSA-P*, where *P* is an abbreviation of the position of the sensor mentioned in Table 1. Ultimately, the number of datasets used amounts to seven.

In the CM dataset, the activities *Standing Up, Walking and Climbing stairs* and *Walking and Talking with Someone* have not been considered as there are very few instances of data.

5.2 Windowing Technique

From [6,10,11], it is observed that the feature computation from the accelerometer data has been efficient by applying windowing. For a data subset (D^α) of size $N_D^\alpha \times N_x$, consisting of data pertaining to a single activity (α) and number of axes (N_x) given by (1), the windowing technique depends on 2 parameters - the window size (w_s) and the percentage of overlap between each window (σ).

$$\begin{aligned} D^\alpha &= \{a_n^i \ \forall \ n \in [1, N_x]\} \ \forall \ i \in [1, N_D^\alpha] \\ &\quad where, \ a_n^i \ \in \mathbb{R} \ are \ accelerometer \\ &\quad signals \ for \ a \ single \ activity \ \alpha \end{aligned} \tag{1}$$

The total number of windows (N_w^α) extracted from D^α is done by discarding values $D_p^\alpha \ldots D_{N_D}^\alpha$ from D^α, where p is given in Eq. 2 and $N_x = 3$, as tri-axial accelerometer data is considered across all datasets.

$$p = w_s * (1 + (N_w^\alpha - 1) * (1 - \frac{\sigma}{100})) + 1 \tag{2}$$

The set of all initial positions of each window and the set of all windows denoted by I^α and W^α are defined in (3) and (4), respectively.

$$I^\alpha = ((i - 1) * w_s) * (1 - \frac{\sigma}{100}) + 1 \ \forall \ i \in [1, N_w^\alpha] \tag{3}$$

$$W^\alpha = \bigcup_{j \in I^\alpha} \{D_i^\alpha \ \forall \ i \in [j, j + w_s - 1]\} \tag{4}$$

Thus, each element in the set W^α consists of a window of shape $w_s \times N_x$. The w_s and the σ of the three datasets was chosen based on the windowing implementations from [3] for DSA, [6] for CM, [4] for HAR and their corresponding values have been tabulated in Table 4. The total number of windows (N_w) extracted from all the activities for each dataset is also shown in Table 4.

5.3 Feature Construction

The following seven features were extracted from each of the N_x axes of a window w: Mean (f_w^1), Standard Deviation (f_w^2), Kurtosis (f_w^3), Skew (f_w^4), Minimum (f_w^5) and Maximum (f_w^6), and Correlation between each pair of axes (f_w^7).

This results in a set of $N_x * 7$ attributes. The importance of the f_w^1 and f_w^7 are discussed in [11] and f_w^3 and f_w^4 have been utilised in [3]. Features f_w^5 and f_w^6 were chosen in order to capture fluctuations between peak values as signified in [12] along with the importance of f_w^2. The feature vector of a window w denoted by F_w is defined in (5).

$$F_w = \left[f_w^1, f_w^2, \ldots, f_w^7\right] \tag{5}$$

Thus, the entire set of feature vectors for W^α, given by F_T^α is expressed in (6).

$$F_T^\alpha = \bigcup_{w \in W^\alpha} F_w \tag{6}$$

Features were extracted for all the activities across all the datasets as mentioned above.

6 Experiment

From [3,4,11,13–15], a set of eight classifiers have been chosen for classification and they have been evaluated for all three datasets using 10-fold cross validation scores tabulated in Table 5. The four classifiers with the highest cross-validation scores across the three datasets have been implemented on a Raspberry Pi Zero

Table 1. Table enumerating sensors with their positions

Dataset	Number of sensors used			Position of sensors	Total number of sensors
	Accelerometer	Gyroscope	Magnetometer		
DSA	5	5	5	Torso (T), Left Arm (LA), Right Arm (RA), Left Leg (LL), Right Leg (RL)	15
CM	1	0	0	Chest	1
HAR	1	1	0	Waist	2

Table 2. Confusion matrix For DSA-RL; Yellow: activity performed, Gray: activity recognized Green: correct classifications, Red: misclassifications

	A1	A2	A3	A4	A5	A6	A7	A8	A9	A10	A11	A12	A13	A14	A15	A16	A17	A18	A19
A1	120	0	0	0	0	0	0	0	0	0	0	0	0	0	0	0	0	0	0
A2	0	116	0	0	0	0	1	3	0	0	0	0	0	0	0	0	0	0	0
A3	0	0	119	0	0	0	0	0	0	0	0	0	0	0	0	0	1	0	0
A4	0	0	0	120	0	0	0	0	0	0	0	0	0	0	0	0	0	0	0
A5	0	0	0	0	118	0	0	0	0	0	0	1	0	0	0	0	0	0	1
A6	0	0	0	0	1	118	0	0	1	0	0	0	0	0	0	0	0	0	0
A7	0	2	0	0	0	0	113	5	0	0	0	0	0	0	0	0	0	0	0
A8	0	4	0	0	1	2	7	93	7	0	0	0	0	0	0	0	0	0	6
A9	0	0	0	0	0	0	0	0	120	0	0	0	0	0	0	0	0	0	0
A10	0	0	0	0	0	0	0	0	0	120	0	0	0	0	0	0	0	0	0
A11	0	0	0	0	0	0	0	0	0	1	119	0	0	0	0	0	0	0	0
A12	0	0	0	0	0	0	0	0	0	0	0	120	0	0	0	0	0	0	0
A13	0	0	0	0	0	0	0	0	0	0	0	0	120	0	0	0	0	0	0
A14	0	0	0	0	1	0	0	0	0	0	0	0	2	116	0	0	0	0	1
A15	0	0	0	0	0	0	0	0	0	0	0	0	0	0	120	0	0	0	0
A16	0	0	0	0	0	0	0	0	0	0	0	0	0	0	0	120	0	0	0
A17	0	0	0	0	0	0	0	0	0	0	0	0	0	0	0	0	120	0	0
A18	0	0	0	0	0	0	0	0	0	0	0	0	0	0	0	0	0	116	4
A19	0	0	0	0	0	1	2	0	8	6	0	0	0	0	0	0	0	1	102

Table 3. Datasets and their attributes

Dataset	Number of activities	Number of participants	N_D
DSA	19	8	1140000
CM	7	15	1923177
HAR	6	30	659136

Table 4. N_w, w_s and σ of each dataset

Dataset	w_s	σ(in %)	N_w
DSA	125	0	9120
CM	52	50	73825
HAR	128	50	10299

with the stock operating system, Raspbian Jessie along with feature computation. The average time taken for computation of a single feature vector (F_w) was found to be 0.01584 s and the time taken for the classification of F_w on a Raspberry Pi Zero has been tabulated in Table 9.

7 Results

7.1 Comparison of Computation and Efficiency

Based on computation and efficiency, the four algorithms chosen in Sect. 6 were ranked as shown in Table 8. The efficiency is based on the 10-fold cross-validation scores given in Table 5 and the computation is based on the time taken to predict the activity of single feature vector shown in Table 9. The best performing classifier was found to be Extra Trees and the most computationally efficient classifier was Decision Trees.

7.2 Classification of Activities Across Datasets

To further evaluate the performance of the proposed model, the classification metrics employed in [3] for DSA, [6] for CM and [4] for HAR along with f1-score, precision and recall were used. All the datasets were randomly partitioned into 75% for training and 25% for testing individually.

DSA Dataset: From Table 5, it can be seen that the data subset corresponding to the accelerometer placed in the right leg (DSA-RL) provided the best scores. Thus, the evaluation metrics that were used: K-Fold Cross Validation Scores, given in Table 5, scores from Repeated Random Sub-Sampling (RRSS) and Leave-One-User-Out cross validation scores (L1OCV) have been shown in Table 6. The maximum K-Fold cross validation scores presented in [3] was 0.99, whereas the model proposed in this paper achieved 0.97 using a single accelerometer. Furthermore, evaluation metrics for the 75%–25% split have been shown in Table 7 for DSA-RL. The confusion matrix for the same across all activities ($A1 \dots A19$) has been portrayed in Table 2.

CM Datase: Table 7 provides the accuracy, precision, recall and f1-scores for the 75%–25% split. The scores for the 10-fold cross validation has been tabulated in Table 5, and the 5-fold cross-validation score that has been implemented in [6] was found to be 0.917 using the Extra Trees classifier.

HAR Dataset: For the HAR dataset, in conjunction with the aforementioned split, a 70%–30% partition over the 30 participants at random was implemented as carried out in [4] and a classification accuracy of 0.83 and an f1-score of 0.84 was achieved for the same using the Extra Trees Classifier while the accuracy achieved in [4] was 0.96. However, the computation of 561 features for one

Table 5. 10-fold cross validation scores

Dataset	Extra Trees	Random Forest	XGBoost	ANN	SVM	Decision Trees	Naive Bayes	AdaBoost
DSA-T	0.9622	0.9533	0.8975	0.9405	0.8955	0.8759	0.6971	0.2681
DSA-RA	0.9551	0.9429	0.8894	0.9149	0.8603	0.8555	0.6120	0.3067
DSA-LA	0.9542	0.9456	0.8905	0.9177	0.8639	0.8832	0.6850	0.3324
DSA-RL	0.9710	0.9666	0.9333	0.9487	0.8652	0.9123	0.8050	0.3999
DSA-LL	0.9697	0.9654	0.9388	0.9411	0.8737	0.9146	0.8148	0.3889
CM	0.9200	0.9183	0.7016	0.8323	0.5434	0.8595	0.5336	0.4452
HAR	0.9839	0.9654	0.8272	0.9402	0.8927	0.9371	0.8312	0.5031

Table 6. Evaluation metrics for DSA-RL

Algorithm	RSSS	LOOCV
Extra Trees	0.9781	0.770671
Random Forest	0.9655	0.7579
ANN	0.93615	0.70012
Decision Trees	0.9157	0.682568

Table 7. Average Precision, Recall, f1-Score and Accuracy for 75%–25% split across all datasets

Metrics	Dataset											
	DSA-RL				CM				HAR			
	Extra trees	Random forest	ANN	Decision tree	Extra trees	Random forest	ANN	Decision tree	Extra trees	Random forest	ANN	Decision tree
Precision	0.962	0.956	0.935	0.891	0.921	0.915	0.804	0.856	0.985	0.968	0.942	0.946
Recall	0.96	0.963	0.939	0.901	0.919	0.91	0.815	0.85	0.985	0.968	0.942	0.945
f1-score	0.969	0.954	0.938	0.902	0.92	0.912	0.804	0.851	0.985	0.968	0.942	0.945
Accuracy	0.968	0.963	0.945	0.905	0.921	0.917	0.815	0.854	0.984	0.967	0.942	0.945

window would be computationally expensive. The results for 10-fold cross validation have been provided in Table 5 and metrics from the 75%–25% split have been presented in Table 7.

Table 8. Rank matrix showcasing efficiency vs computation

Algorithm	Efficiency rank	Computation rank
Extra Trees	1	3
Random Forest	2	4
ANN	3	2
Decision Trees	4	1

Table 9. Time taken for classification of a single feature vector in Raspberry pi Zero in seconds

Dataset	Extra Trees	Decision Tree	ANN	Random Trees
DSA-T	1.026	0.018	0.024	1.014
DSA-RA	1.013	0.011	0.023	1.104
DSA-LA	1.024	0.011	0.023	1.011
DSA-RL	1.016	0.011	0.025	1.015
DSA-LL	1.027	0.011	0.021	1.021
CM	1.060	0.012	0.011	1.012
HAR	1.016	0.011	0.018	1.094

8 Conclusion

The Activity Recognition system designed with the aforementioned modelling of optimization involving the selection of a single accelerometer from an initial set of sensors and nominal feature extraction proved to be computationally viable for classification of activities without compromising on the efficiency. This, in turn, made the system optimized in terms of costs of computation & hardware and energy requirement. The four algorithms chosen in Sect. 6 from an initial set of eight classifiers were analyzed and ranked according to efficiency and computation. It was observed that the four classifiers accurately recognized activities from all the three datasets as shown in Sect. 7, despite skewness in classes exhibited by the CM dataset. It was also found to be feasible for near real-time performance on a Raspberry Pi Zero (5 USD), thus making the system low-cost and affordable for end-users in developing countries.

Acknowledgement. The authors would like to acknowledge Solarillion Foundation for its support and funding of the research work carried out.

References

1. Lara, O.D., Labrador, M.A.: A survey on human activity recognition using wearable sensors. IEEE Commun. Surv. Tutor. **15**(3), 1192–1209 (2013)
2. Lester, J., Choudhury, T., Borriello, G.: A practical approach to recognizing physical activities. In: Pervasive, vol. 3968, pp. 1–16 (2006)
3. Altun, K., Barshan, B., Tunçel, O.: Comparative study on classifying human activities with miniature inertial and magnetic sensors. Pattern Recognition **43**(10), 3605–3620 (2010)
4. Anguita, D., Ghio, A., Oneto, L., Parra, X., Reyes-Ortiz, J.L.: A public domain dataset for human activity recognition using smartphones. In: ESANN (2013)
5. Casale, P., Pujol, O., Radeva, P.: Personalization and user verification in wearable systems using biometric walking patterns. Pers. Ubiquitous Comput. **16**(5), 563–580 (2012)
6. Casale, P., Pujol, O., Radeva, P.: Human activity recognition from accelerometer data using a wearable device. In: IbPRIA, vol. 6669, pp. 289–296 (2011)
7. Ronao, C.A., Cho, S.B.: Human activity recognition using smartphone sensors with two-stage continuous hidden Markov models. In: 2014 10th International Conference on Natural Computation (ICNC), Xiamen, pp. 681–686 (2014)
8. Garcia-Ceja, E., Brena, R.F.: Building personalized activity recognition models with scarce labeled data based on class similarities. In: UCAmI, vol. 9454, pp. 265–276 (2015)
9. Nguyen, L.T., Tague, P., Zeng, M., Zhang, J.: SuperAD: supervised activity discovery. In: UbiComp/ISWC Adjunct, pp. 1463–1472 (2015)
10. Ravi, N., Dandekar, N., Mysore, P., Littman, M.L.: Activity recognition from accelerometer data. In: AAAI, pp. 1541–1546 (2005)
11. Bao, L., Intille, S.S.: Activity recognition from user-annotated acceleration data. In: Pervasive, vol. 3001, pp. 1–17 (2004)
12. Bedogni, L., Di Felice, M., Bononi, L.: By train or by car? Detecting the user's motion type through smartphone sensors data. In: 2012 IFIP Wireless Days, Dublin, pp. 1–6 (2012). https://doi.org/10.1109/WD.2012.6402818
13. Keally, M., Zhou, G., Xing, G., Wu, J., Pyles, A.: PBN: towards practical activity recognition using smartphone-based body sensor networks. In: Proceedings of the 9th ACM Conference on Embedded Networked Sensor Systems, Seattle, Washington, pp. 246–259 (2011)
14. Chen, T., Guestrin, C.: XGBoost: a scalable tree boosting system, CoRR, vol. abs/1603.02754 (2016)
15. Geurts, P., Ernst, D., Wehenkel, L.: Extremely randomized trees. Mach. Learn. **63**(1), 3–42 (2006)

Efficient Polynomial Multiplication via Modified Discrete Galois Transform and Negacyclic Convolution

Ahmad Al Badawi[1(✉)], Bharadwaj Veeravalli[1], and Khin Mi Mi Aung[2]

[1] Department of Electrical and Computer Engineering, National University of Singapore, Singapore 117583, Singapore
ahmad@u.nus.edu, elebv@nus.edu.sg
[2] Data Storage Institute A*Star, Singapore 138634, Singapore
Mi_Mi_AUNG@dsi.a-star.edu

Abstract. Univariate polynomial multiplication in $\mathbb{Z}_q[x]/\langle x^n + 1\rangle$ has brought great attention recently. Thanks to new construction of cryptographic solutions based on lattice and ring-learning with errors problems. A number of software libraries, such as NTL and FLINT, implements fast multiplication algorithms to perform this operation efficiently. The basic notion behind fast polynomial multiplication algorithms is based on the relation between multiplication and convolution which can be computed efficiently via fast Fourier transform (FFT) algorithms. Hence, efficient FFT is crucial to improve fast multiplication performance. An interesting algorithm that cuts FFT length in half is based on the discrete Gaussian transform (DGT). DGT was first proposed to work only with primes that support Gaussian integers arithmetic known as Gaussian primes. We modify this algorithm to work with not necessarily Gaussian primes and show how its parameters can be found efficiently. We introduce an array of optimization techniques to enhance the performance on commodity 64-bit machines. The proposed algorithm is implemented in C++ and compared with mature and highly optimized number theory libraries, namely, NTL and FLINT. The experiments show that our algorithm performs faster than both libraries and achieves speedup factors ranging from 1.01x–1.2x and 1.18x–1.55x compared to NTL and FLINT, respectively.

Keywords: Polynomial multiplication · Discrete Galois transform
Discrete Fourier transform · Negacyclic convolution
Negative wrapped convolution

1 Introduction

Polynomials are essential in several areas of mathematics and science; for instance, in coding theory polynomials can be found in several codes such as Bose, Chaudhuri and Hocquenghem (BCH) codes and error correcting codes

K. Arai et al. (Eds.): FICC 2018, AISC 886, pp. 666–682, 2019.
https://doi.org/10.1007/978-3-030-03402-3_47

(ECC) [1]. In cryptography, some cryptosystems are engineered based on polynomials like the advanced encryption standard (AES) [2], and Learning with Errors (LWE) based cryptosystems [3]. Another cryptographic application that is solely based on polynomial interpolation is the secret sharing. Furthermore, numerical systems employ polynomials for approximate computations using Taylor series and Chebyshev polynomials [4]. On the other hand, computer algebra systems such as Mathematica, Maple, PARI-GP, ... etc., employ polynomial arithmetic for exact symbolic calculations. In fact, polynomial arithmetic is also involved in ordinary numbers arithmetic due to the structural and algorithmic similarities between the two. In other words, an algorithm that multiplies two polynomials can be easily modified to work on numbers instead. The intensive usage of these mathematical expressions demands high performance algorithms to perform their arithmetic. Although algebraists and number theorists may not be greatly interested in how fast the arithmetic can be calculated, efficiency is of paramount importance for computer scientists and engineers. This has led to a huge realm of research for dedicated fast and efficient algorithms for polynomial arithmetic.

Polynomial addition and subtraction are straightforward and simple enough that can be performed in linear time $\mathcal{O}(n)$ where n is the degree of the input polynomials (assuming similar degree for both operands). On the other hand, polynomial multiplication requires more intensive calculations that, surprisingly, no proved lower bound exist for its complexity in the general computational model [5]. The schoolbook algorithm requires $\mathcal{O}(n^2)$ operations. Nevertheless, this complexity can be reduced to $\mathcal{O}(n \log n)$ using more complex algorithms based on Discrete Fourier Transform (DFT). The DFT comes to the scene since multiplication can be calculated via convolution, an operation that can be performed efficiently using DFT. In fact, multiplying two polynomials is equivalent to the convolution of the corresponding two signals where signals are constructed using polynomial coefficients.

We should remark here that the ordinary polynomial multiplication which takes operands of degree less than n and produces a result of degree less than $2n-1$ might be followed by a reduction step, i.e., reducing the result modulo an irreducible polynomial to obtain a polynomial of degree less than n. This multiplication may be referred to as: multiplication in a polynomial ring. One popular ring is $R : \mathbb{Z}[x]/\langle x^n + 1\rangle$; the set of integer polynomials modulo $x^n + 1$. This ring has brought great attention recently especially with the advent of learning with errors (LWE), its ring extension (RLWE) and lattice based cryptosystems which employ the ring to cater for more efficient computations. Examples of these cryptosystems include: Luybashevsky's public-key cryptosystem [6,7], SWIFFT hash functions [8], and some novel homomorphic encryption schemes such as BGV [9], YASHE [10] and FV [11].

The reader should note that a more restricted version of the ring is usually used to limit the size of polynomial coefficients, namely, $\mathbb{Z}_q[x]/\langle x^n + 1\rangle$, where $q \in \mathbb{Z}$, i.e., the set of integer polynomials modulo both $f(x)$ limiting degree growth and q limiting coefficients growth. In this work, we address the problem of polynomial multiplication in the celebrated ring $\mathbb{Z}_q[x]/\langle x^n + 1\rangle$.

The problem has been studied thoroughly in literature and led to various algorithms. Common features in these algorithms are the use of convolution theory and DFT to solve the problem in quasi-linear time $\mathcal{O}(n \log n)$. In this work, we provide a new algorithm based on a modified version of Crandall's Discrete Galois Transform (DGT) that has very appealing features in terms of transform length, precision of calculations, and execution time. We employ several optimization techniques and compare our results with state-of-the-art algorithms. Our experiments show speedup factors ranging from 1.01x to 1.55x over current mature and highly optimized number theory libraries.

The main contributions and scope of the paper can be summarized as follows:

- Presenting a modified version of DGT based multiplication in $\mathbb{Z}_q[x]/\langle x^n + 1\rangle$ in finite field $GF(p^2)$ where p is not necessarily a Gaussian prime.
- Presenting an array of optimization techniques such as choosing special primes and in-place Fourier transform algorithms to improve the overall performance.
- Implementing and comparing the algorithm with mature number theoretic libraries such as NTL [12] and FLINT [13].

1.1 Organization of the Paper

The rest of the paper is organized as follows: in Sect. 2, we give an overview about multiplication algorithms in $\mathbb{Z}_q[x]/\langle x^n + 1\rangle$ by listing monumental milestones that led to efficient algorithms. Next, in Sect. 3 we introduce mathematical material that we believe is crucial to understand the notions presented later on. We propose our algorithm in Sect. 4. In Sect. 5 we provide full details and layout of our implementation and introduce the key optimization techniques used. Our comparison methodology, testing environment, experiments and comparison results are exposed in Sect. 6. Finally, Sect. 7 concludes the work and provides guidelines for potential future work.

2 Literature Review

Fast multiplication algorithms that outperform the quadratic running time of classic algorithms and run in quasi-linear time have emerged by Karatsuba in 1962 [14]. Karatsuba algorithm splits the operands in half and proceeds recursively. The algorithm was then generalized by Toom and Cook [15] where the operands are cut into several parts and multiplication proceeds on the smaller parts. These two algorithms obviously follow the divide and conquer scheme and widely used in several number theory libraries.

Another fast algorithm that has lower complexity than Karatsuba and Toom-Cook is due to Schonhage and Strassen [16]. It employs Fourier transform and convolution theory to multiply large numbers. The algorithm exploits the idea that integer convolution, which can be performed in $\mathcal{O}(n)$ in Fourier domain, can be used to multiply numbers. The algorithm can be adapted smoothly to multiply polynomials as well. The basic concept behind the algorithm is Lagrange interpolation theorem and point-value polynomial representation instead of the canonical coefficient representation. The reader may seek chapter 8 and 10 in [17] for further details.

An interesting work that employs Schonhage-Strassen algorithm is cuHE [18]. It was proposed by Dai and Sunar to implement generic polynomial arithmetic on GPUs using CUDA. They compute the number theoretic transform (NTT) in $GF(p)$ where p =0xFFFFFFFF00000001 is a Solinas prime, after doubling and zero-padding the input polynomials, using Emmart and Weems FFT datapath [19]. The library also provides polynomial reduction API so that ring operations can be supported. Moreover, cuHE supports arithmetic for polynomials with large coefficients using residual number system (RNS) representation and Chinese remainder theorem (CRT) reconstruction. It should be noted that at the time of writing, cuHE supports natively only polynomials of degrees $\{8, 16, 32\} * 2^{10}$.

Chen et al. in [20] employs Schonhage-Strassen algorithm to multiply polynomials in $\mathbb{Z}_p[x]/\langle x^n + 1\rangle$, where p is a "small" prime number of size less than 32 bits, for RLWE and somewhat homomorphic encryption (SWHE) cryptosystems using FPGA hardware accelerators. They employ an adapted version of the NTT known as discrete weight transform (DWT) to multiply in $\mathbb{Z}_p[x]/\langle x^n + 1\rangle$ without an extra reduction step nor doubling and zero-padding polynomials. They also compute the NTT in $GF(p)$ where p is prime. However, the primes supported should satisfy five conditions to guarantee applicability and high performance.

Another similar work that employs Schonhage-Strassen algorithm is due to Akleylek et al. [21] for lattice-based cryptosystems on GPUs using CUDA. They perform the NTT over $GF(p)$, where p is also a small prime number of size less than 32 bits, to multiply two ring elements in $\mathbb{Z}_p[x]/\langle x^n + 1\rangle$. The implementation requires $p \equiv 1 \pmod{2n}$ to guarantee the existence of square root of n-th primitive root of unity modulo p. Although almost arbitrary primes can be used, they did not consider the efficiency of reduction modulo p which can degrade the overall performance due to its intensive involvement in computation.

The interesting feature in Donald and Akleylek's implementations is the exploit of special properties of the ring $\mathbb{Z}_p[x]/\langle x^n + 1\rangle$ to achieve polynomial reduction for free. Furthermore, they do not need to double and zero-pad input polynomials to apply the standard Schonhage-Strassen algorithm. However, other than the limitations we mentioned above, both suffer from computing NTT over a set of primes. This means that one need to optimize DWT for several primes which may not be possible. For instance, one needs to precompute and store powers of primitive roots of unity for several sets of primes if the RNS and CRT needs to be applied to support large coefficients, an approach that may not be desired if memory is scarce.

We propose here an algorithm based on an adapted version of the discrete Galois transform that cuts the transform length into half and works for non-Gaussian primes p in $\mathbb{Z}_p[x]/\langle x^n + 1\rangle$. The only condition required for p is the existence of n-th primitive roots of unity modulo p, i.e., $n|(p-1)$.

The reader should make a distinction between the numbers p and q. Usually, q is a product of several "small"[1] primes $\{p_0, p_1, \ldots, p_{r-1}\}$. Multiplying polynomials in $\mathbb{Z}_q[x]/\langle x^n + 1\rangle$ can be done by several independent multiplications in $\mathbb{Z}_{p_i}[x]/\langle x^n + 1\rangle$. Mathematical transforms such as DFT and NTT are usually done in $GF(p_i)$. However, in some implementations, the transforms are done in $GF(\hat{p}) \gg p_i$, where $\hat{p}$ is a specially chosen prime. Examples of the latter approach are cuHE [18] and Emmart and Weems large integer multiplier [19]. We also follow this approach and use a special Solinas prime number which allows us to optimize DGT computation accordingly in $GF(\hat{p}^2)$.

The special prime $\hat{p}$ is usually chosen to support efficient modulo operation which is heavily involved in transformations. Moreover, single-prime designs need to store precomputed constants, such as primitive roots and twisting factors, for a single prime. This is an appealing feature for platforms with scarce memory resources such as GPUs and FPGAs. More importantly, performing transformations modulo a single prime can lead to more efficient algorithms as one needs to optimize transformations in a single finite field. However, one limitation of single-prime designs is the bound incurred on coefficients magnitude that may lead to increased number of transforms to be performed. This will become clear when we propose our algorithm in Sect. 4.

3 Mathematical Preliminaries

In this section we review basic concepts in convolution theory, DGT, and their relation to polynomial multiplication. We start by a formal representation of our problem statement introducing all the notations used throughout the paper. Our aim is to establish the nomenclature necessary to understand the notions and relations presented in later on.

3.1 Notation

We use capital letters to refer to sets and small letters for elements of a set. As usual, $\mathbb{Z}, \mathbb{Q}, \mathbb{R}$, and $\mathbb{C}$ denote the integers, rationals, reals, and complex numbers, respectively. We use $\lceil . \rceil$, $\lfloor . \rfloor$, $\lfloor . \rceil$ to denote the round up, round down, and round to nearest integer, respectively. We usually represent elements in $\mathbb{Z}_q$ by integers modulo q in the range $\{\lceil -\frac{q}{2} \rceil, ..., \lfloor \frac{q-1}{2} \rfloor\}$ unless stated otherwise. We borrow vector notation to represent polynomials, for instance polynomial $a(x)$ of degree less than n can be represented as: $a = [a_0, a_1, ..., a_{n-1}]$, where a_i is the i-th coefficient of $a(x)$. Following Fourier transformation representation convention, transformed vectors are represented by capitalizing the vectors' symbol.

[1] Small usually means any size that fits in machine word.

3.2 Polynomial Multiplication in $\mathbb{Z}_q[x]/\langle x^n + 1\rangle$

Let $n, q \in \mathbb{Z}^+$ such that n is a power of 2. Let R denote the ring $\mathbb{Z}/\langle x^n + 1\rangle$ be the set of polynomials with integer coefficients of degree less than n, and R_q denote the ring $\mathbb{Z}_q/\langle x^n + 1\rangle$ be the set of polynomials modulo both $\langle x^n + 1\rangle$ and q. Suppose $a(x), b(x) \in R_q$ of degree less than n and we need to compute their product $c(x) = a(x).b(x) \in R_q$ of degree less than n and coefficients in $\mathbb{Z}_q$. We should note here that reduction of a polynomial modulo q can be done in linear time $\mathcal{O}(n)$ by a single pass and reducing each coefficient modulo q. Thus, we may only focus on algorithms that produce the product in R. Hereafter, we also use the word reduction to refer to reducing a polynomial modulo another polynomial.

Classical methods, e.g. schoolbook multiplication followed by polynomial reduction like long division, require $O(n^2)$ to compute $c(x)$. Equation 1 shows how the product can be computed without reduction:

$$\bar{c}(x) = a(x).b(x) = \sum_{i=0}^{n-1}\sum_{j=0}^{n-1} a_i b_j x^{i+j} \tag{1}$$

Still a classical polynomial division algorithm is required to reduce $\bar{c}(x)$ to $c(x)$ in R_q. However, since we are using the special irreducible polynomial $f(x) = x^n + 1$, division can be forgone as follows: Since $x^n + 1 \equiv 0 \ (mod\ x^n + 1)$ then $x^n \equiv -1 \ (mod\ x^n + 1)$. This allows us to replace any x^n by -1 in (1) that results in $\bar{c}(x)$ reduced in R_q for free. This stratagem is known as the negative wrapped [22] or negacyclic convolution [23] and can be formally expressed as shown in (2):

$$c(x) = a(x).b(x) = \sum_{i=0}^{n-1}\sum_{j=0}^{n-1} (-1)^{\lfloor \frac{i+j}{n} \rfloor} a_i b_j x^{(i+j \ mod\ n)} \tag{2}$$

In simpler words, (2) states that polynomial multiplication in R_q can be computed via negacyclic convolution.

3.3 Convolution Theory

As we showed in the previous subsection, multiplication can be performed via convolution. In fact, (1) gives the formal definition of integer convolution of two signals, where a signal is composed using the polynomial coefficients. Furthermore, (2) refers to the definition of negacyclic convolution of a and b.

The convolution theorem states that the Fourier transform of a convolution can be computed via point-wise (dyadic) product of the Fourier transform of the two signals at hand. This is expressed formally in (3)

$$a * b = \mathcal{F}^{-1}(\mathcal{F}(a).\mathcal{F}(b)) \tag{3}$$

where, $*$ denotes the convolution operator, $\mathcal{F}$ and $\mathcal{F}^{-1}$ refer to the Fourier transform and its inverse respectively. In this work, as we are dealing with polynomials of finite coefficients, we only address discrete signals and therefore $\mathcal{F}$ refers to the discrete Fourier transform.

It is a well-known fact that DFT and its inverse can be computed in $\mathcal{O}(n \log n)$ via any fast Fourier transform (FFT)/(IFFT) algorithm. Thus, the overall time complexity of convolving two signals of length n is equal to 2 FFTs and 1 IFFT of length n totaling $\mathcal{O}(n \log n)$. This is an important result that forms the basis of a large class of fast multiplication algorithms.

3.4 Discrete Galois Transform

Discrete Galois transform (DGT) is an adapted version of DFT over the Galois Field $GF(p^2)$, where p is prime, proposed by Crandall in [23]. Although DGT requires that $p \equiv 3 \pmod 4$, i.e. primes that support Gaussian integer arithmetic known as Gaussian primes, we show that the same transform can be applied using primes that are $\equiv 1 \pmod 4$, i.e., non-Gaussian primes. In fact, any prime number belongs to one of those sets.

An element $u \in GF(p^2)$ can be represented as $u_{re} + iu_{im}$ where $u_{re}, u_{im} \in Z_p$ and $i = \sqrt{-1}$ is the imaginary unit. These elements are also known as Gaussian integers. Arithmetic with Gaussian integers are similar to complex number arithmetic with a reduction modulo p for the real and imaginary parts. The DGT and its inverse of signal $x = \{x_0, x_1, \ldots, x_{n-1}\}$ of length n and $x_i \in GF(p^2)$, are shown in (4) and (5), respectively:

$$X_k = \sum_{j=0}^{n-1} x_j g^{-jk} \pmod p \tag{4}$$

$$x_k = n^{-1} \sum_{j=0}^{n-1} X_j g^{jk} \pmod p \tag{5}$$

where, g is a primitive n-th root of unity in $GF(p^2)$. The reader may have noticed the similarity between DGT and DFT to a small degree and DGT and number theoretic transform NTT to greater degree.

DGT enjoys several appealing properties such as: (1) negacyclic convolution of two n-point signals can be computed via $(n/2)$-point DGT/IDGT; (2) some Gaussian primes provides faster modulo operations such as Mersenne primes; (3) DGT supports lengths $\leq p+1$ due to the large size of $GF(p^2)$ multiplicative group and the simplicity of finding primitive roots of unity [24].

Despite these attractive features, DGT requires complex number arithmetic which may hinder its usage over other methods. However, our experiments show that DGT can still achieve good performance and outperform other algorithms.

4 Negacyclic Convolution via DGT

Armed with the notations and basic concepts of the structure of the polynomial ring of interest, convolution, and DGT, we are now ready to introduce polynomial multiplication in $\mathbb{Z}_q[x]/\langle x^n + 1\rangle$ via DGT which is shown in Algorithm 1. It can be easily seen that the algorithm runs asymptotically in time complexity similar to DGT/IDGT. If fast Fourier-like algorithms used to compute them, the overall time complexity of Algorithm 1 is $\mathcal{O}(n \log n)$.

Algorithm 1. Polynomial multiplication in R_q via the DGT

Let R_q be the ring $\mathbb{Z}_q[x]/\langle x^n + 1\rangle$, $q \in \mathbb{Z}^+$ and n is a power of 2. Let $\hat{p} \in \mathbb{Z}^+$ be a prime, g be a primitive $n/2$-th root of unity in $GF(\hat{p}^2)$, and $h \in \mathbb{Z}[i]$ be an $n/2$-th root of i in $GF(\hat{p}^2)$. Let $a(x), b(x), c(x) \in R$ be polynomials of degree less than n with integer coefficients $\in [0, q \leq \lfloor\lfloor\sqrt{\hat{p}/(2n)}\rfloor\rfloor]$.

Input: $a(x), b(x), g, g^{-1}, h, h^{-1}, n, n^{-1}, \hat{p}$
Output: $c(x) = a(x).b(x) \pmod{(q, x^n + 1)}$

Precompute:
$g^j, g^{-j}, h^j, h^{-j} \pmod{\hat{p}}$, where $j = 0, \ldots, n/2 - 1$
Initialize:
fold over input signals:
$a'_j = a_j + ia_{j+n/2}$
$b'_j = b_j + ib_{j+n/2}$
Twist the folded signals:
$\bar{a'}_j = a'_j h^j \pmod{\hat{p}}$
$\bar{b'}_j = b'_j h^j \pmod{\hat{p}}$
Compute $n/2$ **DGT**:
$A = DGT(\bar{a'})$
$B = DGT(\bar{b'})$
Point-wise multiplication:
$C_j = A_j.B_j \pmod{\hat{p}}$
Compute $n/2$ **IDGT**:
$\bar{c'} = IDGT(C)$
Remove twisting factors:
$c'_j = \bar{c'}_j h^{-j} \pmod{\hat{p}}$
Unfold output signal:
$c_j = Re(c'_j)$ ▷ Re returns the real part
$c_{j+n/2} = Im(c'_j)$ ▷ Im returns the imaginary part
return c

Algorithm 1 has some subtle details that should be considered carefully. First, the coefficients of the product signal must be representable $\pmod{\hat{p}}$, i.e. lie in $\{\lceil -\hat{p}/2\rceil, \lfloor \hat{p}/2\rfloor\}$, hence the condition of having the coefficients of input polynomials in $[0, q \leq \lfloor\sqrt{\hat{p}/(2n)}\rfloor]$. If the polynomials of interest have coefficients greater than this bound, one can represent the polynomials in Residual Numbering System (RNS), perform the multiplication on the corresponding polynomial residue using DGT algorithm, and finally reconstructs the final polynomial using CRT reconstruction. This in fact is a common approach used to do arithmetic on large numbers/polynomials. In our implementation, we deal with small coefficients such that CRT is not required, since our main goal is to show a proof-of-concept that the proposed algorithm works flawlessly with non-Gaussian primes.

The reader should also distinguish between q and $\hat{p}$. q is a bound on polynomial coefficients to limit coefficient growth. Also, q needs not be a prime number. On the other hand, $\hat{p}$, which has to be prime, defines the structure $GF(\hat{p}^2)$. It is used to compute DGT/IDGT in $GF(\hat{p}^2)$. In fact, in this work $\hat{p}$ is a 64-bit special Solinas prime that offers efficient modulo reduction. If q happens to be greater than $\hat{p}$, it should be factored into a set of co-prime numbers $\{q_0, q_1, \ldots, q_{l-1}\}$ and RNS representation and CRT reconstruction can be used.

An important implementation issue is the modulo representation. In most programming languages, $(\bmod\, \hat{p})$ operation returns an integer $\in [0, \ldots, \hat{p}-1]$. There is no problem of computing DGT using this range, but an extra treatment step would be required to scale the result back in $\{\lceil -\hat{p}/2 \rceil, \lfloor (\hat{p}-1)/2 \rfloor\}$. This is an important caveat especially if one uses RNS/CRT trick to deal with polynomials with large coefficients. One must first get the negacyclic convolution result in $\{\lceil -\hat{p}/2 \rceil, \lfloor (\hat{p}-1)/2 \rfloor\}$ then reduction modulo the RNS moduli can be used safely. One must also pay attention that the prime $\hat{p}$ used in DGT does not belong to the RNS moduli.

Although DGT algorithm as proposed originally requires that $p \equiv 3 \pmod 4$, i.e., primes that support Gaussian arithmetic known as Gaussian primes, we argue here that it can work with not necessarily Gaussian primes. The discussion below is provided to verify our argument.

First, We believe that the restriction was made to reduce the difficulty of finding primitive roots of unity and i, which are required in the computation. Moreover, working with some Gaussian primes such as Mersenne primes, makes finding primitive roots even simpler due to Creutzburg and Tasche theorem [25] which gives a closed-form formula to find the roots. Secondly, the Galois field $GF(p^2)$ with these primes is well defined and has p^2 elements written in the form $x + iy$.

On the other hand, if we work with $p \equiv 1 \pmod 4$, $GF(p^2)$ is not defined properly since $\sqrt{-1} = i$, is an element of $GF(p)$ and the set we obtain by adjoining Z_p and i has only p elements not p^2. Although the algebraic structure is not well-defined, this has no effect on the DGT multiplication algorithm as long the ordinary complex numbers arithmetic are applied modulo p. Moreover, our experiment results add more empirical guarantees on the correctness of our argument.

The advantage of working with any prime is handy for the following reasons: (1) one does not have to restrict the search space while searching for the primes of interest in computing DGT; (2) some primes provide efficient modulo operations and may not be $\equiv 3 \pmod 4$ such as some Solinas primes [26]; (3) if one chooses to use Mersenne primes, the most suitable prime for current 64-bit machines is $2^{61} - 1$ but it wastes 3 bits which in turns reduces the maximum value of convolution supported. In fact there is a special 64-bit Solinas prime, $p = $ 0xFFFFFFFF00000001, that best serves 64-bit machines in computing DGT and offers efficient modulo operation. This prime has been used in our implementation for another important reason that is presented in Sect. 5.3. In fact, our knowledge of this prime led us to investigate if the DGT can be applied using primes $\equiv 1 \pmod 4$.

The only problem in working with primes that are not $\equiv 3 \pmod 4$ is that finding primitive roots of unity and roots of i requires a trial and error approach. But since this is only being done once, this extra cost cannot be considered a huge obstacle. Nevertheless, we show in Sect. 5.2 how to find these roots efficiently.

Lastly, the reader should note that DGT can be computed over the complex numbers domain. This has an advantage that finding primitive roots of unity or i is trivial. Also, several optimized and mature libraries exist that can compute the FFT efficiently. The only problem in this approach, is that approximations and rounding errors done due to floating point operations in $\mathbb{C}$ can be significant and produce wrong results if the parameters were above certain thresholds.

5 Implementation Layout

In this section, we provide full details of our implementation. All the essential building blocks such as DGT/IDGT, modulo operation, Gaussian integer arithmetic are introduced. We also list the major optimization techniques that we think had led to an efficient implementation.

5.1 DGT and IDGT Computation

DGT operation and its inverse look very similar to DFT and its inverse IDFT. In fact, DFT can be thought of as an abstract transformation that can be applied in any algebraic structure that has well-defined primitive roots of unity and inverse of the transform length. This means that any efficient FFT data-path can be used to implement the DGT/IDGT as long one respects the underlying domain arithmetic.

One important concept that has to be taken into consideration while choosing the forward and inverse DGT algorithms is the bit-reversal scrambling procedure. Several FFT algorithms require data to be reordered in some way. Usually, this reordering is applied to obtain a proper order transform and reduce computation cost. However, since we only use FFT for polynomial evaluation and interpolation, proper order is not important if the inverse FFT took the reordering in consideration. In fact, if a decimation-in-frequency algorithm was used for FFT and a decimation-in-time used for IFFT, bit-reversal can be avoided [24].

We use Gentleman-Sande [27] FFT data-path to compute the forward DGT. Our choice of Gentleman-Sande algorithm can be justified by the following reasons: (1) Gentleman-Sande is a decimation-in-frequency algorithm that requires bit-reversal at the end of computation; (2) the algorithm is in-place, i.e. memory-efficient; (3) data size is a power of 2 that is ideally suitable for Gentleman-Sande algorithm. These properties make Gentleman-Sande ideally suitable for the forward DGT algorithm.

Algorithm 2 shows the basic Gentleman-Sande algorithm. Although, bit reversal is not used, we show it here for completeness. The algorithm is adapted to work in Galois finite field $GF(\hat{p}^2)$.

Algorithm 2. DGT via Gentleman-Sande [27]

Let $\hat{p} \in \mathbb{Z}^+$ be a special prime = 0xFFFFFFFF00000001, $g \in \mathbb{Z}_{\hat{p}}$ be a primitive k-th root of unity. Let x be a folded signal of length $k = 2^d$ composed of the original polynomial (of length 2^{d+1}) coefficients. Elements of x are Gaussian integers $\in GF(\hat{p}^2)$.

Input: $x, g, k, \hat{p}$
Output: $x = DGT(x)$

Precompute:
$g^j \pmod{\hat{p}}$, where $j = 0, \ldots, k-1$

Procedure:
for $m = k/2;\ m \geq 1;\ m = m/2$ **do**
  for $j = 0;\ j < m;\ j{+}{+}$ **do**
    $a = g^{(jk/(2m))}$
    for $i = j;\ i < k;\ i+ = 2m$ **do**
      $x_i = x_i + x_{i+m} \pmod{\hat{p}}$
      $x_{i+m} = a(x_i - x_{i+m}) \pmod{\hat{p}}$
bit-reverse(x) ▷ can be avoided
return x

The IDGT computation is quite similar to forward DGT but requires an extra step of multiplying the resulting vector by a rescaling factor ($k^{-1} \pmod{\hat{p}}$). Furthermore, the inverse of primitive k-th root of unity (g^{-1}) is used instead. It should be noted that powers of g^{-1} can be computed from powers of g since $g^{-i} \equiv g^{(k-i) \pmod{k}} \pmod{\hat{p}}$. For IDGT, we use the famous Cooley-Tukey FFT data-path [28]. It is a decimation-in-time in-place FFT algorithm. We also adapt the algorithm to work in $GF(\hat{p}^2)$. The algorithm is shown in Algorithm 3.

Both Algorithms 2 and 3 are variants of radix-2 in-place fast Fourier transform algorithms and run asymptotically in $\mathcal{O}(n \log n)$ with $\mathcal{O}(n)$ space complexity.

5.2 Gaussian Integers Arithmetic

In this subsection, we show and efficient method to generate the primitive roots of i in $GF(p^2)$, where $p \equiv 1 \bmod 4$. We also list some optimization techniques to improve the arithmetic involved.

(1) Generating k-th Roots of i $\pmod{\hat{p}}$*:* As we have said previously, the DGT was proposed for prime numbers that are also Gaussian integer primes, i.e., congruent to $3 \pmod 4$. However, we showed that this is not a necessary condition and any prime can be used.

One critical challenge in applying the DGT multiplication algorithm is finding k-th primitive roots of i. In general, finding these roots in a field other than $\mathbb{C}$

Algorithm 3. IDGT via Cooley-Tukey [28]

Let $\hat{p} \in \mathbb{Z}^+$ be a special prime = 0xFFFFFFFF00000001, $g^{-1} \in \mathbb{Z}_{\hat{p}}$ be the inverse of k-th primitive root of unity. Let x be a folded signal of length $k = 2^d$ composed of the original polynomial (of length 2^{d+1}) coefficients. Elements of x are Gaussian integers $\in GF(\hat{p}^2)$.

Input: $x, g^{-1}, k, \hat{p}$
Output: $x = IDGT(x)$

Precompute:
$g^{-j} \pmod{\hat{p}}$, where $j = 0, \ldots, k-1$

Procedure:
bit-reverse(x) ▷ can be avoided
for $m = 1;\ m < k;\ m = 2m$ **do**
  for $j = 0;\ j < m;\ j{+}{+}$ **do**
    $a = g^{(-jk/(2m))}$
    for $i = j;\ i < k;\ i{+}= 2m$ **do**
      $x_i = x_i + a x_{i+m} \pmod{\hat{p}}$
      $x_{i+m} = x_i - a x_{i+m} \pmod{\hat{p}}$
return x

is not an easy procedure even for a super computer when the parameters are fairly large. The reason is that a trial and error approach is usually used to find them. If one however is using a Mersenne prime, these roots can be found easily thanks to Creutzburg and Tasche theorem [25] which gives a close form formula to calculate the roots. Since our prime is not a Mersenne prime, we had to use the common trial and error approach to find the roots. Although the prime we are dealing with is not quite large, finding primitive roots of i for large k was not a trivial task. We developed a highly optimized Mathematica script to solve the equation $z^k \equiv i \pmod{\hat{p}}$ where $z \in GF(\hat{p}^2)$. The basic idea is to perform binomial expansion and solve the resulting two equations simultaneously. This approach was not efficient. For instance, finding 2^{14}-th primitive root of $i \pmod{\hat{p}}$ takes 156.45 h and requires 33 GB of memory. Although one may argue that this task is not important as it is only performed once, however, a more efficient solution can be attractive.

A much more efficient approach starts with the fact that $\hat{p}$ can be factored in the Gaussian integer domain into two Gaussian integers; namely, $f_0 = 65536 + 4294967295i$ and $f_1 = 65536 - 4294967295i$. The multiplicative group of Gaussian integers modulo each factor forms a cyclic group of order $\hat{p} - 1$. One, then finds a generator in each group, say ζ_j, where $j \in \{0, 1\}$ by generating a random element and checking if it has an order of $\hat{p} - 1$. Provided that a generator was found, one can find a k-th primitive root of i as follows: $\zeta_j^{(\hat{p}-1)/(4*n)} \pmod{f_j}$. Finally, the CRT can be used to reconstruct the root in $GF(\hat{p}^2)$.

Using the second approach, primitive k-th roots of i modulo $\hat{p}$, where $k \in \{.25, 0.5, 1, 2, \ldots, 64\} * 2^{10}$ can be found in less than a second. For completeness, Table 1 shows instances of k-th primitive roots of $i \pmod{\hat{p}}$ in $GF(\hat{p}^2)$.

Table 1. Primitive k-th Roots of $i \pmod{\hat{p}}$, where $\hat{p} =$ 0xFFFFFFFF00000001

$k/2^{10}$	Root of i
0.25	$1100507988529617178 + 13061373484646814047i$
0.5	$5809945479226292735 + 4344400649288295733i$
1	$1973388244086427726 + 10274180581472164772i$
2	$2796647310976247644 + 10276259027288899473i$
4	$1838446843991 + 11906871093314535013i$
8	$350701160942 + 2376526094797987232i$
16	$6583264258929468914 + 6843633496861054794i$
32	$8792747750079974660 + 5349516358062242429i$
64	$4079625552322729221 + 5648961845066732578i$

(2) Multiplying Gaussian Integers $\pmod{p}$: For multiplication of Gaussian integers in $GF(\hat{p}^2)$, we use Karatsuba [14] algorithm which computes the complex-like multiplication using 3 multiplications and 4 additions/subtractions $\pmod{\hat{p}}$ instead of the schoolbook algorithm that requires 4 multiplications and 2 additions/subtractions $\pmod{\hat{p}}$.

5.3 Reduction $\pmod{\hat{p}}$

The prime we use in our implementation has some beautiful properties that allow one to perform reduction modulo $\hat{p}$ efficiently. This can be shown as follows:

Since size of $\hat{p}$ is 64-bit, an element $\in \mathbb{Z}_{\hat{p}}$ is at most 64-bit number. Hence, we only care about reducing a 128-bit number resulting from multiplying two 64-bit numbers. The prime number we use here is $\hat{p} =$ 0xFFFFFFFF00000001 which can be expressed as $\hat{p} = 2^{64} - 2^{32} + 1$. Following Solinas method [26], starting with $\hat{p} = f(2^k) = t^2 - t + 1$ where $k = 32$, the reduction matrix can be found to equal $X = \begin{bmatrix} -1 & 1 \\ -1 & 0 \end{bmatrix}$. This allows us to compute the reduction of a 128-bit number modulo $\hat{p}$ by performing addition and subtractions of its four 32-bit words. Algorithm 4 depicts how this can be done using only shifts, additions, and subtractions.

6 Experiments and Results

In this section, we compare our implementation with two famous number theory libraries; namely: NTL and FLINT. We first present our comparison methodology and testing environment.

6.1 Methodology

We developed the DGT multiplication algorithm via C++ on a lightly loaded 64-bit machine equipped with 2 CPUs (Intel(R) Xeon(R) CPU E5-2620 v3 at

Algorithm 4. Efficient reduction modulo $\hat{p}$ [26]

Let $a \in \mathbb{Z}^+$ be a 128-bit integer, and prime $\hat{p} = \text{0xFFFFFFFF00000001}$
Input: $a, \hat{p}$
Output: $b = a \pmod{\hat{p}}$

Procedure:
$w_0 = a$ ▷ w_j: 32-bit word
$w_1 = a \gg 32$
$w_2 = a \gg 64$
$w_3 = a \gg 96$
$b = ((w_2 + w_1) \ll 32) - w_3 - w_2 + w_0$
return b ▷ b: 64-bit integer

2.40 GHz), 62 GB RAM and runs ArchLinux (4.8.13-1-ARCH). The compiler used is GCC (6.3.1 20170109).

No explicit parallelism was incorporated during the development such as multi-threading or multi-processors programming. We also did not use any third-party library for our computation. All the algorithms such as DGT computation, modulo operation, and DGT multiplication were all implemented from scratch.

To test the performance of our algorithm, we compare it with NTL [12] version (10.3.0-1) and FLINT version (2.5.2) [13]. Both libraries were compiled from source and built using the same compiler with default configurations.

Coefficients of the polynomials involved in the tests were generated randomly using C++ native (rand) function. We generate the data of a predefined size and prepare NTL and FLINT objects encapsulating the same data.

Any initialization operation that would only be done once was not included in timing figures. Computing powers of roots, data generation, library objects creation, initialization and destruction were not included in the timing figures. Time was measured using Linux native function (gettimeofday). Each experiment were performed 100 times and the average running time is always reported.

6.2 Comparison with NTL and FLINT

We show here performance results of our implementation of Algorithm 1 in the finite field $GF(\hat{p}^2)$ and compare it with NTL and FLINT. We aim here at showing that our implementation is a potential rival for NTL and fairly better than FLINT. Table 2 shows the running times of multiplication in $\mathbb{Z}_p[x]/\langle x^n + 1\rangle$ in milliseconds for different polynomial degrees.

It can be clearly seen that our implementation using DGT is slightly faster than NTL and fairly faster than FLINT. A speedup from 1.01x to 1.2x over NTL and 1.18x to 1.55x over FLINT was achieved. We should note here that our implementation is quite straightforward, i.e., no explicit parallelization, such as multi-threading or vector operations were incorporated, which are used in NTL and FLINT. We expect even further better results if parallel programming techniques are employed.

Table 2. Running time in (millisec) for multiplication in $\mathbb{Z}_p[x]/\langle x^n + 1\rangle$ via NTL, FLINT and DGT

$n/2^{10}$	NTL	FLINT	DGT
0.5	0.4714	0.5206	0.3864
1	0.9683	1.1019	0.8381
2	1.7744	2.0426	1.5795
4	3.7021	4.3349	3.2678
8	8.9681	9.7609	7.8854
16	18.7248	21.8485	18.4732
32	36.2898	44.2422	33.8424
64	77.8016	103.7360	73.2465
128	168.4040	237.3830	152.7910

7 Conclusion

We proposed an efficient algorithm to multiply polynomials in $\mathbb{Z}_p[x]/\langle x^n + 1\rangle$. The algorithm employs an adapted version of Crandall's discrete Galois transform (DGT) that works with not necessarily Gaussian primes. Theoretical and empirical evidences were provided to prove the correctness of our argument.

We implemented the algorithm using C++ in finite fields $GF(p^2)$. An array of optimization techniques were employed to guarantee optimum performance. These optimization techniques include usage of a special prime to compute DGT in $GF(p^2)$. We also compute DGT and its inverse so that bit-reversal is avoided. No explicit parallelization techniques, such as multi-threading or vector instructions were incorporated in the implementation.

The algorithm was compared with NTL and FLINT number theory libraries. Our algorithm outperforms both libraries and achieves 1.01x–1.2x over NTL and 1.18x–1.55x over FLINT.

An immediate and useful extension to the work reported here could be in performance enhancement by employing parallelization techniques such as multi-threading or vector instructions. Moreover, we believe that the algorithm can be ported smoothly to run on hardware accelerators such as FPGAs and GPUs. Furthermore, a slight speedup is expected if higher radix algorithms were used in computing DGT/IDGT instead of the radix-2 algorithms used in this work.

Acknowledgement. This work was supported by Data Storage Institute, A*STAR and the National University of Singapore. The authors would like to thank the stack-exchange users: Ofir and D_S on math and J.M. ◊ on Mathematica.

References

1. Blahut, R.E.: Algebraic codes for data transmission. Cambridge University Press, New York (2003)
2. Daemen, J., Rijmen, V.: The Design of Rijndael: AES-The Advanced Encryption Standard. Springer Science & Business Media (2013)
3. Lyubashevsky, V., Peikert, C., Regev, O.: On ideal lattices and learning with errors over rings. Cryptology ePrint Archive, Report 2012/230 (2012)
4. Hernández, M.: Chebyshev's approximation algorithms and applications. Comput. Math. Appl. **41**(3–4), 433–445 (2001)
5. Ailon, N.: A lower bound for Fourier transform computation in a linear model over 2x2 unitary gates using matrix entropy. arXiv preprint, arXiv:1305.4745 (2013)
6. Lyubashevsky, V., Peikert, C., Regev, O.: On ideal lattices and learning with errors over rings. J. ACM (JACM) **60**(6), 43 (2013)
7. Lyubashevsky, P., Regev, O.: A toolkit for ring-LWE cryptography. In: EUROCRYPT, vol. 7881, pp. 35–54. Springer (2013)
8. Lyubashevsky, V., Micciancio, D., Peikert, C., Rosen, A.: Swifft: a modest proposal for FFT hashing. In: International Workshop on Fast Software Encryption, pp. 54–72. Springer (2008)
9. Brakerski, Z., Gentry, C., Vaikuntanathan, V.: (Leveled) fully homomorphic encryption without bootstrapping. In: Proceedings of the 3rd Innovations in Theoretical Computer Science Conference, pp. 309–325. ACM (2012)
10. Bos, J.W., Lauter, K., Loftus, J., Naehrig, M.: Improved security for a ring-based fully homomorphic encryption scheme. In: IMA International Conference on Cryptography and Coding, pp. 45–64. Springer (2013)
11. Fan, J., Vercauteren, F.: Somewhat practical fully homomorphic encryption. IACR Cryptology ePrint Archive 2012, 144 (2012)
12. Shoup, V., et al.: NTL: a library for doing number theory (2001)
13. Hart, W.B.: Flint: fast library for number theory. Computeralgebra Rundbrief (2013)
14. Karabutsa, A., Ofman, Y.: Multiplication of many-digital numbers by automatic computers. DOKLADY AKADEMII NAUK SSSR, vol. 145, no. 2, p. 293 (1962)
15. Toom, A.L.: The complexity of a scheme of functional elements realizing the multiplication of integers. Soviet Mathematics Doklady **3**(4), 714–716 (1963)
16. Schönhage, A., Strassen, V.: Schnelle multiplikation grosser zahlen. Computing **7**(3–4), 281–292 (1971)
17. Gathen, J.V.Z., Gerhard, J.: Modern Computer Algebra. Cambridge University Press, New York (2013)
18. Dai, W., Sunar, B.: cuHE: a homomorphic encryption accelerator library. In: International Conference on Cryptography and Information Security in the Balkans, pp. 169–186. Springer (2015)
19. Emmart, N., Weems, C.C.: High precision integer multiplication with a GPU using Strassen's algorithm with multiple FFT sizes. Parallel Process. Lett. **21**(03), 359–375 (2011)
20. Chen, D.D., Mentens, N., Vercauteren, F., Roy, S.S., Cheung, R.C., Pao, D., Verbauwhede, I.: High-speed polynomial multiplication architecture for ring-LWE and SHE cryptosystems. IEEE Trans. Circuits Syst. I **62**(1), 157–166 (2015). Regular Paper

21. Akleylek, S., Dağdelen, Ö., Tok, Z.Y.: On the efficiency of polynomial multiplication for lattice-based cryptography on GPUs using CUDA. In: International Conference on Cryptography and Information Security in the Balkans, pp. 155–168. Springer (2015)
22. Winkler, F.: Polynomial algorithms in computer algebra. Springer Science & Business Media (2012)
23. Crandall, R.E.: Integer convolution via split-radix fast Galois transform. Center for Advanced Computation Reed College (1999)
24. Crandall, R., Pomerance, C.: Prime Numbers: A Computational Perspective, vol. 182. Springer Science & Business Media (2006)
25. Creutzburg, R., Tasche, M.: Parameter determination for complex number-theoretic transforms using cyclotomic polynomials. Math. Comput. **52**(185), 189–200 (1989)
26. Solinas, J.A., et al.: Generalized mersenne numbers. University of Waterloo, Faculty of Mathematics (1999)
27. Gentleman, W.M., Sande, G.: Fast Fourier transforms: for fun and profit, pp. 563–578. In: Proceedings of the November 7-10, 1966, Fall Joint Computer Conference. ACM (1966)
28. Cooley, J.W., Tukey, J.W.: An algorithm for the machine calculation of complex fourier series. Math. Comput. **19**(90), 297–301 (1965)

Storage Optimization Using Secure Image Hashing

B. Sri Gurubaran(✉), G. Shailesh Dheep, E. R. S. Subramanian, V. Aishwarya, and A. Umamakeswari

SASTRA University, Thanjavur, India
gurubaran1997@gmail.com, ngsaileshsaran@gmail.com, subramanianramanathan97@gmail.com, aishwaryavasu1997@gmail.com, aum@cse.sastra.edu

Abstract. With the current boom in smartphones and social media, image sharing faces problems, like redundant data. Thereby, compromises the storage of a user's device. In this paper, we design an image hashing and verification technique to check for redundancy. The algorithm creates a hash of the image and is stored along with the image before sending. Upon receiving an image, its hash code is retrieved from the image and is cross referenced with the lookup table in the device. If an entry already exists, the received image is discarded. This reduces the storage of redundant images in the device and thereby avoids wastage of device memory.

Keywords: Hashing · Hash-map · Redundancy · SHA-512 · AES-ECB

1 Introduction

In the current rising trends, the usage of mobile phones has risen considerably, thus social media have shifted towards the mobile environment, where the storage space and computing power are limited. Therefore, having irrelevant or redundant content leads to the risk of using up the phone's entire storage. So, it is best to avoid having redundant data on our mobile devices. And the major cause for presence of redundant data is the storing of the redundant images shared via social media. Though this seems farfetched, the current surge in the popularity social media has led to transfer of huge amounts of data back and forth, mostly images and videos. And most of the times the same image or video is shared by many people so that it leads to the storing up of the same information again and again.

One of the best ways to avoid redundancy is to assign a unique code to every image, if an image with the same code is encountered it can be discarded. This type of unique code is known as a hash. There are many hashing techniques available for text or numeric data, but no such standard algorithms exist for an image. Also, there is the problem of collision where two different images are assigned the same hash code, this is a critical issue with respect to the problem at hand, because if we can't assign unique values then we can't distinguish between the images.

Since, the mobile environment has limited computing power, so it is not prudent to generate the hash for an image again and again. Therefore, the hash value must be

K. Arai et al. (Eds.): FICC 2018, AISC 886, pp. 683–691, 2019.
https://doi.org/10.1007/978-3-030-03402-3_48

stored and must be available easily for quick reference. In that case an efficient lookup method should exist to cross check the hash code of the received image with the hash codes of the image already existing on the device.

To avoid such scenarios an image hashing algorithm is proposed where a unique hash code is generated for the image and is stored along with the image before sending. And on the receiver end, the received image's hash code is retrieved and is checked with the hash codes of the images already existing on the device. These codes are stored in hash table with a reference to the image stored in the memory. If the receiver already has an image with the same hash code the received image is discarded. This is a simple yet effective technique to handle the redundant data in a mobile environment.

2 Related Work

Robust image hashing by Venkatesan et al. [1] introduces an image hashing technique which makes use of various signal processing strategies to generate random binary strings. The proposed technique uses message authentication codes (MACs) thus minimizing the probability of hash collision.

Robust and secure image hashing by Swaminathan et al. [2] developed a novel algorithm to generate an image hash based on Fourier transform and controlled randomization. The proposed hash function is found to be resilient to various kinds of content modifications.

A symmetric color image encryption algorithm using the intrinsic features of bit distributions by Zhang et al. [3] analyzes the intrinsic features of bit distributions and other issues related to the bit information of an image. The expand and shrink strategy is employed to shuffle the image with a reconstructed permuting plane.

A clustering based approach to perceptual image hashing by Monga et al. [4] splits the image hashing technique into feature extraction and data clustering being the intermediate and final hashes respectively. A polynomial-time heuristic clustering algorithm is proposed to determine the final hash length. Randomized clustering algorithms are developed to verify secure image hashing.

Hughes et al. [5] proposed a data matching system to identify redundant flows between a source and a destination received by a communication interface. The flow is stored in the processor as packets. Some part of the packet data is checked for a potential match to the data already present in the storage using hashes. The most likely and the second most likely data matches' sizes are determined. If the match sizes are not too small, a retrieve instruction is generated for the data present in the storage.

Douglis et al. [6] proposed a redundancy elimination mechanism at the block level. The file objects are initially divided into blocks or chunks. Identical blocks are suppressed. Resemblance detection is performed on the remaining chunks to identify redundancy so as to benefit from delta encoding. Any remaining chunks are compressed. Fingerprints of various chunks are merged together to form super fingerprints. Two objects having similar super fingerprints are more likely to be similar. The reduced object is stored and is referenced by its super fingerprint.

Zhang et al. [7] reviews the system architecture and main processing of protocol independent DRE (Data Redundancy Elimination) techniques. The various mechanisms involved in protocol-independent DRE such as fingerprinting mechanism, cache management mechanism and many more are discussed. Additionally, several redundancy elimination systems in wireline, wireless and cellular networks are proposed. Several techniques to enhance the DRE performance, such as DRE bypass techniques, non-uniform sampling and chunk overlap are discussed.

Non-deterministic Image Encryption based on Symmetric Cryptosystem by Sri Gurubaran et al. [8] makes use of a modified AES technique for image encryption. A unique timestamp is used while encryption. Simulation results are provided thus validating the effectiveness of the proposed technique.

3 Proposed Framework

The framework proposed generates hash by using a combination of a few of the pre-existing algorithms like, SHA-512 [9]. Also, other algorithms like, AES-ECB [10, 11] is used for improving the security and the integrity of the hash.

The following are the scenarios where the proposed framework comes into play:

- Obtaining an image
- Sending an image
- Receiving an image

3.1 Obtaining an Image

Whenever a new image is stored in the device, it is necessary to generate the hash for images. From Fig. 1 it can be observed that the raw pixel data of the image is obtained and is used to generate the hash. The pixel data are given as input to the SHA-512 hashing algorithm which generates a 512-bit hash value for the data which is encrypted using an AES-ECB algorithm with the key as the first 128-bits of the pixel data (Dropping last bit in each byte of data). The purpose of this is to maintain the integrity of the message, since it is being communicated between users. And this generated hash value is stored in a hash-map with reference to the location of the image.

3.2 Sending an Image

When an image needs to be sent by the user to another device, the hash value of the image is taken and is stored along with the image. From Fig. 2 we can clearly observe that the all the bits of the hash are stored in the least significant bit of each and every pixel data value [12]. This way, on receiving the receiver need not calculate the hash of the receive image and can just retrieve the hash value from the image itself.

3.3 Receving an Image

The hash value of the received image is obtained by retrieving all the last bits of the pixel data, and that hash value is cross-checked with the hash values already stored in the device's memory. The hash-map is searched using the obtained hash value. And, if a reference to an image exists, then the received image is discarded. Or else, the received image's hash is also stored in the hash-map [13] with a reference to the image.

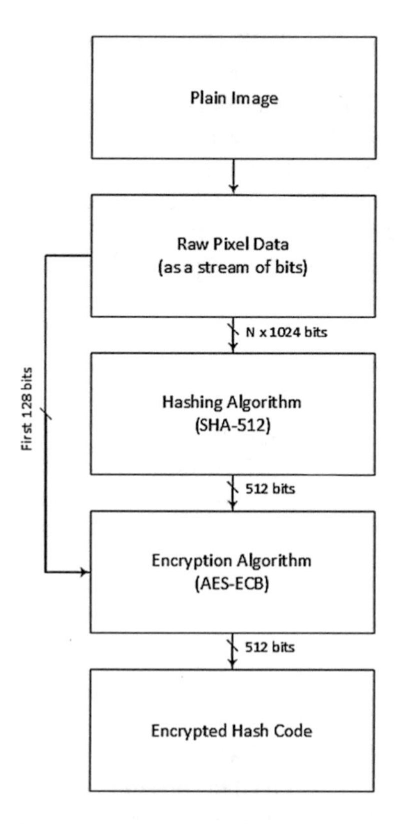

Fig. 1. Hash code generation schemes.

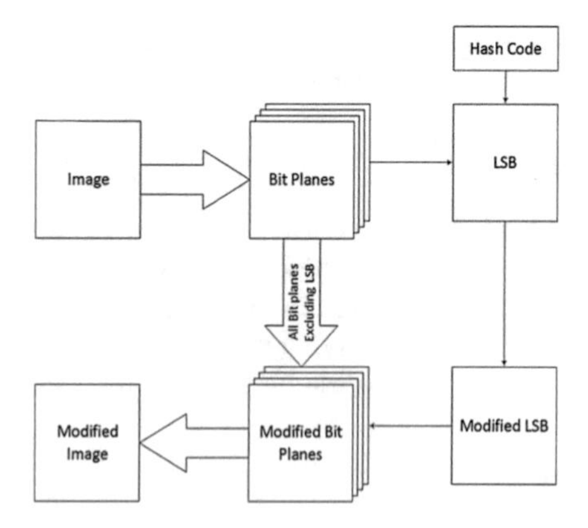

Fig. 2. Hash code mixing with the image.

3.4 Hash Code Generation Process

Step 1: Input image is obtained
Step 2: Pixel data is retrieved from the image
Step 3: Cipher key(CK) is obtained from the first 128 bits of the pixel data (dropping thee LSB in each byte)
Step 4: The pixel data is padded with 0's and appended with the total length(128-bits) to form blocks, each of size 1024 bits.
Step 5: Repeat for every block
Step 5.1: Repeat for 80 times
Step 5.1.1: Obtain 64-bit message processed by a message schedule,

$$W_t = \begin{cases} M_t(i), & 0 \leq t \leq 15 \\ ROTL^1(W_{t-3} \oplus W_{t-8} \oplus W_{t-14} \oplus W_{t-16}), & 16 \leq t \leq 79 \end{cases}$$

$M_t(i)$denotes the first 64bits of the message block i.

$ROTL^1$ denotes Rotate left function by one position.

Step 5.1.2: Generate round constant K_t, which is a sequence of the first 64-bits of the cube roots of the first 80 prime numbers.
Step 5.1.3: Pass the values W_t, K_t, H_{t-1} to the SHA Round Function, where H_{t-1} denotes the intermediate hash values of the previous round.
Step 5.1.4: For every 64-bit input (Round Function)
Step 5.1.4.1: If it is the first round, H_0 (IV) will be the first 64-bits of the fractional parts of the square roots of the first eight prime numbers.
Step 5.1.4.2: Or else,

$$T1 = h + [(e\,AND\,f) \oplus (NOT\,g)] + \sum\nolimits_{1}^{512} e + W_t + K_t$$

$$T2 = \sum\nolimits_{1}^{512} a + [(a\,AND\,b) \oplus (b\,AND\,c) \oplus (c\,AND\,a)]$$

a, b, c, d, e, f, g and h are initialized with intermediate hash values of the previous round.

$$a = T1 + T2$$

$$b = a, c = b, d = e$$

$$e = d + T1$$

$$f = e, g = f, h = g$$

Step 5.2: The intermediate hash code value will be,

$$H_i = SUM_{64}(H_{i-1}, abcdefgh)$$

Step 6: Now the obtained hash code is passed onto an AES-ECB encryption block. Where it is encrypted using the key CK.

Step 7: Thus, the encrypted hash code for the image is generated.

4 Experimental Results

4.1 Histogram Analysis

After comparing the images of the original and hash code mixed image observable differences cannot be seen with the naked eye.

On comparing the histograms from Figs. 3 and 4, it is evident that the images are same in structure but slightly different in content. This is backed by the computation of the correlation coefficient, mean squared error, and structural similarity index between the original and the hashed image, as tabulated in Table 1.

To avoid the overhead of calculating the hash code repeatedly, the hash code will be stored in the least significant bit of the pixel data. Therefore, there may be some minor differences between the original and the hashed image, which affects the mean

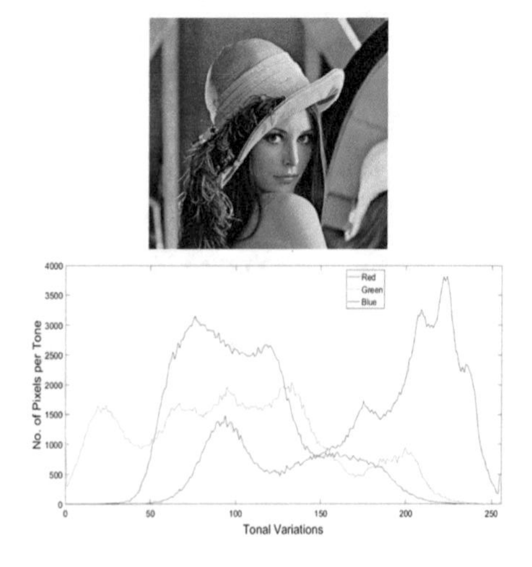

Fig. 3. Original image and its histogram.

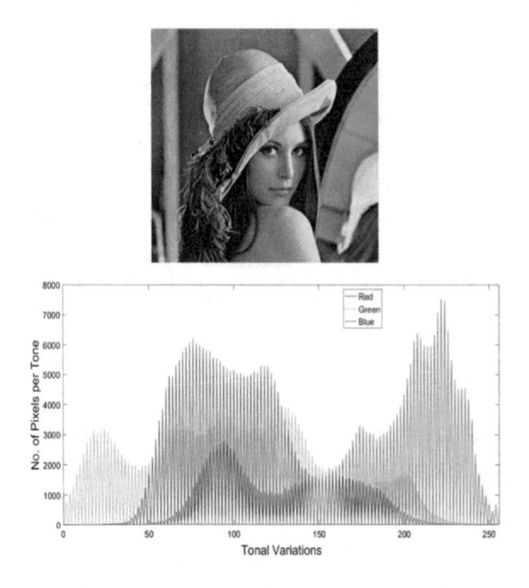

Fig. 4. Hash code mixed image and its histogram.

Table 1. Comparison metrics

	Between original and hashed image		
	Red component	Green component	Blue component
Correlation coefficient	1	1	0.9999
Mean square error (MSE)	0.4999	0.4996	0.5008
Structural similarity index (SSIM)	0.9983	0.9988	0.9990

square error values. If the accuracy of the image needs to be ensured, then the efficiency achieved by the proposed algorithm will be compromised.

4.2 Peformance Metrics

When implemented in a Snapdragon 801 device with 64 GB Hard disk and 3 GB RAM, Running on an Android 6.0.1 (Marshmallow) Operating System. The time taken for a 512×512 image was found and tabulated in Table 2.

Table 2. Time taken for execution

Trial	Generation of hash (ms)	Retrieval and checking of hash (ms)	Searching in a hash-map (ms)	
			1000 entries	*10000 entries*
Average of 10 entries	1059.6	72.4	23.8	32.8
Average of 20 entries	1102.4	81.1	25.2	37.6

The size of the 512 × 512 image was found to be 768 KB. And the space taken to store 20 copies of the same image is 15360 KB. But, the same image when only having a single copy and an entry in the hash-map was found to be 860 KB, i.e., size of image (768 KB) + Size of one entry in a hash-map (92 kB). There is considerable savings of around 94% in even such a simple case.

Though this is an increase in size is disadvantageous, if the image will not be repeated. But, in case of redundancy this reduces the space occupied considerably. This is a reasonable trade-off for reducing the redundant images on the storage.

4.3 Resistance to Attacks

Though the purpose of the hash is to verify the content of the image and therefore there is no need to check the confidentiality and the authenticity of the content received. Still, attacks common to hashing algorithms are addressed by this proposed framework.

(1) *Collision attacks*: Since, the algorithm follows a SHA-512 hashing standard the probability of a single collision to occur at a probability of 75% is as low as one in 1.9×10^{77} [14]. And after a round of encryption the collision rate was found to be still a low value as the normal hashing algorithm.

(2) *Preimage attacks*: The algorithm shows good resistance to differential and meet in the middle attacks, due to the AES encryption process carried out in ECB mode which reduces the risk of exposure by differential attacks.

5 Conclusion

In this paper, an image hashing algorithm is proposed whose primary task is to address the wastage of storage space in mobile devices due to redundant images. Using this algorithm, we can check if an image in the device and the one received are the same by simply comparing their hash codes instead of going through tedious image processing techniques. The robustness of the algorithm to various attacks was verified and was found to be resilient to collision and malicious distortions. The algorithm proposed has derived ideas and algorithms from various fields of computer science.

This algorithm can be extended onto videos and other types of files which are commonly shared and have a high probability of redundancy like documents, audios, etc. And the implementation and analysis further on these types of file will be covered in future work.

References

1. Venkatesan, R., Koon, S.M., Jakubowski, M.H.: Robust image hashing. In: Proceedings of the 2000 International Conference on Image Processing (2000)
2. Swaminathan, A., Mao, Y., Wu, M.: Robust and secure image hashing. IEEE Trans. Inf. Forensics Secur. **1**(2), 215–230

3. Zhang, W., Wong, K., Yu, H., Zhu, Z.: A symmetric color image encryption algorithm using the intrinsic features of bit distributions. Commun. Nonlinear Sci. Numer. Simul. **18**(3), 584–600 (2013)
4. Monga, V., Banerjee, A., Evans, B.L.: A clustering based approach to perceptual image hashing. IEEE Trans. Inf. Forensics Secur. **1**, 68–79 (2006)
5. Hughes, D.A., Burns, J., Yin, Z.: Data matching using flow based packet data storage US8929380 B1 (2015)
6. Douglis, F., Kulkarni, P., LaVoie, J.D., Tracey, J.M.: Method and apparatus for data redundancy elimination at the block level, US8135683 B2 (2012)
7. Zhang, Y., Ansari, N.: On protocol-independent data redundancy elimination. IEEE (2014)
8. Sri Gurubaran, B., Sasikala Devi, N., Subramanian, E.R.S., Geophilus, D.: Non-deterministic image encryption based on symmetric cryptosystem. Procedia Comput. Sci. **93**, 791–798 (2016)
9. Gallagher, P., Director, A.: Secure hash standard (shs). FIPS PUB, 180–183 (1995):
10. Daemen, J., Rijmen, V.: The rijndael algorithm. In: The First Advanced Encryption Standard Candidate Conference 1998
11. Schneier, B.: Algorithm Types and Modes. Applied Cryptography, 2nd edn., 20th Anniversary Edition, pp. 189–211(1996)
12. Kawaguchi, E., Eason, R.O.: Principle and applications of BPCS-steganography. In: Proceedings of SPIE- The International Society for Optical Engineering. vol. 3528 (1999)
13. Liang, Y.D., et al.: Object Oriented Programming with (1987)
14. Girault, M., Cohen, R., Campana, M.: A generalized birthday attack. In: Advances in Cryptology—EUROCRYPT 1988. Springer, Heidelberg (1988)

Author Index

K. Arai et al. (Eds.): FICC 2018, AISC 886, pp. 693–695, 2019.
https://doi.org/10.1007/978-3-030-03402-3

Printed in the United States
By Bookmasters